PERSPECTIVES IN CIVIL ENGINEERING

*Commemorating the
150th Anniversary of the
American Society of Civil Engineers*

American Society of Civil Engineers

1852 – 2002

Building a Better World

EDITED BY
JEFFREY S. RUSSELL

Published by the American Society of Civil Engineers

Library of Congress Cataloging-in-Publication Data

Perspectives in civil engineering : commemorating the 150th anniversary of the American
   Society of Civil Engineers / edited by Jeffrey S. Russell.
       p. cm.
     Includes bibliographical references and index.
     ISBN 0-7844-0686-3
      1. Civil engineering--History. 2. American Society of Civil Engineers--History. I.
   Russell, Jeffery S. II. American Society of Civil Engineers.

   TA19.P45 2003
   624'.09--dc22   2003060160

Published by American Society of Civil Engineers
1801 Alexander Bell Drive
Reston, Virginia 20191
www.asce pubs.asce.org

Any statements expressed in these materials are those of the individual authors and do not necessarily represent the views of ASCE, which takes no responsibility for any statement made herein. No reference made in this publication to any specific method, product, process or service constitutes or implies an endorsement, recommendation, or warranty thereof by ASCE. The materials are for general information only and do not represent a standard of ASCE, nor are they intended as a reference in purchase specifications, contracts, regulations, statutes, or any other legal document.

ASCE makes no representation or warranty of any kind, whether express or implied, concerning the accuracy, completeness, suitability, or utility of any information, apparatus, product, or process discussed in this publication, and assumes no liability therefore. This information should not be used without first securing competent advice with respect to its suitability for any general or specific application. Anyone utilizing this information assumes all liability arising from such use, including but not limited to infringement of any patent or patents.

ASCE and American Society of Civil Engineers—Registered in U.S. Patent and Trademark Office.

Photocopies: Authorization to photocopy material for internal or personal use under circumstances not falling within the fair use provisions of the Copyright Act is granted by ASCE to libraries and other users registered with the Copyright Clearance Center (CCC) Transactional Reporting Service, provided that the base fee of $18.00 per article is paid directly to CCC, 222 Rosewood Drive, Danvers, MA 01923. The identification for ASCE Books is 0-7844-0686-3/03/ $18.00. Requests for special permission or bulk copying should be addressed to Permissions & Copyright Dept., ASCE.

Foreword

I remember my conversations with Bruce Gossett more than 5 years ago as this volume was literally being conceived. What a treasure and benefit for future generations to have a snapshot of the profession at the outset of the 21st century as told by the leading experts in their respective fields. While I had a small part in organizing this publication, there is no denying that this project would never have gotten off the ground if it were not for the dedication of the ASCE staff and the eminent authors themselves who worked tirelessly to bring this volume to life. This volume is the direct result of their work and passion for civil engineering, and is the principal reason that this book is such a success.

As editor of this project, I would like to thank Bruce Gossett, Johanna Reinhart, Charlotte McNaughton, Suzanne Coladonato, and Jackie Perry for their administrative and editorial expertise and support. We also want to thank all the editors of the ASCE journals who considered participating, and especially those editors who helped contact and coordinate expert authors to write these landmark papers, and then facilitated the peer review process of each paper. These editors are listed with the appropriate journal title in the table of contents.

On behalf of ASCE, I would like to thank each of the authors for not only their papers, but their contributions to the civil engineering profession. They are (in alphabetical order): Pedro Albrecht, C. E. Bakis, L. C. Bank, Zdenek P. Ba ant, Haym Benaroya, Arnon Bentur, Leonhard Bernold, Herman Bouwer, Richard Bradshaw, V. L. Brown, Darcy Bullock, David Campbell, Ching-Jen Chen, Koon Meng Chua, E. Cosenza, J. F. Davalos, Scott A. David, Norbert J. Delatte, Marshall J. English, Steven J. Fenves, Kenneth J. Fridley, Mousa Gargari, Francis E. Griggs, Jr., Mark E. Grismer, Terry T. Hall, Jr., Yousef Haik, Chris T. Hendrickson, Glenn J. Hoffman, Milan Jirásek, Futoshi Katsuki, C. G. Keyes, Jr., J. J. Lesko, Herbert S. Levinson, Zongzhi Li, James A. Liggett, Dallas N. Little, Richard W. Lyles, A. Machida, Amir Mirmiran, Sudhir Misra, Louay N. Mohammad, Hiroshi Mutsuyoshi, Thomas M. Petry, A. L. Prasuhn, A. Essam Radwan, William J. Rasdorf, Kevin L. Rens, S. H. Rizkalla, Freddy L. Roberts, Jose M. Roësset, Robert H. Scanlan, Cliff J. Schexnayder, Vijay P. Singh, Kumares C. Sinha, Kenneth H. Solomon, K. K. Tanji, T. C. Triantafillou, Patrick Tripeny, Taketo Uomoto, Wesley W. Wallender, L. B. Wang, Richard Weingardt, David A. Woolhiser, and James T. P. Yao. Regretfully, two of the authors, Neal FitzSimons and Paul R. Wolf, have passed away.

As with all projects of this magnitude, there were countless people involved. To everyone not named who participated in this historic volume, I offer heartfelt thanks and gratitude.

Introduction

Let us consider how significantly and fundamentally the practice of civil engineering has changed over the past 150 years. Gone are the compass, chain, and slide rule, and here to stay are the advanced technologies, including global positioning and virtual 4-D design that are reshaping the very nature of the civil engineering enterprise. Gone are the ox-drawn cart and makeshift building elements. They have been replaced with increasingly automated and adaptable mechanized tools and innovative building materials. Gone too are the hardships of roughing out railroad track in the American wilderness, but here to stay are the systems-level concerns that are challenging the very core of the profession. Finally, gone are the days when the civil engineer could afford to operate in a local context only, without considering the regional and environmental impact of design. Here to stay are the days that the civil engineer must become responsible for offering system-level analysis for clients, the architectural / engineering / construction industry, and the public at large.

The civil engineering enterprise is becoming increasingly infused with nano-, bio-, and information technology. Buildings, structures, and materials are growing "smarter" as wireless sensors, self-healing materials, and integrated systems are providing greater control and information about operating state and performance. That information that can be uploaded into the design and operation of future structures. New construction, especially the grand and symbolic civil works projects of the past, is decreasing while re-construction, refurbishment, and ever-more-intricate additions to the aging infrastructure dominate the public landscape. On most every project, public or private, there has been a shift from a frontier, grass roots philosophy with little awareness for how projects affect environmental or social systems to a new outlook that is global, sustainable, and community-based in scope. Rather than accept as inevitable and then treat pollution and waste, comprehensive efforts are regularly taken to design out inefficiencies through recycling, greener materials, and better reliability and maintainability. The public has become a valued and courted stakeholder, and public hearings and meetings are becoming regular and important forums and milestones in the design process. The stove-piped, segmented project process of the past, with experts passing their knowledge over the fence to other experts downstream, is being replaced with integrated and fluid project processes, not only from design-builders but from the increasing use of cross-functional teams, pre-project planning, asset management, and designing for maintainability, sustainability, life cycle cost, and safety and security.

This is an exciting time for civil engineers, but we find ourselves at a time of transition. As the papers in this book attest, civil engineering has grown in size, complexity, and scope since ASCE was founded in 1852. Civil engineering was once a refined art passed down primarily in the field by seasoned practitioners, in which field-derived solutions were commonplace and in many ways definitive practice. Civil engineering has become a multi-pronged discipline emphasizing planning, scientifically based solutions, and specialized expertise. What has not changed in this transition from a learned art to a studied profession, I believe, is the spirit behind this great endeavor and the Society that unites it. The men who founded ASCE and the men and women who constitute ASCE today share a common interest in building a better world, quite literally. We are engineers because we want to know how things work, and we want to make them work better. We are *civil* engineers because we want to see the orthogonal lines on schematics grow into hospitals, water treatment plants, and safe roads and highways; we want to improve the quality of life for our citizens; and we want to know that we, in our own unique ways, are making a difference.

The vision of ASCE is, after all, "Engineers as global leaders building a better quality of life." It's all there in that one statement—the focus on the international context, the belief that the built environment shapes and contributes to societal well-being, and the desire to serve others. These traits are the spirit of the profession, and in this volume they are discussed and celebrated. What is also in that statement, though we may not recognize it at first sight, is change. The verb *build* implies that we are making and remaking the world, that we are, in essence, agents of change. While change is never easy, we are a profession whose very existence is rooted in transformation, and therefore, we are forever learning, growing, and changing. We find ourselves at a point when we must reflect on where we have been, where we stand, and most importantly, where we are going.

As you will see thumbing through these pages, the papers collected here perform this task quite admirably and excitingly, as they celebrate the history, heritage, and accomplishments of the profession while assessing the state-of-the-art and future challenges. Each paper explores a particular aspect of civil engineering knowledge and practice. The authors, experts who have mastered their fields and helped shape the profession, present the cutting-edge research and the notable people and projects in their respective fields. While these papers collectively provide a snapshot of the profession, each one has an eye towards the future. They offer predictions and likely developments in the years to come, especially regarding the needs of ASCE and the civil engineering profession.

As editor of this volume I have had the unique privilege to watch this important document take shape, and if I may be so bold, I would like to highlight a few of the themes that emerge. We have come a long way since the founding of the Society, but in some ways, perhaps, we have strayed too far in certain directions, and not far enough in others. At the outset of the 21st century, the civil engineering profession is faced with many challenges and opportunities. If the profession is to grow in prestige, efficacy, and reputation, we must confront the challenges before us. Most importantly, we must become better able to accommodate, predict, and embrace change in the complex arena of modern practice.

In response to increased complexity, we have become a highly specialized profession, mastering subsets of knowledge in order to contribute as a team-member to certain aspects of a project. While this strategy has worked quite admirably, it may not be the best strategy to foster sustainability and life cycle costing, while designing to minimize waste and vulnerability. If we are to become leaders of the built and natural environments, we need to broaden our scope from the educational foundation on up to expectations of professional practice.

Time and again these papers implicitly and explicitly call for individuals capable of integrating and synthesizing various domains of knowledge into systems-level understanding. You can see the authors grappling with this concept as they call for merging scales (micro and macro), maximizing benefits, achieving multi-objective optimization, taking a regional view, and designing complete systems as opposed to discrete subsystems and components. Now more than ever, it is paramount for civil engineers to embrace their role as the single profession that makes it their business, their expertise, and their mission to understand, address, and manage the built and natural environments. We must become the profession that excels at system-level planning, despite increasing levels of uncertainty and risk, because if we do not, other professions with leadership and management skills will fulfill this crucial project role.

Which brings me to a second thread running through these papers. As a profession we continue to struggle with the soft side of engineering, and to fulfill our roles as the master integrators and technical leaders of the 21st century, we must embrace a more holistic and realistic definition of what our purpose is and how we can achieve it. The future will demand new knowledge, skills, and attitudes from civil engineers, and to provide a foundation for life long learning and excellence, the profession must continue to set high expectations in education and practice. This includes not only advancing our technical knowledge, but our understanding of leadership, communication, teamwork, and management—what many are calling professional skills. While civil engineers should continue to provide technical expertise and consultation, we must also position ourselves as the team leaders, project leaders, and public policy experts who help our country and the world confront fresh water supplies, irrigation, waste management, an aging infrastructure, and quickly depleting natural resources.

To realize our full potential, we must develop and hone our communication and facilitation skills in order to bring various constituencies together to balance the multiple perspectives and interests facing every project. Technical solutions, by themselves, are no longer the sole measure of project success; optimizing engineering solutions to meet multiple criteria, whether environmental, social, or economic, are increasingly defining professional practice. Accordingly, these papers call again and again for a well-rounded practitioner of the future, with a foundation in liberal arts and professional skills complementing a solid technical background, an understanding of systems, and a personal commitment to life-long learning and continuous improvement.

Finally, these papers themselves serve as models of the well-written argument, embodying a holistic concern for specialties within the larger profession and a wider world beyond technical problems and solutions. Their very existence is proof that civil engineers can and do possess the kind of insight and perspective necessary to confront the challenges of the 21st century. It is our professional duty to embody the perspectives so powerfully articulated in this volume as we strive to uphold the ASCE Mission: "To provide essential value to our members, their careers, our partners, and the public through: Developing Leadership, Advancing Technology, Advocating Lifelong Learning, and Promoting the Profession."

The choice is ours: we create our future and destiny. What will future observers 30 or 100 or 150 years hence think about this volume and our profession? Will civil engineers be the master integrators and technical leaders of the built and natural environments, engaged in the critical discussions related to our planet's resources and management, and thereby enhancing the quality of life? Will we even continue as one group, united to serve the betterment of humankind through well-conceived projects and facilities? We cannot hope to answer these questions for that future reader, but we can answer these questions for ourselves and for our profession today. As we look to the future, let us be mindful of the men and women who have made our profession strong, reliable, and respected. As Albert Einstein said, we are standing on the shoulders of giants. Let each of us continue to improve the profession for all civil engineers—past, present, and future—by seizing the opportunity to make civil engineers leaders of the 21st century and beyond.

Jeffrey S. Russell, P.E., Ph.D.
Madison, WI
June 30, 2003

Contents

Journal of Performance of Constructed Facilities

Kenneth L. Carper, Editor

Journal of Computing in Civil Engineering

Hani G. Melhem, Editor

Journal of Aerospace Engineering

Firdaus E. Udwadia, Editor

Journal of Materials in Civil Engineering

Antonio Nanni, Editor

American Society of Civil Engineers *150th Anniversary Paper*

Building a Better World 1852–2002

Past and Future of Construction Equipment—Part IV

Cliff J. Schexnayder, F.ASCE,[1] and Scott A. David[2]

Abstract: The development of construction equipment has followed the major changes in global transportation. In 1420, Giovanni Fontana was dreaming of and diagramming dredging machines. Development of the steam shovel was driven by a demand for an economical mass excavation machine to support the era of railroad construction. The Cummins diesel engine was developed in the early 1900s as the road-building phase of transportation construction began. In the short term, the basic machine frame will not change, but productivity, accuracy, and utility should improve because of enhancements. Machines will evolve into a mobile counterweight driven by an energy-efficient powerplant. This mobile counterweight will serve as a work platform for an array of hydraulic tools, and it will have synthesized computers that instantly communicate by satellite with distant management teams reporting diagnostics, production, and position.

DOI: 10.1061/(ASCE)0733-9364(2002)128:4(279)

CE Database keywords: Construction equipment; Excavation; High technology.

Foreword

In 1975 the Construction Division of ASCE celebrated its founding 50 years earlier in 1925. At that time a series of papers appeared in the *Journal of the Construction Division* discussing 50 years of engineered construction. That series included three papers [(Douglas 1975; Klump 1975; Larkin and Wood 1975)] discussing the *past and future of construction equipment*: All three papers are discussed in this paper. In honor of ASCEs 150th anniversary, the title and subject matter of this paper follow from these earlier papers, produced 27 years ago.

The Dreams

The development of construction equipment has followed the major changes in global transportation. When travel and commerce were by water systems, builders dreamed of machines to aid in the dredging of ports, rivers, and canals. As early as 1420—72 years before Columbus discovered America—the Venetian Giovanni Fontana was dreaming of and diagramming dredging machines. Leonardo da Vinci designed such a machine in 1503. A few of these machines were actually built, but power was still by the muscle of man: the power source for one of these machines was a lonely runner on a treadmill.

[1]Eminent Scholar, Del E. Webb School of Construction, Box 870204, Arizona State Univ., Tempe, AZ 85287-0204.

[2]Graduate Student, Del E. Webb School of Construction, Box 870204, Arizona State Univ., Tempe, AZ 85287-0204.

Note. Discussion open until January 1, 2003. Separate discussions must be submitted for individual papers. To extend the closing date by one month, a written request must be filed with the ASCE Managing Editor. The manuscript for this paper was submitted for review and possible publication on August 24, 2001; approved on March 19, 2002. This paper is part of the *Journal of Construction Engineering and Management*, Vol. 128, No. 4, August 1, 2002. ©ASCE, ISSN 0733-9364/ 2002/4-279–286/$8.00+$.50 per page.

In 1852, when the American Society of Civil Engineers and Architects (later the ASCE) was incorporated, construction in the United States was changing from canal building to railroad construction. The Middlesex Canal, which connected Boston to the Merrimack River at Lowell, had been in service since 1803, but in 1853 the Boston & Lowell Railroad superseded it. The canal transportation systems, which had played such an important role in the development of the country, were about to be succeeded by the railroads, but construction, be it building canals or railroads, was still achieved by the brawn of man and beast.

In 1843 Caleb Eddy, agent of the Middlesex Canal corporation, wrote. "The inventions and ingenuity of man are ever onward, and a new cheap and more expeditious mode of transportation by steam power has been devised which seems destined to destroy that which was once considered invulnerable." Eddy had a good understanding of how better and cheaper energy sources—technology—affected the canal. The technology of capturing energy for beneficial use is a major driver in equipment change.

Steam Power Machines

The first practical power-shovel excavating machine was built in 1835. William S. Otis, a young civil engineer with the Philadelphia contracting firm of Carmichael & Fairbanks, designed this ingenious machine. The first "Yankee Geologist" machines, as they were called, were put to work in 1838 and 1841, respectively, for railroad work in Massachusetts and New York. Cohrs (1997) reports that only seven of these excavators were ever built, but Anderson (1980) says that the most accurate estimate of the number of machines manufactured is 20. The Anderson number is for the time period from 1840 to 1860; Cohrs does not state a time period. Anderson also adds that every recorded user was a business associate of Otis or his uncle, Daniel Carmichael. It is known that four "Yankee Geologist" machines went to Russia to work on the St. Petersburg–Moscow Railroad. George W. Whistler may have had a hand in this interesting transaction; in 1842 he had moved to Russia after agreeing to manage construction of the St. Petersburg–Moscow Railroad. Whistler was a West Point-educated engineer who surveyed many of the early railroads and actually supervised the first laying of track for a passenger railroad in the United States (Hill 1957).

Another shovel went to England, where it was put to work on a railway in Essex. Anderson leaves the impression that Carmichael had some part in the Russian and English work, but the relationship may really be a connection between Carmichael and Whistler. When working with the Baltimore and Ohio Railroad (B&O) and still a lieutenant in the Army, Whistler was sent to England by the B&O to examine railroad construction there, so he had personal railroad connections in that country. As an interesting note, the second Otis excavator was employed in Canada until 1905.

Otis's patent application stated

> I . . . have invented certain improvements in the apparatus for and mode of excavating earth for the construction of railroads, canals, or other purposes where excavation may be necessary which I effect by means of certain appendages to a scraper of the ordinary construction, which scraper is to be worked by a crane and is to take the earth immediately from the banks from which the excavations are to be made . . . .

He went on to describe how by a system of pulleys to move its arms and bucket the load could be taken up by the scraper, raised by the crane, and turned to be dumped. But he made only two patent claims: (1) the principle of the power thrust of the shovel handle; and (2) a type of friction clutch on the power transmission to the handle. The technology change was related to energy and energy transmission.

Continued development of the steam shovel was driven by a demand for an economical mass excavation machine. An era of major construction projects began in 1880. These projects demanded machines to excavate large quantities of earth and rock. In 1881, Count Ferdinand de Lesseps's French company began work on the Panama Canal. Less than a year earlier, on December 28, 1880, the Bucyrus Foundry and Manufacturing Company, of Bucyrus, Ohio, had been issued its certificate of incorporation. Bucyrus became a leading builder of steam shovels, and 25 years later, when the Americans took over the canal work, the Bucyrus Company would be a major supplier of steam shovels for that effort. Other large projects undertaken in the 1880s that required excavation machines included the Chicago Drainage Canal. The Chicago Canal alone involved moving 30,580,000 m³ (40,000,000 cu yd) of earth and rock excavation.

However, the most important driver in excavator development was the railroad. In the first 50 years of ASCE's existence, about 80% of today's railway network worldwide came into being. Between 1885 and 1897 approximately 112,630 km (70,000 mil) of railway were constructed in the United States. William Otis developed his excavator machine because the construction company Carmichael & Fairbanks, for which he worked and in which his uncle, Daniel Carmichael, was a senior partner, was in the business of building railroads.

The Bucyrus Foundry and Manufacturing Company came into being because Dan P. Eells, a bank president in Cleveland, was associated with the Lake Erie and Western Railway, the Chicago and Saint Louis Railway, and the Ohio Central Railroads. Eells saw an opportunity to profit from a company that produced railroad equipment, and that is what the company did for the first 4 years. But in 1882 the Ohio Central Railroad gave the new com-

Fig. 1. Early 20th century steam shovel (note this machine is mounted on steel traction wheels)

pany its first order for a steam shovel, and sales to the Northern Pacific Railroad (two machines) and to the Savannah, Florida, and Western Railroad soon followed.

The first Bucyrus excavator was shipped on June 3, 1882, to the Northern Pacific Railroad. In May 1883 the company exhibited a shovel at the National Exposition of Railway Appliances in Chicago.

Fig. 1 shows an early 20th century steam shovel mounted on steel traction wheels. Most machines of this period were still mounted on a railway carriage.

Internal Combustion Engines

By 1890 courts of law in Europe had ruled that Nikolaus Otto's patented four-cycle gasoline engine was too valuable an improvement to keep restricted any longer. Following the removal of that legal restraint, many companies began experimenting with gasoline-engine-powered carriages. The Best Manufacturing Company (predecessor to Caterpillar Inc.) demonstrated a gasoline tractor in June of 1893.

Between 1900 and 1911 the Bucyrus Company had conferences with the General Electric Company concerning the use of electrical devices and power for large shovels. The Bucyrus Company came to the conclusion that such a power source was not yet practical. Conversely, Bucyrus did conclude that the only other power source alternative was the internal combustion engine.

The first application of the internal combustion engine to excavating equipment occurred in 1910, when the Monighan Machine Company of Chicago shipped a dragline powered by a 37.2 kW [50 horsepower (hp)] Otto engine to the Mulgrew-Boyce Company of Dubuque, Iowa (Anderson 1980). Bucyrus shipped a gasoline-powered placer dredge to the Yuba Construction Company of Soloman, Alaska, in 1911.

Henry Harnischfeger brought out a gasoline-engine-powered shovel in 1914, and following World War I, the diesel engine was put into excavators as the power source. The Cummins diesel engine had been developed in the early 1900s by a self-taught mechanic named C. L. "Clessie" Cummins. Working out of an old cereal mill in Columbus, Indiana, Cummins's first engine was a Dutch-designed, 4.5 kW (6 hp), farm-type diesel. The Cummins engine soon became popular in other applications, including power shovels. Warren A. Bechtel, who first entered the construc-

Lucky to get on road like this.

Fig. 2. Photograph with Dwight D. Eisenhower's description from personal collection of Dwight D. Eisenhower (Courtesy, Dwight D. Eisenhower Library)

tion field in 1898 in Oklahoma Territory and quickly built a reputation for successful railroad grading, pioneered the use of motorized trucks, tractors, and diesel-powered shovels in construction.

The Erie Steam Shovel Company (later consolidated with the Bucyrus Company) deserves special attention as a machine innovator. Erie saw a need for an excavation machine designed especially for building construction and in 1914 produced a full-revolving steam shovel equipped with a 0.57 m³ (3/4 cu yd) dipper. This was followed by a 0.38 m³ (1/2 cu yd) version in 1916. These machines filled a market niche, but what made them different was that Erie fabricated all parts with jigs and templates. Interchangeable parts were part of Erie's manufacturing plan.

Erie also noted the advantages of other power sources for its machines and first experimented in 1925 with a 0.76 m³ (1 cu yd) gas-plus-air excavator combining a gasoline engine, an air compressor, and air-operated motors. The price of this machine was higher than that of either steam or gasoline models offered by other manufacturers. So in 1926 Erie began changing its complete line of machines to gasoline. The change was difficult, but once the line of four machines from 0.38 m³ (1/2 cu yd) to 0.96 m³ (1 1/4 cu yd) was in place the results were phenomenal for Erie's sales (Anderson 1980).

In the winter of 1922–1923, the first gas-powered shovel was brought to Connecticut, and in the spring of 1923, it was employed on a federally aided road construction project running from Litchfield to Torrington. The third phase of transportation construction had begun. Contractors needed equipment for road building. In 1919 Dwight D. Eisenhower, as a young army officer, took an Army convoy cross-country to *experience* the condition of the nation's roads (Fig. 2). But as the country began to improve its road network, World War II intervened and road building came to a near halt as the war unfolded.

Large Projects as Incubators for Machine Innovation

Large construction projects provide a fertile test bed for equipment innovation. William Mulholland rode on horseback into Los Angeles in 1877 and proceeded both to better himself and to contribute to the city he loved. He passed from ditch tender to straw boss, to foreman, to superintendent learning construction.

Later, as city engineer, Mulholland directed an army of 5,000 men who labored for 5 years to construct the Los Angeles Aqueduct, which stretches 383 km (238 mil) from the Owens River to Los Angeles.

Los Angeles Aqueduct

In 1908 the Holt Manufacturing Company (the other predecessor to Caterpillar Inc.) sold three gas-engined caterpillar tractors to the city of Los Angeles for Mulholland to use in constructing the Los Angeles Aqueduct. Besides crossing several mountain ranges, the aqueduct passed through the Mojave Desert, a severe test site for any machine. The desert and mountains presented a challenging test for the Holt machines, but Benjamin Holt viewed the entire project as an experiment and development exercise (Leffingwell 1997).

Holt found that cast-iron gears wore out very quickly from sand abrasion, so he replaced them with steel castings. The brutal terrain broke suspension springs and burned up the two-speed transmissions in his tractors, and the low gear was simply not low enough for the climbs in the mountains. Holt made modifications to the tractors both at his factory and in the desert. His shop manager Russell Springer set up repair facilities in the project work camps. Mulholland in his final report after completion of the project labeled the Holt tractors the only unsatisfactory purchase of the project, but Holt had developed a much better machine because of the experience.

Boulder Dam

In the years between the two world wars, one particular construction project stands out because of the contributions to equipment design that resulted from the undertaking. Boulder Dam (later named Hoover Dam) was the catalyst for major equipment developments. According to Larkin and Wood (1975), "On that project R. G. LeTourneau, the master of cutting and welding, eliminated the bolted connection and showed the true capabilities of welded equipment with cable operated attachments. This same man, through his numerous inventions and innovations in tractor/scraper design and manufacture, made possible the current earth-moving behemoths in use today." In the 1940s, those machines went on to build airfields around the world in support of the war effort.

Other developments that came from the Boulder Dam project included sophisticated aggregate production plants and improvements in concrete preparation and placement. The project was an enormous proving ground for new equipment and techniques. The use of long flight conveyor systems for material delivery was proven there. Henry Kaiser later put that knowledge to use in constructing Shasta Dam.

Three Significant Developments

After the war roadbuilding roared back, and in 1956 Eisenhower, now president, signed the legislation that established the interstate highway program. To support the roadbuilding effort, scrapers increased in capacity from 7.6 to 23 m³ (10 to 30 cu yd). With the development of the torque converter and the power shift transmission, the front end loader began to displace the old "dipper" stick shovels. Concrete batch and mixing plants changed from slow, manually controlled contraptions to hydraulically operated and electronically controlled equipment.

High-Strength Steels

Up to and through World War II machine frames had been constructed with steels in the 210,000 to 242,000 kN/m^2 (30,000 to 35,000 psi) yield range. After the war, steels in the 280,000 to 311,000 kN/m^2 (40,000 to 45,000 psi) range with proportionally better fatigue properties were introduced (Klump 1975). The new high-strength steel made possible the production of machines having a greatly reduced overall weight. The weight of a 37,000 kg (40 t) off-highway truck's body was reduced from 11,000 to 7,200 kg (25,000 to 16,000 lb) with no change in body reliability.

Nylon Cord Tires

The use of nylon cord material in tire structures made larger tires with increased load capacity and heat resistance a practical reality (Klump 1975). Nylon permitted the actual number of plies to be reduced by as much as 30% with the same effective carcass strength, but with far less bulk or carcass thickness. This allowed tires to run cooler and achieve better traction and improved machine productivity.

High-Output Diesel Engines

Manufacturers developed new ways to coax greater horsepower from a cubic inch of engine displacement. Compression ratios and engine speeds were raised, and " . . . the art of turbocharging was perfected to a practical level, resulting in a 10–15% increase in flywheel horsepower" (Klump 1975).

Big Iron

In 1955 the Caterpillar Tractor Co. introduced its first D9 tractor, a 30,000 kg (33 t) machine with a flywheel rating of 213 kW (286 hp). Euclid followed in 1957 with the TC-12, a 325-flywheel kW (436-flywheel hp) crawler tractor. This iron monster used a separate engine to power each track. When they began developing the site for Dulles International Airport in Virginia, R. G. Le Tourneau was there with a 118,000 kg (130 t) capacity double-bowl scraper. Expansion and size also characterized wheel-loader machines, but the most significant improvement was the introduction of articulation in 1962, which increased the productivity of wheel loaders.

The Next 10 Years

Three professionals offering both contractor and equipment dealer viewpoints were questioned about trends in equipment technology and management. On the dealer side are Tom Bennet of RDO Equipment in Phoenix and Jerry Jung of Michigan CAT, Novi, Michigan. RDO is part of the John Deere dealer network. Mike Monnot of H. B. Zachry, headquartered in San Antonio, presents the contractor viewpoint. Each interviewee was asked the same three questions:

1. Will there be new types of construction equipment or advances in capabilities?
2. How will construction firms obtain the equipment: buy, rent, or lease?
3. How will equipment management change?

Fig. 3. Dozer blade laser control system demonstrated at 2001 Bauma, Munich, Germany

New Types of Construction Equipment

Bennet does not foresee any radically new equipment on the horizon. Everything he sees is merely refinement of the inventions of LeTourneau; earthmoving equipment will remain fundamentally the same.

Jung thinks that new attachments will mean improved utility for the contractor's fleet.

Monnot is in agreement with Bennet and Jung and does not foresee any new classifications of equipment, with the possible exception of equipment for fiberoptic cable installation.

Equipment technology or innovation can be categorized (Goodrum 2001) into three broad categories:

- *Level of control*: equipment advancements that transfer operational control from the human to the machine;
- *Amplification of human energy*: shift of energy requirements from the man to the machine; and
- *Information processing*: gathering and processing of information by the machine.

The responses of our professionals to this question are organized using this system of technology categorization.

Level of Control

Bennet believes safety will continue to be very important. Safety features and operator station improvements will evolve to compensate for the less experienced workforce available today and in the foreseeable future. Related to workforce quality is the proliferation of laser-controlled implements. Laser-guided equipment makes it much easier for inexperienced operators to perform fine-grade activities (Fig. 3).

Monnot says that even the lowly soil compaction roller will see improvement. New technology will allow the roller itself to measure the density of the fill and transmit that data via radio back to the project office. If proven accurate and successful, this development will reduce the number of inspectors required to perform compaction tests.

Further labor-saving improvements will include automated blade controls for finegrade operations and guidance systems for mass grading. The laser and global positioning satellite (GPS) guided systems will become more common and reduce the need for surveyors. All the grader or dozer operator will need to do is load the digital terrain model (DTM) into the on-board computer and then guide the machine where the display indicates. Machine position, along with cut or fill information, will be on a screen in front of the operator at all times. This may turn the operator's job into a video game.

Sound waves can also be used to control machine functions. Ultrasonic sound waves are emitted from a sensor unit and received back to the sensor. The principle is much the same as a ship's sonar. The time the sound wave takes to leave and come back is measured and the distance calculated by the computer. The computer keeps the machine on grade and controls the blade-lift functions. Finegrading a roadbase with a sonic blade control system requires only steering and travel speed input from the operator to keep the sensor over the grade wire or other fixed grade. One equipment group demonstrated a sound wave control system for asphalt pavers at the 2001 Bauma International Trade Fair.

At the Swiss Federal Institute of Technology in Zurich a shotcrete robot for tunnel work has been developed in collaboration with Master Builder Technologies (Girmscheid and Moser 2001). This shotcrete robot is equipped with a laser device that scans the surface of the tunnel on which the shotcrete should be applied. The computer then calculates a virtual surface based on the measured spherical tunnel surface. The virtual plane is located at a distance from the tunnel surface on which the nozzle of the shotcrete robot will be guided and held perpendicular to the rock surface. This makes the optimized application of shotcrete possible, with low rebound and high surface smoothness.

Ultimately, operators sitting in a machine cab may be eliminated altogether. Caterpillar is developing and testing automated rock hauling units for mining. These units are linked by radio to the office and tracked by GPS. The superintendent need only use a laptop to send the start signal, and the trucks do the rest, leaving the lineup at set intervals and following the prescribed course. The superintendent can track the progress of each machine on the computer; if a truck develops a problem, the situation is signaled to the superintendent for corrective action.

Amplification of Human Energy
According to Bennet all machine progress will be in equipment technology. There is a trend to all-hydrostatic drive machinery; transmissions and torque converters will become obsolete. However, Caterpillar is resisting the trend toward hydrostatic drive. As hydraulic systems and electronic controls improve, equipment operation will be simplified, which is a continuing trend that is already being experienced.

There will most likely be a trend to smaller and more versatile equipment. Such equipment is necessary for accomplishing reconstruction projects. Highway reconstruction is often performed within space-restricted single or double lane closures. The ability to work within a narrow space with small but multitooled equipment will be a real advantage.

There may come a time when the base machine is only considered a *mobile counterweight with a hydraulic powerplant*. The

base machine will perform a variety of tasks through multiple attachments. This trend has started with hydraulic excavators and the many attachments available such as hammers, compactors, shears, and material-handling equipment. Wheel-loaders have seen the introduction of the tool-carrier concept. Along with the standard bucket, other attachments such as brooms, forks, and stingers are readily available to perform multitude of tasks on the jobsite. Other attachments will be developed, offering the contractor more versatility from a base investment.

Jung believes that there will not be any radically new equipment but there will be more and improved attachments available for existing equipment. Machines will become multifunctional. He feels that there will continue to be improvements to hydraulic systems and to the hydraulic power at the end of an excavator boom.

Information Processing
Monnot believes there will be continued refinements to on-board machine diagnostics. These will shorten troubleshooting and repair time. There will be expansion of tools such as the Caterpillar Electronic Technician (ET) and Vital Information Management System (VIMS) among all manufacturers and deeper into the product lines. Komatsu, for example, had an impressive display of their information management tool at the 2000 MineExpo in Las Vegas.

Safety is a major concern for these self-guided behemoths. Sensors scan all around the truck for potential problems. If an obstacle is encountered in the roadway larger than a basketball the lead truck stops and sends a signal to all the other trucks to stop. It also signals the superintendent where they are located so that someone can be dispatched to check the problem. It would be unlikely that the automated technology could be adapted to smaller construction equipment in the foreseeable future.

There is general agreement among those interviewed that nothing radical would be forthcoming in the foreseeable future. The basic machine frame will not change, but productivity, accuracy, and utility should improve because of enhancements.

Obtaining Equipment in the Future

Bennet sees equipment leasing becoming more popular now and in the foreseeable future, but it is his view that construction firms will continue to obtain equipment by all three of the traditional methods—buy, rent, and lease—but the breakdown for each is uncertain.

Tax Code
The tax code plays a major role in how a company will obtain and pay for equipment. If it is more advantageous to lease, that is what the contractor will do. Other factors are the company's cash flow and retained earnings. Both affect bonding capacity and the ability to bid new work.

Rental equipment is a very competitive market. Traditionally, a dealership would rent equipment more as a machine sales tool rather than as a supply source for a contractor's short-term equipment needs. It was hoped that the short-term rental would result in a machine purchase if the contractor was satisfied with the machine. Now, with the proliferation of major equipment rental firms, dealers have had to adjust their rental strategies. The rental firms have grown to the point of becoming quasi-dealers themselves. They provide parts and service for their equipment and have the purchasing power to negotiate competitive prices from the manufacturers. This has forced the dealerships to conduct their business more like a rental company.

Dealer Rental Strategy

Jung feels that there will be an increase in equipment rental and rent-to-own plans. Currently in the United States, rental equipment makes up on average 20% of a contractor's fleet. Contrast that number with Europe and Japan where, according to Jung, the rental equipment components of the fleet are 40 and 80% respectively. The current wisdom dictates that the United States will follow suit in the coming years. Dealerships will need to make adjustments to their rental strategy to ensure market share.

Types of Projects

Monnot feels that companies will rely more on lease-to-purchase plans as long as the tax code remains the same. However, companies need to carefully analyze their ownership costs each way: buy, rent, or lease. The proper decision is very dependent on the company's financial situation.

Companies will probably buy less large equipment unless more highway work is planned, bid, and let. New highway construction is essentially complete, and the market for large scrapers may taper off. No one wants to be stuck with a fleet of large and expensive yet obsolete equipment. Renting or leasing these types of equipment may make more sense in the future.

The general consensus seems to be that rental and leasing will increase in popularity, but the degree will be dictated by the U.S. tax code. Dealerships will need to adapt or be left behind by the equipment rental firms that are now in the market. Some manufacturers are already using the rental dealer network as a quasi-distributor network.

Equipment Management

The first aspect that Bennet sees is that equipment management will change based on shorter ownership periods.

Technological Advantage

Bennet believes equipment innovations will come to market at a faster rate, and as a consequence individual machines will experience productivity obsolescence sooner. That does not mean that the equipment will be ready for the scrap yard in 2 or 3 years, but that the technological advantage and therefore the competitive advantage will be gone. These 2- or 3-year-old machines will have a useful place in the secondary market; for example, farmers and small firms that cannot afford first-line equipment. Dealerships will expand guaranteed buy-back programs to allow for a predictable salvage value at the end of the machine's economic life. More accurate financial planning to include the equipment component will improve a company's success.

Maintenance—Technical Training Required

Maintenance will see many changes. There will be a move toward more dealer-performed service rather than self-performed service. Even today, says Bennet, the number of dealer technicians is increasing while the number of contractor mechanics is decreasing. With the equipment changing so rapidly, it is difficult, if not impossible, for a contractor technician to remain current without the technical training available to the dealer technician. Another factor encouraging this trend is the latest array of diagnostic tools and equipment available to dealer technicians. The combination

of up-to-date factory technical training coupled with the latest diagnostic tools usually means shorter down time for the equipment and reduced repair cost.

Guaranteed maintenance programs will help the contractor manage the maintenance budget for the fleet. Under such a program, a set fee is paid up front at the start of the machine's service life and the dealership takes care of the maintenance. Bennet's firm has written some contracts that cover everything but fuel, but most contracts are not that comprehensive. The company does expect this program to expand in the future and to be a major source of revenue.

Similarly, Jung believes that dealerships will offer more "total cost guarantees" for the purchased equipment. This program will cover all repairs and maintenance, with the exception of wear parts, for 10,000 h of machine life.

Customer service will continue to improve with expanded maintenance service. Dealers will expand lube and maintenance services to allow small contractors to better maintain their equipment. Customers will be able to reduce or eliminate their mechanical support staff without sacrificing machine availability.

Power Plant Change

Component replacement is another change that Bennet sees on the horizon. Rebuilding will become less economical, and component exchange will be standard practice. Even engines will become a "throw-away" item, designed to last for a certain number of hours and then be scrapped, at which point a new engine will be installed in the unit.

Bennet feels that engine technology may change the most. Lightweight ceramic diesel engines may soon be available that will become quieter and more fuel-efficient. In an effort to maximize power and fuel efficiency, all diesels will be turbocharged.

Information Processing for Management

Jung thinks that technology improvements will continue to evolve. GPS tracking of equipment will allow the contractor to keep track of the location and hour meter reading of every unit. At some point even machine operating condition will be available in real time for all users.

Monnot also views advancements in the physical tracking of equipment as important. GPS satellites will track and transmit machine location, hour meter reading, and vital operating conditions such as engine temperature and oil pressure. The equipment manager will be able tell at all times where the machines are located and if they are being used. Machines that appear to have left their assigned project can be tracked even if stolen. Machines that indicate low oil pressure or overheating could trigger the dispatch of a service technician to check the problem before the operator or foreman even knows there is a problem.

Monnot says, "where the contractor will see the most changes in equipment will be in fleet management." Costs for the equipment will be more closely monitored and improved equipment rates generated for bidding purposes. A mismanaged equipment fleet has the potential to lose a lot of money. Fortunately, tools are available to aid the tracking of everything put into a machine, how it is being operated, and what productivity it is achieving.

Service diagnostics and real-time communication of data for historical purposes will shorten unscheduled downtime and provide the estimators with equipment cost numbers they can trust. It will be easier for the equipment manager to know the rate of wear for almost any machine in the fleet and to schedule routine maintenance and component replacement.

Fuel Consumption—Equipment Management Tool

Fuel and lubricants sometimes become a black hole on the project. Most companies still rely on the person adding fuel and oil in the field to write the added quantities on a log sheet. Technology is available to aid the tracking of fuel and lubricants. A computer barcode system can be used to identify the fueled machine. Computer records on the fuel or lubrication truck will track the type and quantity for each product dispensed to the machine. The information will then be sent to a computer in the office for processing and machine record updating.

Currently, correlating machine use hours (hour meter reading) and oil sampling are the most popular means by which companies determine service and component replacement intervals. Many in the industry feel that they are not good predictors in cases of extremely heavy or light duty machine use.

Equipment that works in extreme conditions may not receive adequate maintenance if the lubricants are changed at prescribed service meter intervals. Fuel consumption is one way to determine if the conditions are "extreme." Machines engaged in heavy use run their engines at full load most of the time and therefore use more fuel per shift. Once a piece of equipment has used a certain quantity of fuel, it is time to change the oil. This tracking will also indicate if the operator is really working the machine to its full potential.

With equipment used in light duty applications it may be possible to extend the oil change interval. If a piece of equipment spends most of the day at idle, it will not use much fuel. However, the equipment manager may want to look at why the machine is not using fuel. Is the piece really necessary on the project? Corrective action may not be required, but fuel consumption can be another tool to indicate potential job-site productivity problems.

Machine components can also benefit from fuel-consumption tracking. Again, high fuel consumption indicates extreme service conditions, and the equipment manager may need to shorten the planned service life of engines, transmissions, and other components to prevent unplanned catastrophic failures. Using fuel consumption to guide service intervals can help a manager control the equipment fleet more efficiently. Fuel consumption becomes a self-compensating system to make sure the equipment is well maintained.

Crane Operation and Safety

There will be safety improvements on cranes. While computer-monitoring systems are not new, they will have added features: the new systems will record, for every pick, the weight of load, boom angle, and spool speed. These historical data could be helpful for bidding future projects or investigating crane accidents. The systems will be able to diagnose the electrical systems on the unit, allowing shorter repair time and improved machine availability.

Fleet management will become a predictable science for all constructors, not just the very large ones. Dealers will offer more guaranteed pricing and service for all customers, and the large equipment users will continue to refine data collection and planned preventative maintenance programs. Tracking all aspects of equipment use, productivity, efficiency, and rates of component wear will become easier and more affordable with the newer technologies.

The Future

James Douglas (1975) predicted that in the distant future it was conceivable that the discovery of new principles of subatomic structure would enable the excavation contractor to liquefy the cut material and pump it to the fill site. Twenty-seven plus years later it appears more probable that refinement and improvement will rule the future of construction equipment. Those predictions have not yet become reality, but they were presented as a vision of the distant future. The ideas Douglas offered are still valid, but without trying to move into the realm of the scientist as Douglas did, what are the equipment possibilities in the misty future? A vision is created here based on several specific tenets: (1) level of control improvements; (2) amplification of human energy; and (3) information processing. While parts of this vision may seem exotic fiction, it should be remembered that at least one manufacturer has already eliminated on-board operators from a fleet of experimental machines, and the idea of machines as mobile counterweights with a hydraulic powerplant is almost a reality today.

Rebuilding San Francisco in 2025

Let us follow the fictitious Heath Barger as he finishes his lunch and walks into the cellar that makes up his office on the afternoon of November 4, 2025. Heath works from his Oklahoma cellar as an independent equipment operator. Currently, he is working with Mega Inc., a large contractor that specializes in rebuilding urban infrastructure. Heath has worked several times in the past with Mega on projects in New York City and Chicago. Mega is a large multinational engineering/construction firm based in Singapore, which has become a world leader in the execution of difficult reconstruction projects. The current project is in San Francisco, which was recently devastated by a major earthquake.

Heath has a master of science degree in civil engineering from the University of Wisconsin, which he completed on-line from home. He is married with four children and likes to work on projects in time zones to the west of Oklahoma, as that means he can start work later in the day.

Heath greets his workspace as he enters the room. His "computing power" is provided by components that are built into the room itself, all of which communicate wirelessly. Three walls in the workspace act as large-scale display devices. Heath generally operates his equipment as he views images displayed on these surfaces. The wall displays work in conjunction with his glasses to give the images a 3D effect. When not interacting with his "work walls," Heath reviews digital data that help him coordinate machine maintenance for his mobile power-source counterweight.

Heath, who is a Native American, used tribal grant money to lease his first mobile power-source counterweight from Globe Machines in Richardson, Tex. Globe owns most of the power source machines working on this project. Their technicians handle the actual machine maintenance but Heath and Globe's maintenance manager always confer before a component exchange. Shirley, the maintenance manager for Globe, has been working with Heath now for over 5 years.

From her workspace in Texas she sends information and instructions to her on-site team of technicians. Through speech, sketching, and hand gestures she simulates to the technicians how to actually install the component in the power source on the project. When she wishes, she calls upon component manufacturer support directly through linked computer systems. 'SunSpot Energy, the manufacturer, has been monitoring all of the machines' components by computer chips implanted in each component. SunSpot has a large historical database on any item of concern. This database will help Shirley and Heath decide on courses of action when there seems to be a problem.

Heath is a constructor of elevated formwork for bridge structures. His job is to design and construct the forming systems into which the concrete specialist located in Chicago will place their moldable mixture of refuse and epoxy. This project is being built following a virtual project design and analysis. This virtual design helped develop and refine the actual project; project estimates, schedules, and work methods and equipment were determined through the design process. Heath specializes in urban formwork construction, and he is one of the world's leading builders of formwork for bridges. His specially designed appliances, which are fitted on the leased power source, will build most of the formwork. With his appliances it is possible to erect the forms at night without generating more than 20 dBA of sound at 15.2 m (50 ft) from the work site; this is well below the San Francisco ordinance limiting nighttime noise to less than 55 dBA.

Heath's role in the project is to design and construct formwork for the designed structures. He does this in a way that, to a large extent, follows the way that office engineers and equipment operators used to work with the concrete forms in the old days. He will use materials that have already been "delivered" to the project. He must coordinate procurement of those resources with the Mega project manager in San Francisco.

Because unexpected events occur in the real world, he calls upon an extensive "project risks library" to simulate unexpected events, and his work strategy has flexibility for dealing with these risks. Once he has determined the exact sequence of work tasks that will lead to successful completion of the forming operation, he can be quite confident that the formwork will be in place as required by the project manager's master schedule.

As Heath carries out his construction operations, the images in his workspace present the actual construction site, and he performs his tasks by both voice and hand movement commands to his machine. The manufacturers of the formwork materials that are used on the project provide many of the control software components that he uses. Over the years, this software has become much more reliable at working together seamlessly.

Like any construction project, much of his work is highly interdependent with the work of others. Heath and other operators work collaboratively through their connected workspaces, and his machine interacts with theirs. The final constructed project is very much the result of their combined and coordinated efforts.

Conclusion

In the near future, few if any new and radical changes are seen on the horizon. Our existing equipment will become more versatile and efficient, with new attachments made possible by improved hydraulics, electronics, and powertrains.

The U.S. contractors will follow Europe and Japan in the move towards more rental and leasing of equipment rather than outright purchasing. The extent of the shift to rental and leasing will be guided by the U.S. tax code. The dealerships representing the manufacturers will need to shift their emphasis on rental to remain competitive in the equipment market.

Fleet management will become easier as technology makes it possible to collect and analyze data about productivity, efficiency, and maintenance. Problems can be identified and corrected before they become expensive catastrophes. Improper operating techniques that hurt production and possibly shorten machine life will be identified the day they occur and not weeks later. Components could be replaced closer to their failure point without risking costly core damage caused by waiting too long to change the component. Fuel consumption rates will determine maintenance intervals, thereby maximizing the balance between protecting the component life versus wasting money on too-frequent oil changes. The future is construction speed, quality, and lifetime project economics.

Since Douglas's (1975) attempt at predicting the distant future of construction equipment, there have been many advances in computer technology, or maybe more specifically chip technology. Additionally, guidance and positioning technology have all become practical tools. These advances in chips and guidance and positioning technology seem to offer a very different future for the application of mechanical devices to construction processes. The purpose of a machine is to perform a task that is beyond the capability of a human. The machines of the future will amplify human energy, as did those of the past, but control can come from any location.

Acknowledgments

The writers would like to express their gratitude to Tom Bennet, sales manager, RDO Equipment, 2649 N 29th Ave, Phoenix, AZ 85009; Jerry Jung, owner, CEO, Michigan Cat, 24800 Novi Rd., Novi, MI 48375; and Mike Monnot, director of equipment, corporate tools and scaffolding, H.B. Zachry, P.O. Box 240130, San Antonio, TX 78224.

References

Anderson, G. B. (1980). *One hundred booming years*, Bucyrus-Erie, South Milwaukee.

Cohrs, H.-H. (1997). *500 years of earthmoving*, KHL International, East Sussex, U.K.

Douglas, J. (1975). "Past and future of construction equipment—Part III." *J. Constr. Div., Am. Soc. Civ. Eng.*, 101(4), 699–701.

Girmscheid, G., and Moser, S. (2001). "Fully automated shotcrete robot for rock support." *Comput.-Aided Civ. Infrastructure Eng. Mag.*, 16(3), 200–215.

Goodrum, P. M. (2001). "The impact of equipment technology on productivity in the U.S. construction industry." PhD dissertation, Univ. of Texas, Austin, Tex.

Hill, F. G. (1957). *Roads, rails, and waterways: The Army Engineers and early transportation*, Univ. Oklahoma Press, Norman, Okla.

Klump, E. (1975). "Past and future of construction equipment—Part II." *J. Constr. Div., Am. Soc. Civ. Eng.*, 101(4), 689–698.

Larkin, F. J., and Wood, S. (1975). "Past and future of construction equipment—Part I," *J. Constr. Div., Am. Soc. Civ. Eng.*, 101(2), 309–315.

Leffingwell, R. (1997). *Caterpillar dozers and tractors*, Lowe and B. Hould, Ann Arbor, Mich.

American Society of Civil Engineers

Building a Better World

150th Anniversary Paper

Some Historical and Future Aspects of Engineering Mechanics

Ching-Jen Chen[1] and Yousef Haik[2]

Abstract: This paper considers some historical aspects of engineering mechanics with emphasis on the long historical development of mechanics from the evolution of mathematics and tool making before the Common Era to modern genomic and nanoengineering research. To focus, the discussion is divided into five periods of development: (1) engineering mechanics before the Common Era B.C., (2) development between the first and sixteenth centuries; (3) acceleration in the seventeenth and eighteenth centuries; (4) glory of the nineteenth and twentieth centuries; and (5) window in the twenty-first century. The paper also attempts to project a personal view on the future development in the next 50 years in engineering mechanics. This is given in a section of science fiction and fact at the end of the paper.

DOI: 10.1061/(ASCE)0733-9399(2002)128:12(1242)

CE Database keywords: History; Engineering mechanics.

Introduction

During the 2001 Mechanics and Materials Summer Conference in San Diego, Dr. Karl Pister in his general lecture mentioned that he was once interested in a career studying fluid mechanics but was discouraged by a senior faculty member who commented, "fluid mechanics is the study of the Bernoulli equation with 1,000 coefficients." He, therefore, chose a career in solid mechanics. Thus, engineering mechanics in the twentieth century may be viewed to have "hard mechanics" such as solid mechanics, material, structure, earthquake engineering, elasticity, and "soft mechanics" such as fluid mechanics, hydrodynamics, turbulence, thermodynamics, hydraulics, and granular materials.

In nineteenth century and even at the early twentieth century the word "engineering" was not in common use and not a popular word to describe our profession. It was more often that engineering mechanics was considered as a part of physics or applied physics. Indeed, the development of engineering mechanics often takes a path out of phase with the development of sciences. Humans often need engineering applications for survival first before the discovery of underlying scientific laws. Principles of fluid dynamics were not understood before Bernoulli, however, water-

works utilizing the intuition of fluid flows were invented long before humans knew any laws of the mechanics. In this paper, we present a condensed view of the development of engineering mechanics in relation to the whole human development.

In reviewing the development of engineering mechanics, we need to point out that engineering mechanics does not stand alone. It is an integral part of engineering and science development. In discussing some historical aspects, it is convenient to consider the whole period "before Christ (before the Common Era)," or B.C., since the world population was small, approximately one million at 5,000 B.C. and ten million at 1 B.C. The second period is from the first to sixteenth centuries. The population at the end of sixteenth century was approximate 500 million. This was the period of engineering development based on algebra, arithmetic, geometry, and nondifferential calculus. The microscope was not invented yet and humans did not have a microscale view of engineering mechanics. The third period includes just two centuries, the seventeenth and eighteenth centuries. This period differs from the second by the creation of differential calculus, which hastened the development of engineering and science and by the invention of the microscope, which marked the beginning of a microscale view of material and substances. The fourth period is the glory of the nineteenth and twentieth centuries marked by the rapid advance and integration of all sciences and engineering, which is impossible to cover, even briefly, in this paper. The last period is for the twenty-first century, which affords us unlimited creation and fascination. Tables 1–4 summarize the events and notes during the first four periods.

Engineering Mechanics before Common Era B.C.

Tools, Basic Mathematics, and Pi

The word "engineering" is derived from the words engine and ingenious, which are derived from a Latin root, *ingeneare*, which means to create or to contrive. Engineers thus design, create, con-

[1]Dean of Engineering, ASCE Fellow and Professor of Mech. Engineering, College of Engineering, Florida A&M Univ.–Florida State Univ., 2525 Pottsdamer St., Tallahassee, FL 32310-6046. E-mail: cjchen@eng.fsu.edu

[2]Director of Communication and Multimedia Services, Faculty of Mech. Engineering, College of Engineering, Florida A&M Univ.–Florida State Univ., 2525 Pottsdamer St., Tallahassee, FL 32310-6046.

Note. Associate Editor: Stein Sture. Discussion open until May 1, 2003. Separate discussions must be submitted for individual papers. To extend the closing date by one month, a written request must be filed with the ASCE Managing Editor. The manuscript for this paper was submitted for review and possible publication on June 5, 2002; approved on June 5, 2002. This paper is part of the *Journal of Engineering Mechanics*, Vol. 128, No. 12, December 1, 2002. ©ASCE, ISSN 0733-9399/2002/12-1242–1253/$8.00+$.50 per page.

Table 1. Development of Mechanics Before Common Era B.C.

Year (B.C.)	Names	Notes and events	References
6000	World	Population	1 million
5000	Norway	Man on skis, Norway rock carving	Stone Age, tools
4000	Egyptian	Calendar, 360 days a year	Pattern and number
3500	Mesopotamia	Civilization of city, lute, harps	Use of 60 and 10
2900	Egyptian	Sail, siphoning, 365 days a year	Step pyramid
2690	Chinese	Compass, Chinese writing	Bronze Age
2100	Egyptian	Bellows, trumpet, zero	Pi, $\pi = 3(1/8)$
1500	Phoenicians	Alphabet, water clocks	Medicine
1300	Chinese	Astronomy, binomial, decimal	Glass work
1000	Egyptian	Solar and lunar year	Iron Age
572–492	Pythagoras	Sun center, law of string	$A^2 + B^2 = C^2$
478–392	Mo Tzu	Optics of image, kite	Sun spot
400–350	Euclid	Elements	$\pi = 256/81$
384–322	Aristotle	Vacuum impossible	Ether idea
287–212	Archimedes	Idea of hydrostatics, buoyancy	Screw pump
200	Chinese	Use of negative number	$2^{1/2} = 1.414$
100 B.C.	World	Population	10 million

trive, build, and invent devices, machines, and products that are useful to our life and existence. The first engineer known by name and achievement is often considered to be Imhotep of Egypt (James and Thorpe 1994), who designed and built the famous stepped pyramid at Saqqara near Memphis, probably around 2550 B.C. The basic Egyptian technique for transporting heavy stones and giant statues by sliding sledges was also known in Scandinavia around 3,000 B.C. Fig. 1 is a Stone Age rock carving found on a tiny island of Rødøy in the far north of Norway. It illustrates that man was skiing on skis twice his own length and with a single pole. The early man knew how to make tools and understood the mechanics of sliding motion to overcome friction.

In the Stone and Bronze Ages (5,000–2,000 B.C.) the population of the world was just over one million, man's ingenuity mainly dealt with basic tools, construction, forces, overcoming friction, sailing, movement of water, mining, and the sounds of music from drums, lutes, harps, lyres, trumpets, and bells. These ancient engineers were less sophisticated in their knowledge of engineering mechanics. They had not studied electricity, magnetism, or chemistry. However, the mathematics of algebra, geom-

Table 2. Development of Mechanics between First and Sixteenth Centuries

Year (A.D.)	Names	Notes and events	References
1 A.D.	World	Population	30 million
20–62	Hero	Conservation of mass	Turbine
50–119	Tsai Lun	Invent paper	Chinese Grand Canal
78–139	Chang Heng	Seismograph	Chain pump
90–168	C. Ptolemy	Earth center, trigonometry	$\pi = 377/120$
200	Chinese	Abacus	$\pi = 730/232$
130–201	C. Galen	degree of hot and cold	Anatomy
400	Arabs	Mathematics, Arabic writing	$\pi = 3.1416$
430–501	Chinese	$x^3 + 2x^2 + 10x = 20$	Gongs, guitars
700–800	Chinese	Printing, bike, paper money	Water clock
820	Al Khowarizimi	Algebra	Use of number zero
1026	G. d'Arezzo	Music notes and names	Color money
1202–1279	C. S. Chin	Mathematical treatise	High algebra
1214–1292	R. Bacon	Mechanics of flight, balloon	Ban Arabic number
1232	Chinese	Fire arrows; rocket	Binomial expansion
1405	Cheng Ho	Ship building	365.25 days/year
1452–1519	L. daVinci	Hydraulics, parachute	Monaliza
1473–1543	N. Copernicus	Sun center, algebra	Columbus (b 1451)
1548	J. Hasler	Nature degree of man	Temperature
1544–1603	W. Gilbert	Earth as spherical magnet	Violin in use (1555)
1546–1601	T. Brahe	Planet motion data	Astronomy
1564–1642	G. Galileo	Gravity, sun center, sound	Telescope
1571–1630	J. Kepler	Laws of motion	Logarithm
1596–1650	R. Descartes	Theory of vortices	Light
1600	World	Population	500 million

Table 3. Acceleration in Seventeenth and Eighteenth Centuries

Year (A.D.)	Names	Notes and events	References
1650	World	Population	545 million
1578–1657	W. Harvey	Circulation of blood	Logarithms
1581–1626	E. Gunter	Invented slide rule	Wind mill
1623–1662	B. Pascal	Computing machine	Magnetic declination
1627–1691	R. Boyle	Air pressure, chemical element	Turbine
1635–1703	R. Hooke	Hooke law; red spot on Jupiter	Cell, pi to 71 digits
1643–1727	I. Newton	Laws of motion, light spectrum	Barometer
		Jet wagon, Newton temperature	E. Halley (b 1656)
1646–1716	W. Leibniz	Computer; Calculus	G. Fahrenheit (b 1714)
1655–1732	B. Cristofori	Invented piano	
1700–1782	D. Bernoulii	Hydraulics; Ricati equation	J. S. Bach (b 1688)
1706–1790	B. Franklin	Jet ship; bifocal lenses	Pennsylvania rifle
1707–1783	L. Euler	Inviscid equation, mechanics	A. Celsius (b 1742)
1717–1783	J. D'Alembert	Wave equation	J. Haydon (b 1732)
1736–1806	C. Coulomb	Electrical force	F. Herschel (b 1738)
1736–1819	J. Watt	Industrial revolution	Steam engine
1749–1827	P. Laplace	Celestial mechanics	A. Mozart (b 1756)
1768–1830	J. Fourier	Heat equation; Fourier series	Hydrogen, oxygen
1777–1851	H. Oersted	Electrical and magnetic coupling	L. Beethoven (b 1770)
1777–1855	K. Gauss	Planet Ceres	Steamboat
1781–1840	S. Poisson	Potential solution	F. Chopin (b 1810)
1785–1836	L. Navier	Viscous flow equation	J. Braham (b 1833)
1789–1857	A. Cauchy	Equation of string	P.Tchaikovsky (b 1840)
1796–1832	S. Carnot	Thermodynamics	Conservation of mass
1800	World	Population	900 million

etry, arithmetic, and measurements were used in the design and construction of pyramids, ships, tools, houses, chariot wheels, and the charting of the sky. We may think in the modern language that the night sky with fascinating constellations was the only television channel for our early ancestors. Arithmetic numbers based on 10 and 60 were used in astronomy, computing the time of year, and the calendar.

Surprisingly, the concept of pi, π, appeared about 3,000 B.C. and was used in many calculations. However, the symbol π for pi was not used until the eighteenth century (A.D.) (Beckmann 1971). The need for pi perhaps came from the practical need of a constant, pi, that specifies the ratio of the periphery of a circular body, for example a tree or a tool handle, C, over its diameter, D. Pi or C/D was given the value of 3(1/8) or 3.125 by the Babylo-

Table 4. Glory of Nineteenth and Twentieth Centuries

Year (A.D.)	Names	Notes and events	References
1800	World	Population	900 million
1805–1865	W. Hamilton	Reformulate mechanics	Lincoln (b 1809)
1810–1879	W. Froude	Ship modeling	Chopin (b 1810)
1819–1903	G. Stokes	Navier–Stokes equations	Liszt (b 1811)
1821–1894	H. Helmholtz	Conservation of energy	Science of hearing
1824–1887	G. Kirchhoff	Thermal radiation	Siemens (b 1816)
1831–1879	J. Maxwell	Electromagnetism	Strauss (b 1825)
1832–1891	N. Otto	Four stroke engine cycle	Brahms (b 1833)
1833–1896	A. Nobel	Dynamite	P. Tchaikovsky (b 1840)
1842–1912	O. Reynolds	Viscous fluid, turbulence	Dvorak (b 1841)
1844–1906	L. Boltzman	Kinetic theory of gases	E. Grieg (b 1843)
1847–1931	T. Edison	Inventions, light bulb	Pi to 707 places
1858–1947	M. Planck	Quantum theory	Bell telephone (1875)
1867–1912	W. Wright	Kitty Hawk flight	O. Wright (b 1871)
1875–1953	L. Prandtl	Boundary layer theory	S. Rachmaninov (b 1873)
1879–1955	A. Einstein	Relativity, quantum theory	Debussy (b 1862)
1881–1963	V. Karman	Supersonic flow, turbulence	Ford car (1893)
1881–1953	L. Richardson	First weather prediction	Copland (b 1900)
1900	World	Population	1.6 billion
1997	Kanda and Takahashi	World record	Pi to 51 billion digits
2000	World	Population	5.1 billion
2001	Human genes	3 billion genomic letters	~75,000 genes

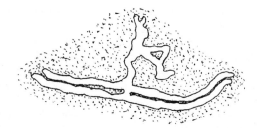

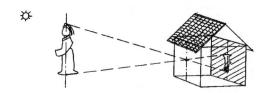

Fig. 2. Principle of simple image by Mo Tzu in 400 B.C.

Fig. 1. Man on skis carved on stone wall in north Norway around 2500 B.C. (James and Thorpe 1994, p. 71, reprinted by permission, all rights reserved)

nians about 2,000 B.C. The value implied from a description of Solomon's Temple (~950 B.C. Hebrew Bible 1 Kings 7:23) edited about 550 B.C. was 30/10 or 3 (Blatner 1997).

The Stone (up to 5,000 B.C.) and Bronze (up to 2,000 B.C.) Ages were also the beginnings of writing. In Asia, the Chinese showed the use of the brush, and Chinese writings were also carved on oracle bone. In other parts of the world, writings in Egypt, Mesopotamia, and the Indus Valley were used to record history, astrology, law, and taxation. One of the major "engineering mechanics" projects during the Stone Age period was the manufacturing of stone tools such as chipped pebbles or flaked stone implements. During the Bronze Age, the engineering mechanic process had expanded to the casting of tools, axes, and pottery, and the construction of various carriers and even boats.

Astronomy, Acoustics, and Optics: Pythagoras and Mo Tzu

With the growth of population, mathematics and technology advanced. This marked the arrival of the Iron Age (1,000 B.C.) and the use of solar and lunar calendars. Writing also was advanced by the introduction of an alphabet by the Phoenicians about 1,500 B.C., which gradually replaced the Egyptian hieroglyphics. Mathematics now extended to include binomial and decimal numbers. Use of the number zero began to appear even though the meaning of zero was not totally understood. Thales of Miletus (born about 624 B.C. in what is now Turkey) correctly predicted a solar eclipse to occur on May 28, 585 B.C. (Calinger 1982; Hellemans and Bunch 1988). Later, Chinese observers reported a supernova in 352 B.C., the earliest known record of such sighting.

Pythagoras of Samos (572–492 B.C.) (Calinger 1982; Ginzburg 1930) was credited with the statement and proof of the theorem that the sum of the squares on the legs of a right triangle equals the squares on the hypotenuse. Also, Pythagoras found that the heaviest hammer ringing out on the anvils by blacksmiths produced the lowest sounds. That led him to further studies on sound from strings. He then showed that the longer the string the deeper the sound and that the note of the octaves lies at fixed points in between. This was the discovery of the Pythagoras basic music law of strings and was one of the early triumphs of systematic mathematical investigation of physical phenomena. The discovery was sufficient to evoke Pythagoras' society to claim that "numbers rule the universe" or "all is number." Pythagoreans speculated that the Earth is a sphere and not as previously believed in the shape of a disk and the Earth revolves around a central fire—the Sun.

In Asia, Mo Tzu (or Mo Ti, 478–392 B.C.) (Dai 1994), a Chinese scientist and philosopher, pioneered the study of optics and proposed the principle of image formation, light rays and

shadow, straight projection of light, and convex and concave reflection. Mo Tzu stated a principle of image formation: that if a person facing the Sun stands in front of a box with a small hole then his inverse image will be seen on the wall inside the box (Dai 1994). Figure 2 illustrates the Mo Tzu principle of photoimaging of a simple camera. In Mo Tzu's writing, he also stated the balance of force and principle of moment. Mo Tzu also spent three years studying the construction and flying of kites (James and Thorpe 1994), perhaps the first known study of kites.

Fundamental Elements and Mathematics

The period from 500 to 600 B.C. marked the rapid advancement of sciences, mathematics, and engineering and their integration. Instead of gods and the supernatural, Thales of Greece (624–546 B.C.) was the first to attempt to propose what the Universe was made of with the idea of elements. Thales considered water as the basic substance and even claimed that the substance of stars is water. Many ideas of basic substances were subsequently proposed by others, such as water, air, fire, earth, rain, and what Aristotle (384–322 B.C.) called "aether" meaning in Greek "blazing." Thales also attempted to explain the invisible power of lodestone that "the magnet has life in it because it moves the iron" (James and Thorpe 1994). The Greek philosopher Leucippus (460–370 B.C.) of Miletus was the first to state a unified idea that all matter was composed of tiny particles so small that nothing smaller was conceivable (Asimov 1989). He called them "atoms," meaning in Greek "indivisible." This is the beginning of the molecular concept.

Around the same time, the Alexandrian mathematician Euclid (400–350 B.C.) authored "the Elements," comprising 13 books of the geometric thought of the Greeks. The accuracy of pi was calculated to 256/81. The Chinese also began to use negative numbers and square roots, giving the square root of 2 as 1.414.

Civil and Hydraulic Engineering: Archimedes

The world population during the time several hundred years before the Common Era had increased to about 10 million, with many slaves available for the heavy labor of construction. This encouraged the Romans to construct buildings, aqueducts, bridges, and roads. There were many engineering mechanical problems. The Romans developed concrete and used it widely in Palestrina (Italy) (Hellemans and Bunch 1988). In China, the first emperor of the Chin dynasty, Shih Huang Ti (259–210 B.C.) burned many books, including many important scientific writings in China, and launched the largest construction project ever carried out, the Great Wall, that outdoes the pyramids. Even astronauts from the Mir space station can see this manmade project (1,400 miles).

Hydraulic engineers during this time were very busy transporting water and applying water flow in many devices. This led to the famous invention of the Archimedean screw by Archimedes (287–212 B.C.), a device for raising water in irrigation systems,

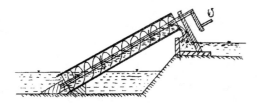

Fig. 3. Archimedes water screw pump invented in 200 B.C.

as shown in Fig. 3. Water clocks and waterwheels for irrigation were commonly in use. Archimedes also found and stated the very important principle of Buoyancy that the buoyant force is equal to the weight of that volume of fluid, which the submerged body displaces (Asimov 1989). Thus, Archimedes is considered to be the creator of statics and hydrostatics. His principle is still found in the elementary books of fluid mechanics and hydraulic engineering today. Archimedes also provided the precise mathematics for the principle of the lever, which was used even in prehistoric times to pry up a heavy rock with a long stick using a smaller rock placed under the stick to give it something to push against. Archimedes estimated the value of pi from geometric constructions to lie between 223/71 and 220/70 (about 3.142) (Asimov 1989). He died during the Roman siege of Syracuse in 212 A.D., stabbed to death by a Roman soldier who resented being told to stand away from a diagram on the sand, which he was studying.

Many engineering devices were in use around the time of the birth of Christ. The Chinese Book of the devil valley master was cited to contain the first clear reference to a lodestone's alignment with Earth's magnetic field; the lodestone device in China is called a "south-pointer" (Hellemans and Bunch 1988). The Chinese also invented cast iron and constructed a 90-mile-long canal from Chang-An to the Yellow River. In Illyria (Yugoslavia and Albania) water-powered mills were used for grinding corn. The Julian calendar began to use three 365-day years followed by one of 366 days. Greek engineers invented the differential gear and constructed lighthouses (Sarton 1993).

In summary, the development of engineering mechanics during the five thousand years B.C. was slow because the population was small. Nevertheless, sciences and engineering had a start. Perhaps since humans were driven by the need for survival so that tools were developed and created. Many wonderful concepts, principles, devices, and machines were invented and proposed, notably, the use of mathematics, use of pi, the calendar, hydraulic engineering, construction, casting, astronomy, and the concept of the atom. Pythagoras, Imhotep, Archimedes, and Mo Tzu were some distinguished engineers in this period. Men gradually were able to expand the use of force by making tools and Pythagoras formulated rational science and mathematics. Imhotep achieved the monumental construction. Mo Tzu understood forces and moments. Archimedes was able to conquer hydrostatics.

Development between First and Sixteenth Centuries

Conservation of Mass, Aqueducts, and Turbines

From the first to sixteenth century, the world population grew from about 10 million to 500 million. Even though the period spans only 1,600 years, the advances in engineering mechanics

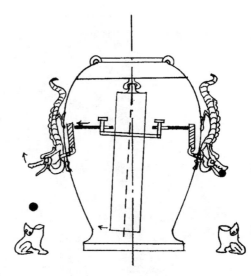

Fig. 4. Earthquake direction detection device in second century

were immense. This period is also marked by great advances in the mathematical fields of geometry, trigonometry, and algebra, and their application to astronomical measurements, celestial mechanics, and engineering design and construction.

In the first century, the population in cities grew. The city of Rome possibly had one million inhabitants and was probably the largest city in the world. Bath houses became very popular for Romans. This led to the construction of many aqueducts for the city. Hero (or Heron) of Alexandria (born in 62–150 A.D.) noted that when water is flowing steadily, the volume of water entering the duct must exit at the end of the duct regardless of the duct size. Even though the statement was restricted to steady flow, it was perhaps the first statement of "The Conservation of Mass Principle." Hero is also known for his study of water fountains, fire engines, siphons, and steam turbines, and the expansion of air caused by heat (Electronic Library 2001). The construction of a coliseum was also made in many cities. In 75 A.D., Emperor Vespasian ordered the building of the coliseum in Rome, which was the largest amphitheater in the world until the construction of the Yale Bowl in 1914 (Hellemans and Bunch 1988).

In 132 A.D., Chang Heng in China invented the world's first seismograph, a device that indicates the direction of an earthquake by dropping a ball from the mouth of a bronze frog (Dai 1994), as shown in Fig. 4. He also constructed a device somewhat like a modern planetarium to keep track of the stars' expected positions in the sky. About the same time, the Chinese also invented the whiffletree, a device that allows two oxen to pull a single cart together.

The Greek astronomer, Ptolemy (Claudius Ptolemaeus) (100–151 A.D.), proposed (Helleman and Bunch 1988) in his writing of "Megale syntaxis tes astronomies," or the astronomical system, that the Earth is the center of the Universe and all the planets revolve around it in combinations of circular motions. Ptolemy worked out mathematical methods for predicting planetary motions, which had a profound influence on human thinking for the next 14 centuries. Ptolemy also wrote Geographia (Geography), which provided an atlas of the known world based on the travels of the Roman legions.

Temperatures and Algebra

Strange as it may seem, the idea of a scale of temperature was familiar to physicians before scientists or engineers had any instruments for measuring it. The great physician, Claudius Galen of Pergamon (129–200) (Lyons and Petrucelli 1987; Knowles Middleton 1966) was the first to introduce, in the second century, the idea of "degrees of heat and cold," four in number each way from a neutral point in the middle. The neutral point was to be a mixture of equal quantities of ice and boiling water (Knowles Middleton 1966). This is the earliest notion of a fixed point or standard of temperature. The engineering use of a temperature scale such as Celsius and Fahrenheit certainly came much later.

Before the turn of the third century, the first writings on alchemy appeared in Egypt (Hellemans and Bunch 1988). In China, Tsai Lun (50–118), a eunuch, was credited with inventing a method for making thin smooth writing paper (Hart 1978), while the archeological evidence in China suggests that paper was in use 250 years earlier, but only for packing and other purposes.

In the fifth to tenth centuries the development of mathematics, particularly algebra, intensified. Historians found that while advances in science and engineering were slow in Europe between 530 and 1000, it flourished in Arabia, India, and China. Uses of the abacus were found in China and India between 200 and 300. About 500, some Indian mathematician who needed to designate an untouched abacus level for the tenth and hundredth used a special symbol for the position number, which we now know as "0." The Arabs picked up the use of zero. Muhammad ibn Al-Khwarizmi (780–850) elucidated this "positional notation" or "0" in his book "Algebra" (810) (Asimov 1989). Pi was calculated to the eighth decimal. The Chinese were also able to solve quadratic and cubic algebraic equations. By the thirteenth century, the use of Arabic numerals was so widespread it was banned in Italy to prevent the decline in using Roman numerals. On the other hand, at the beginning of the fifteenth century, the competition between the abacist who used an abacus for computation

and the algorist who invoked Arabic numerical calculations became quite furious. Eventually, the algorist won acceptance. Fig. 5 shows the scene of a competition in 1503 (Menninger 1969).

By the fifteenth century, many engineering devices and machines were invented such as balloons, fire arrows, rockets, umbrellas, bicycles, gongs, guitars, horseshoes, paddle-wheel boats, and printing machines. The use of hydraulic power became very popular as the prime mover for many machines in grinding, irrigation, wine making, water supply, textiles, and manufacturing. On the other hand, colored paper money began in circulation (Asimov 1989) and the Italian d'Arezzo (995–1050) invented musical notation (Rowley 1977). In the meantime, human knowledge of the world geography advanced with the discovery of America by Christopher Columbus (1451–1506).

Fluid Mechanics and Fluids Engineering

Among many studies of fluid motion in this period, Leonardo da Vinci (1452–1519) provided many details of flow visualization and description of flows. He even attempted to document turbulent flows. As shown in Fig. 6 (Richter 1970) he gave the following description of the characteristics of a drawing he made in 1495: "The clouds scattered and torn. Sand blown up from the seashore. Trees and plants must be bent." In the present day language of turbulence, one may interpret this: "The clouds' motions are irregular and random, and the turbulent eddies are cascading. The sand particles on the seashore are entrained. Trees and plants are bent, subject to large turbulent shearing forces near the ground." Leonardo da Vinci also observed many flow dynamics including river flows and the capillary action of liquids in small-bore tubes. He also designed a horizontal water wheel, which is

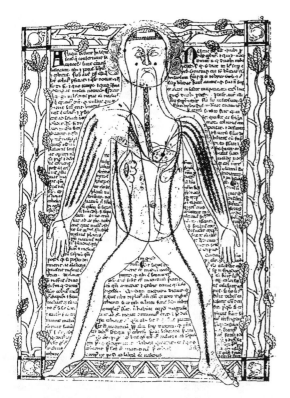

Fig. 7. Anatomy of the circulation system described in thirteenth century (Lyons and Petrucelli 1987, p. 679, reprinted by permission, all rights reserved)

today's turbine. However, even at the end of the sixteenth century, no mathematical model was available to describe flow motions.

In 1327, China's 1100-mile-long Grand Canal was completed, connecting Beijing to the Yangtze River. Water clocks using water as the power source were in popular use in China and Europe. In Holland, windmills were in use in 1408 to carry water from the inland area out to the sea.

Law of Motion and Gravity: Copernicus, Kepler, and Galileo

Population of the world in the fifteenth and sixteenth centuries was over 100 million. During this period, marvelous advances and inventions were made in physics, astronomy, and engineering. Nicholas Copernicus in Polish Prussia (1473–1543) completed the book "On the Revolutions of Celestial Bodies," in 1530, but did not publish it until 1543 for fear of reprisals from the church (Hellemans and Bunch 1988). In it, he offered arguments that the Earth and other planets travel around the Sun. The book astonished the world. The 14 centuries that separate Copernicus from the last great astronomer, Ptolemy, are ones of great scientific achievement in celestial mechanics. However, the book sold slowly and soon went out of print and the second edition was not printed until 1566.

In the meantime, Johannes Kepler of Prague (1571–1630) (Caspar 1993) who assisted Tycho Brahe (1546–1601) and then succeeded him at the Danish observatory published "Astrnomia Nova," (or new astronomy). In the book, he described planets revolving around the Sun in elliptical orbits, which sweep out equal areas in equal time intervals. Kepler introduced the concept of inertia and expounded on the true doctrine of gravity, "Gravity

is a mutual affection between parent bodies which tends to unite them and join them together." Kepler's "Harmonice Mundi" contains his third law of planetary motion that the squares of the periods of revolution of any two planets are proportional to the cubes of their mean distances from the Sun and defended the Copernicus system (Hellemans and Bunch 1988). These observations convinced Galileo Galilei (1564–1642) (Porter 1988; Dugas 1988) to support Copernicus' theory and to publish "Dialogue on the Two Chief World Systems—Ptolemaic and Copernican" in 1632, which raised a storm of controversy with the Church. Kepler's observations and concepts of motion later also contributed to the confirmation of Newton's theory of gravity.

In summary, science and engineering in the period between the first and sixteenth centuries marked the integration of reasoning and experimental verification, notably the observation of planetary motion and the celestial system. Galileo's "De Motu," is a good example of the description of experiments on the dropping of various bodies and recording them accurately with mathematics. Universities were established in many regions to study astronomy, medicine, mathematics, and religion. However, it took 14 centuries to renew the human view of the Universe that the Earth is not its center. Ludolph van Ceulen calculated the value of pi correctly to 20 places in 1597 (Grun 1991). Telescopes and microscopes were invented. Big ships were built and a naval fleet was led by Cheng Ho of China in 1405 to explore the South Pacific Ocean. The compass was used in navigation.

Acceleration in Seventeenth and Eighteenth Centuries

The world population at the beginning of seventeenth century had grown to 600 million. However, the population in the Americas around 1700 was only 300,000. Developments in sciences and technology took place mainly in Europe and elsewhere.

Biomechanics, Computational Machines, and Magnetism

Gunter (1581–1626) invented many measuring instruments, among them the forerunner of the modern slide rule. Gunter's scale is a 2 ft rule with scales of chords, tangents, and logarithmic lines for solving navigational problems. Gunter introduced the words "cosine" and "cotangent" into the language of trigonometry.

In 1628, William Harvey (1578–1657) published a book, "Anatomical treatise on the movement of the heart and blood in animals," (Hellemans and Bunch 1988), revealing for the first time from his experiment that blood circulates in the body. This marked the beginning of biomechanical investigation. Figure 7 (Muir 1994) shows an anatomical sketch of blood vessels before Harvey's discovery. This thirteenth century manuscript of the circulatory system based on Galenic physiology describes the heart as manufacturing a "vital spirit" which resides in the blood and is equivalent to the soul of man. On the contrary, Harvey performed experiments with live animals and was concerned solely with the mechanical flow of blood. He showed that the blood could flow in only one direction both in the heart and in arteries and veins. The heart is clearly a pump and not just an organ that sucked in blood as in Galen's view.

Pascal (1623–1662) (Muir 1994) worked with his father to confirm Torricelli's theory that applied Galileo's laws of motion to fluids, which explained that because of atmospheric pressure water would not rise above 33 ft in a suction pump. This led to the invention of the barometer, the hydraulic press, and the syringe. In 1647 Pascal patented a calculating machine named "Pascaline" to assist in accounting. The machine was a hit since it sold better than one modified by Leibniz (1646–1716).

In 1600, De Magnete written by Gilbert (1544–1603) (Dugas 1988) suggested that Earth is a great spherical magnet and various substances can be used to produce static electricity. Gilbert conjectured that magnetism and electricity produced by rubbing amber were two allied emanations of a single force. This is the beginning of investigation of physical science based on experimentation. Galileo's observation of light reflecting from the moon firmly proved the Copernican system of the Sun's centrality. Observing Jupiter, Galileo (Calinger 1982) also discovered that it has four moons.

Newton Temperature Scale, Newton Wagon, and Franklin Jet Boat

At the beginning of the eighteenth century, precision in scientific measurements was pursued by many scientists. The temperature scale first established by Galen and later by many others was being revised. The great Isaac Newton (1642–1727) (Anon 1701) had a go at thermometry, Newton's temperature scale. Newton chose linseed oil as the thermometric liquid, and for the lower fixed point (0°) "the heat of the air in winter when water begins to freeze." The second fixed point (12°) was the blood heat "the maximum heat that the thermometer can attain by contact with the human body." On Newton's temperature scale, the boiling point of water turns out to be 34°. Newton's temperature scale was never seriously used and was replaced by better thermometers by Daniel Gabriel Fahrenheit (1686–1736) and Anders Celsius (1701–1744) (Lyons and Petrucelli 1987).

Meanwhile, engineers and researchers continued to invent and design new devices. Many were implemented and used. However, many more, even patented, were not used for one reason or another. Following are two not so well known examples because they were not practical and efficient at that time.

Isaac Newton (Encyclopedia Britannica 2000) applied his third law of motion, that for every "action" (force) which one body exerts on another, there is an equal and opposite "reaction" exerted by the second body on the first, to design a Newton wagon, as shown in Fig. 8. The Newton wagon was powered by a steam jet issuing from the spherical boiler to the rear, thus exerting a force on the wagon moving it forward. The steering of the wagon was accomplished by rotating the jet by means of a handle attached to the spherical boiler. Newton's wagon was never commercially successful at his time perhaps due to the fact that the force created by the steam jet is insufficient to move the wagon and that the space needed for the large boiler does not leave room for seating passengers and carrying cargos. However, today's jet engines use exactly the same idea as that of Newton's wagon.

In the mid-eighteenth century, steam driven vessels were built and successfully operated on the Hudson and Mississippi rivers. However, Benjamin Franklin (1706–1790) had no faith in steam operation and suggested a jet-propelled boat. Figure 9 showed a design of Franklin's jet propelled boat (Wilson 1954), to be worked by hand pumping water through a tube and ejecting it from the stern at high speed, the reaction of the expelled jet forcing the boat forward. James Rumsey of Virginia demonstrated the jet-propelled craft before George Washington, who was sufficiently impressed to invest money for further work. However, Rumsey went to England and met only discouragement since James Watt (1736–1819) by this time had built an efficient engine

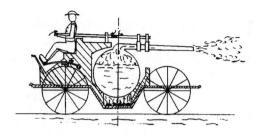

Fig. 8. Newton wagon in seventeenth century

and saw little future in jet boats. Today, jet boats and ships are in popular use with machines and engines replacing hand pumps.

Foundation of Modern Sciences and Engineering

The world population at the beginning of the eighteenth century was approximately 600 million. Progress in the sciences and engineering, and so the quality of life for the ordinary person, had taken a giant step following the late seventeenth century invention of the differential calculus by Newton and Leibniz and studies of air and material properties by Hook, Boyle, and Pascal. Opposite to the telescope used by Galileo in observing planets and stars, a microscope was also invented in 1590 by Dutch spectacle maker, Zacharias Janssen (1580–1638) (Asimov 1989). This marked a new era of microscopic viewing for humans and eventually in 1683 for the Dutch microscopist, Antoni van Leeuwenhoek (1632–1723) the discovery of living organisms in a water pond that he called "animalcules," but we call "micro-organisms" now.

Table 3 lists the notes and events in the eighteenth century. Table 3 shows that by the mid- and late eighteenth century much of the foundation of today's engineering mechanics in fluid mechanics, solid mechanics, and material behavior, including electromagnetic phenomenon, had been laid by D. Bernoulli (1700–1782) for hydraulics; Euler (1707–1783) for mathematics, inviscid flows, and celestial mechanics; D'Alembert (1717–1783) for dynamics and wave motion; Coulomb (1736–1806) for magnetic and electric attraction; Watt (1736–1819) for the steam engine; Laplace (1749–1827) for celestial mechanics and probability; Fourier (1768–1830) for the heat conduction equation and Fourier series; Oersted (1777–1851) for electromagnetic phenomena; Gauss (1777–1855) for magnetism, mechanics, and mathematics; Poisson (1781–1840) for mechanics and materials; Navier (1785–1836) for viscous fluid motion and structure mechanics; Cauchy (1789–1857) for partial differential equations; and Carnot (1796–1832) for the thermal efficiency cycle.

It is also interesting to match contemporary musicians with the cited scientists in Table 3. Therefore, Euler could enjoy the music of Bach. Gauss could enjoy the music of Haydn, Navier to the music of Mozart and Beethoven, while Stokes might listen to the music of Chopin and Liszt, and Carnot to the music of Brahms and Tchaikovsky.

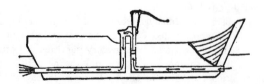

Fig. 9. Franklin jet boat in eighteenth century (Westcott 1857)

In summary, history had never had a period of two hundred years like the seventeenth and particularly the eighteenth century for discovering and creating new knowledge in mathematics, physics, and engineering mechanics. At the end of the eighteenth century pi was calculated in 1794 to 140 decimal places by G. Vega (James and Thorpe 1994).

Glory of Nineteenth and Twentieth Centuries

In the beginning of the nineteenth century, the world population was 900 million. The United States now reached over 5 million and began to advance science, engineering, and technology, and the training of engineers.

Beginning of Engineering Education

Military engineering is the oldest of the engineering skills and was the precursor of the profession of civil engineering. Evidence of the work of the earliest military engineers can be found in the hill forts constructed in Europe during the late Iron Age, and later in the massive fortresses built by the Persians. One epic feat of ancient military engineering was the pontoon bridge built by the engineers of the Persian king Xerxes across the Hellespont (modern Dardanelles) (Encyclopedia Britannica 2002). The military engineers learned the art and practice of designing and building military works and of building and maintaining lines of military transport and communications.

Sebastian le Prestre de Vauban (1633–1707), chief engineer to King Louis XIV, was the person perhaps most responsible for the establishment of military engineering education. He incorporated engineering and the practice of war by establishing the first professionally trained corps of military engineers. Vauban trained engineers according to his own writings and designs, which consisted of elaborations on the bastion with multiple lines of defense and fortifications for small firearms as well as artillery. When Louis XIV's reign ended in 1661, France had a standing army of 300,000, the largest, best trained, and best equipped European fighting force since the Roman legions, thanks to this military engineering education. Vauban's teachings were widely copied and became the authoritative work on military fortification for the eighteenth century.

However, in the second half of the eighteenth century after the invention of the steam engine by James Watt (1736–1819) in 1763, the growth of applied science and the Industrial Revolution was such that there were groups of engineers who were neither military men nor were they concerned with the execution of work exclusively for military purposes. This class of people, then predominantly male, was given the name of civil engineers. In 1815, J. L McAdam (1756–1836) (Asimov 1989) conducted the first paved road around Bristol. This rapidly spread elsewhere to construct "macadamized" highways. It is termed a "highway" because he recommended building the road higher than the fields on either side. By the nineteenth century, the engineering profession had expanded both in Europe and America.

In the early nineteenth century, the engineering curriculum in America by and large was modeled after European systems, namely, English, French, and German. The English mode of engineering is exemplified by the Sheffield Scientific School at Yale, established in 1846, and by the Lawrence Scientific School at Harvard, founded in 1847. The French mode of engineering is modeled after the Ecole Polytechniques founded during the French Revolution primarily for military engineering education and represented by the Rensselaer School in 1825 and the Poly-technic Institute of Brooklyn, established in 1854. The German mode is the Technische Hochschule or Institute of Technology. The Massachusetts Institute of Technology, established in 1861, is of this mode.

In 1862, the Morrill Land Grant Act provided each senator with 30,000 acres of land to establish new land grant colleges such as the Iowa College of Agricultural and Mechanical Arts (now Iowa State Univ.), specifically to teach "such branches of learning as are related to agriculture and mechanic arts." This reflected the technology needed for two major industries, namely, agriculture and machinery, in the United States. Engineers were artists then. The Department of Agriculture was established by President Lincoln in 1862. Lincoln called the department the "people's department" since 90% of the people were farmers. With several land grant acts, eventually each state had one or more land grant colleges. By the late nineteenth century, the three largest engineering concentrations in the U.S. were civil, mechanical, and electrical engineering. In 1848, the American Association for the Advancement of Science was founded. The American Society of Civil Engineers was established in 1852, the American Society of Mechanical Engineers in 1880, and the American Society of Electrical Engineers in 1884.

Engineering Mechanical Marvels

Engineering in the nineteenth century marked the arrival and addition of the application of electromagnetism, chemistry, material science, and atomic physics. M. Faraday (1791–1867) first discovered electromagnetic induction and devised the electrical transformer. Eventually, electricity could be generated cheaply in large quantities. This brought the electrical revolution. For example, light bulbs, electrical elevators, and refrigerators became available. A. G. Bell (1847–1922) (Wilson 1954), born in Scotland, moved to America. He managed to transmit his voice electrically in 1875 and patented the telephone in 1876. T. A. Edison (1847–1931) (Muir 1994), the most prolific inventor with more than 1000 patents in his lifetime, invented the electric vote-recording machine, incandescent light bulb, microphone, kinetoscope, storage battery, gramophone, and talking motion picture. This created the era of consumer industries in electrical appliances, telephone, telegraph, and radio.

On the other hand, steam engines, steamboats, and power generators were in popular use. Beginning in the 1820s, the first railroads in the United States were built. The railroad industry quickly became the largest employer and training ground for American engineers. Engineering work once thought impossible was becoming commonplace. Soon, no project on earth seemed beyond the capability of civil engineers. Among the nineteenth century's most impressive engineering feats was the Brooklyn Bridge, completed in 1883. With successful construction ventures, U.S. engineers were not intimidated by the complexity of designing and building a canal across Panama. The Panama Pacific International celebrated the completion of the canal in 1915. Sewing machines and typewriters were invented and manufactured as consumer products. Large harvesting farm machines, machine guns, and tanks were also designed and used. C. Babbage (1792–1871) invented the modern computer using punched cards.

In fluid mechanical developments, Bell (Wilson 1954) also designed hydrofoils to make ships move faster and kites capable of lifting humans. The screw propeller was designed and built to propel a ship by R. Wilson (1803–1882) (Asimov 1989). Engineers and balloonists were able to make a balloon ascent that reached a height of 4 miles (Asimov 1989) in 1804. The airplane to carry human beings was proposed G. Cayley (1773–1857).

Integration of Sciences and Engineering in Twentieth Century

The twentieth century is the greatest period of achievement of mankind in the integration of sciences and engineering. The National Academy of Engineering and 29 professional engineering societies announced in 2000 a list of the 20 greatest engineering achievements (National Academy of Engineering 2000). They are in the order of achievement:

1. electrification;
2. automobile;
3. airplane;
4. safe and abundant water;
5. electronics;
6. radio and television;
7. agricultural mechanization;
8. computers;
9. telephone;
10. air conditioning and refrigeration;
11. interstate highways;
12. space exploration;
13. internet;
14. imaging technologies;
15. household appliances;
16. health technologies;
17. petroleum and gas technologies;
18. laser and fiber optics;
19. nuclear technologies; and
20. high-performance materials.

Electrification, automobile, highway, air travel, and telephone may be the ultimate symbols of personal freedom. Engineering mechanics is intimately associated with these developments. Electronics, computers, the Internet, radio, and television define the symbol of twentieth century technology. They have transformed businesses, industries, and lives around the world, increased productivity, and opened access to vast amounts of knowledge. Engineering ingenuity will fuel the technology revolution in the twenty-first century.

As a note on music, Maxwell could enjoy the music of Liszt and even Strauss, Reynolds to Brahms and Tchaikovsky. The first attempt to numerically predict weather was made by L. F. Richardson (1881–1953) (Richardson 1965). He failed to predict the weather due to the large grid size and central difference scheme he used on the nonlinear convective term of momentum equations. However, he would be able to enjoy the music of Debussy and Copland. Edison could listen to music of Dvorak and Grieg, and Karman to Rachmaninov. A note on pi, it was calculated to 5.1 billion digits (Blatner 1997) in 1997.

Window in Twenty-First Century

Engineering Mechanics and Education

The engineering achievements of the first half of the twentieth century are marked by the generation of electricity, availability of electrical appliances, pavement of highways, production of internal combustion engines, building of flying machines, invention of photography and movies, and telephones. Television, computers, space exploration, wireless devices, electronics, communication,

transistors, nuclear technologies, prosthetic devices, and the internet distinguished the second half of the twentieth century.

The engineering disciplines in the early twentieth century were mainly classified as civil engineering, mechanical engineering, electrical engineering, and chemical engineering, and in the latter half of the twentieth century the addition of environmental engineering, materials engineering, computer engineering, industrial engineering, and most recently, biomedical engineering. The new engineering disciplines that may be established in the first half of the twenty-first century will be more multidisciplinary in nature such as energy engineering, microsystems engineering and informational engineering.

The newest and most rapidly advancing science is life science. The engineering curriculum in addition to mathematics, physics, and chemistry will definitely need to emphasize biology or the fundamentals of life science. In 2001, scientists were still unsure of how many genes are in the human genome. The long-held notion was that a human has about 100,000 genes. However, the International Human Genome Sequencing Consortium found evidence for 29,691 human transcripts (Launder 2001) and Venter et al. (Venter 2001) of Celera Genomics to be 39,114 genes. The latest estimate (Hollon 2001) gives between 65,000 and 75,000. The 46 human chromosomes (22 pairs of autosomal chromosomes and 2 sex chromosomes) between them house almost 3 billion base pairs of DNA that contain 65,000–75,000 protein-coding genes. This indicated that human genes have only twice as many as the worm or fly model organisms. The dynamics of the uncertainty of human genomes shows the rapid advances of life science research and illustrates that much unknown knowledge remains to be explored. When scientists learn more about genomes and the meaning of each gene, this knowledge will open up a new engineering challenge and new domains for engineering mechanics.

Engineering mechanics in the future will certainly embrace and evolve with the advances of other and new sciences. "Bio," "micro," "nano," and "information" technology mark some characteristics of the new century. Curricula such as nanomechanics, micromechanics, genomechanics, molecular mechanics, and molecular system engineering may very well be offered in engineering mechanics classes in the near future.

Fiction and Fact

Many of the scientific advances and engineering products made today were science fiction to our ancestors and even to our grandparent's generation. To name a few: wireless telephones, jet airplanes, space stations, computers, picture phones, and the internet. So goes the prediction of scientific advances and possible engineering products for the year 2050. The possibility exists that many engineering devices that are fictitious today would become reality in 2050. Thus, science fiction today is likely to become science fact or engineering reality in the future. Here are several fascinations for time capsules to be realized in 2050:

1. disk storage capable of lifetime oral and written history;
2. nanomachines for gene and cell therapy;
3. microchip daily health monitoring systems;
4. memory transfer machines;
5. large scale solar power generation;
6. space station on mars;
7. hydrogen fuel cells to power personal cars;
8. magnetic levitated intelligent vehicles;
9. intelligent and friendly house robots;
10. engineering applications on new fundamental particles;

11. understanding of human genomes;
12. superconductivity materials at high temperature;
13. nanomedical devices;
14. holographic movie and transmission;
15. wireless energy transmission;
16. earthquake prediction and prevention;
17. control of hurricanes;
18. floating cities and airports;
19. space station hotels; and
20. ocean aqua culture and farming.

These engines of tomorrow will create the economical development, improve the quality of our life, and change the culture.

Conclusions

This paper is written at the invitation of the editor, Dr. Stein Store, of the ASCE Journal of Engineering Mechanics, on the 150th anniversary of the ASCE. The suggestion was to write an overview of our personal interest which we deemed important to the profession covering (1) a historical appraisal of how the field evolved; (2) the state of the art of the discipline; and (3) reasonable projections of how the field will evolve in the next 5, 10, or 50 years. Unlike Dr. Pister, the authors had a different and challenging beginning in studying fluid mechanics. Chen was given a NASA project to simulate ablation of the nose cone for a space capsule as his doctoral research. The excitement of achieving an approximate simulation of the high-temperature ablation phenomenon in a low-temperature environment in a laboratory (Chen and Ostrach 1971) provided the author a career interest in fluid mechanics. Haik as his doctoral research studied the magnetic effect on biological fluid mechanics, another different kind of fluid mechanics (Haik et al. 1999). Therefore, the field of engineering mechanics continues to evolve. We are sure that there will be many different developments in twenty-first century engineering mechanics. The authors, in writing this paper, took an overarching view of past, present, and future on engineering mechanics.

In summary, before the common era, engineering mechanics was primarily motivated by the need of human survival and dealt with tool making, construction of habitation, and movement of water, and used arithmetic, algebra, trigonometry, and geometry. From the first to sixteenth century the development of engineering mechanics was without the differential calculus and full comprehension of the laws of motion and gravity. Without the microscope, microscopic views of engineering mechanics were missing. Nevertheless, immense advances were made in celestial mechanics, invention of machines, construction of canals, and irrigation. Also, the integration of common writing with mathematics was achieved by using Arabic numerals based on 10. In the seventeenth and eighteenth centuries, engineering mechanics encompassed military machines, biomechanics of heart circulation, structure mechanics, mechanics of new materials, mechanics of heat and thermodynamics, energy conversion, fluid mechanics, and hydraulics. Entering the nineteenth and twentieth centuries, the field of engineering mechanics had expanded to microelectromechanical systems, transportation, mechatronics, new materials, electronics, semiconductors, aerodynamics, aerospace, and electrodynamics. The twenty-first century is destined to add additional sciences to engineering mechanics. Some examples are nano- and molecular engineering, microsytems engineering, quantum engineering, information engineering, and genomic engineering.

Studying the history of human development in the sciences, engineering, and technologies, one may reach a conclusion that the advances of sciences and engineering stem from human needs for survival before understanding. Humans have abilities in (1) reasoning and logic leading to laws and principles to understand physics, chemistry, and biology; (2) counting and analysis using numbers, arithmetic, trigonometry, algebra, geometry, differential calculus, and digital computation; (3) observation and learning from experience of wars, living in the Earth environment, observation of sky and astronomy, and use of materials; and (4) engineering education to pass on accumulated knowledge. It is impossible to imagine what the field of engineering mechanics will be in another 150 years of the ASCE.

References

Anonymous. (1701). "Scala graduum caloris." *Philos. Trans., London,* 22, 824–829.

Asimov, I. (1989). *Asimov's chronology of science and discovery*, Harper and Row, New York.

Beckmann, P. (1971). *A history of π (Pi)*, St. Martin, New York, 12–16.

Blatner, D. (1997). *The joy of π*, Walker, New York, 12–14.

Calinger, R., ed. (1982). *Classics of mathematics*, Moore, Oak Park, Ill., 50–51.

Caspar, M., translator. (1993). *Kepler*, C. D. Hellman, ed., Dover, New York.

Chen, C. J., and Ostrach, S. (1971). "Low temperature simulation of hypersonic melting ablation and the observed waves." *AIAA J.,* 9(6), 1120–1125.

Dai, N. T. (1994). *Chinese ancient physics*, Tai-Wan San Wu, Taipei, 78, 83–89 (in Chinese).

Dugas, R. (1988). *A history of mechanics*, J. R. Maddox, translator, Dover, New York.

Electronic Library (2001). *Heron of alexandria*, Encyclopedia.com

Encyclopedia Britannica Intermediate. (2000). "Jet propulsion." ⟨http://www.thaitechnics.com/engine/engine intro.html⟩.

Encyclopedia Britannica. (2002). "Military engineering." ⟨http://www.britannica.com⟩.

Ginzburg, B. (1930). *The adventure of science*, Tudor, New York, 1–23.

Grun, B. (1991). *The timetables of history*, Simon and Schuster, New York.

Haik, Y., Pai, V., and Chen, C. J. (1999). "Biomagnetic fluid dynamics." *Fluid dynamics at interfaces*, W. Shyy and R. Narayanan, eds., Cambridge University Press, New York, 34, 439–452.

Hart, M. H. (1978). *The 100—A ranking of the most influential persons in history*, Citadel, New York.

Hellemans, A. and Bunch, B. (1988). *The timetables of science*, Simon and Schuster, New York, 26–35.

Hollon, T. (2001). "Human genes: How many?" *Scientist,* 15(20), 1 and 14–15.

James, P., and Thorpe, N. (1994). *Ancient inventions*, Ballantine, New York, 71.

Knowles Middleton, W. E. (1966). *A History of the thermometer and its use in meteorology*, Johns Hopkins Press, New York, 3–4 and 57–58.

Launder, E. W., et al. (2001). "Initial sequencing and analysis of the human genomes." *Nature (London),* 409, 860–921.

Lyons, A. S., and Petrucelli, R. J. II (1987). *Medicine—An illustrated history*, Abrams, 619.

Menninger, K. (1969). *Number words and number symbols—A cultural history of numbers*, Dover, New York, 350.

Muir, H. ed. (1994). *Larousse dictionary of scientists*, Larousse, Paris.

National Academy of Engineering (2000). "Greatest engineering achievements of the twentieth century." ⟨http://www.greatachievements.org/greatachievements⟩.

Porter, R., ed., (1988). *Man masters nature—twenty-five centuries of science*, Braziller, New York.

Richardson, L. F. (1965). *Weather prediction by numerical process*, Dover, New York.

Richter, J. P. (1970). *The notebooks of Leonardo da Vinci*, Dover, New York, 1 and 2.

Rowley, G., ed. (1977). *The book of music*, Tiger Book International, London.

Sarton, G. (1993). *Hellenistic science and culture in the last three centuries B.C.* Dover, New York, 117–128.

Venter, J. C., et al. (2001). "The sequence of the human genome." *Science,* 291, 1304–1351.

Westcott, T. (1857). *The life of John Fitch, inventor of the steamboat,* Lippincott, Philadelphia.

Wilson, M. (1954). *American science and invention,* Simon and Schuster, New York, 52.

American Society of Civil Engineers — *150th Anniversary Paper*

Nonlocal Integral Formulations of Plasticity and Damage: Survey of Progress

Zdeněk P. Bažant[1] and Milan Jirásek[2]

Abstract: Modeling of the evolution of distributed damage such as microcracking, void formation, and softening frictional slip necessitates strain-softening constitutive models. The nonlocal continuum concept has emerged as an effective means for regularizing the boundary value problems with strain softening, capturing the size effects and avoiding spurious localization that gives rise to pathological mesh sensitivity in numerical computations. A great variety of nonlocal models have appeared during the last two decades. This paper reviews the progress in the nonlocal models of integral type, and discusses their physical justifications, advantages, and numerical applications.

DOI: 10.1061/(ASCE)0733-9399(2002)128:11(1119)

CE Database keywords: Plasticity; Damage; Cracking; Localization.

Introduction

Historical Beginnings

Most standard constitutive models for the mechanical behavior of solids used in engineering applications fall within the category of *simple nonpolar materials* (Noll 1972), for which the stress at a given point uniquely depends on the current values and possibly also the previous history of deformation and temperature *at that point only*. Deformation is in this context characterized by the deformation gradient or by an appropriate strain tensor, i.e., it is fully determined by the first gradient of the displacement field. Intuitively, it seems to be clear that the history of observable variables (strain and temperature) defines the "excitation" of the material point and that the corresponding "response" in terms of stress and entropy evolution should be a unique functional of the local excitation at that point. However, this intuitive feeling tacitly relies on the assumption that the material can be treated as a continuum at an arbitrarily small scale. Only then the finite body can be decomposed into a set of idealized, infinitesimal material volumes, each of which can be described independently as far as the constitutive behavior is concerned. Of course, this does not mean that the individual material points are completely isolated, but their interaction can take place only on the level of balance equations, through the exchange of mass, momentum, energy, and entropy.

In reality, however, no material is an ideal continuum. Both natural and man-made materials have a complicated internal structure, characterized by microstructural details whose size ranges over many orders of magnitude. Some of these details can be described explicitly by spatial variation of the material properties. But this can never be done simultaneously over the entire range of scales. One reason is that such a model would be prohibitively expensive for practical applications. Another, more fundamental reason is that on a small enough scale, the continuum description per se is no longer adequate and needs to be replaced by a discrete mass-point model (or, ultimately, by interatomic potentials based on quantum mechanics).

Constructing a material model, one must select a certain resolution level below which the microstructural details are not explicitly "visible" to the model and need to be taken into account approximately and indirectly, by an appropriate definition of "effective" material properties. Also, one should specify the characteristic wave length of the imposed deformation fields that can be expected for the given type of geometry and loading. Here, the term "wave length" applies not only to dynamics, where its meaning is clear, but also to statics, where it characterizes to the minimum size of the region into which the strain can localize.

If the characteristic wave length of the deformation field remains above the resolution level of the material model, a conventional continuum description can be adequate. On the other hand, if the deformation field is expected to have important components with wave lengths below the resolution level, the model needs to be enriched so as to capture the real processes more adequately. Instead of refining the explicit resolution level, it is often more effective to use various forms of generalized continuum formulations, dealing with materials that are nonsimple or polar, or both.

Some early attempts can be traced back to the 19th century (Voigt 1887, 1894), but the first effective formulation of this kind was proposed by Cosserat and Cosserat (1909). They considered

[1] Walter P. Murphy Professor of Civil Engineering and Materials Science, Northwestern Univ., Evanston, IL 60208. E-mail: z-bazant@northwestern.edu

[2] Research Engineer, Swiss Federal Institute of Technology (EPFL), 1015 Lausanne, Switzerland. E-mail: Milan.Jirasek@epfl.ch

Note. Associate Editor: Stein Sture. Discussion open until April 1, 2003. Separate discussions must be submitted for individual papers. To extend the closing date by one month, a written request must be filed with the ASCE Managing Editor. The manuscript for this paper was submitted for review and possible publication on June 21, 2002; approved on June 24, 2002. This paper is part of the *Journal of Engineering Mechanics*, Vol. 128, No. 11, November 1, 2002. ©ASCE, ISSN 0733-9399/2002/11-1119–1149/$8.00+$.50 per page.

material particles as objects having not only translational but also rotational degrees of freedom, described by the rotation of a rigid frame consisting of three mutually orthogonal unit vectors. A somewhat simpler concept was used by Oseen (1933) and Ericksen (1960) in their work on liquid crystals—they enriched the kinematic description by the rotation of a single vector, characterizing the orientation of the elongated axis of each crystal. After Günther (1958) had reopened the question of an *oriented continuum* and pointed out its relation to the theory of dislocations, the old idea of the Cosserat brothers inspired a rapid development leading to the couple-stress elasticity (Mindlin and Tiersten 1962; Toupin 1962, 1964; Koiter 1964), theory of elasticity with microstructure (Mindlin 1964), micropolar and micromorphic theories (Eringen and Suhubi 1964; Eringen 1964, 1966a,b), and multipolar theory (Green and Rivlin 1964a; Green 1965). Nonlinear extensions were proposed, e.g., by Lippmann (1969) and Besdo (1974).

All these generalized Cosserat theories characterize the motion of a solid body by additional fields that are independent of the displacement field and provide supplementary information on the small-scale kinematics. For example, the continuum with microstructure (Mindlin 1964) uses for this purpose a second-order tensor field that has the meaning of a "microscopic deformation gradient," in general different from the macroscopic deformation gradient evaluated from the displacement field. If the microscopic deformation gradient is restricted to orthogonal tensors, the Cosserat or micropolar continuum is recovered as a special case. On the other hand, a further generalization of the continuum with microstructure leads to the micromorphic continuum (Eringen 1966b). In fact, the continuum with microstructure can be identified with the so-called micromorphic continuum of grade 1 and degree 1.

The micropolar continuum model can be conceived as a continuum approximation of elastic lattices whose members possess a finite bending stiffness. For instance, an orthotropic micropolar model was developed for large regular elastic frames and applied to buckling of tall buildings (Bažant 1971; Bažant and Christensen 1972a,b).

Another important family of enriched continua retains the displacement field as the only independent kinematic field and improves the resolution by incorporating the gradients of strain (i.e., higher gradients of displacement) into the constitutive equations. Interest in such *higher-grade materials* or *gradient theories* was stimulated by Aero and Kuvshinskii (1960), Grioli (1960), Rajagopal (1960), and Truesdell and Toupin (1960). These pioneers took into account only those components of the strain gradient that correspond to curvatures, i.e., to gradients of rotations. This is equivalent to the Cosserat theory with constrained rotations, in which the rotations of the rigid frame associated with each material particle are not independent but are identified with the rotation tensor resulting from the polar decomposition of the macroscopic deformation gradient. Subsequently, the gradient theory was extended by including the effects of the stretch gradients (Toupin 1962), second strain gradients (Mindlin 1965), and gradients of all orders (Green and Rivlin 1964b). Krumhansl (1965) discussed the need for higher-order displacement gradients in continuum-based approximations of discrete lattices.

The last broad family of enriched continuum models is the family to be reviewed here. It consists of *nonlocal models of the integral type*. As early as 1893, Duhem noted that stress at a point should, in principle, depend on the state of the whole body. Nonlocal approaches were exploited in various branches of physical sciences, e.g., in optimization of slider bearings (Rayleigh 1918),

or in modeling of liquid crystals (Oseen 1933), radiative transfer (Chandrasekhar 1950), and electric wave phenomena in the cortex (Hodgkin 1964). Rogula (1965) proposed a nonlocal form of the constitutive law for elastic materials.

Nonlocal elasticity was subsequently refined by Eringen (1966c), Kröner and Datta (1966), Kröner (1966, 1967), Kunin (1966a,b, 1968), Edelen (1969), Edelen and Laws (1971), Edelen (1971), Eringen et al. (1972), Eringen and Edelen (1972), and others. These early studies, frequently motivated by homogenization of the atomic theory of Bravais lattices, aimed at a better description of phenomena taking place in crystals on a scale comparable to the range of interatomic forces. They showed that nonlocal continuum models can approximate the dispersion of short elastic waves and improve the description of interactions between crystal defects such as vacancies, interstitial atoms, and dislocations.

During the last quarter of a century, it has become clear that neither distributed damage in materials nor transitions to discrete microstructural models can be adequately characterized by local constitutive relations between stress and strain tensors. A great variety of nonlocal models, involving either spatial integrals or gradients of strain or internal variables, have been developed. The present paper attempts to review the main existing models, classify them, and compare their properties. Before we focus our attention on models of the integral type, let us discuss nonlocality in a more general context.

Strong and Weak Nonlocality

In solid mechanics, an integral-type nonlocal material model is a model in which the constitutive law at a point of a continuum involves weighted averages of a state variable (or of a thermodynamic force) over a certain neighborhood of that point. Clearly, nonlocality is tantamount to an abandonment of the principle of local action of the classical continuum mechanics. A gradient-type nonlocal model, while adhering to this principle mathematically, takes the field in the immediate vicinity of the point into account by enriching the local constitutive relations with the first or higher gradients of some state variables or thermodynamic forces. A salient characteristic of both the integral- and gradient-type nonlocal models is the presence of a characteristic length (or material length) in the constitutive relation.

The term "nonlocal" has in the past been used with two senses, one narrow and one broad. In the narrow sense, it refers strictly to the models with an averaging integral. In the broad sense, it refers to all the constitutive models that involve a characteristic length (material length), which also includes the gradient models. This broad sense stems from the realization that some gradient models are derived as approximations to the nonlocal averaging integrals, and that for all the gradient models the gradient, in fact, includes a dependence on the immediate (infinitely close) neighborhood of the point under consideration.

A mathematical definition of nonlocality has been given, e.g., by Rogula (1982). The fundamental equations of any physical theory can be written in the abstract form

$$Au = f \qquad (1)$$

where f = given excitation, u = unknown response, and A = corresponding operator (possibly nonlinear) characterizing the system. Typically, u and f are functions or distributions defined over a certain spatial domain V. Operator A is called *local* if it has the following property:

1. If two functions u and v are identical in an open set O, then their images Au and Av are also identical in O.

Equivalently, one could say that whenever $u(\mathbf{x}) = v(\mathbf{x})$ for all $\mathbf{x}$ in a neighborhood of point $\mathbf{x}_0$, then $Au(\mathbf{x}_0) = Av(\mathbf{x}_0)$. It is easily seen that differential operators satisfy this condition, because the derivatives of an arbitrary order do not change if the differentiated function changes only outside a small neighborhood of the point at which the derivatives are taken. For example, standard one-dimensional elasticity is described by the ordinary differential equation

$$-[E(x)u'(x)]' = f(x) \qquad (2)$$

where E = modulus of elasticity, u = displacement, f = body force, and the prime denotes differentiation with respect to the spatial variable x. It is easily verified that the locality condition is satisfied. Equation (2) combines the strain-displacement equation, $\epsilon(x) = u'(x)$, equilibrium equation, $\sigma'(x) + f(x) = 0$, and the (local) elastic constitutive equation, $\sigma(x) = E(x)\epsilon(x)$, where ϵ is the strain and σ is the stress. In nonlocal elasticity, the constitutive equation has the form

$$\sigma(x) = \int_{-\infty}^{\infty} E(x,\xi)\epsilon(\xi)\ d\xi \qquad (3)$$

where $E(x,\xi)$ = kernel of the elastic integral operator, generalizing the notion of the elastic modulus. The corresponding generalization of Eq. (2) then reads

$$-\left[\int_{-\infty}^{\infty} E(x,\xi)u'(\xi)\ d\xi\right]' = f(x) \qquad (4)$$

Due to the presence of a spatial integral, the locality condition is violated [unless the elastic kernel has the degenerate form $E(x,\xi) = E(x)\delta(x - \xi)$, where δ is the Dirac distribution, in which case the local elasticity is recovered].

According to the foregoing definition, one could say that the local theories are those described by differential equations and nonlocal theories are those described by integrodifferential equations. But this refers to nonlocality in the narrow sense. There is another important aspect, related to the presence or absence of a characteristic length. From the mathematical point of view, the absence of a characteristic length is manifested by the invariance of the fundamental equations with respect to scaling of the spatial coordinates [for a precise definition, see Rogula (1982)]. For example, in standard linear elasticity, fundamental Eq. (2) remains valid if x is replaced by $\tilde{x} = sx$, u is replaced by $\tilde{u} = su$, and f is replaced by $\tilde{f} = f/s$, where s is a positive scaling parameter (the prime is then interpreted as the derivative with respect to $\tilde{x}$). This indicates that the theory does not possess any characteristic length. A local theory invariant with respect to spatial scaling is called *strictly local*, while a local theory not invariant with respect to spatial scaling is called *weakly nonlocal*. Weakly nonlocal theories are typically described by differential equations that contain derivatives of different orders. The coefficients multiplying the terms of different orders have different physical dimensions, and from their ratios it is possible to deduce a characteristic length.

Typical examples of weakly nonlocal theories are the Navier–Bernoulli beam on an elastic (Winkler) foundation, or a Timoshenko beam. In the former case, the characteristic length is proportional to the flexural wave length (of spatial oscillations produced by a concentrated force), which is itself proportional to $\sqrt[4]{EI/c}$ where EI = bending cross-sectional stiffness and c = foundation modulus (elastic constant of the foundation).

A Timoshenko beam can be considered as a specific one-dimensional version of a Cosserat continuum. The characteristic length is dictated by the square root of the ratio between the bending stiffness and the shear stiffness of the cross section. For a fixed shape of the cross section, the characteristic length is proportional to the beam depth. Note that, in this one-dimensional description, one "point" of the generalized continuum corresponds to a cross section of the beam. Therefore, only the span, but not the depth, of the beam is an actual geometric dimension of the model in the physical space. The beam depth is a part of the generalized material model, represented by the moment-curvature relation and by the relation between the shear force and the shear distortion. The presence of a characteristic length means that the solutions for different spans cannot be obtained by simple scaling of a reference solution for a given span. Such a scaling would be possible only if the beam depth was also scaled, but this corresponds to a change of the "material."

The foregoing example clearly shows that the solution of a problem can be governed by the ratio of the physical dimensions of a structure to an intrinsic material length. In the present case, this material length arises from the dimensional reduction and has its origin in the geometrical dimension that is no longer explicitly resolved by the model. In analogy to that, the material lengths that are present in various forms of generalized continuum theories arise from the homogenization procedure and have their origin in the characteristics of the heterogeneous microstructure that are no longer explicitly resolved.

To summarize the suggested classification, continuum models for the mechanical behavior of solids (same as other continuum theories) can be divided into

1. strictly local models, which encompass nonpolar simple materials;
2. weakly nonlocal models, exemplified by polar theories and gradient theories (higher-grade materials); and
3. strongly nonlocal models, such as models of the integral type.

It is worth noting that the recently emerged implicit gradient models (Peerlings et al. 1996; Geers et al. 2001; Engelen et al. 2002) are classified as strongly nonlocal, because they are equivalent to integral-type models with special weight functions used for weighted averaging. In the present survey, we restrict our attention exclusively to strongly nonlocal models of the integral type.

Integral-Type Nonlocal Models

Nonlocal Elasticity

The theories of nonlocal elasticity advanced by Eringen and Edelen in the early 1970s (Edelen et al. 1971; Eringen 1972; Eringen and Edelen 1972) attributed a nonlocal character to many fields, e.g., to the body forces, mass, entropy, or internal energy. They were too complicated to be calibrated and experimentally verified, let alone to be applied to any real problems. Later simplifications finally led to a practical formulation in which only the stress–strain relations are treated as nonlocal, while the equilibrium and kinematic equations and the corresponding boundary conditions retain their standard form (Eringen and Kim 1974; Eringen et al. 1977). The related variational principles have recently been developed by Polizzotto (2001). This formulation will now be briefly presented, in order to introduce the basic concepts and prepare a basis for extensions to the inelastic behavior, especially to nonlocal plasticity.

A linear, small-strain, nonlocal theory of elasticity can be derived from the assumption that the elastic energy of a body V is given by the quadratic functional

$$W = \frac{1}{2} \int_V \int_V \boldsymbol{\epsilon}^T(\mathbf{x}) \mathbf{D}_e(\mathbf{x}, \boldsymbol{\xi}) \boldsymbol{\epsilon}(\boldsymbol{\xi}) \mathrm{d}\mathbf{x} \mathrm{d}\boldsymbol{\xi} \qquad (5)$$

where $\boldsymbol{\epsilon}(\mathbf{x}) =$ strain field and $\mathbf{D}_e(\mathbf{x}, \boldsymbol{\xi}) =$ generalized form of the elastic stiffness. The difference from the standard local theory consists in the fact that, in general, it is impossible to express the global energy as a spatial integral of an energy density that would depend only on the local value of strain. Only if $\mathbf{D}_e(\mathbf{x}, \boldsymbol{\xi}) = \mathbf{D}_e(\mathbf{x}) \delta(\mathbf{x} - \boldsymbol{\xi})$ Eq. (5) reduces to

$$W = \frac{1}{2} \int_V \boldsymbol{\epsilon}^T(\mathbf{x}) \mathbf{D}_e(\mathbf{x}) \boldsymbol{\epsilon}(\mathbf{x}) \mathrm{d}\mathbf{x} = \int_V w[\boldsymbol{\epsilon}(\mathbf{x}), \mathbf{x}] \mathrm{d}\mathbf{x} \qquad (6)$$

where $w(\boldsymbol{\epsilon}, \mathbf{x}) = \frac{1}{2} \boldsymbol{\epsilon}^T \mathbf{D}_e(\mathbf{x}) \boldsymbol{\epsilon}$ is the elastic energy density. Physically, the generalized energy expression (5) is needed if the body cannot be decomposed into infinitely small cells that interact only through tractions on their boundaries. This is the case, for instance, in the presence of long-range interactions among material particles, such as atoms or molecules.

Differentiating Eq. (5) with respect to time, we obtain the rate of change of elastic energy

$$\dot{W} = \frac{1}{2} \int_V \int_V \dot{\boldsymbol{\epsilon}}^T(\mathbf{x}) \mathbf{D}_e(\mathbf{x}, \boldsymbol{\xi}) \boldsymbol{\epsilon}(\boldsymbol{\xi}) \mathrm{d}\mathbf{x} \mathrm{d}\boldsymbol{\xi}$$

$$+ \frac{1}{2} \int_V \int_V \boldsymbol{\epsilon}^T(\mathbf{x}) \mathbf{D}_e(\mathbf{x}, \boldsymbol{\xi}) \dot{\boldsymbol{\epsilon}}(\boldsymbol{\xi}) \mathrm{d}\mathbf{x} \mathrm{d}\boldsymbol{\xi} = \frac{1}{2} \int_V \int_V \boldsymbol{\epsilon}^T(\mathbf{x}) [\mathbf{D}_e(\mathbf{x}, \boldsymbol{\xi})$$

$$+ \mathbf{D}_e^T(\boldsymbol{\xi}, \mathbf{x})] \dot{\boldsymbol{\epsilon}}(\boldsymbol{\xi}) \mathrm{d}\mathbf{x} \mathrm{d}\boldsymbol{\xi} = \int_V \int_V \dot{\boldsymbol{\epsilon}}^T(\mathbf{x}) \mathbf{D}_e^{\mathrm{sym}}(\mathbf{x}, \boldsymbol{\xi}) \boldsymbol{\epsilon}(\boldsymbol{\xi}) \mathrm{d}\mathbf{x} \mathrm{d}\boldsymbol{\xi} \quad (7)$$

where

$$\mathbf{D}_e^{\mathrm{sym}}(\mathbf{x}, \boldsymbol{\xi}) = \frac{1}{2} [\mathbf{D}_e(\mathbf{x}, \boldsymbol{\xi}) + \mathbf{D}_e^T(\boldsymbol{\xi}, \mathbf{x})] \qquad (8)$$

is the symmetric part of the generalized stiffness. Note that the symmetrization is carried out simultaneously with respect to the components of the matrix $\mathbf{D}_e$ as well as to the arguments $\mathbf{x}$ and $\boldsymbol{\xi}$. We will assume that the generalized stiffness is right away defined such that it satisfies the symmetry conditions $\mathbf{D}_e = \mathbf{D}_e^T$ and $\mathbf{D}_e(\mathbf{x}, \boldsymbol{\xi}) = \mathbf{D}_e(\boldsymbol{\xi}, \mathbf{x})$, and we will drop the superscript "sym." The last expression in (7) can be written in the form

$$\dot{W} = \int_V \dot{\boldsymbol{\epsilon}}^T(\mathbf{x}) \boldsymbol{\sigma}(\mathbf{x}) \mathrm{d}\mathbf{x} \qquad (9)$$

where

$$\boldsymbol{\sigma}(\mathbf{x}) = \int_V \mathbf{D}_e(\mathbf{x}, \boldsymbol{\xi}) \boldsymbol{\epsilon}(\boldsymbol{\xi}) \mathrm{d}\boldsymbol{\xi} \qquad (10)$$

Eq. (9) is the standard expression for the internal power delivered by stress $\boldsymbol{\sigma}$ at strain rate $\dot{\boldsymbol{\epsilon}}$. Consequently, $\boldsymbol{\sigma}$ is identified as the stress, and Eq. (10) is the constitutive equation of nonlocal elasticity. Since the internal power expression has the standard form, the principle of virtual power leads to exactly the same equilibrium equations and traction boundary conditions as it does in standard (local) elasticity.

For simplicity, we will restrict our attention to macroscopically homogeneous bodies. It is reasonable to assume that the interaction effects decay with distance between the two points $\mathbf{x}$ and $\boldsymbol{\xi}$ (this is sometimes called the "attenuating neighborhood hypoth-

esis") and, unless there is experimental evidence to the contrary, that all stiffness coefficients decay in the same manner. This motivates the commonly assumed form of the generalized stiffness

$$\mathbf{D}_e(\mathbf{x}, \boldsymbol{\xi}) = \mathbf{D}_e \alpha(\mathbf{x}, \boldsymbol{\xi}) \qquad (11)$$

where $\alpha =$ certain attenuation function. The stress-strain law then reads

$$\boldsymbol{\sigma}(\mathbf{x}) = \int_V \mathbf{D}_e \alpha(\mathbf{x}, \boldsymbol{\xi}) \boldsymbol{\epsilon}(\boldsymbol{\xi}) \mathrm{d}\boldsymbol{\xi} = \mathbf{D}_e \int_V \alpha(\mathbf{x}, \boldsymbol{\xi}) \boldsymbol{\epsilon}(\boldsymbol{\xi}) \mathrm{d}\boldsymbol{\xi} = \mathbf{D}_e \bar{\boldsymbol{\epsilon}}(\mathbf{x})$$

$$(12)$$

where

$$\bar{\boldsymbol{\epsilon}}(\mathbf{x}) = \int_V \alpha(\mathbf{x}, \boldsymbol{\xi}) \boldsymbol{\epsilon}(\boldsymbol{\xi}) \mathrm{d}\boldsymbol{\xi} \qquad (13)$$

is the nonlocal strain.

In an infinite isotropic body, the attenuation function depends only on the distance between points $\mathbf{x}$ and $\boldsymbol{\xi}$, and we can write

$$\alpha(\mathbf{x}, \boldsymbol{\xi}) = \alpha_\infty(\|\mathbf{x} - \boldsymbol{\xi}\|) \qquad (14)$$

To remove ambiguity, α_∞ is scaled so as to satisfy the normalizing condition

$$\int_{V_\infty} \alpha_\infty(\|\boldsymbol{\xi}\|) \mathrm{d}\boldsymbol{\xi} = 1 \qquad (15)$$

where $V_\infty =$ entire (one-, two-, or three-dimensional) Euclidean space in which the problem is formulated. Consequently a uniform "local" strain field $\boldsymbol{\epsilon}(\mathbf{x}) = \boldsymbol{\epsilon}_0$ is transformed into a uniform nonlocal strain field

$$\bar{\boldsymbol{\epsilon}}(\mathbf{x}) = \int_{V_\infty} \alpha(\mathbf{x}, \boldsymbol{\xi}) \boldsymbol{\epsilon}(\boldsymbol{\xi}) \mathrm{d}\boldsymbol{\xi} = \boldsymbol{\epsilon}_0 \int_{V_\infty} \alpha(\mathbf{x}, \boldsymbol{\xi}) \mathrm{d}\boldsymbol{\xi} = \boldsymbol{\epsilon}_0 \qquad (16)$$

and the corresponding stress field is given by $\boldsymbol{\sigma}(\mathbf{x}) = \mathbf{D}_e \boldsymbol{\epsilon}(\mathbf{x}) = \mathbf{D}_e \boldsymbol{\epsilon}_0 = \boldsymbol{\sigma}_0$. This gives to $\mathbf{D}_e$ the physical meaning of elastic stiffness under uniform straining.

The attenuation function, also called the nonlocal weight function or the nonlocal averaging function, is often assumed to have the form of the Gauss distribution function

$$\alpha_\infty(r) = (\ell \sqrt{2\pi})^{-N_{\mathrm{dim}}} \exp\left(-\frac{r^2}{2\ell^2}\right) \qquad (17)$$

where $\ell =$ parameter with the dimension of length and $N_{\mathrm{dim}} =$ number of spatial dimensions. Function (17) has an infinite support, which means that nonlocal interaction takes place between any two points, no matter how far from each other they are. For reasons of computational efficiency, it is more advantageous to use attenuation functions with a finite support, e.g., the polynomial bell-shaped function (Bažant and Ožbolt 1990)

$$\alpha_\infty(r) = c \left\langle 1 - \frac{r^2}{R^2} \right\rangle^2 \qquad (18)$$

where the Macauley brackets $\langle \cdots \rangle$ denote the positive part, defined as $\langle x \rangle = \max(0, x)$. Definition (18) contains, again, a parameter with the dimension of length R, which in this case plays the role of the interaction radius, because $\alpha_\infty(r)$ vanishes for $r \geq R$. The scaling factor c is determined from condition (15) and is equal to $15/(16R)$ in one dimension, $3/(\pi R^2)$ in two dimensions, and $105/(32\pi R^3)$ in three dimensions.

There is no unique way of defining the exact form of the attenuation function in a finite body. In nonlocal elasticity, it is usually assumed that $\alpha(\mathbf{x}, \boldsymbol{\xi})$ is still given by Eq. (14), regardless of the presence of boundaries. The integral $\int_V \alpha(\mathbf{x}, \boldsymbol{\xi}) \mathrm{d}\boldsymbol{\xi}$ taken over the finite body V is then smaller than 1 for **x** close to the boundary ∂V, and uniform straining of the finite body does not generate a uniform stress. For materials with long-range elastic forces, this phenomenon has a clear physical explanation. For instance, consider a regular atomic lattice under constant "strain," i.e., under a uniform relative increase of the interatomic distances. For atoms that are sufficiently far from the boundary, the forces generated by the displacement of the surrounding atoms within the interaction distance cancel out due to symmetry. However, if a part of the neighborhood is cut off by the boundary, some of these forces disappear and equilibrium gets disturbed. In a local continuum, it is sufficient to replace the effect of the missing material by tractions on the boundary. However, in a lattice with long-range interactions, there is a boundary layer of thickness R that "feels" the absence of the material behind the boundary. To restore equilibrium, it is not sufficient to apply traction on the boundary, but additional forces are needed in the entire boundary layer.

Motivations of Nonlocality

The aforementioned initial advances were motivated by deviations from the local constitutive models at small scales, caused solely by microstructural *heterogeneity* on the scale of the characteristic length. Recently, a sophisticated explanation of the need for nonlocal terms in homogenized elastic models of random composites has been given by Drugan and Willis (1996) and Luciano and Willis (2001).

An entirely different motivation of nonlocality—the strain-softening character of distributed damage—came to light during the 1970s. It happened as a result of entering the computer era. Finite-element programs made it suddenly feasible to simulate the distributed cracking observed in failure tests of concrete structures. The need to develop concrete vessels and containments for nuclear reactors led to lavish research funding. To approximate the distributed cracking by a continuum, damage models with *strain softening* had to be introduced into finite-element codes. The first among these models was probably the smeared cracking model of Rashid (1968).

The concept of strain softening violated the basic tenets of continuum mechanics as understood at that time, particularly the conditions of stability of material (Drucker's stability) and well posedness of the boundary value problem. Many theoreticians took the firm position that the concept of strain softening in any form was unsound and dismissed its proponents contemptuously as diletants. The controversy, amusing in retrospect but deadly serious at that time, created passionate polemics at conferences, arguments with reviewers, and fights for money at funding agencies (Bažant 2002a). A proposal for a model with strain softening was sure to be rejected if sent for review to these dogmatic theoreticians. But, eventually, it all had a positive effect—a compelling motivation for nonlocal models of a new kind.

The finite-element simulations of failures with distributed (smeared) cracking demonstrated, and the analysis of stability and bifurcation confirmed, that a local inelastic constitutive law with strain-softening damage inevitably leads to spurious localization of damage into a zone of zero volume (Bažant 1976). This causes the numerical solution to become unobjective with respect to the choice of mesh and, upon the mesh refinement, to converge to a solution with a vanishing energy dissipation during structural failure.

Such physically absurd computational results were linked to two problematic features, pointed out already by Hadamard (1903), discussed by Thomas (1961), and emphasized by Sandler (1984), Read and Hegemier (1984), and others. For a dynamic problem in one spatial dimension, they can be described as follows:

1. A material whose tangential stiffness becomes negative has an imaginary wave speed, and thus cannot propagate waves. (Today we know that strain-softening concrete, in fact, can propagate unloading waves and, due to a rate effect on crack growth, also loading waves of a sufficiently steep front.)
2. The dynamic initial-boundary-value problem then changes its type from hyperbolic to elliptic and becomes ill posed, which means that an infinitely small change in the initial conditions can lead to a finite change in the dynamic solution. This was analytically documented for wave propagation in a strain-softening bar by Bažant and Belytschko (1985); see also Bažant and Cedolin (1991), Sec. 13.1.

In multiple spatial dimensions, some wave speeds can remain real even when the tangential stiffness tensor ceases being positive definite, which means that stress waves can still propagate but not in an arbitrary direction. The initial-boundary-value problem is, again, ill posed even though, in general, it does not become elliptic.

Introduction of a characteristic length into the constitutive model (Bažant 1976; Bažant and Cedolin 1979; Cedolin and Bažant 1980; Pietruszczak and Mróz 1981; Bažant and Oh 1983), and formulation of a nonlocal strain-softening model (Bažant et al. 1984) and its second-gradient approximation (Bažant 1984a), were then shown to prevent the spurious localization of strain-softening damage (i.e., to serve as a localization limiter), to regularize the boundary value problem (i.e., make it well posed), and to ensure numerical convergence to physically meaningful solutions. In relation to nonlinear fracture mechanics, the characteristic length in quasibrittle materials with distributed cracking may be physically interpreted as (or related to) the effective size of the fracture process zone at the tip of a macroscopic crack. With respect to homogenization theory, the characteristic length may be taken as equal to (or related to) the size of the representative volume of the material.

The third, practically most compelling, motivation of nonlocality was the *size effect* (in the present context understood as the dependence of the nominal strength on the structure size). The existence of a nonstatistical size effect was brought to light by fracture experiments on concrete (Walsh 1972, 1976; Bažant and Pfeiffer 1987; Bažant and Planas 1998; Bažant 2002b) and discrete numerical simulations using, e.g., the random particle and lattice models (Bažant et al. 1990; Schlangen and van Mier 1992; Schlangen 1993; Jirásek and Bažant 1995; van Mier 1997). In the absence of a characteristic length, the size effect must have the form of a power law. This is the case, for example, in linear elastic fracture mechanics. The incorporation of a characteristic length is needed to describe a transitional type of size effect, in which the scaling according to one power law at scales much smaller than the characteristic length transits to scaling according to another power law at scales much larger than the characteristic length (Bažant 1984b; Bažant and Planas 1998; Bažant 2002b).

Clearly pronounced size effects were also observed in tests of metals on the millimeter and micrometer scales. The results of experiments with bending of thin beams (metallic films) (Richards 1958; Stolken and Evans 1998), torsion of thin wires (Morrison 1939; Fleck et al. 1994), and microindentation (Nix 1989; Ma and Clarke 1995; Poole et al. 1996) cannot be described by the standard plasticity theory that lacks a characteristic length. Even in the elastic range, size effects on the torsional and bending

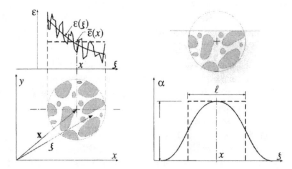

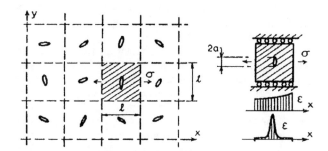

Fig. 2. Idealized crack array and energy release zone of one crack, analyzed in Bažant (1991)

Fig. 1. Left: representative volume of material used for nonlocal strain averaging; the scatter of microstresses and smoothed macrostress profile demonstrate how the stress value at the center of representative volume differs from the stress corresponding to average strain over this volume. Top right: averaging zone when the theoretical nonlocal neighborhood of a point protrudes through the boundary of solid. Bottom right: bell-shaped weight function for nonlocal averaging integral and its relation to characteristic length ℓ.

stiffness contradicting the standard elasticity theory were measured for porous materials such as foams and human bones (Lakes 1986).

To sum up, the development of nonlocal models was historically motivated by
1. the need to capture small-scale deviations from local continuum models caused by material heterogeneity;
2. the need to achieve objective and properly convergent numerical solutions for localized damage;
3. the need to regularize the boundary value problem (prevent ill posedness); and
4. the need to capture size effects observed in experiments and in discrete simulations.

Causes of Nonlocality

The physical causes of nonlocality, some of which have already been mentioned among the historical motivations, may be summarized as follows:
1. Heterogeneity of microstructure and its homogenization on a small scale on which the smoothed strain field cannot be considered as uniform. What matters for the macroscopic stress (averaged over the representative volume) is not the strain value at the center point but the average strain value within the representative volume, which can be very different (Fig. 1, left). The heterogeneity may, for instance, be caused by, and the characteristic length governed by, the grain (inclusion) size, the pore size, or the size of crystals in a metal. In concrete, what is decisive is the maximum aggregate size and spacing.
2. Homogenization of regular or statistically regular lattices or frames, for example, those to be used for large planned space structures or for very tall buildings (this may, of course, be regarded as a special case of heterogeneity).
3. The fact that distributed cracking is physically observed yet impossible to simulate numerically with local continuum models, as already elaborated on.
4. The fact that the growth of a microcrack is not decided by the local stress or strain tensor at the continuum point corresponding to the microcrack center but by the overall energy release from a finite volume surrounding the whole microcrack (Fig. 2). The growth depends on the average deformation of that volume (Bažant 1987, 1991). The size of this volume is determined by the size of the microcracks and their average spacing, which may, but need not, be related to the size of the inhomogeneities in the material.
5. Microcrack interaction, particularly the fact that one microcrack may either amplify the stress intensity factor of another adjacent microcrack or shield that crack, depending on the orientations of the microcracks, the orientation of the vector joining their centers, and the microcrack sizes. This leads to a different kind of nonlocality that is not described explicitly by a spatial averaging integral but implicitly by an integral equation with a kernel of zero average (Bažant 1994; Bažant and Jirásek 1994; Jirásek and Bažant 1994).
6. In the case of metal plasticity, the density of geometrically necessary dislocations in metals, whose effect, after continuum smoothing, naturally leads to a first-gradient model (Fleck and Hutchinson 1993; Gao et al. 1999; Huang et al. 2000; Fleck and Hutchinson 2001).
7. Weibull-type extreme value statistics of quasibrittle failure. As recently realized, without assuming the failure probability at a point of the material to depend on the average strain from a finite neighborhood of the point rather on the continuum stress at that point, a Weibull-type weakest link theory of quasibrittle structural failure runs into some paradoxical situations or incorrect predictions (Bažant and Xi 1991; Bažant and Novák 2000a,b,c), which cannot be avoided without resorting to a nonlocal probabilistic model (see the section on nonlocal probabilistic models of failure).

Nonlocal Averaging Operator

Generally speaking, the nonlocal integral approach consists in replacing a certain variable by its nonlocal counterpart obtained by weighted averaging over a spatial neighborhood of each point under consideration. If $f(\mathbf{x})$ is some "local" field in a solid body occupying a domain V, the corresponding nonlocal field, labeled by an overbar, is defined by

$$\bar{f}(\mathbf{x}) = \int_V \alpha(\mathbf{x},\boldsymbol{\xi}) f(\boldsymbol{\xi}) \, \mathrm{d}\boldsymbol{\xi} \qquad (19)$$

where $\alpha(\mathbf{x},\boldsymbol{\xi}) =$ chosen nonlocal weight function. In applications to softening materials, it is often required that the nonlocal operator should not alter a uniform field, which means that the weight function must satisfy the normalizing condition

$$\int_V \alpha(\mathbf{x},\boldsymbol{\xi}) \, \mathrm{d}\boldsymbol{\xi} = 1 \qquad \forall \mathbf{x} \in V \qquad (20)$$

In an infinite, isotropic, and homogeneous medium, the weight function depends only on the distance $r = \|\mathbf{x} - \boldsymbol{\xi}\|$ between the "source" point, ξ, and the "receiver" point, $\mathbf{x}$. So, we may write $\alpha(\mathbf{x},\xi) = \alpha_\infty(\|\mathbf{x} - \xi\|)$ where $\alpha_\infty(r)$ is typically chosen as a non-negative bell-shaped function (17) or (18), monotonically decreasing for $r \geq 0$ (Fig. 1, bottom right). The smallest distance between points $\mathbf{x}$ and ξ at which the interaction weight $\alpha_\infty(\|\mathbf{x} - \xi\|)$ vanishes (for weight functions with a bounded support) or becomes negligible (for weight functions with an unbounded support) is called the nonlocal interaction radius R. The interval, circle, or sphere of radius R, centered at $\mathbf{x}$, is called the domain of influence of point $\mathbf{x}$.

In the vicinity of the boundary of a finite body, it is simply assumed (without any deep theoretical support) that the averaging is performed only on the part of the domain of influence that lies within the solid (Fig. 1, top right). To satisfy condition (20), the weight function is usually defined as

$$\alpha(\mathbf{x},\xi) = \frac{\alpha_\infty(\|\mathbf{x} - \xi\|)}{\int_V \alpha_\infty(\|\mathbf{x} - \zeta\|)\,d\zeta} \tag{21}$$

However, this modification breaks the symmetry of the weight function with respect to the arguments $\mathbf{x}$ and ξ. In certain types of nonlocal theories it is desirable to work with a symmetric weight function. Polizzotto (2002) and Borino et al. (2002) proposed another modified weight function,

$$\alpha(\mathbf{x},\xi) = \alpha_\infty(\|\mathbf{x} - \xi\|) + \left[1 - \int_V \alpha_\infty(\|\mathbf{x} - \zeta\|)\,d\zeta\right]\delta(\mathbf{x} - \xi) \tag{22}$$

which preserves symmetry and satisfies condition (20).

In computer programs, the nonlocal average at $\mathbf{x}$ is calculated as a weighted sum over the values at all the finite-element integration points ξ lying within the nonlocal interaction radius R. One inevitable penalty of nonlocal averaging is that the bandwidth of the stiffness matrix gets increased. This increases the relative attractiveness of explicit finite-element schemes, which do not necessitate the assembly and decomposition of the stiffness matrix. For nonlocal damage models, it is nevertheless possible to construct the consistent tangent stiffness matrix (Huerta and Pijaudier-Cabot 1994) and use it efficiently in implicit finite-element computations (Jirásek and Patzák 2002).

Original Nonlocal Model for Strain Softening and Its Limitations

In the 1980s, nonlocal models were extended to inelastic materials. Eringen developed his nonlocal formulation of isotropically hardening plasticity in strain space (Eringen 1981), perfect plasticity with associated flow, and deformation theory of plasticity (Eringen 1983). Subsequently, Eringen and Ari (1983) applied these models to simulation of the yielding zone at the fracture front.

The nonlocal averaging concept was also applied to a strain-softening damage model by Bažant et al. (1984) in order to regularize the boundary value problem and prevent spurious dependence of the process zone width and of the energy dissipation on mesh refinement. Initially, the averaging operator was applied to the total strain tensor ϵ. With a uniform weight function, this model could be easily implemented in a finite-element code by imbricating (i.e., overlapping) the finite elements in the manner of roof tiles ("imbrex" in Latin), as shown in Fig. 3. With the condition that the element size be kept constant regardless of mesh refinement, the elements themselves performed the strain averaging. The characteristic length ℓ was then proportional to the fixed element size.

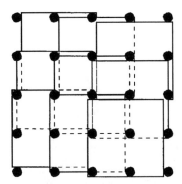

Fig. 3. Imbrication (overlapping) of finite elements, approximately equivalent to averaging of total strain with uniform weight function (Bažant et al. 1984)

However, this "imbricate" concept caused unwanted problems. A nonlocal continuum in which the *total* strain is nonlocal can exhibit zero energy modes of instability (it appears that similar problems could have afflicted also various previous nonlocal elastic models and went undetected, apparently for lack of finite-element implementation). That such instabilities exist becomes clear upon noting that the one-dimensional integral equation $\bar\epsilon(x) \equiv \int_{x-R}^{x+R}\epsilon(s)\,ds = 0$ is satisfied by any function

$$\epsilon(s) = A\sin\frac{\pi n(s + c)}{R} \tag{23}$$

where n = arbitrary positive integer, and A and c = arbitrary real constants. In general, it was shown that such instabilities are avoided if and only if the Fourier transform of the weight function $\alpha_\infty(r)$ is everywhere positive (Bažant and Chang 1984). This is not the case for a uniform weight function (and thus not for the "imbricate" continuum model). Neither is this the case for a triangular weight function or a truncated polynomial function (18).

The Gaussian function (17) has a positive Fourier transform. But, since the tail of its Fourier transform approaches zero, the situation is close to instability. A robust remedy is to add a multiple of the Dirac distribution to the weight function, which is equivalent to an "overlay" (parallel coupling) with an elastic local continuum. But, then, full strain softening down to a zero stress becomes impossible.

So, this original nonlocal model for regularizing strain-softening problems was found usable only for a material that suffers merely partial softening damage followed by rehardening, but not in general situations. Therefore, other remedies were sought, as described next.

Nonlocal Damage and Smeared Cracking

Nonlocal Damage Models

Understanding the source of the aforementioned problems suggested the remedy—the instability modes obviously cannot arise if the nonlocal averaging is applied to variables that can never decrease. Such variables, for example, include the damage variable Ω or the maximum level of damage energy release rate Y_{max} in continuum damage mechanics, or the cumulative plastic strain κ in strain-softening plasticity.

This idea proved successful. It was first applied to continuum damage mechanics, exemplified by the simple isotropic damage model with one scalar damage variable (e.g., Lemaitre and Chaboche 1990). In its local form, this model can be described by the following set of equations:

$$\boldsymbol{\sigma} = (1 - \Omega)\mathbf{D}_e\boldsymbol{\epsilon} \qquad (24)$$

$$\Omega = \omega(Y_{\max}) \qquad (25)$$

$$Y_{\max}(t) = \max_{\tau \le t} Y(\tau) \qquad (26)$$

$$Y = \tfrac{1}{2}\boldsymbol{\epsilon}^T\mathbf{D}_e\boldsymbol{\epsilon} \qquad (27)$$

In the above, $\boldsymbol{\sigma}$ = column matrix of six stress components, $\boldsymbol{\epsilon}$ = column matrix of six engineering strain components, $\mathbf{D}_e$ = elastic material stiffness matrix, and Ω = damage variable that grows from zero (virgin state) to one (fully damaged state) depending on $Y_{\max}$, which is the maximum value of the damage energy release rate Y ever attained in the previous history of the material up to the current state. Of course, Y is not a rate in the sense of a derivative with respect to time. It equals minus the derivative of the free-energy density $\rho\psi(\boldsymbol{\epsilon},\Omega) = (1 - \Omega)\boldsymbol{\epsilon}^T\mathbf{D}_e\boldsymbol{\epsilon}/2$ with respect to the damage variable Ω, and so it represents the "rate" at which energy would be released during (artificially induced) damage growth at constant strain and temperature.

The monotonically increasing function ω in Eq. (25) controls the evolution of damage and thus affects the shape of the stress–strain curve. It is usually designed such that $\Omega = 0$ as long as $Y_{\max}$ remains below a certain threshold value, Y_0. In view of the aforementioned conclusion about the proper nonlocal approach, this local damage evolution law was adapted to a nonlocal form in either of the two following ways (Pijaudier-Cabot and Bažant 1987; Bažant and Pijaudier-Cabot 1988), which represent averaging of the damage energy release rate or of the damage variable:

$$\Omega = \omega(\bar{Y}_{\max}), \quad \text{or} \quad \Omega = \overline{\omega(Y_{\max})} \qquad (28)$$

The convergence and stability of this formulation was verified in various ways; e.g., by refining the mesh for a uniaxially stretched softening bar. The numerical solutions for a progressively increasing number of finite elements are plotted in Fig. 4. As seen, the profiles of strain increment throughout the softening zone converge as the number of elements along the bar is increased from 12 to 100. Also, the averaging of damage and of energy release rate give, in this case, very similar results.

The latter, though, is not true for large postpeak deformations, as demonstrated by Jirásek (1998b). At very large extensions of the bar, a complete fracture must be simulated, which means that the stress must be reduced to zero. Jirásek (1998b) showed that this is not true for some types of nonlocal averaging, and found that averaging of different variables gives rather different responses. The nonlocal damage formulations that he considered are summarized in Table 1. The central column presents in a compact form the nonlocal stress–strain law for the isotropic damage model whose local version is described by Eqs. (24)–(27). The right column shows a possible generalization to anisotropic damage (top part) or to a completely general inelastic model (bottom part). Beside the symbols already defined, the following notations are used: $\gamma = \Omega/(1 - \Omega)$ = compliance variable, $\mathbf{D}_s$ = damaged (secant) stiffness matrix, $\mathbf{D}_u$ = unloading stiffness matrix, $\boldsymbol{\Omega}$ = damage tensor, $\mathbf{Y}$ = tensor of damage energy release rates work-conjugate to $\boldsymbol{\Omega}$, $\mathbf{C}_e = \mathbf{D}_e^{-1}$ = elastic compliance matrix, $\mathbf{C}_i$ = inelastic compliance matrix, and $\mathbf{s} = \mathbf{D}_e\boldsymbol{\epsilon} - \boldsymbol{\sigma}$ = inelastic stress. An overdot denotes differentiation with respect to time.

The load-displacement diagrams generated in a uniaxial tensile test by different nonlocal damage formulations are shown in Fig.

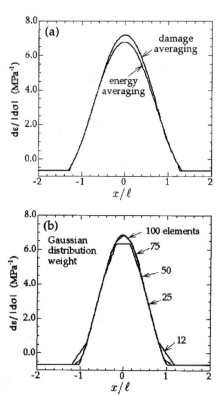

Fig. 4. Profiles of incremental strain per unit stress increment obtained in nonlocal numerical simulation of a tensioned bar: (a) comparison of two types of averaging; (b) convergence as the number of elements is increased (after Bažant and Pijaudier-Cabot 1988).

5. The (local) damage evolution law $\Omega = \omega(Y_{\max})$ is constructed such that the response remains linear up to the peak stress and the softening curve is exponential, asymptotically approaching the horizontal axis. The dashed curve in Fig. 5 corresponds to the unstable solution for which the strain remains uniform. The actual stable solutions are characterized by a nonuniform strain distribution, with strain increments localized into a finite interval, the size of which is controlled by the characteristic length (see Fig. 4). The initial response, right after the onset of localization, is about the same for all the models considered here. At later stages of softening, only the formulations labeled $\omega(\bar{\epsilon})$, $\bar{Y}$, and $\bar{\gamma}$ give a reasonable behavior, with the residual strength approaching zero as the applied elongation is increased [Fig. 5(a)]. The other formulations lead to locking effects and sometimes fail to converge [Fig. 5(b)]. Thus, it can be concluded that the complete fracture is correctly reproduced by models that average the equivalent strain,

Table 1. Overview of Nonlocal Damage Formulations

Formation	Isotropic damage model	General model
$\omega(\bar{\epsilon})$	$\boldsymbol{\sigma} = [1 - \omega(\overline{Y\epsilon}))]\boldsymbol{D}_e\varepsilon$	$\boldsymbol{\sigma} = \boldsymbol{D}_s(\bar{\epsilon})\varepsilon$
$\bar{Y}$	$\boldsymbol{\sigma} = [1 - \omega(\overline{Y\epsilon}))]\boldsymbol{D}_e\varepsilon$	$\boldsymbol{\sigma} = \boldsymbol{D}_s(\Omega(\overline{Y(\epsilon)}))\varepsilon$
$\bar{\epsilon}$	$\boldsymbol{\sigma} = [1 - \omega(Y\epsilon))]\boldsymbol{D}_e\varepsilon$	$\boldsymbol{\sigma} = \overline{\boldsymbol{D}_s(\epsilon)\varepsilon}$
$\bar{\gamma}$	$\boldsymbol{\sigma} = [1 + \overline{\gamma(Y\epsilon))}]^{-1}\boldsymbol{D}_e\varepsilon$	$\boldsymbol{\sigma} = [C_e + \overline{C_e(\epsilon)}]^{-1}\varepsilon$
$\bar{s}$	$\boldsymbol{\sigma} = \boldsymbol{D}_e\varepsilon - \overline{\omega(\epsilon)\boldsymbol{D}_e\varepsilon}$	$\boldsymbol{\sigma} = \boldsymbol{D}_e\varepsilon - \overline{s(\epsilon)}$
$\overline{\Delta s}$	$\dot{\boldsymbol{\sigma}} = (1 - \omega)\boldsymbol{D}_e\dot{\varepsilon} - \bar{\dot{\omega}}\boldsymbol{D}_e\varepsilon$	$\boldsymbol{\sigma} = \boldsymbol{D}_u\dot{\varepsilon} - \overline{\dot{s}(\varepsilon,\dot{\varepsilon})}$
$s(\bar{\epsilon})$	$\boldsymbol{\sigma} = \boldsymbol{D}_e\varepsilon - \omega(\bar{\epsilon})\boldsymbol{D}_e\bar{\varepsilon}$	$\boldsymbol{\sigma} = \boldsymbol{D}_e\varepsilon - s(\bar{\epsilon})$

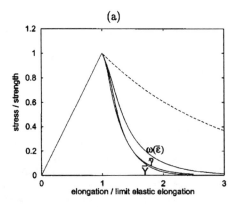
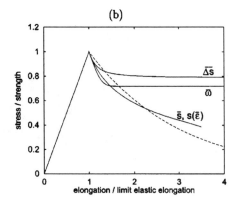

Fig. 5. Comparison of postpeak load-deflection curves of a tensioned bar calculated for nonlocal damage with averaging applied to different variables: (a) nonlocking formulations $\omega(\bar{\epsilon})$, $\bar{Y}$ and $\bar{\gamma}$; and (b) locking formulations $\bar{\omega}$, $\bar{s}$, $\overline{\Delta s}$, and $s(\bar{\epsilon})$.

the energy release rate, or the compliance variable. The evaluation of the inelastic stress from the nonlocal strain, the same as the averaging of the damage variable, inelastic stress, or inelastic stress increment, leads to spurious residual stresses and to an expansion of the softening zone across the entire bar. Models based on averaging of inelastic stress will be addressed in more detail in the section on nonlocal adaptation of general constitutive models.

The basic model with damage evolution driven by the damage energy release rate (27) is simple and appealing from the theoretical point of view, since it can be formulated within the framework of generalized standard materials (Halphen and Nguyen 1975). However, it is not suitable for quasibrittle materials, because it gives the same response in tension and in compression. To emphasize the effect of tension on the propagation of cracks, Mazars (1984) proposed to link the damage to the so-called equivalent strain, defined as the norm of the positive part of the strain tensor. He developed an isotropic damage model for concrete with two damage parameters, ω_t and ω_c, which correspond to uniaxial tension and uniaxial compression, respectively, and both depend on the maximum previously reached value of the equivalent strain. For a general stress state, the actual damage parameter is interpolated according to the current values of principal stresses. The nonlocal formulation of Mazars' model was refined by Saouridis (1988) and Saouridis and Mazars (1992), following the basic idea of Pijaudier-Cabot and Bažant (1987). The averaged quantity was the equivalent strain, which corresponds to a natural generalization of formulation $\bar{Y}$.

A number of nonlocal damage formulations appeared in the literature during the last decade. For instance, di Prisco and Mazars (1996) improved the nonlocal version of Mazars' model by introducing irreversible strains and an additional internal variable controlling the volumetric expansion of concrete in compression. The model was further refined by Ferrara (1998), who considered the characteristic length as a variable parameter that depends on the current level of damage. Valanis started from his anisotropic damage theory based on the integrity tensor (Valanis 1990) and reformulated it as nonlocal, with the damage rate dependent on the current local damage, strain, and positive part of the nonlocal strain rate (Valanis 1991). Kennedy and Nahan (1996) proposed a nonlocal anisotropic damage model (with fixed axes of material orthotropy) for composites and applied it to the failure analysis of laminated shells (Kennedy and Nahan 1997). Their approach is an anisotropic extension of the formulation $\omega(\bar{\epsilon})$. A different extension of this formulation was used by Comi (2001), who defined

two internal lengths leading to two nonlocal averages of the strain tensor, driving the damage evolution in tension and in compression. Jirásek (1999) started from the microplane damage framework established by Carol and Bažant (1997), which exploits the principle of energy equivalence (Cordebois and Sidoroff 1979). He developed an anisotropic damage model for tensile failure of concrete and regularized it by averaging the compliance parameters related to microplanes of different orientations, which corresponds to an anisotropic extension of formulation $\bar{\gamma}$. Fish and Yu (2001) derived a nonlocal damage model for composites by a multiscale asymptotic analysis of the damage phenomena occurring at the micro-, meso-, and macrolevel. Huerta et al. (1998) developed a mesh-adaptive technique for nonlocal damage models, and Rodríguez-Ferran and Huerta (2000) performed adaptive simulations of concrete failure using a residual-type error estimator (Díez et al. 1998).

Nonlocal Smeared Cracking

The nonlocal concept has further been applied to the smeared cracking models widely used for concrete structures and rocks. Such models transform the cumulative effect of microcrack growth and coalescence into additional inelastic strain due to cracking. Adding the smeared cracking strain to the elastic strain, one obtains the total strain. A secant or tangential compliance tensor of the cracked material, which is orthotropic, can be easily obtained. If the damage remains distributed, the cracking strain corresponds to the average of crack opening divided by the spacing between parallel cracks. But, the model can be used even after localization of damage into large macroscopic cracks. To obtain an objective description, it is necessary either to relate the softening part of the stress–strain law to the numerically resolved width of the fracture process zone (which depends on the finite-element size and tends to zero as the mesh is refined), or to reformulate the model as nonlocal and enforce a mesh-independent width of the process zone by introducing a characteristic length.

The smeared cracking model has two variants: (1) the fixed crack model, in which the crack orientation is fixed when the maximum principal stress first attains the strength limit (Rashid 1968; de Borst 1986); and (2) the rotating crack model (Cope et al. 1980; Gupta and Akbar 1984; Rots 1988), in which the crack orientation is rotated so as to always remain perpendicular to the maximum principal strain direction. The latter concept, which usually gives more realistic results, does not mean that the

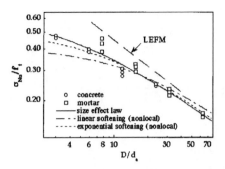

Fig. 6. Size effect obtained with the nonlocal smeared cracking model, for two types of strain-softening law (after Bažant and Lin 1988a)

cracks would actually rotate. Rather, it means that cracks of many orientations exist; cracks of some orientation close and cracks of another orientation open, with the effect that the orientation of the dominant cracks rotates. In the case of the fixed crack model, one needs to cope with the problem of a secondary crack system crossing at some angle the system formed previously (de Borst 1986).

Since smeared cracking models take into account the orientation of cracks, they automatically reflect the crack-induced anisotropy of the material. It is possible to interpret them formally as special anisotropic damage models.

A nonlocal generalization of the rotating crack model, in which the cracking strain is averaged using the nonlocal operator defined in Eq. (19), was investigated by Bažant and Lin (1988a). It was shown that the size effect obtained by finite-element simulations approximately follows Bažant's size effect law (Bažant 1984b). Fig. 6 illustrates the differences in size effect for two different types of the strain-softening law (linear and exponential).

An alternative nonlocal formulation of the rotating crack model can be based on the evaluation of the damaged material stiffness matrix from the nonlocal average of the total strain (Jirásek and Zimmermann 1998b). The stress is then computed as the product of the nonlocally evaluated stiffness with the local strain (this ensures that the response in the elastic range is local).

In terms of the nonlocal damage formulations summarized in Table 1, the Bažant–Lin approach corresponds to formulation $\bar{\gamma}$ while the Jirásek–Zimmermann approach corresponds to formulation $\omega(\bar{\epsilon})$. When applied directly to the standard rotating crack model, the latter formulation leads to instabilities due to negative shear stiffness coefficients. A stable behavior is guaranteed for a modified rotating crack model with transition to scalar damage (Jirásek and Zimmermann 1998b).

The following example, taken from Jirásek and Zimmermann (1997), demonstrates the ability of the nonlocal rotating crack model to reproduce a relatively complex curved fracture pattern. In a series of experiments, Nooru-Mohamed (1992) tested the double-edge-notched (DEN) specimen shown in Fig. 7(a), which can be subjected to a combination of shear and tension (or compression). Nooru-Mohamed performed the experiments for a number of loading paths, some of them even nonproportional. One of the most interesting loading scenarios is path 4c, which produces curved macroscopic cracks [Fig. 7(b)]. The specimen is first loaded by an increasing "shear" force, P_s, while keeping the "normal" force, P, at zero. After reaching the peak of the $P_s - \delta_s$ curve, the type of loading changes. From that moment on, P_s is kept constant and the normal displacement δ is increased, which results into a nonzero reaction P.

The aforementioned experiment was simulated using the nonlocal rotating crack model with transition to scalar damage. The meshes were constructed following a pseudoadaptive technique proposed by Jirásek and Zimmermann (1997), starting from a mesh of uniform density. A sequence of three progressively refined meshes along with the corresponding process zones is shown in Fig. 8, in which the light regions indicate high levels of damage. The simulated process zone matches the experimentally observed cracks very well.

Alleviation of Mesh Orientation Bias for Crack Propagation Direction

In connection with smeared cracking, it is appropriate to point out one advantageous side benefit brought about by the nonlocal averaging concept—the alleviation of mesh orientation bias. Fig. 9 shows a beam analyzed with an aligned mesh and with a deliberately rotated mesh. The local cracking models or cohesive models

(a)   (b)

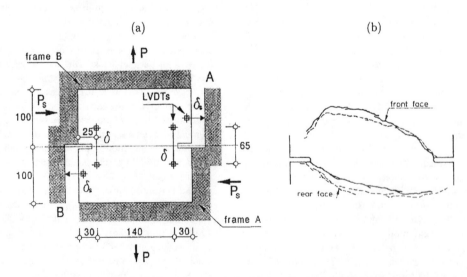

Fig. 7. Double-edge-notched specimen and observed cracks (after Nooru-Mohamed 1992)

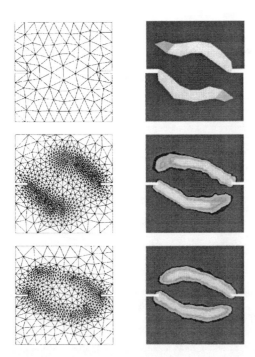

Fig. 8. Simulation of curved crack propagation in a double-edge-notched specimen using a pseudoadaptive technique: Sequence of meshes and fracture process zones (after Jirásek and Zimmermann 1997).

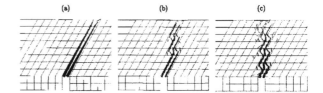

Fig. 10. Simulated crack pattern in the central part of a notched three-point bend specimen: (a) standard rotating crack model (local), (b) rotating crack model with transition to scalar damage (local), and (c) nonlocal rotating crack model with transition to scalar damage (after Jirásek and Zimmermann 1998b).

on a fixed mesh would give an inclined direction of propagation for the rotated mesh. Simply, the cracks prefer to run in the direction of mesh lines. The nonlocal model, on the other hand, gives a nearly vertical propagation of the damage band even for the rotated mesh, provided that the elements are no larger than about $\frac{1}{3}$ of the process zone width (Bažant and Lin 1988a). The $\frac{1}{3}$ rule for avoiding the mesh orientation bias seems to be generally applicable to all the models with nonlocal averaging.

The nonlocal averaging concept seems to be the most effective way to avoid mesh-induced directional bias, provided, of course, that the use of small enough finite elements is feasible. Fig. 10 shows the fracture process zone in a notched three-point bend

specimen simulated with smeared crack models on a skewed mesh. Due to symmetry, the actual crack trajectory should be straight and vertical. Of course, the real crack path is tortuous and, in one single experiment, may deviate from this ideal trajectory. However, the computational simulation is supposed to reproduce the mean trajectory, averaged over a large number of experiments, which is no doubt expected to lie on the axis of symmetry. The local version of the standard rotating crack model exhibits strong directional bias—the crack band propagates along the mesh lines over the entire depth of the specimen [Fig. 10(a)]. It is interesting to note that a partial improvement is achieved already by means of the local version of the modified rotating crack model with transition to scalar damage [Fig. 10(b)]. This can be explained by the fact that, for the standard rotating crack model, the crack trajectory shown in Fig. 10(b) would lead to stress locking (Jirásek and Zimmermann 1998a), which means that it is easier for the macroscopic crack to keep propagating along the mesh lines.

When the modified rotating crack model is reformulated as nonlocal, the results improve further, even for a rather coarse mesh for which the nonlocal interaction radius used in the present example is only 1.25 times the element size. The resulting macroscopic crack trajectory only slightly deviates from the axis of symmetry, despite the skewed character of the mesh [Fig. 10(c)]. The directions of "local cracks" (marked by dark rectangles at individual Gauss points) oscillate and are not aligned with the overall crack trajectory. However, this is quite natural because these local cracks, defined just for the purpose of visualization of

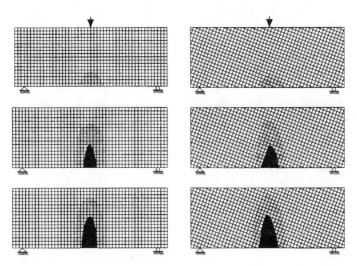

Fig. 9. Study of mesh-orientation bias with nonlocal smeared cracking model of Bažant and Lin (1988a)

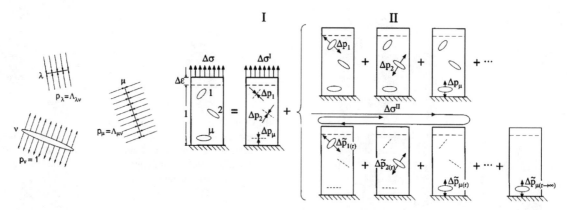

Fig. 11. Microcrack interaction approach: stress on cracks μ and λ caused by unit pressure on the faces of crack ν, and superposition and iteration approach to a system of interacting cracks

the results, are computed from the local inelastic strains. At very late stages of the stiffness degradation process, the local strains localize in a single layer of elements, even though the model is nonlocal. This cannot be judged as a deficiency of the model. What matters is that the load-displacement diagram and the overall direction of the cracking band are reproduced correctly, regardless of the mesh size and orientation. Still better results can be expected for a refined mesh with elements several times smaller than the interaction radius.

Oriented Long-Range Nonlocality Due to Microcrack Interactions

The models discussed so far introduce ad hoc nonlocal averaging, without any physical justification. Bažant (1994) proposed a physical motivation of nonlocal damage, based on a micromechanical analysis of microcrack interactions. He exploited the solution procedure for a statistically homogeneous system of random cracks in an elastic matrix, developed in detail by Kachanov (Kachanov 1985; Kachanov 1987). The solution is a sum of a trivial solution of an elastic solid with all the cracks imagined closed (as if "glued"), and the solution for the case in which the cracks are "unglued" and their faces are loaded by releasing the stresses transmitted across the glued cracks (Fig. 11). Only the second case matters for the stress intensity factors of the cracks. The solution for many cracks can be constructed by an iterative relaxation scheme on the basis of the known solution of the stress transmitted across the plane of an arbitrary closed (glued) crack in

an infinite space containing only one pressurized open crack. This solution is well known, not only for two but also for three dimensions (Fabrikant 1990).

Based on this kind of solution, it was shown (Bažant 1994; see also Bažant and Planas 1998, Sec. 13.3) that the nonlocal incremental constitutive law should have the general form

$$\Delta\boldsymbol{\sigma} = \mathbf{D}_e \Delta\boldsymbol{\epsilon} - \Delta\tilde{\mathbf{s}} \qquad (29)$$

in which the nonlocal inelastic stress increment $\Delta\tilde{\mathbf{s}}$ is the solution of the Fredholm integral equation

$$\Delta\tilde{\mathbf{s}}(\mathbf{x}) = \Delta\tilde{\mathbf{s}}(\mathbf{x}) + \int_V \Lambda(\mathbf{x},\boldsymbol{\xi})\Delta\tilde{\mathbf{s}}(\boldsymbol{\xi})\,d\boldsymbol{\xi} \qquad (30)$$

$$\text{averaging} + \text{crack interactions}$$

The inelastic stress is understood as the stress drop due to the presence of cracks, as compared to the elastic stress that would be induced in an undamaged material by the same strain; see Fig. 12.

Equation (30) is justified by the fact that its approximation by a discrete sum over $\Delta\tilde{\mathbf{s}}$ values at the random crack centers reduces this integral equation to a matrix equation for the solution of the many-crack system in Fig. 11. The first term on the right, which describes the standard nonlocal averaging according to Eq. (19), is justified by material heterogeneity and by the finiteness of the energy release zone of a crack. The second term, which describes microcrack interactions, contains the microcrack interaction function $\Lambda(\mathbf{x},\boldsymbol{\xi})$, which represents statistical continuum smearing of

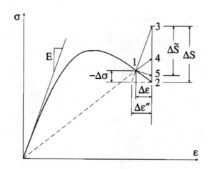

Fig. 12. Stress decomposition into elastic and inelastic parts

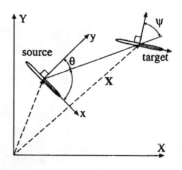

Fig. 13. Parameters describing relative position of two interacting cracks

discrete coefficients $\Lambda_{\mu\nu}$ of the aforementioned matrix equation, each of which is defined as the stress across a glued crack μ produced by a unit pressure applied on the faces of crack ν (Fig. 11). The domain of the integral must be taken as the entire body volume V, because the decay of function Λ with distance is rather slow.

The microcrack interaction function Λ depends not only on the distance $r = \|\mathbf{x} - \boldsymbol{\xi}\|$ between points $\mathbf{x}$ and $\boldsymbol{\xi}$, but also on the orientation of dominant cracks at those points. The dominant crack is assumed to be perpendicular to the maximum principal stress direction. In two dimensions, the interaction factor

$$\Lambda(r,\theta,\psi) \approx -\frac{k(r)}{2\ell^2}[\cos 2\theta + \cos 2\psi + \cos 2(\theta + \psi)] \quad (31)$$

is a function of the distance r, of the angle θ between the radial coordinate axis r and the maximum principal stress direction at the source point $\boldsymbol{\xi}$, and of the angle ψ between the radial coordinate axis r and the maximum principal stress direction at the receiver point $\mathbf{x}$ (Fig. 13). Parameter ℓ is a length parameter reflecting the dominant spacing of interacting microcracks. A similar but more complicated approximate expression was proposed for function Λ in three dimensions (Bažant 1994).

The far-away asymptotic form of radial function $k(r)$ for $r \gg \ell$ was derived from the remote asymptotic field of a single pressurized crack in an infinite space. It turns out that, asymptotically for $r \to \infty$, $k(r)$ is proportional to r^{-d} where d is the number of spatial dimensions (2 or 3). As the simplest match between this asymptotic field and the value 0 required for $r \to 0$, Bažant (1994) proposed to use in two dimensions the expression

$$k(r) = \left(\frac{\kappa \ell r}{r^2 + \ell^2}\right)^2 \quad (32)$$

An interesting point to note is that the asymptotic decay of $k(r)$ with r is very slow, in fact, so slow that the integral in Eq. (30) is at the limit of integrability for the case of an infinite solid. Another point to note is that the integral in Eq. (30) does not represent nonlocal averaging because the weight function Λ does not satisfy the normalizing condition (20). In fact, for a body symmetric with respect to the origin we would get

$$\int_V \Lambda(\mathbf{0}, \boldsymbol{\xi}) d\boldsymbol{\xi} = 0 \quad (33)$$

For angles $|\theta| > 56°$ (in two dimensions), the normal stress component in the direction of the normal to a crack at $\boldsymbol{\xi}$ is tensile and tends to amplify the growth of a crack at $\mathbf{x}$. For $|\theta| \leq 56°$, it tends to shield the other crack from stress and thus inhibit its growth. Thus, a physically important feature of the crack interaction function is that it can distinguish between remote amplification and shielding sectors of a crack, which is ignored in the standard approach.

Eq. (31) was introduced by Ožbolt and Bažant (1996) into a nonlocal finite-element code with the microplane constitutive model for concrete. Eq. (30) was solved at each loading step iteratively, simultaneously with the equilibrium iterations. Ožbolt and Bažant (1996) found that the modified nonlocal averaging based on microcrack interactions makes it easier to reproduce, with the same set of material parameters (including the characteristic length), the results of different types of experiments, especially the failures dominated by tension and those dominated by shear and compression.

Measurement of Material Characteristic Length for Nonlocal Damage

The characteristic length ℓ is a material parameter controlling the spread of the nonlocal weight function. It cannot be directly measured but can be indirectly inferred by inverse analysis of test results. However, a reliable identification procedure can be based only on experiments that are sufficiently sensitive to the nonlocal aspects of the material behavior.

At this point, it is important to note that there is a subtle difference between the notions of an intrinsic material length and a nonlocal characteristic length. The latter is a special case of the former, because nonlocal averaging is only one of the possible enrichments that introduce a length parameter into the continuum theory. Certain other enrichments, e.g., by higher-order gradients or independent microrotation fields, can be classified as weakly nonlocal theories and have similar regularizing effects on the solution. When properly formulated, all the nonlocal theories in the broad sense lead to continuous strain profiles and to process zones (regions of localized strain) of nonzero *width and length*.

On the other hand, there exist special techniques such as the crack band model (Bažant 1982; Bažant and Oh 1983), to be discussed in detail later on, for which the global solution characteristics (such as the load-displacement diagram and the energy dissipated during failure) remain objective with respect to the mesh refinement but the process zone eventually collapses into a surface or curve of vanishing thickness. Since the process zone keeps a finite *length*, such models still carry information about the intrinsic material length and are capable of describing size effects of the transitional type. This material length is, however, not really a *nonlocal* characteristic length. It is present, for instance, in models that work simultaneously with a fracture energy and a material strength (such as the crack band model). A classical example is provided by Irwin's parameter (Irwin 1958),

$$r_p = \frac{K_c^2}{\pi \sigma_0^2} \quad (34)$$

which was introduced to characterize the length of the plastic zone around a crack tip in elastoplastic fracture mechanics [for concrete, it was adopted by Hillerborg et al. (1976)]. Here, K_c is the fracture toughness (critical value of the stress intensity factor) and σ_0 is the yield stress.

The foregoing analysis suggests that the nonlocal models have, in fact, two parameters with the dimension of length, one of them characterizing the length and the other the width of the process zone. They are, in general, independent. Any identification procedure that extracts from the given experimental data only a single length parameter and relates it to the nonlocal characteristic length is in reality based on the tacit assumption that the two length parameters are somehow linked, e.g., that their ratio is fixed. For a limited class of materials, such an assumption can be reasonable, but it can hardly be considered as a general rule. A reliable and fully general identification procedure should be capable of extracting two independent length parameters, but for this it is necessary to have a sufficiently rich set of experimental data (rich in terms of the variety of information, not just in terms of the number of specimens tested).

A typical example is provided by the size effect method. Sometimes it is assumed that the nonlocal characteristic length can be determined by fitting the size effect on the nominal strength of specimens, in other words, by trying to reproduce the dependence of the peak load on the structure size for a set of geometrically similar fracture specimens covering a sufficiently

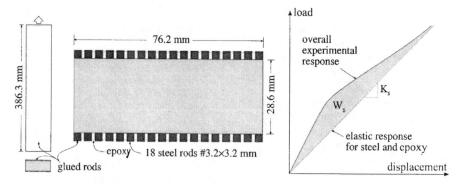

Fig. 14. Left: cross section of concrete specimen restrained by elastic rods; right: load-displacement curves of restrained specimen and area W_s

wide range of sizes (or, more generally, brittleness numbers, in the case of dissimilar specimens). In reality, this procedure can give only one material length parameter, which is related (through a geometry factor determined from fracture mechanics) to the length of the process zone rather than to its width.

A rigorous determination of the nonlocal characteristic length merely from the size effect on the peak loads for one test geometry is, therefore, impossible. Only if one takes into account the fact that both dimensions of the process zone are related to the same microstructural property, e.g., to the maximum aggregate size and spacing in concrete, it is possible to assume that the ratio between the two length parameters is equal to a certain constant and relate the length provided by the size effect method to the nonlocal characteristic length. In practice, this assumption can be hidden in fixing certain other parameters, e.g., those that define the local softening law.

For a gradient damage model, the optimal estimation of the characteristic length of cement mortar reinforced by cellulose fibers was studied in detail by Carmeliet (1999). His conclusions remain valid for nonlocal models of the integral type. Carmeliet used the Markov estimation method, which gives not only the mean values of the material parameters, but also their standard deviations and correlation coefficients, provided that a sufficient number of test results are available. The identification procedure exploited the results of uniaxial tensile tests on a double-notched specimen, and of four-point bending tests on single-notched specimens of different sizes in the range 1:8 and with two different relative notch depths. Young's modulus was determined directly from the tensile tests, and the Markov method aimed at determining the tensile strength, the nonlocal characteristic length, and a parameter that appears in the damage law and controls the area under the local stress–strain curve, i.e., the dissipation density per unit volume in the absence of localization. When the identification was based on the peak loads from all the tests, Carmeliet found quite a high correlation between the strength and the characteristic length (correlation coefficient 0.88). This indicates a strong dependence between these parameters, meaning that the procedure just outlined cannot determine them independently—they are somehow constrained by an implicit hypothesis. Even if the identification procedure is refined by taking into account the entire load-displacement curves (not only the peak loads), the correlation coefficient remains between 0.84 and 0.95, depending on which experiments are taken into account. Correlation coefficient 0.95 is obtained when the parameters are evaluated only from the geometrically similar specimens with the same relative notch size, which is the typical case considered by

the size effect method. This again confirms that the nonlocal characteristic length cannot be uniquely determined from the size effect alone.

For a reliable identification of the nonlocal length parameter, one needs the results of tests that are sensitive to the nonlocal aspects of the material behavior. One possible solution is to combine two kinds of tests: (1) those leading to localized damage patterns with (2) those where the damage remains distributed. The former kind is sufficiently represented by standard fracture tests on notched specimens. The latter kind is represented, for instance, by the elastically restrained tensile test, proposed by Bažant and Pijaudier-Cabot (1989) and later modified by Mazars and Berthaud (1989), and Mazars et al. (1990). In this kind of experiment, tension is applied to a long prismatic specimen of concrete, restrained on the surface by gluing to it many thin longitudinal aluminum rods (Fig. 14). The cross section of the rods must be sufficient to prevent the tangential stiffness of the specimen from ever being negative, despite softening damage in concrete. Furthermore, to permit simple analysis, the concrete thickness must be small enough (only a few aggregate sizes) to force the deformation to be essentially uniform across the thickness. The area of the load-displacement curve lying above the line that corresponds to the stiffness of the elastic rods alone (without the concrete) represents the energy W_s dissipated by strain-softening damage within the specimen. Dividing W_s by the volume V of tensioned concrete in the specimen, one gets an estimate of the dissipation density per unit volume, $g_F = W_s/V$, which is valid in the absence of localization and should correspond to the area under the local stress–strain diagram. Once the parameters of the local stress–strain law are fixed, the nonlocal characteristic length remains the only undefined parameter. It can be determined from the ratio G_F/g_F where G_F is the fracture energy (dissipation per unit area), which can be determined from fracture tests. In fact, it is not even necessary to use the size effect method; the average peak load for a single specimen size would suffice, provided that the specimen is large enough to make the process zone substantially smaller than the whole specimen.

It must, nevertheless, be noted that the aforementioned experimental procedure would be perfect only if the elastically restrained tensile test provided reliable information on the complete stress–strain law, including the value of the tensile strength. However, a refined analysis by Boudon-Cussac et al. (1999) revealed that this technique can have a large error due to partial debonding at the interface between the concrete and the glued rods. If any debonding occurs, the true behavior of concrete is not easy to deduce, because the stress state in concrete ceases to be uniform, even in the central part of the specimen. Instead of a straightforward interpretation of the test results, one would have to resort to inverse analysis. Another spoiling effect, albeit no doubt less serious, is that a surface restraint by homogeneously deforming rods does not exactly replicate the ideal microstructural deformation at a cross section of a large concrete specimen with statistically homogeneous deformation.

An alternative identification procedure can be based on local measurements in or around the process zone. Knowing the details of the local strain distribution, one can evaluate the nonlocal characteristic length by looking for the best match between the experimental and numerical strain profiles. Geers et al. (1996a) applied this idea with success to short glass-fiber-reinforced polypropylene, based on the experimental results reported by Geers et al. (1996b), who measured the displacements at a number of points arranged in a grid on the surface of the compact tension (wedge splitting) specimen. For concrete, Denarié et al. (2001) tried to measure strains inside the process zone using optical fibers with Bragg gratings, but the accuracy of this technique in the presence of cracking and the correct interpretation of the results are not completely clear. Nevertheless, these experiments confirmed that the process zone has a finite width, because the measured strains were way above the elastic limit, even though the fibers were not crossed by the final macroscopic crack. Another interesting point is that the residual strains measured at complete failure were rather high. This confirms the fact, well known from fracture testing of concrete, that the assumption of unloading to the origin, characteristic of pure damage models, is simplistic. Realistic models should incorporate permanent strains, even under tension.

Although detailed information on the precise distribution of strains in the process zone is highly valuable, it is difficult to obtain. For identification of the nonlocal characteristic length, it is nevertheless sufficient to know at least the width of the process zone. An estimate can be obtained by monitoring and evaluating the acoustic emissions produced by cracking, or by x-ray inspection. Such experimental techniques have been developed among others by Mihashi and Nomura (1996), Landis (1999), Otsuka and Date (2000), or Labuz et al. (2001).

Nonlocal Adaptation of General Constitutive Models

The section on nonlocal damage models dealt with various nonlocal formulations of the simple isotropic damage model with one scalar damage variable. It would be very useful to develop a unified nonlocal formulation applicable to any inelastic constitutive model with softening and acting as a reliable localization limiter, insensitive to mesh bias and free of locking effects. However, such a formulation is not available.

The first four formulations from Table 1 rely on the concept of damage and, consequently, they can be generalized only to anisotropic damage models or to the closely related smeared crack models. For other constitutive frameworks, such as plasticity or microplane theory, a straightforward generalization is not available.

The formulations based on the notion of inelastic stress may seem to be ideal candidates for a unifying averaging scheme, because the inelastic stress s can be defined for any type of constitutive model, simply as the difference between the elastically evaluated stress, $\mathbf{D}_e\boldsymbol{\epsilon}$, and the actual stress, $\boldsymbol{\sigma}$; see Fig. 12. In the elastic range, the inelastic stress by definition vanishes, and so the model response remains local. In the present context, this is a desirable feature, because we are interested mainly in strain lo-

calization due to softening and the deviations from locality in the elastic range are not expected to play an important role.

In the inelastic range, the nonlocal constitutive law could be defined as (Bažant et al. 1996)

$$\boldsymbol{\sigma} = \mathbf{D}_e\boldsymbol{\epsilon} - \overline{\mathbf{s}(\boldsymbol{\epsilon})} \qquad (35)$$

where $\overline{\mathbf{s}(\boldsymbol{\epsilon})}$ = nonlocal average of the inelastic stress $\mathbf{s}$ evaluated from the local strain $\boldsymbol{\epsilon}$. Unfortunately, this averaging scheme is suitable only at the early stages of softening but later inevitably leads to spurious locking effects. This has already been documented in Fig. 5(b) for the case when the local constitutive law is based on damage mechanics. In a general case, the reason for locking can be explained as follows.

Under uniaxial tension with uniform strain, the actual stress $\sigma = E\epsilon - s$ is expected to be non-negative. Consequently, in the general case with nonhomogeneous strain, the inelastic stress evaluated from the local stress–strain law cannot exceed the elastic stress, which can be written as $s(x) \leqslant E\epsilon(x)$ for all x. Now, let x_{max} be the spatial coordinate of the point at which the strain attains its maximum value ϵ_{max}. It is easy to show that the nonlocal inelastic stress at x_{max} must satisfy the inequality

$$\bar{s}(x_{max}) = \int_L \alpha(x_{max},\xi)s(\xi)\mathrm{d}\xi \leqslant \int_L \alpha(x_{max},\xi)E\epsilon(\xi)\mathrm{d}\xi$$

$$\leqslant E\epsilon_{max}\int_L \alpha(x_{max},\xi)\mathrm{d}\xi = E\epsilon_{max} \qquad (36)$$

The derivation exploits the basic properties of the nonlocal weight function, namely, the fact that $\alpha(x,\xi)$ is non-negative and normalized according to Eq. (20). From the one-dimensional version of Eq. (35) it is clear that the stress at x_{max} can vanish only if $E\epsilon(x_{max}) = \bar{s}(x_{max})$, i.e., if Eq. (36) holds with equality signs. But, this is possible only if $\epsilon(\xi) = \epsilon_{max}$ for all the points within the domain of influence of point x_{max}, i.e., for all ξ such that $\alpha(x_{max},\xi) > 0$. Note that under uniaxial tension with negligible body forces the stress is uniform along the entire bar. The foregoing analysis means that the stress in the bar can vanish only if the strain distribution is uniform along the entire bar. In other words, the strain distribution cannot remain localized until complete failure. This is true independently of the type of constitutive law, as long as this law does not generate compressive stress under monotonic uniaxial tension.

Consequently, the nonlocal formulation with averaging of the inelastic stress is inherently incapable of describing complete failure, and so it cannot be used as a unifying concept. The same holds for the modifications that define the inelastic stress incrementally or evaluate the inelastic stress from the nonlocal strain.

Nonlocal Plasticity

Associated Plasticity with Isotropic Hardening or Softening

In the standard (local) version of the flow theory of plasticity with isotropic hardening or softening, the yield function, typically, has the form

$$f(\boldsymbol{\sigma},\kappa) = F(\boldsymbol{\sigma}) - \sigma_Y(\kappa) \qquad (37)$$

where $\boldsymbol{\sigma}$ = stress, κ = scalar hardening-softening variable, $F(\boldsymbol{\sigma})$ = equivalent stress (e.g., the von Mises equivalent stress for J_2 plasticity, or the maximum principal stress for Rankine plasticity), and σ_Y = current yield stress. The evolution of the yield stress as a function of the hardening variable is described by the hardening law. This law is, for future use, conveniently written as

$$\sigma_Y(\kappa) = \sigma_0 + h(\kappa) \qquad (38)$$

where σ_0 = initial yield stress and h = hardening function. The derivative $H = \mathrm{d}h/\mathrm{d}\kappa$ is called the plastic modulus. The fundamental equations of associated elastoplasticity include also the elastic stress–strain law,

$$\boldsymbol{\sigma} = \mathbf{D}_e(\boldsymbol{\epsilon} - \boldsymbol{\epsilon}_p) \qquad (39)$$

the flow rule,

$$\dot{\boldsymbol{\epsilon}}_p = \dot{\lambda}\mathbf{f}(\boldsymbol{\sigma}) \qquad (40)$$

and the loading–unloading conditions

$$\dot{\lambda} \geq 0, \quad f(\boldsymbol{\sigma},\kappa) \leq 0, \quad \dot{\lambda}f(\boldsymbol{\sigma},\kappa) = 0 \qquad (41)$$

Here, $\boldsymbol{\epsilon}$ = (total) strain, $\boldsymbol{\epsilon}_p$ = plastic strain, $\mathbf{D}_e$ = elastic material stiffness, $\dot{\lambda}$ = rate of the plastic multiplier, and $\mathbf{f} \equiv \partial f/\partial\boldsymbol{\sigma}$ = gradient of the yield function, defining the direction of plastic flow.

The hardening variable κ reflects the changes in the microstructure induced by plastic flow. Its rate is usually related to the plastic strain rate by the strain-hardening hypothesis

$$\dot{\kappa} = \|\dot{\boldsymbol{\epsilon}}_p\| \qquad (42)$$

or by the work-hardening hypothesis

$$\dot{\kappa} = \frac{\boldsymbol{\sigma}^T\dot{\boldsymbol{\epsilon}}_p}{F(\boldsymbol{\sigma})}. \qquad (43)$$

Substituting the plastic strain rate (40) into Eq. (42) or (43), we can write

$$\dot{\kappa} = k\dot{\lambda} \qquad (44)$$

where $k = \|\mathbf{f}(\boldsymbol{\sigma})\|$ for strain hardening and $k = \boldsymbol{\sigma}^T\mathbf{f}(\boldsymbol{\sigma})/F(\boldsymbol{\sigma})$ for work hardening.

During plastic yielding, the yield function must remain equal to zero, and so the rates of the basic variables must satisfy the consistency condition $\dot{f} = 0$. This makes it possible to derive the well-known expression for the rate of the plastic multiplier,

$$\dot{\lambda} = \frac{\langle \mathbf{f}^T\mathbf{D}_e\dot{\boldsymbol{\epsilon}}\rangle}{\mathbf{f}^T\mathbf{D}_e\mathbf{f} + kH} \qquad (45)$$

where the Macauley brackets $\langle\cdots\rangle$ denote the positive part; for more details, see, e.g., Jirásek and Bažant (2001), Sec. 20.1.

In the presence of softening, characterized by a negative value of the plastic modulus H, the boundary value problem becomes ill posed and must be regularized. This can be achieved by a suitable nonlocal formulation.

Eringen's Nonlocal Plasticity

Historically, the first nonlocal formulations of plasticity were proposed by Eringen in the early 1980s. He set up the framework for three classes of nonlocal plasticity models, based on the deformation theory, flow theory, and strain–space plasticity. Eringen did not consider the case of softening and did not intend his models to serve as localization limiters. Rather, he was interested in the continuum-based description of interacting dislocations and in the nonlocal effects on the distribution of stress around the crack tip in elastoplastic fracture mechanics. Nevertheless, for the sake of comparison it is useful to briefly describe his approach.

Eringen (1981) started from the plasticity theory formulated in the strain space by Green and Naghdi (1965) and Naghdi and Trapp (1975). The yield function is here written in terms of the strain, plastic strain, and hardening variable(s). The original theory was developed for large-strain applications, but for the present purpose we could write the yield condition as

$$g(\boldsymbol{\epsilon},\boldsymbol{\epsilon}_p,\kappa) = 0 \qquad (46)$$

This condition is equivalent to the classical yield condition in the stress space, $f(\boldsymbol{\sigma}) = \sigma_Y(\kappa)$, if the function g is defined as

$$g(\boldsymbol{\epsilon},\boldsymbol{\epsilon}_p,\kappa) = F[\mathbf{D}_e(\boldsymbol{\epsilon} - \boldsymbol{\epsilon}_p)] - \sigma_Y(\kappa) \qquad (47)$$

The evolution laws for the internal variables and the loading–unloading conditions are postulated in the general form

$$\dot{\boldsymbol{\epsilon}}_p = \dot{\lambda}\boldsymbol{\rho}(\boldsymbol{\epsilon},\boldsymbol{\epsilon}_p,\kappa) \qquad (48)$$

$$\dot{\kappa} = \dot{\lambda}k(\boldsymbol{\epsilon},\boldsymbol{\epsilon}_p,\kappa) \qquad (49)$$

$$g(\boldsymbol{\epsilon},\boldsymbol{\epsilon}_p,\kappa) \leq 0, \quad \dot{\lambda} \geq 0, \quad \dot{\lambda}g(\boldsymbol{\epsilon},\boldsymbol{\epsilon}_p,\kappa) = 0 \qquad (50)$$

where $\boldsymbol{\rho}$ and k are given functions. In analogy to the stress–space formulation, the rate of the plastic multiplier

$$\dot{\lambda} = -\frac{\left\langle \left(\dfrac{\partial g}{\partial\boldsymbol{\epsilon}}\right)^T\dot{\boldsymbol{\epsilon}}\right\rangle}{\left(\dfrac{\partial g}{\partial\boldsymbol{\epsilon}_p}\right)^T\boldsymbol{\rho} + \dfrac{\partial g}{\partial\kappa}k} \qquad (51)$$

can be evaluated from the consistency condition $\dot{g} = 0$, valid during the plastic flow.

In the nonlocal plasticity theory of Eringen (1981), the stress is computed by averaging the local stress that would be obtained from the local model. This is equivalent to rewriting the stress–strain law (39) in the form

$$\boldsymbol{\sigma} = \overline{\mathbf{D}_e(\boldsymbol{\epsilon} - \boldsymbol{\epsilon}_p)} \qquad (52)$$

If the elastic stiffness is constant in space, Eq. (52) is equivalent to

$$\boldsymbol{\sigma} = \mathbf{D}_e\overline{(\boldsymbol{\epsilon} - \boldsymbol{\epsilon}_p)} \qquad (53)$$

which is a straightforward extension of the nonlocal elastic law (12). Thus, the nonlocal approach of Eringen can be characterized as averaging of the stress or of the elastic strain. The response of the model is nonlocal already in the elastic range, which is not a desirable feature in applications to problems with strain localization due to softening. Stress averaging does not act as a localization limiter anyway. For instance, in the one-dimensional tensile test, all the solutions obtained with the corresponding local model would remain admissible, since the stress is constant due to the equilibrium condition and the constant field is not modified by the averaging operator.

As a further step, Eringen (1983) formulated nonlocal theories of plasticity in the stress space. In the flow theory he considered only perfect von Mises plasticity with an associated flow rule, but the logic he followed would, in a general case, lead to the stress–strain law

$$\boldsymbol{\sigma} = \overline{\mathbf{D}_e\boldsymbol{\epsilon}} - \mathbf{D}_e\boldsymbol{\epsilon}_p \qquad (54)$$

which is in the case of a spatially constant elastic stiffness equivalent to

$$\boldsymbol{\sigma} = \mathbf{D}_e(\overline{\boldsymbol{\epsilon}} - \boldsymbol{\epsilon}_p) \qquad (55)$$

Thus, one could say that the averaged quantity is the elastic stress or the total strain. Again, the response is nonlocal already in the elastic range, and the formulation cannot be used as a localization limiter, but this time the reason is different. While the previously mentioned model with nonlocal stress would not prevent localization into a set of zero measure, the present model with nonlocal strain would not allow any localization at all (in the one-dimensional test problem). This becomes clear if one realizes that, at the bifurcation from a uniform strain state, the nonlocal strain rate in the elastically unloading region would have to be equal to a negative constant while in the plastically softening region it would have to be equal to a positive constant. However, the non-local strain obtained by applying an integral operator with a continuous weight function is always continuous, even in situations when the local strain has the character of a Dirac distribution. Consequently, the nonlocal strain cannot have a jump at the elastoplastic boundary, and the plastic strain cannot localize into any region smaller than the entire bar.

Nonlocal Plasticity Model of Bažant and Lin

The first nonlocal formulation of *softening* plasticity was proposed by Bažant and Lin (1988b); it was applied in finite-element analysis of the stability of unlined excavation of a subway tunnel in a grouted soil. The underlying cohesive–frictional plasticity model was based on the Mohr–Coulomb yield condition and a linear softening law with the work-softening hypothesis. Bažant and Lin (1988b) proposed to replace the plastic strain in the stress–strain law (39) by its nonlocal average,

$$\overline{\boldsymbol{\epsilon}_p}(\mathbf{x}) = \int_V \alpha(\mathbf{x},\boldsymbol{\xi})\boldsymbol{\epsilon}_p(\boldsymbol{\xi})\,d\boldsymbol{\xi} \qquad (56)$$

where the local plastic strain $\boldsymbol{\epsilon}_p$ is obtained by integrating in time the rate $\dot{\boldsymbol{\epsilon}}_p$ evaluated from the standard expressions (45) and (40). As an alternative, Bažant and Lin (1988b) also proposed to average the rate of the plastic multiplier (45) and then substitute the nonlocal average

$$\overline{\dot{\lambda}}(\mathbf{x}) = \int_V \alpha(\mathbf{x},\boldsymbol{\xi})\dot{\lambda}(\boldsymbol{\xi})\,d\boldsymbol{\xi} \qquad (57)$$

into Eq. (40) instead of $\dot{\lambda}$. They noted that this modification is computationally more efficient (since the averaged quantity is a scalar and not a tensor) and that it leads to very similar numerical results as the formulation with averaging of the plastic strain. Fig. 15(a) shows four successive mesh refinements used in the tunnel excavation analysis, and the (exaggerated) deformation of the mesh caused by the excavation (Bažant and Lin 1988b) is plotted in Fig. 15(b). The contours of the strain-softening zone given in Fig. 15(c) demonstrate negligible differences, which confirm that averaging of the plastic strain or of the plastic multiplier indeed acts as a localization limiter. However, this particular model with

$$\boldsymbol{\sigma} = \mathbf{D}_e(\boldsymbol{\epsilon} - \overline{\boldsymbol{\epsilon}_p}) \qquad (58)$$

turns out to be a special case of the nonlocal formulation with averaging of the inelastic stress, already discussed in the section on nonlocal adaptation of general constitutive models. In local plasticity, the expression for the inelastic stress reads

$$\mathbf{s} = \mathbf{D}_e\boldsymbol{\epsilon} - \boldsymbol{\sigma} = \mathbf{D}_e\boldsymbol{\epsilon} - [\mathbf{D}_e(\boldsymbol{\epsilon} - \boldsymbol{\epsilon}_p)] = \mathbf{D}_e\boldsymbol{\epsilon}_p \qquad (59)$$

and the corresponding nonlocal inelastic stress is

$$\overline{\mathbf{s}} = \overline{\mathbf{D}_e\boldsymbol{\epsilon}_p} = \mathbf{D}_e\overline{\boldsymbol{\epsilon}_p} \qquad (60)$$

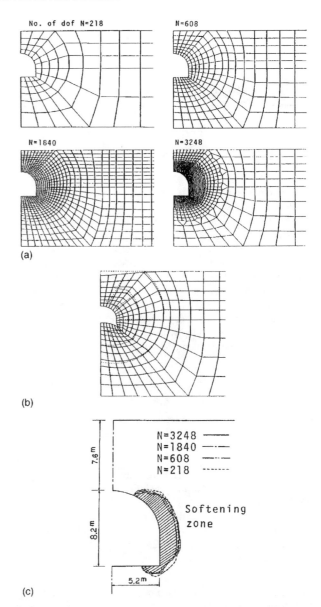

(a)

(b)

(c)

Fig. 15. Nonlocal plastic analysis of subway tunnel excavation in a grouted soil (after Bažant and Lin 1988b): (a) finite-element meshes, (b) deformed mesh with $N = 608$ degrees of freedom, and (c) contours of the strain-softening zone.

provided that the elastic stiffness $\mathbf{D}_e$ is constant in space. Consequently, this nonlocal plasticity model is expected to exhibit locking at later stages of the process of strain softening.

Plasticity with Nonlocal Softening Variable

Perhaps the simplest nonlocal formulation of plasticity with isotropic strain softening is obtained if the current yield stress is computed from the nonlocal average of the softening variable. The softening law (38) is replaced by

$$\sigma_Y = \sigma_0 + h(\bar{\kappa}) \qquad (61)$$

where

$$\bar{\kappa}(\mathbf{x}) = \int_V \alpha(\mathbf{x},\boldsymbol{\xi})\kappa(\boldsymbol{\xi})\,d\boldsymbol{\xi} \qquad (62)$$

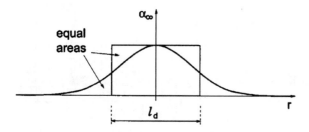

Fig. 16. Geometrical meaning of dissipation length ℓ_d

is the nonlocal softening variable. The flow rule and the evolution law for the local softening variable κ are kept unchanged.

As shown by Planas et al. (1993), this simple nonlocal plasticity model is, in fact, equivalent to a cohesive crack model (Hillerborg et al. 1976). The plastic strain localizes into a set of zero measure but, in contrast to the local model, the energy dissipation does not vanish. The total work G_F spent by the complete failure (material separation) over a unit area of the cohesive crack under uniaxial tension is equal to the product $g_F \ell_d$, where g_F is the area under the local stress–strain curve (work per unit volume in a local continuum) and l_d is the equivalent dissipation length imposed by the nonlocal model. It can be shown that, for the present model,

$$\ell_d = \frac{1}{\alpha(\mathbf{x}_s,\mathbf{x}_s)} \qquad (63)$$

where $\mathbf{x}_s =$ point at which the localized cohesive crack forms. Far from the boundary, $\ell_d = 1/\alpha_\infty(0)$ is a constant proportional to the internal length parameter of the nonlocal weight function. It has the geometrical meaning of the width of a rectangle that has the same area as the nonlocal weight function and the same height at the origin, see Fig. 16. In the proximity of a physical boundary, the dissipation length (63) decreases, which means that the local fracture energy is smaller than in an infinite body. It has been suggested (Hu and Wittmann 2000) that this could partially explain the size effect on the fracture energy observed in experiments with concrete and mortar.

Formulation Combining Nonlocal and Local Softening Variables

The singular character of the strain distribution generated by the simple nonlocal plasticity model from the previous section is an exception. As a rule, nonlocal models lead to solutions character-

ized by a high degree of regularity. This is the case, for instance, for nonlocal damage and smeared crack models. For nonlocal softening plasticity, regularity can be achieved by a modification of the variable driving the yield stress degradation. Vermeer and Brinkgreve (1994) defined the variable $\bar{\kappa}$ that enters the softening law (61) as a linear combination of the local cumulative plastic strain and its nonlocal average:

$$\bar{\kappa}(\mathbf{x}) = (1-m)\kappa(\mathbf{x}) + m\int_V \alpha(\mathbf{x},\xi)\kappa(\xi)\ d\xi \qquad (64)$$

The same definition of the softening variable was also considered by Planas et al. (1996) and by Strömberg and Ristinmaa (1996).

Note that Eq. (62) is obtained as a special case with $m=1$. Another special case with $m=0$ corresponds to the local softening plasticity model. One might expect the typical values of m used by the generalized model to lie between these two cases but, interestingly, the plastic region has a nonzero width only for $m > 1$ (which may be called an overnonlocal formulation).

The localized strain distributions in a one-dimensional tensile test obtained with various values of m are plotted in Fig. 17. Far from the boundary, the distribution of plastic strain inside the plastic region is symmetric [Fig. 17(a)]. If the plastic region is adjacent to the physical boundary of the bar, the largest plastic strain is attained right on the boundary [Fig. 17(b)]. The width of the plastic region is an increasing function of parameter m and tends to zero for $m \to 1^+$. For a given m, the width of the plastic region is proportional to the internal length imposed by the nonlocal averaging function. It is worth noting that the shape of the plastic strain profile does not depend on the softening modulus, and that the width of the plastic region does not change during the softening process (in the uniaxial test).

Since the yield stress in Eq. (61) depends on the nonlocal softening variable, the yield function and the loading–unloading conditions have a nonlocal character, and the stress evaluation cannot be performed at each Gauss integration point separately. This is the price to pay for the enhancement of the standard continuum description, which restores the well posedness of the boundary value problem and leads to numerical solutions that exhibit no pathological sensitivity to finite-element discretization.

Thermodynamics-Based Symmetric Nonlocal Plasticity

One noteworthy aspect of the nonlocal models presented so far has not yet been mentioned—the fact that they yield nonsymmetric tangential stiffness matrices. This point was studied, and the cause explained for the nonlocal damage model (Bažant and

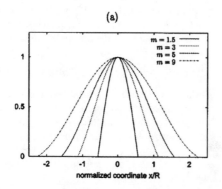

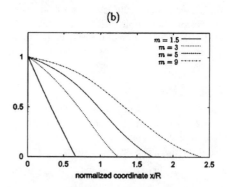

Fig. 17. Plasticity model combining local and nonlocal softening variables: Effect of parameter m on the plastic strain profiles localized (a) inside the bar, and (b) at the boundary.

Pijaudier-Cabot 1988). The structure of the nonlocal stiffness matrix was described by Huerta and Pijaudier-Cabot (1994), and the numerical implementation and convergence rates were studied by Jirásek and Patzák (2002). No convergence problems attributed to nonsymmetry have so far been encountered in computations with the aforementioned models.

Modifications achieving symmetry would nevertheless be highly desirable, not only for theoretical reasons but also for improving computational efficiency of implicit solution schemes (through the use of more efficient equation solvers). A symmetric structure of nonlocal models can be obtained by an appropriate generalization of the postulate of maximum plastic dissipation (understood here in the global sense). The thermodynamic framework established for nonlocal elasticity by Edelen and Laws (1971) and Edelen et al. (1971) served as a basis for thermodynamically consistent formulations of nonlocal softening plasticity (Svedberg 1996; Svedberg and Runesson 1998; Polizzotto et al. 1998).

From the thermodynamic point of view, the elastic stress–strain law and the plastic hardening law are state laws that can be derived from a suitably defined free-energy potential. For models formulated within the framework of generalized standard materials (complying with the postulate of maximum plastic dissipation), the flow rule, the definition of the hardening variable and the loading–unloading conditions are complementary laws that can be derived from the dual dissipation potential defined as the indicator function of the set of plastically admissible states; for a detailed discussion, see, e.g., Jirásek and Bažant (2001, Chap. 23). The postulate of maximum plastic dissipation restricts the class of models covered by this framework to those that satisfy the conditions of convexity (of the yield function) and normality (of the plastic flow and of the evolution of hardening variables). For instance, models with nonassociated flow rules remain outside the class of generalized standard materials, even though they can still be thermodynamically admissible (Bažant and Cedolin 1991). Nevertheless, the postulate of maximum dissipation leads to a symmetric structure of the constitutive equations, and it is interesting to explore its possible extensions to nonlocal material models.

For plasticity with isotropic softening, Borino et al. (1999) considered the free-energy density in the form

$$\rho\psi(\boldsymbol{\epsilon},\boldsymbol{\epsilon}_p,\bar{\kappa}) = \rho\psi_e(\boldsymbol{\epsilon}-\boldsymbol{\epsilon}_p) + \rho\psi_p(\bar{\kappa}) \qquad (65)$$

where $\rho\psi_e(\boldsymbol{\epsilon}-\boldsymbol{\epsilon}_p) = \frac{1}{2}(\boldsymbol{\epsilon}-\boldsymbol{\epsilon}_p)^T\mathbf{D}_e(\boldsymbol{\epsilon}-\boldsymbol{\epsilon}_p) =$ elastically stored energy and $\rho\psi_p(\bar{\kappa}) =$ plastic part of the free energy, usually interpreted as the energy stored in microstructural changes. The postulate of maximum plastic dissipation then leads to an associated nonlocal plasticity model with the softening law (38) replaced by

$$\sigma_Y = \sigma_0 + \widetilde{h(\bar{\kappa})} \qquad (66)$$

where $\bar{\kappa} =$ (usual) nonlocal average of the softening variable κ (cumulative plastic strain), and the tilde over $h(\bar{\kappa})$ means that the thermodynamic force $q = \rho\partial\psi/\partial\bar{\kappa} = \rho d\psi_p/d\bar{\kappa} = h(\bar{\kappa})$ is subjected to the so-called dual averaging. The dual averaging operator is defined implicitly by the identity

$$\int_V \tilde{f}g\,dV = \int_V f\bar{g}\,dV \qquad (67)$$

which must hold for all functions f and g for which the right-hand side makes sense. Since

$$\int_V f(\mathbf{x})\bar{g}(\mathbf{x})\ d\mathbf{x} = \int_V f(\mathbf{x}) \int_V \alpha(\mathbf{x},\boldsymbol{\xi})g(\boldsymbol{\xi})d\boldsymbol{\xi}d\mathbf{x}$$

$$= \int_V \int_V \alpha(\mathbf{x},\boldsymbol{\xi})f(\mathbf{x})d\mathbf{x}\ g(\boldsymbol{\xi})d\boldsymbol{\xi} \qquad (68)$$

we obtain $\tilde{f}(\boldsymbol{\xi}) = \int_V \alpha(\mathbf{x},\boldsymbol{\xi})f(\mathbf{x})\ d\mathbf{x}$, or, equivalently,

$$\tilde{f}(\mathbf{x}) = \int_V \alpha(\boldsymbol{\xi},\mathbf{x})f(\boldsymbol{\xi})d\boldsymbol{\xi} \qquad (69)$$

This formula indicates that the dual averaging operator differs from the original (primal) one only in the order of arguments of the weight function α. For an infinite domain V_∞, the weight function is symmetric (because it depends only on $\|\mathbf{x}-\boldsymbol{\xi}\|$), and so there is no difference between the primal and dual averaging (thus the nonlocal operator is self-adjoint). For a finite domain V, the weight function is nonsymmetric in the proximity of the boundary, and $\tilde{f}$ is, in general, different from $\bar{f}$. In particular, the dual averaging operator does not satisfy the normalizing condition that a constant field f is transformed into a constant field $\tilde{f} \equiv f$.

As pointed out by Rolshoven and Jirásek (2001), the nonlocal plasticity formulation with softening law (66) does not provide full regularization, similar to the simpler formulation with softening law (61). Plastic strain still localizes into a set of zero measure. This is not possible in the nonlocal plasticity theory proposed by Svedberg and Runesson (1998), who considered the free energy as a function of both the local and nonlocal internal variables. A particular model of this type defines the plastic part of the free energy in the form $\psi_p(\kappa,\bar{\kappa}) = \psi_{p1}(\kappa) + \psi_{p2}(\bar{\kappa}-\kappa)$, i.e., as a sum of two terms that depend on the local softening variable and on the difference between the nonlocal and local softening variable, respectively. The consistent thermodynamic approach then leads to the hardening–softening law

$$\sigma_Y = \sigma_0 + h_1(\kappa) + \widetilde{h_2(\bar{\kappa}-\kappa)} - h_2(\bar{\kappa}-\kappa) \qquad (70)$$

where function $h_1 \equiv \rho\psi'_{p1}$ describes softening under homogeneous conditions (uniform distribution of plastic strain) and function $h_2 \equiv \rho\psi'_{p2}$ is the correction due to nonlocal effects. This model has been shown to give a localized plastic zone of a nonzero size (Borino and Failla 2000). However, its scope of application is limited; while it reasonably captures the initial localization profile, it cannot describe complete failure with zero residual resistance of the material. This can be easily explained for the uniaxial tensile test. If the material completely loses cohesion at late stages of the deformation process, the local softening law must be such that $\sigma_0 + h_1(\kappa) = 0$ for κ exceeding a critical value κ_c. In the zone where the plastic strain exceeds κ_c, the actual yield stress (70) can vanish only if the nonlocal correction vanishes, i.e., if

$$\widetilde{h_2(\bar{\kappa}-\kappa)} = h_2(\bar{\kappa}-\kappa) \qquad (71)$$

Sufficiently far from the boundary, the dual averaging operator is identical with the primal one. Arguments similar to those used by Jirásek and Rolshoven (2002) in their analysis of the Vermeer–Brinkgreve nonlocal plasticity model (previous section) lead to the conclusion that Eq. (71) is satisfied only if $h_2(\bar{\kappa}-\kappa)$ is constant, and this is possible only if $\bar{\kappa}-\kappa$ is constant, provided that the hardening function h_2 is monotonic. Applying the same reasoning once again, it can be shown that the difference $\bar{\kappa}-\kappa$ is constant across the plastic zone of a nonzero size only if κ is constant along the entire bar. This means that when the plastic strain at the center of the localized plastic zone reaches κ_c, the plastic zone starts expanding and the strength degradation gets delayed, which is perceived as a stress locking effect. For nonlinear softening laws, this effect builds up gradually.

The foregoing thermodynamically based nonlocal plasticity has apparently not yet been applied in multidimensional finite-element codes. The same general idea was adapted to nonlocal damage by Benvenuti et al. (2000). Comi and Perego (2001) proposed a simpler formulation that permits an explicit evaluation of nonlocal averages, without the need for solving a nonlocal consistency condition, which has the character of an integral equation. However, since they used a primal averaging operator based on the scaled weight function (21), the dual operator did not preserve a uniform field, which caused certain problems in the vicinity of physical boundaries. A remedy was found by Borino et al. (2002), who constructed a self-adjoint operator that preserves a uniform field. The resulting thermodynamically based nonlocal damage model seems to perform well, but it still requires a two-fold nonlocal averaging, and so it is more computationally expensive than the simple nonlocal damage theory of Pijaudier-Cabot and Bažant (1987).

The thermodynamically based formulation is very attractive from the theoretical viewpoint, and the fact that it leads to symmetric stiffness matrices is an asset from the computational viewpoint. The dissipation inequality is automatically satisfied. However, there are two penalties to pay: (1) a greater complexity of the constitutive model; and (2) a broadened bandwidth in the stiffness matrix, caused by the double application of the nonlocal operator. Furthermore, while a very appealing theoretical framework has been developed, much of its physical foundation remains clouded.

The way to symmetrize the nonlocal model based on crack interactions is not known at present. It is even unclear whether an effort in that direction would be physically justified.

Nilsson's Model

In his thesis (Nilsson 1994) and journal article (Nilsson 1997), Nilsson proposed a thermodynamically motivated nonlocal plasticity model that shares many similar features with the model of Borino et al. (1999). However, there are also some important differences. In the most general version of his model, Nilsson considered all the arguments of the free-energy potential as nonlocal. Since this would lead to a stress–strain law that is not easily invertible, Nilsson focused his attention on what he called the model with restricted nonlocality, which deals with total strain only in its local form. For isotropic softening, the free-energy density is written as

$$\psi(\boldsymbol{\epsilon},\bar{\boldsymbol{\epsilon}}_p,\bar{\kappa})=\psi_e(\boldsymbol{\epsilon}-\bar{\boldsymbol{\epsilon}}_p)+\psi_p(\bar{\kappa}) \qquad (72)$$

In contrast to Eq. (65), the plastic strain that appears in the elastic part of the free energy is nonlocal. Consequently, the elastic part of the stress–strain law (39) is now replaced by

$$\boldsymbol{\sigma}=\mathbf{D}_e(\boldsymbol{\epsilon}-\bar{\boldsymbol{\epsilon}}_p) \qquad (73)$$

The softening law (38) is the same as for the simple formulation of nonlocal plasticity, and the flow rule (40) and loading–unloading conditions (41) remain standard. However, this is not consistent with the postulate of maximum plastic dissipation, as pointed out by Borino and Polizzotto (1999) and admitted by Nilsson (1999). It even turns out that, under certain circumstances, the global dissipation could become negative (Jirásek and Rolshoven 2002), which means that the model is not really ther-

modynamically consistent. Moreover, the plastic strain again localizes into a set of zero measure, similar to the model with a nonlocal softening variable from the section on plasticity with nonlocal softening variable.

Nonlocal Version of Dislocation-Based Gradient Plasticity for Micrometer Scale

While some numerical experts assert that the gradient models are more friendly to a computer programer than the nonlocal averaging models, Gao and Huang (2001) recently took the opposite view. For the purpose of finite-element analysis, they converted the dislocation-based gradient plasticity model for metals on the micrometer scale (Gao et al. 1999; Huang et al. 2000) to a form in which the strain gradient is approximated by a nonlocal integral over a representative volume. A generalized form of their idea is as follows.

Within a certain representative volume V_c surrounding a given point $\mathbf{x}$, the strain tensor may be approximated as $\boldsymbol{\epsilon}(\mathbf{x}+\mathbf{s})\approx\boldsymbol{\epsilon}(\mathbf{x})+\boldsymbol{\epsilon}_{,m}(\mathbf{x})s_m$, where the subscript preceded by a comma denotes a partial derivative and repeated indices imply summation. Multiplying this equation by s_k and integrating over V_c, one gets

$$\int_{V_c}\boldsymbol{\epsilon}(\mathbf{x}+\mathbf{s})s_k\mathrm{d}\mathbf{s}\approx\boldsymbol{\epsilon}(\mathbf{x})\int_{V_c}s_k\mathrm{d}\mathbf{s}+\boldsymbol{\epsilon}_{,m}(\mathbf{x})\int_{V_c}s_ms_k\mathrm{d}\mathbf{s} \qquad (74)$$

The integrals on the right-hand side characterize the geometry of the representative cell. If point $\mathbf{x}$ is located at the center of gravity of the cell, the first-order moments $\int_{V_c}s_k\,\mathrm{d}s$ vanish. The second-order moments

$$I_{mk}=\int_{V_c}s_ms_k\mathrm{d}\mathbf{s} \qquad (75)$$

are components of the inertia tensor, which is always regular, and thus invertible. In the simplest case of a cubic or spherical cell, I_{mk} is a multiple of the unit tensor. Multiplying Eq. (74) from the right by the inverse tensor I_{kl}^{-1}, we obtain for the strain gradient the approximation

$$\boldsymbol{\epsilon}_{,l}(\mathbf{x})\approx\mathbf{I}_{kl}^{-1}\int_{V_c}\boldsymbol{\epsilon}(\mathbf{x}+\mathbf{s})s_k\mathrm{d}\mathbf{s} \qquad (76)$$

If the strain gradient terms in a gradient plasticity model are replaced by this approximation, the numerical solution can be programed in the same manner as the nonlocal averaging models.

In strain-gradient plasticity, the size of the representative cell is dictated by the characteristic length

$$\ell_g=\alpha_g(G/\sigma_0)^2b \qquad (77)$$

where $G=$ elastic shear modulus, $\sigma_0=$ yield stress, $b=$ Burgers vector of edge dislocation, and $\alpha_g=$ dimensionless semiempirical constant of the order of 1 (Gao et al. 1999). Alternatively, one could consider the approximation of the gradient by an integral as a purely numerical procedure and select V_c as the smallest neighborhood of the given point $\mathbf{x}$ that contains a sufficient number of material points traced by the finite-element program. Eq. (76) can then be interpreted as a generalized finite-difference scheme. In one dimension, the minimum required number of integration points would be 2, and if these points are located symmetrically with respect to x, Eq. (76) yields the standard finite-difference formula

$$\epsilon'(x)\approx\frac{\epsilon(x+h)-\epsilon(x-h)}{2h} \qquad (78)$$

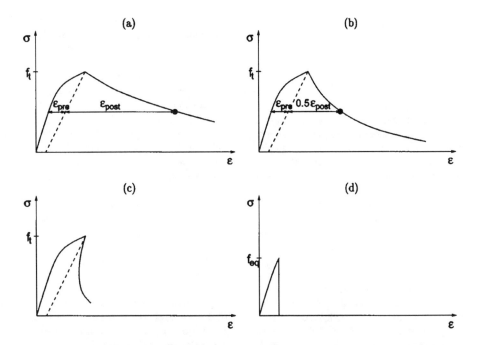

Fig. 18. Adjustment of the stress–strain diagram according to the crack band approach: (a) master curve and decomposition of strain into the prelocalization and postlocalization parts, (b) horizontal scaling of the postlocalization part for an element twice as large as the physical process zone, (c) snap back occurring for a very large element, and (d) conservation of fracture energy by strength reduction.

However, in a general case (with more integration points or irregular point arrangement), this approach does not provide the optimal accuracy, and it would be preferable to replace s_k in Eq. (74) by a suitably modified weight function.

Partial Regularization and Substitutes for Nonlocality

Crack Band Model

The crack band model (Bažant 1982; Bažant and Oh 1983) is (aside from viscosity) practically the simplest, and computationally most effective, way to avoid the pathological sensitivity to mesh refinement. Since the beginning, it has been the most widely used model for distributed cracking of concrete and geomaterials and has been incorporated in a number of commercial finite-element codes. The basic idea of the crack band model is very similar to the mesh-adjusted softening modulus technique, proposed in the context of softening plastic shear bands by Pietruszczak and Mróz (1981). Essentially, the same approach was developed by Willam et al. (1986) under the name of the composite fracture model.

This simple model is endowed with some, but not all, of the characteristics of nonlocal models. It ensures the correct energy dissipation in a localized damage band (or equivalent fracture), and it gives the correct transitional size effect. The main difference from the nonlocal models is that the nonlocal spatial averaging is replaced by an energy-based rescaling of the postlocalization part of the stress–strain relation, which takes into account the density of the computational mesh (size of the finite elements). The global load–displacement diagram can be captured correctly, but the width of the numerically resolved fracture process zone depends on the element size and tends to zero as the

mesh is refined. This is why the crack band model (or fracture energy approach) cannot be considered as a true localization limiter. It provides only a partial regularization of the problem, in the sense that the global response characteristics do not exhibit spurious mesh sensitivity. However, the mesh-induced directional bias is still present.

The crack band approach is based on the observation that, if a local softening model is used, the numerically resolved process zone typically localizes into one layer of finite elements, called the crack band. The effective width of the crack band, h_{ef}, can be estimated using the rules proposed by Bažant (1985b) and refined by Rots (1988). A reasonable estimate is obtained by projecting one element onto the direction normal to the band, which is (for quasibrittle failure) usually close to the direction of maximum principal strain. The user specifies the stress–strain relation for the basic case when h_{ef} is equal to the actual effective width of the physical fracture process zone, L_s, which is proportional to the intrinsic material length ℓ. As far as possible, the size of the finite elements in the strain-softening distributed damage zone is kept equal to L_s. But if, for reasons of computational efficiency, the finite elements need to be larger, then the postlocalization portion of the master constitutive diagram (on which the tensorial constitutive law is based) is scaled horizontally (i.e., in the direction of the strain axis) by the factor L_s/h_{ef}; see Fig. 18(b). This makes the strain softening steeper and a proper scaling ratio achieves that the energy dissipated by the cracking band per its unit advance remains the same. Note that the postlocalization portion is the difference, for the same stress, of the strain at the softening curve from the strain at the unloading diagram emanating from the point at which localization occurs (typically the peak), not from a vertical line dropping down from that point [Fig. 18(a)]. In the simple case of a triangular stress–strain diagram with linear strain softening in uniaxial tension characterized by the tangent modulus $E_t < 0$, the adjusted postpeak tangent modulus $\tilde{E}_t$ must satisfy the condition (Bažant 1982; Bažant and Oh 1983)

$$-\frac{1}{\tilde{E}_t} = \frac{2G_f}{h_{ef}f_t^2} - \frac{1}{E} \qquad (79)$$

where E = initial Young's modulus and $G_f = L_s(E^{-1} - E_t^{-1})f_t^2/2$.

The increase of finite-element size, however, has a limit. For a certain large enough critical element size, h_{crit}, the adjusted postpeak diagram develops a snapback [Fig. 18(c)], which is inadmissible (the stress would cease to be a unique functional of the strain history). To avoid the spurious snapback, one can use a vertical stress drop and scale down the strength limit f_t to some equivalent value f_{eq} such that h_{ef} times the area under the stress–strain diagram with the vertical drop would remain equal to G_f [Fig. 18(d)]. It turns out that, generally, $f_{eq} \propto 1/\sqrt{h_{ef}}$, and for the special case of a triangular stress–strain diagram

$$f_{eq} = f_t\sqrt{2l_0/h_{ef}} \qquad (80)$$

where $l_0 = EG_f/f_t^2$ = Irwin-type characteristic length of the material. This expedient was proposed by Bažant and Cedolin (1979) and Cedolin and Bažant (1980), and good practical results with finite elements much larger than L_s were obtained in situations where failure is driven by fracture.

To model a very narrow cracking band, or to approximate a line fracture, finite elements smaller than L_s may be used in the crack band model. The same scaling of the postpeak diagram can again be applied, making the strain softening less steep. As the mesh is refined, the band of cracking elements converges to a layer of zero thickness, i.e., to a line (in two dimensions) or surface (in three dimensions). Thus, the limit corresponds to the solution of a cohesive crack model. The mathematical aspects of this correspondence were studied by Simo et al. (1993).

Scaling of the stress–strain diagram in the above sense is straightforward only for models that explicitly control the evolution of inelastic strain, e.g., for softening plasticity or smeared crack models. In that case, the desired scaling effect is achieved by a modification of the hardening modulus (derivative of stress with respect to inelastic strain). In continuum damage mechanics, nonlinearity and softening are controled by the damage evolution law, and the reduction factor $1 - \omega$ multiplies the total strain. It is, therefore, not easy to scale only the postlocalization part of strain while keeping the unloading part unaffected. An exact scaling procedure leads to a nonlinear equation (except for the special case of linear softening), which must be solved iteratively during each stress evaluation. It is also difficult to extend the crack band approach to mixed-mode failure or failure under complex three-dimensional stress states. In such general cases, the standard fracture energy is not sufficient to characterize the dissipation in localized failure modes. Generalized forms of the crack band approach that cover a variety of failure modes have been proposed, e.g., for softening plasticity models (Etse and Willam 1994; Kang 1997) and for the microplane constitutive model M4 for concrete (paper in preparation by Bažant, Zi, Jendele, and Novák).

Whereas the nonlocal models with at least three finite elements across the width of the localization band are the best known way to avoid mesh orientation bias for the propagation direction (in fact, better than the known mesh-adaptive schemes), the crack band model is poor in this regard. It performs best if the path of the cracking band (or the fracture to be approximated by it) is known in advance and if the mesh is laid out so that a mesh line would coincide with this path. If the crack path is not known, one should solve the problem with meshes of different inclinations. The most reasonable solution is usually that providing the smallest peak load or the steepest postpeak load deflection diagram.

A caveat needs to be mentioned with respect to those infrequent situations where the strain-softening damage does not localize, being stabilized, for example, by a heavy enough reinforcement net or by an adjacent layer of compressed material (see the examples mentioned later in the section on limitations of cohesive models). In such situations, the postpeak strain-softening behavior must not be rescaled. So, in using the crack band approach, the user must separately assess whether it is reasonable to expect localization. However, in some cases, diffuse softening damage patterns in some parts of the structure can coexist with localized cracks in other parts, and these parts may even change during the loading process. In such cases it is then next to impossible to define a reasonable rule for the adjustment of the stress–strain diagram according to the element size. A consistent solution can be obtained only when finite-element sizes $h = L_s$ can be used (in which case no rescaling is needed for the crack band approach), or with a fully regularized nonlocal model.

As explained before, the crack band model uses a modification of the postpeak part of the constitutive law enforcing the correct energy dissipation by a localized crack band. When large finite elements need to be used (larger than the actual width of the fracture process zone), the resolution of the localization process can be improved by special enrichments of the finite-element approximation that allow capturing narrow bands of highly localized strain inside the elements. The pioneering paper by Ortiz et al. (1987) paved the way to special elements with embedded localization bands (Belytschko et al. 1988; Sluys 1997; Oliver et al. 1998). These elements can describe a layer of softening material separated from the surrounding, elastically unloading material by two parallel weak discontinuity planes across which the displacement remains continuous but the out-of-plane strain components can have a jump. The width of the band can be specified independently of the mesh (provided that the elements are not too small), and it can be set equal to the actual width of the process zone, L_s, considered as a material parameter. A one-dimensional element of this kind would give exactly the same response (in terms of the relationship between the nodal displacements and nodal forces) as a standard element with the softening modulus adjusted according to the element size, as described for the crack band model. For two- and three-dimensional elements, the formulation with embedded discontinuities can better reproduce the kinematics of highly localized deformation modes and thus avoid certain locking effects (Jirásek 2000a).

In a recent study (Bažant et al. 2001; Červenka et al. 2002), a new model for a damage localization layer within a finite element has been developed. For the special case of an orthogonal brick element (whose nodal displacements are compatible with uniform strain), the model reduces to a combination of one brick element for the strain-softening localization band and another brick element for the parallel layer of elastically unloading material, such that the interface conditions of stress equilibrium and of kinematic compatibility are exactly satisfied. In the general case of an arbitrary finite element containing a localization layer of arbitrary orientation, the incremental stress–strain relation is assumed to be the same as in an "equivalent" orthogonal brick element that has the same total volume, and the same volume and width of the localization layer. The interface conditions (numbering 6) are solved within each loading step or iteration, together with iterations for the nonlinear constitutive law. Implementation of this localization element together with microplane model M4 for concrete (Bažant et al. 2000) led to good results for various localization problems.

Cohesive Crack and Cohesive Zone Models

Traction-Separation Law

The cohesive crack model for concrete, formulated by Hillerborg et al. (1976) under the name of fictitious crack model, represents a refinement of the classical cohesive models, with the main difference that a cohesive crack is not considered to initiate and propagate along a predetermined path but is assumed to initiate anywhere in the structure if the tensile stress reaches the strength limit. Although this approach is not really a nonlocal model, it provides an alternative to achieve one objective of the nonlocal models—a mesh-size independent response in the presence of strain-softening damage or fracture. The early applications in the field of concrete fracture were limited to mode-I situations. An extension to cracks and interfaces opening under mixed-mode conditions was developed, e.g., by Cervenka (1994), who exploited a loading function in the space of normal and shear tractions, originally proposed by Carol and Prat (1990) in the context of statically constrained microplane models. Scaling properties and asymptotic solutions of cohesive crack models were studied by Planas and Elices (1992, 1993) and Bažant and Li (1995).

Phenomena such as debonding, delamination, or intergranular damage in composites and metals are often described by cohesive zone models, going back to the ideas of Dugdale (1960) and Barenblatt (1962) and recently developed in a modern computational framework by Needleman (1987), Tvergaard and Hutchinson (1992), Tvergaard and Hutchinson (1993), and Ortiz and coworkers (Camacho and Ortiz 1996; Pandolfi et al. 1999; Ortiz and Pandolfi 1999).

Cohesive models describe highly localized inelastic processes by traction-separation laws that link the cohesive traction transmitted by a discontinuity line or surface to the displacement jump, characterized by the separation vector. The physical interpretation and the numerical implementation of such models can be manifold. The discontinuity surface can be an internal boundary between two different materials (e.g., a matrix-inclusion interface), a plane of weakness (e.g., a rock joint), or a simplified representation of a narrow localization zone (e.g., of a fracture process zone or a shear band). Some formulations assume a nonzero initial elastic compliance of the cohesive zone, which may be either physically motivated by the reduced stiffness of the interface layer as compared to a perfect bond between two materials, or considered as a purely numerical artifact corresponding to a penalty-type enforcement of displacement continuity in the elastic range. Other formulations enforce displacement continuity directly, and allow a nonzero separation only after a certain initiation condition has been met.

Cohesive models for which the traction-separation law begins with a positive slope (i.e., models with a nonzero elastic compliance, or models with softening preceded by hardening) must be used with care. If the stress increases above the initiation limit f_0 (which is equal to zero for models with a nonzero elastic compliance), it must also increase above f_0 at some material points sufficiently close to the crack face, yet a new cohesive crack cannot initiate at that point. Because of this inconsistency, cracks described by such cohesive models cannot be assumed to initiate at an arbitrary location. Their path must be specified in advance or restricted to a predefined finite number of segments.

Needleman's original model is based on an elastic potential without any internal variables, i.e., it has the character of a nonlinear elastic model with a path-independent work of separation. Such a simple approach is suitable only if the crack opening grows monotonically, but it fails to give reasonable results for (even partially) closing cracks. Moreover, the model has a nonzero initial compliance, which means that the displacement jump starts growing whenever nonzero tractions are applied, no matter how small they are. This is appropriate for simulations of preexisting material interfaces (grain boundaries, matrix-inclusion interfaces, etc.) with a well-defined geometrical structure that can be taken into account by the finite-element mesh. When the aim is to simulate a crack propagating along an arbitrary path that is not known in advance, potential surfaces (or lines, in two dimensions) of decohesion must be interspersed throughout the material (Xu and Needleman 1994). Their high initial stiffness has an adverse effect on the conditioning of the global stiffness for implicit methods, or on the critical time step for explicit methods. Moreover, due to the finite number of potential discontinuity segments, the crack propagation path is locally always constrained to a discrete set of directions, typically spaced by 45 or 60°.

A cohesive zone model with a nonzero initial compliance was also used by Tvergaard and Hutchinson (1993) and Wei and Hutchinson (1999), who studied the interplay between plastic yielding in a small process zone and separation processes at an interface between two materials.

Ortiz (1988) suggested a derivation of the traction-separation law for mode I from a micromechanical model based on an array of collinear microcracks. Camacho and Ortiz (1996) studied fracture and fragmentation in brittle materials using a cohesive model with possible crack initiation under mixed mode. The model was then reformulated within a thermodynamic framework (Pandolfi et al. 1999) and extended to the area of finite opening displacements (Ortiz and Pandolfi 1999). Due to insistence on a simple symmetric formulation with a potential, the resultant of the shear and normal stresses in the cohesive crack has in these models the same direction as the relative (normal and shear) displacement across the crack. For shear with a nearly vanishing normal stress, the relative displacement is predicted to be nearly parallel to the crack plane. This means that the dilatancy due to shear, observed in most nonmetallic materials, is not taken into account (at least not on the element level, although some dilatancy gets produced globally if the sliding interelement crack has a zig-zag path).

Discretization Techniques for Cohesive Models

Discretization techniques used in conjunction with cohesive models can be divided into two broad categories, depending on whether the displacement discontinuity can occur only between adjacent elements, or can run across the finite-element mesh along an arbitrary trajectory.

In the first category, one can distinguish methods that assume an a priori given trajectory of the discontinuity, methods that trace the trajectory by continuous remeshing, and methods that admit a potential discontinuity between any two adjacent finite elements.

The second category consists of finite-element formulations with embedded discontinuities, based on the enhanced assumed strain method, and extended finite elements based on the partition-of-unity method.

In analogy to elements with embedded localization bands, which use discontinuous strain approximations, it is possible to construct elements with embedded localization lines (in 2D) or planes (in 3D), across which the displacement field is discontinuous (Dvorkin et al. 1990; Klisinski et al. 1991; Olofsson et al. 1994; Simo and Oliver 1994; Oliver 1996). For this numerical scheme, the trajectory of the cohesive crack or cohesive zone is independent of the finite-element mesh. This technique is much more flexible than the standard scheme with discontinuities allowed only at element interfaces, and it eliminates the need for continuous remeshing.

A vast majority of embedded crack formulations developed in the 1990s use a nonconforming approximation of the displacement jump. On one hand, this is convenient, because the approximation of the displacement jump in one element can be completely decoupled from the other elements and eliminated on the local level. However, the elements then become sensitive to the orientation of the discontinuity line with respect to the nodes, and in some unfavorable situations the response of the element ceases to be unique (Jirásek 2000b).

A much more robust implementation can be achieved if the discontinuous enrichment is based on the partition-of-unity concept (Melenk and Babuška 1996; Duarte and Oden 1996), with a conforming approximation of the displacement jump and completely independent strain components in the two parts of the element separated by the crack (Wells 2001). This formulation is intimately related to the manifold method, developed by Shi (1991) for preexisting material discontinuities such as rock joints, and to the extended finite-element method (XFEM), developed by Belytschko and co-workers for stress-free cracks and material interfaces (Moës et al. 1999; Sukumar et al. 2000; Daux et al. 2000) and adapted by Moës and Belytschko (2002) for cohesive cracks.

The traction-separation laws for embedded discontinuities are usually postulated in a plasticity format (Simo and Oliver 1994; Oliver 1996; Olofsson et al. 1994; Armero and Garikipati 1995; Larsson et al. 1996; Ohlsson and Olofsson 1997) or constructed as extensions of the cohesive crack model to mixed-mode situations (Dvorkin et al. 1990; Klisinski et al. 1991; Lotfi and Shing 1995) using concepts similar to fixed crack versions of smeared crack models. However, the plasticity-based models do not properly describe unloading of a brittle material. This problem becomes especially severe at late stages of the degradation process when the crack is stress free and, according to the plasticity theory, a reversal of the opening rate immediately generates a compressive traction, which is not physically realistic. The theories inspired by the cohesive crack model for concrete are close to interface damage mechanics (Simo and Oliver 1994; Oliver 1996; Armero 1997), which provides a natural description of the gradual loss of integrity. In this framework, a difficult issue is the proper treatment of stiffness recovery upon complete crack closure, with possible frictional sliding. Such effects were consistently taken into account by Cangemi et al. (1996) and by Chaboche and coworkers (Chaboche et al. 1997a,b) in the context of interface models for delamination and debonding of fiber–matrix composites. For cracks, a similar approach was proposed by Jirásek (1998a) and Jirásek and Zimmermann (2001).

Limitations of Cohesive Models
The cohesive crack or cohesive zone models provide an objective description of fully localized failure. In this case, the cohesive traction-separation law with softening does not need any adjustment for the element size, because mesh refinement does not change the resolved crack pattern. However, the mesh independence is questionable if the cracking pattern is diffuse. This can happen, e.g.,
1. in concrete with a sufficiently dense and strong reinforcing net;

2. on the tensile side of a reinforced concrete beam, where distributed cracking is stabilized by the tensile reinforcement;
3. in a system of cooling (or drying) cracks propagating into a halfspace, especially if the temperature profile has a steep front (Bažant and Ohtsubo 1977; Bažant and Cedolin 1991, Sec. 12.5);
4. in dynamic problems when stress waves travel across the structure; and
5. if heterogeneity of the material is explicitly taken into account as the mesh is refined.

Mesh refinement can cause parallel potential cohesive cracks to become arbitrarily close. Thus, if they are under the same stress, their cumulative opening displacements per unit volume of material (if uniformly stressed) can become arbitrarily large in such situations. Such behavior was illustrated by Bažant's example (Bažant 1985a,b, 1986) of a concrete bar under uniaxial tension, with strong enough axial steel reinforcement such that the overall tangential stiffness of the bar never becomes negative, despite softening in the cracks (same as Fig. 14).

Consequently, cohesive models with an assumed mesh-independent softening law are applicable only when the strain-softening damage is a priori known to localize. In a certain sense, they are complementary to continuum-based models, for which a mesh-independent stress–strain law with softening can cover only the cases when cracking remains perfectly distributed. In a general case with regions of diffuse and localized cracking (and perhaps with gradual transition from diffuse damage to localized fracture), both classes of models would need an adjustment taking into account the element size, shape, and orientation.

For example, the model developed at LCPC Paris (Rossi and Wu 1992) works with elastic elements separated by cohesive interfaces, and the material parameters describing those interfaces are considered as random variables. To obtain mesh-independent results, at least in the global sense, it is necessary to adjust the statistical characteristics of the random strength distribution to the size of the elastic elements. For small elements, the strength characterizes the properties on a small scale, and it has a large variation. For large elements, the strength should characterize the overall behavior on a larger scale, resulting from the combined effect of small-scale processes that cannot be captured explicitly on the given resolution level. Experimental size effect investigations indicate that both the mean value and the variation of strength decrease as the size of the sample or specimen increases. The empirically derived rule for the size dependence of the statistical characteristics of random strength distribution developed at LCPC seems to give globally objective results when used in simulations on different meshes.

Another limitation of the cohesive cracks is that they cannot capture stress multiaxiality in the fracture process zone. The usual cohesive crack or cohesive zone models postulate a traction-separation relation that takes into account only the out-of-plane components of the stress tensor at the discontinuity surface. This approach suits just fine all the notched fracture specimens in common use but is not representative of many applications. The effect of the compressive normal stresses acting parallel to the crack plane (called sometimes the "T stresses") is ignored. The fact that this cannot be universally correct is clear from the observation that if one of the T stresses approaches the uniaxial compressive strength f_c of the material, the strength limit of the cohesive crack model should approach zero. Thus, for general situations, a multiaxial cohesive crack model would have to be developed. Even though this would, in principle, be possible, the multiaxial behavior in the fracture process zone can be captured more easily by using a nonlocal model or crack band model with a good triaxial constitutive law.

Finally, since the softening damage in the early stage of opening of a cohesive crack occurs within a band of a certain width, the cohesive stress at location x should depend not merely on the crack opening at x but on the average of crack opening taken over a certain neighborhood of x. In other words, the cohesive law itself may need to be considered as nonlocal (Bažant 2001, 2002b), which was taken into account in the gradient-enhanced formulation of the cohesive crack model proposed by van Gils (1997).

Regularization by Real or Artificial Viscosity

In dynamics, nonlocal averaging can sometimes be avoided, as a convenient approximation, by considering material viscosity, rate effect, or damping. Introducing artificial viscosity may serve as an expedient substitute for nonlocality.

Indeed, since the dimension of viscosity η is kg/(m s), the dimension of Young's modulus E (or strength) is kg/(m s^2), and the dimension of mass density ρ is kg/m^3, there exists a material length associated with viscosity, given by

$$\ell_v = \frac{\eta}{v\rho} = \frac{\eta}{\sqrt{E\rho}} \qquad (81)$$

where $v = \sqrt{E/\rho}$ = longitudinal wave propagation velocity. Consequently, any rate dependence in the constitutive law combined with inertial effects introduces a length scale. This effect was exploited as a localization limiter regularizing the boundary value problem, for example, by Needleman (1988). In a similar fashion, a length scale is introduced by the delay effects in the damage evolution law (Ladevèze 1992, 1995; Ladevèze et al. 2000).

There is, however, an important difference from the nonlocal models—the viscosity-induced nonlocality gradually disappears with the passage of time. For load durations much larger than the relaxation (or retardation) time associated with the type of viscosity used, the modeling is not completely objective. Thus, the viscosity or rate effect can be used as a substitute for a nonlocal model only within a narrow range of time delays and rates, generally not spreading more than one order of magnitude. Therefore, if an artificial viscosity is used, cautious insight is needed.

If a characteristic length, ℓ_v, is considered in dynamic problems, then a characteristic time τ_0 is automatically implied as the time of passage of a wave front over the distance ℓ_v:

$$\tau_0 = \frac{\ell_v}{v} = \frac{\eta}{v^2\rho} = \frac{\eta}{E} \qquad (82)$$

In problems of impact of missiles on concrete walls, for example, the duration of the dynamic event is not too long compared to τ_0. Therefore, nonlocal averaging is generally not needed. Due to inertia effects, there is not enough time for spurious localization of strain-softening damage to develop, whether or not any viscosity or damping is taken into account.

If the viscosity is introduced in the constitutive law in such a manner that there is no softening for fast enough loading [e.g., Bažant and Li's (1997) model for the rate dependence or cohesive cracks in concrete], then the boundary value problem is regular (well posed) for fast enough loading even in the absence of inertia forces (i.e., in quasistatics), simply because there is no strain softening in the structure.

Nonlocal Probabilistic Models of Failure

It is impossible to do justice to the subject without pointing out the importance of nonlocality in probabilistic modeling of failure. The evolution of the classical probabilistic theory of strength, begun in qualitative terms by Mariotte (1686), culminated with the studies on extreme value statistics and discovery of Weibull distribution by Fischer and Tippett (1928), and the application of this distribution to fracture by Weibull (1939). This theory rests on the hypothesis that the structure fails if a material element that is infinitesimal compared to structure size D fails. The theory works just fine for fatigue-embrittled metals and fine-grained ceramics. It is noteworthy that since the size effect of this statistical theory is a power law, self-similarity is implied, and no characteristic structure size D_0 exists. The model possesses no characteristic length. This is not realistic for quasibrittle materials.

This deficiency gets manifested in structures that do not fail immediately after the failure of one infinitesimal element. It requires modification for quasibrittle structures, consisting of materials such as concrete, composites, or ice, which are characterized by a sizable fracture process zone (FPZ), not negligible compared to the structure size D. Two types of statistical behavior must be distinguished:

1. The structure fails only after a long fracture or a long damage band develops, which is typical of reinforced concrete, fiber composites, and compression dominated problems. The path of the fracture or band and the location of the FPZ are dictated mainly by mechanics on the macroscale and are little sensitive to material randomness when the structure is large. Within the FPZ the material behaves randomly, but if the size of the FPZ, which is of the same order as the intrinsic material length ℓ, is negligible compared to structure size D (i.e., if $D \gg \ell$), the statistical influence disappears and the size effect is purely deterministic (Bažant and Xi 1991). The classical Weibull theory cannot treat the large size limit for which the fracture or band front, in relative coordinates, becomes perfectly sharp. The reason is that the Weibull integral, see Eq. (83) below, diverges for a singular stress distribution, which is approached as $D/\ell \to \infty$.

2. The structure fails at the initiation of fracture from the surface, which is exemplified by flexural failure of an unreinforced beam. If the structure is not large compared to ℓ, the stress redistribution in the cross section engendered by the FPZ is important and dominates the failure load. The statistical effects are then negligible, and the classical Weibull statistical theory does not apply. However, for $D \gg \ell$, the FPZ relative to D becomes a point, and the structure fails as soon as the first point fails. Since this point can have many random locations, the statistical effect on failure is important, and in this limit case the Weibull theory does apply. The larger the structure, the smaller is the strength that can be encountered by this randomly located point.

The nonlocal concept has been shown to provide a remedy to the abovementioned problems. In the simplest theory, which deals only with the failure load and ignores the randomness of the deformations and stresses prior to failure, the probability of structural failure, p_f, is given by the Weibull integral whose integrand involves the (local) maximum principal stress $\sigma_I(x)$. The nonlocal remedy simply consists in replacing $\sigma_I(x)$ with the nonlocal maximum principal stress $\bar{\sigma}(x)$. Thus, the Weibull integral over structure volume (in the case of tensile failures) takes the form (Bažant and Xi 1991; Bažant and Novák 2000b):

$$p_f = 1 - \exp\left\{ -\int_V \left\langle \frac{\bar{\sigma}_I(x)}{\sigma_0} \right\rangle^m \frac{dx}{V_r} \right\} \qquad (83)$$

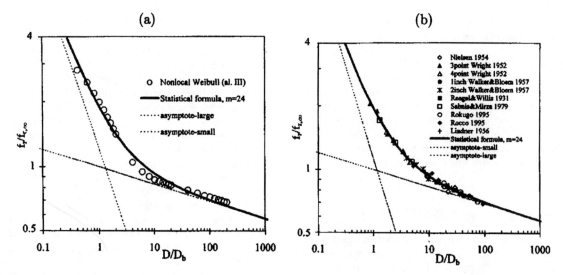

Fig. 19. Logarithmic size effect plots of (a) the numerical results obtained with the nonlocal Weibull-type probabilistic theory, and (b) previously published test data on the modulus of rupture of concrete, f_r (i.e., flexural strength of unreinforced beams), and their fit by an asymptotic formula [in (b), the optimum fit is obtained first for each data set separately and then the data are plotted in relative coordinates]; after Bažant and Novák (2000a).

where m and σ_0 are material constants; m is the Weibull modulus ($m \approx 24$ for concrete), and σ_0 is the Weibull scale parameter, with the dimension of stress. The nonlocal stress cannot be defined simply as the average stress. Rather, it must be defined in terms of the nonlocal inelastic stress or nonlocal inelastic strain. Choosing the latter, one has

$$\tilde{\sigma} = \mathbf{D}_e(\epsilon - \bar{\epsilon}_i) \qquad (84)$$

where ϵ_i is the inelastic strain and $\bar{\epsilon}_i$ is its nonlocal average. For details, consult Bažant and Novák (2000b,c), who showed that the theory gives good agreement with extensive test data regarding the combined probabilistic-energetic size effect on the modulus of rupture of unreinforced concrete beams (Bažant and Novák 2000a). The logarithmic size effect plots of the numerical results obtained with the nonlocal Weibull-type probabilistic theory are shown in Fig. 19(a). Note that a deterministic nonlocal theory gives almost the same results for the left half of the plot but terminates with a horizontal asymptote. The inclined large-size asymptote corresponds to the classical Weibull statistical size effect without nonlocality. The combination of nonlocality and statistics is needed for the transition, through the intermediate sizes. Evidently, amalgamation of the statistical and nonlocal theories is crucial for correct extrapolation from small-scale laboratory tests to large structures (such as the vertical bending fracture of an arch dam).

A much more difficult proposition is the development of a nonlocal probabilistic theory that would give not only the failure probabilities, independent of the loss of positive definiteness of the structural stiffness matrix, but would also describe the probability distributions of deflections and stresses prior to failure. The theory must yield the probability distribution of the first eigenvalue λ_1 of the structural stiffness matrix as the failure is approached. The main difficulty is that what matters is the distribution tail, with probabilities of the order of 10^{-7}. For $D \to \infty$, the tail of the distribution of λ_1 as $\lambda_1 \to 0$ must have the form of Weibull distribution when the relative structure size D/ℓ is approaching infinity, because in that limit the classical Weibull theory must be recovered. The existing stochastic finite-element methods do not meet this requirement, and thus their use for predicting failure loads of extremely small probability is doubtful. Formulation of a realistic general statistical nonlocal theory is a challenge for the future.

Concluding Thoughts and Future Path

While two decades ago strain-softening damage models were regarded as controversial, the nonlocal concept has by now rendered them respectable and their use realistic. Nonlocality is now generally accepted as the proper approach for regularizing the boundary problems of continuum damage mechanics, for capturing the size effect, and for avoiding spurious localization, giving rise to pathological mesh sensitivity.

The last two decades of research gave birth to a wide variety of nonlocal models, with differences that are not quite justified by diversity in the types of materials and practical application. One may now expect a period of crystalization in which many artificially complex or oversimplified models will fade, being recognized as superfluous, and only a few will gain a permanent pedestal in the pantheon of knowledge.

Speculations though could be made in this regard, they are better left to future scrutiny. The strength of each model's pedestal will be judged by the physics of the microstructure, and the permanence of that pedestal will be decided by passage of each model through the sieve of practical applications. Doubtless much further research lies ahead, and polemics will enliven the path into the future.

Acknowledgments

Partial financial support of the first writer has been obtained under Grant No. N00014-91-J-1109 from the Office of Naval Research (program director Y. D. S. Rajapakse) to Northwestern University, and financial support of the second writer has been provided by the Swiss Commission for Technology and Innovation under Grant Nos. CTI 4424.1 and CTI 5501.1.

References

Aero, E. L., and Kuvshinskii, E. V. (1960). "Fundamental equations of the theory of elastic materials with rotationally interacting particles." *Fiz. Tverd. Tela (S.-Peterburg)*, 2, 1399–1409.

Armero, F. (1997). "Localized anisotropic damage of brittle materials." *Computational plasticity: Fundamentals and applications*, D. R. J. Owen, E. Oñate, and E. Hinton, eds., International Center for Numerical Methods in Engineering, Barcelona, Spain, 635–640.

Armero, F., and Garikipati, K. (1995). "Recent advances in the analysis and numerical simulation of strain localization in inelastic solids." *Computational plasticity: Fundamentals and applications*, D. R. J. Owen and E. Oñate, eds., International Center for Numerical Methods in Engineering, Barcelona, Spain, 547–561.

Barenblatt, G. I. (1962). "The mathematical theory of equilibrium of cracks in brittle fracture." *Adv. Appl. Mech.*, 7, 55–129.

Bažant, Z. P. (1971). "Micropolar medium as a model for buckling of grid frameworks. *Developments in mechanics.*" *Proc., 12th Midwestern Mechanics Conference*, Univ. of Notre Dame, 587–593.

Bažant, Z. P. (1976). "Instability, ductility, and size effect in strain-softening concrete." *J. Eng. Mech. Div.*, 102(2), 331–344.

Bažant, Z. P. (1982). "Crack band model for fracture of geomaterials." *Proc., 4th Int. Conf. on Numerical Methods in Geomechanics*, Z. Eisenstein, ed., Univ. of Alberta, Edmonton, 3, 1137–1152.

Bažant, Z. P. (1984a). "Imbricate continuum and its variational derivation." *J. Eng. Mech.*, 110(12), 1693–1712.

Bažant, Z. P. (1984b). "Size effect in blunt fracture: Concrete, rock, and metal." *J. Eng. Mech.*, 110(4), 518–535.

Bažant, Z. P. (1985a). "Fracture in concrete and reinforced concrete." *Mechanics of geomaterials: Rocks, concretes, and soils*, Z. P. Bažant, ed., Wiley, London, 259–303.

Bažant, Z. P. (1985b). "Mechanics of fracture and progressive cracking concrete structures." *Fracture mechanics of concrete: Structural application and numerical calculation*, G. C. Sih and A. DiTommaso, eds., Martinus Nijhoff, Dordrecht, The Netherlands, 1, 1–94.

Bažant, Z. P. (1986). "Mechanics of distributed cracking." *Appl. Mech. Rev.*, 39, 675–705.

Bažant, Z. P. (1987). "Why continuum damage is nonlocal: Justification by quasiperiodic microcrack array." *Mech. Res. Commun.*, 14, 407–419.

Bažant, Z. P. (1991). "Why continuum damage is nonlocal: Micromechanics arguments." *J. Eng. Mech.*, 117(5), 1070–1087.

Bažant, Z. P. (1994). "Nonlocal damage theory based on micromechanics of crack interactions." *J. Eng. Mech.*, 120(3), 593–617.

Bažant, Z. P. (2001). "Concrete fracture models: Testing and practice." *Eng. Fract. Mech.*, 69, 165–206.

Bažant, Z. P. (2002a). "Reminiscences on four decades of struggle and progress in softening damage and size effect." *Concr. J. (Japan Concr. Inst.)*, 40, 16–28.

Bažant, Z. P. (2002b). *Scaling of structural strength*, Hermes-Penton, London.

Bažant, Z. P., and Belytschko, T. B. (1985). "Wave propagation in a strain-softening bar: Exact solution." *J. Eng. Mech.*, 111(3), 381–389.

Bažant, Z. P., Belytschko, T. B., and Chang, T.-P. (1984). "Continuum model for strain softening." *J. Eng. Mech.*, 110(12), 1666–1692.

Bažant, Z. P., Caner, F. C., Carol, I., Adley, M. D., and Akers, S. (2000). "Microplane model M4 for concrete: I. Formulation with work-conjugate deviatoric stress." *J. Eng. Mech.*, 126(9), 944–953.

Bažant, Z. P., and Cedolin, L. (1979). "Blunt crack band propagation in finite element analysis." *J. Eng. Mech.*, 105(2), 297–315.

Bažant, Z. P., and Cedolin, L. (1991). *Stability of structures*, Oxford University Press, New York, Chap. 10.

Bažant, Z. P., Červenka, J., and Wierer, M. (2001). "Equivalent localization element for crack band model and as alternative to elements with embedded discontinuities." *Fracture mechanics of concrete structures*, R. de Borst et al., ed., Balkema, Lisse, The Netherlands, 765–772.

Bažant, Z. P., and Chang, T.-P. (1984). "Instability of nonlocal continuum and strain averaging." *J. Eng. Mech.*, 110(10), 1441–1450.

Bažant, Z. P., and Christensen, M. (1972a). "Analogy between micropolar continuum and grid frameworks under initial stress." *Int. J. Solids Struct.*, 8, 327–346.

Bažant, Z. P., and Christensen, M. (1972b). "Long-wave extensional buckling of large regular frames." *J. Struct. Div., ASCE*, 98(10), 2269–2289.

Bažant, Z. P., and Jirásek, M. (1994). "Nonlocal model based on crack interactions: A localization study." *J. Eng. Mater. Technol.*, 116, 256–259.

Bažant, Z. P., and Li, Y.-N. (1995). "Stability of cohesive crack model." *J. Appl. Mech.*, 62, 959–969.

Bažant, Z. P., and Li, Y.-N. (1997). "Cohesive crack model with rate-dependent crack opening and viscoelasticity: I. Mathematical model and scaling." *Int. J. Fract.* 86, 247–265.

Bažant, Z. P., and Lin, F.-B. (1988a). "Nonlocal smeared cracking model for concrete fracture." *J. Struct. Eng.*, 114(11), 2493–2510.

Bažant, Z. P., and Lin, F.-B. (1988b). "Nonlocal yield-limit degradation." *Int. J. Numer. Methods Eng.*, 26, 1805–1823.

Bažant, Z. P., and Novák, D. (2000a). "Energetic-statistical size effect in quasibrittle failure at crack initiation." *ACI Mater. J.*, 97, 381–392.

Bažant, Z. P., and Novák, D. (2000b). "Probabilistic nonlocal theory for quasibrittle fracture initiation and size effect: I. Theory." *J. Eng. Mech.*, 126(2), 166–174.

Bažant, Z. P., and Novák, D. (2000c). "Probabilistic nonlocal theory for quasibrittle fracture initiation and size effect: II. Application." *J. Eng. Mech.*, 126(2), 175–185.

Bažant, Z. P., and Oh, B.-H. (1983). "Crack band theory for fracture of concrete." *Mater. Struct.*, 16, 155–177.

Bažant, Z. P., and Ohtsubo, H. (1977). "Stability conditions for propagation of a system of cracks in a brittle solid." *Mech. Res. Commun.*, 4, 353–366.

Bažant, Z. P., and Ožbolt, J. (1990). "Nonlocal microplane model for fracture, damage, and size effect in structures." *J. Eng. Mech.*, 116(11), 2485–2505.

Bažant, Z. P., and Pfeiffer, P. A. (1987). "Determination of fracture energy from size effect and brittleness number." *ACI Mater. J.*, 84, 463–480.

Bažant, Z. P., and Pijaudier-Cabot, G. (1988). "Nonlocal continuum damage, localization instability and convergence." *J. Appl. Mech.*, 55, 287–293.

Bažant, Z. P., and Pijaudier-Cabot, G. (1989). "Measurement of characteristic length of nonlocal continuum." *J. Eng. Mech.*, 115(4), 755–767.

Bažant, Z. P., and Planas, J. (1998). *Fracture and size effect in concrete and other quasibrittle materials*, CRC, Boca Raton, Fla.

Bažant, Z. P., Tabbara, M. R., Kazemi, M. T., and Pijaudier-Cabot, G. (1990). "Random particle model for fracture of aggregate or fiber composites." *J. Eng. Mech.*, 116(8), 1686–1705.

Bažant, Z. P., and Xi, Y. (1991). "Statistical size effect in quasibrittle structures: II. Nonlocal theory." *J. Eng. Mech.*, 117(11), 2623–2640.

Bažant, Z. P., Xiang, Y., Adley, M., Prat, P. C., and Akers, S. A. (1996). "Microplane model for concrete: II. Data delocalization and verification." *J. Eng. Mech.*, 122(3), 255–262.

Belytschko, T., Fish, J., and Engelmann, B. E. (1988). "A finite-element with embedded localization zones." *Comput. Methods Appl. Mech. Eng.*, 70, 59–89.

Benvenuti, E., Borino, G., and Tralli, A. (2000). "A thermodynamically consistent nonlocal formulation for elastodamaging materials: Theory and computations." *Proc., ECCOMAS 2000*, Barcelona, Spain, CD-ROM.

Besdo, D. (1974). "Ein Beitrag zur nichtlinearen Theorie des Cosserat Kontinuums." *Acta Mech.*, 20, 105–131.

Borino, G., and Failla, B. (2000). "Thermodynamically consistent plasticity models with local and nonlocal internal variables." *Proc., European Congr. on Computational Methods in Applied Sciences and Engineering, ECCOMAS*, Barcelona, Spain, CD-ROM.

Borino, G., Failla, B., and Parrinello, F. (2002). "A symmetric formulation for nonlocal damage models." H. A. Mang, F. G. Rammerstorfer, and J. Eberhardsteiner, eds., *Proc., 5th World Congress on Computa-

tional Mechanics (WCCM V), Vienna Univ. of Technology, Vienna, Austria, ISBN 3-9501554-0-6, ⟨http://wccm.tuwien.ac.at⟩.

Borino, G., Fuschi, P., and Polizzotto, C. (1999). "A thermodynamic approach to nonlocal plasticity and related variational approaches." *J. Appl. Mech.,* 66, 952–963.

Borino, G., and Polizzotto, C. (1999). "Comments on 'Nonlocal bar revisited by Christer Nilsson'." *Int. J. Solids Struct.,* 36, 3085–3091.

Boudon-Cussac, D., Hild, F., and Pijaudier-Cabot, G. (1999). "Tensile damage in concrete: Analysis of experimental technique." *J. Eng. Mech.,* 125(8), 906–913.

Camacho, G. T., and Ortiz, M. (1996). "Computational modeling of impact damage in brittle materials." *Int. J. Solids Struct.,* 33, 2899–2938.

Cangemi, L., Cocu, M., and Raous, M. (1996). "Adhesion and friction model for the fiber/matrix interface of a composite." *Proc., 1996 Engineering Systems Design and Analysis Conf.,* 157–163.

Carmeliet, J. (1999). "Optimal estimation of gradient damage parameters from localization phenomena in quasibrittle materials." *Mech. Cohesive-Frict. Mater.,* 4, 1–16.

Carol, I., and Bažant, Z. P. (1997). "Damage and plasticity in microplane theory." *Int. J. Solids Struct.,* 34, 3807–3835.

Carol, I., and Prat, P. C. (1990). "A statically constrained microplane model for the smeared analysis of concrete cracking." in *Computer aided analysis and design of concrete structures*, N. Bićanić and H. Mang, eds., Pineridge, Swansea, U.K., 919–930.

Cedolin, L., and Bažant, Z. P. (1980). "Effect of finite-element choice in blunt crack band analysis." *Comput. Methods Appl. Mech. Eng.,* 24, 305–316.

Červenka, J. (1994). "Discrete crack modeling in concrete structures." PhD thesis, Univ. of Colorado, Boulder, Colo.

Červenka, J., Bažant, Z. P., and Wierer, M. (2002). "Equivalent localization element for crack band approach to mesh-size sensitivity in microplane model." *Int. J. Numer. Methods Eng.,* in press.

Chaboche, J. L., Girard, R., and Levasseur, P. (1997a). "On the interface debonding models." *Int. J. Damage Mech.,* 6, 220–257.

Chaboche, J. L., Girard, R., and Schaff, A. (1997b). "Numerical analysis of composite systems by using interphase/interface models." *Computational Mech., Berlin,* 20, 3–11.

Chandrasekhar, S. (1950). *Radiation transfer,* Oxford University Press, London.

Comi, C. (2001). "A nonlocal model with tension and compression damage mechanisms." *Eur. J. Mech. A/Solids,* 20, 1–22.

Comi, C., and Perego, U. (2001). "Numerical aspects of nonlocal damage analyses." *Rev. Europ. Elements Finis,* 10, 227–242.

Cope, R. J., Rao, P. V., Clark, L. A., and Norris, P. (1980). "Modeling of reinforced concrete behavior for finite-element analysis of bridge slabs." *Numerical methods for nonlinear problems,* Pineridge, Swansea, 1, 457–470.

Cordebois, J. P., and Sidoroff, F. (1979). "Anisotropie élastique induite par endommagement." *Comportement mécanique des solides anisotropes, No. 295, Colloques internationaux du CNRS,* Editions du CNRS, Grenoble, France, 761–774.

Cosserat, E., and Cosserat, F. (1909). *Théorie des corps déformables,* Herrman, Paris.

Daux, C., Moës, N., Dolbow, J., Sukumar, N., and Belytschko, T. (2000). "Arbitrary branched and intersecting cracks with the extended finite-element method." *Int. J. Numer. Methods Eng.,* 48, 1741–1760.

de Borst, R. (1986). "Nonlinear analysis of frictional materials." PhD thesis, Delft Univ. of Technology, Delft, The Netherlands.

Denarié, E., Saouma, V. E., Iocco, A., and Varelas, D. (2001). "Concrete fracture process zone characterization with fiber optics." *J. Eng. Mech.,* 127(5), 494–502.

di Prisco, M., and Mazars, J. (1996). "Crush–crack: A nonlocal damage model for concrete." *J. Mech. Cohesive Frict. Mater.,* 1, 321–347.

Díez, P., Egozcue, J. J., and Huerta, A. (1998). "A posteriori error estimation for standard finite-element analysis." *Comput. Methods Appl. Mech. Eng.,* 163, 141–157.

Drugan, W. J., and Willis, J. R. (1996). "A micromechanics-based nonlocal constitutive equation and estimates of representative volume element size for elastic composites." *J. Mech. Phys. Solids,* 44, 497–524.

Duarte, C. A., and Oden, J. T. (1996). "H–p clouds—An h–p meshless method." *Numer. Methods Partial Diff. Eq.,* 12, 673–705.

Dugdale, D. S. (1960). "Yielding of steel sheets containing slits." *J. Mech. Phys. Solids,* 8, 100–108.

Duhem, P. (1893). "Le potentiel thermodynamique et la pression hydrostatique." *Ann. Sci. Ec. Normale Super.,* 10, 183–230.

Dvorkin, E. N., Cuitiño, A. M., and Gioia, G. (1990). "Finite elements with displacement interpolated embedded localization lines insensitive to mesh size and distortions." *Comput. Methods Appl. Mech. Eng.,* 90, 829–844.

Edelen, D. G. B. (1969). "Protoelastic bodies with large deformation." *Arch. Ration. Mech. Anal.,* 34, 283–300.

Edelen, D. G. B., Green, A. E., and Laws, N. (1971). "Nonlocal continuum mechanics." *Arch. Ration. Mech. Anal.,* 43, 36–44.

Edelen, D. G. B., and Laws, N. (1971). "On the thermodynamics of systems with nonlocality." *Arch. Ration. Mech. Anal.,* 43, 24–35.

Engelen, R. A. B., Geers, M. G. D., and Baaijens, F. P. T. (2002). "Nonlocal implicit gradient-enhanced elastoplasticity for the modeling of softening behavior." *Int. J. Plast.,* in press.

Ericksen, J. L. (1960). "Anisotropic fluids." *Arch. Ration. Mech. Anal.,* 4, 231–237.

Eringen, A. C. (1964). "Simple microfluids." *Int. J. Eng. Sci.,* 2, 205–217.

Eringen, A. C. (1966a). "Linear theory of micropolar elasticity." *J. Math. Mech.,* 15, 909–924.

Eringen, A. C. (1966b). "Mechanics of micromorphic materials." *Proc., 11th Int. Congr. of Applied Mechanics,* Springer, Berlin, 131–138.

Eringen, A. C. (1966c). "A unified theory of thermomechanical materials." *Int. J. Eng. Sci.,* 4, 179–202.

Eringen, A. C. (1972). "Linear theory of nonlocal elasticity and dispersion of plane waves." *Int. J. Eng. Sci.,* 10, 425–435.

Eringen, A. C. (1981). "On nonlocal plasticity." *Int. J. Eng. Sci.,* 19, 1461–1474.

Eringen, A. C. (1983). "Theories of nonlocal plasticity." *Int. J. Eng. Sci.,* 21, 741–751.

Eringen, A. C., and Ari, N. (1983). "Nonlocal stress field at a Griffith crack." *Cryst. Lattice Defects Amorphous Mater.,* 10, 33–38.

Eringen, A. C., and Edelen, D. G. B. (1972). "On nonlocal elasticity." *Int. J. Eng. Sci.,* 10, 233–248.

Eringen, A. C., and Kim, B. S. (1974). "Stress concentration at the tip of a crack." *Mech. Res. Commun.,* 1, 233–237.

Eringen, A. C., Speziale, C. G., and Kim, B. S. (1977). "Crack-tip problem in nonlocal elasticity." *J. Mech. Phys. Solids,* 25, 339–355.

Eringen, A. C., and Suhubi, E. S. (1964). "Nonlinear theory of simple microelastic solids." *Int. J. Eng. Sci.,* 2, 189–203, 389–404.

Etse, G., and Willam, K. J. (1994). "Fracture-energy formulation for inelastic behavior of plain concrete." *J. Eng. Mech.,* 120(9), 1983–2011.

Fabrikant, V. I. (1990). "Complete solutions to some mixed boundary value problems in elasticity." *Adv. Appl. Mech.,* 27, 153–223.

Ferrara, L. (1998). "A contribution to the modeling of mixed-mode fracture and shear transfer in plain and reinforced concrete." PhD thesis, Politecnico di Milano, Milano, Italy.

Fischer, R. A., and Tippett, L. H. C. (1928). "Limiting forms of the frequency distribution of the largest and smallest member of a sample." *Proc. Cambridge Philos. Soc.,* 24, 180–190.

Fish, J., and Yu, Q. (2001). "Multiscale damage modeling for composite materials: Theory and computational framework." *Int. J. Numer. Methods Eng.,* 52, 161–191.

Fleck, N. A., and Hutchinson, J. W. (1993). "A phenomenological theory for strain gradient effects in plasticity." *J. Mech. Phys. Solids,* 41, 1825–1857.

Fleck, N. A., and Hutchinson, J. W. (2001). "A reformulation of strain gradient plasticity." *J. Mech. Phys. Solids,* 49, 2245–2271.

Fleck, N. A., Muller, G. M., Ashby, M. F., and Hutchinson, J. W. (1994). "Strain gradient plasticity: Theory and experiment." *Acta Metall. Mater.,* 42, 475–487.

Gao, H., and Huang, Y. (2001). "Taylor-based nonlocal theory of plasticity." *Int. J. Solids Struct.*, 38, 2615–2637.

Gao, H., Huang, Y., Nix, W. D., and Hutchinson, J. W. (1999). "Mechanism-based strain gradient plasticity. I. Theory." *J. Mech. Phys. Solids*, 47, 1239–1263.

Geers, M. G. D., de Borst, R., and Brekelmans, W. A. M. (1996a). "Computing strain fields from discrete displacement fields in 2D solids." *Int. J. Solids Struct.*, 33, 4293–4307.

Geers, M. G. D., Engelen, R. A. B., and Ubachs, R. J. M. (2001). "On the numerical modeling of ductile damage with an implicit gradient-enhanced formulation." *Rev. Euro. Eléments finis*, 10, 173–191.

Geers, M. G. D., Peijs, T., Brekelmans, W. A. M., and de Borst, R. (1996b). "Experimental monitoring of strain localization and failure behavior of composite materials." *Compos. Sci. Technol.*, 56, 1283–1290.

Green, A. E. (1965). "Micromaterials and multipolar continuum mechanics." *Int. J. Eng. Sci.*, 3, 533–537.

Green, A. E., and Naghdi, P. M. (1965). "A general theory of an elastic-plastic continuum." *Arch. Ration. Mech. Anal.*, 18, 251–281.

Green, A. E., and Rivlin, R. S. (1964a). "Multipolar continuum mechanics." *Arch. Ration. Mech. Anal.*, 17, 113–147.

Green, A. E., and Rivlin, R. S. (1964b). "Simple force and stress multipoles." *Arch. Ration. Mech. Anal.*, 16, 325–353.

Grioli, G. (1960). "Elasticità asimmetrica." *Annali di matematica pura ed applicata*, Ser. IV, 50, 389–417.

Günther, W. (1958). "Zur Statik und Kinematik des Cosseratschen Kontinuum." *Abh. Braunschweigischen Wissenschaftlichen Gesellschaft*, 10, 195–213.

Gupta, A. K., and Akbar, H. (1984). "Cracking in reinforced concrete analysis." *J. Struct. Eng.*, 110(8), 1735–1746.

Hadamard, J. (1903). *Leçons sur la propagation des ondes*, Hermann, Paris.

Halphen, B., and Nguyen, Q. S. (1975). "Sur les matériaux standards généralisés." *J. Mec.*, 14, 39–63.

Hillerborg, A., Modéer, M., and Peterson, P. E. (1976). "Analysis of crack propagation and crack growth in concrete by means of fracture mechanics and finite elements." *Cem. Concr. Res.*, 6, 773–782.

Hodgkin, A. L. (1964). *The conduction of nervous impulse*, Thomas, Springfield, Ill.

Hu, X. Z., and Wittmann, F. H. (2000). "Size effect on toughness induced by cracks close to free surface." *Eng. Fract. Mech.*, 65, 209–211.

Huang, Y., Gao, H., Nix, W. D., and Hutchinson, J. W. (2000). "Mechanism-based strain gradient plasticity. II. Analysis." *J. Mech. Phys. Solids*, 48, 99–128.

Huerta, A., Díez, P., Rodríguez-Ferran, A., and Pijaudier-Cabot, G. (1998). "Error estimation and finite-element analysis of softening solids." *Advances in adaptive computational methods in mechanics*, P. Ladevèze and J. T. Oden, eds., Elsevier, Oxford, U.K., 333–348.

Huerta, A., and Pijaudier-Cabot, G. (1994). "Discretization influence on regularization by two localisation limiters." *J. Eng. Mech.*, 120(6), 1198–1218.

Irwin, G. R. (1958). "Fracture." in *Handbuch der Physik*, S. Flügge, ed., Springer, Berlin, VI, 551–590.

Jirásek, M. (1998a). "Finite elements with embedded cracks." *LSC Internal Rep. No. 98/01*, Swiss Federal Institute of Technology, Lausanne, Switzerland.

Jirásek, M. (1998b). "Nonlocal models for damage and fracture: Comparison of approaches." *Int. J. Solids Struct.*, 35, 4133–4145.

Jirásek, M. (1999). "Comments on microplane theory." *Mechanics of quasibrittle materials and structures*, G. Pijaudier-Cabot, Z. Bittnar, and B. Gérard, eds., Hermès Science, Paris, 55–77.

Jirásek, M. (2000a). "Comparative study on finite elements with embedded cracks." *Comput. Methods Appl. Mech. Eng.*, 188, 307–330.

Jirásek, M. (2000b). "Conditions of uniqueness for finite elements with embedded cracks." *Proc., 6th Int. Conf. on Computational Plasticity*, Barcelona. CD-ROM.

Jirásek, M., and Bažant, Z. P. (1994). "Localization analysis of nonlocal model based on crack interactions." *J. Eng. Mech.*, 120(7), 1521–1542.

Jirásek, M., and Bažant, Z. P. (1995). "Macroscopic fracture characteristics of random particle systems." *Int. J. Fract.*, 69, 201–228.

Jirásek, M., and Bažant, Z. P. (2001). *Inelastic analysis of structures*, Wiley, Chichester, U.K.

Jirásek, M., and Patzák, B. (2002). "Consistent tangent stiffness for nonlocal damage models." *Comput. Struct.*, in press.

Jirásek, M., and Rolshoven, S. (2002). "Comparison of integral-type nonlocal plasticity models for strain-softening materials." *Int. J. Eng. Sci.*, in press.

Jirásek, M., and Zimmermann, T. (1997). "Rotating crack model with transition to scalar damage: I. Local formulation, II. Nonlocal formulation and adaptivity." *LSC Internal Rep. No. 97/01*, Swiss Federal Institute of Technology, Lausanne, Switzerland.

Jirásek, M., and Zimmermann, T. (1998a). "Analysis of rotating crack model." *J. Eng. Mech.*, 124(8), 842–851.

Jirásek, M., and Zimmermann, T. (1998b). "Rotating crack model with transition to scalar damage." *J. Eng. Mech.*, 124(3), 277–284.

Jirásek, M., and Zimmermann, T. (2001). "Embedded crack model: I. Basic formulation." *Int. J. Numer. Methods Eng.*, 50, 1269–1290.

Kachanov, M. (1985). "A simple technique of stress analysis in elastic solids with many cracks." *Int. J. Fract.*, 28, R11–R19.

Kachanov, M. (1987). "Elastic solids with many cracks: A simple method of analysis." *Int. J. Solids Struct.*, 23, 23–43.

Kang, H. D. (1997). "Triaxial constitutive model for plain and reinforced concrete behavior." PhD thesis, Univ. of Colorado, Boulder, Colo.

Kennedy, T. C., and Nahan, M. F. (1996). "A simple nonlocal damage model for predicting failure of notched laminates." *Compos. Struct.*, 35, 229–236.

Kennedy, T. C., and Nahan, M. F. (1997). "A simple nonlocal damage model for predicting failure in a composite shell containing a crack." *Compos. Struct.*, 39, 85–91.

Klisinski, M., Runesson, K., and Sture, S. (1991). "Finite element with inner softening band." *J. Eng. Mech.*, 117(3), 575–587.

Koiter, W. T. (1964). "Couple stresses in the theory of elasticity." *Proc. K. Ned. Akad. Wet., Ser. B: Phys. Sci.*, 67, 17–44.

Kröner, E. (1966). "Continuum mechanics and range of atomic cohesion forces." *Proc., 1st Int. Conf. on Fracture*, T. Yokobori, T. Kawasaki, and J. Swedlow, eds., Japanese Society for Strength and Fracture of Materials, Sendai, Japan, 27.

Kröner, E. (1967). "Elasticity theory of materials with long range cohesive forces." *Int. J. Solids Struct.*, 3, 731–742.

Kröner, E., and Datta, B. K. (1966). "Nichtlokale Elastostatik: Ableitung aus der Gittertheorie." *Z. Phys.*, 196, 203–211.

Krumhansl, J. A. (1965). "Generalized continuum field representation for lattice vibrations." *Lattice dynamics*, R. F. Wallis, ed., Pergamon, London, 627–634.

Kunin, I. A. (1966a). "Model of elastic medium simple structure with spatial dispersion." *Prikl. Mat. Mekh.*, 30, 942 (in Russian).

Kunin, I. A. (1966b). "Theory of elasticity with spatial dispersion." *Prikl. Mat. Mekh.*, 30, 866 (in Russian).

Kunin, I. A. (1968). "The theory of elastic media with microstructure and the theory of dislocations." *Mechanics of generalized continua*, E. Kröner, ed., Springer, Heidelberg, 321–329.

Labuz, J. F., Cattaneo, S., and Chen, L.-H. (2001). "Acoustic emission at failure in quasibrittle materials." *Constr. Build. Mater.*, 15, 225–233.

Ladevèze, P. (1992). "A damage computational method for composite structures." *Comput. Struct.*, 44, 79–87.

Ladevèze, P. (1995). "A damage computational method for composites: Basic aspects and micromechanical relations." *Computational Mech., Berlin*, 17, 142–150.

Ladevèze, P., Allix, O., Deü, J.-F., and Lévêque, D. (2000). "A meso-model for localization and damage computation in laminates." *Comput. Methods Appl. Mech. Eng.*, 183, 105–122.

Lakes, R. S. (1986). "Experimental microelasticity of two porous solids." *Int. J. Solids Struct.*, 22, 55–63.

Landis, E. N. (1999). "Micro–macro-fracture relationships and acoustic emissions in concrete." *Constr. Build. Mater.*, 13, 65–72.

Larsson, R., Runesson, K., and Sture, S. (1996). "Embedded localization band in undrained soil based on regularized strong discontinuity—

Theory and FE analysis." *Int. J. Solids Struct.,* 33, 3081–3101.

Lemaitre, J., and Chaboche, J.-L. (1990). *Mechanics of solid materials,* Cambridge University Press, Cambridge, U.K.

Lippmann, H. (1969). "Eine Cosserat Theorie des plastischen Fließens," *Acta Mech.,* 8, 255–284.

Lotfi, H. R., and Shing, P. B. (1995). "Embedded representation of fracture in concrete with mixed finite elements." *Int. J. Numer. Methods Eng.,* 38, 1307–1325.

Luciano, R., and Willis, J. R. (2001). "Nonlocal constitutive response of a random laminate subjected to configuration-dependent body force." *J. Mech. Phys. Solids,* 49, 431–444.

Ma, Q., and Clarke, D. R. (1995). "Size-dependent hardness in silver single crystals." *J. Mater. Res.,* 10, 853–863.

Mariotte, E. (1686). *Traité du mouvement des eaux,* M. de la Hire, ed., English translation by J. T. Desvaguliers, London (1718).

Mazars, J. (1984). "Application de la mécanique de l'endommagement au comportement nonlinéaire et à la rupture du béton de structure." *Thèse de Doctorat d'Etat,* Univ. Paris VI, France.

Mazars, J., and Berthaud, Y. (1989). "Une technique expérimentale appliquée au béton pour créer un endommagement diffus et mettre en évidence son caractère unilateral." *Compt. Rendus Acad. Sci., Paris,* 308, 579–584.

Mazars, J., Berthaud, Y., and Ramtani, S. (1990). "The unilateral behavior of damaged concrete." *Eng. Fract. Mech.,* 35, 629–635.

Melenk, J. M., and Babuška, I. (1996). "The partition of unity finite-element method: Basic theory and applications." *Comput. Methods Appl. Mech. Eng.,* 39, 289–314.

Mihashi, H., and Nomura, N. (1996). "Correlation between characteristics of fracture process zone and tension-softening properties of concrete." *Nucl. Eng. Des.,* 165, 359–376.

Mindlin, R. D. (1964). "Microstructure in linear elasticity." *Arch. Ration. Mech. Anal.,* 16, 51–78.

Mindlin, R. D. (1965). "Second gradient of strain and surface tension in linear elasticity." *Int. J. Solids Struct.,* 1, 417–438.

Mindlin, R. D., and Tiersten, H. F. (1962). "Effects of couple stresses in linear elasticity." *Arch. Ration. Mech. Anal.,* 11, 415–448.

Moës, N., and Belytschko, T. (2002). "Extended finite-element method for cohesive crack growth." *Eng. Fract. Mech.,* 69, 813–833.

Moës, N., Dolbow, J., and Belytschko, T. (1999). "A finite-element method for crack growth without remeshing." *Int. J. Numer. Methods Eng.,* 46, 131–150.

Morrison, J. L. M. (1939). "The yield of mild steel with particular reference to the effect of size of specimen." *Proc. Inst. Mech. Eng.,* 142, 193–223.

Naghdi, P. M., and Trapp, J. A. (1975). "The significance of formulating plasticity theory with reference to loading surfaces in strain space." *Int. J. Eng. Sci.,* 13, 785–797.

Needleman, A. (1987). "A continuum model for void nucleation by inclusion debonding." *J. Appl. Mech.,* 54, 525–531.

Needleman, A. (1988). "Material rate dependence and mesh sensitivity in localization problems." *Comput. Methods Appl. Mech. Eng.,* 67, 69–85.

Nilsson, C. (1994). "On nonlocal plasticity, strain softening, and localization." *Rep. No. TVSM-1007,* Division of Structural Mechanics, Lund Institute of Technology, Lund, Sweden.

Nilsson, C. (1997). "Nonlocal strain softening bar revisited." *Int. J. Solids Struct.,* 34, 4399–4419.

Nilsson, C. (1999). "Author's closure." *Int. J. Solids Struct.,* 36, 3093–3100.

Nix, W. D. (1989). "Mechanical properties of thin films." *Metall. Trans. A,* 20A, 2217–2245.

Noll, W. (1972). "A new mathematical theory of simple materials." *Arch. Ration. Mech. Anal.,* 48, 1–50.

Nooru-Mohamed, M. B. (1992). "Mixed-mode fracture of concrete: An experimental approach." PhD thesis, Delft Univ. of Technology, Delft, The Netherlands.

Ohlsson, U., and Olofsson, T. (1997). "Mixed-mode fracture and anchor bolts in concrete: Analysis with inner softening bands." *J. Eng. Mech.,* 123(10), 1027–1033.

Oliver, J. (1996). "Modeling strong discontinuities in solid mechanics via strain softening constitutive equations. Part 1: Fundamentals. Part 2: Numerical simulation." *Int. J. Numer. Methods Eng.,* 39, 3575–3624.

Oliver, J., Cervera, M., and Manzoli, O. (1998). "On the use of strain-softening models for the simulation of strong discontinuities in solids." *Material instabilities in solids,* R. de Borst and E. van der Giessen, eds., Wiley, Chichester, U.K., 107–123.

Olofsson, T., Klisinski, M., and Nedar, P. (1994). "Inner softening bands: A new approach to localization in finite elements." *Computational modeling of concrete structures,* H. Mang, N. Bićanić, and R. de Borst, eds., Pineridge, Swansea, U.K., 373–382.

Ortiz, M. (1988). "Microcrack coalescence and macroscopic crack growth initiation in brittle solids." *Int. J. Solids Struct.,* 24, 231–250.

Ortiz, M., Leroy, Y., and Needleman, A. (1987). "A finite-element method for localized failure analysis." *Comput. Methods Appl. Mech. Eng.,* 61, 189–214.

Ortiz, M., and Pandolfi, A. (1999). "Finite-deformation irreversible cohesive elements for three-dimensional crack-propagation analysis." *Int. J. Numer. Methods Eng.,* 44, 1267–1282.

Oseen, C. W. (1933). "The theory of liquid crystals." *Trans. Faraday Soc.,* 29, 883–899.

Otsuka, K., and Date, H. (2000). "Fracture process zone in concrete tension specimen." *Eng. Fract. Mech.,* 65, 111–131.

Ožbolt, J., and Bažant, Z. P. (1996). "Numerical smeared fracture analysis: Nonlocal microcrack interaction approach." *Int. J. Numer. Methods Eng.,* 39, 635–661.

Pandolfi, A., Krysl, P., and Ortiz, M. (1999). "Finite-element simulation of ring expansion and fragmentation: The capturing of length and time scales through cohesive models of fracture." *Int. J. Fract.,* 95, 279–297.

Peerlings, R. H. J., de Borst, R., Brekelmans, W. A. M., and de Vree, J. H. P. (1996). "Gradient-enhanced damage for quasibrittle materials." *Int. J. Numer. Methods Eng.,* 39, 3391–3403.

Pietruszczak, S., and Mróz, Z. (1981). "Finite-element analysis of deformation of strain-softening materials." *Int. J. Numer. Methods Eng.,* 17, 327–334.

Pijaudier-Cabot, G., and Bažant, Z. P. (1987). "Nonlocal damage theory." *J. Eng. Mech.,* 113(10), 1512–1533.

Planas, J., and Elices, M. (1992). "Asymptotic analysis of a cohesive crack: 1. Theoretical background." *Int. J. Fract.,* 55, 153–177.

Planas, J., and Elices, M. (1993). "Asymptotic analysis of a cohesive crack: 2. Influence of the softening curve." *Int. J. Fract.,* 64, 221–237.

Planas, J., Elices, M., and Guinea, G. V. (1993). "Cohesive cracks versus nonlocal models: Closing the gap." *Int. J. Fract.,* 63, 173–187.

Planas, J., Guinea, G. V., and Elices, M. (1996). "Basic issues on nonlocal models: Uniaxial modeling." *Tech. Rep. No. 96-jp03,* Departamento de Ciencia de Materiales, ETS de Ingenieros de Caminos, Univ. Politécnica de Madrid, Ciudad Univ. sn., 28040 Madrid, Spain.

Polizzotto, C. (2001). "Nonlocal elasticity and related variational principles." *Int. J. Solids Struct.,* 38, 7359–7380.

Polizzotto, C. (2002). "Remarks on some aspects of nonlocal theories in solid mechanics." *Proc., 6th National Congr. SIMAI,* Chia Laguna, Italy, CD-ROM.

Polizzotto, C., Borino, G., and Fuschi, P. (1998). "A thermodynamic consistent formulation of nonlocal and gradient plasticity." *Mech. Res. Commun.,* 25, 75–82.

Poole, W. J., Ashby, M. F., and Fleck, N. A. (1996). "Microhardness of annealed and work-hardened copper polycrystals." *Scr. Mater.,* 34, 559–564.

Rajagopal, E. S. (1960). "The existence of interfacial couples in infinitesimal elasticity." *Ann. Phys. (Leipzig),* 6, 192–201.

Rashid, Y. R. (1968). "Analysis of prestressed concrete pressure vessels." *Nucl. Eng. Des.,* 7, 334–344.

Rayleigh, O. M. (1918). "Notes on the theory of lubrication." *Philos. Mag.,* 35, 1–12.

Read, H. E., and Hegemier, G. A. (1984). "Strain softening of rock, soil, and concrete—A review article." *Mech. Mater.,* 3, 271–294.

Richards, C. W. (1958). "Effects of size on the yielding of mild steel beams." *Proc. Am. Soc. Test. Mater.*, 58, 955–970.

Rodríguez-Ferran, A., and Huerta, A. (2000). "Error estimation and adaptivity for nonlocal damage models." *Int. J. Solids Struct.*, 37, 7501–7528.

Rogula, D. (1965). "Influence of spatial acoustic dispersion on dynamical properties of dislocations. I." *Bull. Acad. Pol. Sci., Ser. Sci. Tech.*, 13, 337–343.

Rogula, D. (1982). "Introduction to nonlocal theory of material media." *Nonlocal theory of material media, CISM courses and lectures*, D. Rogula, ed., Springer, Wien, 268, 125–222.

Rolshoven, S., and Jirásek, M. (2001). "On regularized plasticity models for strain-softening materials." *Fracture mechanics of concrete structures*, R. de Borst, J. Mazars, G. Pijaudier-Cabot, and J. G. M. van Mier, eds., Balkema, Lisse, 617–624.

Rossi, P., and Wu, X. (1992). "Probabilistic model for material behavior analysis and appraisement of concrete structures." *Mag. Concrete Res.*, 44, 271–280.

Rots, J. G. (1988). "Computational modeling of concrete fracture." PhD thesis, Delft Univ. of Technol., Delft, The Netherlands.

Sandler, I. S. (1984). "Strain-softening for static and dynamic problems." *Proc., Symp. on Constitutive Equations*, ASME, Winter Annual Meeting, New Orleans, 217–231.

Saouridis, C. (1988). "Identification et numérisation objectives des comportements adoucissants: Une approche multiéchelle de l'endommagement du béton." PhD thesis, Univ. Paris VI.

Saouridis, C., and Mazars, J. (1992). "Prediction of the failure and size effect in concrete via a biscale damage approach." *Eng. Comput.*, 9, 329–344.

Schlangen, E. (1993). "Experimental and numerical analysis of fracture processes in concrete." PhD thesis, Delft Univ. of Technology, Delft, The Netherlands.

Schlangen, E., and van Mier, J. G. M. (1992). "Simple lattice model for numerical simulation of fracture of concrete materials and structures." *Mater. Struct.*, 25, 534–542.

Shi, G. H. (1991). "Manifold method of material analysis." *Trans., 9th Army Conf. on Applied Mathematics and Computing*, Minneapolis, 57–76.

Simo, J. C., and Oliver, J. (1994). "A new approach to the analysis and simulation of strain softening in solids." *Fracture and damage in quasibrittle structures*, Z. P. Bažant, Z. Bittnar, M. Jirásek, and J. Mazars, eds., E&FN Spon, London, 25–39.

Simo, J. C., Oliver, J., and Armero, F. (1993). "An analysis of strong discontinuities induced by strain softening in rate-independent inelastic solids." *Computational Mech., Berlin*, 12, 277–296.

Sluys, L. J. (1997). "Discontinuous modeling of shear banding." *Computational plasticity: Fundamentals and applications*, D. R. J. Owen, E. Oñate, and E. Hinton, eds., Int. Center for Numerical Methods in Eng., Barcelona, Spain, 735–744.

Stolken, J. S., and Evans, A. G. (1998). "A microbend test method for measuring the plasticity length scale." *Acta Mater.*, 46, 5109–5115.

Strömberg, L., and Ristinmaa, M. (1996). "FE formulation of a nonlocal plasticity theory." *Comput. Methods Appl. Mech. Eng.*, 136, 127–144.

Sukumar, N., Moës, N., Moran, B., and Belytschko, T. (2000). "Extended finite-element method for three-dimensional crack modeling." *Int. J. Numer. Methods Eng.*, 48, 1549–1570.

Svedberg, T. (1996). "A thermodynamically consistent theory of gradient-regularized plasticity coupled to damage." *Licentiate thesis*, Chalmers Univ. of Technology.

Svedberg, T., and Runesson, K. (1998). "Thermodynamically consistent nonlocal and gradient formulations of plasticity." *Nonlocal aspects in solid mechanics*, A. Brillard and J. F. Ganghoffer, eds., EUROMECH Colloquium 378, Mulhouse, France, 32–37.

Thomas, T. Y. (1961). *Plastic flow and fracture of solids*, Academic, New York.

Toupin, R. A. (1962). "Elastic materials with couple stresses." *Arch. Ration. Mech. Anal.*, 11, 385–414.

Toupin, R. A. (1964). "Theories of elasticity with couple stress." *Arch. Ration. Mech. Anal.*, 17, 85–112.

Truesdell, C., and Toupin, R. A. (1960). "Classical field theories of mechanics." *Handbuch der physik*, Springer, Berlin, III, 1.

Tvergaard, V., and Hutchinson, J. W. (1992). "The relation between crack growth resistance and fracture process parameters in elastic–plastic solids." *J. Mech. Phys. Solids*, 40, 1377–1397.

Tvergaard, V., and Hutchinson, J. W. (1993). "The influence of plasticity on mixed mode interface toughness." *J. Mech. Phys. Solids*, 41, 1119–1135.

Valanis, K. C. (1990). "A theory of damage in brittle materials." *Eng. Fract. Mech.*, 36, 403–416.

Valanis, K. C. (1991). "A global damage theory and the hyperbolicity of the wave problem." *J. Appl. Mech.*, 58, 311–316.

van Gils, M. A. J. (1997). "Quasibrittle fracture of ceramics." PhD thesis, Eindhoven Univ. of Technology, The Netherlands.

van Mier, J. G. M. (1997). *Fracture processes of concrete*, CRC, Boca Raton, Fla.

Vermeer, P. A., and Brinkgreve, R. B. J. (1994). "A new effective nonlocal strain measure for softening plasticity." *Localization and bifurcation theory for soils and rocks*, R. Chambon, J. Desrues, and I. Vardoulakis, eds., Balkema, Rotterdam, The Netherlands, 89–100.

Voigt, W. (1887). "Theoretische Studien über die Elastizitätsverhältnisse der Krystalle." Abh. Königlichen Gesellschaft Wiss. Göttingen.

Voigt, W. (1894). "Über Medien ohne innere Kräfte und eine durch sie gelieferte mechanische Deutung der Maxwell-Hertzschen Gleichungen." *Abh. Königlichen Gesellschaft Wiss. Göttingen*, 40, 72–79.

Walsh, P. F. (1972). "Fracture of plain concrete." *Indian Concr. J.*, 46, 469–470 and 476.

Walsh, P. F. (1976). "Crack initiation in plain concrete." *Mag. Concrete Res.*, 28, 37–41.

Wei, Y., and Hutchinson, J. W. (1999). "Models of interface separation accompanied by plastic dissipation at multiple scales." *Int. J. Fract.*, 95, 1–17.

Weibull, W. (1939). "The phenomenon of rupture in solids." *Proc. R. Swedish Inst. Eng. Res. (Ing. Akad. Handl.)*, 153, 1–55.

Wells, G. N. (2001). "Discontinuous modeling of strain localization and failure." PhD thesis, Delft Univ. of Technol., Delft, The Netherlands.

Willam, K., Bićanić, N., and Sture, S. (1986). "Composite fracture model for strain-softening and localized failure of concrete." *Computational modeling of reinforced concrete structures*, E. Hinton and D. R. J. Owen, eds., Pineridge, Swansea, 122–153.

Xu, X.-P., and Needleman, A. (1994). "Numerical simulations of fast crack growth in brittle solids." *J. Mech. Phys. Solids*, 42, 1397–1434.

American Society of   *150th Anniversary*
Civil Engineers 1852–2002 *Paper*
Building a Better World

Observations on Low-Speed Aeroelasticity

Robert H. Scanlan, Hon.M.ASCE[1]

Abstract: A brief history of developments in the field of low-speed aeroelasticity is provided in the context of application to long-span bridge structures. The paper begins with summary of some of the significant developments in aeroelasticity for low-speed aeronautical applications in the early 20th century. The role of the pivotal Tacoma Narrows failure and subsequent investigation is introduced. The development of formal experimental and analytical tools for the prediction of long-span bridge response to wind is presented, and their roots in—but differences from—classical low-speed aeroelasticity are presented and discussed (e.g., aerodynamic admittance). The important issue of Reynolds number scaling is discussed and posed as a problem that requires resolution in future research. Details of analytical and experimental techniques are not provided herein; readers are referred to the references for developments in these areas. The intent of the paper is to emphasize the parallels between these two strongly related fields, and in so doing, highlight the role of classical aeronautical engineering in the development of state-of-the-art bridge wind engineering.

DOI: 10.1061/(ASCE)0733-9399(2002)128:12(1254)

CE Database keywords: Aeroelasticity; Bridges; Wind loads.

Introduction

The vortex-excited Aeolian harp of the ancient Greeks is one of the earliest historical examples of an aeroelastic or fluid-structure interactive phenomenon. Its characteristics were closely studied in the 1870s by Strouhal (1878). Among much later examples might be cited the striking cases of wind-induced flutter of the truss wires, wings, and tails of stick-and-wire aircraft of World War I vintage. Most of the early aircraft examples of aeroelastic effects were physically observed phenomena unaccompanied by theory, but by the 1920s attempts at theoretical descriptions of aircraft-related flight-induced oscillatory phenomena had advanced considerably with the work of Frazer (Frazer and Duncan 1928), Collar, Pugsley, and Glauert (Glauert 1928) in England and Birnbaum (1924), Wagner (1925), and others in Germany. Den Hartog's (1932) transmission-line galloping analysis is a representative nonaeronautical aeroelastic example. His simple steady-flow "galloping criterion" associated with a negative lift gradient has served to elucidate incipient galloping conditions in numerous practical cases over some seven decades.

By the mid- and latter 1930s firm analytical descriptions of the flutter of thin airfoils in incompressible flow had become available through the work of individuals like Küssner (1929) and

Schwarz in Germany and Theodorsen (Theodorsen 1935; Theodorsen and Garrick 1941), Garrick (1939), von Karman Sears (1941) in the United States. From December 1941 through 1945 the United States was at war, and engineers, including the writer, applied their skills to aircraft design, analysis, and production. Technical problems in aeroelasticity were studied intensely during this period. Together with others in different locations in the United States the writer worked on problems of aircraft structural dynamics and flutter in 1942–1946. A body of structural dynamic theory and practice was thus accumulated in those technically productive years. Particularly with the great impetus given aircraft development during World War II, studies of aeroelastic effects progressed by substantial leaps. By now, such studies have ranged throughout all flight speed regimes, from incompressible through supersonic and beyond. The present account will focus on some of the low speed effects only.

In the 1950s three American textbooks on aeroelasticity appeared: Scanlan and Rosenbaum (1951) *Aircraft Vibration and Flutter*, Fung (1955) *The Theory of Aeroelasticity*, and Bisplinghoff et al. (1955) *Aeroelasticity*. These texts set forth state-of-the-art precepts in the field of aircraft aeroelasticity that subsequently played useful roles in practical aircraft design and parallel academic background studies. The discussion of bluff-body aeroelasticity offered at a later point in this paper would be lacking in perspective without at least mention of these important summaries of pioneering studies.

With the preceding remarks as a brief introduction, the present review will proceed from these historical concerns to examine a selected few of the low-speed aeroelastic phenomena associated with bluff structures—particularly bridges—in the wind. While throughout the 20th century aircraft design advanced by concentrating on high-speed flow around streamlined objects, in more recent decades the field of civil engineering has benefited from increasing attention to the many modest yet recondite concerns of slower flows around bluff bodies.

[1]Deceased, formerly Homewood Professor, Dept. of Civil Engineering, The Johns Hopkins Univ., Baltimore, MD 21218; this paper was almost complete at the time of Robert Scanlan's death, and was completed by Robert N. Scanlan (son), Elizabeth C. Scanlan (spouse), and Nicholas P. Jones (colleague).

Note. Associate Editor: Stein Sture. Discussion open until May 1, 2003. Separate discussions must be submitted for individual papers. To extend the closing date by one month, a written request must be filed with the ASCE Managing Editor. The manuscript for this paper was submitted for review and possible publication on June 25, 2002; approved on June 25, 2002. This paper is part of the *Journal of Engineering Mechanics*, Vol. 128, No. 12, December 1, 2002. ©ASCE, ISSN 0733-9399/2002/12-1254–1258/$8.00+$.50 per page.

Few nonaeronautical aeroelastic examples have received the broad publicity given the dramatic wind-induced collapse of the Tacoma Narrows Bridge that occurred on November 7, 1940. This took place in a cross wind of some 42 mi/h under the direct observation (and filmed documentation) of a few eyewitnesses. Chief among these eyewitnesses was Professor F. B. Farquharson of the University of Washington, who was preoccupied with that event over many months, both before and after the collapse. The Tacoma Narrows Bridge disaster is widely and correctly cited as the critical triggering episode of modern bridge aeroelasticity and may now be viewed as the key motivating point of departure for studies in that field.

The initial, most important, and most accurate review of the event occurred in a detailed report (Ammanh et al. 1941) addressed to the U.S. Federal Works Agency. In that report a key section outlined the experimentally demonstrated evolution of the aerodynamic damping of a torsionally oscillating section model of the Tacoma Narrows Bridge, examined in the wind tunnel by Professor Dunn of Caltech.

In spite of the clarity of the 1941 report, it did not succeed in putting the matter to rest in the public mind. Perhaps it was read by too few. In any event, a series of Tacoma Narrows tales, begun in 1941, has continued sporadically ever since—some 60 years to date. Many of these tales veered substantially—even inventively—wide of the known facts.

Billah and Scanlan (1991) undertook, in a paper in the American Journal of Physics to redress some of the varying accounts of the destructive Tacoma Narrows event. The deck of that bridge girder consisted of a squat, H-section profile with two outstanding characteristics: a shape strongly susceptible of tripping a leading edge vortex, and a relatively weak structural stiffness in torsion that enabled twisting oscillation. These characteristics combined in strong wind to engender flutter instability. Objecting to the oft-cited but over-simplistic "resonance" characterization of the associated bridge motion found in numerous textbooks, we described the Tacoma phenomenon instead as an interactive wind-induced, self-excited torsional oscillation. The bridge response just prior to collapse was *not* simply a case of the well-known Karman "vortex street" phenomenon.

During 1944–1954, Farquharson published a set of five comprehensive reports on the original and replacement Tacoma Narrows designs (Farquharson 1949). In the period of the late 1940s and early 1950s, Prof. F. Bleich of Columbia University offered a theoretical analysis of the Tacoma phenomenon based, unfortunately, on the Theodorsen airfoil theory (clearly misapplied in that context) (Bleich 1948). During the 1950s Vincent and Farquharson, on a more practical tack, created and tested dynamic wind tunnel section models of several bridges, including the original Tacoma Narrows and Golden Gate spans. Both of the associated section models demonstrated strong torsional instability in the wind tunnel. To the present day, the Golden Gate span evidences only marginally acceptable stability under cross wind, this in spite of the addition to the deck girder of a lower-chord horizontal "wind truss" installed around 1955. In the early 1990s Raggett and Scanlan made wind tunnel models and performed theoretical studies that demonstrated the possibility of substantially increasing the flutter speed of the Golden Gate Bridge by adding certain deck and railing fairings.

In the late 1960s and early 1970s me and my students Sabzevari and Tomko (Scanlan and Sabzevari 1967, 1969; Scanlan and Tomko 1971) established early analytic models to parallel the action of several of the physical wind tunnel models created and studied by Vincent (1952) and Farquharson (1949) in the United States and Scruton (1952) in England. Our analytical modeling followed the natural course of first imitating the *style* of the theoretical Theodorsen airfoil model, but we soon recognized that the typical *bluff body* represented by most bridge deck sections then extant could not be expected to establish detailed flow characteristics and structural responses simulating those involving an airfoil. In fact, the net oscillatory forces engendered around the typical bluff body needed to be viewed characteristically as the net of a complex of *experimentally* determined physical forces—often associated with separated flow, rather than being closed-form smooth-flow analytical descriptions, like those portrayed in Theodorsen's theory. This situation led to the eventual experimental identification of flutter derivatives. To date, such motion-linked aerodynamic derivatives have been extensively identified, and once identified, they have been successfully applied to bridge deck model studies in many venues internationally. Their use has been efficient and valuable in numerous design studies.

Retracing the evolutionary steps involving the development of flutter derivatives, it was first observed that in spite of complex wind-structure interactions, the net motion of most bluff bridge sections, possessed of substantial inertia, displayed almost pure sinusoidal or damped sinusoidal response. This characteristic led naturally to simple means for identification of the associated oscillating wind force coefficients. Since the introduction of the free-oscillation methods of Scanlan and Tomko (1971) for bridge section models, many techniques have been exploited over the years demonstrating different alternate experimental methods for obtaining flutter derivatives (e.g., Sarkar et al. 1994).

Following the early style of airfoil aeroelastic theory, linear, first order, sectional lift, and moment effects were initially represented in the form

$$L = \frac{1}{2}\rho U^2 B(a_1 \dot{h} + a_2 \dot{\alpha} + a_3 \alpha)$$

$$M = \frac{1}{2}\rho U^2 B^2(b_1 \dot{h} + b_2 \dot{\alpha} + b_3 \alpha)$$

where ρ = air density; U = cross-wind velocity; B = model width, along wind; a_i, b_i = appropriate constants; and h, α = vertical and torsion coordinates, respectively.

Scanlan and Tomko (1971) reorganized these expressions into a format that incorporated the dimensionless flutter derivatives. Arrays of similar coefficients have, by the present time, appeared in several other contexts. To date at least two dozen doctoral theses from various places around the world have been written on the subject of bridge deck aeroelasticity, featuring flutter derivatives. At a somewhat more recent stage of development, in which three freedoms—h (lift), α (twist), and p (drag)—were included, a full set of 18 dimensionless flutter derivatives—H_i^*, A_i^*, P_i^* ($i = 1, 2, \ldots, 6$)—were developed and given the following form (Singh et al. 1994):

$$L_{ae} = \frac{1}{2}\rho U^2 B \left[KH_1^*\left(\frac{\dot{h}}{U}\right) + KH_2^* B\left(\frac{\dot{\alpha}}{U}\right) + K^2 H_3^* \alpha + K^2 H_4^* \frac{h}{B} \right.$$

$$\left. + KH_5^*\left(\frac{\dot{p}}{U}\right) + K^2 H_6^*\left(\frac{p}{B}\right) \right]$$

$$D_{ae} = \frac{1}{2}\rho U^2 B \left[KP_1^*\left(\frac{\dot{p}}{U}\right) + KP_2^* B\left(\frac{\dot{\alpha}}{U}\right) + K^2 P_3^* \alpha + K^2 P_4^* \frac{p}{B} \right.$$

$$\left. + KP_5^*\left(\frac{\dot{h}}{U}\right) + K^2 P_6^*\left(\frac{h}{B}\right) \right]$$

$$M_{ae}=\frac{1}{2}\rho U^2 B^2\left[KA_1^*\left(\frac{\dot{h}}{U}\right)+KA_2^* B\left(\frac{\dot{\alpha}}{U}\right)+K^2 A_3^*\alpha+K^2 A_4^*\frac{h}{B}\right.$$

$$\left.+KA_5^*\left(\frac{\dot{p}}{U}\right)+K^2 A_6^*\left(\frac{p}{B}\right)\right]$$

where H_i^*, A_i^*, P_i^* = dimensionless flutter derivatives, functions of $K = B\omega/U$, and where ω = circular flutter frequency. In this formulation, the sway or drag degree of freedom (p), not generally present for airfoils, but important for some flexible bridge spans, has been included. This effect was demonstrated to be important, for example, in the case of the current world's longest suspension span, the Akashi-Kaikyo, in Japan. In the present paper it will be sufficient to focus only on lift and moment terms involving h and α.

It has been observed that the type of experimental flutter derivatives obtained by Scanlan and Tomko (1971) have repeatedly permitted calculations that support full-bridge dynamic predictions. The paradigm for such implementations has been a close parallelism with the style and format of the Theodorsen thin-airfoil flutter theory. In this theory the concept of linear superposition of small effects, such as airfoil angle of attack, is justified both analytically and physically. In the context of bridge decks, the equivalent step of analysis must be considered an approximation only, as it is throughout the present discussion. If in this process anomalous effects appear they may be considered subjects for further experimental investigation. This paper reviews some elements of thin airfoil theory and their reinterpretation in the form of linearized bluff-body theory.

For typical full-bridge flutter analysis, acquired sectional flutter derivative forces are integrated spanwise over a large number (typically dozens) of natural structural modes. Solution of the corresponding eigenproblem then yields the wind speed of flutter. In this, the procedures, though presently substantially enhanced by computer aid, follow theoretical methods already outlined in the 1940s and 1950s or before. Details of the flutter eigenvalue problem have, as always in either bridge or aircraft cases, been computationally demanding.

The employment of aeroelastic formulations based on flutter derivatives follows the original style of aircraft flutter, in which flutter analyses written in the time domain are based on flutter derivatives expressed in the frequency domain. This "hybrid" formulation served well for the flutter case alone but did not provide for arbitrary structural motions described wholly in the time domain, such as occur under wind buffeting action.

The Wagner (1925) lift-growth or "indicial" function $\varphi(s)$ associated with a theoretical step change in airfoil angle of attack α provided a basis for time-domain lift force development under arbitrary time-dependent (small) angle of attack change α as follows:

$$L(s)=\frac{1}{2}\rho U^2 B C_L'\left\{\varphi(0)\alpha(s)+\int_0^\infty \varphi'(\sigma)\alpha(s-\sigma)d\sigma\right\}$$

where

$$s=\frac{2Ut}{B}$$

$$\varphi'(s)=\frac{d\varphi}{ds}$$

$$C_L'=\frac{dC_L}{d\alpha}$$

Garrick (1939) demonstrated the following Fourier transform identity between the classic Theodorsen circulation function $C(k)$ and Wagner function $\varphi(s)$:

$$C(k)=\varphi(0)+\overline{\varphi'(s)}$$

where

$$\overline{\varphi'(s)}=\int_0^\infty \varphi'(s)e^{-iks}ds$$

This link permits writing the Theodorsen airfoil flutter theory alternately in either the time or frequency domains. A single lift-growth function $\varphi(s)$ can be shown to be sufficient to represent lift or moment expressions in the Theodorsen theory. This single function must later be generalized to several functions in bluff-body theory.

Jones (1940) offered the following excellent approximation for the Wagner lift-growth function:

$$\varphi(s)=1-ae^{-bs}-ce^{-ds}$$

with

$$\varphi'(s)=abe^{-bs}+cde^{-ds}$$

where $a = 0.165$, $b = 0.0455$, $c = 0.335$, and $d = 0.300$. A wide variety of choices for the constants a, b, c, and d is clearly available to fit other indicial functions. The Theodorsen function

$$C(k)=F(k)+iG(k)$$

can be approximated by

$$F(k)=1-a-c+\frac{ab^2}{b^2+k^2}+\frac{cd^2}{d^2+k^2}$$

$$G(k)=-k\left[\frac{ab}{b^2+k^2}+\frac{cd}{d^2+k^2}\right]$$

Flutter lift and moment components are each associated with two forms of effective wind vertical angle of attack, i.e. α and $\dot{h}/U$, so that, at flutter in complex notation, the following four force components may be written.

Lift:

$$L_h=\frac{1}{2}\rho U^2 BK[H_1^*-iH_4^*]\frac{\dot{h}}{U}$$

$$L_\alpha=\frac{1}{2}\rho U^2 BK^2[iH_2^*+H_3^*]\alpha$$

Moment:

$$M_h=\frac{1}{2}\rho U^2 B^2 K[A_1^*-iA_4^*]\frac{\dot{h}}{U}$$

$$M_\alpha=\frac{1}{2}\rho U^2 B^2 K^2[iA_2^*+A_3^*]\alpha$$

Each of these four functions may be associated with a separate "circulation" function $F(\ )+iG(\ )$. Assuming (via implicit linearization) that the flutter derivative and associated notation $H_i^*(k)$ and $A_i^*(k)$ hold for either airfoil or bridge deck sections, the following set of equivalences can be demonstrated:

$$K[H_1^*-iH_4^*]=C_L'[F_{Lh}+iG_{Lh}]=C_L'[\varphi_h(0)+\overline{\varphi_h'}]$$

$$K^2[iH_2^*+H_3^*]=C_L'[F_{L\alpha}+iG_{L\alpha}]=C_L'[\varphi_\alpha(0)+\overline{\varphi_\alpha'}]$$

$$K[A_1^* - iA_4^*] = C_M'[F_{Mh} + iG_{Mh}] = C_M'[\psi_h(0) + \overline{\psi_h'}]$$

$$K^2[iA_2^* + A_3^*] = C_M'[F_{M\alpha} + iG_{M\alpha}] = C_M'[\psi_\alpha(0) + \overline{\psi_\alpha'}]$$

where the following definitions and observations are in order: $C_L', C_M' =$ slope of associated lift or moment curve; φ_h, φ_α = indicial lift functions; $\psi_h, \psi_\alpha =$ indicial moment functions; $F(\ ), G(\ ) =$ associated circulation functions.

This array of theoretical results groups together both frequency- and time-domain aspects of the linearized flutter problem, either for thin airfoil or bluff-body (bridge section).

A further observation is that *aerodynamic admittances* (χ) can be derived from these relationships. In this context four such functions will be available. For example (Scanlan and Jones 1999)

$$|\chi|^2 = K^2[H_1^{*2} + H_4^{*2}]/C_L'^2$$

which emphasizes a typical form of the link between admittances and the flutter derivatives. Such forms occur in expressions for structural buffeting by wind components u, w. For example, lift can be written

$$L = \frac{1}{2}\rho U^2 B\left[2C_L\chi_u\frac{u}{U} + C_L'\chi_w\frac{w}{U}\right]$$

for specific complex admittances χ_u, χ_w.

For quasisteady (static) lift, admittances are equal to unity. It is worth commenting that in the literature over a number of years, bridge deck bluff-body nonairfoil admittance was often incorrectly identified with the Sears airfoil admittance. The present review should correct this long-standing misinterpretation.

Summary

In the analytic theory of thin airfoil flutter, terms involving small angles of attack, such as α or $\dot{h}/U$, are linearly additive both theoretically and experimentally. This situation does not carry over in the strictest sense to bluff bodies like bridge deck sections around which flow effects are generally nonlinear. In practice, however, if the relevant structural displacements remain small, as with low-amplitude vibrations, most studies involving flutter derivatives yield reasonably accurate results for depicting overall system dynamics. Katsuchi et al. (1999) and Miyata and Yamada (1990) separately demonstrated response predictions well within 10% for a 1:100 scale model of the full Akashi-Kaikyo span.

Bridge structures with inertia large compared to the air, and with linear elastic structural characteristics, tend to respond like simple linear oscillators close to their natural frequencies, or in alternate formulation, the structural model under study may expressly be driven in prescribed sinusoidal cycles. Exploitation of like circumstances permits identification of flutter derivatives from net integrated aerodynamic force components. A particular note should be added. Some bluff bodies with box-like sections evidence flow regimes in which self-organized, periodic vortex shedding, and associated periodic body motion occur. In such situations, special care is required to interpret flutter derivatives appropriately (Scanlan 1998).

A continuing source of uncertainty in experimental bridge deck aeroelasticity is the low value of Reynolds number inherent in typical small-scale bridge models tested at low speed in atmospheric wind tunnels. Analogous circumstances also hold for building models in boundary layer wind tunnels. In such circumstances Reynolds number can be low by several orders of magnitude. This avenue of wind engineering research has—for obvious practical reasons—rarely been explored in the civil engineering context because of the sparse and inadequate means available to exploit it. The usual argument adduced with objects having "sharp" edges is that these permit definitive flow separations and thus ostensibly a near-Reynolds number equivalence of scaled force characteristics between small- and full-scale structural forms. This mostly unverified assumption merits thorough research or at least the establishment of appropriate means to *interpret* ostensibly analogous aerodynamic force equivalence when actual physical similarity cannot be achieved. Models of extremely small scale present a veritable kaleidoscope of flow/pressure situations (separation, reattachment, etc.) that differ from their prototype equivalents.

In low-speed atmospheric wind tunnels, proper duplication of full-scale Reynolds number effects is a practical impossibility, as is the proper realization of equivalent full-scale turbulence. Overall, these shortcomings may be even more egregious with aeroelastic phenomena. These questions—commonly dismissed or summarily treated at present—remain open for serious future resolution.

Farquharson (1949) listed a fair number of bridges that failed under wind—some as much as a century before the Tacoma Narrows episode. The Tacoma Narrows warning came at a juncture at which danger signs began to be heeded and responded to in technical depth. While numerous bridges have continued to exhibit disturbances under wind, no comparable catastrophes have since been reported. This reflects the fact that since 1940 bridge aeroelasticity, of both new and existing designs, has been seriously developed into an effective engineering art.

References

Ammann, O. H., von Karman, T. and Woodruff, G. B. (1941). "The failure of the Tacoma Narrows Bridge." *Rep. to the Federal Works Agency*, Washington, D.C., March 28.

Billah, K. Y., and Scanlan, R. H. (1991). "Resonance, Tacoma Narrows Bridge failure, and undergraduate physics textbooks." *Am. J. Phys.*, 59(2), 118–124.

Birnbaum, W. (1924). "Das ebene problem des schlagenden flügels." *Zeitschrift für angewandte Mathematik und Mechanik*, 4(277) (in German).

Bisplinghoff, R. L., Ashley, H., and Halfman, R. L. (1955). *Aeroelasticity*, Addison-Wesley, Cambridge, Mass.

Bleich, F. (1948). "Dynamic instability of truss-stiffened suspension bridges under wind action." *Proc. ASCE*, 74(8), 1269–1314.

Den Hartog, J. P. (1932). "Transmission line vibration due to sleet." *Trans., AIEE*, 1074–1076.

Farquharson, F. B., Ed. (1949). *Aerodynamic stability of suspension bridges, Bull. No. 116*, Univ. of Washington Engineering Experimental Station, Parts 1–5.

Frazer, R. A., and Duncan, W. J. (1928). "The flutter of airplane wings." *Reports and Memoranda, 1155*, Aeronautical Research Committee, London.

Fung, Y. C. (1955). *The theory of aeroelasticity*, Wiley, New York.

Garrick, I. E. (1939). "On some fourier transforms in the theory of non-stationary flows." *Proc., 5th Int. Congress for Applied Mechanics*, Wiley, New York, 590–593.

Glauert, H. (1928). "The force and moment of an oscillating airfoil." *Br. ARC, R&M* 1216, November.

Jones, R. T. (1940). "The unsteady lift on a wing of finite aspect ratio." *NACA Rep. 681*, U.S. National Advisory Committee for Aeronautics, Langley, Va.

Katsuchi, H., Jones, N. P., and Scanlan, R. H. (1999). "Multimode coupled flutter and buffeting analysis of the Akashi-Kaikyo Bridge." *J. Struct. Eng.*, 125(1), 60–70.

Küssner, H. G. (1929). "Schwingungen von flugzeugflügeln." *DVL Jahrbuch*, 313–334 (in German).

Miyata, T., and Yamada, H. (1990). "Coupled flutter estimate of a suspension bridge." *J. Wind. Eng. Ind. Aerodyn.*, 33(1&2), 341–348.

Sarkar, P. P., Jones, N. P., and Scanlan, R. H. (1994). "Identification of aeroelastic parameters of flexible bridges." *J. Eng. Mech.*, 120(8), 1718–1742.

Scanlan, R. H. (1998). "Bridge flutter derivatives at vortex lock-in." *J. Struct. Eng.*, 124(4), 450–458.

Scanlan, R. H., and Jones, N. P. (1999). "A form of aerodynamic admittance in bridge aeroelastic analysis." *J. Fluids Struct.*, 13(7-8), 1017–1027.

Scanlan, R. H., and Rosenbaum, R. (1951). *Aircraft vibration and flutter*, MacMillan, New York.

Scanlan, R. H., and Sabzevari, A. (1967). "Suspension bridge flutter revisited." *ASCE Structural Engineering Conf.*, Preprint 468, ASCE, New York.

Scanlan, R. H., and Sabzevari, A. (1969). "Experimental aerodynamic coefficients in the analytical study of suspension bridge flutter." *J. Mech. Eng. Sci.*, 11(3), 234–242.

Scanlan, R. H., and Tomko, J. J. (1971). "Airfoil and bridge deck flutter derivatives." *J. Eng. Mech. Div.*, 97(6), 1717–1737.

Scruton, C. (1952). "Experimental investigation of aerodynamic stability of suspension bridge with special reference to seven bridges." *Proc., 5th Civil Engineering Conf.*, 1 Part 1, No. 2, 189–222.

Sears, W. R. (1941). "Some aspects of non-stationary airfoil theory and its practical application." *J. Aeronaut. Sci.*, 8(3), 104–108.

Singh, L., Jones, N. P., Scanlan, R. H., and Lorendeaux, O. (1994). "Simultaneous identification of 3-DOF aeroelastic parameters." *Wind engineering, retrospect and prospect*, Vol. 2, Wiley, New York, 972–981.

Strouhal, V. (1878). "Uber eine besondere Art der Tonerregung." *Annalen der Physik* (in German).

Theodorsen, T. (1935). "General theory of aerodynamic instability and the mechanism of flutter." *T.R. No. 496*, National Advisory Committee for Aeronautics, Langley, Va.

Theodorsen, T., and Garrick, E. (1941). "Flutter calculations in three degrees of freedom." *T.R. No. 741*, National Advisory Committee for Aeronautics, Langley, Va.

Vincent, G. S. (1952). "Mathematical prediction of suspension bridge behavior in wind from dynamic section model tests." *Int. Assoc. for Bridge Struct. Eng.*, 12, 303–321.

Wagner, H. (1925). "Uber die Enstehung des dynamischen Auftriebes von Trag-flügeln." *Zeitschrift für angewandte Mathematik und Mechanik*, 5(17) (in German).

American Society of Civil Engineers 1852 – 2002 Building a Better World

150th Anniversary Paper

What Is Hydraulic Engineering?

James A. Liggett[1]

Abstract: This paper, written to mark ASCE's 150th anniversary, traces the role of hydraulic engineering from early or mid-twentieth-century to the beginning of the twenty-first century. A half-century ago hydraulic engineering was central in building the economies of the United States and many other countries by designing small and large water works. That process entailed a concentrated effort in research that ranged from the minute details of fluid flow to a general study of economics and ecology. Gradually over the last half-century, hydraulic engineering has evolved from a focus on large construction projects to now include the role of conservation and preservation. Although the hydraulic engineer has traditionally had to interface with other disciplines, that aspect of the profession has taken on a new urgency and, fortunately, is supported by exciting new technological developments. He/she must acquire new skills, in addition to retaining and improving the traditional skills, and form close partnerships with such fields as ecology, economics, social science, and humanities.

DOI: 10.1061/(ASCE)0733-9429(2002)128:1(10)

CE Database keywords: Hydraulic engineering; History.

Introduction

The answer to the title question will be framed by the experience of the individual reader. Hydraulic engineering is a broad field that ranges from the builder to the academic researcher. Without such a range it would not be the dynamic field that it is and, more importantly, it could not have contributed to society in the positive way that it has over the past century, and it would not continue to be a viable, challenging, and important profession. To illustrate the historical perspective to this question, and in so doing illustrate the evolution of hydraulic engineering, the present work uses one of the more visible activities involving hydraulic engineers—large water projects and especially dams in the United States. The reader, though, should not be misled into neglecting the myriad of other activities in which hydraulic engineers engage, some—individually or in combination—equally important to dams. The huge increase over the past 150 years in understanding of flow processes, especially those that occur in nature, and the associated ability to quantify these processes for analysis, design, and prediction is especially important.

However, the direct answer to what is hydraulic engineering does not lie solely in its history. The profession has always been a leader in the use of the latest technology; thus, technological innovation plays a vital part in the modern practice of hydraulic engineering. Innovations include modern computation, including techniques to make detailed flow processes and their complex interactions with other processes easily understandable. They also include the use of modern electronics for data gathering in the laboratory and the field and a myriad of other tools such as satellite photography, data transmission, global-positioning satellites, geographical data systems, lasers for laboratory and field measurement, radar, lidar, and sonar. Most importantly, they include the hydraulic engineer's interaction with the natural environment and ecology, an interaction that holds great promise and challenge. Indeed, the challenges of the last century, brilliantly solved by the collaboration of academics, small and large private companies, and government action agencies, are being replaced by new demands that will require even more interchange.

That interchange—not a new theme, but one that is beginning to dominate the future of hydraulic engineering—is the primary focus of this paper. First, however, we take a look at where hydraulics has been. A half-century ago the answer to the title question was obvious. The decades at mid-twentieth-century constituted the heydays of hydraulic engineering. It was the big-dam era, the time of large irrigation projects, large power projects, large flood-control projects, large navigation projects—large projects! Strangely, that era was short lived, at least in the United States, because of economic and ecological considerations. It lasted only about a half-century. Where does that leave hydraulic engineering at the beginning of the 21st century? What is hydraulic engineering now in an era of substantially increasing interdisciplinary developments?

A Time of Construction

Fig. 1 shows the history of dam construction in the United States from 1902 to 1987 in five-year periods. Immediately after World War II, dam construction surged, but it tapered off to very little by the late 1980s. Table 1 shows the largest U.S. dam projects, approximately 10 by height of dam and approximately 10 by reservoir size. All the projects on the list were completed between

[1]Professor Emeritus, School of Civil and Environmental Engineering, Cornell Univ., Ithaca, NY 14853. E-mail: jal8@cornell.edu

Note. Discussion open until June 1, 2002. Separate discussions must be submitted for individual papers. To extend the closing date by one month, a written request must be filed with the ASCE Managing Editor. The manuscript for this paper was submitted for review and possible publication on October 8, 2001; approved on October 8, 2001. This paper is part of the *Journal of Hydraulic Engineering*, Vol. 128, No. 1, January 1, 2002. ©ASCE, ISSN 0733-9429/2002/1-10−19/$8.00+$.50 per page.

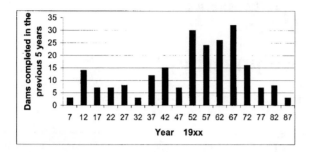

Fig. 1. A summary of dam building in the United States, 1902–1987 (Redrawn from Rhone 1988)

1936 and 1979. It is remarkable that dam building—or at least the largest dams—in the United States occurred in about a half-century.

The prevailing philosophy, both in society and in engineering, in the early and midparts of the century was that we could conquer nature and put it to the use of mankind. In particular, the case for dam building was compelling: dams provided flood control; they provided storage for irrigation and water supply, especially important in the arid west; they provided (what was then) a substantial amount of electrical power; and they provided recreation. The success of Hoover Dam (1936) as one of the nation's monumental construction projects—the largest attempted up to that time in the United States and the world—seemed to prove the point. As Reisner (1986) said, Hoover Dam's "turbines would power the aircraft industry that helped defeat Hitler, would light up downtown Los Angeles and 100 other cities. Hoover Dam proved it could be done." (Appendix I, Economics and War) The total number of dams in the United States grew to 75,000. Hydraulic engineers were building the U.S. economy (with a bit of credit to the structural engineers who designed and built the struc-

tural aspects of water resource projects) and, perhaps, we suffered a bit from the "monument syndrome" (Hirshleifer et al. 1960).

Of course it was not all dams. Flood control, irrigation, water supply, groundwater, and many other areas of civil engineering were the subjects of hydraulic engineering, and activity in those areas was equally vigorous. Although structural engineering would continue to employ more than any other specialty of civil engineering, hydraulic engineering was the glamour specialty and was surging. The big-dam era is symbolic, but hydraulic engineering in the twentieth century was about much more. The following were some other notable projects, to name a few.

The California Water Projects. Southern California, a region with a large population and little water, began its search for water in 1904 with the Owens Valley Project (completed 1913). After the construction of Hoover Dam and Parker Dam, the Colorado River supplied water to California to supplement that from Owens Valley. In an insatiable search for more, the California State Water Project was begun in 1960 to bring water from Northern California (Oroville Dam on the Feather River) to Southern California (Fig. 2). All of these projects have generated controversy, but they have enabled Southern California to grow and have opened the region, especially the Central Valley, to supply fruits and vegetables that feed the nation.

The Central Arizona Project. Dams along the Salt River, primarily Theodore Roosevelt Dam (1911, the first multipurpose project constructed by the Bureau of Reclamation), supplied water for irrigation and domestic use to the Salt River Valley. The Central Arizona Project (completed in the 1980s) transfers water to the cities of Phoenix and Tucson from the Colorado River. It consists of an aqueduct (336 miles long) from the southern end of Lake Havasu (Parker Dam) and includes 15 pumping plants, 3 tunnels, and a dam with storage reservoir (New Waddell Dam and Lake Pleasant).

The Arkansas River Project. The Arkansas River Navigation System was approved by Congress in 1946 and completed in

Table 1. Largest Dams and Reservoirs in the United States

Dam	River	State	Type	Height (m)	Reservoir Cap ($m^3 \times 10^9$)	Year
Oroville	Feather	Calif.	Earthfill	230	4.30	1968
Hoover	Colorado	Ariz.-Nev.	Arch	221	34.85	1936
Dworshak	North Fork Clearwater	Id.	Gravity	219	4.26	1973
Glen Canyon	Colorado	Ariz.	Arch	216	33.30	1963
New Bullards Bar	North Yuba	Calif.	Arch	194	1.18	1970
New Melones	Stanislaus	Calif.	Earthfill	191	2.96	1979
Swift	Lewis	Wash.	Earthfill	186	0.93	1958
Mossyrock	Cowlitz	Wash.	Arch	185	1.60	1968
Shasta	Sacramento	Calif.	Gravity	183	5.61	1945
New Don Pedro	Tuolumne	Calif.		178	2.5	1971
Hungry Horse	South Fork Flathead	Mont.	Arch	172	4.28	1953
Grand Coulee	Columbia	Wash.-Ore.	Gravity	168	11.79	1942
Ross	Skagit	Wash.	Arch	165	1.90	1949
Fort Peck	Missouri	Mont.	Earthfill	76	22.12	1940
Oahe	Missouri	S.D.	Earthfill	74	27.43	1962
Garrison	Missouri	N.D.	Earthfill	64	27.92	1953
Wolf Creek	Cumberland	Ky.			4.93	1951
Fort Randall	Missouri	S.D.	Earthfill	50	5.70	1953
Flaming Gorge	Green	Utah	Arch	153	4.67	1964
Toleda Bend	Sabine	La.-Tex.			5.52	1968
Libby	Columbia	Mont.	Gravity	129	7.17	1973

Fig. 2. The Wind Gap pumps. A part of the water delivery system to Southern California, the A. D. Edmonston Pumping Plant lifts water nearly 2,000 feet up the Tehachapi Mountains where it then crosses through a series of tunnels to the Los Angeles Basin. It, along with other projects, enables the cities of Southern California to grow and prosper in an arid climate.

1971. It controls flooding and provides a navigable waterway for shipment of agricultural products, lumber, petroleum, and coal by means of 17 dams and locks along the waterway.

The Mississippi River Navigation and Flood Control Projects. Work on the Mississippi River has been continuing for so long that it seems almost forgotten as a major hydraulic engineering feat. The Mississippi River Commission, created by act of Congress in 1879, is responsible for flood control and navigation along the river (Fig. 3). The main stages of the navigation improvement program included a channel 9-feet deep and 250-feet wide at low water between Cairo, Illinois, and Baton Rouge, Louisiana (authorized in 1896), widening of the channel to 300

Fig. 3. The Mississippi River near Muscatine, Iowa. The photo illustrates the barge traffic on the river ("tows," although the barges are actually pushed by the tug). The series of pools is a fish hatchery. This site is near the Iowa Institute of Hydraulic Research Mississippi Riverside Environment Research Station, which is intended to study ecology and environmental considerations along the river.

feet (1928), and deepening to 12 feet (1944). Channel improvement and maintenance are still under way along with a ship channel 45-feet deep from Baton Rouge to the Gulf of Mexico (authorized 1945). In the upper part of the river, 29 locks and dams have been constructed to create a 9-foot-deep channel to Minneapolis-St. Paul. It has become part of vast inland waterway from the Gulf of Mexico and Florida to Canada, the Great Lakes, and the St. Lawrence Seaway. After the flood of 1927, the Corps of Engineers began the process of levee construction. From Cape Girardeau, Missouri, to the Gulf of Mexico, the Mississippi is encased in levees and sea walls, as is much of the river to the north. The Mississippi projects have enabled the city of New Orleans to exist, have opened the central United States to the economic transportation of goods, and have enabled agricultural production unparalleled in the history of the world.

The Tennessee River. In 1933, the Tennessee Valley Authority (TVA) was established for the multiple purposes of flood control, navigation, electrical power, water supply, and, importantly, for the economic development of a previously depressed region. TVA has made the Tennessee one of the most controlled rivers in the world.

This small sample indicates the importance of hydraulic engineering in mid-twentieth-century. All projects mentioned herein are in the United States, but similar activity took place throughout much of the world. Although the construction of big dams has ceased in the United States [Seven Oaks Dam (Southern California, completed in 1999) would not have made the list in Table 1 at 168-m high, but it is of substantial size, and was constructed for flood control], it continues in some parts of the world. These and other projects graphically illustrate the paradigm of controlling nature for the benefit of mankind. There is no question that they have brought great economic benefit to the entire nation and, regionally, to the areas in which they were constructed. Indeed, the first half of the twentieth century was a little Dark Age in the United States marked by the great depression and two world wars. Those who might criticize the engineering accomplishments of that time from a distance have not had to live under such conditions. For example, the TVA has transformed a poor, underdeveloped area of the country into one rich in energy resources and agricultural opportunities. If the title question on this paper had been "What *was* hydraulic engineering?" these, along with many other large projects and innumerable small ones, such as municipal water supply and groundwater management, certainly supply the answer.

A Time of Enlightenment

The compelling promise of large water-control projects and other hydraulic works was fulfilled completely. That activity was accompanied by a sort of revolution in knowledge and rational analysis that took place in engineering in the 1950s and 1960s. First, engineering had discovered its scientific basis. In hydraulic engineering, the landmark events were the publication of Rouse's (1938) book *Fluid Mechanics for Hydraulic Engineers* and Vennard's (1940) book *Elementary Fluid Mechanics*. These books and their followers set apart the teaching of hydraulics, a mostly empirical subject, from fluid mechanics, a subject based on mathematical analysis. Other branches of engineering were showing a parallel change. Rational analysis had become popular. This development created an optimism that with the proper mathematical analysis we could solve many nagging problems that were holding back progress.

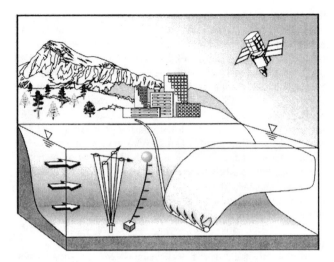

Fig. 4. Plume dynamics. The use of a multitude of sensors for velocity and concentration of substances coupled with satellite data transmission and used in 3D numerical modeling to solve pollution problems in waterways, lakes, and oceans is illustrated. Adapted from Roberts (1999).

Second, the computer became a practical tool for engineers in the latter half of the century. First used as a research tool in the late 1950s, their use spread to the engineering office in the 1960s and 1970s. Although numerical methods had been a sophisticated subject long before automatic computation, it now took on practical importance and held the promise to solve those equations that were presented in elementary and advanced fluid mechanics courses. Now, we believed, we really were on the verge of solving all the practical and relevant hydraulic engineering problems.

The first such solutions were those that we had been taught in the classroom but were laborious. Examples included the steady-state solution of pipe network problems and the calculation of open-channel flow profiles. Finally, hydraulic engineers had gained the ability to solve such problems as unsteady open-channel flow (Isaacson et al. 1954), but we learned from this development that simply plugging the equations into the computer was not an easy process. In fact the solution by Isaacson et al. (1954), the mathematicians, was largely a failure, and we had to await the advancement by Preissmann (1961)–a mathematician working for an engineering consultant, Sogreah—to show the way. The devil was in the details; it was not simply a mathematical exercise but required engineering judgement to determine which of the details were important and which could be ignored.

For the first time, our multidimensional and time-dependent problems seemed within our grasp (Fig. 4). The dimensional approximation (i.e., approximating a fundamentally 3D problem in two dimensions or a 2D problem in one dimension) was not always necessary. These developments led us to believe that it was only a matter of (a short) time before hydraulic engineering became a science almost as rational as physics. The world was filled with meaningful, interesting, and economically important problems, and we were gaining the means to solve them. It was a great time!

Time of New Challenges

If there was a single turning point it was probably the construction of Glen Canyon Dam on the Colorado River in northern Arizona. Environmentalists, primarily the Sierra Club, had criticized the dam since its inception. Glen Canyon is essentially the uppermost part of the Grand Canyon (Appendix, Glen Canyon), one of the jewels in the system of national monuments. To build power dams in the Grand Canyon seemed rather like harnessing the thermal energy of Old Faithful in Yellowstone National Park or using Yosemite and Bridalveil Falls for electrical power. Although such projects would be rejected by society today, it is interesting to recall a long-since forgotten plan proposed at the turn of the twentieth century by the English physicist and hydropower consultant Lord Kelvin (Burton 1992) to turn Niagara Falls into a grand hydropower plant (and, indeed, hydropower is currently being produced at that site).

Other dams have been proposed for the Grand Canyon area. The two most notable are Marble Canyon Dam (abandoned in the 1960s) and Bridge Canyon Dam (sometimes called Hualapai Dam as it is on the Hualapai Indian Reservation. It was officially canceled in 1984 but still shown as a dam site on many Arizona maps). Considerable exploratory work was done on these sites, especially at the Marble Canyon site. There are tunnels deep into the rock and innumerable places where core drilling took place [Figs. 5(a and b)]. Obviously, these dams were serious projects, and construction was almost begun. The Sierra Club takes credit for blocking the construction of these dams (although it once favored Bridge Canyon Dam in a resolution of November 12, 1949). However, economics and the realities of construction probably paid a significant role in the fact that they were never built. Marble Canyon Dam was to be placed in a limestone formation, bringing into question its long-term safety. Access to the dam site is difficult and would have necessitated costly construction of roads. Evaporation from the water-short Colorado River from the lake surface was a negative factor. A strong argument at the time was that hydropower was unnecessary because nuclear energy was to supply abundant electricity, so cheap that we would not have to meter it. (The Sierra Club was initially a promoter of nuclear power, but it has since changed its view.)

Everyone recognized that Lake Powell (behind Glen Canyon Dam) would flood the wild river, covering picturesque rock formations—said by many to be better than the Grand Canyon—and some archeological sites. However, there seemed to be little recognition at the time of the downstream changes. The Colorado River carries a heavy sediment load that is now being retained in Lake Powell. The river through the Grand Canyon has changed from a muddy stream to one that is more or less clear and carries significant sediment from the tributaries and side canyons only during the summer rainy season. The sediment no longer nourishes the beaches in the Grand Canyon and has changed the habitat for fish. Previously, the temperature of the water ranged from near freezing to the mid-seventies (degrees Fahrenheit). Now it is a near-constant 49 °F near Glen Canyon Dam and increases somewhat in the summer to the headwaters of Lake Mead. Flooding occurred on an annual basis; whereas, now the flow is relatively constant (Appendix, Flood) with the result that some of the larger sediment brought in by tributaries is not moved by the main river. The downstream ecology has changed forever, or at least as long as Glen Canyon Dam exists. The endangered natural fish will never be fully restored.

It is not clear, however, that the ecological changes are undesirable. The Colorado is now a cold-water trout stream; the fish seem more desirable. Those who use the river for recreation can now make use of the water (and are not constantly covered by mud in the rapids), which was impossible under former condi-

Fig. 5. Marble Canyon. Although not officially part of the Grand Canyon, most people consider it such, and it is in the Grand Canyon National Park: (a) The Colorado River consists of riffles and pools. The pool shown in the photo is natural, not the product of a dam, and reflects the colors of the Redwall limestone. Is this the perfect place for a dam or should it remain in its natural state for the enjoyment of future generations?; (b) Marble Canyon dam site. The hole in the cliff of the upper part of the picture is an exploratory tunnel in the Red-wall limestone. The light colored material is tailings from that tunnel. There are several such tunnels on both sides of the river and numerous core drills branch from each tunnel.

tions. The issue raises the question: Is a change in ecology always undesirable?

Large dams have been attacked as inefficient, citing, for example, Lake Mead, which loses to evaporation 10% of the flow of the Colorado River, enough to supply Los Angeles. Some who initially favored the construction of Glen Canyon Dam have since changed their minds. Most notable was Barry Goldwater, former

Fig. 6. A billboard that has been put up around Arizona. The Glen Canyon controversy invokes strong emotions on both sides. Many of these billboards have been vandalized, most commonly by painting out the line "Don't let the Sierra Club" so that it reads "drain Lake Powell."

senator from Arizona, Republican presidential candidate in 1964, and a leader of the conservative movement in the United States. Senator Goldwater stated that the one vote in the Senate that he regretted was his vote for Glen Canyon Dam. Stewart Udall, three-term congressman from Arizona (first elected in 1954) and 8 years as Secretary of the Interior in the Kennedy and Johnson administrations, went from favoring three dams in the Grand Canyon to two dams to one dam to no dam (Fradkin 1981). Secretary of Interior Bruce Babbitt in the Clinton administration and former governor of Arizona now champions the cause of dam removal (although his efforts have been confined to relatively small dams). It is indeed notable that many who once favored the dam projects as economic pluses, and who ran for office on platforms advocating such projects, have now changed their tune. Secretary Babbitt, especially, has proposed eliminating a large number of dams in the United States and has been campaigning hard—with a symbolic sledgehammer in hand—to that end. Although Secretary Babbitt has stated that he is opposed to the destruction of Glen Canyon Dam, the campaign to remove the dam is in full swing, with the opposition vigorously defending the dam (Fig. 6). It is an emotional issue in the Southwest, with both sides holding demonstrations and generating a large amount of newsprint.

Also controversial is the situation on the Snake and Columbia Rivers where the salmon are endangered in part by the dams. (Fishing is, perhaps, the largest factor in the decrease of salmon, and agricultural pollution plays an important part). Unfortunately, the construction of fish ladders, guide barriers [Figs. 7(a and b)] and transportation of fish over the dams (primarily in the downstream direction because the dams are destructive to the smelt) have not solved the problem. Additionally, hatcheries have not been able to restore the natural cycle of salmon breeding and spawning. [The Corps of Engineers has spent more than $50 million per year on the Columbia River Salmon Program (USACOE 2000)]. Aside from the environmentalists' objections, those dams pit one economic interest against another—those who profit by the dams and lakes against who that profit by the fish and other features of the natural rivers.

The present situation has been the consequence of two factors. First, neither the public, lawmakers, nor the designers and planners of many of the projects appeared to realize the consequences

(a)

FISH
GUIDEWALL

(b)

Fig. 7. A guide barrier for directing salmon smelt to a by-pass channel around hydropower turbines. The images are from a study by Weber et al. (2001); (a) Although such systems are at least partially successful, the early dams were constructed with no provision for migratory fish: (b) Numerical simulation of fish response to flow conditions along the guide barrier. Numerical codes not only simulate flow fields, they are being developed to simulate fish response to such flow fields.

of the projects. Those who opposed the dams were termed extremists. Benefit-cost studies were questioned at the time by eminent economists, but prodam interests always made sure that they indicated a favorable benefit/cost ratio. Most important was the intense lobbying by politicians of the areas in which the projects were to be built and the optimism expressed by local leaders of the economic benefits of the projects. Second, a change of attitude apparently has occurred among the citizenry. The new attitude is one of conservation of nature rather than conquering nature for our benefit. Loucks et al. (2000) ask, "What will the desires of future generations be?" and answer pessimistically, "Clearly, our guesses about those future desires, even the educated guesses, will be wrong." [Despite such abject pessimism, Loucks et al. (2000) go on to discuss how "sustainable water resource management" might be accomplished.]

The "irrefutable" arguments in favor of dams have not just been set aside; they have been reversed, at least for the moment.

A Time of Reality

The promise of mathematically based solutions to the problems of hydraulic engineering was indeed fulfilled to a great extent, espe-

cially through the use of numerical methods. That is not to say that traditional engineering judgement became unimportant. In fact, the role of the engineer became even more crucial when interpreting computational results that might portray poor solutions. Before computer-based computation reached its ultimate conclusion, however, it ran into a serious roadblock, that of scale. Many of the solutions that we want occur at problem scales of a few meters to thousands of kilometers. Unfortunately, such solutions often depend on what occurs at the sub-millimeter scale (Liggett 1996). That fact makes the necessary discretization of equations clearly unachievable, now and in the foreseeable future. The example of long-term (say, 6 month) weather prediction is often used. In that case, the 3D, time-dependent equations would have to describe every blade of grass, every pebble on the beach, and the movement of all animals as well as the large-scale motion of the atmosphere. Not only is there no computer at present or contemplated that could handle such a problem, the task of describing initial and boundary conditions is beyond comprehension. The vast majority of hydraulic engineering problems suffer a similar difficulty. (Numerical analysts can point to considerable advances in computation with turbulence, e.g., large eddy simulation. Such advances will continue and form, perhaps, the most important area of modern fluid mechanics research.)

The result is that although numerical computations have not reached the end of their potential, future advances will depend on the speed of the machines and the power of the algorithms while advancing at a slower pace than once predicted and seemed to be obtainable 40 years ago.

A Time of Quandary

The pessimism expressed by Loucks et al. (2000) about forecasting the wishes of future generations might well be applied to determining the wishes of the present generation. The debate over the production and use of energy is symptomatic, and it is related to hydraulic engineering. Most water projects are either large users or producers (or both) of power. Moreover, hydraulic engineers have been active researchers on alternate sources such as ocean thermal and wave power and wind power. How should society address the problem of sufficient electrical energy?

Renewable sources include wind, direct solar, water, renewable biomass, ocean thermal, and tidal power. As attractive as some of these are, they all—or in combination—have serious difficulties. Wind and direct solar power can only furnish a fraction of the energy that we now use and both require an enormous area of land surface. Water power depends on dams and, even if the public favored more dam construction, economical sites are scarce. Renewable biomass is a form of solar power that also takes much area, has pollution consequences (Appendix, Pollution), and also does not seem capable of producing sufficient power. Ocean power is too widely dispersed to be economic.

Nonrenewable sources have the obvious disadvantage that they cannot be sustained over time. They consist of fossil fuels, nonrenewable biomass, and nuclear energy. The first two have pollution consequences. Nuclear energy is currently regarded as too hazardous in the United States, although much of the world is embracing it as a major source of energy. Breeder reactors, which cannot be built in the United States by law, largely because they can produce weapons-grade plutonium, can go a long way toward making nuclear power renewable. The problem of disposal of waste (largely a false problem in the opinion of the writer) is also greatly mitigated by breeder technology.

Everyone favors conservation. Indeed the United States does seem to be profligate in use of energy, and could conserve a very large amount without a great change in the standard of living, although many life styles would have to undergo considerable change. However, even drastic conservation measures would only be a blip on the curve of increasing energy use. Assuming, say, a 20% decrease due to conservation, in a few years we would find ourselves in the same dilemma. That is not to say that conservation is not worthwhile, only that it is not a long-term solution.

Indicative of the quandary for hydraulic engineering are the changes occurring to the city of Phoenix. Phoenix recently adopted a flood-control policy in which the small streams (usually dry) would be allowed to flow freely and more or less naturally, instead of past policy in which they were encased in a concrete floodway. Such a change was, of course, applauded by conservationists. But the consequence is that the natural streams require much more space than the concrete floodways, space that could be used for parks or housing. Parks may be compatible with the space, but a restriction on housing decreases density, forcing building further out into the surrounding area and creating more urban sprawl, a current hot-button criticism of many cities. The Corps of Engineers has proposed a similar solution to parts of the Napa River (California) in which flood plains and marshes would be used to convey floods in place of concrete channelization.

There are many cases in which it seems impossible to have our cake and eat it. Do we really want rapid economic development or conservation? preservation of ecology or individual space? pristine forests or recreation? unspoiled national parks or universal access? freedom from traffic and destruction of environment by highways or freedom of individual movement? (many indicators point to a period of deconstruction for highways that may be comparable to the deconstruction of dams), more industry or less pollution? These and many more issues are at the nub of the dilemma of a sustainable economy in general and sustainable (Appendix, Sustainable) water-resource planning in particular. The nature versus growth question is illustrated by a sign in the agriculturally rich Mississippi River flood plain and flyway for migratory birds that asks, "Food for folk or fowl?"

Thus, the twentieth century, so certain of the path to prosperity at midcentury, has ended in a quandary. A discussion of the question "Where do we go from here?" will most likely dominate the first half of the twenty-first century. In an increasingly integrated world, even the goals are not clear. Barrett and Odum (2000) state " . . . politicians used to talk about 'the greatest good for the greatest numbers' as a goal for society. But this slogan is rarely heard now because society is finding out by experience that the greatest good, in terms of quality of life for the individual, comes when the numbers are not as high as they can possibly be—and when the per capita impacts are not maximized, either."

A Cornell University ecologist has estimated that the resources of the earth could sustain a population of about 2 billion with a standard of living slightly lower than that of the present-day United States. That number is about one-third of the current population of the earth. The implication is that anything greater than 2 billion will mean a lower average standard of living. If the estimate is anywhere near correct, there are few or no "developing" countries; most of the earth's people are destined to spend their lives in poverty, and the pressure on ecosystems will only increase. Indeed, it is difficult to think of a major problem that is not either caused or exacerbated by population pressure. This issue is such a political hot potato with cultural and religious overtones that politicians avoid it. Its impact on every facet of modern life is such that it deserves an open and rational debate,

including the choices that are available and their consequences. At present, such a debate apparently takes place mostly in the ecology and economics literature (Arrow et al. 1996; Barrett and Odum 2000). That issue, together with global climate change, will have a large impact on the future of hydraulic engineering as we attempt to supply an increasingly crowded world with water resources and energy. It constitutes a challenge equivalent to any that the profession has faced and, as usual, challenge equals opportunity.

A Time for Consilience and Opportunity

To say that hydraulic engineers must become more a part of a team approach to the solution of modern-day problems is somewhat trite, but it is a fact. Solutions, or at least best shots at solutions, to multifaceted problems lie in rational analysis and good design involving a variety of expertise. If construction required hydraulic expertise, so does preservation and restoration.

The professional responsibilities of the hydraulic engineer demand, in turn, that an effort be made to integrate disciplines involved in water-resource projects. Indeed, life itself does not exist without water, the substance that also gives life to our profession. Wilson (1998) has used the word "consilience" (Appendix, Consilience) to express the fact that knowledge in engineering, science, *and* humanities is interconnected. He states " . . . true reform will aim at the consilience of science with the social sciences and humanities . . . Every college student should be able to answer the following question: What is the relation between science and the humanities, and how is it important for human welfare? Every public intellectual and political leader should be able to answer that question as well" (Wilson 1998).

By virtue of the many issues they raise, water projects commonly merge engineering, science, humanities, and societal desires. However, integration can be difficult, especially when societal desires evolve and change. Even if it is difficult to predict the desires of the next generation, we have not been very conscientious in studying the impact of what we have been doing. Glen Canyon Dam was completed in 1963, but only in 1982 was the Glen Canyon Environmental Study initiated to determine if an environmental impact statement (EIS) on the operation of the dam was warranted [The National Environmental Policy Act (NEPA) of 1969 required Environmental Impact Statements. Glen Canyon was completed in 1963, several years before NEPA]. Seven years later, in 1989, that committee determined that an EIS was appropriate and such a study was undertaken. Another 7 years went by before the EIS was produced. [Perhaps it is symptomatic that a study to control temperatures downstream of Glen Canyon Dam for the preservation of native and endangered warm water fish has the late date of January, 1999 (USBR 1999).] Perhaps it is only in hindsight to say that the EIS process should have been completed *before* construction began. Thus, the view of Loucks et al. (2000) that planners cannot predict the wishes of future generations missed the point in this case. Planners did not even *know* what future generations were going to get.

It is probably not an exaggeration to say that in the 1960s such studies would have been unwelcome. The prevailing attitude among hydraulic engineers was "Let's quit studying the project to death and get on with the job." That attitude was not without merit. Nonessential investigations and the resulting increases in cost could easily doom a major project and justify the critics' charges of cost overruns and schedule delays. Had many of those projects required present-day justification, much of their economic benefit would have been lost. From the point of view of the year 2001 with the critics campaigning to remove dams, many would say that such studies should have been considered crucial.

A Language Problem

Hydraulic engineers have always worked with nonengineering disciplines (economists, meteorologists, biologist, and others), especially in the planning of large projects. Outstanding examples include the salmon problem in the northwest [Figs. 7(a and b)], ecological studies along the Missouri River, and the operation of Glen Canyon dam. Such cooperation has steadily increased in the latter part of the twentieth century as its necessity has become more apparent. However, the research community has focused primarily on technical progress rather than on "interdisciplinary studies." Interdisciplinary cooperation is not easy and is frequently frustrating.

First, we have to learn a new language—for example, what is "nature's capital" as opposed to economic capital (Barrett and Odum 2000) or "gross natural product" versus gross national product (Naudascher 1996a)? Even more frustrating is the fact that ecological and economic concepts are difficult to express quantitatively, and generally impossible to analyze using the methods of physics. When we bring in social science and humanities, that difficulty is multiplied many times. But a quantitative understanding of such disciplines is no more difficult than explaining the concepts of fluid mechanics, hydraulics, and hydrology to people uninitiated in hydraulic engineering. Finally, researchers who work on such projects are likely to receive little credit. Papers with no mathematics or only simple formula are considered "soft" to fellow engineers, and certainly engineers will have difficulty gaining a foothold in other areas. In research universities, tenure and promotion committees are likely to be suspicious of such interdisciplinary activities. Indeed, such suspicion is well founded and appropriate.

Although we have readily adapted advances in technologically based disciplines—computing, instrumentation, mathematics, numerical methods—into hydraulic engineering, the current challenge is different in that these areas speak the language of engineering and science; they are quantitative. In adapting technology to technological problems, productive and unproductive paths are quickly discovered. In attempting to integrate hydraulic engineering with ecology and social science, it is easy to become bogged down in largely unproductive rhetoric. When the goals are not well defined, the path to those goals cannot be clear.

Liquids in Nature

Hydraulic engineering is the application of fluid mechanics to the liquid earth. Although some applications involve man-made systems (pipe networks, for example), many deal with the complexities of nature. Those latter applications include river engineering; sediment transport; groundwater movement; lake, ocean, and reservoir dynamics (including the complications of stratification); waves; surface flow; and the alteration of natural flows by man, including pollution. Hydraulic engineering is clearly a field for those who love nature and who are comfortable in applying the laws of fluid mechanics for the betterment of mankind while preserving nature. It is a field that has changed in the last half-century, but the challenges were never greater than are those of the present day.

The technical (mechanical) challenges in hydraulic engineering are immense. No one could accurately predict the results of the artificial flood release of Glen Canyon in 1996 toward the

objective of partial restoration of nature and natural habitat. Nor is this problem one of traditional hydraulic engineering because "restoration" is not simple; it contains multiple implications in the fields of ecology and biology as well as engineering. The immensity of the problems requires the instrumentation, simulation, and communications tools that increasingly are at our disposal (Fig. 4).

A Crossroads

In a real sense hydraulic engineering is at a crossroads. The midcentury challenges were met and largely conquered, albeit with inadequate foresight in some cases. That lack of foresight is only a minor factor in the problems of the twenty first century. In the United States, society has the luxury of debate whether to conquer nature for our economic benefit or to preserve nature for the enjoyment of future generations (and, perhaps, for their economic benefit). Hydraulic engineers can sit on the sidelines and simply do the bidding of the politician or we can influence the debate. Certainly the professional attitude *at the very least* requires us to lay out the alternatives as we know them and to perform the research to know the alternatives and their consequences as best we can.

The words "water shortage" are often heard. In the United States there is no permanent water shortage, anywhere (although temporary shortages may exist). The country is bordered by oceans that have an unlimited supply of water. Water is a commodity and, as such, it can be priced according to its supply and demand. With sufficient engineering works we can supply water anywhere, but of course at a price. Traditional hydraulic engineering is only a part of the determination of that price; the other part is the effect on ecology of the source and of the region that receives the water. Outside of the United States permanent water shortages occur in regions where the resources do not exist for its acquisition, either locally or by importation. Naudascher (1996a) argues that in such areas public works projects such as big dams have not helped the really needy. However, they have, without doubt, often contributed to the economy and the general welfare of several nations and to their political stability. Neither blanket condemnation nor blanket acceptance of such projects is a reasonable stance. The displacements of peasants by large dams and their inability to benefit from the irrigation, power, and recreation are well known and documented, but such displacements must be balanced against the destruction and forced relocations caused by flooding. According to Naudascher (1996b), flood control may contribute to impoverishment in that it eliminates the natural fertilization of the land through the deposition of silt and eliminates the flushing of salt. He also points out that these deprivations may actually be counted as benefits because the fertilizer industry increases sales and irrigation, drainage, and desalinization schemes—necessary due to isolation from the river—are added to the gross national product of the nation. Unfortunately, political power often rests in those that see only one side of the problem. Fortunately, the hydraulic engineering literature contains a much more balanced perspective than is common in the popular media.

A Time for Education

At midcentury, few thought that such considerations were the responsibility of the hydraulic engineer. Now it seems that they cannot be ignored if hydraulic engineers are to be professionals. The educational and research burden at midcentury was technical—fluid mechanics, mathematics, and a bit of economics along with subjects in hydraulic engineering. Now in addition to the traditional role we must be much broader, studying ecology, biology, resource management, a smattering of systems analysis, and related items plus humanities and social science. Although every engineering curriculum contains these latter two items, they are loosely required with only the vaguest of goals and no thoughts toward the unity of knowledge in the sense of Wilson (1998). In other words, the modern hydraulic engineer must be able to speak "ecology" in the broadest sense of the word. We must be team players with a variety of disciplines. To be part of a team, courses in the language and culture of ecology, biology, economics, social studies, and the humanities have to be a part of the education, including the continuing education of hydraulic engineers (Liggett and Ettema 2001). Our universities must do a better job of integrating disciplines—of consilience—than they have up to the present. Courses in these subjects should not be individual and unconnected hurdles on the path to a degree (Ettema 2000).

However, this approach contains its own hazards. "Environmentalist" is all too often a buzzword and signifies someone who cares about the environment but knows little science or engineering and is likely to embrace the latest "green" fad. One who calls him/herself an environmentalist is frequently regarded as a refugee from academia who cannot make it in science or engineering. Thus, the educational requirements for hydraulic engineers should not be relaxed. *No one should be able to call him/herself a hydraulic engineer until he/she has mastered the science, mathematics, mechanics, and engineering.* When dealing with environmental issues we must speak from a solid background, not repeat the dogma of the Sierra Club or other groups. Although the position of such groups often stems from expert knowledge and is the best that we know at the time, it is too often a knee-jerk reaction of those who seem to believe that everything man-made, especially a large engineering project, is bad.

A New Time of Hydraulic Engineering

The challenges of hydraulic engineering of the last half century remain. They can be stated as familiar questions: How can we better predict and calculate sediment transport? ice effects? open channel hydraulics? water supply for irrigation and municipalities? groundwater flow and groundwater remediation? How can we better link hydraulics, hydrology, and weather forecasting? How can we better characterize turbulence so that it does not defeat our calculations of diffusion, boundary friction, transport, and fluid flow in general? How can we better use computational fluid mechanics to study the complex problems that nature has given the hydraulic engineer? How can we design better and more efficient structures? All these questions and more are crucial not only to the traditional role of hydraulic engineers, but also to our emerging responsibility as a partner in society's decisions for what is best for sustaining human development and environmental well being.

Only if we remain knowledgeable in these matters can we enter the debate as experts on specific questions such as: Should Bridge Canyon Dam be built? Should Glen Canyon Dam be removed? Should the Snake River Dams be removed? Should flood control projects be constructed with higher dikes and levees or should we restore flood plains and marshes for relief? And we should provide expertise on mankind's role in preserving nature while attempting to provide a decent standard of living for the people of an overpopulated earth. Hydraulic engineering must go

far beyond the realm of applied fluid mechanics while retaining a base deeply rooted in fluid mechanics. These questions (and those regarding less developed countries, only briefly mentioned herein) can be answered only by the consilience of hydraulic engineering with the humanities and social sciences while being especially careful to maintain the quality and integrity of hydraulic engineering. Such a goal may be as difficult as the characterization of turbulence, but it is as important.

The challenges of the twenty-first century may not contain the same machismo of the twentieth century, but they are certainly as important and even more challenging. It is still a great profession!

Parting Comments

In an attempt to address its title question, this paper considers the role of hydraulic engineering in the development of large water-control projects in the twentieth century. Although the dams associated with those projects are symbolic, highly visible, useful, and sometimes controversial, they are, of course, only a part of hydraulic engineering activities. This paper also is largely about hydraulic engineering in the United States. The development of large water projects, including dams, continues in many other countries and in some cases appears essential to their development. Attitudes and conditions in many countries may differ considerably from those in the United States; therefore, it is not appropriate to judge them in the light of the United States experience. The account given in this work is intended to be broad—and intended to make the point that our profession is becoming broader—in terms of hydraulic engineering's place amidst human endeavors. Obviously, no one answer to the title question is entirely satisfactory. Readers should apply their own perspectives and answers to that question.

Acknowledgments

Robert Ettema, editor of *JHE*, invited me to write this paper, and he read each draft. His suggestions proved to be valuable and most are incorporated herein. He called my attention to Wilson (1998), which brought into closer focus some of the ideas of the original draft. Others that have contributed markedly are Jacob Odgaard, Philip Burgi, Henry Falvey, and Carlos Alonso. However, none of these people has entirely agreed with the writer's opinions as expressed herein. Disagreement and the resulting debates can only make us stronger.

Appendix

Notes

Economics and War. Of course, many other projects contributed to the war effort and the economic development of the west. The United States was especially fortunate to have the huge electrical resources of the Columbia River come on line with the completion of Grand Coulee at the beginning of the war. Power from Grand Coulee and other Columbia River dams supplied the bauxite furnaces that were a cornerstone in aircraft production. The United States may have won the war without Grand Coulee, Hoover, and Bonneville, but it would have been a longer war with more casualties. An extensive (211 pages plus annexes) analysis of the project can be found in Ortolano et al. (2000). They study the economics, the projected and actual impacts, what went right and what went wrong, and they identify the winners and losers from the dam construction. The analyses of Ortolano et al. (2000) apply much more broadly than the Grand Coulee project.

Glen Canyon. Glen Canyon is separated from the Grand Canyon by Marble Canyon and is outside of the Grand Canyon Recreational Area. However, it is one long canyon with a wide point near the confluence of the Little Colorado River and low walls in the vicinity of Lee's Ferry and the confluence or the Paria River, a few miles downstream of the dam.

Flood. There was an artificial, 16-day-long "flood" in 1996 to mimic part of the natural cycle of the Colorado River. That "flood" was small compared to natural floods. The USBR with a great fanfare of national publicity declared the flood a success in restoring beaches and natural habitat to the Grand Canyon. Conversations with boatmen who direct commercial trips through the Canyon indicate that its success was very short lived, perhaps as little as a month and no more than a year. The flood has not been repeated, at least up to the time of this writing (but a smaller, two-day flow of power plant capacity took place in 1997 and a previous testing of reconstructed spillways discharged more water for a longer period of time).

Pollution. Pollution consists of the chemicals and ash produced as biomass is burned and CO_2 is emitted. However, technology can make it a mostly clean process and even net CO_2 emissions might be close to zero when capturing of CO_2 by growth is considered. In the case of some biomass, for example, ethanol from corn, more energy goes into the production than is extracted from the fuel.

Sustainable. The definition of "sustainable" is vague. Consider the following paragraph from Barrett and Odum (2000): "Much has been written in recent years regarding the need to live within a society that sustains its resources for the future, a goal that requires rating plans for the future based on the concept of sustainable development (e.g., Lubchenco et al.1991, Huntley et al.1991, NCR 1991, Heinen 1994, Goodland 1995). A forum on 'Perspectives on Sustainability,' which appeared in Ecological Applications (November 1993), attempted to summarize many of the earlier perspectives surrounding this topic. Unfortunately, considerable confusion remains, especially among the citizenry, as to what is meant by sustainable development. Dictionaries define 'to sustain' as 'to hold,' 'to keep in existence,' 'to support,' 'to endorse without failing or yielding,' 'to maintain,' or 'to supply with necessities or nourishment to prevent from falling below a given threshold of health or vitality.' Given these definitions, the businessperson often views sustainability as sustaining profits based on ever increasing consumption of limited natural resources or sustaining rapid economic growth forever! At the other extreme, the definition in the widely cited Brundtland report (WCED 1987)—namely, that 'sustainable development is development that meets the needs of the present without compromising the ability of future generations to meet their own needs' (p. 8)—is so vague as to be impossible to quantify or implement." (References not included herein.)

Consilience. From the *Oxford English Dictionary*: "consilience konsi.liens. [f. next: see -ence.] The fact of 'jumping together' or agreeing; coincidence, concurrence; said of the accordance of two or more inductions drawn from different groups of phenomena." Consilience is the title of the book by Wilson (1998) that treats the unity of all knowledge.

References

Arrow, K., et al. (1996). "Economic growth, carrying capacity and the environment." *Ecological Applications,* 6(1), 13–15.

Barrett, G. W., and Odum, E. P. (2000). "The twenty-first century: The world at carrying capacity." *BioScience,* 50(4), 363–368.

Burton, P. (1992). *Niagara: A history of the falls,* McClelland and Stewart, Toronto.

Ettema, R. (2000). "Adrift in the curriculum." *J. Prof. Issues Eng. Educ. Pract.,* 126(1), 21–26.

Fradkin, P. (1981). *A river no more: The Colorado River and the West,* Knoph, New York.

Hirshleifer, J., DeHaven, J. C., and Milliman, J. W. (1960). *Water supply: Economics, technology, and policy,* University of Chicago Press, Chicago.

Isaacson, E., Stoker, J. J., and Troesch, B. A. (1954). "Numerical solution of flood protection and river regulation problems (Ohio-Mississippi floods)." *Rep. II, Rep. No. IMM-NYU-205,* Institute of Mathematical Sciences, New York Univ.

Liggett, J. A. (1996). "Fundamental issues in numerical hydraulics," *Issues and directions in hydraulics,* T. Nakato and R. Ettema eds., Balkema, Rotterdam, The Netherlands.

Liggett, J. A., and Ettema, R. (2001). "Civil engineering education: The past, present and future." J. Hydraul. Eng., 127(12).

Loucks, D. P., Stakhiv, E. Z., and Martin, L. R. (2000). "Sustainable water resources management." *J. Water Resour. Plan. Manage. Lcd.,* 126(2), 43–47.

Naudascher, E. (1996a). "Turning towards the needy: The unexplored direction," *Issues and Directions in Hydraulics,* Nakato and Ettema, eds., Balkema, Rotterdam, The Netherlands.

Naudascher, E. (1996b). "Impact of the High Aswan Dam on Egypt— Discussion," *Issues and Directions in Hydraulics,* Nakato and Ettema, eds., Balkema, Rotterdam, The Netherlands.

Ortolano, L., et al. (2000). "Grand Coulee Dam and the Columbia Basin Project, USA," case study report prepared as an input to the World Commission on Dams, ⟨www.dams.org⟩

Preissmann, A. (1961). "Propagation de intumescences dans les canaux et rivieres," 1st Congres de l'Assoc. Francaise de Calcul, Grenoble, 443–442.

Reisner, M. (1986). *Cadillac desert: The American West and its disappearing water,* Viking, New York.

Rhone, T. J. (1988). "Development of hydraulic structures," Paper presented at the ASCE Conference on Hydraulic Engineering, Colorado Springs, Colo. August, 8–12.

Roberts, P. J. W. (1999). "Modeling Mamala Bay outfall plumes. I: Near-Field." *J. Hydraul. Eng.,* 125(6), 564–573.

Rouse, H. (1938). *Fluid mechanics for hydraulic engineers,* McGraw-Hill, New York.

USACOE. (2000). "Civil works environmental initiative: Value to the nation," ⟨http://www.wrsc.usace.army.mil/iwrnew/pdf/brochures/ 3_99EnvBroch_P.pdf⟩

USBR. (1999). "Glen Canyon dam modifications to control downstream temperatures: Plan and draft environmental assessment," ⟨http:// www.uc.usbr.gov/environment/pdfs/gcdtc.pdf⟩

Vennard, J. K. (1940). *Elementary fluid mechanics,* Wiley, New York.

Weber, L. J., Nestler, J., Lai, Y. G., and Goodwin, R. A. (2001). "Simulating fish movement behavior using CFD and water quality models for improved fish passage, protection and impact assessment," *NATO advanced research workshop: New paradigms in river and estuary Management,*

Wilson, E. O. (1998). *Consilience: The unity of knowledge,* Vantage, New York.

American Society of Civil Engineers *150th Anniversary Paper*

Irrigation Hydrology: Crossing Scales

Wesley W. Wallender[1] and Mark E. Grismer[2]

Abstract: Hydrology is the science concerned with distribution, circulation, and properties of water of the earth and its atmosphere, across the full range of time and space scales. Subject matter ranges widely from chemical and physical properties to the relation of water to living things. Irrigation hydrology is constrained to analysis of irrigated ecosystems in which water storage, applications, or drainage volumes are artificially controlled in the landscape and the spatial domain of processes varies from micrometers to tens of kilometers while the temporal domain spans from seconds to centuries. The continuum science of irrigation hydrology includes the surface, subsurface (unsaturated and groundwater systems), atmospheric, and plant subsystems. How do we scale up highly nonlinear physical, chemical, and biological processes understood at natural scales to macro- and mega-scales at which we measure and manage irrigated agroecosystems? How do we measure, characterize, and include natural heterogeneity in scaling nonlinear processes? In this paper, we discuss scaling issues and related research opportunities in irrigation hydrology with the hope of helping the irrigation-drainage engineering/ science profession better address scaling problems in formulating designs affecting irrigated ecosystems.

CE Database keywords: Irrigation; Hydrology; Drainage; Nonlinear systems; Heterogeneity; Scale, ratio.

DOI: 10.1061/(ASCE)0733-9437(2002)128:4(203)

Introduction

While irrigation has dramatically increased agricultural productivity in the short term, subsequent water resource depletion as well as soil and water contamination threaten long-term sustainability of irrigated ecosystems. Improved understanding of processes crossing space and time scales sets the foundation for quantification of both positive and negative impacts needed to manage irrigation-drainage projects for long-term sustainability. Hydrology is the science of choice to better understand complex ecosystems because water, a ubiquitous and life-supporting substance, links physical and biological systems at multiple temporal and spatial scales.

Hydrology is the science concerned with distribution, circulation, and properties of water of the earth and its atmosphere, over the full range of time and space scales. Subject matter ranges widely from chemical and physical properties to the relation of water to living things. Irrigation hydrology is constrained to the water-related science of irrigated ecosystems in which the spatial domain of processes varies from micrometers to hundreds of kilometers while the temporal domain spans from seconds to cen-

turies. The continuum science of irrigation hydrology includes the surface, subsurface (unsaturated and groundwater systems), atmospheric, and plant subsystems in which at least one part of the water storage, application, or drainage systems is artificially controlled in the landscape. In this paper, we discuss scaling issues and related research opportunities in irrigation hydrology with the hope of helping the irrigation-drainage engineering/science profession better address scaling problems in formulating designs affecting irrigated ecosystems.

Continuum, Scales, and Scaling

Viewing irrigation hydrology within the broader context of hydrologic sciences is in the spirit of the National Research Council (NRC 1990) report emphasizing analysis of "cycling of water at all scales from microprocesses of soil water to the global processes of hydroclimatology." It remains a challenge to precisely define appropriate space and time scales of seemingly simple processes many of which are in fact nonlinear and occur across heterogeneous systems. Fig. 1 schematically illustrates how water influences and is influenced by transport and transformation of air, solutes, soil, and biota (plants and animals) that are vital for life and spans the continuum from groundwater through the vadose zone to the atmosphere.

Irrigation hydrology falls within the three-dimensional domain shown in Fig. 1 and is studied at several time and space scales listed in Table 1. The microscale is the smallest scale at which a system can be considered a continuum. Below this scale, materials are viewed as discrete particles and properties such as water content and flow velocity are not continuous functions. At this scale gas, liquid, and solid phases are considered as independent yet interacting continua. The macroscale is larger than the microscale but smaller than the scale of the entire system of interest. Gas, liquid, and solid phases are commonly considered as over-

[1] Professor, Depts. of Land, Air and Water Resources (Hydrology) and Biological and Agricultural Engineering, Univ. of California–Davis, Davis, CA 95616.

[2] Professor, Depts. Of Land, Air and Water Resources (Hydrology) and Biological and Agricultural Engineering, Univ. of California–Davis, Davis, CA 95616.

Note. Discussion open until January 1, 2003. Separate discussions must be submitted for individual papers. To extend the closing date by one month, a written request must be filed with the ASCE Managing Editor. The manuscript for this paper was submitted for review and possible publication on September 12, 2001; approved on January 3, 2002. This paper is part of the *Journal of Irrigation and Drainage Engineering*, Vol. 128, No. 4, August 1, 2002. ©ASCE, ISSN 0733-9437/2002/4-203–211/$8.00+$.50 per page.

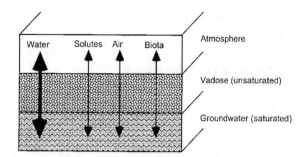

Fig. 1. The water continuum in hydrology

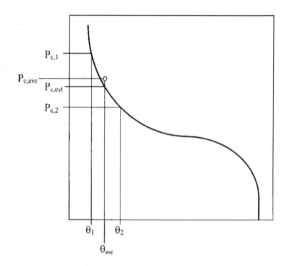

Fig. 2. Conceptual nonlinear function between soil-water content θ and average soil-water capillary pressure P_c

lapping (spatially averaged) continua. At the megascale and above, spatial variations are ignored and the system is usually described using temporal and spatial averages.

The "natural" scale may be defined as a "characteristic" dimension in time and space associated with an observation of a variable or a process while "scaling" refers to concepts, or paradigms used to adjust relationships between variables or processes as the scale of observation differs from the natural scale. Often, a more intuitive characteristic measure is the representative elementary volume (REV) used by Bear (1972). For homogeneous systems REV is the minimum volume of observation such that the volume average of a property, such as soil porosity, does not change as the averaging volume increases. In this case, the corresponding natural length scale is several pore diameters, or at least 3–5 grain diameters (Clausnitzer and Hopmans 1999) for glass beads. For nonstationary properties (e.g., trend in variable mean) and inhomogeneous (heterogeneous) systems the lower limit of the REV for porosity is related to the particle size while the upper limit is related to the distance across which average porosity trends.

Even in homogeneous systems, differences between natural and measurement length scales may compromise validity of nonlinear constitutive relationships commonly used in irrigation hydrology such as that between soil-water content and capillary pressure, or matric suction potential. This occurs despite resolution of conflicts (Corey and Klute 1985) between definition of potentials based on the aqueous phase and the water molecule (Nitao and Bear 1996). Measurement of soil-water content requires a sample large enough to include the entire pore-size distribution, or approximately equivalent to that required for the porosity REV. However, neglecting adsorptive force fields, capillary pressure is defined only across an air-water interface in a particular interconnected pore space. As an analogy for soil, if a bundle of capillary tubes having a range of diameters that are initially full of water (saturated) is subjected to a particular suction that exceeds the capillary rise value of some of the tubes, it will result in only those tubes having a diameter greater than that value draining; the smaller tubes remain "saturated." The bundle, or "bulk water content" includes both the "saturated" and empty tubes, while the applied capillary pressure applies to only those tubes that drained. That is, the natural scale of capillary pressure is at the pore scale, while that for water content is several times larger. As the relationship between capillary pressure (P_c) and pore hydraulic radius (R) is hyperbolic (i.e., $P_c \propto 1/R$), the relationship between water content and capillary pressure is a nonlinear step function. Similarly, for a homogeneous soil, the natural length scale of capillary pressure is microscopic (Nitao and Bear, 1996) and the constitutive relationship is highly nonlinear. Pressure, which varies within the soil-water content-based REV, is aver-

aged across the fluid phase and the resulting function (Fig. 2) depends on the averaging method. Conceptually, an average capillary pressure versus soil-water content relationship could be scaled up. For example, combining two adjacent REVs having water contents θ_1 and θ_2, yields an average water content $\theta_{ave} = [\theta_1 + \theta_2]/2$. Entering the nonlinear function shown in Fig. 2, θ_{ave} for the larger sample volume leads to an "average" pressure $P_{c,ave} = [P_{c,1} + P_{c,2}]/2$ which may fall on an upscaled function. The difference between $P_{c,est}$ and $P_{c,ave}$ is the error if the unscaled function is used with the data from a device measuring the larger volume of soil. Furthermore, there are multiple combinations of adjacent water contents each giving a unique "averaged" pressure and these pairs do not fall on a single scaled-up curve. Hence, the original and the upscaled functions depend on the method of averaging. While conceptually an average capillary pressure versus soil-water content relationship can be established for the REV, microscopic capillary pressure cannot be measured nondestructively. Furthermore, most laboratory measurements of soil-water content are at sample sizes considerably larger than the soil-water content REV. Instead, some "averaged" capillary pressure determined from a wick, ceramic candle, or ceramic plate is assigned to the measured soil-water content to give volume-dependent, constitutive relations. If the relationship between capillary pressure and soil-water content were fundamentally linear, no scaling issue between the different measurement scales would result. Measured relationships between unsaturated soil hydraulic conductivity (K) and capillary pressure, or soil-water content encounter the same difference in scale problem combined with even greater nonlinearity (i.e., $K \propto R^2$).

For heterogeneous porous media, hydraulic function(s) vary by location within the medium and this also causes scaling problems. When adjacent volumes are averaged, the functions (linear or nonlinear) are averaged and, as above, there is a cloud of points rather than a single upscaled curve. Together, heterogeneity of the medium and nonlinearity of processes, or relations cause problems in scaling processes across the irrigated ecosystem. Beven and Fisher (1996) describe the "scaling problem" as a set of concepts that enable information or models developed at one particular scale to be used in making predictions at another scale.

Harvey (1997) and Buggman (1997) suggest six causes of scale and scaling problems for hydrologic systems that also apply to the irrigated ecosystem.

Table 1. Space and Time Scales for Water Properties and Processes in Irrigation Hydrology

Time	Space			
	Microscale (10^{-5} 10^{-2} m)	Macroscale (10^{-2} 1 m)	Megascale (1 10^4 m)	Global Scale (10^4 10^8 m)
Seconds	Water properties Pore-water processes Geochemical processes	Porous media flow Pipe flow	Infiltration Transpiration Precipitation Surface flow Drainage	Atmospheric processes
Hour-day Years Centuries Millennia	Time average ↓	Space average →		

1. Spatial heterogeneity in surface/subsurface processes;
2. Nonlinearity in system responses at different scales;
3. Processes may require threshold scales to occur (related to nonlinearity);
4. Dominant processes of concern may change with different scales;
5. Development of emerging processes resulting from mutual interaction of small-scale processes; and
6. Disturbance regimes (e.g., dams, canals, etc.) superimposed on natural systems.

Fundamentally, the latter four causes are special cases of or a combination of those associated with process nonlinearity and system heterogeneity. Focusing also on the last four causes, Wagenet (1998) noted that scale translation efforts have often failed due to either a key controlling process or characteristic that has been overlooked, or when multiple factors interact/combine to create unique phenomena heretofore not considered/studied. One possible opportunity in analysis of scaling problems associated with irrigation hydrology is measurement of integrative water application or drainage processes at several scales from the plot to the region.

Scaling in Irrigation and Drainage Systems

Irrigation-drainage engineering has focused on analyses of processes occurring at several spatial/temporal scales ranging from microscopic pore-scale phenomena to regional, or water-district scale issues associated with water use and delivery. In practice length, area- or volume-dependent measurements are taken of independent and dependent variables resulting in length scale-dependent constitutive models. For example, Wallender (1987) developed a model to predict yield from soil-water content and found a model dependence on water content sample size. Likewise, the scale "jump" in agricultural hydrology from that at which individual processes such as infiltration, drainage, evapotranspiration, soil loss, or crop yield have been studied, to the global scale at which climate change impacts manifest themselves has presented conceptual as well as practical problems of scales and scaling (Schulze 2000). Thus, in order to predict human impacts on water resources, air quality, and agricultural production irrigation hydrologists need to consider "upscaling" of nonlinear processes in the presence of heterogeneity.

Pore to Laboratory Column Scales

In recent hydrologic research, several schemes and methods have been developed to upscale, or volume average (e.g., Boe 1994;

Whitaker 1999; Ekrann and Aasen 2000) physical laws, (e.g., the momentum equation) applicable to pore-scale processes in order to describe processes, or phenomena at the column, or bench scale. In irrigation hydrology, the obvious processes of concern that are upscaled from the pore scale include assessment of soil-water hydraulic properties (e.g., Desbarats 1995; Or and Tuller 1999; Held and Celia 2001) plant-water uptake (Cermak and Nadezhdina 1998) and water-quality related processes (e.g., Macdonald et al. 1999). At microscales, water and its constituents flow and transform as well as interact with soil surfaces in the rhizosphere (i.e., root-, soil-, and soil-water interfaces), the unsaturated zone below the root zone and the groundwater system at time scales from seconds to millenia (Table 1). Both measurement support size limitations as well as developing realistic solutions of continuity-type equations at microscales have led to measurement and modeling at macroscales. For example, the microscale linear momentum (Navier-Stokes) equations are simplified to the Stokes' equations and space-averaged to derive Darcy's equation (Whitaker, 1999). Combining the Darcy equation with mass conservation to model spatially averaged velocity or water potential results in the Richard's equation. Similarly, the Navier-Stokes equations are simplified for the case of free-surface (e.g., open-channel) flow in border or furrow irrigation to yield the St. Venant equations. These cross-section averaged equations are solved for one- or two-dimensional flow velocities and depths in furrow, border, and basin irrigation systems.

When considering micro-scale heterogeneities across soil pores, or from soil cores to plot scales, Hopmans et al. (2002) summarize application of "aggregation" methods (Wheatcraft and Tyler 1986) to scale up pore-scale measurements of large and small hydraulic conductivity "textures" to an average value for the core as a whole.

When considering microbial degradation of contaminants or nutrients in soils, models of bacterial populations often do not resolve pore-scale variability problems, rather, substrate and biomass concentrations are taken as bulk averages. These models unrealistically predict monotonically increasing biomass growth everywhere except where the limiting substrate concentration is very small. Macdonald et al. (1999) examines the possibility of biofilm mass-transfer limitations at the pore scale using both the traditional biofilm model as well as previously published results from an upscaling model. Results from the biofilm model suggest that limitations on biofilm growth due to mass-transfer resistance could be significant in coarse-grained soils with adequate substrate availability. The upscaling approach also indicated that the degree of mass transfer resistance is reduced at higher-groundwater velocities.

When considering crop consumptive-use processes, the work of Cermak and Nadezhdina (1998) in upscaling plant-water uptake measurements from individual trees to stands may be valuable. They estimated sapwood depth according to xylem water content and more precisely according to radial patterns of sap flow rate in five coniferous and four broad-leaved species of different diameter, age, and site conditions and compared it to sapwood cross-sectional area, a simple biometric parameter widely used for scaling up the transpiration data between trees and forest stands. Sapwood area estimated by the two methods was almost equal in some species (e.g., Cupressus arizonica), but differed significantly in other species (e.g., Olea europaea, Pinus pinea). Radial pattern of sap flow rate was a more reliable indicator of sapwood then xylem water content for sap flow scaling purposes. Sapwood area could be used for upscaling transpiration rates from measuring points (e.g., stems) to whole trees and from trees to stands only for the same measurement periods. That is, averaging sapwood cross-sectional areas within the tree stand provided a reasonable estimate of the stand transpiration rate as a whole illustrating how heterogeneity effects (i.e., variable soils, nutrients, tree growth, shading, etc.) within the tree stand on transpiration rates appear to diminish when transitioning between the tree and stand scales.

There are unexplored processes at the microscale deserving upscaling research that will lead to improved water management. For example, a common, though not yet well-understood phenomenon is soil shrink/swelling and cracking during and between irrigation events (Waller and Wallender 1993; Lima and Grismer 1994; Nichols and Grismer 1997). Solution of mass, momentum, energy, and entropy balance equations of continuum mechanics at the microscale is one opportunity to understand and simulate this behavior and build upon these early attempts. The prediction of water and solute transport is complicated by simultaneous strain and its causal stress in the solid phase. With improved theoretical understanding at the microscopic scale leading to superior upscaled models which are consistent with measurement support, development of more efficient salt management strategies for commonly irrigated and drained cracking soils will evolve.

In addition to saturated and multiphase (unsaturated) flow in irrigated porous media, free-surface flow in canals as well as flow and erosion in borders and furrows is governed by the microscopic-scale Navier-Stokes equations. Imagine water flow at the leading edge of the advancing front in furrow irrigation where flow is fully three dimensional. The flow domain evolves with time as soil aggregates disintegrate. Fluid density changes as the soil particles are entrained in the water. Recommendations for erosion control as well as pesticide and herbicide runoff can be guided via the understanding of these processes at the microscale. Thus, there are numerous research opportunities to combine the solution of the balance equations with process theory from physics, chemistry, and biology at microscales to better understand and predict sediment/pollutant flow and transport but, again, research on upscaling to measurement support is needed.

Appropriate measurement technology is required to calibrate and validate microscale models. Remote sensing methods such as nuclear magnetic resonance are currently available to measure zero-, first-, and second-order tensor fields such as water content (zero order or scalar), flow velocity (first order or vector) and velocity gradient (second order or informally called tensor) at the pore scale and may assist in developing a greater understanding of microscopic processes and space averaging methods for improved management and design of irrigation-drainage systems in the future.

Laboratory to Plot Scales

Modifying the characterization of soil-water properties from laboratory columns (order of 10^{-1} m) to plots (order of 10 m) involves scaling of hydraulic functions, or aggregation of soil textures (Hopmans et al. 2002). For example, Tidwell and Wilson (1997) describe procedures for upscaling permeability estimates to the plot and field scales. Such an analysis is also critical to evaluation of soil moisture variability and recharge rates across a landscape due to small changes in topography. Green (1995) developed a method of upscaling both water content and hydraulic conductivity on the assumption of uniform matric potential for upscaling water content, and use of van Genuchten's function to represent the upscaled water-retention curve. He used numerical experiments to address how the combination of undulating topography and moisture-dependent anisotropy affected soil-water redistribution and distribution of recharge rates in both space and time. He found that small-scale soil layers created anisotropy in upscaled (spatially averaged over many small layers) hydraulic conductivity. Values of anisotropy at moderately low saturations were generally much higher than at saturation or near saturation. Moreover, zones of soil-water accumulation acted as preferential recharge areas. Different combinations of landscape topography, anisotropy, and steady recharge rates affected the spatial distributions of both Darcy flux and upscaled soil-water contents. Darcy flux varied by three or four orders of magnitude, from a minimum beneath topographic crests to a maximum beneath swales such that vertical Darcy flux at a depth of 2 m was strongly correlated ($R^2 = 0.96$) with topographic curvature. This latter observation again suggests that larger-scale measurements (e.g., topography, tree stands) that incorporate smaller-scale heterogeneities may provide better insight into processes of concern (e.g., seepage, transpiration) than when modeled at the smaller scales.

At this scale range, again there are ample opportunities for upscaling research. For example, as cited above, upscaled simulated flow patterns have potentially significant ramifications for root zone leaching of pesticides and solutes to groundwater. In the latter case, Buyuktas and Wallender (2001) modeled three-dimensional flow and transport to subsurface drains for a spatially variable soil and found that upscaled soil hydraulic properties could be used for this case to represent heterogeneous soil properties.

Plot to Field Scales

Upscaling from plot (order 10^1 m) to the field scale (order $10^2 - 10^3$ m) is again complicated by nonlinearity and variability of process parameters such that some process descriptions are revised, or recharacterized so as to include heterogeneities via a range of statistical methods (e.g., kriging, renormalization, and aggregation). For the irrigation hydrologist, such analyses are usually applied to soil-water transport processes (i.e., soil-water retention and permeability) that have been studied extensively in the past decade by engineers, geologists, hydrologists, and soil scientists and is available in their respective journals (e.g., Rubin and Gomez-Hernandez 1990; Russo 1992; Indelman 1993; Indelman and Dagan 1993a,b,c; Sanchezvila et al. 1995; King 1996; Wen and Gomez-Hernandez 1996; Ewing 1997; Hristopulos and Christakos 1997; Bierkens and vanderGaast 1998; Hunt 1998; Tegnander and Gimse 1998; Wen and Gomez-Hernandez 1998; Hristopulos and Christakos 1999; Sahimi and Mehrabi 1999; Tidwell and Wilson 1999; Sanchez-Vila et al. 1999). On the other hand, in ecological engineering studies, Ahn and Mitsch (2002) found that taking a scaling perspective based on wetland function and "performance" (e.g., water use, nutrient uptake/transformation, plant coverage, hydraulic residence times) to be useful towards comparison of 1 m^2 versus 1 ha systems. In comparing the small and large wetland systems, they found nearly the same performance during time periods of one to two years despite obvious heterogeneities (e.g., preferential flow) in the 1 ha system.

There is also some question as to whether commonly accepted methods for estimating evapotranspiration (e.g., Penman-Monteith equation) by irrigation hydrologists that are validated from lysimeter-scale analyses apply at larger field/region scales despite their mathematical elegance (Schulze 2000). Perhaps in an effort to better upscale from the plant to canopy gaseous exchange processes in a manner that may be useful to us, Wohlfahrt et al. (2000) develop a model that quantifies CO_2 and H_2O gas exchange of whole plants and the canopy. The model combines leaf (gas exchange, energy balance) and canopy (radiative transfer, wind attenuation) scale simulations. Net photosynthesis and stomatal conductance are modeled using a nitrogen-sensitive model of leaf gas exchange. An analytical solution to the energy balance equation is adopted to calculate leaf temperatures. Radiative transfer, considered separately for the wave bands of photosynthetically active, near-infrared and long-wave radiation, is simulated by means of a model that accounts for multiple scattering of radiation using detailed information on canopy structure as input data. Partial pressures of CO_2 and H_2O, as well as air temperatures within the canopy are not modeled, but instead, measured values are used as input data to the model.

Also of potential significant importance to irrigation hydrologists is the upscaling of nonlinear surface erosion processes measured at the plot scale up to field and watershed scales associated with hillslope agriculture [e.g., vineyards and orchards; see Battany and Grismer (2000)]. Le Bissonnais et al. (1998) measured crust formation, runoff, and erosion during two seasons from 1-m^2, 20-m^2, and 500-m^2 plots and up to a small 70 ha catchment. Runoff fraction of total rainfall, sediment concentration and net soil loss from the 20-m^2 and 500-m^2 plots were practically equivalent and several times greater than that obtained from the 1-m^2 plots and 70 ha catchment. Interestingly, peak sediment concentration and soil loss from the 1-m^2 plots and 70 ha catchment were quite similar, despite different processes (e.g., preferential flow) occurring at each scale. Overall, ground cover was the primary factor controlling erosion, though the variation in results at the different scales underscored the need for additional evaluation. King et al. (1998) incorporates this research into upscaling of an erosion model from small areas to regions.

Next consider upscaling processes and measurements, and their implications for irrigation and drainage system design using economic sampling principles. First we discuss processes. Generally, irrigation design is based on results of time and area-averaged mass balance equations. For overland flow processes, solution of the momentum equation is approximated or ignored. In the case of surface irrigation modeling, a predetermined shape is assumed for the surface flow profile as a replacement for the St. Venant equation. Likewise, infiltration, an input to the volume balance equation of surface flow, is measured over areas ranging from 1–100 m^2 and infiltration rate and cumulative depth functions are generally based on area-averaged measurements. For sprinkle irrigation performance evaluation, the momentum equation is ignored and application volume depends on the projected area of the catch can and on the duration of the test. Thus, the average and spatial variation of water application depths are area-and time-averaged measurements.

Rather than using scales of measurement that optimize constitutive relations and are consistent with the processes of interest, they have been largely determined by practical considerations. For sprinkle irrigation assessments, readily available oil cans drove their use in measuring uniformity. Fischer and Wallender (1988) recommended larger collectors to enhance accuracy and precision and to reduce the duration and number of tests. For surface irrigation system evaluation, although variability of infiltration measurements decreases as the area of the infiltrometer increases (Wallender 1987), the area of the infiltrometer is constrained by practical considerations such as water, the ability to construct and install large infiltrometers, or the ability to solve the inverse problem of determining the infiltration function from measurements of flow geometry, slope, hydraulic roughness, and water advance. Similarly, the construction and function of gypsum blocks, tensiometers, neutron probes determines the space- and time-averaging of soil-moisture measurements. Likewise, drainage and groundwater measurements, using piezometers or wells, are area or volume as well as duration dependent. Such measurement scales may be incongruent with that optimal for the irrigation-drainage process of interest (Wallender 1987).

Last, economic principles are used to determine optimal sampling for design. Optimal sampling minimizes the sum of design error cost and sampling cost (Ito et al. 1999). They calculate the cost of a design error (or cost information) as the difference between optimal returns to water in which infiltration function information is complete and incomplete. The optimal sampling density minimizes the sum of unit information and infiltration measurement cost. The use of suboptimal space and time scales of processes and constitutive relations carry into suboptimal sampling strategies and system designs.

A water budget analysis serves to illustrate this compromised process. To control sampling cost by decreasing sampling density, several fields are aggregated and all but one component [e.g., evapotranspiration (ET), surface water delivery, drainage, tailwater, change in vadose zone storage] is measured or assumed and the remaining term is selected as the closure term (e.g., deep percolation). To further control sampling cost, in addition to decreasing sampling density, the duration between samples is increased. Less frequent measurements are taken because it is difficult (i.e., numerous simultaneous measurements over a large area) and expensive to measure changes in storage as well as flux of water and solutes using small sample areas spaced close together over large areas. However, rarely is the cost of the lack of information included in the analysis. Sampling cost is constrained by arbitrary budgets instead of the sum of sampling and suboptimal management.

Field to Region-Watershed Scale

Upscaling field-scale measurements, processes and simulation to the regional or watershed scale involves aggregation of data/processes collected at multiple levels into a coherent framework for subsequent analyses of impacts at the larger scales. This data management is crucial to later application and is often accomplished with the assistance of Geographic Information Systems (GIS). Multispectral remote sensing data are often analyzed and incorporated to produce a vegetative cover thematic layer, that may be used to provide data for ET and infiltration estimation. In irrigation hydrology, agricultural fields become a part, perhaps the largest fraction, of the overall landscape or watershed to be considered. For example, Purkey and Wallender (2001) modeled the effect of removing land parcels from agricultural production located up- and down-gradient of shallow groundwater on subsurface drainage. For single parcels, the 31% drainage reduction potential of downgradient land retirement was more attractive than the 16% reduction associated with the upgradient land retirement. Regulatory and water management agencies are increasingly interested in the overall impact of irrigated agricultural production on the landscape (including groundwater) as the health of water supplies for local communities and environment are intertwined with agricultural practices in the region. Regional scale agricultural economic-type models and evaluations have also been developed (e.g., Lee and Howitt 1996), but we retain our focus here on physically based models and the processes associated with upscaling (e.g., Davy and Crave 2000). We consider upscaling from the field to region and underscore the importance of data management and inclusion of impacts on ground and surface water systems within the agricultural production region.

Integrated modeling of interdependent surface, soil-water and groundwater processes requires multiple scales of investigation, or analysis, to achieve objectives associated with improved water management, or environmental protection. There have been problems in some efforts to upscale land-use cover information from field to regional scales. For example, Yang (1997) investigated the effects of changing spatial scale on the representation of land cover and found that significant errors may be introduced in heterogeneous areas where different land cover types exhibit strong spectral contrasts. The extent of agreement between spatially aggregated coarse resolution land cover datasets and full resolution datasets changed depending on the properties of the original datasets, including the pixel size and class definition. To assist in identification and resolution of such problems, Geurink (1999) designed a database structure that includes elements for organization and characterization of field and laboratory observations with utilities for both data management and derivation of extensive, integrated model parameters. Selection and resolution of physical properties, stored and maintained in the database, are governed by temporal and spatial sensitivity of hydrologic processes of the region. The broadest range of applications for data acquired from the database can be attained when data are independent of scale, structure, and process algorithm of a model. Data management utilities have been outlined to query and acquire data, evaluate and correct data errors and omissions, characterize data statistically, and develop data sets for integrated model calibration. His results suggest that uncertainty and sensitivity analyses of simulation results can be used to refine observation resolution requirements for physical properties and to guide identification of additional locations or physical properties for which to acquire data. Moreover, centralized, hierarchal accessibility to data appears to promote multidisciplinary application of common data while maintaining protection protocols.

At the surface, where crop growth and water use (or ET) are of primary importance practically, Olesen et al. (2000) developed a crop simulation model (*CLIMCROP*) for investigating different methods of aggregating simulated county and national crop yields for winter wheat in Denmark with and without irrigation for a range of soil types and climatic conditions. While the model captured most of the spatial variation in observed yields, except at the coarsest resolutions of soil data, the finest resolution of soil and climate data gave the best fits between simulated and observed spatial autocorrelation in yield. The model better represented interannual yield variability on loamy soils as compared to that for sandy soils. The results indicated that upscaling of simulated yields for Danish conditions required a spatial resolution of soil data of 10 km^2 or finer and that more climate (i.e., ET)

stations are required if yields at larger scales (i.e., regional) are to be estimated. Crop simulation models have also been developed for rice production and associated evolution of methane from the paddies (e.g., Matthews et al. 2000; van der Gon et al. 2000).

Irrigation hydrologists may take some direction in developing agricultural landscape models that incorporate landscape impacts from those developed for ecological systems or forested watersheds. Cernusca et al. (1998) used a combined approach of field measurements and process-based modeling to assess effects of land-use changes in the Eastern Alps, the Swiss Alps, the Spanish Pyrenees, and the Scottish Highlands ecosystems. An analysis of structures and processes in the context of land-use changes is performed, scaling from the leaf to the landscape level including analysis of vegetation and soil-type spatial distribution, canopy structure, ecosystem water relations, catchment area hydrology, ecosystem microclimate and energy budgets, gas exchange of single plants to canopies, and potential risks associated with land-use changes. At each level of integration, modeling results were validated by direct field measurements. For example, with gas exchange processes, the up-scaling began with experimental measurements of gas exchange by single leaves using CO^2/H_2O porometers, proceeds to the canopy level using Bowen-ratio and Eddy-correlation techniques, and finally to the gas exchange of composite landscapes using a scintillation anemometer system/differential optical absorption spectroscopy system equipment in combination with a instrumented aircraft flying at a constant height above the landscape. A hierarchy of process-based models was used to link the gas exchange processes at the different scales: a single leaf model based on a biochemical model of photosynthesis' a newly developed multispecies canopy model taking into account multiple light scattering and nonuniform distribution of leaves, and a landscape model providing a pixel-by-pixel integration. In an analogous fashion, Llorens and Gallart (2000) upscaled interception by individual pine needles and bunches to determine stem flow, throughfall, and eventually canopy and catchment scale forest water storage through combined field measurements and modeling. They found that modeled estimates of forest canopy water storage were 30% greater than that determined from more commonly used indirect estimates based on rainfall-throughfall regressions.

Lumped process and distributed models used to upscale field observations of processes have also been evaluated in forest and watershed hydrology. For example, Yu (2000) evaluated three types of forest hydrologic models, "black-box," lumped parameter, and physically based distributed models (PBDMs). Black-box models are essentially statistical input-output watershed models, while lumped-parameter and PBDM models are empirically derived models based on hydrological processes in the watershed. Relative to black-box and lumped-parameter models, PBDMs considered watershed spatial heterogeneity through different cells and flow pathways and can express hydrological processes in both spatial and temporal dimensions. The writer suggested that landscape ecology knowledge combined with GIS could support PBDMs to solve the upscaling problem. Birkinshaw and Ewen (2000) also successfully use a physically based, spatially distributed river catchment model combined with "point-based" equations for transformation of carbon and nitrogen pools in the soil and subsequent advection dispersion of these nutrients to ground and surface waters in the watershed. Soutter and Loague (2000) also found that application of a "point-scale" pesticide leaching model coupled with variation in soil-type data and land/water-use conditions was adequate to evaluate regional dibromo-3-chloropropane (DBCP) leaching in the Fresno, Calif. area. Previously, but and in a similar fashion, Kersebaum and Wenkel (1998) evaluated the effect of different data qualities and aggregation levels on results of an integrated soil-water and nitrogen dynamics model for an area of three districts that were part of a former large-scale study on nitrogen losses to groundwater. Measurements and simulations of groundwater recharge and nitrogen leaching indicated that aggregation of soil map units had a relative small effect as compared to averaging of weather data and errors of land-use characterization. Use of long-term average weather data lead to underestimation of groundwater recharge rates. Nitrogen uptake by plants in the regional study was often less than that simulated with actual weather data and observed crop rotations. Refsgaard et al. (1999) uses a combined data aggregation and distributed modeling approach to evaluate regional-scale nitrate leaching and groundwater contamination. Readily available European level data from standard databases were supplemented by selected data from national sources. Model parameters were obtained from these data by use of "transfer functions" and upscaled from point to field to catchment scales using effective parameters with a statistically based data aggregation procedures that preserved areal distribution of soil types, vegetation types, and agricultural practices on a catchment basis. In comparing simulation results with measurements for a pair of small Danish watersheds, the upscaling/aggregation procedure appeared to be applicable in many areas with regard to root zone processes such as runoff generation and nitrate leaching, but had important limitations with regard to stream hydrograph shape due to failure to account for scale effects in relation to stream-aquifer interactions. Groundwater contamination issues also pose human-health risks that require combined physical molding and health-risk assessments at the regional scale (e.g., Pelmulder 1994; Pelmulder et al. 1996) and such analyses may need to be included with regional-scale economic assessments (e.g., Lee and Howitt 1996) of irrigated agricultural impacts to formulate regional water quality policies and management of agricultural practices.

Ultimately, watershed contamination, or landscape water-use models developed to assess effects of changing land use, or management conditions need to incorporate contamination effects on the overall ecological system. Habersack (2000) discusses the river-scaling concept as a basis for ecological assessments and notes that since river morphology results from transport of water and sediments, the size of project areas and the analysis procedures are critical factors. Restricting the assessment of abiotic and biotic river components and its variability to a certain scale may neglect the fact that ecological integrity depends on the stream process scale. He uses a two-step procedure for assessing the ecological integrity at various temporal and spatial scales. During the so-called downscaling phase, abiotic and biotic components are analyzed at the regional, catchment, sectional, local, and point scales and illustrate the importance of mass balance analyses, and application of fractals and self-similarity studies for channel development. As the scale varies dramatically, results obtained from various analysis tools are significantly different but interdependent. Biotic analyses are also performed at these scales, so that the interrelations between channel morphodynamics and habitat quality can be derived at the end of this first phase. In the second phase, upscaling integrates and aggregates the results from the first step in order to yield overall conclusions on ecological integrity.

Finally, assessment of farm and regional water use efficiency in agriculture is essential towards rational allocation of limited water resources (Grismer 2001a,b). For example, Tuong and Bhuiyan (1999) evaluate the water losses associated with rice production and identify that water used during land preparation, and seepage and percolation during crop growth are the primary limitations and discuss strategies for increasing farm-level water-use efficiency and upscaling from on-farm to system-level water savings.

Region-Watershed to Global Scale

Upscaling from regional to global conditions requires remote sensing measurements and incorporation of global circulation models coupled with changing land-use and anthropomorphic impacts on climate change and variability. This scale of research or application is typically beyond the scope normally considered by the irrigation hydrologist, but may soon come to be an issue of concern as land-use impacts and agricultural policies and practices in one country often affect neighboring nations as well as more distant nations across the ocean (e.g., dust storms in China/ Mongolia appearing in California in spring 2001). In addition to global climate change considerations are aspects of population growth and adequate nutrition (Renault and Wallender 2000). Harvey (2000) discusses general aspects and problems associated with upscaling to the global scale, while De Grandi et al. (2000) illustrate use of satellite imagery and data to evaluate ecosystem changes in Africa.

Opportunities

As a profession, irrigation engineers/hydrologists are developing creative measurement and sampling technology as well as analytical and statistical theory to improve monitoring and assessment of water quantity and quality impact from irrigated agriculture. Though a daunting task, in gaining a better understanding of irrigation hydrology processes through measurement and modeling across broad space and time scales, we hope to discover linkages which transform results from one time and space scale to another with no or limited additional measurement.

This approach to field and laboratory research combined with modeling leading to optimal irrigation and drainage management is recognized as an important strength in irrigation hydrology. As such, several research/application opportunities remain including: (1) scaling up fundamental physical and biological processes in time and space to better represent the agricultural production system as part of the broader landscape; (2) developing indicators to identify system function (malfunction) at all scales; and (3) manage production systems, at appropriate time and space scales, from input (e.g., water and nutrient application) as well as output (e.g., productivity and environmental impact) perspectives. Clearly, new scale-appropriate measurement methods are needed to facilitate such studies, an opportunity in its own right.

In reviewing the ASCE articles database, it was interesting to note that in the past three decades there have been only a handful of articles published in symposia or journals that consider in their keywords, or titles the combination of irrigation or drainage with hydrology. Irrigation-drainage engineering scaling issues have not been overtly considered though they are apparent in many research articles, perhaps as a result of combined empiricism and "judgement" often required of the engineering discipline. In contrast, scale and scaling issues have been a dominant research theme in our sister fields of hydrology, soil science, and agricultural economics during the past decade. In hydrology, for example, books by Stewart et al. (1996) and Sposito (1998) provide excellent reviews of the subject, while in soil science, entire journal issues (e.g., see Finke et al. 1998) have been devoted to the problems of upscaling processes. Perhaps we should consider that our role as engineers/hydrologists of the agricultural landscape is less to develop new techniques of information management, i.e., new software, but more to better apply what is already available to solve the "real world" problems we face. For example, Geurink (1999) in working with the multiple scales associated with an integrated surface/ground water model developed a database structure that facilitates analyses and scaling between levels. However, Dumanski et al. (1998) noted "Innovations in the informational sciences are so rapid and constant that we often have to run hard just to avoid being left behind." Consequently, our challenge is to develop the information paradigms necessary to develop a better understanding of the major issues and policy changes related to our science, and then apply those new techniques that suit these requirements.

References

Ahn, C., and Mitsch, W. J. (2002). "Scaling considerations of mesocosm wetlands in simulating large created freshwater marshes." *Ecol. Eng.*, 18(3), 327–342.

Battany, M. C., and Grismer, M. E. (2000). "Rainfall runoff, infiltration and erosion in hillside vineyards: Effects of slope, cover and surface roughness." *Hydrolog. Process.*, 14, 1289–1304.

Bear, J. (1972). *Dynamics of fluids in porous media*, Dover, New York.

Beven, K. J. and Fisher, J. (1996). "Remote sensing and scaling in hydrology." *Scaling up hydrology using remote sensing*, Stewart, J. B., Engman, E. T., Fedes, R. A. and Kerr, Y., eds., Wiley, Chichester, U.K., 1–18.

Bierkens, M. F. P., and vanderGaast, J. W. J. (1998). "Upscaling hydraulic conductivity: Theory and examples from geohydrological studies." *Nutrient Cycl. Agroecosyst.*, 50(1–3), 193–207.

Birkinshaw, S. J., and Ewen, J. (2000). "Nitrogen transformation component for SHETRAN catchment nitrate transport modeling." *J. Hydrol.*, 230(1-2), 1–17.

Boe, O. (1994). "Analysis of an upscaling method based on conservation of dissipation." *Transp. Porous Media*, 17(1), 77–86.

Bugmann, H. (1997). "Scaling issues in forest succession modeling." *Elements of Change 1997 - Session One: Scaling from site-specific observations to global model grids*, S. J. Hassol and J. Katzenberger, eds., Aspen Global Change Institute, Aspen, Colo. 45–57.

Buyuktas, D., and Wallender, W. W. (2001). "Numerical simulation of water flow and solute transport to tile drains." *J. Irrig. Drain. Eng.*, in press.

Clausnitzer, V., and Hopmans, J. W. (1999). "Determination of phase-volume fractions from tomographic measurements in two-phase systems." *Adv. Water Resour.*, 22(6), 577–584.

Cermak, J., and Nadezhdina, N. (1998). "Sapwood as the scaling parameter: Defining according to xylem water content or radial pattern of sap flow?" *Ann. Sci. Forest. (Paris)*, 55(5), 509–521.

Cernusca, A., Bahn, M., Chemini, C., Graber, W., Siegwolf, R., Tappeiner, U., and Tenhunen, J. (1998). "Ecomont: A combined approach of field measurements and process-based modeling for assessing effects of land-use changes in mountain landscapes." *Ecol. Modell.*, 113(1–3), 167–178.

Corey, A. T., and Klute, A. (1985). "Application of the potential concept to soil water equilibrium and transport." *Soil Sci. Soc. Am. J.*, 49(1), 3–11.

Davy, P., and Crave, A. (2000). "Upscaling local-scale transport processes in large-scale relief dynamics." *Phys. Chem. Earth Part A—Sold Earth Geodesy*, 25(6–7), 533–541.

De Grandi, G. F., Mayaux, P., Malingreau, J. P., Rosenqvist, A., Saatchi, S., and Simard, M. (2000). "New perspectives on global ecosystems from wide-area radar mosaics: Flooded forest mapping in the tropics." *Int. J. Remote Sens.*, 21(6–7), 1235–1249.

Desbarats, A. J. (1995). "Upscaling capillary pressure-saturation curves in heterogeneous porous media." *Water Resour. Res.*, 31(2), 281–288.

Dumanski, J., Pettapiece, W. W., and McGregor, R. J. (1998). "Relevance of scale dependent approaches for integrating biophysical and socio-economic information and development of agroecological indicators." *Nutrient Cycl. Agroecosyst.*, 50, 13–22.

Ekrann, S., and Aasen, J. O. (2000). "Steady-state upscaling." *Transp. Porous Media*, 41(3), 245–262.

Ewing, R. E. (1997). "Aspects of upscaling in simulation of flow in porous media." *Adv. Water Resour.*, 20(5–6), 349–458.

Finke, P. A., Bouma, J., and Hoosbeek, M. R. (1998). "Preface." *Nutrient Cycl. Agroecosyst.*, 50, 1.

Fischer, G. R., and Wallender, W. W. (1988). "Collector size and test duration effects on sprinkler water distribution measurement." *Trans. ASAE*, 31(2), 538–542.

Geurink, J. S. (1999). "Concepts of data management and assessment for multi-scale integrated hydrologic modeling." PhD Civil Engineer, Univ. of South Florida, Tampa, Fla., 355.

Green, T. R. (1995). "The roles of moisture-dependent anisotropy and landscape topography in soil-water flow and groundwater recharge." PhD Civil Engineering, Stanford Univ., Palo Alto, Calif., 327.

Grismer, M. E. (2001). "Regional alfalfa yield, ET_c and water value in western states." *J. Irrig. Drain. Eng.*, 127(3), 131–139.

Grismer, M. E. (2002). "Regional cotton lint yield, ET_c and water value in Arizona and California." *Agric. Water Manage.*, 54(3), 227–242.

Habersack, H. M. (2000). "The river-scaling concept (RSC): A basis for ecological assessments." *Hydrobiologia*, 422, 49–60.

Harvey, L. D. D. (1997). "Upscaling in global change research." *Elements of Change 1997 – Session One: Scaling from site-specific observations to global model grids*, S. J. Hassol and J. Katzenberger, eds., Aspen Global Change Institute, Aspen, Colo. 14–33.

Harvey, L. D. D. (2000). "Upscaling in global change research." *Clim. Change*, 44(3), 223.

Held, R. J., and Celia, M. A. (2001). "Pore-scale modeling and upscaling of nonaqueous phase liquid mass transfer." *Water Resour. Res.*, 37(3), 539–549.

Hopmans, J. W., Nielsen, D. R., and Bristow, K. L. (2002). "How useful are small-scale soil hydraulic property measurements for large-scale vadose zone modeling?" *Heat and Mass Transfer in Natural Environment*, D. Smiles, P. A. C. Raats, and A. Warrick, eds., American Geophysical Union Geophysical Monograph Series, in press.

Hristopulos, D. T., and Christakos, G. (1997). "An analysis of hydraulic conductivity upscaling." *Nonlin. Anal.-Theory Methods Appl.*, 30(8), 4979–4984.

Hristopulos, D. T., and Christakos, G. (1999). "Renormalization group analysis of permeability upscaling." *Stoch. Environ. Res. Risk Assess.*, 13(1–2), 131–160.

Hunt, A. G. (1998). "Upscaling in subsurface transport using cluster statistics of percolation." *Transp. Porous Media*, 30(2), 177–198.

Indelman, P. (1993). "Upscaling of permeability of anisotropic heterogeneous formations 3. Applications." *Water Resour. Res.*, 29(4), 935–943.

Indelman, P., and Dagan, G. (1993a). "Upscaling of conductivity of heterogeneous formations - General approach and application to isotropic media." *Transp. Porous Media*, 12(2), 161–183.

Indelman, P., and Dagan, G. (1993b). "Upscaling of permeability of anisotropic heterogeneous formations 1. The general framework." *Water Resour. Res.*, 24(4), 917–923.

Indelman, P., and Dagan, G. (1993c). "Upscaling of permeability of anisotropic heterogeneous formations 2. General structure and small perturbation analysis." *Water Resour. Res.*, 29(4), 925–933.

Ito, H., Raghuwanshi, N. S., and Wallender, W. W. (1999). "Economics of furrow irrigation under partial infiltration information." *J. Irrig. Drain. Eng.*, 125(3), 105–111.

Kersebaum, K. C., and Wenkel, K.-O. (1998). "Modeling water and nitrogen dynamics at three different spatial scales: Influence of different data aggregation levels on simulation results." *Nutrient Cycl. Agroecosyst.*, 50(1–3), 313–319.

King, P. R. (1996). "Upscaling permeability - Error analysis for renormalization." *Transp. Porous Media*, 23(3), 337–354.

King, D., Fox, D. M., Daroussin, J., LeBissonnais, Y., and Danneels, V.

(1998). "Upscaling a simple erosion model from small areas to a large region." *Nutrient Cycl. Agroecosyst.*, 50(1–3), 143–149.

Le Bissonnais, Y., Benkhadra, H., Chaplot, V., Fox, D., King, D., and Daroussin, J. (1998). "Crusting, runoff and sheet erosion on silty loamy soils at various scales and upscaling from m² to small catchments." *Soil Tillage Res.*, 46(1–2), 69–80.

Lee, D. J., and Howitt, R. E. (1996). "Modeling regional agricultural production and salinity control alternatives for water quality policy analysis." *Am. J. Agric. Econom.* 78(1), 41–53.

Lima, L. A., and Grismer, M. E. (1994). "Application of fracture mechanics to cracking of saline soils." *Soil Sci.*, 158(2), 86–96.

Llorens, P., and Gallart, F. A. (2000). "A simplified method for forest water storage capacity measurement." *J. Hydrol.*, 240(1–2), 131–144.

MacDonald, T. R., Kitanidis, P. K., McCarty, P. L., and Roberts, P. V. (1999). "Mass-transfer limitations for macroscale bioremediation modeling and implications on aquifer clogging." *Ground Water*, 37(4), 523–531.

Matthews, R. B., Wassmann, R., Knox, J. W., and Buendia, L. V. (2000). "Using a crop/soil simulation model and GIS techniques to assess methane emissions from rice fields in Asia. IV. Upscaling to national levels." *Nutrient Cycl. Agroecosyst.*, 58(1–3), 201–217.

National Research Council (NRC). (1990). "Opportunities in the Hydrologic Sciences." *Committee on Opportunities in the Hydrologic Sciences, Water Science and Technology Board*, National Academy Press, Washington, D.C., 368.

Nichols, J. R., and Grismer, M. E. (1997). "Measurement of fracture mechanics parameters in silty-clay soils." *Soil Sci.*, 162(5), 309–322.

Nitao, J. J., and Bear, J. (1996). "Potentials and their role in transport in porous media." *Water Resour. Res.*, 32(2), 225–250.

Olesen, J. E., Bocher, P. K., and Jensen, T. (2000). "Comparison of scales of climate and soil data for aggregating simulated yields of winter wheat in Denmark." *Agr. Ecosyst. Environ.*, 82(1–3), 213–228.

Or, D., and Tuller, M. (1999). "Liquid retention and interfacial area in variably saturated porous media: Upscaling from single-pore to sample-scale model." *Water Resour. Res.*, 35(12), 3591–3605.

Pelmulder, S. D. (1994). " A framework for regional scale modeling of water resources management and health risk problems with a case study of exposure sensitivity to aquifer parameters." PhD thesis, Univ. of California, Los Angeles.

Pelmulder, S. D., Yeh, W. W. G., and Kastenberg, W. E. (1996). "Regional scale framework for modeling water resources and health risk problems." *Water Resour. Res.*, 32(6), 1851–1861.

Purkey, D. R., and Wallender, W. W. (2001). "Drainage reduction under land retirement over shallow water table." *J. Irrig. Drain. Eng.*, 127(1), 1–7.

Refsgaard, J. C., Thorsen, M., Jensen, J. B., Kleeschulte, S., and Hansen, S. (1999). "Large scale modeling of groundwater contamination from nitrate leaching." *J. Hydrol.*, 221(3–4), 117–140.

Renault, D., and Wallender, W. W. (2000). "Nutritional water productivity and diets." *Agric. Water Manage.*, 45(2000), 275–296.

Rubin, Y., and Gomez-Hernandez, J. J. (1990). "A stochastic approach to the problem of upscaling of conductivity in disordered media — theory and unconditional numerical simulations." *Water Resour. Res.*, 26(4), 691–701.

Russo, D. (1992). "Upscaling of hydraulic conductivity in partially saturated heterogeneous porous formation." *Water Resour. Res.*, 28(2), 397–409.

Sahimi, M., and Mehrabi, A. R. (1999). "Percolation and flow in geological formations: upscaling from microscopic to megascopic scales." *Physica A*, 266(1–4), 136–152.

Sanchez-Vila, X., Axness, C. L., and Carrera, J. (1999). "Upscaling transmissivity under radially convergent flow in heterogeneous media." *Water Resour. Res.*, 35(3), 613–621.

Sanchez-Vila, X., Girardi, J. P., and Carrera, J. (1995). "A synthesis of approaches to upscaling of hydraulic conductivities." *Water Resour. Res.*, 31(4), 867–882.

Schulze, R. (2000). "Transcending scales of space and time in impact studies of climate and climate change on agrohydrological responses." *Agric. Ecosyst. Environ.*, 82(1–3), 185–212.

Soutter, L. A., and Loague, K. (2000). "Revisiting the Fresno DBCP case study simulations: The effect of upscaling." *J. Environ. Qual.*, 29(6), 1794–1805.

Sposito, G. (1998). *Scale dependence and scale invariance in hydrology*, Cambridge Univ. Press, Cambridge, U.K.

Stewart, J. B., Engman, E. T., Fedes, R. A., and Kerr, Y. (1996). "Scaling up hydrology using remote sensing." Wiley, Chichester, U.K.

Tegnander, C., and Gimse, T. (1998). "Flow simulations to evaluate upscaling of permeability." *Math. Geol.*, 30(6), 717–731.

Tidwell, V. C., and Wilson, J. L. (1997). "Laboratory method for investigating permeability upscaling." *Water Resour. Res.*, 33(7), 1607–1616.

Tidwell, V. C., and Wilson, J. L. (1999). "Permeability upscaling measured on a block of Berea Sandstone: Results and interpretation." *Math. Geol.*, 31(7), 749–769.

Tuong, T. P., and Bhuiyan, S. I. (1999). "Increasing water-use efficiency in rice production: Farm-level perspectives." *Agric. Water Manage.*, 40(1), 117–122.

van der Gon, H. A. C. D., van Bodegom, P. M., Houweling, S., Verburg, P. H., and van Breemen, N. (2000). "Combining upscaling and downscaling of methane emissions from rice fields: Methodologies and preliminary results." *Nutrient Cycl. Agroecosyst.*, 58(1–3), 285–301.

Wagenet, R. J. (1998). "Scale issues in agroecological research chains." *Nutrient Cycl. Agroecosyst.*, 50, 23–34.

Wallender, W. W. (1987). "Sample volume statistical relations for water content, infiltration and yield." *Trans. ASAE*, 30(4), 1043–1050.

Waller, P. M., and Wallender, W. W. (1993). "Changes in cracking, water content and bulk density of salinized swelling clay field soils." *Soil Sci.*, 156(6), 414–423.

Wen, X. H., and Gomez-Hernandez, J. J. (1996). "Upscaling hydraulic conductivities in heterogeneous media - An overview." *J. Hydrol.*, 183(1–2), R9–R32.

Wen, X. H., and Gomez-Hernandez, J. J. (1998). "Upscaling hydraulic conductivities in cross-bedded formations." *Math. Geol.*, 30(2), 181–211.

Wheatcraft, S. W., and Tyler, S. W. (1988). "An explanation of scale-dependent dispersivity in heterogeneous aquifers using concepts of fractal geometry." *Water Resour. Res.*, 24, 566–578.

Whitaker, S. (1999). *Theory and applications of transport in porous media: The method of volume averaging*, Kluwer-Academic, Dordrecht, The Netherlands, 219.

Wohlfahrt, G., Bahn, M., Tappeiner, U., and Cernusca, A. (2000). "A model of whole plant gas exchange for herbaceous species from mountain grassland sites differing in land use." *Ecol. Modell.*, 125(2–3), 173–201.

Yang, W. (1997). "Effects of spatial resolution and landscape structure on land cover characterization." PhD thesis, Univ. of Nebraska–Lincoln, Lincoln, Neb.

Yu, P.-T. (2000). "Application of physically-based distributed models in forest hydrology." *Forest Res.*, 13(4), 431–438.

American Society of Civil Engineers 1852—2002 Building a Better World

150th Anniversary Paper

Integrated Water Management for the 21st Century: Problems and Solutions

Herman Bouwer[1]

Abstract: Most of the projected global population increases will take place in third world countries that already suffer from water, food, and health problems. Increasingly, the various water uses (municipal, industrial, and agricultural) must be coordinated with, and integrated into, the overall water management of the region. Sustainability, public health, environmental protection, and economics are key factors. More storage of water behind dams and especially in aquifers via artificial recharge is necessary to save water in times of water surplus for use in times of water shortage. Municipal wastewater can be an important water resource but its use must be carefully planned and regulated to prevent adverse health effects and, in the case of irrigation, undue contamination of groundwater. While almost all liquid fresh water of the planet occurs underground as groundwater, its long-term suitability as a source of water is threatened by nonpoint source pollution from agriculture and other sources and by aquifer depletion due to groundwater withdrawals in excess of groundwater recharge. In irrigated areas, groundwater levels may have to be controlled with drainage or pumped well systems to prevent waterlogging and salinization of soil. Salty drainage waters must then be handled in an ecologically responsible way. Water short countries can save water by importing most of their food and electric power from other countries with more water, so that in essence they also get the water that was necessary to produce these commodities and, hence, is virtually embedded in the commodities. This "virtual" water tends to be a lot cheaper for the receiving country than developing its own water resources. Local water can then be used for purposes with higher social, ecological, or economic returns or saved for the future. Climate changes in response to global warming caused by carbon dioxide emission are difficult to predict in space and time. Resulting uncertainties require flexible and integrated water management to handle water surpluses, water shortages, and weather extremes. Long-term storage behind dams and in aquifers may be required. Rising sea levels will present problems in coastal areas.

DOI: 10.1061/(ASCE)0733-9437(2002)128:4(193)

CE Database keywords: Water management; Integrated systems; Population; Water pollution; Water reuse; Ground-water supply; Developing countries.

Introduction

Population growth and higher living standards will cause ever increasing demands for good quality municipal and industrial water, and ever increasing sewage flows. At the same time, more and more irrigation water will be needed to meet increasing demands for food for growing populations. Also, more and more water will be required for environmental concerns such as aquatic life, wildlife refuges, recreation, scenic values, and riparian habitats. Thus, increased competition for water can be expected. This will require intensive management and international cooperation. Since almost all liquid fresh water on the planet occurs under-

ground, groundwater will be used more and more and, hence, must be protected against depletion and contamination, especially from nonpoint sources like intensive agriculture.

While growing populations and increasing water requirements are a certainty, a big uncertainty is how climates will change and how they will be affected by man's activities like increasing emissions of CO_2 and other greenhouse gases, particulate matter, and other contaminants like ozone and nitrous oxides. There still is no agreement among scientists how and when the climate will change, and what changes will occur where. The main conclusion so far seems to be that climate changes (natural and anthropogenic) are likely, that they are essentially unpredictable on a local scale, and that, therefore, water resources management should be flexible so as to be able to cope with changes in availability and demands for water (McClurg 1998). This calls for integrated water management where all pertinent factors are considered in the decision making process. Such a holistic approach requires not only supply management, but also demand management (e.g., water conservation and transfer of water to uses with higher economic returns), water quality management, recycling and reuse of water, economics, conflict resolution, public involvement, public health, environmental and ecological aspects, socio-cultural aspects, water storage (including long-term storage or water "banking"), conjunctive use of surface water and groundwater, water

[1]Agricultural Research Service, U.S. Water Conservation Laboratory, 4331 E. Broadway Rd., Phoenix, AZ. 85040. E-mail: hbouwer@uswcl.ars.ag.gov

Note. Discussion open until January 1, 2003. Separate discussions must be submitted for individual papers. To extend the closing date by one month, a written request must be filed with the ASCE Managing Editor. The manuscript for this paper was submitted for review and possible publication on November 21, 2001; approved on February 28, 2002. This paper is part of the *Journal of Irrigation and Drainage Engineering*, Vol. 128, No. 4, August 1, 2002. ©ASCE, ISSN 0733-9437/2002/4-193–202/$8.00+$.50 per page.

pollution control, flexibility, regional approaches, weather modification, and sustainability. Agricultural water management increasingly must be integrated with other water management and environmental objectives.

Global Population and Water Supplies

The present world population of about six billion is projected to almost double in this century. Almost all of this population increase will be in the third world, where there are already plenty of water and sanitation problems and where about 1,400 people (mostly children) die every hour due to waterborne diseases (Bouwer 1994, and references therein). Also, there will be more and more migration of people from rural areas to cities, creating many large cities including megacities with more than 20 million people that will have megawater needs, produce megasewage flows, and have megaproblems. Already, there is talk that people in these megacities should have little gardens where they can grow their own food and recycle their own waste. There would then be little difference between megacities with a lot of small gardenlike farming and rural areas with dense populations, especially in the suburban fringes of the cities. All of these people and their animals living closely together could present serious health problems as viruses and other pathogens that normally affect only animals can be transferred to humans. This could cause epidemics of potentially global proportions because of lack of immunity and vaccines, much like the ebola and acquired immune deficiency syndrome viruses and the various flu outbreaks caused by swine or chicken viruses. If the animals are also given regular doses of antibiotics to promote faster growth, antibiotic resistant strains of pathogens could be created which could cause serious human pandemics.

For adequate living standards as in western and industrialized countries, a renewable water supply of at least 2,000 m^3 per person per year is necessary (Postel 1992). If only 1,000–2,000 m^3 is available, the country is water stressed, while below 500 cubic meters per person per year, it is water scarce. Nomadic desert people can subsist on only a few m^3 per person per year (not including their animals). The global renewable water supply is about 7,000 m^3 per person per year (present population). Thus, there is enough water for at least three times the present world population. Hence, water shortages are due to imbalances between population and precipitation distributions.

Almost all of the water of our planet (97%) occurs as salt water in the oceans (Bouwer 1978 and references therein). Of the remaining 3%, two-thirds occur as snow and ice in polar and mountainous regions, which leave only about 1% of the global water as liquid fresh water. Almost all of this (more than 98%) occurs as groundwater, while less than 2% occurs in the more visible form of streams and lakes which often are fed by groundwater. Groundwater is formed by excess rainfall (total precipitation minus surface runoff and evapotranspiration) that infiltrates deeper into the ground and eventually percolates down to the ground water formations (aquifers). For temperate, humid climates, about 40% of the precipitation ends up in the ground water. For mediterranean-type climates, it is more like 10 to 20%, and for dry climates it can be as little as 1% or even less (Bouwer 1989; Tyler et al. 1996). These natural recharge rates give an idea of the safe or sustainable yields of aquifers that can be pumped from wells without depleting the groundwater resource. In many areas of the world, especially the drier ones, groundwater is the main water resource. Natural recharge rates are difficult to predict with any accuracy (Stone et al. 2001), and often pumping greatly exceeds recharge, so that groundwater levels are declining. It is frightening to consider what will happen in these areas when the wells go dry and no other water resources are available.

Water Storage via Dams

Future climatic changes may also include more weather extremes, like more periods with excessive rainfall and more periods with low rainfall that cause droughts. Also, in relatively dry climates, small changes in precipitation can cause significant changes in natural recharge of groundwater. To protect water supplies against these extremes and changes, more storage of water is needed, including long-term storage (years to decades) to build water reserves during times of water surplus for use in times of water shortage. Traditionally, such storage has been achieved with dams and surface reservoirs. However, good dam sites are getting scarce and dams have a number of disadvantages like interfering with the stream ecology, adverse environmental effects, displacement of people for new dam reservoirs, loss of scenic aspects and recreational uses of the river, increased waterborne diseases and other public health problems, evaporation losses (especially undesirable for long-term storage), high costs, potential for structural problems and failure, and no sustainability since all dams eventually lose their capacity as they fill up with sediments (Pearce 1992; Jobin 1999; Postel 1999, and references therein). For these reasons, new dams are increasingly difficult to construct, except in some countries (mostly third world) where the advantages of abundant and cheap hydroelectric power are considered to outweigh the disadvantages of dams. One of the advantages of dams is that they can be operated to even out the flow in the downstream river, regardless of seasonal or longer-term variations in rainfall. On the other hand, the ease to turn the turbines off and on to meet peaking power or other short-term fluctuations in electricity demands can adversely affect the downstream ecology. For example, as stated by Newcom (2001): "Dams in California have been blamed by scientists and many in the environmental community as being one of the major catalysts responsible for moving salmon species in California—and through the Northwest—to the endangered and threatened lists of the federal Endangered Species Act. The reasons for the demise of these anadromous fish because of dams are varied and include limited spawning habitat; decreased downstream flows that limit backwater habitats serving as rearing areas for fry and juveniles to mature; increased predation by non-native fish species; entrainment from pumps and turbines; varying water temperatures; reduced nutrient-rich sediment and spring migration flows; and dissolved gases." One way to make dam operation for generation of hydropower environmentally more acceptable and in compliance with environmental laws is to increase the capacity for generation of thermal power so that hydropower that produces undesirable extremes in flows and temperatures of the water below dams can be avoided. For California, such laws are the Endangered Species Act, the Californian Environmental Quality Act, the National Environmental Policy Act, and the Water Quality Act. Under federal law, nonfederal hydropower facilities must be relicensed every 40–50 years—a process that can take years to complete (Newcom 2001). The relicensing process also can mean the end for older, often obsolete dams where modifications to meet new regulations would be so expensive that destroying the dam is the best solution. However, dam decommissioning and demolition often is not a simple process. It can be very complex and expensive (Tatro 1999) and it has been the subject of special short courses (Univ. of Wisconsin 2001). Dams on international rivers require intensive cooperation among the countries involved, so that countries downstream from the dam are not adversely affected and have a voice in the location, design, and operation of the dam. New dam projects require careful planning to minimize adverse environmental, public health, and sociocultural effects.

Water Storage via Artificial Recharge of Groundwater

If water cannot be stored above ground, it must be stored underground, via artificial recharge of groundwater. Already, more than 98% of the fresh liquid water supplies of the world occurs underground (Bouwer 1978, and references therein), and there is plenty of room for more. Artificial recharge is achieved by putting water on the land surface where it infiltrates into the soil and moves downward to underlying groundwater (Bouwer 1997, 1999). Such systems require permeable soils (sands and gravels are preferred) and unconfined aquifers with freely moving groundwater tables. Infiltration rates typically range from 0.5 to 3 m/day during flooding. With continued flooding, however, suspended particles in the water accumulate on the soil surface to form a clogging layer that reduces infiltration rates. Biological, chemical, and physical actions further aggravate the clogging. Thus, infiltration basins must be periodically taken out of service to allow drying, cracking, and, if necessary, mechanical removal of the clogging layer. Taking drying periods into account, long-term infiltration rates for year round operation of surface recharge systems may be in the range of 100–400 m/year.

Artificial recharge may be implemented with in-channel and off-channel infiltration systems. In-channel systems consist of low dams across the streambed or of T- or L-shaped levees in the streambed to back up and spread the water so as to increase the wetted area and, hence, infiltration in the streambed. Off-channel systems consist of specially constructed shallow ponds or basins that are flooded for infiltration and recharge. Where streamflows are highly variable, upstream storage dams or deep basins may be necessary to capture short-duration high-flow events for subsequent gradual release into recharge systems. Also, recharge systems can be designed and managed to enhance environmental benefits (e.g., aquatic parks, trees and other vegetation, and wildlife refuges).

Since sand and gravel soils are not always available, less permeable soils like loamy sands, sandy loams, and light loams are increasingly used for surface infiltration recharge systems. Such systems may have infiltration rates of only 30–60 m/year for year round operation. Thus, relative evaporation losses are higher and in warm, dry climates could be about 3–6% of the water applied, as compared to about 1% for basins in more permeable soils. Systems in finer textured soils also require more land for infiltration basins. However, the larger land requirements enhance the opportunity for combining the recharge project with environmental and recreational amenities.

Where sufficiently permeable soils are not available or surface soils are contaminated, artificial recharge also can be achieved via infiltration trenches or recharge pits or shafts (Bouwer 1997, 1999). If the aquifers are confined, i.e., between layers of low permeability, artificial recharge can be achieved only with recharge or "injection" wells drilled into the aquifer. The cost of such recharge often is much higher than the cost of infiltration with basins because wells can be expensive and the water must

first be treated to essentially remove all suspended solids, nutrients, and organic carbon to minimize clogging of the well–aquifer interface. Since such clogging is difficult to remove, prevention of clogging by adequate pretreatment of the water and frequent pumping of the well is better than complete well remediation. Increasingly, recharge wells are constructed as dual purpose wells for both recharge and extraction to allow recharge when water demands are low and surplus water is available (i.e., during the winter), and pumping when water demands are high like in the summer. Such storage and recovery wells are used for municipal water supplies so that water treatment plants do not have to meet peak demands but can be designed and operated for a lower average demand, which is financially attractive (Pyne 1995).

The big advantage of underground storage is that there are no evaporation losses from the groundwater. Evaporation losses from the basins themselves in continuously operated systems may range from 0.5 m/year for temperate humid climates to 2.5 m/year for hot dry climates. Groundwater recharge systems are sustainable, economical, and do not have the ecoenvironmental problems that dams have. In addition, algae which can give water quality problems in water stored in open reservoirs do not grow in groundwater. Because the underground formations act like natural filters, recharge systems also can be used to clean water of impaired quality. This principle is extensively used as an effective low-technology and inexpensive method to clean up effluent from sewage treatment plants to enable unrestricted and more aesthetically acceptable water reuse (see "Water Reuse" section). The systems then are no longer called recharge systems but soil–aquifer treatment (SAT) or geopurification systems.

Conjunctive Use and Water Banking

Nature's way of storing water is underground, where about 98% of all the liquid and fresh water of the world occurs (Bouwer 1978, and references therein). The other 2% mostly occurs in streams and lakes, which often are fed by groundwater. Groundwater is a dependable source of water and less affected by the vagaries of climate than surface water. Often, surface water and groundwater are used conjunctively, surface water when available, and groundwater when the streams or lakes are low or dry. Where water requirements have been increasing, there often has been a tendency to pump more groundwater with all the undesirable effects such as aquifer depletion, land subsidence, salt water intrusion, and higher pumping costs. The solution then is either to build more dams for surface storage, or to store more water underground via artificial recharge of groundwater. Underground storage is preferred where dams are not feasible and also when the water may have to be stored for long periods (years to decades) and evaporation losses from the dam reservoirs are not acceptable. Such long-term underground storage is often called water banking (McClurg 2001). Some of the issues in groundwater banking have to do with water rights, especially where surface water and groundwater are governed by different water right systems. For example, surface water may be governed by prior appropriation or the riparian principle, whereas groundwater rights maybe in the hands of the owner of the overlying land (Bouwer 1978 and references therein). Thus, when surface water is used for groundwater recharge, the question is who owns the water after it has joined the aquifer? Also, after long-term storage (decades, for example), is the recharge water still recoverable from the aquifer or has it moved laterally away from the region over which the recharging entity has jurisdiction? There may also be water quality issues, for example, where the groundwater is of better quality than the surface water used for recharge, or where the recharge water is of good quality and picks up undesirable chemicals from the aquifer such as arsenic, boron, and dissolved salts. Effects on groundwater levels must also be considered to avoid undue groundwater rises during recharge and undue declines during extraction. Some states (California, for example) allow extraction of groundwater in excess of the amount put into the aquifer by artificial recharge. This excess would then consist of the natural recharge. However, such natural recharge is difficult to predict, especially in dry climates where recharge may only be a small percentage (1% for example) of an already very small precipitation (Stone et al. 2001). Other states like Arizona, where natural recharge is very low, require groundwater extractions to be no more than 95% of the artificial recharge inputs, thus leaving 5% of the recharge in the aquifer. The best approach is to monitor groundwater levels in the area of water banking and groundwater pumping so that pumping rates can be increased where groundwater levels are rising, and decreased where they are falling.

Groundwater and Salinity Control for Sustainable Irrigation

There are many serious cases of pollution of surface water and groundwater by point sources (e.g., sewage and industrial wastewater discharges, leaking ponds or tanks, and waste disposal areas). However, point source pollution is, at least in principle, relatively simple to control and prevent. A much greater threat to the liquid fresh water resources of the planet is nonpoint source pollution of groundwater. A significant nonpoint source of groundwater pollution is agriculture, with its use of fertilizer, pesticides, and salt containing irrigation water that contaminate the drainage water as it moves from the root zone to the underlying groundwater. The problem can be expected to get worse in the future as agriculture must intensify (including use of more agricultural chemicals) to keep up with the demands for more food and fiber by increasing populations. Pollution of groundwater also causes pollution of surface water wherever the contaminated groundwater moves into streams where it maintains the base flow, and also into lakes and coastal waters.

In humid areas with rainfed agriculture, the main contaminants in the drainage water from the root zone are nitrate and pesticide residues (Bouwer 1990). In irrigated areas, the drainage water also contains the salts that were brought in with the irrigation water. To avoid accumulation of salts in the root zone, excess irrigation water must be applied to leach the salts out of the soil so as to maintain a salt balance in the root zone. For efficient irrigation systems, the excess water may be about 20% of the total irrigation water applied. In dry climates, this means that the salt concentrations in the drainage water are about five times higher than in the irrigation water, which often is much too high for drinking and for irrigation of all but the most salt tolerant crops. For more efficient irrigation, the salt concentrations in the drainage water will even be higher. For less efficient irrigation, and also where there is significant rainfall, the salt concentrations in the drainage water will be lower. Recent successes in genetically altering plants to make them more salt tolerant offers hope for widening the choice of crops that can be grown with salty water (Apse et al. 1999).

Where groundwater levels are high, drains need to be installed to remove the drainage water from the soil and to avoid waterlogging and salinization of the soil. Discharges from the drains then contain salt and residues of agricultural chemicals and, hence, they are a source of water pollution. The least undesirable ultimate disposal of this water may be in the oceans. Inland disposal can degrade surface water. Disposal in evaporation ponds requires considerable land for "salt lakes" that could eventually become environmental hazards. Use of the salty drainage water for sequential irrigation of increasingly salt tolerant plants (including trees like tamarisk, eucalyptus, and salt tolerant poplars) and ending with halophytes like salicornia and certain grasses will concentrate the salts in small volumes of water (Shannon et al. 1997). The volume of the final drainage water may then only be a few percent of the original irrigation water so that salt concentrations could be 20–100 times higher than that in the original irrigation water. Disposal into evaporation ponds will then require much less land. Another alternative is desalination of the drainage water by, for example, reverse osmosis. The desalted water can then be used for potable and other purposes, but the process still leaves a reject brine that requires disposal. Concentrating the salts into smaller volumes of water by sequential irrigation, evaporation ponds, or membrane filtration also reduces the cost of transporting the salty water to oceans, salt lakes, or other places for "final" disposal. Leaving the water in evaporation ponds will eventually cause the salts to crystallize, which can then be disposed as solid waste in designated landfills.

Where groundwater levels are deeper (often due to prior groundwater pumping), the drainage water will move down to the groundwater and reduce its quality to the point where it becomes useless for drinking and general irrigation. Without desalting of groundwater, the further use and pumping of groundwater will stop. If irrigation is continued, groundwater levels then will rise (typically about 0.3–2 m/year) and eventually threaten underground pipe lines, basements, gravel pits, landfills, cemeteries, deep-rooted old trees, etc. Finally, they can cause waterlogging and salinization of the soil, so that nothing will grow anymore and the areas become salt flats. Inability to control groundwater below irrigated land has caused the demise of old civilizations and is still the reason why so much irrigated land in the world is losing productivity or is even being abandoned today (Postel 1999). To prevent this waterlogging and salinization, groundwater pumping must be resumed or deep agricultural drains must be installed to keep groundwater levels at safe depths. The salty, contaminated water from these wells or drains must then be managed as discussed in the previous paragraph. Irrigation without groundwater control ultimately causes waterlogging and salinity problems, and irrigation can only be sustainable if salts and drainage water are adequately removed from the underground environment and managed for minimum environmental damage.

An intriguing possibility is to use the evaporation ponds as solar ponds to produce hot water for heating and/or electric power generation. In an experimental solar pond project in El Paso, Texas, the pond is 3 m deep with a 1 m layer of low salinity water on top, a 1 m layer of medium salinity in the middle, and a 1 m layer of high salinity (brine) at the bottom (Xu 1993). Sun energy is then trapped as heat in the bottom layer while the lighter top layers prevent thermal convection currents and act as insulators. The hot brine from the bottom layer is pumped to a heat exchanger where a working fluid like isobutane or freon is vaporized which then goes through a turbine to generate power. The working fluid is condensed in another heat exchanger that is cooled with normal water which is recirculated through a cooling tower. The working fluid then returns to the brine heat exchanger where it is preheated by the brine return flow from the heat exchanger to the pond before it is vaporized again. The El Paso pond has a surface area of 0.3 ha and generates 60–70 kW. At this rate, a solar pond system of about 5,000 ha could generate about 1,000 mW of electricity, which is typical of a good sized power plant. There is enough heat stored in the hot brine layer to also generate power at night. Sequential irrigation, membrane filtration, and solar ponds for power generation have the advantage that they treat the salty water as a revenue producing resource that helps offset the cost of final disposal of the salts.

Where sewage effluent is used for irrigation, a whole new spectrum of pollutants can be added to the soil (Bouwer et al. 1999). If not attenuated in the root zone, these pollutants can show up in the drainage water at much higher concentrations than in the effluent (about five times higher for efficient irrigation in dry climates, less for inefficient irrigation and/or areas with significant rainfall). Thus, in addition to the usual nitrates and salts, the drainage water could also contain disinfection byproducts (DBPs) like trihalomethanes (THMs) and haloacetic acids (HAAs) that were formed in the drinking water when it was chlorinated for public water supply and the chlorine reacted with natural dissolved organic carbon in the water to form chloroform, bromodichloromethane, and other DBPs (McCann 1999). Then when it became sewage effluent and was chlorinated again and this time with high chlorine doses and long contact times to kill all the pathogens, a whole new suite of DBPs could be formed. There is great concern about cancer, adverse pregnancy outcomes, and other health effects of DBPs in drinking water. The U.S. Environmental Protection Agency will lower the maximum contaminant level for THMs from 100 to 80 μg/L, and for HAAs to 60 μg/L, with further reductions being expected (McCann 1999). A recently discovered DBP is N-nitrosodimethylamine (NDMA), which is an extremely carcinogenic compound formed by the reaction of chlorine with dimethylamine (DMA). The California Department of Health Services has set a NDMA drinking water action level of 20 ng/L (California State Department of Health Services 1998) and has recently lowered it to 10 ng/L. However, adequate dose-response relations for humans are not available. Thus, while chlorination effectively kills bacteria and viruses to avoid infectious disease outbreaks from sewage irrigation, it also creates chemicals that may have adverse long-term health effects. Alternative disinfection procedures that do not use chlorine, like ultraviolet irradiation, soil–aquifer treatment, or "time" should be considered.

In addition to DBPs, the treated sewage effluent and, hence, the waters into which it is discharged can also be expected to contain pharmaceuticals, industrial chemicals like polychlorinated biphenyls (PCB) and others that may have biological effects, and personal care products. These chemicals enter the wastewater with discharges from pharmaceutical and other industries, hospitals and other medical facilities, households where unused medicines are flushed down the toilets, and human excreta which contain incompletely metabolized medicines (Richardson and Bowron 1985; Daughton and Ternes 1999; Daughton and Jones-Lepp 2001; Kolpin et al. 2002). Pharmaceutically active chemicals also include certain industrial chemicals like dioxin, pesticides, and chlorinated organic compounds. While not directly toxic or carcinogenic, these chemicals may produce adverse health effects by interfering with hormone production (endocrine disruptors), by weakening immune systems, and by other biological responses. So far, most studies of pharmaceuticals and pharmaceutically active chemicals have been carried out on aquatic animals where adverse effects on hormone production and reproductive processes, including feminization of the males, have been observed (Goodbred et al. 1997). Since their long-term and syn-

ergistic effects on humans are not known, pharmaceuticals and similar chemicals should be kept out of the water environment as much as possible (Zullei-Seibert 1998). Farm animals with their ingestion of hormones, antibiotics, and veterinary medicines, can also be a source of pharmaceuticals in water as their manures and wastewater from animal feeding operations are spread on land from where they can run off into surface water or percolate down to groundwater (Daughton and Jones-Lepp 2001).

Other potential contaminants in the drainage water from sewage irrigated crops and plants are humic substances like humic and fulvic acids. These are known precursors of DBPs when the water is chlorinated. The humic substances are formed as stable endproducts wherever organic matter is biodegraded. Since effluent with its nutrients can be expected to produce lush vegetation when used for irrigation, there will be more biomass on and in the soil which, upon biodegradation, could produce increased levels of humic substances in the drainage water and, eventually, in the underlying groundwater. When this water is pumped from wells and chlorinated for potable use, increased levels of DBPs can then be expected in the drinking water. Thus, where sewage effluent is used or planned to be used for irrigation, careful studies should be made of the potential effects of groundwater, especially where the groundwater is, or will be, used for drinking.

Water Pollution and Total Maximum Daily Loads

Pollution of natural waters will become increasingly serious as growing populations demand more high-quality water while at the same time producing more wastes that will often be returned to those waters. Until recently, the focus in the USA has been mostly on point sources of water pollution (discharges of sewage effluent and industrial water) which are controlled through discharge permits under the authority of the 1972 Clean Water Act and specified in the National Pollutant Discharge Elimination System. While this program has led to considerable improvement in surface water quality, fishable and swimmable conditions have not always been met. As a matter of fact, the report "Assessing the TMDL (total maximum daily loads) Approach to Water Quality Management" recently published by the National Research Council mentions that the USA still has about 21,000 polluted river segments, lakes, and estuaries making up over 300,000 river and shore miles and 5 million lake acres (National Research Council 2001). Thus, whereas until now, pollution control has been based on controlling effluent discharges, the next phase will also control nonpoint sources of pollution, mainly due to urban and agricultural runoff, drainage of groundwater into surface water, and atmospheric fallout. The main pollutants of concern are nutrients and sediment, but they could also include certain pesticides, pharmaceuticals, and other chemicals of emerging concerns. Control of these contaminants will be based on entire watersheds and it will be achieved by establishing TMDLs for the entire systems. The TMDL approach was already included in the 1972 Clean Water Act as Section 303d. However, it was largely overlooked until the Environmental Protection Agency (EPA) in response to lawsuits and other pressures from environmental groups developed TMDL regulations that were promulgated on 13 July 2000. Cost estimates by the EPA for implementing the TMDL program range from $900 million to $4.3 billion per year (Gray 2001), which primarily would be borne by dischargers. TMDLs could also be developed for groundwater, especially where it drains into surface water such as gaining streams or groundwater-fed lakes.

The TMDL concept is a dramatic switch from effluent based standards to ambient water standards, and from controlling point sources to controlling entire watersheds. In view of the high cost of implementing the program, attainment of the desired water quality may be questionable, particularly since the underlying scientific principles may not be fully understood. However, while attainment may be an issue, there is interest in moving ahead with the program while practicing adaptive management to make adjustments in the program where the results are not as expected Christen 2001; National Research Council 2001; Wagner 2001), and to minimize administrative complications (Smith 2002). Others, especially those who will be financially affected by TMDLs, favor delay or modification of TMDLs and their implementation (Christen 2001). For agriculture, this may mean more use of best management practices for control of erosion, and of nutrients and pesticides in runoff water. Vegetated buffer strips on a watershed scale also may be effective in controlling nonpoint source pollution of surface water (Schultz et al. 1995; Isenhart et al. 1998; Lee et al. 1999).

Global Change

Few issues have received so much attention and have generated so much controversy as the effects of increasing concentrations of CO_2 and other greenhouse gases in the atmosphere on temperature and climate. Predictions range from serious effects on ecosystems and our health (Office of Science and Technology 1997), increased flooding, and desertification (Hulme and Kelly 1993) to everything is normal and just part of the natural climate fluctuations that have been going on for ages as a result of the dynamic nature of planet earth. Sometimes it appears that conclusions are based primarily on consensus and majority opinions. What all this controversy shows, however, is that it is not known to a sufficient degree of accuracy what is going to happen in space and in time. Thus, it is difficult to make adequate plans. In addition to gradual, long-term climate changes, more abrupt changes within the span of a human generation may also happen (Showstack 2001, and references therein).

Models for predicting global precipitations are based on models for predicting global temperatures in response to increasing CO_2 concentrations in the atmosphere. However, the temperature predictions are fraught with uncertainties (Kimball 2002), which makes precipitation prediction very difficult. The models do not even do well in predicting present precipitation patterns (Kimball 2002). However, because temperatures are projected to rise globally, average evaporation from oceans and other bodies of water will also increase, and therefore, globally averaged precipitation will also increase. However, the precipitation patterns may change (Albritton et al. 2001). Over the higher latitudes, precipitation is predicted to increase. Decreases are projected for Central America in the summer and for South Africa and Australia. Over much of the United States, projections are inconsistent, but with small increases indicated for winter in both Western and Eastern North America. Albritton et al. (2001) also note that a strong correlation exists between interannual variability and mean precipitation. Consequently, future increases in mean precipitation are likely to lead to increases in precipitation variability.

It is not surprising that some countries, especially small ones with little geographic and hydrologic diversity, are concerned about future water resources management and have tried to make some predictions as to what may happen to them in the long term. Such countries include The Netherlands which is concerned about increased flooding caused by the Rhine due to larger peak flows and rising sea water levels, and Israel which is concerned about water resources. The Dutch predictions (de Jong et al. 2001) are based on estimated average temperature increases (4°C by 2100), from which they estimate precipitation increases (4% in summer and 25% in winter) which then go in their hydrological model to predict flood flows. These predictions are useful for long-range planning and they indicate that for the next 20 years, flood control dikes will still be feasible. As time proceeds, climate and climate science will develop further so that more detailed and reliable climate scenarios can be formulated. Sea level rises by the year 2100 are predicted to be in the range of 20–110 cm. Analyses such as these are useful for long-range planning for other river basins. If, indeed, increasing flood flows are expected, raising levees ultimately may no longer be feasible and construction of parallel flood ways may be the best approach. Normally, these flood ways would be farmed and there would be no expensive structures, so that when they are used for flood control and the "green" rivers become real rivers, there is minimum damage.

The green river concept can also be applied to small rivers or streams. An example is the Indian Bend Wash in Scottsdale Ariz., which drains a watershed of about 500 km^2 of urban and mountainous areas with short concentration times. Rainfall averages about 20 cm/year with occasional downpours of 2–5 cm in a few days. For the last 15 km, the wash runs through urban Scottsdale before it discharges into the Salt River. This normally dry wash is about 150 m wide and was an eyesore of weeds, old tires, discarded washing machines, etc. that flooded every few years or so and made level street crossings unpassable. In the 1960s, flood control plans were developed which started with the usual approach of a concrete channel with levees on each side and houses and other urban developments right up to the levees. However, this plan was opposed by the public who did not want a Los Angeles-type "concrete canyon" and instead opted for a greener solution with a soft edge channel and recreational facilities. The result was a green river about 150 m wide with levees along the outer edges, a meandering low-flow channel a few m wide in the middle, and lakes, golf courses, sports fields, picnic areas, playgrounds and hiking and biking trails in the rest of the wash. The area now is a prime, high-density, and very popular recreation facility, and an example of what can be achieved with normally dry stream beds in urban settings. Small floods occur every few years, and large floods that cover most of the green river about every 20 years. The 100-year flood is 850 m^3/s. The flood of record so far is 570 m^3/s, which occurred in 1972 and has a recurrence interval of 70 years. Floods are short lived. They reduce or interrupt recreation for only a few days to about a week, and cause little or no damage.

Israel also has made predictions of future climates for various scenarios based on local climatic trends and on national and regional climatological research and models (Gabbay 2001). The projected changes between now and for the year 2100 are

- mean temperature increase 1.6–1.8°C,
- reduction in precipitation 4–8%,
- increase in evapotranspiration 10%,
- delayed winter rains,
- increased rain intensity and shortening of the rainy season,
- greater seasonal temperature variability,
- increased frequency and severity of extreme climatic events, and
- greater spatial and temporal climatic uncertainty.

Because of the uncertainties in global change predictions, especially in space and in time, the best policy for water resources management is flexibility so as to be able to handle floods and droughts, and surpluses and shortages. This is best achieved through integrated water management, as defined earlier. Global change may also affect infectious disease outbreaks. The occurrence of such diseases already shows distinct geographical distributions and dependency on seasonal variations. Thus, prediction of climate changes and their effects on disease outbreaks will be useful in developing appropriate public health programs to prevent or control such outbreaks (National Academy of Sciences 2001).

For irrigation, the effect of climatic changes on water supplies must also be considered in relation to the effects of increasing CO_2 concentrations in the atmosphere on crop water use efficiencies and yields. As stated by Kimball:

"The degree of influence of global change on future water resources is difficult to predict because various components are likely to be affected in opposing ways. Global warming [surface temperature projected to increase 1.2 to 5.8°C (mean of 3.5°) by 2100, depending on CO_2 emissions scenario and on the particular general circulation model (GCM) used for the projection] would tend to increase evapotranspiration (ET) rates and irrigation water requirements. At the same time, precipitation is projected to increase globally, which would both decrease irrigation water requirements and increase water supplies, although regional pattern changes are very uncertain. The direct effects of elevated CO_2 (projected to reach 540 to 950 μmol mol^{-1}, depending on CO_2 emissions scenario) on plants likely will cause increases in stomatal resistance (about 20–40% for a 350 μmol mol^{-1} increase in CO_2 concentration for most herbaceous plants, with woody plants affected less), which will also tend to reduce ET. But at the same time, the elevated CO_2 will stimulate increases in plant leaf area (probably on the order of 10% in peak leaf area index for a 350 μmol mol^{-1} increase in CO_2 concentration for C_3 plants with C_4 plants responding less) and canopy temperature, both of which increase ET.

The sensitivity of "reference" ET for alfalfa to the several opposing future influences was examined using a form of the Penman–Monteith equation that is under consideration for adoption as a standard by the American Society of Civil Engineers. For constant future relative humidity, annual reference ET at Maricopa, Ariz., would increase 2.1%/°C (or 7.1% for the projected mean 3.5°C rise in global temperature). Increasing stomatal resistance reduces ET 0.16%/% (or 3.0 to 5.7% for a 20 to 40% increase in resistance), whereas increasing leaf area increases ET 0.16%/% (or 1.6% for a 10% increase in leaf area). The combined effect of these three influences would be a net increase of 2.7 to 5.7% in ET.

However, irrigation requirement is the difference between seasonal ET for a well-watered crop and the amount of water available from precipitation and soil storage, and the latter two likely will also be affected by global change. Modeling studies which have been done using scenarios of future weather projected by various GCMs predict that irrigation requirements will increase substantially, on the order of 35% for the U.S. overall but with wide variability depending on GCM, region, and crop.

Fortunately, overall precipitation is also projected to increase, which will have favorable effects on runoff or streamflow or irrigation water supply. One study projected global runoff to increase 10%. However, the regional variability is large and uncertain, and another study predicted water yields for the year 2030 to decrease in the Southern U.S. and the Great Plains and to increase in the East and Far West, whereas for the year 2095, they predicted no change to substantial increases for much of the U.S. Another aspect of global warming is that greater proportions of annual precipitation will fall as rain rather than snow and that snowpacks will melt faster, which means that some important agricultural regions in the U.S. may lose a substantial part of a huge free snowpack "reservoir" that presently stores winter precipitation at higher elevations for summer irrigation at lower elevations.

Both the projected climate changes and the direct physiological effects of elevated CO_2 on plants likely will cause shifts in optimal production regions for many crops. Further, human economic and social factors likely also will cause changes in land use and associated demands for irrigation water. In addition, there likely will be shifts in natural vegetation on the upstream watersheds, which may change the supplies of water available for irrigation in the future.

In conclusion, global change very likely will affect future irrigation and water resources. The effects of climate and CO_2 on seasonal crop water use are relatively well understood—slight increases (2–6%) are predicted for plausible scenarios of future temperature and CO_2. The effects on irrigation requirements and on water supplies are much more uncertain due to the uncertainties in projected precipitation patterns. It behooves future water resource planners and future growers to try to be as flexible as possible."

Most of the studies of the effects of elevated CO_2 concentration in the atmosphere on crop yield and water requirements have been done with pure CO_2. In reality, however, concentrations of other gases in the atmosphere may also increase. Some of these could have adverse effects on crop yields. For example, levels of ozone have more than doubled in the past 100 years and are predicted to continue rising at an even faster rate in the future (Hough and Derwent 1990). In one experiment, yield increases in potatoes induced by elevated CO_2 levels in the air were substantially reduced by the presence of elevated ozone (Finnan et al. 2002).

Another consequence of increasing temperatures and global change is rising seawater levels, primarily due to melting of polar ice sheets and thermal expansion of oceans (Warrick et al. 1996). As stated by Anderson et al. (2001): "Coastal change occurs in response to natural processes that operate across a wide range of spatial and temporal scales. Long-term, century-scale impacts of climate change that will affect coastal environments include decimeter-scale sea-level rises; shifts in sea–surface temperatures, which will likely influence tropical storm tracts, as well as storm frequency and magnitude; and precipitation variations that may impact sediment flux to coastal areas. Other effects may include changes in coastal and ocean currents and wave regimes."

In many parts of the world, people have been migrating toward coastal cities, which already causes serious stresses on coastal environments that can only get worse as rates of global sea-level rise increase. In addition to direct flooding of low areas, additional backing up and flooding problems can be expected where surface water and wastewater (pipelines) are discharged into the ocean. Groundwater levels in coastal areas will also rise as natural discharge of groundwater in the ocean is reduced. Salt water intrusion into coastal aquifers can also increase, especially where groundwater is pumped from wells.

Carbon emissions can be reduced by conservation and efficient use of energy, by using nonfossil energy sources (hydropower, wind, solar, nuclear, and ethanol or other biofuels) and by growing more plants for carbon sequestration in biomass and soil (Schlesinger 1999). Biofuels still emit carbon into the atmosphere but, unlike carbon from fossil fuels, it is recycled carbon via photosynthesis. Oceans also hold considerable amounts of carbon (Siegenthaler and Sarmiento 1993).

Water Reuse

All water is recycled through the global hydrologic cycle. However, planned local water reuse is becoming increasingly important for two reasons (Bouwer 1993). One is that the discharge of sewage effluent into surface water is becoming increasingly difficult and expensive as treatment requirements become more and more stringent to protect the quality of the receiving water for aquatic life, recreation, and downstream users. The cost of the stringent treatment may be so high that it becomes financially attractive for municipalities to treat their water for local reuse rather than for discharge. The second reason is that municipal wastewater often is a significant water resource that can be used for a number of purposes, especially in water short areas. The most logical reuse is for nonpotable purposes like agricultural and urban irrigation, industrial uses (cooling, processing), environmental enhancement (wetlands, wildlife refuges, riparian habitats, and urban lakes), fire fighting, dust control, and toilet flushing. This requires treatment of the effluent so that it meets the quality requirements for the intended use. Adequate infrastructures like storage reservoirs, canals, pipelines, and dual distribution systems are also necessary so that waters of different qualities can be transported to different destinations. Aesthetics and public acceptance are important aspects of water reuse, especially where the public is directly affected.

Treatment plant processes for unrestricted nonpotable reuse are primary and secondary treatment followed by tertiary treatment consisting of flocculation, sand filtration, and disinfection (ultraviolet irradiation or chlorination) to make sure that the effluent is free from pathogens (viruses, bacteria, and parasites). Such tertiary effluent can then be used for agricultural irrigation of crops consumed raw by people or brought raw into the kitchen, urban irrigation of parks, playgrounds, sports fields, golf courses, road plantings, etc., and urban lakes, fire fighting, toilet flushing, industrial uses, and other purposes. The tertiary treatment requirement was developed in California and is followed by most industrialized countries (Bouwer 1993; Asano 1998).

The California tertiary treatment is relatively high technology and expensive and is, therefore, often not feasible in third world countries. To avoid use of raw sewage for irrigation, and to still make such irrigation reasonably safe from a public health standpoint, the World Health Organization (1989) has developed guidelines that are based on epidemiological analyses of documented disease outbreaks and that are achievable with low-technology treatment such as in-series lagooning with long detection times (about one month). While this treatment does not produce pathogen-free effluent, epidemiological studies have indicated that use of such effluent for the irrigation of crops consumed raw greatly reduces health risks compared to untreated sewage. As a precaution, however, the vegetables and fruit grown with such effluent should only be consumed by the local people that hopefully have developed some immunity to certain pathogens. Tourists and other visitors from the outside should not eat the local raw fruits and vegetables, and the produce should not be exported to other markets. Also, the lagooning treatment must be viewed as a temporary solution and full tertiary treatment plants should be built as soon as possible, especially when the lagoons become overloaded, detention times become too short for adequate pathogen removal, and the lagoon system cannot be deepened or expanded.

Additional treatment of secondary or tertiary effluent and lagoon effluent can also be obtained by using the effluent for artificial recharge of groundwater where underground formations function as natural filters that can significantly reduce concentrations of suspended solids, nitrogen, phosphorus, organic carbon, trace elements, and micro-organisms (Bouwer 1993; 1997; 1999). The resulting SAT greatly enhances the aesthetics of water reuse because the purified water comes from wells and not from sewage treatment plants and, hence, has lost its identity as "treated sewage." Water after SAT also is clear and odorless. SAT is especially important in countries where there are social or religious taboos against direct use of "unclean" water (Ishaq and Khan 1997; Warner 2000) or where expensive advanced treatment plants are not feasible.

Potable use of sewage effluent basically is a practice of last resort, although unplanned or incidental potable reuse occurs all over the world where sewage effluent is discharged into streams and lakes that are also used for public water supplies (Crook et al. 1999), and where cess pits, latrines, septic tanks, and sewage irrigation systems leak effluent to underlying groundwater that is pumped up again for drinking. In-plant sewage treatment for direct potable reuse requires advanced processes that include nitrogen and phosphorous removal (nitrification/dentrification and lime precipitation), removal of organic carbon compounds (activated carbon adsorption), removal of dissolved organic and inorganic compounds and pathogens by membrane filtration (microfiltration and reverse osmosis), and disinfection. Even when all these treatment steps are used and the water meets all drinking water quality standards, direct potable reuse where the treated effluent goes directly from the advanced treatment plant into the public water supply system (pipe-to-pipe connection) may never be practiced. People see this as a "toilet-to-tap" connection and public acceptance will be very difficult to obtain. Rather, to protect against accidental failures in the treatment plant and to enhance the aesthetics and public acceptance of potable water reuse, the potable reuse should be indirect, meaning that the effluent should first go through surface water (streams or lakes) or groundwater (via artificial recharge) before it can be delivered to public water supply systems. The surface water route has several disadvantages, including algae growth that can cause taste and health problems since some algal metabolites are toxic. To minimize algae growth, the wastewater may then have to be treated to remove nitrogen and phosphorous, which increases the reuse costs. Also, water is lost by evaporation and the water is vulnerable to recontamination by animals and human activities. These disadvantages do not exist with the groundwater route, where the water also receives SAT benefits. Groundwater recharge also enables seasonal or longer storage of the water to absorb differences between water supply and demand, and mixing of the effluent water with native groundwater when it is pumped from wells. Water reuse basically compresses the hydrologic cycle from an uncontrolled global scale to a controlled local scale. Since all water is recycled in one way or another, the quality of the water at its point of use is much more important than its history.

Virtual Water

Water-short areas can minimize their use of water by importing commodities that take a lot of water to produce like food and electric power, from other areas or countries that are blessed with more water. The receiving areas then are not only getting the commodities, but also the water that was necessary to produce them. Since this water is "virtually" embedded in the commodity, it is called virtual water (Allan 1998). For example, for every kilogram of wheat imported, the country also gets about 1 m^3 of virtual water at much less cost than the price or value of local water resources, if available, in the country itself. Using a lot of water just to satisfy a national pride of being self-sufficient in food production (especially staple foods) will then not be economical if these foods can be imported much cheaper from water rich countries (Wichelns 2001). More and more areas in the world face serious water shortages with little prospect of having adequate water for their inhabitants, even by trying to move more water to people or more people to water. Imports of virtual water embedded in food and other commodities may then economically and politically be a very good solution, and probably the easiest way to achieve peaceful solutions to water conflicts.

As economies and trade become more and more global in scope, global movement of food from water rich to water poor countries should be just as feasible as moving petroleum products from oil rich to oil poor countries. To ensure that global distribution of food will not be used as political weapons, it should be internationally controlled with representation of the importing countries. Other opportunities for saving local water resources by importing virtual water include import of electric power from areas with more abundant water for cooling of thermal power plants, with dams for hydroelectric power production, or with coastal areas that provide ocean water for cooling. The virtual water concept could also be useful in protecting wetlands of international ecological significance against water diversions and drying up to produce more irrigation water, such as the Sudd wetlands in the Sudan and the Okovanggo Basin in Botswana (Postel 1999). International cooperation could then be established to develop ecotourism in these areas that will provide revenues for the import of staple foods and the virtual water therein.

Moving virtual water will be much cheaper than moving the water itself, which is also being considered. Proposals range from building huge pipelines or aqueducts to hauling water in tankers and towing icebergs from polar regions or large rafts with fresh water from river discharges into oceans (McCann 2000; Handelman 2001). For water rich countries, such water exports can be a significant source of revenue.

Conclusions

Increasing populations and uncertain climatic changes will pose heavy demands on water resources in the future. Holistic approaches, integrated water management principles, and international cooperation will be necessary to develop sustainable systems and prevent catastrophes. Agricultural water management must be integrated with other water management practices, since the actions of one user group will affect the water interests of others. More research needs to be done to make sure that management of water and other resources is based on sound science and engineering. Much greater local, national, and international efforts, cooperation, and expenditures are needed to meet future food and water requirements in sustainable, peaceful, and environmentally responsible ways. The challenges are there.

References

Albritton, D. L., et al. (2001). "Climate Change 2001: The Scientific Basis, Contribution from Working Group I to the Third Assessment Report, Intergovernmental Panel for Climate Change." *Technical Summary*, Cambridge Univ. Press, Cambridge, U.K., 21–83.

Allan, J. J. (1998). "Virtual water: a strategic resource, global solutions to regional deficits." *Ground Water*, 36, 545–546.

Anderson, J., Rodriguez, A., Fletcher, C., and Fitzgerald, D. (2001). "Researchers focus attention on coastal response to climate change." *Trans., Am. Geophys. Union*, 82(44), 513–520.

Apse, M. P., Akaron, G. S., Snedden, W. A., and Blumwald, E. (1999). "Salt tolerance conferred by overexpression of a vacuolar Na^+/H^+ antiport in Aribidopsis." *Science*, 285(31), 1256–1258.

Asano, T., ed. (1998). *Wastewater reclamation and reuse*, Technomic, Lancaster, Penn.

Bouwer, H. (1978). *Groundwater hydrology*, McGraw–Hill, New York, 480.

Bouwer, H. (1989). "Estimating and enhancing groundwater recharge." *Groundwater recharge*, M. L. Sharma, ed., Balkema, Rotterdam, The Netherlands, 1–10.

Bouwer, H. (1990). "Agricultural chemicals and groundwater quality," *J. Soil Water Conservat.*, 45(2), 184–189.

Bouwer, H. (1993). "From sewage farm to zero discharge." *Eur. Water Pollution Control*, 3(1), 9–16.

Bouwer, H. (1994). "Irrigation and global water outlook." *Agric. Water Manage.*, 25, 221–231.

Bouwer, H. (1997)."Role of groundwater recharge and water reuse in integrated water management." *Arabian J. Sci. Eng.*, 22, 123–131.

Bouwer, H. (1999). "Chapter 24: Artificial recharge of groundwater: systems, design, and management." *Hydraulic design handbook*, L. W. Mays, ed., McGraw–Hill, New York, 24.1–24.44.

Bouwer, H., Fox, P., Westerhoff, P., and Drewes, J. E. (1999). "Integrating water management and reuse: causes for concern?" *Water. Qual. Internat.*, 1999, 19–22.

California State Department of Health Services (CSDHS). 1998. NDMA in drinking water, California State Department of Health Services, Sacramento, Calif.

Christen, K. (2001). "TMDL program broken but fixable, NRC report finds." *Water Env. Technol.*. 13(9), 31–36.

Crook, J., MacDonald, J. A., and Trussell, R. R. (1999). "Potable use of reclaimed water." *J. Am. Water Works Assoc.*, 91(8), 40–49.

Daughton, C. G., and Ternes, T. A. (1999). "Pharmaceuticals and personal care products in the environment: agents of subtle change?" *Environ. Health Perspect.*, 107, Supplement 6:907–938.

Daughton, C. B., and Jones-Lepp, T. L., eds. (2001). *Pharmaceuticals and personal care products in the environment*, ACS Symposium Series 791, American Chemical Society Washington, D.C.

de Jong, J., Können, G., and Kattenberg, S. (2001). "Climate changes in the Rhine basin." Special report, Royal Dutch Meteorological Institute, P. O. Box 201, 3730 AE DeBilt, The Netherlands.

Finnan, J. M., Donnelly, A., Burke, J. I., and Jones, M. B. (2002). "The effects of elevated concentrations of carbon dioxide and ozone on potato (*Solanum tuberosum L.*) yield." *Agric., Ecosyst. Environ.*, 88, 11–22.

Gabbay, S. (2001). "Vulnerability and adaptation to climate change." *Israel Env. Bull.*, 24(1), 11–14.

Goodbred, S. L., et al., "Reconnaissance of 17B-estradiol, 11-ketotestosterone, vitellogenian, and gonad histopathology in common carp of United States Streams: potential for contaminant-induced endocrine disruption." U.S. Geological Survey Open File Rep. No. 96-627. Sacramento, California.

Gray, R. (2001). "EPA sets cost estimate on TMDLs." *Water Eng. Manage.*, 148(10), 8.

Handelman, S. (2001). "Exporting fresh water." *Time Mag.*, August B14–B15.

Hough, A. M., and Derwent, R. G. (1990). "Changes in the global concentration of tropospheric ozone due to human activities." *Nature (London)*, 344, 645–648.

Hulme, M., and Kelly, M. (1993). "Exploring the links between desertification and climate change." *Environment,* 35:4, 39–11,45.

Isenhart, T. M., Schultz, R. C., and Colletti, J. P. (1998). "Chapter 19: Watershed restoration and agricultural practices in the midwest: Bear Creek of Iowa." *Watershed restoration: Principles and practices,* J. E. William, C. A. Wood, and M. P. Dombeck, eds., 318–334.

Ishaq, A. M., and Khan, A. A. (1997). "Recharge of aquifers with reclaimed wastewater: A case for Saudi Arabia." *Arabian J. Sci. Eng.,* 22(1C), 133–141.

Jobin, W. (1999). *Dams and disease,* Taylor and Francis, Florence, Kentucky, 544.

Kimball, B. A. (2002). "Global change and water resources." *Irrigation of Agricultural Crops Monograph,* R. J. Lascano and R. E. Sojka, eds., American Society of Agronomy, Madison, Wis., in press.

Kolpin, D. W., et al. (2002). "Pharmaceuticals, hormones, and other organic wastewater in U.S. streams, 1999–2000." *Environ. Sci. Technol.,* 36(6), 1202–1211.

Lee, K., Isenhart, T. M., Schultz, R. C., and Mickelson, S. K. (1999). "Sediment and nutrient trapping abilities of switchgrass and bromegrass buffer strips." *Agroforestry Syst.,* 44, 121–132.

McCann, B. (1999). "By-product blues." *Water,* 21, 15–18.

McCann, B. (2000). "Oceanic answer." *Water,* 21, 26–28.

McClurg, S. (1998). "Climate change and water: what might the future hold?" *Western Water,* 1998, 4–13.

McClurg, S. (2001). "Conjunctive use: banking for a dry day." *Western Water,* (Jul/Aug), , 4–13.

National Academy of Sciences (2001). *Under the weather: Climate, ecosystems, and infectious disease,* National Academy Press, Washington, D.C.

National Research Council (NAS), (2001). *Assessing the TMDL approach to water quality management,* K. Reckhow, Chair., National Academy Press, Washington, D.C.

Newcom, J. S. (2001). "Dealing with the shock." *Western Water* (Sep/Oct), , 4–13.

Office of Science and Technology, 1997. "Climate change: State of knowledge." Washington, DC.

Pearce, F. (1992). *The dammed,* The Bodley Head, London, 276.

Postel, S. (1992). *Last oasis,* Worldwatch Institute, Washington, D.C.

Postel, S. (1999). *Pillar of sand,* Worldwatch Institute, Washington, D.C.

Pyne, R. D. G. (1995). *Groundwater recharge and wells: A guide to aquifer storage and recovery,* Lewis Publishers, Boca Raton, Fla.

Richardson, M. L., and Bowron, J. M. (1985). "The fate of pharmaceutical chemicals in the aquatic environment." *J. Pharmacol.,* 37, 1–12.

Schlesinger, W. H. (1999). "Carbon sequestration in soils." *Science,* 284, 2095.

Schultz, R. C., et al. (1995). "Design and placement of a multi-species riparian bugger strip system." *Agroforestry Syst.,* 31, 117–132.

Shannon, M., Cervinka, V., and Daniel, D. A. (1997). "Chapter 4: Drainage water reuse." *Management of agricultural drainage water quality,* C. A. Madromootoo, W. R. Johnston, and L. S. Willardson, eds. Water Rep. No. 13, Food and Agricultural Organization of the United Nations, Rome, Italy, 29–40.

Showstack, R. (2001). "Panel urges measures to minimize effects of future abrupt climate changes." *Trans., Am. Geophys. Union,* 82(52), 653–654.

Siegenthaler, U., and Sarmiento, J. L. (1993). "Atmospheric carbon dioxide and the ocean." *Nature (London),* 365, 119–225.

Smith, J. D. (2002). "U.S. EPA's new rule confusing, delays TMDL program; should be scrapped." *Water Environ. Technol.,* 14(2), 6–7.

Stone, D. B., Moomaw, C. L., and Davis, A. (2001). "Estimating recharge distribution by incorporating runoff from mountain areas in an alluvial basin in the Great Basin region of the southwestern United States." *Ground Water,* 39(6), 807–818.

Tatro, S. B. (1999). "Dam breaching, the rest of the story." *Civ. Eng. (N.Y.),* 69(4), 50–55.

Tyler, S. W. et al. (1996). "Soil-water flux in the southern Great Basin, United States: temporal and spatial variations over the last 120,000 years." *Water Resour. Res.,* 32, 1481–1499.

University of Wisconsin (2001). "Succeeding with a dam decommissioning project." Short course offered by Department of Engineering Professional Development, Madison Wis.

Wagner, E. (2001). "There is no perfect time to issue a TMDL rule." *Water Environ. Technol.,* 13(9), 8–10.

Warner, W. S. (2000). "The influence of religion on wastewater treatment." *Water,* 21, 11–13.

Warrick, R. A., LeProvost, C., Meier, M. F., Oerlemans, J., and Woodworth, P. L. (1996). "Changes in sea level," in Climate Change 1995. *The Science of Climate Change,* J. T. Houghton, L. G. Meira Filho, B. A. Callander, N. Harris, A. Kattenberg, and K. Maskell, eds., Cambridge Univ. Press, New York, 361–405.

Wichelns, D. (2001). "The role of 'virtual water' in efforts to achieve food security and other national goals, with an example from Egypt." *Agric. Water Manage.,* 49(2), 135–155.

World Health Organization. (1989). "Health guidelines for the use of wastewater in agriculture and aquaculture." *Tech. Bull. Ser. 77,* WHO, Geneva, Switzerland.

Xu, H., ed. (1993). *Salinity gradient solar ponds—a practical manual.* Vol. 1 (Solar pond design and construction) and Vol. 2 (Solar pond operation and maintenance). Dept. of Industrial and Mechanical Engineering, Univ. of Texas, El Paso, Tex.

Zullei-Seibert, N. (1998). "Your daily "drugs" in drinking water? State of the art for artificial groundwater recharge." *Proc., 3rd Int. Symposium on Artificial Recharge of Groundwater,* Amsterdam, The Netherlands, 405–407.

American Society of Civil Engineers

150th Anniversary Paper

Building a Better World

A Paradigm Shift in Irrigation Management

Marshall J. English, M.ASCE[1]; Kenneth H. Solomon, M.ASCE[2]; and Glenn J. Hoffman[3]

Abstract: In coming decades, irrigated agriculture will be called upon to produce up to two thirds of the increased food supply needed by an expanding world population. But the increasing dependence on irrigation will coincide with accelerating competition for water and rising concern about the environmental effects of irrigation. These converging pressures will force irrigators to reconsider what is perhaps the most fundamental precept of conventional irrigation practice; that crop water demands should be satisfied in order to achieve maximum crop yields per unit of land. Ultimately, irrigated agriculture will need to adopt a new management paradigm based on an economic objective—the maximization of net benefits—rather than the biological objective of maximizing yields. Irrigation to meet crop water demand is a relatively simple and clearly defined problem with a singular objective. Irrigation to maximize benefits is a substantially more complex and challenging problem. Identifying optimum irrigation strategies will require more detailed models of the relationships between applied water, crop production, and irrigation efficiency. Economic factors, particularly the opportunity costs of water, will need to be explicitly incorporated into the analysis. In some cases the analysis may involve multi-objective optimization. The increased complexity of the analysis will necessitate the use of more sophisticated analytical tools. This paper examines the underlying logic of this alternative approach to irrigation management, explores the factors that will compel its adoption, and examines its economic and environmental implications. Two important concerns, sustainability and risk, are discussed in some depth. Operational practices for implementing the new approach are contrasted with current, conventional irrigation practices. Some of the analytical tools that might be employed in the search for optimum irrigation strategies are reviewed. Finally, the limited and largely intuitive efforts that have already been made to implement this new paradigm are discussed.

DOI: 10.1061/(ASCE)0733-9437(2002)128:5(267)

CE Database keywords: Irrigation practices; Management; Economic factors; Environmental issues; Risk

Introduction

A fundamental shift in irrigation practice is likely to evolve over the next few decades. Economic pressures on farms, increasing competition for water, and the adverse environmental impacts of irrigation will motivate a new approach to irrigation based on economic efficiency rather than crop water demand (Kirda and Kanber 1999). This new approach, which might be described simply as "optimization," has been characterized as a new paradigm (Perry 1999). Here we argue that this fundamental change in irrigation management is desirable and inevitable, and that it will be a significant departure from current practice.

Irrigation optimization should not be confused with scientific irrigation scheduling, the systematic tracking of soil moisture or crop water status to determine when and how much to irrigate. As

[1]Professor, Bioengineering Dept., Oregon State Univ., Corvallis, OR 97330. E-mail: englishm@engr.orst.edu

[2]Professor and Head, Bioresource and Agricultural Engineering Dept., California Polytechnic Univ., 1 Grand Ave., San Luis Obispo, CA 93407.

[3]Professor and Head, Biological Systems Engineering Dept., Univ. of Nebraska-Lincoln, 223 LW Chase Hall, Lincoln, NE 68583-0726.

Note. Discussion open until March 1, 2003. Separate discussions must be submitted for individual papers. To extend the closing date by one month, a written request must be filed with the ASCE Managing Editor. The manuscript for this paper was submitted for review and possible publication on December 17, 2001; approved on May 19, 2002. This paper is part of the *Journal of Irrigation and Drainage Engineering*, Vol. 128, No. 5, October 1, 2002. ©ASCE, ISSN 0733-9437/2002/5-267–277/$8.00+$.50 per page.

originally conceived (Gear et al. 1976) and as generally practiced, irrigation scheduling is predicated on maximizing yields, hence current scheduling procedures do not explicitly account for costs and revenues. Optimization, on the other hand, explicitly accounts for these economic factors. That is not to say that scientific scheduling is not part of the optimization approach. On the contrary, new and more sophisticated irrigation scheduling techniques will be needed to implement optimal irrigation plans.

This paper is in three parts. The first ("The Nature of Irrigation Optimization") deals with maximizing farm profits, the simplest optimization problem. Optimum irrigation implies lower levels of applied water than conventional irrigation. Reduced water applications, in turn, imply potentially greater irrigation efficiencies and reduced environmental impacts. The linkages between profit, water use, efficiency, and environmental impacts are outlined briefly.

The second part ("A Broader View of Irrigation Optimization") deals with irrigation strategies designed for broader objectives such as increased food security, regional and national economic growth, and mitigation of nonpoint source pollution. These broader issues may involve multi-objective optimization for benefits that are not, strictly speaking, economic in nature, which implies more complex analyses than the relatively simple, clearly defined problem of profit maximization. Two other important issues are dealt with in this part of the paper. One is risk, a critical concern for many farmers. Inherent uncertainties in the relationships between water use, crop yield, and net income imply eco-

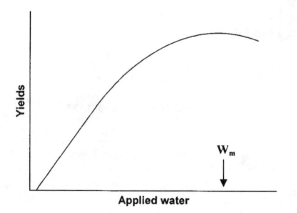

Fig. 1. General form of relationship between applied water and crop yield

nomic risk. Some optimization strategies may amplify that risk. A truly optimum irrigation strategy would therefore need to account not only for expected profit but also for the farmer's tolerance for risk. The second issue is management of salinity. Always a concern in irrigated agriculture, the management of salinity can be more critical and more challenging under an optimum irrigation regime.

The third part of the paper ("A Practical View of Irrigation Optimization") explores some of the practical implications of an optimization approach. Economic forces and resource constraints that will motivate adoption of this new approach are outlined. Optimum practices will be contrasted with current, conventional practices. The role of Operations Research, the technical discipline that deals with analysis of optimal strategies, will be reviewed. Finally, limited initial efforts to apply the principles of optimization in practice on production farms will be discussed.

It is hoped this paper will offer leaders in the irrigation industry a clearer understanding of an approach to irrigation design and management that is altogether different from current conventional practices. For some the paper may reveal research needs or business opportunities. For others it could be a catalyst for change in management practices. In short, our objective is to motivate, accelerate, and facilitate a fundamental and positive change in irrigation management.

The Nature of Irrigation Optimization

We will illustrate the essential features of irrigation optimization by focusing first on a few simple analyses. Though they generally disregard many of the complexities of real-world farming, these analyses will illustrate the essential nature of an optimum approach to irrigation.

Maximizing Net Income When Water is not Limited

We first consider the simplest case, optimum seasonal water use by a farmer whose sole objective is maximizing net income. Let us assume that the farmer has limited land, but that inputs of water, fertilizers, and other factors are not constrained. A general relationship between applied water and crop yield per unit of land is shown in Fig. 1. Crop yield functions are commonly based on evapotranspiration (ET) rather than applied water because such functions are more or less independent of the irrigation system,

soils and other local factors that influence the shape of the curve in Fig. 1. However, in practice, irrigators control the application of water, not ET. In order to develop optimal irrigation strategies it is therefore necessary, in one way or another, to work with the relationship between applied water and yield, the relationship shown in Fig. 1. At low levels of applied water, up to about 50% of full irrigation, yields tend to increase linearly with applied water (Doorenbos and Kassam 1979; Vaux and Pruitt 1983; Hargreaves and Samani 1984). As more water is applied the relationship becomes curvilinear, a consequence of accelerating losses from increased surface evaporation, runoff, and deep percolation. Beyond the point of maximum yield the curve turns downward, reflecting yield losses from anaerobic root zone conditions, diseases, and leaching of nutrients associated with excessive water use.

Let us represent this crop yield function as $y(w)$, where y indicates yield per unit of land (Mg/ha) and w is the depth of applied water (mm). What level of applied water would constitute "full" irrigation in Fig. 1? The nominal irrigation requirement is generally defined as the amount of water needed to achieve "full production potential" (Doorenbos and Pruitt 1992). But given the spatial variability of soil and applied water, it is not possible to irrigate an entire field with perfect uniformity, and consequently achieving full production potential at every point in a field is not feasible. So, for purposes of this paper, let us interpret "full irrigation" to mean the level of applied water that achieves the highest *field average* yield, indicated as W_m in Fig. 1. If our objective is to maximize yield per unit of land, we need only set the derivative of the yield function to zero and solve for w. In other words, the objective of conventional irrigation is defined by

$$\frac{\partial y(w)}{\partial w} = 0 \qquad (1)$$

Now let us turn to the general relationship between applied water and production costs. Relevant production costs include: (i) variable costs of water, energy, labor, and maintenance; (ii) capital and other fixed costs associated with land preparation, irrigation system capacity, and other costs not directly associated with irrigation or crop yield; and (iii) the costs of chemical applications, harvest, and other production costs that vary with yields (and therefore with water use). Note that all of these costs are relevant to the optimization problem. This was demonstrated in a case study by English and Nuss (1982) in which only 41% of the cost reductions associated with optimum irrigation would be linked to reduced costs of irrigation *per se*. The remaining 59% of savings would derive from reductions in capital investment, chemical use, seeding and harvest, all of which would decline with reduced water use. Their analysis did not consider subsidies and tax issues, but these may also be relevant.

We will represent production costs as a function of applied water, $c(w)$ ($/m^3$). Using that cost function and the earlier yield function, and representing the unit price paid for the crop as p_c ($/Mg), the relationship between applied water and net income can be written

$$i(w) = p_c \cdot y(w) - c(w) \qquad (2)$$

where $i(w)$ represents net income per unit of land ($/ha).

If our objective is maximum net income per unit of land, rather than maximum yield, the optimum level of irrigation can be determined by setting the derivative of the net income equation to zero

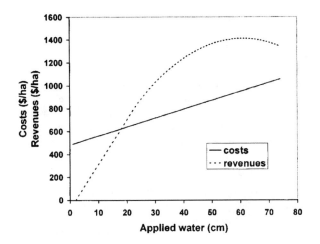

Fig. 2. Costs and revenue as function of applied water

$$\frac{\partial i(w)}{\partial w} = p_c \cdot \frac{\partial y(w)}{\partial w} - \frac{\partial c(w)}{\partial w} = 0 \qquad (3)$$

After rearranging, optimum irrigation is then defined by

$$\frac{\partial y(w)}{\partial w} = \frac{1}{p_c} \frac{\partial c(w)}{\partial w} \qquad (4)$$

The terms on the right, the inverse of crop price and the derivative of the cost function, will both be positive and nonzero. Equation (4) therefore indicates that at the optimum point the derivative of the yield function will be positive and nonzero. That is, the profit maximizing point will be found in the rising part of the yield curve, which is to the left of the yield-maximizing point. Thus a profit-maximizing strategy will use less water per unit of land than a conventional, yield-maximizing strategy. As an example, English and Raja (1996) analyzed optimum irrigation of winter wheat on an Oregon farm, assuming a crop price of $147/Mg and using the following locally derived yield and cost functions:

$$y(w) = -0.5348 + 0.03326 \cdot w - 0.0000273 \cdot w^2 \qquad (5)$$

$$c(w) = 482.30 + 0.779 \cdot w \qquad (6)$$

The resulting cost and revenue curves are shown in Fig. 2. Using this yield function in Eq. (1) and solving for w, the yield-maximizing level of applied water would be 610 mm; net farm income would be $453.70 per hectare. On the other hand, the profit-maximizing level of irrigation, from Eq. (4), would be 510 mm, 16% less than full irrigation, and the net income would be $491.51/ha, an increase of 8.3%.

Maximizing Net Income When Water is the Limiting Resource

When irrigation is constrained by limited water availability or limited irrigation system capacity, the water saved by reducing the depth of irrigation might be used to irrigate additional land. The problem then is to determine the optimum trade-off between the depth of applied water and the area to be irrigated. That is, we wish to maximize total income, $I(w)$, where

$$I(w) = [p_c y(w) - c(w)] \cdot A \qquad (7)$$

The irrigated area, A (ha) is determined by the total volume of available water, W_T (m³), and the depth of applied water, w. That is

$$A = \frac{W_T}{w} \qquad (8)$$

Combining Eqs. (7) and (8), optimum water use will be defined by the value of w that satisfies the equation (English 1990)

$$-A \cdot \frac{\partial i(w)}{\partial w} = i(w) \cdot \frac{\partial A}{\partial w} \qquad (9)$$

Using the earlier yield and cost functions [Eqs. (5) and (6)] and crop price, but assuming a constrained water supply, English and Raja (1996) estimated that the profit maximizing depth of irrigation would be 370 mm, about 39% less than the nominal irrigation requirement of 610 mm mentioned earlier. The water thus saved would be used to increase the area irrigated by 64% and the net income derived from the limited water supply would be increased by 49%.

Optimization for Multiple Fields and Crops

If a farm has sufficient water, optimum irrigation depths can be determined on a field-by-field basis. The optimization problem becomes more complex when multiple fields and crops are to be irrigated with a limited water supply. Since limited water implies an opportunity cost for water, the decision maker must consider all fields and crops and any alternative uses of water simultaneously, allocating a larger share of the water to more profitable crops, or perhaps marketing water to off-farm uses. Analyses of such cases typically rely on mathematical programming techniques (e.g., linear programming or dynamic programming) to optimize for the entire farm as a single planning unit.

The added complexity of such whole-farm analyses is illustrated in a paper by Martin and vanBrocklin (1989) who used dynamic programming to determine optimal planting and irrigation strategies for a mix of unirrigated and irrigated crops. Areas planted in three crops were optimized for the farm as a whole, within limits imposed by governmental production constraints. Irrigation water, drawn primarily from a limited ground water source, was rationed under a water-bank program with water allocations contracted in five-year blocks. The analysis also accounted for seasonal weather variability. The optimal strategies varied by season within the five-year contract period. Estimates of the optimal proportion of land to irrigate varied from about 50 to 100% of available land, with the optimal depth of applied water varying from 65 to 85% of nominal full irrigation.

Efficiency, Adequacy, and Leaching

Since the application of water is always nonuniform, some fraction of an irrigated field will generally be under irrigated while the remainder is fully irrigated or over irrigated. The percentage of a field that is fully irrigated or over irrigated is referred to as the irrigation adequacy. It was pointed out earlier that NRCS guidelines call for 90% adequacy; that is, 90% of the field should receive enough water to avoid loss of crop yield or quality. If the depth of applied water is less than the nominal requirement the adequacy will be less than 90%. Yields are likely to be reduced as a result, as noted earlier, but that will be accompanied by reduced percolation, increased application efficiency (Hart and Reynolds 1965; Shearer 1978) and reduced leaching of soluble chemicals.

An analysis by Stewart et al. (1974) provides some perspective on changes in deep percolation that might be expected as applied water is reduced. Using experimental data relating corn yield to applied water in Davis, CA, they estimated that a 5.8%

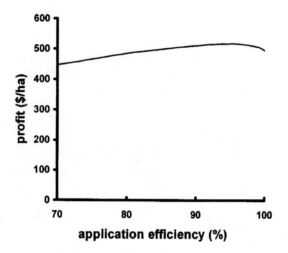

Fig. 3. Model analysis of profit versus application efficiency (from English 1999)

reduction in total applied water would be accompanied by a 40% reduction in percolation. They estimated that yields would be reduced by 1.1%, though with no resulting reduction in profits.

The implication is that reduced water use can lead to simultaneous increases in farm profits and application efficiencies. Fig. 3 illustrates an analysis of the relationship between application efficiency and profit for irrigated winter wheat for a Columbia Basin farm (from English 1999). The nominal application efficiency (the application efficiency that would be realized if the NRCS standard of 90% adequacy is achieved) in this case would be 82%. As water use is reduced, both efficiency and net income would increase until an application efficiency of 95% is reached at the profit-maximizing level of irrigation. Beyond that point profits decline as efficiency increases.

The two studies cited above are examples of a larger body of field research and other model studies supporting the conclusions that: (1) reduced irrigation is associated with increased application efficiency; (2) yield reductions may be small compared to the reductions in water use, hence net farm income may increase up to a point; and (3) the reduced irrigation may be accompanied by reduced leaching.

In some situations the increased application efficiency from optimum management may save water that would otherwise be irretrievably lost. For example, Kelly and Ayer (1982) estimated that if corn and cotton in California were irrigated at profit-maximizing levels rather than yield-maximizing levels, the water saved could supply a city of over 600,000 people. Much of that water would otherwise be lost to the saline evaporation ponds and aquifers of the southern San Joaquin Valley.

On the other hand, there are cases where increased application efficiency is not desirable, for example, when runoff from one farm is an essential source of irrigation water for another farm (Sekler 1999). And, in fact, situations where virtually all return flows are recaptured and reused, as evidenced by basin-wide efficiencies approaching 100%, are common. But even in those situations, increased efficiencies may increase the net benefits derived from limited water supplies by reducing production costs (as outlined earlier), improving the distribution of water among farmers and other users, and reducing the negative, off-farm effects of irrigation (Wichelns 2001).

A Broader View of Irrigation Optimization

The benefits of irrigation optimization may be greater for society as a whole than for individual irrigators. The larger community may be concerned with broader objectives, such as maximizing regional income or food security, or minimizing resource use or environmental impacts. The limiting resource may be something other than the water supply. For example, the pollution assimilation capacity of water that receives return flows may be more significant than the water supply *per se*.

Food Security

Conventional irrigation management is predicated on maximizing the rate of food production per unit of land (Mg per hectare). Maximization of *total* food production can be a more important concern. The problem of maximizing food production with a limited supply of water can be stated as

$$\text{Maximize:} \quad Y_{\text{total}} = y(w) \cdot A \tag{10}$$

where Y_{total} (Mg) represents total food production derived from a given water supply; and A represents the total irrigated area, as determined by the total water supply and depth applied [see Eq. (8)]. Taking the derivative with respect to w and rearranging terms leads to the defining equation for maximizing total food production

$$\frac{\partial y(w)}{\partial w} = -\left[\frac{1}{A} y(w) \frac{\partial A}{\partial w} \right] \tag{11}$$

The inverse of area $(1/A)$ and the yield $[y(w)]$ are both positive and nonzero. If the total water supply is constrained, and if water saved by reducing the depth of irrigation (w) is used to put additional land under irrigation, A will increase as w decreases. Since the derivative of A is therefore negative and nonzero, the right side of the equation must be nonzero and positive. The left side of the equation must therefore also be positive. If the derivative of $y(w)$ is positive at the point of maximum total yield, that point must be on the rising part of the yield curve. In short, optimum total food production with a limited water supply implies less than full irrigation.

To illustrate, let us consider maize production in southern Africa. Maize, a drought sensitive crop, supplies about 80% of the food supply for northeastern Zimbabwe. It is a region where, as David Livingstone noted more than 150 years ago, droughts are the rule rather than the exception. We will use the following generic maize yield function, derived from 23 published yield functions by Solomon (1985), to develop an irrigation strategy for maximizing maize production per unit of water:

$$y(w) = 6.0(-0.84 + 4.43W_R - 3.52W_R^2 + 1.11W_R^3 - 0.18W_R^4) \tag{12}$$

where W_R, the relative water supply, is defined as

$$W_R = \frac{(w + \text{rain})}{W_{\text{MAX}}} \tag{13}$$

where w = depth of applied water; rain = in-season rainfall (assumed to be nil); and W_{MAX} = amount of water needed for maximum yield (m³/ha). For northeastern Zimbabwe, W_{MAX} is estimated to be 525 m³/ha and the estimated maximum attainable yield is 6.0 Mg/ha.

From Eqs. (8) and (10)

$$Y_{\text{total}} = y(w) \cdot \frac{W_T}{w} \tag{14}$$

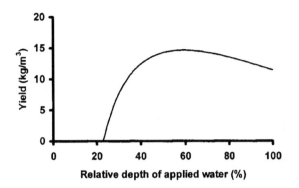

Fig. 4. Maize production versus relative applied water (W_R) in northeastern Zimbabwe

Dividing both sides of this equation by 525 m³ of water (the yield-maximizing volume per hectare) gives us the yield per cubic meter, plotted in Fig. 4 as a function of relative applied water, W_R. Under full irrigation ($W_R = 1.0$) total yield would be the maximum attainable 6.0 Mg/ha, or 11.43 kg/m³. As W_R is reduced the productivity increases, reaching a maximum when W_R is 0.593 (59.3% of full irrigation, or 311 m³/ha). The calculated yield rate per unit of water at that point reaches a maximum of 14.61 kg/m³, approximately 28% greater than the total maize production that would be achieved with nominal, full irrigation.

Regional and National Issues

Other regional or national objectives may bring an even wider range of issues and constraints into the analysis. Kazakhstan offers an interesting case in point. Development and renovation of irrigation projects to increase food production is being undertaken with World Bank funding, with the stipulation that the bank's loans be matched by Kazakhstan in hard foreign currency. Given the very unsteady national currency, hard currency is a severely limited resource. The essential criterion in irrigation planning is therefore maximization of food production per unit of hard currency.

As another example, recent irrigation projects in Egypt have been designed in part to provide for resettlement to reduce urban population pressures (El-Hafez 2002). The national government has subsidized these new farms in order to maximize their probability of economic success. Optimization in this case involves maximizing the total area of new land under cultivation, consistent with an acceptable level of profitability, while minimizing risk to individual farmers, a complex set of objectives indeed.

These broader perspectives may also extend to opportunity costs and/or long term costs. For example, the Kazakhstan government, in evaluating candidate irrigation projects, has chosen to base energy costs on world energy prices rather than local energy prices, reflecting the opportunity costs of energy that the country could otherwise export. Another example, the cost of aluminum irrigation pipe in the United States illustrates the longer term perspective. The cost of aluminum pipe is generally based on local pipe prices, a large component of which is due to the energy cost of aluminum production. However, some of those energy costs could be recaptured if the aluminum were recycled. From a long term, national perspective, a more complete analysis might need to consider not only the initial aluminum pipe cost but also the eventual cost recovery from recycling, and the collection and hauling costs of that recycling.

Irrigation Efficiency and Environmental Impacts

Environmental damages attributed to irrigation include salinization and alkalinization of soils, contamination of ground water with nitrates and residues of pesticides, sediment loading and contamination of surface waters by fertilizers and biocides, decreased assimilative capacity of streams from reduced in-stream flows, and contamination of return flows from salinity and sometimes toxic levels of selenium and boron. These are all linked to the intensity of irrigation water use. Because optimum irrigation generally implies reduced water use and increased application efficiencies, it offers a way to mitigate these adverse effects. For example, Miller and Anderson (1993) and Frazier et al. (1999) demonstrated an inverse relationship between net income and nitrate contamination of ground water, and concluded that economically optimum irrigation would improve ground water quality. Similarly, Johnson et al. (1991) found that optimizing irrigation would reduce nitrate leaching in central Oregon. It should be noted that reduced leaching to ground water does not guarantee improved ground water quality. If fertilizer use is not adjusted to account for the reduced leaching, irrigation optimization may actually increase contaminant concentrations in ground water by reducing the volume of ground water and its corresponding dilution capacity (Campagna et al. 1999; Connor and Perry 1999). But if fertilizer applications are well managed, reducing applied water may be a more cost effective way to mitigate ground water pollution than reducing fertilizer use (Frazier et al. 1999). Studies such as these have led one economist to argue that development of economically optimum irrigation strategies can be regarded as a useful strategy for reducing adverse impacts of irrigation (Mjelde et al. 1990).

Uncertainty and Risk

Uncertainty adds another dimension to the optimization problem. Given the unpredictable effects of weather, disease, and various other factors, the crop yields that will be produced with a given level of applied water are inherently uncertain. This is graphically illustrated by the histogram in Fig. 5 showing variability of bean yields with a given level (300 mm) of applied water in southern Idaho (English 1981). This histogram was generated using a statistical model, calibrated with observed bean yields from 53 fields over a five-year period, that simulated (1) potentially attainable yields in response to applied water, and (2) yield losses associated with adverse weather, poor management practices, disease, pest infestations, and other hazards. This representation of yield variability offers a perspective on the intrinsic uncertainty of the yield function, $y(w)$. Production costs and crop prices, the $c(w)$ and p_c terms used earlier, are likewise uncertain. An analysis based on these quantities is therefore also uncertain. The relationship between net income and crop yield, crop price and production costs is still represented by Eq. (1), but the factors embedded in that equation may need to be treated as random variables, hence net income is itself a random variable. We might estimate its probability distribution, then base irrigation decisions on the expected net income, as calculated by

$$E(I) = \int i \cdot f(i)\, di \qquad (15)$$

where i represents random net income; $E(I)$ denotes its expected value; and $f(i)$ its probability density function.

For many managers the irrigation strategy of choice will be that which maximizes $E(I)$. However, the decision problem may

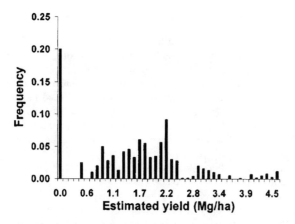

Fig. 5. Simulated statistical distribution of bean yields with 300 mm of water applied in southern Idaho

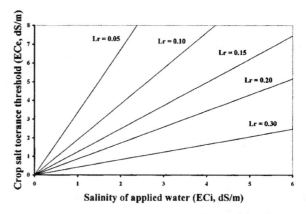

Fig. 6. Graphical solution for leaching requirement (LR) as function of salinity of applied water and crop salt tolerance threshold (adapted from Hoffman and van Genuchten 1983)

be complicated by the fact that other decision makers are risk averse. Irrigation strategies that offer the highest expected net income may also entail a greater risk of loss. That is particularly true when water is in short supply. If the optimum plan calls for simultaneously reducing depths of applied water and increasing irrigated area, any losses incurred may be amplified by the increased area under irrigation.

Given a choice between a high-risk strategy with a high statistical expectation of profit and an alternative strategy with less potential profit but less likelihood of loss, a risk-averse farmer may prefer the latter option. The significance of risk aversion was demonstrated by English and Orlob (1978) who employed Bayesian Decision Theory to study irrigation optimization under conditions of uncertainty. Using empirical utility functions (mathematical expressions that can be used to predict which of two risks a decision maker will prefer) derived by Lin (1973) and Buccolla (1976) for a group of six California farmers whose attitudes ranged from highly risk averse to risk indifferent, they estimated that the most risk-averse manager would prefer an irrigation strategy offering 40% lower expected profit than that preferred by the most risk tolerant manager. The more risk averse farmers would tend to use more water per unit of land.

Salinity and Sustainability

The threat of salt-affected soils and saline irrigation waters is a major concern in irrigated agriculture. The problem encompasses total dissolved solids (salinity), excessive sodium (sodicity), and toxic elements (toxicity). Any of these are a threat to sustainable irrigated agriculture, especially in arid areas. In humid regions, in semiarid regions with Mediterranean-type climates and heavy winter rainfall, or in monsoon-type climates with heavy summer rains, salinity may not be a problem. In these climates, precipitation will normally provide ample soil water to leach the potentially hazardous salts below the crop root zone, though low permeability of soil layers or the hydrogeology may sometimes prevent adequate leaching, resulting in localized areas of excessive salinity.

Where salinity is a hazard, water must drain through the crop root zone to prevent solutes from accumulating to levels detrimental to crop production. The minimum fraction of applied water that must pass through the root zone, termed the leaching requirement (L_r), has been established by Hoffman and van Genuchten (1983) and can be estimated from Fig. 6.

All soils have the ability to transmit soil water provided hydraulic conductivity is adequate. If natural drainage is not sufficient to leach salts from the root zone, artificial drainage to an appropriate outlet must be installed to sustain crop productivity. The loss of productivity from excess salinity may take decades to develop or just one crop season, depending upon the severity of the problem.

The problem of managing salinity is exacerbated when irrigation managers strive to maximize farm profits by applying less water, since less water may then be available to satisfy the leaching requirement. If the leaching requirement is not satisfied everywhere in the field, soil salinity will increase in those parts of a field where the applied water is less than evapotranspiration plus leaching requirement. Because of nonuniformity in irrigation applications, one must decide whether to apply copious amounts of water to assure the leaching requirement is satisfied throughout the field or to accept some reduction in yield in parts of the field rather than over irrigate most of the field. Advances in irrigation technology and management are needed to approach the goal of matching the leaching requirement.

How frequently the leaching requirement must be met is a very complex issue that depends upon the salt tolerance of the crop, the variation of salt tolerance during plant growth, the salinity of the irrigation water, the amount of effective rainfall, and soil drainage. For rotations containing both sensitive and tolerant crops, it may be possible to grow a salt sensitive crop followed by a tolerant crop before soil salinity reaches detrimental levels and leaching must occur. When both nonsaline and saline irrigation waters are available, it may be possible to control salinity by cycling these waters for several years before leaching is required.

Given the spatial variabilities of soils and applied water, the unequal sensitivities of different crops at different stages of growth and the complex aquatic chemistry involved, the problem of irrigation optimization under saline conditions is challenging indeed. Nevertheless, the economic principles of optimization still apply where salinity is a concern. Knapp and Dinar (1987) used a dynamic programming model to study alternative strategies for irrigation using a high cost, primary source of water (salinity of 0.67 dS/m) and low cost reused drainage water (salinity of 7.96 dS/m) on a soil with an initial salinity (EC_e) of 7 dS/m. The optimal level of irrigation was found to be sensitive to production costs, but in any event would be less than the yield maximizing level of irrigation. Soil salinity would be reduced in the course of the irrigation season, indicating that the optimal strategy would be sustainable.

A Practical View of Irrigation Optimization

Pressures for Change

The narrowing of farm profit margins in recent years can be expected to motivate a new approach to irrigation based on the principle of optimization. The need to minimize environmental impacts from irrigation will be another factor driving adoption of the new paradigm. But perhaps the most compelling pressure will be escalating competition for water. Irretrievable water use worldwide has increased more than 400% since 1950, largely in response to increasing populations (Kirda and Kanber 1999). It has been argued that by 2025 the world demand for fresh water may be approaching the limits of readily accessible supplies (Postel et al. 1996). While some do not agree with that assessment, local shortages are already a reality. Fredericksen (1997) has made a compelling argument that when climatic variability and the inertia of political and social institutions are considered, the world is already facing a water crisis. Major water supplies that have been taken for granted are nearing their exploitable limits, becoming the focus of intense development and potential conflict. The Tigres and Euphrates rivers, recently dammed near their sources in Turkey, are now being controlled and heavily exploited by a country that, historically, had been a minor stake holder. The Indus river, supporting one of the oldest and largest irrigated areas in the world, is already over subscribed, with water deliveries averaging an estimated 35% less than nominal crop water requirements (Trimmer 1990). The Nile, the very emblem of irrigated agriculture, has supplied Egypt with abundant irrigation water for seven thousand years. As recently as 1990 the per capita water supply for Egypt was 922 m^3/year, almost double the internationally accepted "water poverty limit" of 500 m^3/year. But by 2025 this figure will have declined to 337 m^3/year, and a large fraction of Egyptian agricultural land could be un-irrigated (El-Hafez 2002). Water supplies on the scale of these examples cannot be moved around the globe to deal with local shortages, but populations will continue to grow for some decades at least. Maximizing food production or national income with limited water will therefore be particularly important in water-short countries.

Contrasting Conventional and Optimum Practices

The optimization paradigm has clear implications for both design and management of irrigation systems. Design procedures will necessarily become more involved, with the designer relying less on design rules of thumb and assuming more responsibility for formulating system specifications and developing case-specific design solutions. Management will require more sophisticated analysis, greater technical competence, and investment in new technologies.

Conventional irrigation practices are based on two key specifications, the crop water requirement and the nominal application efficiency. The crop water requirement was defined long ago as that which will prevent crop water stress that would cause loss of yield or quality (Haise and Hagan 1967). The nominal application efficiency, which derives from an NRCS standard noted earlier, is defined as that which can be attained when irrigation adequacy is

90% for high or medium valued crops, or 75% for low valued crops. Conventional irrigation is therefore defined in terms of the amount of applied water required to prevent crop water stress in 90% of the field. Though dated, these stipulations are still the foundation of standard irrigation practice worldwide. A 1992 revision of the Food and Agricultural Organization of the United Nations Irrigation and Drainage Paper 24 ("Crop water requirements") defines crop water requirements in terms of full production potential, and a relatively recent NRCS revision to the National Engineering Handbook assumes 90% adequacy in deriving nominal values of low-quarter and low-half application efficiencies (NRCS 1993).

When the optimization paradigm is adopted, irrigation design and management can no longer be based on such *a priori* specifications. To determine the most economical level of irrigation, the designer will need to employ crop production models and operations research, two altogether new elements in system design, and will need to plan for some optimal degree of crop stress. Rather than regarding application efficiency as a predetermined constant for a given irrigation system, the efficiency will be a decision variable influenced by the management strategy chosen (English 1999). Just as sprinkler design guidelines derived from years of experience became obsolete with the advent of buried PVC solid-set systems, other well-established rules of thumb will become obsolete as economic considerations are explicitly addressed, and formulation of new, fixed guidelines will be difficult at best.

Irrigation scheduling will also involve new challenges. As water use is reduced and the margin for error narrows, more precise field determinations of crop water uptake, soil moisture conditions, and salinity status will be needed. New techniques for soil moisture measurements and progressively more accurate models of evapotranspiration, used in tandem, will be critical in this effort. Regardless of the performance of these new models and technologies, however, the inescapable spatial and temporal variability of soil moisture measurements may necessitate mathematical filtering (English et al. 1981; Aboitiz-Uriarte 1983).

While an irrigation system is commonly thought of in terms of its hardware—its distribution system, pumps, land improvements, etc.—the technology also includes information in the form of theoretical relationships, analytical tools, and the education to utilize them. The optimization paradigm outlined here involves much greater analytical complexity and technical sophistication than currently needed for conventional irrigation. Accordingly, we should anticipate that the irrigation manager will need a higher level of technical competence and better sources of information and advice. The cooperative irrigation advisory services (e.g., CIMIS, Agrimet) and commercial irrigation scheduling companies that have emerged over the past three decades to assist farmers with conventional irrigation scheduling indicate an existing need for such specialized assistance, even without the added complexity of optimization. A broader and more sophisticated menu of information and advice may be demanded of these services in coming decades as farmers adopt an economic approach to irrigation management.

While it is beyond the scope of this paper, we note also that the pressure to obtain the best economic use of irrigation water is likely to change water law as well. Increasing competition for water has already had an effect, for example, with the development of water-rights markets in certain areas to serve specific objectives.

Operations Research

One distinguishing feature of the new irrigation management paradigm will be an increased reliance on operations research, the branch of applied mathematics concerned with formulation and solution of optimization problems (Simmons 1975). Operations Research involves: (i) modeling of real-world processes and phenomena, and (ii) development of algorithms for identifying optimal strategies (Hillier and Lieberman 1980).

With regard to the first of these, modeling capabilities may be a critical issue in efforts to optimize irrigation management. In recent decades the research community has made important progress in modeling interactions between irrigation timing and amount, crop yields, the disposition of water as ET and percolation, and the quantity and quality of return flows. But some of the most widely used models of today, when applied to the problem of irrigation optimization, are found to have significant shortcomings. Some, for example, do not realistically simulate efficiencies and return flows, spatial variability or uncertainty. With accelerating use of models for real-world operations, and diminishing reliance on guidelines and standard operating procedures, much more will be demanded of these models. To illustrate this point, consider the Center d'Etudes, de Recherches et d'Essais Scientifiques (CERES) models (Jones and Kiniry 1986) and similar models developed specifically for such crops as maize, potatoes and cotton simulate crop growth and development in response to primary factors: nutrient availability, leaf area, availability of soil water, sunlight, and other climatic conditions. These widely used models do not explicitly account for spatial variability in the field, and therefore cannot adequately simulate the disposition of applied water. In effect, they cannot simulate efficiency. The CERES models require the user to stipulate the application efficiency, which is then treated as a constant. Since the stipulated efficiency determines the amount of water applied, and that in turn determines the volume of losses, the efficiency assumed by the user becomes self-fulfilling (English et al. 1985). [There have been experiments with a hybrid CERES model that combines the plant response simulation capabilities of the CERES models with elements of a spatially variable irrigation efficiency model developed by English et al. (1992).]

Another widely used line of empirical models is based on the relationship between evapotranspiration and crop yield (Doorenbos and Kassam 1979). Hoffman and van Genuchten (1983) have extended these ET-based models to predict yield as a function of average root zone salinity and leaching fraction, given steady-state conditions in the root zone (see Fig. 6). Letey (1991) has reviewed and characterized such models for use in irrigation management. However, these ET-yield models do not account for losses from runoff, deep percolation, or accelerated evaporation, and therefore do not provide any information about efficiency. It is left to the analyst to estimate efficiency independently. A comprehensive simulation model of application efficiency developed by the New Zealand Ministry of Agriculture and Fisheries and Oregon State University (English et al. 1992) has been developed for that purpose. Alternatively, Warrick and Yates (1987) derived formulas for estimating yields based on three different functional forms of yield-water use relationships and three different distribution models of spatial variability.

With regard to analytical techniques for identifying optimum strategies, a wide variety of algorithms have been proposed specifically for irrigation optimization over the past 40 years. Where input factors (e.g., water and nitrogen) can be treated as continuous variables and the availability of these inputs is not constrained in any way, simple calculus can be used to optimize for each field individually (as was done in the first example of this paper). When inputs are constrained, for example, when water supply is limited, and where a number of alternative strategies can be considered, a variety of classical operations research techniques might be employed. One of the most widely used of these is linear programming. This technique has been adapted to overcome difficulties that are specifically associated with irrigation optimization; convex linear programming has been used to deal with the nonlinear relationship between applied water and crop yield; chance constrained linear programming has been used to accommodate a stochastic water supply (Maji and Heady 1978).

Dynamic programming is another frequently used algorithm for optimization [see earlier discussion of the work by Martin and van Brocklin (1989)]. Though not as computationally efficient as linear programming, this more flexible technique allows greater freedom in the use of discrete and stochastic variables and nonlinear functional relationships. Linear programming and dynamic programming have also been combined in a hybrid analysis of inter-seasonal and intra-seasonal water allocations (e.g., Matanga and Marino 1977).

Simulation modeling, though relatively inefficient, allows the greatest flexibility in dealing with the complexities of the real world (e.g., English et al. 1992). Analysis based on simulation modeling implies that a search technique of some sort must be used to identify optimal strategies. Search algorithms may be computationally intensive and can involve difficulties in distinguishing between local and global optima. A promising technique developed by Canpolat (1997) involved the use of genetic algorithms to substantially increase the efficiency of the search procedure for evaluating complex sets of alternative seasonal water use plans.

Analytical algorithms for dealing with uncertainty and risk in irrigation are less straightforward than the techniques discussed above and generally do not produce singular, clearly defined conclusions. Nevertheless, practical ways of accounting for risk have been proposed. These often involve characterizing alternative irrigation strategies in terms of the mean and variance of profits. Lin (1973) concluded that most farmers' personal value systems can be quantified in terms of living standard, ownership of land, leisure time, and risk acceptance/aversion. Mean and variance of farm income are regarded as indirect measures of these intrinsic values, and therefore have been used by some as indices of value in decision analysis (Mjelde et al. 1990).

A variety of algorithms have been proposed for choosing between alternative strategies where risk is a concern. Perhaps the simplest way to do this is akin to the advice a stock broker might offer a client, simply presenting the estimated mean and variance of outcomes associated with each alternative strategy and leaving the choice to the client. In this regard, techniques have been developed to at least narrow the range of alternatives to those most likely to conform to the client's preferences. Stochastic dominance techniques identify quasioptimal strategies that account for risk in a general way (Mjelde et al. 1990). Applications of stochastic dominance methods have been explored by Cochrane and Mjelde (1989) and Harris and Mapp (1980).

A study by Martin and VanBrocklin (1989) cited earlier (see above; "Optimization for Multiple Fields and Crops") illustrates incorporation of risk and uncertainty into irrigation planning. They considered two alternative objectives: (i) to maximize expected net income over a five-year planning period; or (ii) to maximize the minimum return over a five-year period. The second of these is a classical risk-avoidance strategy. These alternative objectives led to substantially different irrigation strategies, both in terms of allocations of land and water within a season and intra-seasonal allocation of water.

As indicated, the analytical techniques proposed for decision analysis under uncertainty are quite diverse. The reader is referred to Mjelde et al. (1990) for a comprehensive review of the subject. The spectrum of irrigation strategies may also extend to defensive strategies to mitigate risk, for example, buying insurance and accepting a lower guaranteed profit in exchange for guaranteed avoidance of serious loss. The irrigation manager might also use a portfolio approach, pursuing several different irrigation strategies as a way of reducing variance without substantially reducing expected net income (English and Orlob 1978).

Optimization in Practice

It was observed more than 20 years ago that considerable research had been done on irrigation optimization on a theoretical level, but little concentrated thought had been given to the question of optimization at the operational level (English and Orlob 1978). That is still true today; the principles of optimization are not being systematically applied under field conditions. When water is limited farmers have often developed intuitive irrigation strategies to maximize net returns to water (BPA 1987), but they have done so without the benefit of rigorous economic, engineering, and scientific analyses. Limited and imperfect rules of thumb have been adopted by some in the irrigation industry to improve the economic efficiency of irrigation. But while these may be based on economics in a very general way, they have been limited in scope and too inflexible to deal with the varied circumstances of working farms, and may therefore lead to irrigation strategies that are suboptimal. Keller and Bliesner (1990) suggest under-irrigating by 20% when water supplies are limited. This recommendation may be appropriate as a first approximation, but it does not explicitly account for economic factors and may therefore miss the optimum by a wide margin. For example, English and Raja (1996) estimated that optimal levels of irrigation in water-limited situations would be between 30% and 50% less than full irrigation for three very different sets of circumstances (wheat in the Columbia Basin, cotton in California's San Joaquin Valley, and maize in sub-Saharan Africa).

Irrigation optimization has been a subject of research for at least four decades, but to our knowledge no rigorous and systematic optimization procedures are being used in production agriculture today. A literature search to verify this point turned up many theory-based articles but no examples of systematic optimization under field conditions. An informal survey of 42 key irrigation professionals was also conducted by one of the authors, inquiring whether they were aware of any systematic applications of the principles of optimization on production farms. Persons contacted included state extension specialists, NRCS and ARS personnel, managers of regional irrigation advisory services (CIMIS and Agrimet) and international research leaders in irrigation management. The response was consistent: none could cite any examples of rigorous, science-based optimization strategies anywhere in the world, though several described intuitive strategies devised by individual farmers.

Summary

During coming decades, the fundamental objective of irrigation is expected to shift from the physiological objective of maximizing crop yields per unit of land to a new economic objective, maximizing net returns to irrigation. In simplest terms, this might be referred to as an "optimization" paradigm. Optimization will generally imply reduced irrigation depths and reduced yields per unit of irrigated land. But as operating costs are reduced and water is freed for other productive uses, farm profits will be increased. Additional water may become available to other off-farm interests in those situations where return flows from irrigation would otherwise be irretrievably lost. Even where irrigation return flows are almost fully recaptured, optimum management can still be expected to benefit farmers and other water users through reduced production costs, improved distribution of water, and reduced environmental impacts.

From a societal perspective, optimum irrigation may be defined more broadly as maximization of overall benefits, including such nonmonetary benefits as water quality protection, food security, increased employment, and resettlement of populations.

The optimization approach will be more challenging than conventional irrigation practice of today. Irrigation planning will need to incorporate crop-water production functions and detailed cost functions that are not normally considered in irrigation planning or scheduling today. Salinity will often be a significant complicating factor. In the general case, analysts will need to deal with multiple objectives and a wide spectrum of alternative strategies, and may need to account for uncertainty and the possibility of greater financial risk. Such complex analyses will call for more sophisticated physical models and will depend upon analytical tools from the domain of Operations Research.

Acknowledgments

The writers gratefully acknowledge the important contributions of John Guitjens, Terry Howell, and William Ritter whose critical reviews and suggestions materially improved this paper.

Notation

The following symbols are used in this paper:

A = total irrigated area;
$c(w)$ = costs of production per unit area as function of depth of applied water;
$E(I)$ = expected value of net income per unit area;
$f(i)$ = probability density function of net income per unit area;
$i(w)$ = net income as function of depth of applied water;
P_c = crop price per unit weight of product;
W_T = total volume of available water;
W_R = relative depth of applied water, as proportion of yield-maximizing depth;
w = depth of applied water;
Y_{total} = total crop production from area under irrigation (A); and
$y(w)$ = crop yield as function of depth of applied water.

References

Aboitiz-Uriarte, M. (1983). "Stochastic soil moisture estimation and forecasting for irrigated fields." MS thesis, Dept. of Civil Engineering, Colorado State Univ., Fort Collins, Colo.

BPA. (1987). "Partial irrigation feasibility study and demonstration project." *Bonneville Power Administration, Portland, Ore.*, by Northwest Economic Associates, Vancouver, Wash.

Buccola, S. T. (1976). "Portfolio evaluation of long-term marketing contracts for U.S. farmer cooperatives." PhD dissertation, Dept. of Agricultural Economics, Univ. of California, Davis, Davis, Calif.

Campagna, P. M., Faux, R. N., and English, M. J. (1999). "Estimating ground water quality as a function of surface irrigation practices." *Proc., Symp. on Irrigation Efficiency, Int. Water Resources Engineering Conf.*, ASCE, Reston, Va.

Canpolat, N. (1997). "Optimization of seasonal irrigation scheduling by Genetic Algorithms." Doctoral dissertation, Dept. of Bioresource Engineering, Oregon State Univ., Corvallis, Ore.

Cochran, M. J., and Mjelde, J. W. (1989). "Estimating the value of information with stochastic dominance. An application from agricultural crop management." *Annual Meeting of the Southwest Region of the Decision Sciences Institute*, Charleston, S.C.

Connor, J. D., and Perry, G. M. (1999). "Analyzing the potential for water quality externalities as the result of market water transfers." *Water Resour. Res.*, 35(9), 2833–2839.

Doorenbos, J., and Kassam, A. H. (1979). "Yield response to water." *FAO Irrigation and Drainage Paper No. 33*. Food and Agriculture Organization of the United Nations, Rome.

Doorenbos, J., and Pruitt, W. H. (1992). "Crop water requirements." *FAO Irrigation and Drainage Paper No. 24*. Food and Agriculture Organization of the United Nations, Rome.

El-Hafez, Sayed Ahmed Abd. (2002). "Irrigation advisory services in Egypt." *Expert consultation on irrigation advisory services in the near east region; Hammamet, Tunisia, 13–16 May, 2002*, Food and Agriculture Organization of the United Nations, Rome, in press.

English, M. J. (1981). "The uncertainty of crop models in irrigation optimization." *Trans. ASAE*, 20(4), 917–928.

English, M. J. (1990). "Deficit irrigation: I. analytical framework." *J. Irrig. Drain. Eng.*, 116(3), 399–412.

English, M. J. (1999). "Determining optimum application efficiencies." *Proc., Symp. on Irrigation Efficiency, International Water Resources Engineering Conf.*, ASCE, Reston, Va.

English, M. J., Glenn, M. J., and VanSickle, J. (1981). "Irrigation scheduling for optimum water use." *Proc., ASAE Symposium on Irrigation Scheduling for Water and Energy Conservation in the 80s*, Chicago, December 14–15, 61–72.

English, M. J., and Nuss, G. S. (1982). "Designing for deficit irrigation." *J. Irrig. Drain. Div.*, 108(2), 91–106.

English, M. J., and Orlob, G. T. (1978). "Decision theory applications and irrigation optimization." *Contribution No. 174*, California Water Resources Center.

English, M. J., and Raja, S. N. (1996). "Perspectives on deficit irrigation." *Agric. Water Manage.*, 32, 1–14.

English, M. J., Taylor, A. R., and Abdelli, S. (1992). "A sprinkler efficiency model." *ICID Bull.*, 41(2), 153–162.

English, M. J., Taylor, A., and John, P. (1985). "Evaluating sprinkler system performance." *New Zealand Agric. Sci.*, 19, 32–38.

Frazier, M. W., Whittlesey, N. K., and English, M. J. (1999). "Economic impacts of irrigation application uniformity in controlling nitrate leaching." *Proc., Symp. on Irr. Efficiency, Int. Water Resources Engineering Conf.*, Seattle.

Frederiksen, H. D. (1997). "Are current water management policies adequate to the task?" *Proc., Theme A, International Association for Hydraulic Research (IAHR) 27th Congress*, San Francisco, August 10–15, M. English and A. Zsollosi-Nagy, eds.

Gear, R. D., Dransfield, A. S., and Campbell, M. (1976). "Effects of irrigation scheduling and coordinated delivery on irrigation and drainage systems." *ASCE National Resources and Ocean Engineering Convention*, San Diego, April 5–8.

Haise, W. R., and Hagan, R. M. (1967). "Soil, plant and evaporative measurements as criteria for scheduling irrigations." *Irrigation of agricultural lands: Agronomy 11*, American Society of Agronomy, Madison, Wis., 577.

Hargreaves, G. H., and Samani, Z. A. (1984). "Economic consideration of deficit irrigation." *J. Irrig. Drain. Div.*, 110(4), 343–358.

Harris, T. R., and Mapp, Jr., H. P. (1980). "A control theory to optimal irrigation scheduling in the Oklahoma Panhandle." *South. J. Agric. Econ.*, 12, 165–171.

Hart, W. E., and Reynolds, W. N. (1965). "Analytical design of sprinkler systems." *Trans. ASAE*, 8(1), 83–85,89.

Hillier, F. S., and Lieberman, G. J. (1980). *Introduction to operations research*, Holland–Day, San Francisco.

Hoffman, G. J., and van Genuchten, M. H. (1983). "Soil properties and efficient water use: Management for salinity control." *Limitations to efficient water use in crop production*, Am. Soc. Agronomy, Madison, Wis., 73–85.

Johnson, S. L., Adams, R. M., and Perry, G. M. (1991). "Assessing the on-farm costs of reducing ground water pollution: A modeling approach and case study." *Am. J. Agric. Econom.*, 73, 1063–1073.

Jones, C. A., and Kiniry, J. R. (1986). *Ceres-maize: A simulation model of maize growth and development*, Texas A&M University Press, College Station, Tex.

Keller, J., and Bliesner, R. D. (1990). *Sprinkle and trickle irrigation*, Chapman–Hall, New York.

Kelly, S., and Ayer, H. (1982). "Water conservation alternatives for California: A micro-economic analysis." *ERS Staff Report No. AGES820417*, Natural Resource Economics Division, Economics Research Service, USDA, Washington, D.C.

Kirda, C., and Kanber, R. (1999). "Water, no longer a plentiful resource, should be used sparingly in irrigated agriculture." *Developments in plant and soil sciences, Vol. 84: Crop yield response to deficit irrigation*, Kirda, Mouteonnet, Hera and Nielsen, eds., Kluwer, London, 1–20.

Knapp, K. C., and Dinar, A. (1987). "Optimum irrigation under saline conditions." *Irrigation Systems for the 21st Century; Proc., Spec. Conf. of the Irrigation and Drainage Division*, ASCE, Portland Ore., July.

Letey, J. (1991). *Crop water production functions and the problems of drainage and salinity*, Kluwer, Norwell, Mass.

Lin, W. R. (1973). "Decisions under uncertainty: An empirical application and test of decision theory in agriculture." PhD dissertation, Dept. of Agricultural Economics, Univ. of California, Davis, Davis, Calif.

Maji, C. C., and Heady, E. O. (1978). "Intertemporal allocation of irrigation water in the Mayarashi Project (India): An application of Chance-constrained linear programming." *Water Resour. Res.*, 14(2), 190–196.

Martin, D., and vanBrocklin, J. (1989). "Operating rules for deficit irrigation management." *Trans. ASAE*, 32(4).

Matanga, G. B., and Marino, M. A. (1977). "Application of optimization and simulation techniques to irrigation management." *Water science and engineering papers*, No. 5003, Dept. of Land, Air and Water Resource, Univ. of California, Davis, Davis, Calif.

Miller, G. D., and Anderson, J. C. (1993). "Farmers' incentives to reduce ground water nitrates." *Management of Irrigation and Drainage Systems; Integrated Perspectives. Specialty Conference*, ASCE-Irrigation and Drainage Division. Park City, Utah, 707–714.

Mjelde, J. W., Lacewell, R. D., Talpaz, H., and Taylor, C. R. (1990). "Economics of irrigation management." *Management of farm irrigation systems*; ASAE, St. Joseph, Mich. Monograph No. 9, 461–493.

NRCS. (1993). "Irrigation water requirements." *National engineering handbook*, Natural Resources Conservation Service, USDA, part 623, Chap. 2.

Perry, C. J. (1999). "The IWMI water resources paradigm—definitions and implications." *Agric. Water Manage.*, 40, 45–50.

Postel, S. L., Daily, G. C., and Erlich, P. R. (1996). "Human appropriation of renewable fresh water." *Science*, 271(9), 785–788.

Sekler, D. (1999). "Revisiting the IWMI paradigm: Increasing the efficiency and productivity of water use." *IWMI Water Brief2*. International Water Management Institute, Colombo, Sri Lanka.

Shearer, M. N. (1978). "Comparative efficiency of irrigation systems." *Proc., Annual Technical Conference*, The Irrigation Association, Arlington, Va., 1883–188.

Simmons, D. M. (1975). *Nonlinear programming for operations research*, Prentice–Hall, Englewood Cliffs, N.J.

Solomon, K. H. (1985). "Typical crop water production functions." *Paper No. 85-2596*, ASAE Winter Meeting, Chicago, December 17–20, American Society of Agricultural Engineers. St. Joseph, Mo.

Stewart, J. I., Hagan, R. M., and Pruitt, W. O. (1974). "Functions to predict optimal irrigation programs." *J. Irrig. Drain. Div.*, 100(2), 179–199.

Trimmer, W. L. (1990). "Applying partial irrigation in Pakistan." *J. Irrig.*

Drain. Div.*, 116(3), 342–353.

Vaux, H. J., and Pruitt, W. O. (1983). "Crop water production functions." *Advances in Irrigation*, D. Hillel, ed., Academic, New York, Vol. 2.

Warrick, A. W., and Yates, S. R. (1987). "Crop yield as influenced by irrigation uniformity." *Advances in Irrigation*, D. Hillel, ed., Academic, New York, Vol. 4.

Wichelns, D. (2001). "An economic perspective on the potential gains from improvements in irrigation water management." *Agric. Water Manage.*, 52, 233–248.

American Society of
Civil Engineers

150th Anniversary
Paper

Building a Better World

1852 – 2002

Water Quality Aspects of Irrigation and Drainage: Past History and Future Challenges for Civil Engineers

K. K. Tanji, M.ASCE,[1] and C. G. Keyes Jr., F.ASCE[2]

Abstract: This paper presents a brief history of irrigation and drainage related to ASCE activities on its Jubilee. This paper discusses legislation and policies that affect irrigation and drainage practices, water quality constituents of increasing concern in irrigation and drainage practices, and presents a prognosis on the future of declining freshwater resources available for irrigated agriculture and growing water quality problems in irrigation and drainage. Civil engineers in ASCE's Irrigation and Drainage Division have compiled an 80-year history of highly meritorious service and accomplishments. In the next millennium, civil engineers will face a formidable challenge in managing and protecting the precious freshwater resources in the U.S.

DOI: 10.1061/(ASCE)0733-9437(2002)128:6(332)

CE Database keywords: Irrigation; Drainage; History; Water resources management.

Introduction

The American Society of Civil Engineers (initially chartered as the American Society of Civil Engineers and Architects) was founded October 23, 1852. Prior to and since that time, civil engineers have contributed significantly toward the development of irrigation and drainage in the United States. For example, in 1763 George Washington surveyed the Dismal Swamp area in Virginia and North Carolina for reclamation that led in 1778 to the charter of the Dismal Swamp Canal Company (Ritter and Shirmohammadi 2001). In 2001 ASCE selected Hoover Dam on the Colorado River, designed by civil engineers and built in the 1930s, as one of the Civil Engineering Monuments of the millennium. Hoover Dam, at the time of construction, was the highest dam, the costliest water project, and contained the largest power plant. Lake Mead, created by Hoover Dam, serves as the water supply for about 18 million people and over 0.6 million ha of irrigated land in the U.S. and an additional 0.2 million ha in Mexico (ASCE News 2001).

This paper presents a brief history of irrigation and drainage related to ASCE activities on its Jubilee. This paper discusses legislation and policies that affect irrigation and drainage practices, water quality constituents of increasing concern in irrigation

and drainage practices, and presents a prognosis on the future of declining freshwater resources available for irrigated agriculture and growing water quality problems in irrigation and drainage. Table 1 presents a chronology of the history of irrigation and drainage and related milestone events.

Brief History of Irrigation and Drainage in the U.S.

Drainage and Irrigation in the East

In the humid eastern United States, drainage was practiced in Colonial America in the 1600s and 1700s to reclaim swamplands (now called wetlands) for agriculture and other land use (Beauchamp 1987). The European settlers installed drainage networks that consisted of open drains and buried subsurface drains utilizing drainage experience from their homelands in England, France, and The Netherlands. The materials used for buried drains before the development of clay tile pipes ranged from tree logs, limbs and brush to stones, bricks, and straw (Ritter and Shirmohammadi 2001). Supplemental irrigation practices in the humid East including water table control for irrigation is a comparatively recent development that started in the early 1950s (Table 1).

Irrigation and Drainage in the West

In the semi-arid West, irrigation was initially practiced about 300 BCE by the Hohokam Indians along the Salt River in Arizona (Tanji 1990) and in the 1700s by the Spanish missionaries. One of the oldest preserved irrigation systems is the Mission Espada Aqueduct along the San Antonio River in Texas. This aqueduct system consisted of a dam built in 1740 with an acequia madre (mother ditch) that was 25 km in length to irrigate about 1,420 ha of crop land (brochure on San Antonio Missions, National Park Service). Modern large-scale irrigation in the West began with the Mormon settlement in the Great Salt Lake Basin in 1847 (NRC 1996), with the Miller–Lux Canal in the San Joaquin Valley of California in 1860 (Tanji et al. 1986), and the passage of the 1902

[1]Professor Emeritus, Dept. of Land, Air and Water Resources–Hydrology, Univ. of California, Davis, CA 95616. E-mail: kktanjii@ucdavis.edu

[2]F.NSPE, Past President, Environmental and Water Resources Institute, ASCE; formerly, Principal Engineer, U.S. International Boundary and Water Commission; and Professor Emeritus and Dept. Head, New Mexico State Univ., Las Cruces, NM 88003.

Note. Discussion open until May 1, 2003. Separate discussions must be submitted for individual papers. To extend the closing date by one month, a written request must be filed with the ASCE Managing Editor. The manuscript for this paper was submitted for review and possible publication on December 20, 2001; approved on July 8, 2002. This paper is part of the *Journal of Irrigation and Drainage Engineering*, Vol. 128, No. 6, December 1, 2002. ©ASCE, ISSN 0733-9437/2002/6-332–340/$8.00+$.50 per page.

101

Table 1. Selected History and Milestones Related to Irrigation and Drainage in the U.S.

Year	Activity
	Before the formation of ASCE
300 BCE	Hohokam Indians practiced irrigated agriculture along the Salt River in Arizona
1600s	Settlements in Colonial America reclaim swamplands with drainage systems
1700s	Spanish missionaries in the West develop acequias to irrigate crop lands
1803–1848	U.S. greatly expands with the Louisiana Purchase (1803), Florida and eastern Louisiana ceded by Spain (1819), annexation of Texas (1845), the Oregon Compromise in the Pacific Northwest (1846), and Western lands ceded by Mexico (1848)
1849, 1850, 1860	Swamp Land Acts transfer federal lands to states for potential land reclamation
1852	American Society of Civil Engineers and Architects formed by 12 founding members, later renamed American Society of Civil Engineers
1862	U.S. Department of Agriculture established
1862	Mortill Land Grant Act passed to form land grant universities and agricultural experimental stations in each state
1890	Professor Hilgard recommends drainage of San Joaquin Valley land in California for excess water and salts
1902	Congress passes Reclamation Act of 1902 for water development projects in the arid West
1919	Congress passes Drainage District Improvement Act to provide drainage in local districts
1922	Irrigation Engineering (IR) Division authorized for ASCE
1935	Soil Conservation Service (predecessor of NRCS) established by Public Law 46
1948	Wilcox recommends irrigation water quality classification scheme
1950	ASCE Committee on Drainage of Irrigated Lands formed
1953	U.S. Committee on Irrigation and Drainage (of ICID) invited to ASCE and Irrigation Engineering (IR) Division became Irrigation and Drainage (I&D) Division
1953	ASCE Committee on Irrigation and Drainage in Humid Lands formed
1954	U.S. Salinity Laboratory publishes Agricultural Handbook No. 60
1962	ASCE Committee on Reuse of Wastewater formed at IR Division
1965	Water Quality Act of 1965 establishes Federal Water Pollution Control Administration
1972	ASCE Technical Committee on Water Quality formed and active to date
1972	ASCE Technical Committee on Drainage formed and active to date
1972	Federal Water Pollution Control Act amended by Public Law 92-500, renamed Clean Water Act in 1977
1974	Congress passes Colorado River Basin Salinity Control Act
1976	FAO publishes Water Quality Guidelines for Agriculture, revised edition in 1985

Table 1. (*Continued*)

Year	Activity
1982	Drainwater selenium found toxic to waterbirds in Kesterson Reservoir, Calif.
1983	Chesapeake Bay Program initiated to restore water quality and wildlife habitats
1985	NRC appoints scientific advisory committee on Irrigation-Induced Water Quality Problems for Kesterson and later DOI's National Irrigation Water Quality Program (NIWQP) in the West
1986	USGS launches 10-yr National Water Quality Assessment Program (NAWQA)
1990	ASCE produces Manual No. 71, Agricultural Salinity Assessment and Management
1994	CALFED Bay–Delta Program initiated to fix quality and quantity problems in Calif.
1995	Irrigation and Drainage Division and Hydraulics Division amalgamated into Water Resources Engineering Division
1999	Water Resources Engineering Division amalgamated into Environmental and Water Research Institute (EWRI)
1999	EPA requests states to implement TMDL on water quality-impaired water bodies
2000	Water Quality and Drainage Technical Committees combined by the Irrigation and Drainage Council of EWRI
2002	ASCE celebrates 150 years of service

Reclamation/Newlands Act. The Reclamation Act established the U.S. Reclamation Service under the egis of the U.S. Geological Service (USGS) to develop water resources for irrigation of federal lands in the western states. The first project was the Newlands Reclamation Project in what is now the Truckee–Carson Irrigation District in Nevada. In 1907 the U.S. Reclamation Service was separated from the USGS, with about 30 ongoing projects in progress, and in 1923 was renamed the U.S. Bureau of Reclamation (USBR) (USBR 2000). In the western United States drainage is designed to remove excess salts in addition to excess water while in the eastern United States drainage is to remove ponded surface water and excess soil water (Skaggs and van Schilfgaarde 1999).

Expansion of Croplands and Drainage

The original states in the Union prior to 1800 were located along the East Coast, with the exclusion of Florida, to east of the Mississippi River. From 1803 to 1848, the U.S. acquired much of what is now the conterminous U.S. (Table 1). The 1850 federal census records show that there were about 45.6 million ha of improved crop land, mainly in the eastern U.S. and in the Ohio and Mississippi Valleys. In 1997 there were 71.2 million ha of farmland of which 20.2 million ha was irrigated, mainly in the West (Fig. 1). In 1950 about 41.6 million ha of land had been drained in the U.S., much of it located in the Ohio and Mississippi River Basins and along coastal areas (Fig. 2, a more recent updated map was not available). Water development for irrigation was a driving force in the settlement of the American West (NRC 1996).

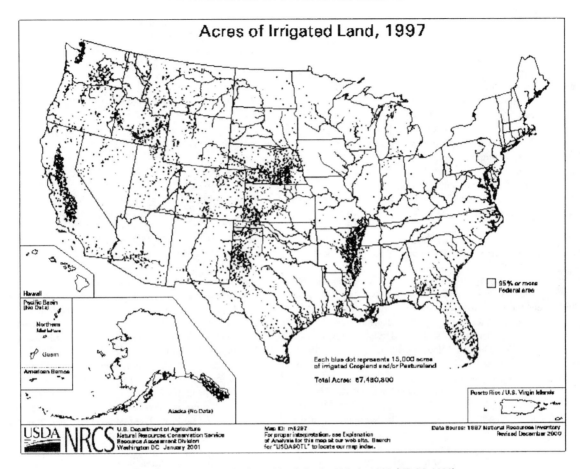

Fig. 1. Distribution of irrigated lands in the U.S. in 1997 (NRCS 1998)

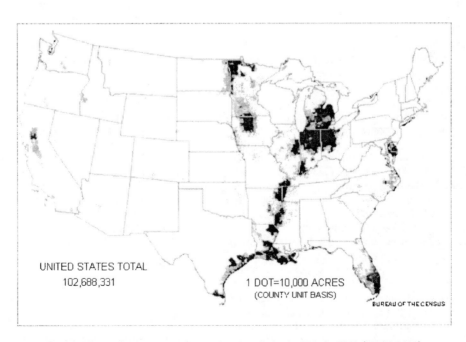

Fig. 2. Distribution of drained agricultural lands in the U.S. in 1950 (USGS 1998)

Inventions

A number of industrial inventions greatly accelerated the development of irrigation and drainage. The invention of the steam shovel and dredge in 1795 made it easier to construct drains for waterlogged lands. The invention of the internal combustion engine in 1826 led to tillage, harvesting, and construction equipment including wheeled tractors and caterpillar type tractors. Clay tile pipes for use in subsurface drainage were first manufactured in Illinois by 1838. The invention of the turbine pump allowed farmers in the 1920s to pump groundwater and to utilize sprinkler irrigation. The invention of plastics in the 1940s led to development of corrugated-wall polyethylene plastic tubing in 1967 for subsurface drainage and smooth-plastic tubing for drip irrigation lines and plastic drip emitters.

Role of the Irrigation and Drainage Division in ASCE

The Irrigation Engineering (IR) Division of ASCE was authorized on June 20, 1922, with 23 members and a 5-member Executive Committee elected by members of the Division (C. G. Keyes Jr., ASCE internal memorandum 1991). The constitution of this Division was fully established by 1925. The Division grew to 645 members and affiliates by 1927. Participants in this Division included engineers from federal and state water-related governmental agencies as well as others from irrigation, drainage and reclamation districts, universities, and consulting firms. The IR Division was renamed the Irrigation and Drainage (I&D) Division in 1953 (Table 1). The early action committees organized in the Division included committees on National Reclamation Policy, Cooperation between Federal and State Agencies, Drainage of Irrigated Lands, Duty of Water, and Interstate Water Rights (C. G. Keyes, Jr., ASCE internal memorandum 1991). Over time the Division's activities extended to water conservation, the ability of owners of irrigated lands to pay irrigation costs, sedimentation in reservoirs, irrigation conveyance losses, drainage of irrigated lands, irrigation and drainage practices in humid areas, drought, reuse of water for irrigation, on-farm irrigation, and water quality. Many of these topics are still relevant today. The Technical Committee on Water Quality and the Technical Committee on Drainage (Table 1) have both been continuously active since 1972 sponsoring numerous specific Task Committees. The water-quality related Task Committees included water quality for irrigation, erosion and sediment control, Colorado River salinity control and management, agricultural evaporation ponds, managing trace elements, agricultural nonpoint source water quality models, and guidelines for wetland restoration of converted crop lands. One of the principal accomplishments of the Task Committee system was the 1990 publication of Manual No. 71, "Agricultural Salinity Assessment and Management" (Tanji 1990). Other equally significant IR and I&D contributions not mentioned herein relate to on-farm irrigation management, crop evapotranspiration, and hydraulics of surface irrigation and drainage. The Irrigation and Drainage Division has had an 80-year history of meritorious service and accomplishment to ASCE and the world.

The 1995 ASCE Strategic Plan called for the creation of semi-autonomous institutes for groups of engineering specialties within ASCE. The Irrigation and Drainage Division and the Hydraulics Division were amalgamated into the Water Resources Engineering Division in 1995. In 1999 the Environmental and Water Resources Institute (EWRI) was formed from the former Technical Divisions of Environmental (Sanitary) Engineering, Water Resources, and Water Resources Planning and Management as was

the Water and Environmental Standards Council, with a membership of about 20,000 (Table 1). The present Irrigation and Drainage Council of EWRI of ASCE oversees the activities of technical committees such as the Water Quality and Drainage Committee.

Legislation and Policies Affecting Irrigation and Drainage

Early Legislation

For nearly two centuries, swamp and overflow land in the U.S. was considered to be wasteland and not of high value unless reclaimed for agriculture or other land use. In the mid-1800s, the Swamp Land Acts (Table 1) was enacted to convert about 5.2 million ha of federally owned swamp lands into state-owned land for potential agricultural development (USGS 1998). It was the intent of these Swamp Land Acts for the states to reclaim them but most states did not reclaim these lands immediately. A Drainage District Improvement Act was passed in 1919 (Table 1) to provide drainage for storm water overflow in swamp and overflow areas. This Act resulted in the formation of local drainage districts within counties and cities. Land was assessed and general obligation bonds were issued to construct and install drain outlets, open drains, conduits, and pipelines.

In 1862 the U.S. Department of Agriculture (USDA) was formed in a nation-wide effort to increase crop and animal production. Also in 1862 the Morrill Land Grant Act was passed to form land grant universities and agricultural experiment stations in each state (Table 1). The first academic department of irrigation was established in 1905 on what is now the Davis campus of the University of California.

The Soil Conservation Service (SCS), the predecessor of the Natural Resources Conservation Service (NRCS), was created in 1935 by Public Law 46 (Table 1) when Congress declared that "soil erosion was a menace to the national welfare" (NRCS 2002). The SCS, in turn, was preceded by the Soil Erosion Service which was formed in 1933 through the efforts of Hugh Hammon Bennett. He was a soil scientist at the USDA and based on his seminal studies on soil erosion in the 1920s and 1930s he crusaded for governmental assistance to landowners to adopt conservation practices to prevent soil erosion utilizing demonstration projects. The Soil Conservation Act of 1935 (Public Law 46) established SCS offices in each state as well as resource conservation districts within the states. The SCS expanded their responsibilities from on-farm conservation practices to reduce soil erosion to interaction of soil and water on a watershed scale. SCS became increasingly involved in drainage with the passage of the Flood Control Acts of 1936 and 1944 as well as the Watershed Protection and Flood Prevention Act of 1954 (NRCS 2002). The SCS and now the NRCS have expanded their activities to water quality aspects with the Clean Water Act of 1972, which will be discussed next. The NRCS works directly with landowners to help them apply new or improved irrigation, drainage, and water management technologies.

Water Quality Acts

The history of water quality management in the U.S. began with the 1899 Rivers and Harbors Act that made it unlawful to discharge any refuse into a navigable waterway unless a permit was obtained from the U.S. Army Corps of Engineers. Federal management of water quality was slow to develop although some streams were known to carry excessive sediment loads and sedimentation in reservoirs was a problem; the waters of the Rio Grande and Colorado Rivers, to name but a couple, became salinized with downstream travel from natural and agricultural drainage. The Water Quality Act of 1965 established the Federal Water Pollution Control Administration (FWPCA) and called for creation of water quality standards for all interstate and coastal waters (Table 1). The 1972 Federal Water Pollution Control Act's objective was to restore and maintain the integrity of the nation's water. Point sources of pollution required NPDES permits prior to discharge and had to meet water quality standards of the receiving water body. Nonpoint sources such as runoff from agricultural and forested lands were excluded from being required to have NPDES permits. One of the objectives was to achieve "fishable and swimable" river water by 1983 but that goal was not achieved. The 1972 Federal Water Pollution Control Act was amended in 1977 and became the Clean Water Act that was further amended in 1987. In 1999 the U.S. Environmental Protection Agency (EPA) proposed regulations to clarify and strengthen the total maximum daily load (TMDL) program under Sec. 303(d) of the Clean Water Act (Table 1). Each states' TMDL process identifies bodies of water in which water quality standards are not being attained and then establishes appropriate waste load allocations or TMDLs for all sources in the watershed, including point and nonpoint sources. Pollution prevention and maintaining the integrity of the nation's bodies of water still remains a formidable engineering challenge.

Colorado River Basin

The Colorado River Basin Salinity Control Project was passed in 1974 (Table 1). Prior to irrigation development, the headwaters of the Colorado River system contained about 50 ppm total dissolved solids (TDS) and after 2,300 km downstream travel into Mexico the salinity was about 400 ppm TDS. But with the construction of over 20 reservoirs and associated irrigation projects and irrigation return flow from these irrigation projects, the salinity in the Colorado River as it entered Mexico rose to as high as 1,200 ppm TDS in the 1960s. Mexico, which had a 1944 international treaty to annually receive about 1.85 billion cubic meters of Colorado River water, expressed concern that crop production in the Mexicali Valley was suffering from an elevated amount of salinity. This treaty made no mention of water quality, but after lengthy negotiations from 1961 to 1973, the 1944 treaty was amended so that water to Mexico at Morales Dam did not exceed 115 ppm more than the salinity at Imperial Dam, the last irrigation diversion from the River in the U.S. To achieve agreement, the USBR and SCS [now the Natural Resources Conservation Service (NRCS)] launched on-farm salinity control measures in the upper basin states. Deep percolation through the Mancos shale formation was identified as one of the principal sources of salinity and on-farm water conservation and canal linings were implemented. The USBR constructed the Yuma Desalting Plant to treat saline drainwater from the Wellton–Mohawk Irrigation and drainage District. Today, the salinity of the Colorado River at Imperial Dam averages about 850 ppm TDS and the Colorado River Water entering Mexico is slightly greater than 850 ppm TDS.

Selenium in the San Joaquin Valley and the West

As part of the Central Valley Project in California, the USBR constructed the San Luis Drain to convey drainage effluents from a 17,000 ha tile-drained site in the Westlands Water District, near Fresno. The saline drainwater was to be temporarily disposed of in the Kesterson Ponds, located in the Kesterson National Wildlife Refuge, because construction of the drain outlet to the Sacramento–San Joaquin Delta was halted due to environmental concerns. After a mere 6 years of operation, selenium in the drainwater was implicated as the toxin that was causing bird deformity and reduced reproduction rates and fish kills in the Kesterson Ponds (Table 1) (Tanji et al. 1986). The gruesomely deformed embryos of waterbirds made national and international news. The Kesterson Ponds were shut down in 1986 and the subsurface drains were plugged in the Westlands Water District. As a result, the Westlands Water District does not even have a surface drainage outlet to dispose of its irrigation return flow. A $50 million, 6 year investigation was carried out in 1985 by the joint federal/ state San Joaquin Valley Drainage Program (SJVDP) (SJVDP 1990) to recommend alternative management options to the drainage, salinity, and selenium problems in about 0.9 million ha of irrigated lands on the west side of the San Joaquin Valley. The new irrigation and drainage management options proposed were drainwater treatment, drainwater reuse, and land retirement since implementation of improved irrigation water management will still produced some residual subsurface drainwater. A committee of the National Research Council that provided scientific oversight to the drainage investigation (NRC 1989) concluded that drainage water quality problems are inevitable wherever irrigation is practiced. This NRC committee stated that there was a need not only for technical solutions but also for the development of policies to minimize and ameliorate drainage problems. A second NRC committee was appointed to help in DOI's National Irrigation Water Quality Program (NIWQP) (Engberg and Sylvester 1993) to investigate and manage other sites in the West that have selenium contamination problems. A number of sites are currently undergoing remediation for selenium contamination.

Coastal Zones and Estuarines

The Coastal Zone Management Act of 1972 established a partnership between the federal and state government to protect and conserve natural resources and biological diversity along U.S. coastal zones. The coastal zone, including estuaries and nearshore marine areas, comprises about 17% of the land mass but contains over half of the U.S. population, 141 million in the 1998 census (NRC 2000). Eutrophication is a major problem in coastal zones and is caused by excess nutrients (nitrogen and phosphorus). The sources of nutrients are natural, such as flood runoffs, and anthropogenic, such as agricultural runoff and discharges from municipalities and industries. Some of these nutrient sources originate in watersheds far from the point of impact, such as hypoxia in the Gulf of Mexico, from nitrogen contaminated runoff in the Mississippi River Basin.

Two major estuarine preservation and restoration programs are noteworthy of mention. The Chesapeake Bay Program was initiated in 1983 to restore the living resources of this Bay (Table 1). This program is a cooperative effort between the U.S. EPA and the states of Maryland, Pennsylvania, and Virginia and the District of Columbia. Nonpoint sources of agricultural pollutants are a major source of water quality degradation in the Chesapeake Bay. Some progress in restorating the water quality and ecosystem is being achieved (Chesapeake Bay Program 2000), including a decrease in the runoff of contaminated agricultural runoff. The present CALFED (California/federal) Bay–Delta Program in California (Table 1) had its beginnings in the mid-1960s as the San Francisco Bay–Delta Program and was jointly conducted by state and federal agencies. This effort later became the CALFED

Program in 1994 (CALFED Bay–Delta Program 2001). The objectives of this program are similar to those of the Chesapeake Bay Program with the exception that waters in the delta are also being exported to provide drinking water to about 23 million Californians and to irrigate about 1.6 million ha of productive agricultural land. Water quality standards formulation and ecosystem restoration are two of the more important goals of the Bay–Delta Program. To date, CALFED has been involved mainly in planning and research issues. However, some significant accomplishments have transpired such as the Central Valley Project Improvement Act, which transferred about 1 billion cubic meters of water for the environment, and the purchase of several islands in the delta by the state for expansion of wetlands and possible freshwater storage. These actions have either reduced farmland or decreased the availability of water for agriculture. The latter has resulted in less than 100% water delivery to Westlands Water District.

Farm Bills

Several recent Food Securities Acts have been passed by Congress to promulgate conservation measures for natural resources. As part of the 1985 Farm Bill, Congress authorized the Conservation Reserve Program (CRP) under which landowners contract with the U.S. Department of Agriculture to convert crop land to perennial grass or tree cover and receive payments for 10 or 15 years U.S. (Congress 1985). These "sodbuster" and "swampbuster" provisions were intended to restore wetlands, prairies, and other natural habitats to reduce soil erosion and commodity surpluses. The Food Security Act of 1985 was amended in 1990 and 1996. The 1990 amendments included replacement of the 1985 CRP with a new Environmental Conservation Acreage Reserve Program that included various water quality and wildlife related amendments to emphasize environmental benefits. The 1996 amendments included a voluntary Wetland Reserve Program for landowners to receive financial incentives to enhance wetlands in exchange for retiring marginal agricultural lands.

NAWQA Program

In 1985, the U.S. Geological Survey embarked on a 10-year investigation of the National Water-Quality Assessment (NAWQA) Program (NRC 2001). The objectives of NAWQA (USGS 1999a) were to ascertain the current water quality conditions for a large part of the nation's freshwater streams, rivers, and aquifers, to describe how water quality is changing, and to obtain improved understanding of natural and human factors that affect water quality conditions (Table 1). Significant findings on nutrients and pesticides in the first 20 study units were reported (USGS 1999b).

Civil engineers from governmental agencies, universities, the agricultural industry, and consultants have been involved in all of the above water quality related legislation, programs, and activities. Many of these activities were featured in ASCE Specialty Conferences sponsored by I&D and published in the *Journal of Irrigation and Drainage Engineering*.

Agricultural Water Quality Constituents of Concern

The principal constituents of concern in irrigation and drainage include salinity, sodium, nitrates, trace elements of boron, selenium, and arsenic, pesticide residue, and sediment. A historical perspective on agricultural water quality to the present is offered.

Humid East Versus Arid West

In the humid East where drainage of crop lands was initiated during Colonial America, water quality of agricultural runoff was not of concern until recent times. Erosion and sediment production were of concern but they were attributed to rainfall runoff. In contrast, some of the crop lands in the arid West whether cultivated or irrigated were affected by salt even before they were irrigated. As early as 1890 Professor Hilgard (1890) of the University of California at Berkeley recommended subsurface drainage be required for the waterlogged, salt-affected lands in the San Joaquin Valley of California (Table 1). The 1902 Reclamation Act resulted in massive irrigation development. Some of areas contained salt-affected land, and much effort was directed toward land reclamation of saline and sodic soils (Table 1). Leaching and drainage were essential to maintain crop production. Saline irrigation drainwater was discharged into streams and groundwater basins with little or no constraint. As irrigation water was extensively diverted upstream and irrigation drainage was disposed of downstream, serious concerns about downstream water quality arose. It is of interest to note that for the San Joaquin River in California salinity and boron were constituents of concern to downstream water users prior to the 1950s, then nitrates became important in the 1960s, pesticide residue in the 1970s and again in the 2000s, selenium in the 1980–1990s, and arsenic in 2000. Similar concerns about downstream degradation in river water quality occurred elsewhere like in the Rio Grande, Colorado, and Pecos River systems.

Salinity

Salinity is a major constituent of water quality concerns in the arid West. The primary source of dissolved mineral salts comprised of Na, Ca, Mg, Cl, SO_4, and HCO_3 ions are derived from natural chemical weathering of earth materials (Hilgard 1907). The deep percolation of irrigation water moves salts into drains and shallow groundwater, and eventually into streams. In regions with low rainfall and high evaporation rates, salts introduced with irrigation water need to be leached out of the crop root zone to maintain crop production. The U.S. Salinity Laboratory (1954) introduced the term leaching requirement (LR), defined as the fraction of infiltrated water that needs to be drained out of the root zone to maintain a salt balance and prevent root zone salt accumulation. Salinity was frequently treated as a conservative parameter in the past but now is considered a reactive parameter that participates in mineral dissolution and precipitation, cation exchange, and ion association (Tanji 1990). Salinity in irrigation water as well as in drainage water continues to be a major water quality problem in the West.

Irrigation Water Quality Guidelines

Because some of the irrigation water used in the West was marginal in quality, water quality guidelines for irrigation came into being. Hilgard (1907) noted that dissolved mineral salts found in the ocean were also found in soil, albeit at much lower concentrations in most instances. He also noted high concentrations of nitrates of geologic origin in some San Joaquin Valley soil. Soluble boron was also found by Hilgard at low concentrations but appeared to be toxic to certain plants. Wilcox of the U.S. Salinity Laboratory was in 1948 one of the first to suggest a classification scheme for irrigation water quality that included salinity, sodicity, boron, and chloride (Table 1). Wilcox evaluated salinity in terms of electrical conductivity (EC) because EC makes soil water less available to plants due to its contribution to a solute (osmotic) potential. Sodium was evaluated in terms of the percent of sodium relative to the total cations (Na,Ca,Mg), recognizing that excess sodium results in poor soil permeability. The percent of sodium was later replaced by the sodium adsorption ratio (SAR) (USSL 1954). Boron and chloride were found to be important because excessive concentrations can cause toxicity to plants and were considered specific ion effects in contrast to relatively benign salinity. Other researchers proposed various classification schemes for irrigation water. Many disagreements were aired at a University of California Water Resources Center (WRC) specialty conference (WRC 1963). Although a generalized salinity classification was proposed, the LR was made crop-specific based upon crop salt tolerance data. The permeability impacts of sodium were found to be a combination of EC and SAR, i.e., low EC and moderate SAR caused permeability problems and a high SAR can be partially compensated for by high EC as far as soil permeability is concerned. The specific ion effects of boron, chloride, and sodium were also made crop-specific. In 1976 FAO published "Water Quality Guidelines for Agriculture" based on the findings of the U.S. Salinity Laboratory that were modified in 1974 by the University of California Committee of Consultants (Table 1). The FAO guideline has been adopted internationally and widely used; it was revised slightly in 1985 (Ayers and Westcot 1985).

Trace Elements

In addition to dissolved mineral salts, saline soils and geologic formations contain trace elements of boron, selenium, molybdenum, and arsenic. Kelley and Brown (1928) conclusively identified boron as a trace element that is toxic to crop plants. Selenium was known to be toxic to animals in the Rocky Mountain states in the 1930s (Rosenfeld and Beath 1962) and became a nationwide concern after the selenium poisoning of waterbirds in Kesterson Reservoir in California. Arsenic has been known to be a common trace element in ground water of Western states and the maximum contaminant level (MCL) of arsenic in drinking water was established at 50 ppb. Since the finding of extensive poisoning from arsenic in drinking water in Bangladesh in the 1990s (British Geological Survey 1999), the MCL for arsenic in the U.S. was re-evaluated and will be reduced to 10 ppb in early 2002 (EPA 2001). For irrigation water, boron and molybdenum appears to be the trace elements of most concern. Boron is toxic to many crop plants while accumulation of molybdenum in forage crops is detrimental to ruminants. The presence of boron, selenium, and molybdenum in irrigation drainage is of concern because of their impact on plants and aquatic biota.

Nitrates

Nitrates in groundwater underlying crop lands became a concern since some drinking water from wells far exceeded public health limits of 10-ppm nitrate/nitrogen. One of the earliest regional studies on nitrate pollution of groundwater was conducted in the late 1960s in the Upper Santa Ana River Basin in California (Ayers and Branson 1973). The sources of nitrates included excessive application of commercial nitrogen fertilizers in citrus orchards beginning in the 1930s, excessive leaching of manure-laden dairy wastewater beginning in the 1950s, and intensive land disposal of municipal sewage water as the urban population began to increase exponentially in the 1960s. The long travel time to deep ground bodies of water required assessment of land use and nitrate loading decades earlier. Extensive studies on nitrate pollution from crop lands in the Midwest began in the 1970s and later were expanded nationwide. The density of feedlots, dairies, and hog farms in certain locales make nitrate contamination of surface and groundwater difficult to avoid.

Pesticide Residue

Pest control using naturally occurring chemicals such as arsenic and nicotine had been practiced in agriculture for more than 1,500 years but since the discovery of DDT in 1939 synthetic organics have become the preferred pest control agents (NRC 1993). The presence of DDT residue in the environment and its ecotoxicological impact were brought to attention in 1962 by Rachel Carson's book *Silent Spring*. Today, dozens of synthetic insecticides, herbicides, and fungicides are used in agriculture including organochlorines, organophosphates, carbamates, and phenoxyacetic acids. The fate and transfer pathways of pesticides applied to crop lands are complex, and require knowledge of their chemical properties such as water solubility, volatility, and sorption to clays and organic matter, their transformations such as chemical hydrolysis and microbial degradation, and their physical transport in gaseous, liquid and solid form. The last mode of movement is typically as sediment-bound transport. Once a pesticide enters the soil, its fate is largely dependent upon sorption and its persistence (NRC 1993). Sorption is commonly evaluated by the use of a sorption coefficient based on the organic carbon content of the soil whereas persistence is evaluated in terms of half-life. Pesticides with low sorption coefficients are likely to leach in soil and accumulate in groundwater. Pesticides with long half-lives could be persistent in soil for decades. The pesticide residues most commonly found in groundwater include alachor, aldicarb, atrazine, carbofuran DBCP, DCPA, and simazine (NRC 1993). Those pesticides strongly sorbed to clay and organic matter may be subjected to removal by surface runoff and sediment, and they include DDT and its metabolites, dieldrin, endosulfan, lindane, heptachlor, and difocal. Other pesticides that have high water solubility and that are weakly sorbed may be lost from crop land in the dissolved state such as 2,4-D, dicamba, dinoseb, molinate, and other herbicides (NRC 1993). Intensive studies are underway to reduce the runoff and deep percolation of pesticide residue.

Sediment

Sediment is produced by erosion of exposed land surfaces by rain and sprinklers as well as by surface water runoff from rainfall and irrigation. The sedimentation process initially involves detachment or dislodging of soil particles by erosive agents and then transport or entrainment by water movement and finally deposition downstream when the water energy is insufficient to continue entrainment (NRC 1993). Research on soil erosion was probably started in the 1930s by Hugh Bennett of the Soil Conservation Service and ultimately led to the universal soil loss equation (USLE) in 1978. The USLE is based on rain-related factors (rainfall erosivity and soil erodibility), soil-related factors (slope length and slope gradient), and land management factors (cover and cultivation management as well as erosion control practices) (NRC 1993). The USLE has been updated and computerized in the 1990s into a revised universal soil loss equation (RUSLE). Excessive sediment concentrations in irrigation water may plug sprinkler jets and drip irrigation emitters. Excessive sediment concentrations in runoff water may suffocate aquatic biota and when the sediment is deposited may reduce the carrying capacity of streams, channels, and reservoirs. The sediment-bound nutrients (nitrogen and phosphorus) and pesticides in runoff water are of much concern to water quality due to their potential eutrophication and toxicity to aquatic life, respectively.

Water Quality Modeling

In the late 1950s mathematical and simulation models became a new tool for engineers. Some of the earliest computer models were of the hydrologic-water quality type applied to river basins (Tanji 1981). In the early models, salinity was treated as a conservative parameter but by the mid-1960s the reactivity of dissolved mineral salts was accounted for by such reactions as mineral solubility, cation exchange, and ion pairing. In the late 1960s and early 1970s, simulation models were extended to include nitrogen transformation and transport in crop lands (Tanji 1982). This development was followed in the 1980s by pesticide transformation and transport models (NRC 1993). Today, simulation models for constituents of concern in agricultural water are continuously being refined and updated (Ritter and Shirmohammadi 2001). As TMDL regulations are implemented, water quality simulation models will be increasingly utilized.

Prognosis on Water Quality Aspects of Irrigation and Drainage

The availability of developed freshwater resources in this nation is finite, and there is increasing competition for not only water quantity but also for water quality by various water users including agricultural, municipal, recreation, power generation, transportation, American Indian water rights, and the environment. The reader is referred to an in-depth analysis of "A New Era for Irrigation" (NRC 1996). The report states, "No one can say with any degree of certainty how irrigated agriculture will change in the near or far term. We can, however, assert with considerable confidence that it will change." This NRC report presents the following conclusions that have been paraphrased by the authors of this paper

1. Irrigation will continue to play an important role in food and fiber production but with some changes including a reduction in acreage.
2. The amount of water dedicated to irrigation will decline given the changing societal values and increasing economic competition for water.
3. The economic force driving irrigated agriculture will be increasingly determined by the ability of U.S. agriculture to compete in the global market.
4. Irrigated agriculture will continue to shift toward large, well financed, diversified farming operations.
5. Changes in federal, state, and local policies that affect irrigated agriculture are expected since the present policies were established in a different era and no longer meet today's societal needs.
6. The term irrigation is being broadened from crop irrigation to irrigation of urban landscaping and golf courses.
7. Advances in irrigation technology are necessary at the same time the federal government is reducing its support of irrigation development.
8. Some portion of the water used by irrigated agriculture will, over time, be shifted to satisfy environmental goals.
9. Irrigation emerged as an individual and collective effort at

the watershed level and its future will be determined at the local watershed level in concert with and accommodating other water users, especially those downstream.

Agriculture utilizes more than 70% of the water supply in the West. Recent competition for water has led to water transfer from agriculture to urban areas and for preservation/restoration of fish and wildlife, e.g., water transfer from the Imperial Irrigation District in southeastern California to the Metropolitan Water District in southwestern California (NRC 1996). Clearly, there will be less freshwater resources used by agriculture in the next millennium. Moreover, the discharge of poor quality agricultural drainage water is increasingly being constrained by regulations. This means agriculture will need to use water more conservatively than at present to reduce the volume of irrigation return flow. Agriculture will need to use and reuse water of marginal quality. Agriculture may face severe quality restrictions on the discharge of drain water that may eventually require some kind of treatment before discharge into public water systems. These and other projections will increase public concern about water quality aspects of irrigation and drainage.

With respect to on-farm water conservation, much can be accomplished with current irrigation technology and improved water management practices if they are fully implemented. Additional improvements may be obtained with new reliable soil and plant sensors linked to computerized irrigation scheduling. Use of marginal quality water by agriculture will require more intensive monitoring of soil and plants. Some changes into crop patterns may be needed when using poorer quality water. For instance, water containing elevated levels of salinity and/or trace elements may be more usable for salt tolerant crops, forage, and halophytes. To meet more stringent waste discharge requirements for irrigation drainage water it may be necessary to treat drainwater using methods ranging from sediment retention ponds and grass strips to constructed flow through wetlands to remove phosphorus and trace elements. Under severe water shortfalls, saline irrigation may need to be desalted to the salinity levels of usable irrigation water by utilizing new membranes and microfilters that require less energy. There will be a need for increased use of renewable energy such as solar, wind, and biomass power. Real-time monitoring of flow and water quality constituents of concern in river systems may be used to guide when and how much irrigation drainage can be discharged and still meet the water quality objectives of the stream. As irrigation projects mature, the vadose zone may become filled with deep percolation water and shallow water tables will need to be managed by root water extraction, subsurface drainage, or ground water management. The design parameters for subsurface drainage systems will necessarily include considerations for water quality, too.

The above conclusions and prognosis will be a formidable challenge to irrigated agriculture. Clearly, there will be problems associated with loss of agricultural water to urban demands and ones for fish and wildlife. Agriculture will need to utilize available freshwater supplies judiciously. Additionally, agriculture will have to utilize marginal quality water in greater quantities than now. New approaches and new inventions will need to be developed. Civil engineers will need to be more creative and innovative, and will need to interact more with other professionals in soil and crop sciences, hydrologists and watershed planners, regulators and environmentalists, and economists and policy makers to successfully provide solutions to water availability and water quality problems.

References

ASCE News. (2001). "ASCE names the Hoover Dam a civil engineering monument of the Millennium." September 27.

Ayers, R. S., and Branson, R. L. (1973). "Nitrates in the Upper Santa Ana River Basin in relation to groundwater pollution." University of California, California Agricultural Experiment Station Bull., 861, 1–60.

Ayers, R. S., and Westcot, D. W. (1985). "Water quality for agriculture, FAO irrigation and drainage paper No. 29 (rev. 1).

Beauchamp, K. H. (1987). "A history of drainage and drainage methods," Farm drainage in the United States—History, status and prospects, G. A. Pavelis, ed., USDA ERS, USDA Misc. Pub. No. 1455, 13–29.

British Geological Survey. (1999). "Groundwater studies for arsenic contamination in Bangladesh." ⟨http://www.bgs.ac.uk/arsenic/bphase1/B find.htm⟩ (December 20, 2001).

CalFed. (2001). "CALFED Bay-Delta Program." ⟨http://www.calfed.ca.gov⟩ (December 20, 2001).

Chesapeake Bay Program. (2000). "Chesapeake Bay Program: How's it doing?" ⟨http://www.chesapeakebay.net/restrtn.htm⟩ (December 20, 2001).

Engberg, R. A., and Sylvester, M. A., (1993). "Concentrations, distribution, and sources of selenium from irrigated lands in western United States." J. Irrig. Drain. Eng. 119(3), 522–536.

Hilgard, E. W. (1890). "Alkali lands, irrigation and drainage in their natural relations." California Agricultural Experiment Station Annual Report, Appendix, 7–56.

Hilgard, E. W. (1907). Soils, their formation, properties, composition, and relation to climate and plant growth in humid and arid regions, Macmillan, New York.

Kelley, W. P., and Brown, S. M. (1928). "Boron in soils and irrigation waters of Southern California and its relation to citrus and walnut culture." California Agriculture Experiment Station, Hilgardia, Calif., 3, 445–458.

National Research Council (NRC). (1989). Irrigation-induced water quality problems, what can be learned from the San Joaquin Valley experience, Committee on Irrigation-Induced Water Quality Problems, J. van Schilfgaarde, Chair, National Academy Press.

National Research Council (NRC). (1993). Soil and water quality, an agenda for agriculture, Committee on Long-Range Soil and Water Conservation, S. Batie, Chair, National Academy Press.

National Research Council (NRC). (1996). A new era for irrigation, Committee on the Future of Irrigation in the Face of Competing Demands, W. R. Gardner, Chair, National Academy Press.

National Research Council (NRC). (2000). Clean coastal waters, understanding and reducing the effects of nutrient pollution, Committee on the Causes and Management of Coastal Eutrophication, R. W. Howarth, Chair, National Academy Press.

National Research Council (NRC). (2001). Opportunities to improve the USGS national water quality assessment program, Committee on Opportunities to Improve the USGS National Water Quality Assessment Program, G. Hallberg, Chair, National Academy Press, in press.

Natural Resources Conservation Service (NRCS). (1998). "Natural resources inventory." ⟨http://www.nhq.nrcs.usda.gov/land/pubs⟩ (December 20, 2001).

Natural Resources Conservation Service (NRCS). (2002). "Natural Resources Conservation Service brief history." ⟨htpp://www.nrcs.usda.gov/history/articles/brief history.html⟩ (May 2, 2002).

Ritter, W. F., and Shirmohammadi, A., eds. (2001). Agricultural nonpoint source pollution, watershed management and hydrology, Lewis.

Rosenfeld, I., and Beath, O. A. (1962). Selenium: Geobotany, biochemistry, toxicity and nutrition, Academic, New York.

San Joaquin Valley Drainage Program (SJVDP). (1990). "A management plan for agricultural subsurface drainage and related problems on the westside San Joaquin Valley." Sacramento, Calif.

Skaggs, R. W., and J. van Schilfgaarde, eds. (1999). "Agricultural drainage." Agronomy No. 38, American Society of Agronomy.

Tanji, K. K. (1981). "River basin hydrosalinity modeling." Agric. Water Manage., 4, 207–225.

Tanji, K. K. (1982). "Chapter 19: Modeling of the soil nitrogen cycle." Nitrogen in agricultural soils, F. J. Stevenson, ed., Agronomy Monograph No. 22, American Society of Agronomy, 721–772.

Tanji, K. K. (1990). "Chapter 1: Nature and extent of agricultural salinity." Agricultural salinity assessment and management, K. K. Tanji, ed., ASCE Manual No. 71, 2–17.

Tanji, K., Lauchli, A., and Meyer, J. (1986). "Selenium in the San Joaquin Valley." Environment, 28(6), 6–11, 34–36.

U.S. Bureau of Reclamation (USBR). (2000). "The Bureau of Reclamation—A brief history." ⟨http://www.usbr.gov/history/borhist.htm⟩.

U.S. Congress. (1985). Food Security Act of 1985. ⟨http://agriculture.house.gov/food security-act-of-1985.htm⟩ (May 2, 2002).

U.S. Environmental Protection Agency (EPA). (2001). "Drinking water priority rulemaking: Arsenic." ⟨http://www.epa.gov/safewater/ars/arsenic.html⟩ (December 20, 2001).

U.S. Geological Service (USGS). (1998). "Wetlands of the United States, their extent and their value to waterfowl and other wildlife and a century of wetland exploitation." Northern Prairie Wildlife Research Center.

U.S. Geological Service (USGS). (1999a). "Review of the phosphorus control measures in the United States and their effects on water quality." Water Resources Investigations Rep. No. 99-4007, NAWQA Program.

U.S. Geological Service (USGS). (1999b). "The quality of our nation's waters, nutrients and pesticides." USGS Circular No. 1225, NAWQA Program.

U.S. Salinity Laboratory (USSL). (1954). "Diagnosis and improvement of saline and alkali soils," USDA handbook No. 60, L. A. Richards, ed.

Water Resources Center (WRC). (1963). Agricultural water quality research conference, University of California Water Resources Center, O. Lunt, ed.

American Society of
Civil Engineers
1852 – 2002
Building a Better World

150th Anniversary
Paper

1852–2002: 150 YEARS OF CIVIL ENGINEERING IN THE UNITED STATES OF AMERICA

By Francis E. Griggs Jr.,[1] Fellow, ASCE

ABSTRACT: On the 150th anniversary of the founding of the American Society of Civil Engineers it is important to review the growth of the profession and the country it serves. This paper introduces the major projects and people who changed the practice of civil engineering from almost an art form in the 1850s to the science that it is today. In the opinion of the writer, those included are the ones that had the greatest impact on the country. Others favoring other projects and people should prepare similar papers for the society.

INTRODUCTION

The American Society of Civil Engineers (ASCE) was founded in 1852 by a small group of engineers meeting in offices of the Croton Aqueduct, in New York City. Their goal was to form a national civil engineering society based upon the Institution of Civil Engineers, in London. They used the constitution of the Boston Society of Civil Engineers, founded four years earlier, as a framework. After approving a constitution and set of by-laws, they prepared a list of potential members who, from their personal knowledge, were practicing civil engineering in the United States. That list of 229 names was a "Who's Who" of American civil engineering. It included the Honorary Members of the Society: S. H. Long, Col. Albert, Prof. Bache, Henry Burden, Moncure Robinson, and Dennis Hart Mahan. In addition, 46 were listed as members. Railroad engineers predominated, as that was what most were practicing at the time. Most were from main cities along the eastern seaboard, and some from as far west as St. Louis. Listed (some with their current job titles) were civil engineers, mechanical engineers, and architects. Many of the men unknown to us went on to design and build the United States just prior to and after the Civil War.

UNITED STATES IN THE 1850s

The enlargement of the Erie Canal, which opened in 1825, was well underway. The Delaware and Hudson

[1]Dir. of Historic Bridge Prog., Clough, Harbour & Assoc. LLP, III Winners Cir., Albany, NY 12205. E-mail: fgriggs@nycap.rr.com

Note. Discussion open until March 1, 2002. To extend the closing date one month, a written request must be filed with the ASCE Manager of Journals. The manuscript for this paper was submitted for review and possible publication on June 15, 2001; revised July 3, 2001. This paper is part of the *Journal of Professional Issues in Engineering Education and Practice*, Vol. 127, No. 4, October, 2001. ©ASCE, ISSN 1052-3928/01/0004-0148–0159/$8.00 + $.50 per page. Paper No. 22244.

Canal and the Lehigh Coal Canal were delivering increasing amounts of anthracite coal to the cities of New York and Philadelphia. The canal-building mania of the earlier part of the century gave way to railroad-building mania in the middle part of the century. After over 25 years, by 1852, the Baltimore and Ohio Railroad made it to the Ohio River at Wheeling. The New York and Erie Railroad was completed to Lake Erie from the Hudson River at Piermont in 1851. The New York Central was being formed by combining many of the short lines that developed between the cities of the state to create a complete line from New York City to Buffalo. In the west, rail lines were laid westerly from Chicago to the Mississippi. By the end of the decade, rail service had spread throughout the land east of the Mississippi and reached as far west as St. Joseph, Missouri. Between 1840 and 1850 the number of railroad miles increased from 2,818 to 9,021 (320%); in the decade of the 1850s from 9,021 to 30,635 miles, or an increase of over 340%.

Bridge building, soon to be a major activity of civil engineers, was still in its infancy. Men like Squire Whipple, Wendell Bollman, and Albert Fink were building iron bridges for the B&O Railroad and the Erie Canal, but most bridges were still built of wood or stone. Charles Ellet built the Fairmount wire cable suspension bridge across the Schuylkill River in 1847. He followed this with his 1,010-foot bridge at Wheeling, in 1849, and had erected a small suspension bridge across the Niagara Gorge in the early 1850s. In Pittsburgh and on the Delaware and Hudson Canal, John A. Roebling built several short suspension bridges for carriages and canal boats. He was working on his double-deck railroad suspension bridge, the successor bridge to Ellet's, at Niagara. His major bridges, however, were still in the future. The early wooden-bridge builders like Palmer, Burr, Wernwag, Towne, and Howe were either dead or past their prime.

Squire Whipple's book on bridge building was published in 1847, and Herman Haupt's in 1851, but the use of science and mathematics to design bridges more efficiently was still in its infancy.

As the railroad opened up more and more of the country, cities and towns grew rapidly. But the technology of designing and constructing buildings had changed little from ancient times; most buildings were wood-frame or masonry-bearing wall construction. In the larger cities of the east a high building would still be only around six stories.

Municipal water works developed as early as the late seventeenth century in Boston. Other cities followed throughout the eighteenth century. These waterworks, however, were suited only for small populations and were inadequate as cities grew. The Croton Aqueduct, the primary water supply for New York City, was opened in 1842 under the engineering genius of John B. Jervis. In the 1850s it was enlarged by some of the founders of ASCE, primarily A. W. Craven and George S. Greene. Boston opened its Chocituate Aqueduct in 1850 under the guidance of Ellis Chesbrough. Other large cities—Philadelphia, Albany, etc.—followed and provided their citizens with a clean, ample, untreated water supply. But disease and subsequent death plagued the major cities of the United States well into the twentieth century. In New Orleans, for example, almost 8,000 citizens died from yellow fever in 1853, while in 1854 in Chicago 1,424 died from cholera. Many cities were virtual ghost towns during the summer months, when these diseases struck fear into the population and caused mass migrations into the countryside.

The collection and treatment of sewage was in its infancy, and typhoid and cholera were common in many cities. It wasn't until 1858 that the same Ellis Chesbrough developed a plan to collect the sewage of Chicago and discharge it into Lake Michigan. This plan relied primarily on dilution as the solution to pollution. It can be said that the beginnings of what was to become the specialty of sanitary engineering began in Chicago and then spread throughout the cities of the United States.

The roadway system was virtually nonexistent, as roads built in the turnpike boom of the early part of the century were not maintained. The Cumberland (National) Road, from Washington to the Ohio River, was completed earlier but had also not been well maintained.

To meet the growing demand for trained engineers, more engineering schools were needed. Prior to 1845 only Rensselaer Polytechnic Institute in Troy, N.Y., and Union College, in Schenectady, N.Y., were offering degrees in civil engineering. The USMA at West Point had been training men moving into the profession for many years, based upon their knowledge of topographic engineering, drafting, and mathematics. For many years their professor had, and would be, Dennis Hart Mahan, one of the first Honorary Members of ASCE. Harvard and Yale established engineering branches in 1847. It wasn't until the passage of the Morrill Land Act, in 1862, that the number of engineering schools in the country began to increase significantly.

When the Boston Society of Engineers was formed in 1848, only three of its members had any formal education and only one of them, Eben Horsford, had a civil engineering degree (1838) from Rensselaer. Most practicing civil engineers were apprentices to men like Benjamin Wright, Charles Ellet, Benjamin LaTrobe, Strickland Kneass, and Stephen H. Long. Foreign emigration of trained engineers, including an influx from Germany in the late 1840s, helped to fill the need.

With a population in 1860 of over 31,000,000 (up over 8,300,000 from 1850) the country was growing rapidly and developing the lands of the Ohio and Mississippi Valleys. The gold rush in the late 1840s brought more people west, and after the Civil War that migration increased.

In 1852, when ASCE was founded, the country was growing rapidly. The railroad system was expanding but the roadway system was virtually nonexistent. The supply of trained civil engineers was limited and the collection and treatment of sewage was in its infancy.

PROJECTS THAT SHAPED THE GROWTH OF THE UNITED STATES

To properly describe the growth of civil engineering over this 150-year period requires difficult decisions as to what projects truly made a difference and who the men were behind them.

In 1955, ASCE named its *Seven Modern Wonders*. Later, in 1994, ASCE updated this list as follows:

1955

- Empire State Building
- Panama Canal
- Hoover Dam
- San Francisco-Oakland Bay Bridge
- Chicago Sewage Disposal System
- Colorado River Aqueduct
- Grand Coulee Dam/Columbia Basin Project

1994

- World Trade Center
- Panama Canal
- Hoover Dam
- Golden Gate Bridge
- Kennedy Space Flight Center
- Interstate Highway System
- Trans-Alaskan Pipeline

Later in 1998, an international committee named the Seven Modern Civil Engineering Wonders of the World. They were the following:

- Panama Canal
- Itaipu Dam in Brazil

- World Trade Center (New York City)
- Chunnel under the English Channel
- CN Tower (Toronto)
- Netherlands North Sea Protection Works
- Golden Gate Bridge

I have chosen the projects based upon my experience (primarily in the structures, transportation, and geotechnical areas) and the literature of the period. They are in approximately chronological order as follows:

- The Chicago Water and Sewage Treatment System 1858–1869
- The Transcontinental Railroad 1863–1869
- The Brooklyn Bridge 1869–1883
- New York City Interborough Rapid Transit (IRT) System 1900–1904
- The Panama Canal 1904–1914
- The Empire State Building 1930–1931
- The Golden Gate Bridge 1931–1936
- The Interstate Highway System 1956–1990

CHICAGO WATER AND SEWAGE TREATMENT SYSTEM, 1858–1869

It has been said that of all the accomplishments of civil engineers, its contribution to public health has saved more lives and contributed more to our standard of living than any other effort. A review of the number of deaths from cholera in Chicago in the years between 1850 and 1854 indicates that almost 3,000 people died. In 1854, the worst year, there were 1,424 deaths and a rate of 2,162 deaths per 100,000 inhabitants.

Contaminated drinking water and the decomposition of human waste were the causes of many deaths not only from cholera and typhoid but also yellow fever. The actual link between typhoid and the decomposition of organic (and chiefly vegetable) substances was not discovered until the 1870s.

If Chicago was to continue to grow and prosper it was necessary that steps be taken to clean up the city. Ellis S. Chesbrough (Fig. 1), the man who would clean up Chicago, came to the city in 1855 as Chief Engineer to the newly formed Chicago Sewerage Commission. When he arrived the city disposed of its sewage directly into the Chicago River, which flowed slowly into Lake Michigan. The residents relied on wells for most drinking water and those wells were near the river. The proximity of the two resulted in the pollution of the wells and in the periods of disease noted. Before coming to Chicago, Chesbrough had experience as a railroad engineer and served as one of the Chief Engineers of the Cochituate Aqueduct in Boston, under the guidance of John B. Jervis. He remained in Chicago with the Commission for the next 24 years.

In 1856 one of the first acts of the Commission was to send Chesbrough to Europe to study and examine sewage disposal systems. In 1858 he published his findings in a *Report of the Results of Examination Made in Relation to*

FIG. 1. Ellis S. Chesbrough

Sewerage in Several European Cities in the Winter of 1856–57. Based upon that experience he greatly modified the design he promoted prior to his trip abroad. His revised plan was the first comprehensive sewerage system in any U.S. city. It entailed raising the entire city so sewage could flow more rapidly into Lake Michigan through the Chicago River. The sewage outfall points close to the shore contaminated the water intake from the lake, requiring him to design a tunnel extending two miles into Lake Michigan to a water intake, therefore minimizing pollution of the drinking water. This project, started in 1864, opened to public acclaim in 1869, and for the first time Chicago had a source of clean water that could not be polluted by discharge of sewage into the lake. The 1866 outbreak of cholera was the last for Chicago since the city now had a water supply system and sewage collection and disposal system, both designed and built by the same man, that enabled the city to grow. Chesbrough was sought after as a consultant by many other cities to design or consult on their systems, and most major United States cities built their systems based upon that of Chicago. Between 1890 and 1900 the city reversed the flow of the Chicago River and dumped its sewage into the Des Plains River, using water from Lake Michigan to flush the river and it sewage downstream. Still later, in 1938, a major treatment program was implemented to minimize the negative effects downstream as far as St. Louis. It was this project that was named one of the Civil Engineering Wonders in 1955.

TRANSCONTINENTAL RAILROAD, 1863–1869

Railroading in the United States is said to have begun with the chartering of the Baltimore and Ohio Railroad and to a lesser degree with the Mohawk and Hudson Railroad. The number of miles spread rapidly between 1831 and the founding of ASCE. It would continue to grow rapidly until the Civil War. During the war, expansion of the system slowed, as most of the country's resources went to the war effort.

The nation had been debating the construction of a transcontinental railroad starting in 1853, when Congress authorized Jefferson Davis, then secretary of war, to make a survey of possible routes. Visionaries like Edwin Ferry Johnson and A. C. Whitney were promoting such a railroad as early as 1837, but it required action by the federal government to move the road from a dream to reality. Three lines were surveyed. The north route was primarily along the 49th parallel, the Buffalo route along the 38th to 39th parallels, and the south route along the 32nd to 35th parallels. By 1856 both the northern and southern states favored a Pacific railroad, but with tensions high over the issue of slavery and states rights, they couldn't agree on the route to select. In 1862, after the secession of the southern states and the beginning of the Civil War, Congress, with the support of President Lincoln, passed the Pacific Railroad Bill.

The war didn't stop Theodore Judah (Fig. 2) from thinking big. In 1854 he went west to design the first railroad in California, the Sacramento Valley Railroad. As early as 1856 he became so obsessed with the idea of a transcontinental railroad that he was frequently called "mad Judah" or "crazy Judah." In 1857 he published a pamphlet titled "A Practical Plan for Building the Pacific Railroad." For the next several years he promoted the idea in Congress and sought backing from California entrepreneurs. He finally convinced Collis Huntington, Leland Stanford, Charles Crocker, and Mark Hopkins, known in history as the "Big Four," to support his efforts and create the Central Pacific Railroad. In 1860 he surveyed the route over the Sierras, which proved his idea was feasible. It wouldn't be easy, but it was possible. In the east Thomas Durant and the Ames Brothers were forming the Union Pacific (UP) Railroad to match Judah's Central Pacific (CP) Railroad. In 1864 Congress passed the Pacific Railroad Act, amending the 1862 Act, which designated the Union Pacific and Central Pacific, both privately owned organizations, to build the transcontinental railroad, providing them with land grants and subsidies. The CP actually started east as early as October 1863, but tension was developing between Judah and the Big Four. Judah went east to find backers to buy out the Big Four in late 1863. He died in the same year from yellow fever contracted while crossing the Isthmus of Panama on the Panama Railroad, which could be called the first transcontinental railroad of the Americas.

The construction of the UP line began from the east with Peter Dey as chief engineer. He surveyed three routes, but the route along the North Fork of Platte River was an early favorite and would be the designated route. General Grenville Dodge (Fig. 3), then fighting in the Civil War, discovered a route through the mountains earlier. After the war Dodge became the UP's Chief Engineer and he and his surveyors laid out the route seeking the

FIG. 2. Theodore Judah

FIG. 3. Grenville Dodge

shortest, straightest line with the lowest grade. Moneymen like Durant, however, did not always agree with his routes and actually wanted the line to be longer than it had to be, as their subsidy was by the mile.

The story of the building of the two converging lines continues to fascinate as new books are constantly printed describing the epic adventure. From an engineering, construction, and supply standpoint, the project marked the evolution from small eastern railroads near their points of supply and labor to remote regions with no natural resources, and little labor, water, or wood. In addition, the UP line had to contend with the Plains Indians, who generally did not want the "iron horse" coming across their lands.

All supplies for the UP were shipped from the east up the Mississippi and Missouri Rivers to Omaha. There they were shipped along the completed railroad to the end of the line. The CP shipped all supplies, such as rail and locomotives, by sea through the Isthmus of Panama on the Panama Railroad or around the Cape. This project required greater coordination of purchasing, shipping, and delivery to point of use than any project in the history of the world. An additional constraint was the fact that backers of both lines were frequently cash poor and relied on prompt payment of government subsidies to continue. These subsidies were not issued until each 20-mile increment of line was approved. Surveyors would be out in front staking the line. They were followed by the grading crews, who in turn were followed by the tracklayers. Add to all these constraints the long bitter winters, a shortage of laborers, and primitive labor-saving equipment. Consider what a great accomplishment it was to build the entire 1,776 miles. The CP relied primarily on Chinese to build its line over the Sierras by means of major tunnels, and thence across Nevada to Utah. The UP relied on veterans of the Civil War and Irish emigrants to cross the plains and the Wastach Mountains. The Union Pacific built 1,086 miles between late 1864 and May 1869, and the Central Pacific built 690 miles of line between January 1863 and May 1869. They connected at Promontory Point on May 10, 1869, with the driving of the Golden Spike.

The total number of miles (1,776) is an appropriate number as the line gave a new freedom of mobility that was lacking prior to the opening of the Railroad. The country would never be the same again. It was now possible to travel its length in seven days for $70. Previously the trip by land across the country could take as much as six months, by ship and the Panama Railroad upwards to three months, and seven months by sea around the cape. It was the Big Four of the CP, Thomas C. Durant, and Oakes and Oliver Ames of the UP who financed the project. Engineers like Peter Dey, Grenville Dodge, S. B. Reed of the UP, and Theodore Judah, James Strobridge, and Samuel Montague of the CP who made it happen. Dodge and Judah, in addition, were the main men involved in getting congressional and presidential approval of the railroad in the late 1850s and early 1860s. They created a railroad compared to which there was "nothing like it in the world."

BROOKLYN BRIDGE 1869–1883

In 1867 John A. Roebling was selected as chief engineer of the New York and Brooklyn Bridge, after a charter for the bridge was issued by the New York State Legislature. Roebling had been advocating a bridge at this site for almost a decade. What Roebling proposed was a suspension bridge with a span of almost 1,600 feet. This was almost 50% longer than his Cincinnati Bridge, which opened earlier that year. Between the founding of ASCE and 1867, suspension bridges had also been built across the Niagara, 821 feet by Roebling and 1,268 feet by Sam-

uel Keefer. James Eads was ready to start his bridge across the Mississippi at St. Louis. Both Roebling and Eads were thinking about using steel instead of wrought iron and both faced difficult foundation problems.

Roebling knew his bridge would be something special as he told Bridge Directors, "the contemplated work, when constructed in accordance with my designs, will not only be the greatest bridge in existence, but it will be the great engineering work of the continent and of the age." His son Washington A. Roebling (Fig. 4), after graduating from Rensselaer in 1857, worked with his father on the Allegheny Bridge, in Pittsburgh, and the Ohio River Bridge. He enlisted as a private in the Civil War and was, after almost four years, discharged as a Major with a Brevet rank of Colonel.

After the war the Roeblings finished the Ohio River Bridge, and in 1867, as a wedding present from his father, Washington took his wife, Emily Warren, on a tour of Europe. On this trip, however, John wanted Washington to look into the latest engineering works on the continent, particularly the use of pneumatic caissons by the French. The elder Roebling knew that the success of his dream demanded a stable foundation for his towers. The borings indicated that bedrock was at 90 feet on the Brooklyn side and 76 feet on the New York side. On July 6, 1869, John Roebling injured his foot in a collision between a ferryboat and the pier on which he was standing while surveying for the Brooklyn Tower. In 16 days he died because of complications from lockjaw. Washington became Chief Engineer and for the next 14 years was dedicated to the fulfillment of his father's dream.

Pneumatic caissons were a new construction technique in the United States at that time. James Eads was sinking very deep foundations at St. Louis and William J. McAlpine sunk pneumatic caissons for the Dry Dock at the Brooklyn Navy Yard in the late 1840s. Roebling's caissons were much larger than either Eads' or McAlpine's

FIG. 4.   Washington A. Roebling

and would carry a much larger loading. He sank his Brooklyn caisson to a depth of only 45 feet, resting it on a tight soil with boulders. He encountered many problems sinking the caisson, including a fire that badly damaged the wooden roof. He finished the caisson and, based upon his experience, greatly modified his design for the much larger New York caisson. The changes, along with sandy soil, greatly accelerated the sinking, but he was going to a much greater depth, and his men started to experience the same caisson disease that he himself suffered on the Brooklyn caisson. With the problem getting worse each foot the caisson sank, he decided to stop the caisson at 78 feet, resting it on the upper cusps of bedrock with dense sand filling the spaces between the rock outcroppings. Shortly thereafter Roebling became incapacitated from the effects of caisson disease and never visited the bridge site again during construction. He observed construction from his home on Brooklyn Heights and used reports from his assistant engineers and wife Emily to direct construction of the anchorages, spinning of the cables, and the hanging of the deck structure. He followed his father's plan in his use of steel wire to spin the cables and later built the entire suspended structure with steel rather than wrought iron.

Cable spinning began in 1876 and was completed in October 1878. It required almost five years to hang the decking and complete the approaches. On May 24, 1883, the bridge opened with a huge ceremony attended by all the local dignitaries as well as by the governor of New York and the president of the United States, Chester A. Arthur. The bridge, designed by John A. and Washington A. Roebling and built by Washington, became an icon of the City and to this day remains one of the most recognized and beloved bridges in the country. It became an example of the daring of the American civil engineer and would remain the longest bridge in the world until the Williamsburg Bridge opened 20 years later. Its majestic stone towers and lace-work of suspender cables and cable stays are perhaps the most photographed elements of any bridge in the country. Roebling's assistant engineers traveled across the country to build bridges of their own and became leaders of the engineering profession. The American engineer, based upon example of the Roeblings, became known around the world for this ability to accomplish the difficult, and even impossible, task. The Brooklyn Bridge has been called the Great Bridge and has fully lived up to the elder Roebling's dream.

NEW YORK CITY RAPID TRANSIT LINES, 1900–1904

New York City grew rapidly in the mid to late nineteenth century, and its necessary expansion to the north and east resulted in major transportation difficulties in moving its people from their homes to their places of work. The horse-drawn surface trolley system addressed the problem for a period, and the advent of the steam-powered train and elevated track helped in the 1880s and

1890s. Those locomotives were dirty and discharged ash and smoke, creating a further negative impact on surrounding dwellings. The elevated structures also created a sense of clutter and darkness in the streets below. The speed on the elevateds rarely exceeded 12 miles per hour and, while an improvement, still was not fast. There was insufficient room for the structure of elevated tracks and all the surface carriage, pedestrian, and wagon traffic on the streets. Proponents of a subsurface rapid transit system began their efforts to build a modern system.

The entrenched elevated companies had considerable political muscle, and they stymied the subway proponents for many years. A committee of the ASCE, chaired by Octave Chanute, recommended the elevated system over the subway in 1875. Its report cited six issues in support of its recommendations, including the problem of clearing tunnels of smoke generated by locomotives and the difficulty of building the subway with such large numbers of utilities buried in city streets.

In the late 1890s the electric motor was developed to power subway trains, thus solving the smoke problem— but the political problem remained. In 1897 the Tremont Street Subway was built in Boston and even though it was a short line, it gave American engineers experience in this type of construction. In 1900 the transportation problem worsened in New York City, and finally a plan was developed to build the IRT (Interborough Rapid Transit) subway. The financing plan, originally proposed by Abram Hewitt, son-in-law of well-known Peter Cooper, required the city to fund construction by a private organization that would build and operate the system, with ownership being retained by the city.

Construction began in March 1900 under Chief Engi-

FIG. 5. William B. Parsons

neer William Barclay Parsons (Fig. 5). Parsons was associated with the subway effort since 1886, when he was appointed engineer for the New York District Railway. The IRT would connect Manhattan, the Bronx, and Brooklyn. The project was to be cut-and-cover between City Hall and 34th Street, with tunneling required between 34th and 42nd Streets and under Central Park. Above 60th Street the contractor could select his own method as long as it did not interfere with street traffic. In cut-and-cover sections the contractor could not open more than 400 feet unless he covered this with a structure on which vehicles and pedestrians could ride or walk. The cut-and-cover method required meticulous planning to disconnect, relocate, and reconnect all buried utilities while maintaining surface and elevated traffic. It was also necessary to ensure that adjacent buildings were not undermined.

The tunnels under the Harlem, to the north, and East Rivers were major projects in themselves. The Harlem River Tunnel was built with twin 14-foot-diameter cast-iron tubes covered with 2 1/2 feet of concrete on the top and a minimum of 1 foot on the sides and bottom. The contractor built this tunnel in a uniquely designed wooden box consisting of wooden sheet piles for walls and a 40-inch-thick wooden top that was anchored to the walls. The water was then forced from the box by pneumatic pressure and the tunnel built under pressure to mid-river. The northerly half of the tunnel was built in a similar fashion with the exception that the top half consisted of cast-iron rings and the required concrete cover was cast off-site, floated to the site, and sunk onto the sheet piles forming the walls. This connection was sealed and the area under the roof dewatered by pneumatic pressure, and the bottom half of the tunnel was completed. The East River Tunnel went through rock on the Manhattan side and was advanced by standard tunneling techniques. The tunnel from Brooklyn, however, was through sand and required shields that were pushed through the sand. The space behind the shield was pressurized to permit soil to be removed and cast-iron rings to be placed. Concrete was then injected between the rings and the bored wall of the tunnel.

Parsons, working with the contractor, successfully implemented this project between March 24, 1900, and October 24, 1904, and the first paying customer rode on the line between City Hall and 125th Street. In his 1906 report Parsons concluded, "the years 1905 and 1906 may be regarded as an epoch in the history of rapid transit, looking to construction of future subways on so extensive a scale as to have been hardly conceivable a few years ago, or even contemplated within the past decade." It was said that it was "one of the great public 'improvements of the twentieth century, but also as an indispensable element in the life of America's largest city.'" Without it, and its extensions, the growth of New York City would have been severely limited. It was conceived in a time of great political instability and lack of direction by political leaders. Leaders like William B. Parsons, the engineer, and August Belmont, the financier, made it happen. Lessons learned

on the IRT were helpful as many other cities grew and needed their own subway systems.

PANAMA CANAL, 1904–1914

The United States' interest in a canal connecting the Atlantic and Pacific Oceans started under President James Polk, who in 1846 signed the Bidlack Treaty with Columbia. This treaty stated that "the right of way or transit across the Isthmus of Panama, upon any modes of communication that now exist, or that may be hereafter constructed, shall be open and free to the government and citizens of the United States. . . ." The Panama Railroad was built by private enterprise across the Isthmus between 1850 and 1855. For a time this eliminated the pressure for building a canal across the Isthmus. In 1875 President Ulysses S. Grant authorized the survey of proposed routes starting very close to the South American continent and extending as far north as Mexico. The two routes that received the most attention were the Panama Route, which followed very closely the route of the railroad, and the Nicaragua Route, which made use of Lake Nicaragua. The latter route became the American Route while the French, under the guidance of Ferdinand DeLessups, in 1879 selected the Panama Route.

Using private monies, the French began preliminary work in early 1880 on a sea-level canal similar to that which DeLessups built at Suez in 1869. Many American and some French engineers expressed their opinions that a sea-level canal was not possible with the funding available and the time frame DeLessups proposed. They pointed out the difficulty of controlling the Chagres River during construction and operation and the problems of diseases like yellow fever and malaria rampant in that part of the world. By force of personality and his ability at Suez to do just what many engineers said couldn't be done, DeLessups was able to obtain large sums of money to progress with the work. But between 1880 and 1886 it became clear that he could not complete a sea-level canal. He then adopted the recommendation of Phillipe Bunau Varilla to build a temporary lock canal that would have ships lock up to an artificial lake at the summit level and then lock down to sea level. While the canal was in operation, constant excavation would continue at the summit, dropping the level until such time as a set of locks could be removed. This excavation would continue until they arrived at a sea-level canal. But by this time disease, lack of money, and a growing realization that he still did not have a plan to control the Chagres River resulted in the company passing into receivership. A new company fored to continue work to the extent needed to maintain the concession that the French received from Columbia in 1879.

In 1903 the United States agreed to buy all equipment on the Isthmus and the title to the work completed dependent upon the signing of a treaty with Columbia. This treaty would give the United States the right to build and run a canal on the route selected by the French. Columbia, at least in the eyes of President Theodore Roosevelt, was slowing down his effort to build the canal. After Columbia rejected the treaty, the citizens of Panama revolted and declared their independence from Columbia. A new treaty, much more in the interest of the United States, was signed in 1904 with the new country of Panama. John Findlay Wallace was appointed the first chief engineer and was effectively told to go and "make the earth fly." Arriving at Panama in 1904 and surveying the work completed by the French, he was very complimentary of their efforts. Between 1904 and 1905 it became clear that organization of the enterprise was unwieldy and resulted in very slow decision-making. At the same time yellow fever returned and killed many American workers. In addition, Wallace in his effort to make the dirt fly did not provide the housing and food required for his workers. All these factors resulted in a very poor start by the United States. Wallace resigned in 1905 after only one year on the largest construction job in the history of the world.

Roosevelt appointed John Stevens (Fig. 6), a leading railroad engineer, to take over the project. Stevens observed that the United States did not have a plan, and he knew that the success of its effort rested on making the Isthmus free of yellow fever and at least minimizing malaria. He saw that housing and food distribution needed to be upgraded to attract and keep the men necessary to dig the canal. He decided to stop most excavation and join with Col. William Gorgas to eliminate the disease-carrying mosquitoes and clean up the canal route. At the same time he devised a material-hauling system to move earth from the Culebra Cut to places where it could be used or successfully wasted. His knowledge of railroading enabled him to set up a very efficient system of steam shovels and dirt cars to keep the dirt/rock moving. He proposed the largest man-made lake in the world at an elevation of 85 feet above sea level to be fed by the Chagres River. This made the river his friend as contrasted to the problem it was for the French effort. He needed to build a large (Gatun) dam at the northerly end of the canal with smaller dams and locks at the southerly end to create this lake.

After a Commission of Engineers expressed the opinion that the United States should build a sea-level canal, Stevens convinced Roosevelt they were wrong. With Roosevelt's support the decision was made to continue with a lock canal. The plan had ships come from the Caribbean lock up to Lake Gatun through a series of three locks and then pass through Gatun Lake and the Culebra Cut (now the Gaillard Cut). They would then be locked down through another set of three locks to the Pacific Ocean. This required the construction of the largest earth fill dam in the world at the time, the Gatun Dam, and the largest excavation project in history.

After making all these decisions and initiating his earth-handling system, Stevens resigned in 1907, much to the consternation of Theodore Roosevelt. Roosevelt decided that he needed to send someone to the Isthmus who would

FIG. 6. John Stevens

FIG. 7. George Goethals

not easily resign, so he appointed Col. George Goethals (Fig. 7) as chief engineer. Goethals came to this new position with almost dictatorial powers, as Roosevelt believed that everything on the Isthmus should be under the control of one man. Goethals was not only chief engineer but was also in charge of the Panama Railroad and the governance of Colon, Panama City, and all lands controlled by the United States along the route of the canal. He had other military men as assistants who would be at their posts as long as the president and Goethals required. While Stevens made most of the key decisions, Goethals made the system work in the most efficient way possible. He did this with Col. Gaillard assigned to the Culebra Cut and Col. Seibert to the Gatun Locks. For the next seven years Goethals and his "Army of Panama" fought huge earth slides (some as large as 3,000,000 cubic yards) in the Culebra Cut, poured more concrete for the locks than had ever been poured on a single project, built the largest earth-fill dam in the world, relocated the Panama Railroad, built port facilities at both ends of the canal, and placed into service a lock control system that was fail-safe and ensured the success of the canal's operation using the power of water dammed up in Gatun Lake. This water worked the locks and generated all the electricity needed to run the support facilities.

In 1914, after seven years of American effort and 34 years after DeLessups broke ground, the canal opened with little fanfare, due to the beginning of WWI. Slides continued to interrupt service years after completion. The United States indeed built the canal and that opened a path between the seas, bringing to a close the greatest project in the history of the young country since the completion of the Transcontinental Railroad in 1869. Ships could now pass between our east and west coasts much faster and carry goods at a much lower cost. Like the Suez Canal 45 years earlier, the Panama Canal changed the world.

EMPIRE STATE BUILDING, 1930–1931

The erection of tall structures can be traced back to the Pyramids of ancient Egypt (height 481 feet), the Pharos of Alexandria (lighthouse with a height of 350 feet) and the Colossus of Rhodes. Masonry church spires became the highest structures in most cities and villages and reached as high as 404 feet at the Salisbury Cathedral. The Washington Monument at a height of 555 feet was completed in masonry in 1884 and Gustave Eiffel built his famous tower in 1889 to a height of 1,000 feet. It wasn't until late in the 19th century that high structures were considered to house large numbers of people. Residential and commercial buildings were limited in height by the number of flights of stairs, usually six flights so that the occupants could reasonably be able to climb to apartments or offices. As elevators developed in the middle to later part of the century, the limiting factor shifted to the efficient height that masonry-bearing walls could be built without taking up excessive amounts of floor space. Exterior bearing walls became excessively wide as the height of the building increased, even when cast-iron columns were used for interior support of floor systems.

With the advent of inexpensive cast and wrought iron, and later steel, it became possible to design and build a steel frame that held up floors and wall-covering materials, thus making bearing walls unnecessary. In the 1870s iron—cast for columns and wrought for beams—framed buildings, one being the 10-story Western Union Building in New York City. In 1884 the first building built with a steel frame was Chicago's 10-story Home Insurance Building, designed by William Jenney. In 1902 in New York, the famous 20-story, 285-foot Flat Iron Building at Times Square was built with a steel frame and six elevators.

The construction of the Woolworth Building in 1913 began the rapid increase in height of what were called skyscrapers. It had a height of 792 feet and an average floor height of 12 feet for its 60 stories. It was the first truly tall building requiring new elevator systems and ways of moving people in a vertical direction. It was an instant success and retained the title of the tallest building in the world for almost 20 years. In the boom times of the late 1920s the Bank of Manhattan was built on Wall Street, a 71-story, 925-foot-high building with a very tall flagpole. Walter Chrysler, the automobile maker, planned his famous building. When the final height of the Bank building was set, he added its famous top, made up of the curving and overlapping metal plates that are so recognizable. It surpassed the Manhattan Bank by 105 feet and set a new record of 1,030 feet when it opened in 1930.

In the race to be the tallest, the Empire State Building was conceived. Officers of DuPont and General Motors were the prime movers behind this project to build a massive office tower on the former site of the Waldorf Astoria Hotel. They elected a former governor of New York and presidential candidate, Al Smith, as president of the board. The plan was to build the tallest structure in the world and to do it in the early part of the Great Depression. A steel frame was selected with a cladding of aluminum, stone, and stainless steel to speed up construction and cut down on weight. The contractor demolished the hotel in less than five months and had the foundation in place in another three. By late March 1930 the building was ready for the first steel. What followed was perhaps one of the greatest construction projects in United States history. They were constructing an 86-floor building in the middle of the busiest city in the world. The erection of the steel proceeded at an average rate of one story per day. In one week it actually grew by 14 stories. It took just 161 days to erect and rivet the entire steel frame. Like clockwork the other trades followed. Scheduling of material delivery was critical, as there was little room on site to store it. Elevator construction followed at an equally rapid pace. The building as erected was 1,050 feet above street level. With the installation of a 200-foot mooring mast (for dirigibles), the building topped out at an elevation of 1,250 feet. (Later, in 1951, it increased to 1,472 feet with the installation of a 222-foot TV antenna.) The building opened on May 1, 1931, ahead of schedule after a total construction time of 1 year and 45 days—20% below the original estimates.

The Empire State Building (Fig. 8), even though surpassed in height by many other buildings in the city and around the world, it is still one of the most recognized and visited. As an engineering achievement it makes every list of landmark structures.

GOLDEN GATE BRIDGE, 1931–1936

After the opening of the Brooklyn Bridge in 1883, the next key suspension bridge built was the Williamsburg Bridge by Leffert L. Buck in 1903. This bridge was

FIG. 8. Empire State Building

FIG. 9. Joseph B. Strauss

longer, wider, and carried a much higher load than the Brooklyn Bridge. It was the first to use steel for its towers in addition to cables and suspended structure. The next major design was the George Washington Bridge, built in 1931 by Othmar Amman, with its 3,500-foot span. Other major suspension bridges in that period were the Bear Mountain Bridge over the Hudson River, the Manhattan Bridge adjacent to the Brooklyn Bridge, and the Ambassador Bridge in Detroit, Michigan. Contemporary with the Golden Gate was the twin-span San Francisco–Oakland Bay Bridge that was noteworthy as a complete bridge project. It has not received the attention and accolades of its neighbor, partially because of its setting but also the fact that its central suspended spans are significantly shorter.

A bridge across the Golden Gate was the vision of Michael O'Shaughenessy, City Engineer of San Francisco, and Joseph B. Strauss (Fig. 9) as early as 1919. O'Shaughenessy saw the need to open up the Marina Peninsula north of the city to vehicular traffic with a bridge rather than the slow ferry system. He asked Strauss if he could build a bridge across the Gate and told him that other engineers had given him a price of $250 million. Strauss indicated that he thought he could build a bridge for $25 million. After borings were taken and a survey of the site completed, Strauss determined that the total length across the Golden Gate was over 6,700 feet and that foundations on rock at a depth of only 50 feet could be built with a central span of 4,000 feet. In 1921 Strauss submitted a plan that combined the suspension and cantilever bridge types. The suspended cable span was 2,640 feet and each cantilever arm 680 feet. Strauss believed that structurally the bridge was technically feasible and within his earlier cost estimate. O'Shaughenessy also asked Gustave Lindenthal and Robert McMath to submit proposals, but their estimated costs were far beyond the City's ability to finance. Strauss' design was, in reality, the only one seriously considered by O'Shaughenessy. In the eye of many observers it was ugly and not in keeping with the natural beauty of the Golden Gate.

Over the next eight years this was the bridge studied as O'Shaughenessy and Strauss tried to convince the voters that a bridge was possible and within their ability to pay. It wasn't until 1929, however, that a new suspension bridge was proposed with a 4,200-foot central span. This was possible by the addition of Charles Ellis to Strauss' staff and the assistance of Othmar Amman (then working on the George Washington Bridge), Charles Derleth, and Leon Moisseiff. Moisseiff developed a method of distributing wind load between the cables and stiffening trusses. Prior to this, trusses were assumed to carry the entire wind load. As a result, as span length increased, the trusses increased significantly in size. With Moisseiff's analysis, the trusses and the look of the bridge could be much lighter as well as less expensive to build. Using the new theory, Ellis then designed the towers, anchorage, cables, and suspended spans.

The design was completed in 1930. It was recognized that, while Strauss had been the promoter of the bridge, the design was the work of Charles Ellis, using the theories of Moisseiff. Strauss, however, was not a man who could share his project with others: he fired Ellis in late 1931. The design was completed, but Ellis continued on his own to work on the tower design over the next nine months. In the fall of 1932 he believed he had completely solved the problem to his satisfaction. He submitted his design to the other consulting engineers and went over Strauss' head to ask for an interview with the bridge's directors. Nothing came of this request and the bridge, as originally designed by Ellis, was ready for construction. Work started on the anchorages and piers in early 1933. The anchorages, although huge, created no major problems, nor did the northerly pier foundation that was close to shore. The southerly pier being well out into the Gate, created major problems in sinking the caisson because of the turbulent nature of the bay. After several major setbacks, the piers were placed and the steelwork was ready to begin.

In the midst of the Great Depression, men and materials were available at bargain prices. The towers, over 748 feet high, went up without problems. Then the John A. Roebling Company came in to spin the 36-in. diameter cables. Using new techniques, it spun faster than on any previous bridge and finished two months ahead of schedule. The suspended structure followed, and by April 1936 the last of the concrete was placed on the deck. The bridge opened officially on May 27, 1936, for pedestrians and on the following day for automobiles.

The bridge, with its central span of 4,200 feet, retained its status as the longest bridge in the world until the Verrazano Narrows bridge opened, in 1964, with its main span of 4,260 feet. It was designed by Othmar Amman, the consultant on the Golden Gate and designer of the George Washington Bridge. Even though the Japanese have recently completed a 6,527-foot bridge, the Golden Gate and the Brooklyn Bridge are perhaps the two most recognized bridges in the world. The Golden Gate Bridge, with its unique color and magnificent setting, is currently undergoing a renovation to increase its resistance to earthquake loading. It has been recognized not only for its beauty but for its contribution to the design and construction of suspension bridges around the world.

INTERSTATE HIGHWAY SYSTEM, 1956–1990

The United States highway system prior to WWII was incomplete and in poor condition. The rapid expansion in the number of automobiles and trucks placed great pressure on government, at all levels, to add to and improve the system. While steps had been taken, the expansion of the network was very slow, and few of what we would call modern highways had been built. The Pennsylvania Turnpike was an exception. When it opened in 1940 it was the most modern, divided, and controlled access road in the country. From Harrisburg to Pittsburgh, it covered over 160 miles. It was financed by bonds and an influx of money from the WPA. The coming of World War II set back highway construction and expansion for the next five to six years. After the war there was a rapid increase of automobiles and trucks on the inadequate network. Several turnpike authorities were created to build roads, as government at the state and federal levels could not find a way to finance the needed improvements. President Truman, with the pressures of the cold war, the financing

of the Marshall plan, and the hot war in Korea, did not have the resources to commit to road building. His successor Dwight Eisenhower, however, was very interested in highways; he began the process that led to the passage of the Federal-Aid Highway Act of 1956 and created what is known as the Interstate and Defense Highway System.

The Federal-Aid Highway Act called for a system of 40,000 miles of multilane divided highways to connect most of the major cities in the United States as well as highways around and through those cities. Its $25 billion cost was to be expended over a 12-year period, and the system would be completed by 1972. It was to be financed through gasoline taxes that would go from 2 to 3 cents per gallon with all tax increase being placed in a Highway Trust Fund. The Federal Government would pay 90% of the cost, and the states 10%. The system was designed to carry the estimated traffic projected for 1972.

Once the project began it became clear that there would not be enough money generated by the gasoline tax to complete it in 12 years. President Eisenhower recommended that additional monies be appropriated from the general revenue fund. Even with this influx of money, the project would take far longer and cost far more than anyone had anticipated earlier in the planning process.

The man most involved in the conception of the program was Thomas MacDonald, who first proposed a similar highway system. MacDonald was chief of the Bureau of Public Roads since 1919 and had overseen and promoted the federal involvement in highway construction. After 34 years in the position, Eisenhower chose not to renew his contract in 1953. He was replaced by Francis DuPont, who immediately proposed that the system be

FIG. 10. Francis Turner

financed entirely by the federal government. His assistant, Francis Turner (Fig. 10), was appointed federal highway administrator in 1969 and became the face of the Interstate Highway System. These men were the guiding force behind the system, but the states did the actual designing and construction of the system. They needed to expand their highway departments and colleges had to increase their production of engineers to design and build the system. Construction companies expanded to complete the ambitious schedule of road construction, and organizations like the American Association of State Highway Officials developed guidelines and standards for all to follow. It was an exciting time to be a civil engineer.

In the early days of the program's implementation, the bulldozer mentality of many engineers resulted in a public uprising against placing highways in cities and the encouragement of alternate means of meeting the transportation needs of the community. Prime examples of this were in New Orleans, Boston, and San Francisco. As their primary charges, highway engineers considered the safety and convenience of the motorists and their need to get as close as possible to their destinations. The fact that the highway might disrupt the social and economic fabric of a community was far less important to the engineer than was the efficiency of the system. Over the 1970s and into the 1980s that approach was modified, and engineers began to take into account other factors than cost per mile and accessibility. The system continued its expansion well into the 1990s, when it was officially completed 37 years after its creation. In reality, however, the system will never be completed—new mileage is being added and the cost of maintaining and upgrading the road continues to the present day. The program was followed in 1991 by ISTEA (the Intermodal Surface Transportation Efficiency Act), and in 1999 by TEA21 (the Transportation Equity Act of the Twenty-First Century). These programs recognize that highways are important but are only one part of the overall transportation needs of the country.

As had the Intercontinental Railroad over a century earlier, the highway system had changed the way Americans lived and traveled. It was now possible to cross the United States in three to four days in your own automobile without stopping except for gasoline, food, and breaks. Cross-country trucking became economical and competed with the railroad for its market. It was now possible to live miles away from your place of employment and make the commute in a short period of time. City dwellers moved to the suburbs to live and retailers followed, building malls at interchanges of the system. Manufacturers and distributors located their new facilities at or near interchanges to cut down on their costs.

The Interstate Highway System, with its approximately 42,000 miles of roadway, comprises only 1% of the total highway mileage in the country but carries 24% of the passenger traffic and 45% of truck traffic in a safe and fast manner. On the downside, people, retailers, manufacturers, etc., left the cities for the suburbs and downtowns became decimated. The interstates directed traffic away from many small towns and cities that relied upon passing traffic for part of their business, and these communities suffered the consequences. In summary, even with these negative effects, engineers can be proud that they created a system that enhanced people's mobility and freedom to work and play at far greater distances from their homes. Companies can move freight and goods much faster and cheaper than at any time in the history of the country. So it is clear that the Interstate Highway System changed the way Americans live, work, and dream in much the same way that the Intercontinental Railroad did a century earlier.

CONCLUSION

Over the past 150 years the American Society of Civil Engineers has witnessed the United States progress rapidly, at a pace far exceeding that of any other country in the world. Some of that progress was due to the fact that in 1852 the country was just starting to flex its muscle and expand its view of what was possible in a free society with abundant natural resources. All these opportunities still needed a native-born citizenry and large numbers of naturalized citizens who had the vision and energy to develop the land of the free to its fullest potential. The Civil Engineer was the one who made much of this happen as he/she developed the waterway, railway, and highway systems, designed and built sewage and water treatment facilities, and built higher and higher buildings and longer and longer bridges. As we look back on this century-and-a-half of progress, we can be proud of what we as a profession have created. Would we have done certain things differently if we had them to do over again? Certainly, but that is from the perspective of hindsight, and it must be recognized that every major project was a creature of its time. Conditions usually are never exactly the same again. From our current vantage point we should not be overly critical of our predecessors, but instead rejoice in their accomplishments and our vision as we plan to create the world of the 21st century.

American Society of Civil Engineers   150th Anniversary Paper

Building a Better World

ASCE History and Heritage Programs

A. L. Prasuhn, F.ASCE,[1] and Neal FitzSimons, F.ASCE[2]

Abstract: Concerned that neither the general public nor civil engineers themselves were fully appreciative of the contributions of civil engineers to the development and standard of living of the country, a few dedicated civil engineers persuaded the ASCE Board of Direction to establish a history and heritage committee and a formal program in 1964. The committee has more or less flourished since that date, with an ever-expanding number of programs. The purpose of this paper is to provide an overview and description of some of these programs, with some consideration of what has worked well, and what has been less successful. It is hoped that this overview will encourage local ASCE units to reenergize their history and heritage efforts. It may also be of use to engineering groups, or perhaps others, contemplating an increased effort in the areas of history, heritage, or preservation. The primary purposes for the program were (1) to create an awareness and pride in civil engineers for the rich history and heritage of their chosen profession; and (2) inform the general public of the role that civil engineering has played in the development of the country and in improving the quality of life. These purposes are accomplished through the following activities: landmark programs, conference sessions and congresses, historic publications, recognition of civil engineers, cooperative efforts, slide shows and videos, tours, historic preservation, and other activities.

DOI: 10.1061/(ASCE)1052-3928(2003)129:1(14)

CE Database keywords: Civil engineering; Engineers; History; ASCE activities.

Introduction

Prior to 1964, ASCE had no organized history and heritage program. This is not to say that civil engineering history was ignored. Rather its overall importance was not recognized by the society. Certainly, the centennial year of 1952, and the centennial issue of *Civil Engineering*, demonstrated that ASCE was willing to support civil engineering history. For the serious reader the *Transactions of ASCE* provide a venerable record of the history and heritage of American civil engineering. In addition to the documentation of the development of civil engineering practice over the first 100 years of ASCE, the papers and ensuing discussion provide enormous insight into the leading civil engineers and their innovative solutions to problems.

Outside of ASCE, the Society for the History of Technology (SHOT) in the United States and the Newcomen Society in Britain provided some opportunities for publishing of technical articles, but other activities were limited or not readily available for American civil engineers. Other examples of individual efforts to promote civil engineering history abound as well. What was clearing lacking was a structured effort by ASCE, as the lead

society, to promote this history and heritage. This paper discusses that structure.

The ASCE history and heritage programs are under the purview of the ASCE Committee on the History and Heritage of American Civil Engineering (CHHACE), the name of which was recently changed to the History and Heritage Committee (HHC). The ASCE Board of Direction needed little encouragement to approve the establishment of a history program in 1964. That little encouragement was provided primarily by Neal FitzSimons, the second writer. The process may be summed up in his own words as follows (FitzSimons 1996):

In my personal view, the present History & Heritage program of the American Society of Civil Engineers began in the study of Dean S.C. Hollister, at Cornell University, Ithaca, New York in the Fall of 1947. I was among a group of Civil Engineering students meeting with the Dean to discuss the first post-war 'Engineers' Day'. As with most meetings, it became a bit tedious and my eyes wandered to the books on the shelves of his den library which surrounded us. A set of four large tomes caught my eye and during coffee-time, I browsed through them—*D'Architecture Hydraulique* by Bernard Forest Belidor. They were fascinating! It was this incident at Cornell that sparked my interest in engineering history.

It was not until October 1963 after my early construction career overseas and an assignment in the field of protection from nuclear weapons that I again became active in ASCE and found that there was no program which focused on history. With some diffidence I appeared before the Board of Direction and proposed a history and heritage program. At their suggestion I wrote a formal letter to William 'Pete' Wisely the Executive Director of the Society. The response came in April, 1964 from President Waldo Bowman in the form of my appointment to a Task Com-

[1]Professor Emeritus, Lawrence Technological Univ., 21000 West Ten Mile Rd., Southfield, MI 48075.

[2]Deceased March 23, 2000; formerly, Engineering Counsel, Kensington, Md.

Note. Discussion open until June 1, 2003. Separate discussions must be submitted for individual papers. To extend the closing date by one month, a written request must be filed with the ASCE Managing Editor. The manuscript for this paper was submitted for review and possible publication on October 8, 2001; approved on September 23, 2002. This paper is part of the *Journal of Professional Issues in Engineering Education and Practice*, Vol. 129, No. 1, January 1, 2003. ©ASCE. ISSN 1052-3928/2003/1-14-20/$18.00.

Fig. 1. Although seemingly out of place in California, the Bridgeport Bridge, completed in 1862, provided a major link for heavy freight between Marysville, Calif., and Virginia City, Nev. It was designated a NHCEL in 1970.

Fig. 2. Site of one of the 23 incline planes on the Morris Canal (with slack-water towpath bridge below) that were required to overcome the cumulative rise and fall of 1,674 ft as the canal crossed New Jersey. The Morris Canal was designated a NHCEL in 1980.

mittee on the History and Heritage of American Civil Engineering (CHHACE) to be chaired by Past-President Gail Hathaway. A few years later he passed the torch to me.

FitzSimons, with a rotating corps of dedicated and prominent civil engineers, lead the creation and operation of the committee for nearly 25 years. He was ably assisted throughout this entire period by staff contact Herbert R. Hands. The first writer chaired the committee from 1990 until 1997. He was followed in turn by Jerry Rogers who more recently turned over the leadership to Henry Petroski in 2000.

CHHACE flourished and has received the blessing of ASCE for over 38 years. As a clear measure of this success, other engineering societies have virtually duplicated portions of the ASCE program whereas other organizations have used the ASCE activities as a guide in setting up their own successful programs. For example, very effective programs operate in Australia, Canada, and Great Britain. Various components of the program are discussed in the following pages.

Civil Engineering Landmark Programs

The historic landmark program is the oldest, the most visible, and probably the most successful activity sponsored by the committee. The Bollman Truss Bridge at Savage, Maryland, was selected in 1966 as ASCE's first National Historic Civil Engineering Landmark (NHCEL). The program has been so successful, that in spite of very stringent qualifications which must be met before ASCE will grant National Landmark status to a project, there are well over 190 National Historic Civil Engineering Landmarks extending across the United States from border to border. Although many of the landmarks have immediate world-wide recognition (e.g., the Brooklyn Bridge, Golden Gate Bridge, or Hoover Dam), others are much more low key, such as the first United States stream gauging station at Embudo, New Mexico, or the pioneer Folsom Hydroelectric Power House, now a state park in California (Figs. 1–3).

In order to qualify as a NHCEL, the project must be at least 50 years old, and have made a significant contribution to both the civil engineering profession and at least a major region of the United States. It was decided at the outset that the NHCEL nominations must come from the local ASCE sections and branches.

Thus, local engineers must be involved; in fact, they have the responsibility for developing not only the nomination, but also a plaque presentation ceremony and the associated public relations efforts. At the recognition ceremony the plaque is presented to the project owner by the president of ASCE or another high-ranking official, and an effort is made to invite the general public, newspapers, and television stations. Historic societies and other local groups with an interest in history are frequently involved, providing help with promotion of the event and increasing the attendance.

On occasion, hundreds of people have attended the presentation ceremonies. Countless others have read about the landmark ceremonies, heard about them on radio or television, or simply stumbled upon a landmark plaque and thereby associated the visible project with civil engineering. Certainly, millions of people have been made more aware of the proud history and heritage of American civil engineering as a result of this program.

The nominations provide documentation that becomes part of a permanent database. Because of the nomination process, the records are somewhat uneven, but still are useful for research and

Fig. 3. Then ASCE President Tom Sawyer presents a plaque designating the St. Claire Railroad Tunnel between Port Huron, Mich., and Sarnia, Ontario, as a NHCEL in 1991. This was the first subaqueous tunnel in North America.

Fig. 4. ASCE's second International Historic Civil Engineering Landmark, the Zuiderzee Enclosure Dam (constructed 1927–1932) near Amsterdam was so recognized in 1983.

Fig. 6. An aerial view of the 1905 Victoria Falls Bridge in Zimbabwe. An ASCE delegation led by then President Stafford Thornton participated in the IHCEL plaque dedication in 1995.

archival purposes. Nominations that the committee feels are not of national caliber are usually designated as local or state historic civil engineering landmarks with a similar plaque and (usually a local) presentation ceremony. They become part of a parallel local landmark program. An unknown number of local or state historic civil engineering landmarks have been so designated. Some states such as California and Texas have designated a large number of local landmarks, others may have very few. Some have been recognized by a landmark plaque similar to the national landmarks, others are only recognized in a booklet or local guide. ASCE has only limited information on the local landmarks, but they add to the overall recognition of our civil engineering heritage just as surely as do the national landmarks.

Some obvious NHCELs have never been nominated. At the head of this list is the Empire State Building. However, its recent recognition by ASCE as a Civil Engineering Monument of the Millennium (ASCE 2001) rightfully celebrates its contribution to civil engineering.

The International Historic Civil Engineering Landmark (IHCEL) program is younger, but the criteria are even stricter. To

be designated an IHCEL, the project must be among the most significant civil engineering landmarks in the world. The first IHCEL, designated in 1979, was the Iron Bridge in England (the oldest extant metal bridge in the world). What a tremendous role this bridge played in the history of bridge engineering! There are now over 31 IHCELs including such diverse projects as the Forth Railway Bridge in Scotland, the Eiffel Tower in Paris, the Panama Canal, Smeaton's Eddystone Lighthouse, the Zuiderzee Enclosure Dam in the Netherlands, the Ecole National des Ponts et Chaussées in France, and the Victoria Falls Bridge in Zimbabwe.

As with the NHCELs, many of the IHCEL recognition ceremonies have been very well attended. Local color has included Morris dancers in England, a bagpipe band in Scotland, alpine horns in Switzerland, strings of firecrackers in China, and a marimba band in Zimbabwe. In addition to providing a showcase for the rich heritage of civil engineering worldwide, the IHCEL presentations have provided a unique opportunity for ASCE to work closely with engineering societies around the world. Often the U.S. delegation and the local civil engineers have an opportunity to participate in a number of social and professional activities in conjunction with the dedication ceremony (Figs. 4–6).

History and Heritage Conferences, Sessions, and Congresses

The CHHACE program has to compete with the many other professional and technical committees of ASCE. Initially it was very difficult to arrange for sessions on history and heritage themes, particularly at national conferences. The first session devoted to history at a national convention was in October 1972 at Houston, Texs. The topic, not too surprisingly, was "National Historic Civil Engineering Landmarks" and it was well attended. The next year, when the meeting was held in Washington, D.C., the local Section prepared, and ASCE published, a *Guide to the Civil Engineering Landmarks of the National Capital Area*, but no sessions were authorized. The very important local publications such as this *Guide* will be discussed below.

Probably the biggest year for CHHACE sessions during its early years was in 1976, the American Bicentennial. At the ASCE national meeting in Philadelphia almost every technical division included papers with historic themes and there was a special CHHACE-sponsored history session. Also in that year there were

Fig. 5. The Thames Tunnel in London—first subaqueous tunnel in the world—was dedicated as an IHCEL in 1991 to recognize the collective engineering genius of Marc and Isambard Kingdom Brunel. Tunnel and tunneling shield are shown in this drawing by Marc Brunel.

a number of specialty conferences that held history sessions. For example, the ASCE "Rivers '76" conference at Colorado State University had a session on the history of American waterways.

The first international history session was held at the Boston convention in 1979. A group of Swiss engineers presented a program on the history of long-span bridges emphasizing the contributions of the Swiss-American civil engineer, Othmar Amman. The following year, CHHACE sponsored a session on "Civil Engineering and Historic Preservation." This led to ASCE participation in the National Academy of Sciences' conference on the "Conservation of Historic Stone Buildings and Monuments" in 1981.

CHHACE sponsored sessions at two national conventions in 1982: in the Spring at Las Vegas, "Civil Engineers Protecting Their Heritage," and in the Fall at New Orleans, "Milestones in the Practice of Civil Engineering." Two sessions were also sponsored in 1983. At the Spring convention in Philadelphia there was a session on "Research Needs for the Rehabilitation of Structures" and in the Fall, in Houston a session was titled "Engineering Classics." At this latter session Ralph Peck spoke on his experiences in working with Karl Terzaghi to a standing room only crowd. The San Francisco ASCE meeting in October 1984 was also a very special one since it included a recognition of the 20th anniversary of CHHACE and the 15th anniversary of Historic American Engineering Record (HAER). The session's theme was "Historic Bridges and their Builders."

The history and heritage sessions have always been of high quality and well attended. On the other hand, CHHACE has not been equally successful in setting up stand-alone history and heritage conferences, as ASCE has persistently maintained, perhaps with good reason, that they would not be financially profitable. When ASCE cut back to one convention a year, CHHACE sessions were reduced accordingly.

However, the recent restructuring of ASCE has permitted a resurgence of history sessions. Beginning in 1996 at the ASCE annual convention in Washington, D.C., there was an extremely successful, parallel National Congress on History and Heritage with as many sessions as in 1976. The congress and ASCE proceedings were a joint effort between the national and local history and heritage committees. This model was repeated at the annual convention in Boston in 1998. Proceedings, edited by Jerry Rogers who was also responsible for the organization of the congresses, were published for both congresses (Rogers et al. 1996, Rogers 1998). It appears that this type of event will become a frequent practice at future annual conferences as it did at the 2001 Annual Conference in Houston, which featured international civil engineering speakers, historic bridge rehabilitation sessions, section history sessions, civil engineering history posters, and a myriad of history topics (Rogers et al. 2001).

Commencing in the early 1990s, CHHACE began a series of history and heritage breakfasts at the annual conventions. Although these have included a number of excellent presentations, attendance has been limited by the cost and the competing events.

Historic Publications

Immediately after CHHACE was authorized, *Civil Engineering* began publishing a series of articles by FitzSimons on early American civil engineers. Although the selected individuals were leaders in the development of early civil engineering, they were largely forgotten by contemporary engineers. From March 1965 through November 1966, these articles were in a "Who am I?"

format. Later the series was widened in scope to include the broader subject of civil engineering history. These continued from 1965 through 1973.

In 1966, ASCE reached an agreement with the Smithsonian Institution to establish a biographical archive of American civil engineering. One of the first results of this was the *Biographical Dictionary of American Civil Engineers* published by ASCE (1972). Under a grant from ASCE, FitzSimons and a graduate student in history wrote 170 biographical sketches of American civil engineers born prior to 1861. These were included in this volume of the dictionary. The material collected was placed in the archive. Fifteen years later, under the direction of Frank Griggs, ASCE published a second volume of this popular reference work with 218 entries born prior to 1900 (Griggs 1991).

CHHACE encouraged ASCE to begin publishing books with a history theme. The first was *The Civil Engineer: His Origins* (ASCE 1970). The biographical dictionary mentioned above was published in 1972, and the following year a joint effort of ASCE, HAER, and Texas Tech University titled *Water for the Southwest* was published (Baker et al. 1973). The next CHHACE publication, a collection of classic papers, titled *American Wooden Bridges* followed in 1976 (ASCE 1976). A similar collection of papers on water supply and wastewater treatment, *Pure and Wholesome*, was published in 1981 (ASCE 1981). These collections were primarily reprints of landmark papers from the ASCE *Transactions* of the previous 120 years. The second volume of the above-mentioned *Biographical Dictionary* was published in 1991. While these publications were reasonably well accepted, ASCE pricing policy limited their circulation, and ASCE has not encouraged subsequent committee publications.

However, ASCE and/or CHHACE have also been involved in a number of other successful publications that include *Engineering Classics* (1978), a collection of articles by the late James Kip Finch that was edited by FitzSimons; *Historic Preservation of Engineering Works* edited by Emory Kemp and Theodore Sande also published in 1978, and the result of a Engineering Foundation conference on preservation; Daniel L. Schodek's book *Landmarks in American Civil Engineering* published by the MIT Press, Cambridge, Massachusetts in 1987 describing with considerable clarity the first 100 NHCELs; *Sons of Martha* (1989), a series of readings collected and compiled by A.J. Fredrich mostly dealing with history and heritage (the title of which is taken from the poem of the same name by Rudyard Kipling in which he refers to engineers and builders as the biblical sons of Martha); *Landmark American Bridges* by Eric DeLony (1993); and *Historic American Covered Bridges* (1997) by Brian J. McKee copublished with Oxford University Press. (It might be noted in passing that Kipling in poem after poem pays tribute to the toil of civil engineers and builders.)

Cooperative agreements with the British Institution of Civil Engineers have lead to an exchange of titles and marketing between ASCE and ICE Thomas Telford, Ltd. This has resulted in a number of outstanding British references on civil engineering history and heritage becoming readily available to American civil engineers. This includes an excellent series of regional books on British civil engineering history titled *Civil Engineering Heritage*, and prepared by the ICE Panel for Historical Engineering Works (PHEW).

The American Society of Mechanical Engineers published a book on National Historic Mechanical Engineering Landmarks in 1979. Although not a civil engineering publication, it is worth mentioning to indicate the broad influence of CHHACE on engineering history.

ASCE Guide to History and Heritage Programs and Local Publications

Possibly CHHACE's most important and influential publication is the *ASCE Guide to History and Heritage Programs* (ASCE 1998). This booklet was originally intended to aid local sections in developing and operating their own programs. It has also had a much wider use by other historical groups in the United States and by many foreign engineers and societies. First distributed in 1968 as a pamphlet, it is now in its sixth major edition. The current 56-page booklet continues to serve both as a "how to do it" and a reference source on civil engineering history. As a direct result of this guide, many dozens of our local ASCE sections have formed committees similar to CHHACE.

With the encouragement of CHHACE, the local units have embarked successfully on their own history and heritage activities. These include designating local landmarks, publishing local landmark guides, developing walking tours of civil engineering landmarks in urban areas, publishing of local engineering histories, providing speakers on engineering history, and a variety of other programs. The first such local publication was the aforementioned *Guide to the Civil Engineering Landmarks of the National Capital* produced in 1973. For the American Bicentennial in 1976, the Los Angeles, Sacramento, and San Francisco sections of ASCE each published high-quality booklets on the civil engineering landmarks in their respective regions of California. The Florida section published a similar high-standard bicentennial booklet, as did many other sections. Some of these booklets remain in circulation today. Many more local publications in a similar vein have followed. Perhaps the largest single effort was a publication by the Texas section titled *The First Eighty Years, a History of the Texas Section* (Wagner and Santry 1993).

Recognition of Civil Engineers

The nomination form requests information on the engineer(s) involved in any historic project to be nominated as a NHCEL. This provides the opportunity for the landmark recognition ceremonies to properly emphasize the civil engineers that were involved in the project and their specific roles that lead to the completion of the successful project. On occasion, the names of the engineers have appeared on the plaque.

In addition to the recognition associated with the landmarks, programs have been established to specifically provide recognition and historic documentation on famous civil engineers themselves. The two volumes of the *Biographical Dictionary of American Civil Engineers* discussed previously provide a very real example. Both volumes have proven to be very popular and useful reference sources. The "Who am I?" articles provide another. A third volume of the series, an international biographical dictionary, is currently under development.

The committee also set up an oral history program. Excellent examples were provided originally by FitzSimons, and a kit was prepared to give local ASCE sections instruction in the techniques of oral history. All tapes and manuscripts are copied and placed in the Smithsonian Institution. This ASCE program has been emulated by other groups such as the American Public Works Historical Society, which has a significant oral history program. Because of the press of other activities, the ASCE oral history program has not flourished and, in fact, has become dormant of late.

In 1966, to reward ongoing efforts, CHHACE member Trent R. Dames endowed an ASCE Civil Engineering History and Heritage Award "to recognize those persons who through their writing, research or other efforts have made outstanding contributions toward a better knowledge of, or appreciation for, the history and heritage of civil engineering" (ASCE Official Register 2002). This award is not just for Americans or even confined to civil engineers. To date, two British engineers, Stanley B. Hamilton and John James, one Canadian engineer, Robert Leggett, and two non-engineers including, David McCullough, author of *The Great Bridge: The Epic Story of the Building of the Brooklyn Bridge* and *Path Between the Seas: the Creation of the Panama Canal, 1870–1914*, have been honored by this award. (More recently the same Trent Dames endowed a Fund for Civil Engineering History at the Huntington Library in San Marino, California.)

A few years ago the ASCE Board of Direction approved an American Civil Engineering Hall of Fame. This is intended to recognize pioneering civil engineers that have been deceased for at least 50 years. The Hall of Fame has not yet been put into effect, but promises to bring deserved attention to our forebears, the civil engineers that have led the way in the United States. The Hall of Fame was intended to complement existing society programs such as the Honorary Membership grade that recognizes the achievements of living civil engineers.

Cooperative Efforts

The committee's longest, and arguably most successful cooperative effort is with the National Park Service (under the U.S. Department of the Interior). Shortly after the inception of CHHACE, talks were begun with the National Park Service and the Library of Congress to establish a national program to document engineering structures and sites. After a few years of effort, a formal agreement was reached and in 1968, Congress approved funding for the Historic American Engineering Record (HAER). This was despite one Senator's incredulity that "engineers could be interested in history."

HAER was modeled after the similar Historic American Buildings Survey (HABS), which was established during the U.S. depression in the 1930s to help provide employment for architects. Today, through the work of HAER, which has closely followed the traditions and standards of HABS, the collections of the Library of Congress includes thousands of drawings, photographs, and documents dealing with all facets of civil engineering structures. Recently, the Shell Oil Foundation donated $500,000 to the Library of Congress to scan photographs and drawings for a new National Digital Library. HAER materials are the first to be scanned and should be available on the Internet.

A related effort of HAER, which also touches on preservation, involves creating an inventory of historic bridges in conjunction with federal and state departments of transportation. Although not without controversy, this program has helped to provide a basis to determine which bridges should be rehabilitated for their historic significance, which should be bypassed and preserved intact, and which should be replaced (but with possible HAER documentation). HAER has also consulted with CHHACE to develop engineering world heritage site nominations within the United States. The first site to be proposed is likely to be the Brooklyn Bridge; however, this project is on hold for political reasons.

By their very nature, the IHCEL recognition ceremonies are always cooperative efforts between the respective engineering societies. Within the United States, a number of NHCEL ceremonies have provided cooperative opportunities because of the complex nature of the landmark. For example, mechanical, electrical, and civil engineers all participated through their professional societies in the ceremonies at the first (in the United States) commercial hydroelectric plant in Appleton, Wisconsin.

CHHACE through ASCE has provided seed money and/or technical assistance for publications, television productions, movies, and other interesting ventures as they come along. One request was for financial help to aid in the cataloging of the John A. Roebling papers. Roebling was of course the engineer of the Brooklyn Bridge. Another excellent example is the television production of *The Great Bridge* (Brooklyn Bridge) by the then relatively unknown producer, Ken Burns. Budget limitations certainly impact CHHACE's ability to accomplish all of its goals and potential activities. For some time, it has not been possible to support these efforts, no matter how worthy.

In the early 1990s CHHACE, the History Committee of the American Society of Mechanical Engineers (ASME), and the Public Works Historic Society (of the American Public Works Association) set up a committee to determine whether society history committees could jointly accomplish more than was apparently possible by the individual societies acting alone. If some success could be demonstrated, it was felt that other societies would be willing to join the effort. After some initial success, and one national workshop in Washington, D.C., the effort floundered, partly because of the lack of strong ASCE backing.

However, the effort was restarted in 1998 with a Consortium Engineering History Workshop involving CHHACE, ASME, and the National Park Service. This has evolved into an annual summer event with the different engineering societies (IEEE, ASME, etc.) organizing the one-day workshops. Attendance has ranged from 30 to 50 over the past 3 years. It remains to see if this will prove more successful than the first attempt.

Finally, CHHACE through it members and ex-members, has frequently responded to all types of queries from the press, television stations, other engineers, other societies, and the general public. Topics have ranged from the Golden Gate Bridge, to IHCELs, to the Seven Modern Engineering Wonders of the World.

Historic Preservation

CHHACE has not been specifically involved in the preservation of engineering works on a regular basis. However, it encourages involvement of engineers in preservation projects, and many civil engineers, committee members or otherwise, have been so involved. Committee involvement has usually been on an ad hoc basis, and has usually taken the form of responses to requests, e.g. how to preserve an old bridge, or should an old bridge be preserved. However, the landmark program has surely had some influence on the preservation of engineering structures. The best and biggest example of CHHACE involvement is perhaps the Statue of Liberty in New York Harbour. When rehabilitation became essential, ASCE, with the assistance of CHHACE, raised $250,000 in the early 1980s to permanently record the structural details of the Statue of Liberty.

The committee position has always been to approach preservation on a realistic basis. While preservation is strongly encouraged and is frequently the desired end result, it has always been felt appropriate to include the economic benefits of rehabilitation and/or preservation when considering the future of engineering structures. Adaptive reuse is also strongly encouraged when preservation intact is not feasible. If actual preservation is not possible, then documentation (e.g., by HAER) is encouraged, if appropriate.

Tours

On a number of occasions, particularly under the leadership of the first writer, CHHACE has cosponsored tours to the Netherlands and Great Britain in conjunction with IHCEL presentations. The tours have ranged in size from a party as small as approximately 16 for a visit in 1983 to the Netherlands to recognize ASCE's second IHCEL, the Zuiderzee Enclosure Dam, up to about 48 participants in 1985 on a visit to England and Scotland to recognize the Forth Bridge as an IHCEL. The tours have generally been 12–14 days in length, and covered a wide range of historic and modern civil engineering works. Participation by the local engineering societies, particularly individual members of the ICE Panel for Historical Engineering Works (PHEW) in Great Britain who have joined the group on a day-by-day basis, made the tours truly memorable experiences.

As presidential international visits became a more formal process, frequently with relatively large delegations, they replaced the CHHACE tours. However, committee members have continued to provide support to ASCE on presidential visits both across the United States and around the world. This has occurred most often when historic landmark presentations have been involved; in which case CHHACE members have assisted with background information and material for presidential addresses.

Slide Shows and Videos

ASCE has produced numerous slide shows and several have had an historical theme. The first produced by the History and Heritage Committee was "Five Thousand Years of Civil Engineering" (1970) based upon a book by Captain Charles Merdinger, USN, who was a member of the first committee. In 1974 after ASCE recognized its first 100 NHCELs, a popular slide show was developed emphasizing the national landmark program. Slide shows have always been well received at local section meetings and public presentations by ASCE members; and many individuals have personal collections based on the landmarks and other historic projects. In 1976 for the American Bicentennial, ASCE, in cooperation with other engineering societies, produced a slide set which illustrated the history of American engineering by using slides of commemorative U.S. postage stamps. Another slide show was "Agents of Progress" with a similar theme, but using Currier and Ives prints.

Videos, CDs, PowerPoint presentations, and use of the Internet, have now become the more popular mediums for presentations of this type. A recently produced video is discussed under "Other Activities." ASCE, with some CHHACE input, has developed a history perspective on it Web site. There is an enormous opportunity for future projects in these areas.

Other Activities

In 1994 CHHACE organized a board of consultants with worldwide representation to denominate the seven modern civil engineering wonders of the world. The globally circulated magazine *Popular Mechanics* included the resulting consensus as the featured article in its December 1995 issue (Pope 1995). The selected seven modern civil engineering wonders are the Channel Tunnel between England and France, the CN Tower in Toronto, the Empire State Building in New York City, the Golden Gate Bridge in San Francisco, Itaipu Dam on the Brazil/Paraguay border, The Netherlands North Sea Protection Works, and the Panama Canal. These projects are now included on ASCE's Web site (see www.asce.org) and queries are still received concerning these projects.

At the request of CHHACE, ASCE recently approved a policy statement on history and heritage. This puts the Society officially on record as recognizing the overall importance of civil engineering history and heritage activities. Among other intents, the policy encourages civil engineers to be individually involved in history and heritage activities, and urges civil engineering faculty and university civil engineering departments to include a historic presence in their curricula.

Another recent activity is the creation of "History Modules," an effort lead by former committee member, Frank Griggs. The idea is to provide, at modest cost, an off-the-shelf product pertaining to the history of civil engineering that busy engineering professors, or others, can make use of without in-depth research on their own part. Initially, a total of four articles were written on various facets of civil engineering history of which three were published by ASCE and one by the Boston Society of Civil Engineers Section. The first three were on the flexural formula, the development of the truss, and the Manning equation and the final one was on the history of geotechnical engineering. It was originally proposed that these topics could be converted into short slide shows, but it may now become necessary to create the somewhat more expensive videos. To lead off, a 40 min video on the history of the truss bridge was produced by Frank Griggs, and distributed by ASCE. It is intended that more would follow, but financial and time constraints have slowed progress on the project.

From time to time, the committee has also been asked to cooperate with other societies, such as the Society for the History of Technology (SHOT) and the Society for Industrial Archaeology (SIA). This has involved joint sponsorship of meetings, help with meeting publicity, and other projects. Some of the committee members have also been active or have had leadership positions in these other organizations.

Closing Remarks

The participation of engineers in history and heritage activities has come a long way since the ASCE Committee on the History and Heritage of American Civil Engineering was established in 1964. There are many similar programs in countries around the world. The CHHACE feels that it has played an important role in the progress that has been made. In the words of the late FitzSimons, the programs "seek to educate both our own profession and the public at large, in all facets of the history of Civil Engineering from the monumental to the mundane. In this way, we will all come to better understand the role engineers have had,

and continue to have in transforming society. Our profession, even before it was formally recognized as such, has always responded to society's changing needs and demands and has therefore had an enormous impact on our economy, the appearance of our landscape, and the health of our people" (FitzSimons 1996).

This has become even more true in these troubled times. Many challenges remain. As population increases, and there are more and more demands for new construction, both in the developed and developing countries, preservation of the engineering heritage continues to demand constant vigilance and an active voice from knowledgeable engineers and related professionals. As to the proper recognition of the contributions of civil engineers, we have hardly scratched the surface.

Acknowledgments

This paper is the outgrowth of some 42 years of collective experience working with the ASCE History and Heritage Program. Some parts of the paper have been presented previously by one or the other of the two writers. The landmark photographs are all taken by the first writer.

References

ASCE. (1970). *The civil engineer: His origins*, ASCE, New York.

ASCE. (1972). *A biographical dictionary of American civil engineers*, ASCE, New York.

ASCE. (1976). *American wooden bridges*, ASCE, New York.

ASCE. (1981). *Pure and wholesome*, ASCE, New York.

ASCE. (1998). *History and heritage of civil engineering*, ASCE, Reston, Va.

ASCE. (2001). "ASCE designates Empire State Building as civil engineering monument of the millenium." *ASCE News*, 26(8), 1 and 6.

ASCE. (2002). *Official Register 2002*, ASCE, Reston, Va.

Baker, T. L., Rae, S. R., Minor, J. E., and Conner, S. V. (1973). *Water for the southwest*, ASCE, New York.

FitzSimons, N. (1996). "History and heritage programs of the American Society of Civil Engineers." *Proc., Annual Conf., Canadian Society for Civil Engineering*, Vol. 1, Montreal.

Griggs, F. E. (1991). *Biographical dictionary of American civil engineers*, Vol. II, ASCE, New York.

Pope, G. T. (1995). "The seven wonders of the modern world," *Popular Mech.*, 172(Dec) 48–56.

Rogers, J. R., ed. (1998). *Engineering history and heritage*, ASCE, Reston, Va.

Rogers, J. R., and Fredrich, A. J., eds. (2001). *International engineering history and heritage: Improving bridges to ASCE's 150th anniversary*, ASCE, Reston, Va.

Rogers, J. R., Griggs, F. E., Jaske, R. T., and Kennon, D., eds. (1996). *Civil engineering history: Engineers make history*, ASCE, New York.

Wagner, F. P., and Santry, I. W. (1993). *The first eighty years, a history of the Texas Section*, 2nd Ed., Texas Section, ASCE, Austin, Tex.

American Society of Civil Engineers

150th Anniversary Paper

1852 - 2002

Building a Better World

State of the Art of Structural Engineering

Jose M. Roësset, Hon.M.ASCE,[1] and James T. P. Yao, Hon.M.ASCE[2]

Abstract: The objective of this paper is to provide an overview of the developments in structural engineering that took place during the past century. This overview includes (1) some of the major structural accomplishments as selected by the writers, (2) the advances in mechanics as the basis of structural analysis, (3) the development of new materials, (4) new fields of research and practice, and (5) the changes in the way design projects are performed. In addition, the writers' personal predictions for future developments during the 21st century are also presented. One of the main features affecting the evolution of structural engineering over the last part of the 20th century has been the advent and rapid development of digital computers as engineering tools. Computers can be used to perform complex and cumbersome computations and to enhance worldwide communications, both with great speed and reliability. This has already had an important effect on the way we design structures and educate civil engineers, but the impact on structural analysis and design as well as on construction planning and management is still in progress. We believe that this impact will be fully felt in the 21st century. Computers will liberate engineers from tedious and routine computations, allowing them to concentrate on more creative and important endeavors. They will facilitate the design of constructed facilities as complete systems rather than by considering each subsystem (such as structure and foundation) separately. They will lead finally to the needed integration of the design and construction processes.

DOI: 10.1061/(ASCE)0733-9445(2002)128:8(965)

CE Database keywords: Structural engineering; History.

Introduction

As we proceed into the 21st century and a new millennium it is worthwhile to reflect on what the 20th century brought us, on where we stand today and where we should be going in the future. As stated once by the late Charles L. Miller at Massachusetts Institute of Technology (MIT), rather than attempting to predict the future, it is more important to decide what the future should be and to try to influence change in that direction. Even this is a personal matter and what anybody sees as desirable features may be quite different from what others would select. The material presented here represents the writers' personal opinion, which is inevitably biased by their own backgrounds and education. This is even more so in selecting names of civil engineers that have made an impact on their profession. In attempting to select only a few, one has to be influenced again by personal feelings and background. No attempt was made to present a comprehensive or exhaustive list. The names mentioned are intended only as examples. Equally incomplete is the list of possible references

dealing with structural engineering, structures, structural models, or the built environment, all terms used by Grigg et al. (2001) in their 264-page treatise entitled *Civil engineering practice in the twenty-first century.*

In this paper the writers attempt to look at some of the accomplishments of the 20th century in structural engineering and to put them in perspective with respect to earlier work. One can look then at some of the new developments that are likely to continue during the 21st century and some of the perceived needs. The main changes in structural engineering during the 20th century and in years to come are due to the developments in digital computers both as powerful tools to perform cumbersome computations and as new means of communication between members of design teams, professors and students, or any other persons. These changes affect both the practice of civil engineering and engineering education.

The computational capabilities provided by today's computers liberate the structural engineer from the laborious task of performing detailed stress analyses and allow the designer to concentrate on the more creative parts of the design process. This implies exploring alternatives, accounting for uncertainties, and integrating properly all the different components of the system (e.g., structure, foundation, and equipment) to be designed. In education they allow students to acquire experience in structural behavior by conducting simulations, looking at various alternatives, and observing visually the structure's response to different excitations.

The new ease in worldwide communications facilitates the concurrent work of many different teams in geographically dispersed regions of the world. This changes significantly the way the design of large projects is conducted. From the educational viewpoint, the new communication tools such as the Internet and e-mail complement the more traditional methods of teaching.

[1]Professor of Civil and Ocean Engineering, and Holder of Wofford Cain Chair in the Dept. of Civil Engineering, Texas A&M Univ., College Station, TX 77843-3136.

[2]Professor of Civil Engineering, and Holder of Lohman Professorship in Engineering Education, Texas A&M Univ., College of Engineering, College Station, TX 77842-3136.

Note. Associate Editor: C. Dale Buckner. Discussion open until January 1, 2003. Separate discussions must be submitted for individual papers. To extend the closing date by one month, a written request must be filed with the ASCE Managing Editor. The manuscript for this paper was submitted for review and possible publication on September 25, 2001; approved on April 2, 2002. This paper is part of the ***Journal of Structural Engineering***, Vol. 128, No. 8, August 1, 2002. ©ASCE, ISSN 0733-9445/2002/8-965-975/$8.00+$.50 per page.

Combined with the computational capabilities, they allow the creation of virtual laboratories where students can carry out virtual experiments for those phenomena that are well known and thus can be simulated, as would be the case for most classroom demonstrations. Clearly, research work dealing with the discovery of new, as yet unknown, phenomena will still require physical experiments. The advantage of virtual experiments or computer simulations for educational purposes is that they can be repeated at will, at the convenience of each student, easily changing geometry and materials to observe immediately the effects of these changes, and learning by induction from direct and visual observation of the results.

The various structural engineering handbooks published during the 20th century, from the Kidder-Parker Architects' and Builders' Handbook (first published in 1884) (see Parker 1931), to Gaylord and Gaylord (1968, 1990) or Chen (1997), provide through the titles of their chapters and their contents an excellent and detailed history of the evolution of the field during that time. Although some topics are common, these handbooks have different chapters treating various subjects. It is thus necessary to read all of the recent ones in order to obtain a complete view of structural engineering in general. Equally instructive are the lists of courses offered, and their descriptions, in the catalogs of Civil Engineering Departments of the leading universities (particularly the most progressive ones that are willing to incorporate new subjects early on), as well as the lists of research projects in progress at these institutions at any given time. In this paper no attempt is made to present an exhaustive list of all the major developments that have taken place. The discussion is limited to those with which the writers are most familiar. Thus, their own education and background once again bias the selection.

Background

L'École Centrale des Ponts et Chaussees, the first Civil Engineering School in the world, was established in Paris, France, in 1747. In the late 18th century, John Smeaton in England coined the name of "civil engineer" as being distinct from the military engineer. The Smeatonian Society can be considered as the precursor of our civil engineering societies. While there were a number of prestigious engineers during this time (Vauban, Coulomb, Smeaton), in France and in Great Britain, civil engineering really took off as a technical profession during the 19th century, with most schools created in the early part of the century imitating the French model. Famous engineers like Castigliano, Cauchy, Navier, Rankine, or Saint-Venant were not only theoreticians but also practicing engineers who designed well-known and important structures. Other illustrious practicing structural designers included Gustave Eiffel, designer of many steel bridges (the Garabit viaduct, for instance) in addition to his famous tower in Paris; Thomas Telford, who designed an early iron chainlink suspension bridge (the Menai suspension bridge in 1826) in addition to many other bridges and aqueducts; and John Roebling, who is considered the father of modern suspension bridges and designed the Brooklyn Bridge. It can be said in fact that most of the structural concepts of the 20th century were a continuation of the accomplishments of the 19th century.

This rich history of civil engineering as a technical profession (without even accounting for all the magnificent civil works conducted by master builders from antiquity) is both a matter of pride and a reason for concern. Because so many of the basic theoretical developments in mechanics took place over a century ago (starting with Hooke's law in 1660), newer engineering professions tend to look at civil engineering as an established discipline without significantly new and revolutionary advances. Yet the 20th century saw not only new frontiers in the size of the structures built (bridge spans, heights of buildings and dams) but also the development of new analytical and numerical tools, new structural concepts, new materials, new construction techniques, and even new subdisciplines. The 25 longest span (suspension) bridges in the world, with spans exceeding 700 m, were all designed and built in the 20th century. So were all the steel cantilever truss bridges with spans exceeding 350 m, all steel arch bridges with spans longer than 300 m, and all concrete arch bridges with spans over 240 m. Prestressed concrete, high-strength concrete, composite construction, light-gauge steel construction, and, more recently and still in the incipient stages, composite and smart materials were all creations of the last century. So were cable-stayed bridges, shell roofs, double curvature arch dams, high-rise buildings with tube action, and offshore platforms. To the names of the famous engineers of the 19th century we can add in the 20th among many others those of Othmar Amman (George Washington and Verrazano Narrows Bridges), David Steinman (Mackinac Bridge), and Joseph Strauss (Golden Gate Bridge) in suspension bridges; Eugene Freyssinet (Plougastel Bridge), Robert Maillart (Salginatobel Bridge), and T. Y. Lin in prestressed and reinforced concrete structures; Fazlur Khan and Les Robertson in high-rise buildings; Eduardo Torroja, Pier Luigi Nervi, and Santiago Calatrava as designers of shell roofs and exciting new types of structures; Ray Clough, Hardy Cross, Fred (A. M.) Freudenthal, Wilhelm Flugge, Gaspar Kani, Eric Reissner, Stephen Timoshenko, and Olgierd Zienkewickz as analysts; and George Housner and Nathan Newmark as the fathers of earthquake engineering among other accomplishments.

Structural Accomplishments

General Comments

A number of excellent books such as those by Kirkham (1914, 1933), Sheiry (1938), Husband and Harby (1947), Goldberger (1981), Billington (1983), Collins (1983), Southworth and Southworth (1984), Westerbrook (1984), Nakamura (1988), Billington (1996), Jackson (1997), Berlow (1998), Stoller (2000), and Abramson (2001) provide detailed descriptions and photographs of various engineering landmarks of the 20th century. The reader is referred to these publications or similar ones for a more complete coverage of the topic and pictures of outstanding structures.

Tall Buildings

Tall buildings have always challenged the imagination of engineers and fascinated the general public. Even if the Ingalls Office Building in Cincinnati, completed in 1902, has been credited as the first skyscraper, tall buildings are emblematic of 20th century structural engineering achievement. Goldberger (1981) reviewed the history of skyscrapers and discussed several of them in some detail. As building height increased, a number of new problems, some requiring important research efforts, had to be faced. Traditional methods of analysis that ignored the axial deformation of the columns were no longer valid, for instance, to determine the lateral displacements due to wind loads. The economics of the project, associated with the amount of usable space per floor, dictated the need for structural solutions different from the standard moment-resisting frames. Wind vibrations became a potential discomfort problem for the tallest structures, and methods to control these motions had to be devised (e.g., tuned mass dampers and viscous dampers). New types of foundations and new measures had to be conceived in some cases.

Each skyscraper is a unique project, incorporating new knowledge and experience. Skyscrapers like the Petronas Towers (445 m, 1997) and the Plaza Rakyat (376 m, 1998) in Kuala Lumpur, Malaysia, the Sears Tower (435 m, 1974) in Chicago, the Jin Mao Building (414 m, 1998) in Shanghai, China, or the World Trade Center Towers (410 m, 1973, destroyed during the attack on America of September 11, 2001) and the Empire State Building (375 m, 1931) in New York City have seen a significant evolution in their structural configurations and are not mere replicas of each other. On July 18, 2001, the Empire State Building was designated as a Civil Engineering Monument of the Millennium (ASCE 2001).

Long-Span Bridges

Long-span bridges have also attracted young people to the structural engineering profession. ENR (1999) listed the 1936 San Francisco–Oakland Bay Bridge, the 1937 Golden Gate Bridge, the 1957 Mackinac Bridge, the 1964 Chesapeake Bay Bridge-Tunnel, the 1966 Severn Bridge, the 1986 Sunshine Skyway Bridge, the 1995 Normandy Bridge, the 1998 Akashi Kaiko Bridge, and the 1999 Tatara Bridge as outstanding achievements of the 20th century. Cable-stayed bridges such as the Tatara Bridge in Japan, the Sunshine Skyway Bridge in Tampa Bay, Fla., or the El Alamillo Bridge in Sevilla, Spain, are also very aesthetic solutions developed during this last century. The Golden Gate Bridge was designated by ASCE as the Civil Engineering Monument of the Millennium in the long-span bridge category (ASCE 2001).

Dams

Although dams (either small diversion dams or larger dams for water storage) have been built since antiquity (e.g., Smith 1975), large dams are another kind of structure that fascinates the public. The Grand Coulee Dam and the Hoover Dam (Jackson 1997; ENR 1999) are two notable examples. The Hoover Dam was the tallest curved gravity dam upon completion in 1936, and continues to be a major tourism attraction up to today. Essentially all dams over 60 m high were designed and built during the last century, the only two exceptions being the 62 m high Goufre D'Enfer gravity dam built in France in 1866 and the 71 m high Puentes Dam built in Spain in 1884.

Shell Structures

Thin shell structures used for long-span roofs, dams, tanks, or cooling towers are creations of the 20th century. The Dorton Arena in Raleigh, N.C., the large domes of auditoria and arenas, the beautiful hyperbolic paraboloids of Felix Candela, the parabolic shells of the St. Louis Priory Chapel, are all innovative and impressive structures. So are the tension structures of Frei Otto or membrane roofs supported by air pressure such as the Pontiac Dome. While shell structures have unfortunately lost some of their appeal in civil engineering, many structural engineers have found challenging jobs in the aeronautics and aerospace industries where shell type structures are used and needed.

Offshore Platforms

Offshore platforms are another creation of the 20th century. The first platform for oil drilling out of the sight of land was installed in 1947 off the coast of Louisiana, in 6 m of water. The water depths at which platforms were installed increased steadily and at a fast pace. By 1955 they had reached 30 m, by 1965 67 m, by 1976 255 m (the Hondo platform, off the coast of California, which had to be fabricated in two pieces and then welded on site), by 1978 300 m (the Cognac platform, in the Gulf of Mexico, fabricated in three modules), and by 1988 405 m (the Bullwinkle platform in the Gulf of Mexico fabricated now in a single piece). Both Cognac and Bullwinkle won awards as the civil engineering achievements of the year. These were all steel jacket structures. In the North Sea the difficulties in accessing the platforms during severe and frequent storms required larger storage capabilities, which, coupled with the European preference for reinforced concrete, led to the design of gravity platforms. Neither of these solutions was feasible, however, as the oil industry was forced to venture into deeper and deeper waters in search of new reservoirs, because their natural periods would become too close to those of the design waves. It was thus necessary to conceive new structural types of much more flexible structures whose natural periods would be again far removed from those of the waves but on the other side of the spectrum.

The structures that have been and are being designed for water depths of the order of 1,000 m and more consist basically of a large floating body tied to the sea bottom by vertical tethers, prestressed by the hull's buoyancy, or mooring lines (catenary or taut moorings). The possible solutions are tension leg platforms (TLPs), spars, floating production systems (FPSs or semisubmersibles), and tanker based floating production, storage, and offloading systems (FPSOs). The Auger platform, a TLP, was installed in the Gulf of Mexico in 1994 in 858 m of water and again won the civil engineering achievement of the year award. It was followed by Mars (887 m in 1995), Ram Powell (900 m in 1997), and Ursa (1,140 m in 1998), all TLPs, as well as the Neptune Spar (570 m in 1997) and the Genesis Spar (840 m in 1998). The industry is looking now at water depths of the order of 3,000 m. The uncertainties in the loads, the need to deal with the interaction between the structure, the surrounding sea, and the bottom soil, and the difficulties of working in such extreme water depths have made offshore structures the new exciting challenge for structural engineers over the last quarter of the century.

Concluding Remarks

We have attempted in the previous sections to provide an overview of some of the developments in different types of structures during the 20th century since the subject of this paper was structural engineering. We must realize, however, that no civil structure can survive without an adequate foundation and that designs become meaningful only when constructed. One cannot ignore entirely therefore the progress made in soil mechanics and in construction methods.

Although Coulomb, Rankine, and others had already been involved with important problems related to soils, the field of soil mechanics as a separate discipline emerged in the 20th century and expanded rapidly due to the contributions of Karl Terzaghi and a large number of outstanding followers worldwide. The progress in our understanding of soil behavior under different conditions and in our ability to numerically predict their behavior has been considerable. It has led to the safer and more reliable design of a number of different types of foundations, the design and construction of a large number of earth dams, and the understanding of how soils affect the characteristics of earthquake motions and the dynamic response of structures. Unfortunately, the needed integration between the design of a structure and its foundation is still lacking. The expansion of the field of soil mechanics with a change in name to geotechnical engineering, the creation of a number of subdisciplines or areas (e.g., geotechnical earthquake engineering or geoenvironmental engineering), and the widening gap at the professional level (ASCE, for instance) between structural and geotechnical engineers are aggravating the situation. It is necessary to consider a structure and its foundation as a single system and to do so it may be necessary for structural engineers to learn more about soils and to design their own foundations.

The 20th century has seen also substantial advances and innovation in construction methods and techniques, which once again influence the designs. Different forms of formwork, precast construction, and segmental construction of bridges are but a few examples. As construction and project management have evolved and become separate programs within civil engineering departments at universities, the gap between structural and construction engineers has increased. We consider this gap a deplorable situation. An effort must be made to integrate design and construction as well as the design of the various subsystems (e.g., structure and foundation).

Structural Analysis

General Comments

Girvin (1948) gave a historical appraisal of mechanics beginning with the first logical proof of Archimedes' principle in approximately 300 B.C., following with the early Greek developments, the medieval period (500–1500 A.D.), the Moorish culture in Spain, the contributions of Roger and Francis Bacon, the Renaissance (1400–1600 A.D.), and the modern period when the principles of statics, dynamics, strength of materials, and theory of elasticity were established.

The bases for structural analysis were set by Hooke and Mariotte in the 17th, by Coulomb, Euler, and Lagrange in the 18th, and by Airy, Betti, Boussinesq, Castigliano, Cauchy, Green, Kirchhoff, Lamb, Muller-Breslau, Navier, Poisson, Rankine, Ritter, Saint Venant, Stokes, Voigt, and Young among others in the 19th century. In the first half of the 20th century Hardy Cross in the United States and Gaspar Kani in Europe developed schemes based on the mathematical solution of simultaneous equations by iteration or relaxation that enabled the computation with a slide rule of the bending moments in large frames with negligible axial deformations. These developments had a major impact in structural analysis and the method of moment distribution (or the method of Cross) in particular was adopted worldwide. The first half of the century saw also the development of the elastic theory of shells with the work of such pioneers as Dischinger, Krauss, Flugge, Mushtari, Novozhilov, Pfluger, Reissner, and Vlasov.

Computer Methods

The main changes and innovations in structural analysis occurred in the second half of the century and were due to the advent of the digital computer. The formulation of matrix structural analysis of assemblies of linear members (plane and space trusses and frames as well as plane grids) was followed by the development and implementation in general purpose computer programs of the finite element and the boundary element (or boundary integral equation) methods for two- and three-dimensional continua. This allowed solving complex structures such as plates and shells of arbitrary shapes with their actual boundary conditions. Previously it was necessary in many cases to introduce approximations and/or to simplify the support conditions in order to find analytical, closed form, solutions. This represented also an important change in emphasis from continuous to discrete mathematical models, and made the routine structural analyst obsolete. Yet continuous solutions, where they exist, remain valuable as a check on the accuracy of the discrete formulations and on the validity of the model or discretization selected. Although analog computation was popular in the 1960s, the fast development of digital computers has dominated the field. Recently, because of the experimental usage of biomaterials in computers, hybrid computers again appear promising.

The advantages and potential pitfalls of excessive dependence on computers is a topic that deserves serious discussion but it falls beyond the scope of this paper.

Nonlinear Analysis

Our understanding of the effect of changes in geometry, due to the deformation and displacements caused by loads, on the stiffness and stability of structures has seen significant progress in the 20th century, beyond the simple concepts of Euler buckling. The work of Timoshenko and Bleich provided designers with the tools to estimate the buckling loads for structural members subjected to a variety of loading conditions, plates, shells, and pipes. This work was extended by a large number of researchers, with more rigorous mathematical formulations, with experimental work, and with simplified procedures at the more practical level. The distinction between bifurcation and limit point buckling and the possibility of predicting both from a single formulation were clarified. The ability to predict the buckling load of a complete assembly of members instead of considering each column as a separate entity with a different buckling load represented another important improvement. Much progress was achieved in the study of shell buckling and nonlinear instability. Yet many of the simplified models and analogies used in practice have at times confused practicing engineers, making them think of stability as a strength rather than a stiffness consideration and leading to improper estimates of buckling effects under dynamic loads.

Plastic analysis was also developed during the 20th century as a means to estimate ultimate loads and failure conditions with relatively simple models, predicting in some cases upper or lower bounds. So were fatigue analysis and fracture mechanics. Once again the finite element method changed drastically the way nonlinear behavior due either to nonlinear material properties or to changes in geometry (stability considerations) was studied. Instead of being able to predict only the ultimate load and failure mechanism for relatively simple assemblies of members or to estimate buckling loads under a number of simplifying assumptions, one can now follow the behavior of a complex structure as the loads increase and it undergoes inelastic deformations, until a limiting condition is reached. This is a very important capability, particularly when considering extreme loads, such as the maximum credible earthquake, for which it might be too expensive to maintain the structure in the linear elastic range, or when considering nonlinear instability or progressive buckling of pipelines. It is an area in which much work remains to be done, to gain confidence in the existing nonlinear constitutive models for various materials, to incorporate three-dimensional effects, and to account for nonstructural components when dealing with actual buildings (rather than idealized bare plane frames).

Structural Dynamics

Structural dynamics is another area that has seen a substantial development in the 20th century, dealing with the problems of impacts and moving loads, wind effects on very flexible structures, design of machine foundations, and earthquake and off-shore engineering. These last areas have brought with them the need to account for the interaction between different media, such as the structure and the underlying (or in some cases surrounding) soil, the structure and the surrounding water, or the combination of water, structure, and soil as a single system. Soil-structure, fluid-structure, and fluid-soil-structure interactions have become important areas of research. During the 20th century, many significant advances have been made in these fields.

Concluding Remarks

Mechanics are the basis for structural analysis and should be taught to all engineering students (Roësset and Yao 1988). Unfortunately, many practicing structural engineers do not appreciate their education in mechanics or the many credit hours devoted to mechanics courses that they took. Educators need to examine how mechanics can be taught in universities so that future engineers will appreciate its role and importance and be able to see its practical usefulness.

Materials

General Comments

We tend to think of iron first, and steel next, as the materials of the 19th century, and attribute to the 20th century reinforced concrete. We should remember, however, that reinforced concrete, first used for flowerpots in 1857, was already extensively used for structures in Europe by the 1880s and that Mörsch's book on the design of reinforced concrete structures was published in 1902. The last century, however, has seen important changes in the types and properties of these two materials. We have seen new alloys for steel, higher-strength steel, and high-performance steels. Meanwhile, additives for concrete were developed along with a greatly enhanced understanding of their chemical and physical characteristics. Dams and pavements have seen the use of roller compacted concrete. The combined use of concrete and steel members in composite construction has proven to be a popular and interesting solution for buildings. The strength of both materials has increased continuously with the major changes in concrete and during the last quarter of the century. It is possible to have today concretes with compressive strengths of the same order of magnitude as those of steel and with similar tensile strength, as well as flowable and self-compacting concrete.

New Materials

The second half of the 20th century has seen the development of materials science as a separate discipline. Plastics, fiberglass, and more recently composite materials consisting of a resin (thermoplastic or thermoset) base with glass or graphite fibers (or a combination of both) have found their way in a number of important applications in the automotive, aerospace, and naval industries. It is possible today to design materials, just as one designs a structure, so as to obtain any desired combination of strength, stiffness, toughness, and ductility. It is also possible to design "smart" materials whose properties change following a desired pattern depending on various conditions (e.g., states of stress or strain, temperature, humidity, and electric current). Unfortunately, civil engineers have not played a major role in this effort because of the high costs of these newer materials.

To date, composite materials (fiber-reinforced polymers) have found application in traditional civil engineering structures for seismic strengthening and retrofit, structural repair, and new forms of bridge decks. Different types of fibers have also been used as reinforcement in reinforced or prestressed concrete structures.

In the offshore field, weight is an important factor, making the use of composites very attractive. It has been reported that (1) a pound saved in the weight of floating structures such as tension leg platforms can represent a saving of about $4 if it is properly accounted for in the design, and (2) the use of phenolic compounds for the grate floors and stairs of the Mars platform resulted in total savings of some $25 million. Even so, the civil engineering applications have not yet reached the volume of these in the aeronautical, naval, and automotive industries. In some cases more research is necessary to understand the long-term behavior of these materials in potentially aggressive environments under different states of stress. Whether the use of these new materials in civil structures will expand will depend primarily on their unit cost and their availability in large quantities with a reliable supply.

Nondestructive Evaluation

At the same time that new materials are being developed, techniques to test these materials (as well as the conventional ones) in place (in situ) and in a nondestructive way have been established. This is essential for quality control, particularly for new materials, and to assess the condition of existing structures, particularly old ones, to evaluate their load resistance capacity and to identify potential damage in these existing structures. The use of nondestructive evaluation (NDE) techniques to determine material properties or structural behavior requires, for a proper interpretation of the data, the use of system identification and damage assessment methodologies that are also the result of research conducted primarily during the last quarter of the century.

Dealing with Uncertainties

General Comments

The desire to account in a more rational way for the uncertainties that exist in the prediction of the loads acting on a structure, in the properties of the materials used, and in the accuracy of the methods of analysis, led engineers in Europe to modify the form of their codes and to make use of probabilistic concepts by the middle of the 20th century. The interest in applying the theory of probabilities to real civil engineering problems started shortly after in the United States, at the academic level, in institutions such as Columbia University, MIT, Stanford, and the University of Illinois, among others. Earthquake engineering, particularly when dealing with the design of important facilities such as nuclear power plants, and offshore engineering, among other fields, are fertile grounds for the application of probabilistic methods. Risk analyses have started to become standard requirements in these fields. Yet the introduction in practice of probability concepts has been slow. Even when the probability-based load and resistance factor design (LRFD) specifications were introduced in design codes, they were used without explicit mention of probabilities. The load and resistance factors were selected through calibration, in order to obtain results similar to those of the working stress design and past experience, rather than on the basis of the existing uncertainties.

Uncertainty Analysis

Almost all civil engineering curricula have now a basic introductory course on statistics/probabilities. In some instances this is purely a mathematical treatment. In others the course emphasizes applications to real civil engineering problems. Unfortunately, this course usually is not followed by other courses. Unless the material is applied again in following design and analysis courses so that the student can see its practical importance, it will be only an additional requirement that can be forgotten once the course is over. Much remains to be done to make probabilities and reliability an integral part of civil engineering education and practice. Yet one must remember that this is just a way to account rationally for uncertainties in predicting the performance of a structure during its life. The uncertainties can only be reduced through a better understanding of the actual physical processes involved. The existence of uncertainties should not be construed as an excuse to introduce systematic errors through the use of either inadequate models or methods of analysis.

Structural Reliability

Structural reliability has been traditionally defined as the probability of the useful life of a given structure exceeding a certain time period. This is a good measure of the level of safety of a structure but it is more meaningful to base reliability of existing structures on symptoms that can be related to structural damage. Cempel (1991) in Poland introduced the concept of symptom-based reliability first, in connection with testing of diesel engines based on their noise level. The same principle could be applied to civil engineering structures. Natke and Cempel (1997) and Wong and Yao (2001) attempted to apply it to civil infrastructure systems. However, the symptoms indicative of structural damage in these structures (damage states, the variables that characterize them, and the values of these variables corresponding to each state) are yet to be defined.

Fuzzy Logic

While academic researchers apply structural reliability principles and develop new methodologies assuming that the required probability distributions (including the tails of these distributions that characterize the rare, extreme events) are perfectly known, engineers in practice find that in many cases it is nearly impossible to select with accuracy a probability distribution or its parameters. The best that can be done in many cases is to define vaguely the probability of an event occurring as low, medium, or high. This has led to the theory of fuzzy sets. Zadeh (1965) published the first paper on fuzzy sets. Basically, the theory of fuzzy sets deals with those events that are meaningful but not well defined. For example, a damaged structure might be classified as collapsed, severely damaged, lightly damaged, and not damaged. With the

exception of the category of "collapsed," the other classifications are meaningful but not clearly defined and thus they are fuzzy events. Although civil engineers were among the first to apply the theory of fuzzy sets (e.g., Wong et al. 1999), there have been very few practical applications to date. Yet in structural reliability studies, there are many situations where fuzzy logic (e.g., Yen and Langari 1999) is potentially applicable.

Earthquake and Wind Engineering

General Comments

Earthquake engineering started as a technical discipline after the 1906 San Francisco earthquake. Thanks to the considerable amount of research that has been conducted in this field over the last 100 years, we have made tremendous progress in our understanding of the nature of earthquakes and earthquake mechanisms. For example, we now understand better the effects of magnitude, distance, geology, topography, and local soil properties on the characteristics (amplitude and frequency content) of the seismic motions that are expected at a specific site, and the behavior of soils and structures when subjected to seismic excitations. Earthquake related research still represents a major fraction of the funds allocated by the National Science Foundation to structural or geotechnical engineering, in addition to the funds provided by this and other government agencies for seismological work. This funding led first to the creation of a National Earthquake Engineering Center at the State University of New York in Buffalo, to the creation later of three national centers, and more recently to the investment of a very large amount of funds to upgrade and create new experimental capabilities, to connect them, and to create a national network of laboratories (NEES).

Earthquake Engineering Research

Structural research in earthquake engineering has dealt with (1) the development of improved dynamic analysis techniques, and (2) many experiments on isolated members, joints, and small assemblies of elements or scaled models of frames, to better understand various failure modes, to fit curves to the measured data in order to obtain design formulas, or to improve structural detailing. As a result of this work, buildings can be designed today, using present codes, to resist earthquakes much more safely than 50 years ago. Although the increase in safety may be attributed in large part to improvements in the supervision of the construction process, the numerous changes particularly in reinforcement of concrete members to provide continuity and ductility, and in the details of the connections, have also played a key role in adding to the safety. The lack of quality control of materials and the construction process are still, however, major causes of catastrophic failures in some countries.

A major shift in seismic research as applied to structures occurred in the last quarter of the century, when the emphasis moved from the design of new buildings to the retrofitting and strengthening of existing structures and the repair of structures damaged during earthquakes. Given the large inventory all over the world of buildings that were designed without appropriate seismic considerations (early building codes or codes in certain regions without any seismic provisions) this represented a logical move. It is ironic, however, that in a highly publicized case of repair and retrofitting, a 10-story building was reduced to seven stories because of lack of information on the condition or capacity of its foundation. This illustrates the importance of damage identification through nondestructive damage evaluation techniques, a field that has been developing over the last quarter of the century, as well as the limitations of looking at only one component of a building (the structure) instead of following the systems approach and integrating it with its foundation and underlying soils supporting it.

Research on the more creative, conceptual, phase of the design process, exploring alternative structural configurations which may be better suited to resist the loading resulting from earthquake excitation, as well as new mechanisms of energy dissipation, base isolation or, in general active and passive control systems, is only relatively recent in spite of the pioneering efforts of Frank Lloyd Wright in the design of the Imperial Palace Hotel in Tokyo. It is to be hoped that research in the 21st century will look more at these topics rather than just continuing forever to test standard configurations. It is also hoped that the earthquake engineering community will look at the overall problem as one involving many important factors instead of trying to look at each component in isolation as a one-dimensional problem. At present, we try to characterize each earthquake by a single value, soil effects by a single descriptive or numerical parameter, and damage to a building by a single measure of ductility.

The National Center for Earthquake Engineering Research at Buffalo also started a coordinated research program involving social scientists and engineers. They looked not only at technical issues but also at the social and economic implications of earthquakes. This was important pioneering work. The cooperation of social scientists and earthquake engineers has been continued by the three succeeding centers: the Pacific Earthquake Engineering Research (PEER) Center on the West Coast, the Mid-America Earthquake (MAE) Center in the Midwest, and the Multidisciplinary Earthquake Engineering Research (MCEER) Center in the East.

Seismic Design Codes

It is fair to say that regular buildings can be designed at present to perform satisfactorily under the potential earthquakes to which they may be subjected during their lifetime. According to the code's philosophy, some nonstructural or even structural damage may occur depending on the severity of the motions but collapse and loss of life should be avoided. Particularly important is the evolution of the codes from working stress design based on linear elastic analyses that are meaningless when substantial inelastic deformations are accepted in the structure. These design codes have now changed to load/resistance factor design based on ultimate conditions, and the profession is expected to finally adopt performance-based design. In a performance-based design code, one would design for an expected or desired level of damage in the case of extreme loads, accounting for the uncertainties present in the process, assuming that one could in fact predict the expected damage accurately. This represents a major improvement in the code philosophy. The main limitation of the approach is in the last assumption. Present analysis procedures cannot predict yet with accuracy the amount of damage that would occur in a complete building, including all its components, under a given earthquake.

The main source of uncertainty remains still the characterization of the design earthquake, due to lack of sufficient historical data on real earthquakes. Starting with the famous 1940 El Centro earthquake record, there are only slightly more than 60 years of real earthquake records. In the long history of the earth, 60 years

of data collection is simply not sufficient. Thus the need to take into account the uncertainty of the excitation in earthquake engineering is continuing.

Concluding Remarks

The economic losses due to earthquakes are very large and seem to increase continuously. Thus the large amount of funding that has been made available for research in earthquake engineering, particularly after a large and damaging earthquake, is justified. The losses due to wind (hurricanes or tornadoes) are also very significant but the funding for wind related research has always been considerably smaller. Wind engineering is, however, another very important area in need of research. In addition to the study of the damage caused by hurricanes or tornadoes and the development of design measures to reduce this damage, there are important problems associated with wind induced vibrations. In tall and flexible buildings wind loads tend to control the design, even in seismic areas, and can result in serious discomfort problems. In suspension and cable-stayed bridges wind can result in vibrations of the cables, as well as serious aerodynamic instabilities. While the failure of the Tacoma Bridge is well known and documented, other types of damage caused by wind are less well understood.

Maintenance, Repair, and Retrofit

General Comments

The need to assess the condition of existing structures, to repair them if they are damaged, or to strengthen them if they do not meet the requirements of modern codes is not just limited to earthquake engineering. It is much more general and related to the upkeep of our vast civil infrastructure. The decision of whether to demolish large amounts of structures once they reach their design lives in order to replace them or whether to maintain and repair them needs to be made. Meanwhile major landmarks throughout the world that had withstood the passage of time for many centuries (some times with periodic repairs) are beginning to suffer more serious, accelerated, deterioration, due to age and environmental conditions aggravated by atmospheric pollution. Conservation of these historic monuments has become a theme of major interest at the international scale and the subject of an increasing number of conferences.

Note that the question is not just how to strengthen existing structures but also how to assess their condition and the need for strengthening. This involves the use of nondestructive testing techniques and the application of new methodologies for damage assessment and identification. Academic programs on evaluation and rehabilitation of buildings have already been created at a number of universities, ranging from the introduction of one or two courses within an existing curriculum, to the development of a set of courses leading to a diploma, and more recently the creation of a complete degree granting program.

Forensic Engineering

Structural engineers have been involved for many years in the rating of bridges and old buildings on the basis primarily of visual inspection, following sets of established procedures. The occurrence of major collapses has led in general to the creation of blue ribbon panels of experts charged with the investigation and determination of the causes of the failure. Forensic engineering was developed as a specialty within structural engineering dealing with these two issues as well as quality control in construction. The new emphasis on maintenance, repair, and retrofit can thus be considered to some extent as a broadening of its scope.

Structural Control

It is equally important to consider how new structures can be designed and built to facilitate their future maintenance as they age, the continuous monitoring of their condition, the early detection of potential problems, and the identification of damage after an extreme loading event such as an earthquake. One could thus talk about design for maintainability and repairability, as well as design for durability. Research in other fields such as electronics, with the development of a large variety of new sensors, health monitoring, fiber optics, "smart" materials, and control mechanisms should play a key role in this effort (e.g., see Housner et al. 1997).

Structural Design

General Comments

The objective of a civil engineer involved in the design of a specific structure is to obtain a system that satisfies a given set of functional requirements both aesthetically and economically. In addition, this system must perform its intended use safely under all the potential loads and environmental actions to which it may be subjected during its lifetime. The complexity of the computations and the effort required to perform structural analyses, to determine the stresses in the members due to a specified set of loads, and to compare them to allowable values provided by the codes, exaggerated for many years the importance of this phase of the design, at the expense of other considerations. In many cases the dimensioning process, where member sizes are selected on the basis of the computed stresses (or strength) and the code requirements, has been considered synonymous with design (particularly in so-called design courses in typical civil engineering curricula), whereas it is in fact just the last step of the analysis. Codes required initially that the computed stresses remain in the linear elastic range of the materials and applied a factor of safety with respect to the onset of yielding to account for potential variations in the loads or material properties. This approach (working stress design) made sense when dealing with performance under normal service loads but did not provide a reliable indication of the margin of safety with respect to collapse. The use of different factors for the loads and the material strength to account for uncertainties and the consideration of the limiting or collapse condition led to the load and resistance factor design codes. When dealing with extreme loads (such as earthquakes) and accepting the possibility of nonlinear behavior, it is no longer sufficient to know the safety or reliability index with respect to collapse. One must be able to predict the amount of expected damage (and the economic losses) for different load levels. This has led finally to performance-based design codes as already discussed in relation to earthquake engineering.

Economy and Computer Applications

It seems that in some cases designers might also have forgotten the fact that their structures had to be built and at a reasonable cost in order to achieve the goal of economy (clear exceptions were offshore structures where the fabrication, transportation, and installation procedures controlled the design). This was probably a logical consequence of separating the design and construction planning processes instead of integrating them. As a result it was necessary in the last quarter of the century to create a new word and to talk about "constructability" as an attribute of the design. Equally important is the need to look not only at the original cost of the structure but also at the costs of maintenance and repairs during its intended lifetime. One could thus add as mentioned above two new words and attributes to the structural design: "maintainability" and "repairability," both closely associated with the possibility of using instrumentation to monitor the performance of the structure on a continuous basis, and to obtain early diagnoses of potential troubles or malfunctions. And all this must be performed within a probabilistic framework.

The reduction in time and cost of the structural analyses brought by the availability of verified software packages allows the designer to concentrate on other issues, such as a better estimation of the potential loads, the investigation of alternatives looking for an optimum solution, the integration of the design of the different components of the system, and the coordination of the design and construction. In the second half of the 20th century there were a number of research efforts on structural optimization. Unfortunately, the application of mathematical optimization techniques to structural design required the selection of an objective function. Total weight seemed a logical choice and the simplest one. Yet weight, which is very important for aeronautical or aerospace applications, as well as for some types of offshore structures, is not a significant contributor to the cost of buildings or most civil engineering structures. New and promising methodologies to perform structural optimization in a much more practical sense were beginning to be developed at the end of the century.

Analysis-Design Integration: Computer Applications

In the late 1950s, a young civil engineer named Charles L. Miller pointed out that the computer should not be just a research instrument in the departments of mathematics or electrical engineering of universities but an everyday tool for practicing engineers. This was the time when we were still laboring with IBM 1620s and experts had predicted that a handful of IBM 7040s would saturate the computation market until the year 2000. He also indicated that to achieve this goal it would be necessary to facilitate the communication between man and machine and proposed the creation of problem oriented languages to replace the cumbersome fixed format data input forms and make the computer more user friendly. This extraordinary vision resulted in his becoming head of the Civil Engineering Department at MIT while in his early 30s. In this capacity he oversaw the development of the first major structural analysis package, the *STRESS* program, and he conceived next the creation of an integrated civil engineering system (ICES) with problem oriented verbal input, dynamic memory allocation, and a common data base, which could be used by structural, geotechnical, construction, mechanical, and electrical engineers to integrate the complete design of a building.

One of the first components of this system was the *STRUDL* package representing an extension of *STRESS* with more sophisticated analysis capabilities and a design orientation. Unfortunately the success of *STRESS* and *STRUDL* led other universities to develop faster, more efficient, and more sophisticated analysis programs, forgetting about the user friendly features or any design considerations. As a result, what was an important advance in computational analysis capabilities represented a significant step backward on the communications, design, and integration fronts. It is interesting to notice that the present analysis packages that take advantage of more user friendly input-output capabilities (with computer-aided design, graphical displays, etc.) have not been developed in academia but by industry. Programs that perform real design, rather than simply checking stresses with code formulas, or integrate the design of the structure and its foundation or the design and the construction phases are scarce and mostly of a proprietary nature. To a large extent the dream of Charles Miller remains yet to be fulfilled. Reinschmidt (1991) discussed this topic in more detail.

Education

General Comments

In the 1950s engineering education in the United States experienced a major shift in emphasis from a very pragmatic know-how and can-do approach to a much more rigorous theoretical treatment of basic and engineering sciences. While the strengthening of the scientific basis was desirable, unfortunately, it was done at the expense of the more practical engineering subjects. At the same time, there has been an increased emphasis on more basic research, not only as an important component of the educational process (particularly for graduate students), but also as an end by itself (and eventually as a main source of funding). Up to that time most engineering professors had a substantial amount of practical experience and maintained themselves in touch with the practice of engineering as did the great engineers of the 19th century. At that time, research was often motivated by real problems encountered in practice, rather than being dictated by funding agencies with the assistance of government panels, often consisting of other academic researchers.

Civil Engineering Education

The composition and background of engineering faculties has changed substantially. At present most faculty members in the United States are hired upon completion of their PhDs without any exposure to practice. After five (or fewer) years of service faculty members will undergo a tenure review that will require their having (1) generated a certain amount of research funding, and (2) published a substantial number of papers. Exposure to practice usually will not count for promotion/tenure considerations and therefore cannot be a serious consideration for a young person trying to make it in the present academic environment. As a result, a situation is reached where many universities have trouble finding faculty members who can teach realistic design courses that are required for accreditation. It seems that many faculty members at research universities are trying to recreate themselves, producing more researchers and faculty members rather than competent, top level professional engineers. This is a dangerous situation in need of remedy.

Some enlightened institutions foresaw the problem and started many years ago hiring practicing engineers as adjunct professors to teach design courses. A better solution is to have prominent engineers retire early and join full time university faculties, participating in the teaching and research efforts in combination and close collaboration with the other faculty members, coordinating and integrating their contributions so as to incorporate realistic examples in all the courses. The time has come for the case study approach, commonly used in other disciplines, to become more widely used in civil engineering education.

Reinstating education, and education of professional engineers in particular, as a major goal of universities will require major changes in their philosophy (Roësset and Yao 2000). It will require also a new type of closer industry-university cooperation. Final decisions on curriculum content should rest with the faculty but industry should receive clear and well-defined benefits from this cooperation. Persons involved in planning civil engineering curricula would do well to read carefully the little jewel by Cross (1952), *Engineers and ivory towers*.

Predictions

On the basis of their own knowledge and of their perceived needs and hopes, reluctantly the writers are willing to make the following predictions.

- There will be taller buildings, deeper platforms in the ocean, and longer bridges built. There will also be new structures built in space and perhaps on other planets, as well as structures under water for an increased number of applications. It is interesting to notice that while Jules Verne's futuristic predictions for space have been greatly exceeded by reality, his vision of submarine life remains to be fulfilled, perhaps because the sky exerts a stronger appeal on human beings than the ocean bed. Yet the importance of the ocean and its resources for human life will continue to increase.

- There will continue to be important progress in our analysis capabilities, allowing us to predict better the behavior of structures under static and dynamic loads, particularly on two fronts: (1) the three-dimensional analysis of complete structures, including nonstructural as well as structural components, considering both the structure and its foundation; (2) the nonlinear analysis of structures with realistic constitutive models and the ability to predict the location and extent of damage that they might suffer. As the methods of analysis advance so will our knowledge and understanding of the physical processes that cause the loads on the structures, such as wind, earthquakes, or sea states, allowing us to better model the excitation. To achieve this, we must continue to collect data on these loads.

- There will be an increased use of uncertainty and risk analyses in structural engineering. Although there have been reliability-based design codes and much progress has been achieved already in structural reliability, most people working in these fields are analysts. Experiments must be performed in order to make further progress. In addition to probabilistic methods, uncertainty analyses will include fuzzy sets if they are to be applied in practice to cases where the probabilities can only be estimated in vague terms. Soft computation including fuzzy logic, genetic algorithms, and neural networks will be applied to more practical problems in structural engineering. Hybrid computers combining the respective advantages of analog and digital computations may enable structural engineers to deal with more practical problems.

- One can foresee buildings and bridges that are instrumented so as to be able to monitor their performance and diagnose potential troubles easily, and structures conceived so that they can be easily maintained, repaired, or replaced. One can foresee also increased use of passive and active control systems as means to respond more effectively to different types of external excitations. There will be increased practical applications of symptom-based reliability and health monitoring of existing structures. Symptom-based reliability will become more useful once we are able to define variables characterizing the different damage states that can be obtained from field measurements and nonlinear analyses, as well as their limiting values. This will require more experimental and analytical research. Future developments in damage assessment and measurement technologies (especially nondestructive evaluation and in situ testing techniques) will also help to increase its practical applications.

- It is believed that the liberating aspects of computers on the demands on structural designers will see their full impact in this century. The time should come to see finally a fully integrated design process in which the engineer can look at the complete system and the interactions of its different components, while technicians carry out on computers the analysis of the various subsystems. The designer will be able to devote more time to seeking optimum solutions (materials and topology) in relation to the functional use of the structure and its overall cost (including initial cost of materials and construction, cost of money, and cost of maintenance and repairs). Life-cycle costs will be considered at the beginning of each project, and visualization will be more commonly used.

- The writers see a more intensive use in civil engineering of the design tools that have been developed in the automobile and aeronautics industries, with three-dimensional graphical models of the structures in the computer that can be updated as the design progresses and modifications are introduced by different teams in concurrent design, or as the simulated or actual construction process goes on. There will also be increased use of simulation and consideration of the construction process during the design in order to guarantee the constructability of the project. More research is necessary, however, to guarantee that concurrent interactive design will proceed smoothly.

- Designers will have a tremendous variety of materials to choose from, but whether ultrahigh-strength and ductile concrete or composite materials consisting of a resin matrix with glass or carbon fibers will see extended use in civil construction, as compared to aerospace or automobile applications, will depend on their cost and commercial availability in large quantities.

- Performance-based design codes will be further developed. More methods of analyses will be acceptable and used in analysis and design. As long as the structure will perform according to the specified usage, engineers will have more freedom in their designs.

- There should be a much larger concern of the structural engineer for issues that are not directly related to the resistance of the structure or the distribution of forces in the members, but which are essential for the functional, economic, or aesthetic viability of the work, or for the acceptance of the project by the owners or the public at large. This will necessitate important changes in the way we educate structural engineers (or civil engineers in general).

- There will be significant changes in engineering education affecting both form (the way in which we teach) and substance (the content of the curriculum). The new multimedia and simulation capabilities available will complement and enhance the effectiveness of the more traditional forms of teaching without replacing them, for education is not merely making information available. There will be as a result an increase in visual and inductive learning. Curricula will pay more attention to sociopolitical and economic issues affecting civil engineering projects and to imparting communication and team working skills to the students. Whether one can teach leadership (in contrast to management or bureaucratic skills) is not clear to the writers, but leadership skills can be further enhanced.

- There will be an increase of virtual congresses and conferences starting with the CE World, the 2002–2003 ASCE virtual congress celebrating the 150th anniversary of ASCE.

Concluding Remarks

The role of design engineers and the practice of structural engineering will change substantially due to the impact of computers in our society. There should not be any reason for concern since these changes will make design work much more interesting and exciting. We may not need as many engineers practicing at a professional level as we are producing today, but hopefully they will have better jobs.

The writers hope also that the 21st century will see a return to valuing substance over appearance. They realize that the best project has no value unless it is sold to the stakeholders and they believe in the importance of communication skills in engineers. Marketing ability is indeed very important. Yet a book should not be judged only by its cover, and, contrary to what was written on the wall of a particular building at a university, marketing is not everything.

Acknowledgments

The writers wish to thank Jeff Russell of the University of Wisconsin at Madison who invited them to write this paper. They also wish to acknowledge the reviewers and the journal editor, Professor Dale C. Buckner, for further improving this manuscript. In addition, they acknowledge the Cain Chair and the Lohman Professorship at Texas A&M University for providing the financial support necessary for the completion of this paper.

References

Abramson, D. M. (2001). *Skyscraper rivals*, Princeton Architectural Press, New York.

ASCE. (2001). "ASCE designates Empire State Building as a Civil Engineering Monument of the Millennium." *ASCE News.*, 26(8), 1.

Berlow, L. H. (1998). *The reference guide to famous landmarks of the world*, Oryx Press, Phoenix.

Billington, D. P. (1983). *The tower and the bridge*, Basic Books, New York.

Billington, D. P. (1996). *The innovators*, Wiley, New York.

Cempel, C. (1991). *Vibroacoustical condition monitoring*, Ellis Horwood, Chichester, England.

Chen, W. F., ed. (1997). *Handbook of structural engineering*, CRC Press, Boca Raton, Fla.

Collins, A. R., ed. (1983). *Structural engineering: Two centuries of British achievements*, Institution of Structural Engineers Anniversary Publication, London.

Cross, H. (1952). *Engineers and ivory towers*, McGraw-Hill, New York.

ENR. (1999). "125 top projects," *ENR*, ⟨http://www.enr.com/new/125topproj2.asp⟩.

Gaylord, E. H., and Gaylord, C. N., eds. (1968). *Structural engineering handbook*, McGraw-Hill, New York.

Gaylord, E. H., and Gaylord, C. N., eds. (1990). *Structural engineering handbook*, 3rd Ed., McGraw-Hill, New York.

Girvin, H. F. (1948). *A historical appraisal of mechanics*, International Textbook Company, Scranton, Pa.

Goldberger, P. (1981). *The skyscraper*, Alfred A. Knopf, New York.

Grigg, N. S., Criswell, M. E., Fontane, D. G., and Siller, T. J. (2001). *Civil engineering practice in the twenty-first century: Knowledge and skills for design and management*, ASCE, New York.

Housner, G. W. et al. (1997). "Structural control: Past, present, and future," special issue, *J. Eng. Mech.,* 123(9), 897–971.

Husband, J., and Harby, W. (1947). *Structural engineering*, 5th Ed., Longman's, New York.

Jackson, D. C., ed. (1997). *Dams, studies in the history of civil engineering*, Vol. 4, Ashgate Vriorium, Brookfield, Mass.

Kirkham, J. E. (1914). *Structural engineering*, Myron C. Clark Publishing Company, Chicago.

Kirkham, J. E. (1933). *Structural engineering*, McGraw-Hill, New York.

Nakamura, T., ed. (1988). *American high-rise buildings*, Japan Architect Company, Tokyo.

Natke, H. G., and Cempel, C. (1997). *Model-aided diagnosis of mechanical systems—fundamentals, detection, localization, and assessment*, Springer, New York.

Parker, H. E., ed. in chief. (1931). *Kidder-Parker architects' and builders' handbook*, 18th Ed. (1st Ed. 1884), Wiley, New York.

Reinschmidt, K. F. (1991). "Integration of engineering design and construction." *J. Constr. Eng. Manage.,* 117(4), 756–772.

Roësset, J. M., and Yao, J. T. P. (1988). "Civil engineering needs in the 21st century," *J. Prof. Issues Eng.,* 114(3), 248–255.

Roësset, J. M., and Yao, J. T. P. (2000). "Roles of civil engineering faculty," *J. Prof. Issues Eng. Educ. Pract.,* 126(1), 8–15.

Sheiry, E. S. (1938). *Elements of structural engineering*, International Textbook Company, Scranton, Pa.

Smith, N. (1975). *Man and water*, Peter Davies, London.

Southworth, M., and Southworth, S. (1984). *A.I.A. guide to Boston*, Boston Society of Architects, Boston.

Stoller, E. (2000). *The John Hancock Center*, Princeton Architectural Press, New York.

Westerbrook, R. (1984). *Structural engineering design in practice*, Construction, New York.

Wong, F. S., Chou, K. C., and Yao, J. T. P. (1999). "Chapter 6: Civil engineering including earthquake engineering." *Practical applications of fuzzy technologies*, H.-J. Zimmermann, ed., Kluwer Academic, Boston, 207–245.

Wong, F. S., and Yao, J. T. P. (2001). "Health monitoring and structural reliability as a value chain." *Comput.-Aided Civ. Infrastruct. Eng.* 16, 71–78.

Yen, J., and Langari, R. (1999). *Fuzzy logic*, Prentice-Hall, Upper Saddle River, N.J.

Zadeh, L. A. (1965). "Fuzzy sets." *Information Control,* 8, 338–353.

American Society of Civil Engineers

150th Anniversary Paper

1852 – 2002

Building a Better World

Special Structures: Past, Present, and Future

Richard Bradshaw[1]; David Campbell[2]; Mousa Gargari[3]; Amir Mirmiran[4]; and Patrick Tripeny[5]

Abstract: *Special structures* are landmarks and testimonials to the achievements of the structural engineering profession. They are true three-dimensional representations of our equilibrium equations and affirmations of our analytical techniques, design standards and construction practices. They include many types of structures, such as: space frames or grids; cable-and-strut and tensegrity; air-supported or air-inflated; self-erecting and deployable; cable net; tension membrane; lightweight geodesic domes; folded plates; and thin shells. This work celebrates the ASCE's sesquicentennial by providing a historical perspective on how *special structures* have evolved, their state-of-practice in the dawn of the 21st century, and a projection of their potential trends and evolution into the future.

DOI: 10.1061/(ASCE)0733-9445(2002)128:6(691)

CE Database keywords: Domes, structural; Fabrics; Grid Systems; Membranes; Spacing; Plates; State-of-the-art reviews.

Introduction

The sesquicentennial of ASCE provides a great opportunity to assess the role of *special structures* in the structural engineering profession with a historical perspective on how these types of structures have evolved, their state-of-practice in the dawn of the 21st century, and a projection of their potential trends and evolution into the future.

From the Georgia Dome in Atlanta, the Livestock Pavilion in Raleigh, and Madison Square Garden in New York to the Olympic Stadium in Munich, and from the Pontiac Silverdome in Michigan to the Sydney Opera House in Australia and the Haj terminal in Saudi Arabia, *special structures* are landmarks and testimonials to the achievements of the structural engineering profession (Fig. 1). They are what makes us most interested in and proud of our profession and what binds us together with the architects and architectural and construction engineers in appreciation of the art of structural design and construction. They are true 3D representations of our static and dynamic equilibrium equations, and affirmations of our analytical techniques, design standards, and construction practices. Each special structure is a pro-

totype by itself, rather than a duplicate produced on an assembly line.

Yet, it is not easy to qualify the term *special structure*, as perhaps loosely used in this paper. For the purpose of this paper, "special structure" refers to innovative long-span structural systems, primarily roofs and enclosures to house human activities. More specifically, they include many types of structures, such as: space frames or grids; cable-and-strut and tensegrity; air-supported or air-inflated; self-erecting and deployable; cable net; tension membrane; geodesic domes; folded plates; and thin shells. We exclude tall buildings and long-span bridges, both of which are addressed separately.

Thin shells and tension membranes are considered form-resistant structures, as they resist loads by virtue of their shape. Neither will function if flat, and both carry loads predominantly through in-plane stresses rather than by bending, granted that thin shells bend as well as compress. Other special structures resist loads mostly in flexure. The typical flat space frame or grid supported by columns or walls acts primarily in flexure even though its individual members behave axially (Fig. 2). Depending on the loads, top and bottom chords will be in tension or compression, similar to the flanges of an I-beam, and the diagonals (acting in tension or compression) carry the shear, much like the web of an I-beam (Cuoco 1997).

Structures that resist loads by bending may be categorized using span-to-depth ratio. For example, this ratio is about 20 for a typical wood joist; whereas, a timber beam with larger loads will have a ratio around 12. Steel bar joists usually run about 24, as do many wide-flange beams and reinforced concrete waffle slabs. Space frame is remarkable with as high a ratio as 35–40. On the other hand, span-to-depth ratio has no significance for form-resistant structures. A more useful measure for an arch is the span-to-rise or the span-to-thickness ratio (Fig. 3). Efficient arches have span-to-rise ratios of 2–3 and span-to-thickness ratios of about 40. Larger span-to-rise ratios generally result in larger axial forces and require a smaller span-to-thickness ratio.

Tensile structures are more efficient than arches because they do not buckle. The Verrazano Narrows Bridge in New York, for example, has a span-to-sag ratio of 10 and a span-to-thickness

[1]Consultant, Richard R. Bradshaw, Inc., 17300 Ballinger, Northridge, CA 91325.

[2]Principal and Chief Executive Officer, Geiger Engineers, 2 Executive Blvd., Ste. 410, Suffern, NY 10901.

[3]Assistant Professor, Dept. of Construction Sciences, Univ. of Cincinnati, Cincinnati, OH 45206.

[4]Editor, Professor, Dept. of Civil Engineering, North Carolina State Univ., Raleigh, NC 27695.

[5]Assistant Professor, Graduate School of Architecture, Univ. of Utah, 375 South 1530 East, Salt Lake City, UT 84112.

Note. Associate Editor: C. Dale Buckner. Discussion open until November 1, 2002. Separate discussions must be submitted for individual papers. To extend the closing date by one month, a written request must be filed with the ASCE Managing Editor. The manuscript for this paper was submitted for review and possible publication on November 8, 2001; approved on March 7, 2002. This paper is part of the *Journal of Structural Engineering*, Vol. 128, No. 6, June 1, 2002. ©ASCE, ISSN 0733-9445/2002/6-691–709/$8.00+$.50 per page.

Fig. 1. (a) Sydney Opera House; (b) Haj Terminal (Geiger Engineers); (c) Millenium Dome (Birdair); (d) Georgia Dome (Geiger Engineers); (e) Pontiac Silver Dome (Geiger Engineers)

ratio of about 400 for its cables (Madugula 2002). With span-to-thickness ratios near 300,000, large air-supported membranes are undoubtedly the most efficient structures, although one may argue that they have a zero span length continuously supported on columns of air.

Although efficient in material use, tensile structures generate large pull forces at their base. For example, the concrete compres-

sion ring encircling the Georgia Dome is 7.9 m (26 ft) across to take such large forces. While large horizontal thrusts are also present in low-rise arches, they can be more easily resisted than the large "pulls" developed by tension membranes and cable domes.

In this work, special structures are categorized into three groups based on the method by which they resist the loads: com-

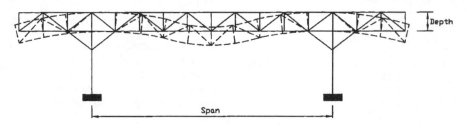

Fig. 2. Space frame acting in bending

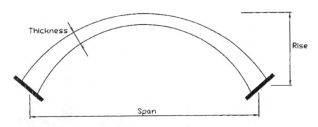

Fig. 3. Ratios in a simple arch

pression structures (shells); tension structures (tension fabric, air-supported, air-inflated, tensegrity, and cable-net structures); and tension/compression reticulated structures (space grids or frames and geodesic domes). The emphasis is placed on the history of special structures (how they came about and why they are unique), their structural behavior (how they withstand the loads), their advantages and disadvantages, their methods of analysis and design, and the noteworthy structures of each type around the world.

Compression Structures: Shells

Brief History of Thin Shells

Architectural thin shells discussed in this work are a modern development. The domes and cylindrical shaped structures of antiq-

uity and the Middle Ages were thick and could only resist compressive loads. The first modern architectural shell is generally credited to that built by the Zeiss optical company in Austria in the 1920s. In the United States, shells were extensively studied by the aircraft industry in the 1930s. In 1933, Donnell, an aeronautical engineer, formulated the general equations for cylindrical shells, including both bending and membrane actions.

While Eduardo Torroja of Spain is credited for the systematic engineering study of *architectural* shells in the 1930s, it was the work of Felix Candela in Mexico that ignited the sudden surge of popularity of shells in the 1950s. His shells were spectacular both for appearance and for bold engineering. At a time when a 75-mm (3 in.) thick shell was considered daring, Candela built a hyperbolic paraboloid shell with less than 16 mm (5/8 in.) thickness for the Cosmic Ray Pavilion at Ciudad Univ. in Mexico City. Fig. 4 and 5 show examples of how Candela skillfully created different shells from the same hyperbolic paraboloid geometry.

It was an article in *Progressive Architecture* (1955) on the shells of Candela that launched the modern shell era by attracting the attention of architects. Figs. 6–12 show some of the remarkable early shells for the air terminal in St. Louis, MIT auditorium in Boston, TWA terminal in New York, Sports Palace in Italy, and Exhibition Hall in Paris.

The latter, designed by Esquillan, is one of the engineering marvels of the 20th century, whose statistics define its uniqueness. In plan, it is an equilateral triangle; 218 m (715 ft) long on a side

(a)

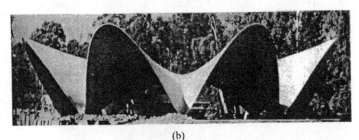

(b)

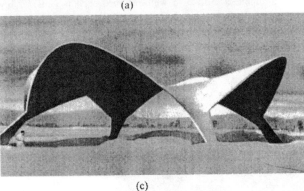

(c)

(d)

Fig. 4. (a) 16-mm (5/8 in.) thick hyperbolic paraboloid shell of Cosmic Ray Pavilion, Mexico City, (Faber 1963); (b) 61-mm (2.4 in.) thick, 30 m (100 ft) span hyperbolic paraboloid shell of a restaurant, Mexico City, (Faber 1963); (c) 61-mm (2.4 in.) thick, 20 m (64 ft) span hyperbolic paraboloid shell of a sales office, Guadalajara, Mexico (Faber 1963); (d) 38-mm (1.5 in.) thick hyperbolic paraboloid shell of a church tilted on edge, Narvarte, Mexico (Faber 1963).

Fig. 5. 83 mm (3.25 in.) thick, 8 m (26 ft) square, 0.6 m (2 ft) rise hyperbolic paraboloid shell, Vallejo, Mexico (Faber 1963)

Fig. 6. 114–216 mm (4.5–8.5 in.) thick, 37 m (120 ft) span intersecting cylindrical shells with ribs at edges and at groin, air terminal, St. Louis, Robert and Schaefer Engineers (Joedicke 1963)

Fig. 7. 89 mm (3.5 in.) thick, 49 m (160 ft) span spherical shell with hinges at abutments, MIT Auditorium, Boston, Amman and Whitney Engineers (Joedicke 1963)

Fig. 9. 59 m (192 ft) span Sports Palace, Rome, Nervi Engineer (Joedicke 1963)

with a rise of 48.8 m (160 ft). If a circular dome were circumscribed about the equilateral triangle, it would be 251 m (825 ft) in diameter, beyond the span of any building today. However, it would be simpler to design the full dome than a triangular piece cut out of it due to the instability of the free edge, which creates a potential for buckling. This problem was prevented using a two-layer shell spread apart by vertical walls. The overall depth of the system is 1.9 m (6.25 ft) at the crown and 2.7 m (9 ft) at the spring line. The thickness of each layer is 60 mm (2.38 in.) at the crown and 120 mm (4.75 in.) at the spring line. The interior precast cross walls are 59 mm (2.33 in.) thick. Thus, no part of this immense shell, the largest ever built, is thicker than 120 mm (4.75 in.). Remarkably, this was all in 1957, before the use of computers.

The history of civil engineering has repeatedly shown that new types of structures have been built before their behavior was fully understood. This is as true of modern shells as it was of the cathedrals of the Middle Ages; that rational explanation for their success was found only after the persistence of their existence forced their recognition. The early practitioners had to rely on intuition and courage rather than on written knowledge. It can be certain that a great deal of anxiety took place before Candela built his 16 mm (5/8 in.) thick Cosmic Ray Pavilion. One could only imagine the fortitude it took to remove the forms from under the 218 m (715 ft) span of Esquillan's Exhibition Hall.

The structures of the skilled practitioners of the art, such as Candela, Esquillan, Torroja, and Nervi are distinguished by their elegance in minimizing the thickness, eliminating the ribs, and avoiding the hinges at the abutments. It suffices to note that the span-to-thickness ratio of a well-designed shell is considerably larger than that of an eggshell. Ribs are used to carry the shear forces from the shell to the abutments and to prevent buckling of

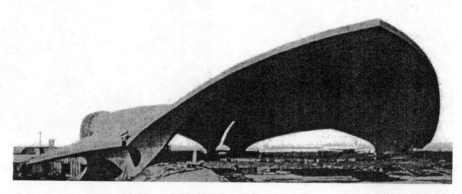

Fig. 8. 178–610 mm (7–24 in.) thick, 37 m (120 ft) span arbitrary shape shell with ribs at edges and in interior, MIT Auditorium, Boston, Amman and Whitney Engineers (Joedicke 1963)

Fig. 10. Inside view of Sports Palace, Rome (coffered ceiling made by pouring a thin layer of concrete over precast concrete boxes which then become part of the structure) (Joedicke 1963)

the edges. However, it is possible to eliminate many ribs by making the shell itself act as the rib. This requires skilled analysis, which test the knowledge and nerve of the designer. Hinges between the shell and the abutments reduce the capacity of the structure and serve only to simplify the design.

Shells and Geometry

There is no type of structure that has so intimate a relationship with space geometry as a shell. There are two important yet simple geometrical observations in shells: all constructed shells are only fragments of a more complete geometrical shape; and all geometric surfaces would either continue to infinity or intersect with themselves.

The shell in Fig. 13 is derived from two intersecting tori or "doughnut" shapes, as shown in Fig. 14. In this example, the doughnut has a pinhole-sized hole. The shell of Fig. 13 is shown

Fig. 11. 218 m (715 ft) side span exhibition hall, Paris, Esquillan Engineer (note size of other buildings in background for scale) (Joedicke 1963)

Fig. 12. Interior precast partition walls inside of exhibition hall, Paris (Joedicke 1963)

Fig. 13. 64–127 mm (2.5–5 in.) thick, 37 m (120 ft) span intersecting tori with no ribs or hinges (Richard Bradshaw)

Fig. 15. Cooling tower, generated by straight lines (Gould 1988)

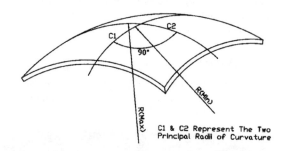

Fig. 16. Conoid, generated by straight line traveling along another straight line at one end and curved line at other end (Joedicke 1963)

using the heavy lines in the figures. By adjusting the parameters of the tori, any desired rise could be obtained. There are advantages in using mathematically defined geometrical shapes as opposed to arbitrary forms. The input into the finite-element (FE) model will require guesswork, unless the surfaces are described mathematically. Also, the formwork of arbitrary shapes is more expensive. Yet, there are famous shells that are not mathematically defined, such as the Sydney opera house and the TWA building in St. Louis.

Shells can be singly curved (e.g., cylinders and cones) or doubly curved (e.g., sphere or hyperbolic paraboloid). Paraboloid is a shell of revolution made by revolving a parabola about its axis. A hyperbola produces a hyperboloid of two sheets when rotated about its axis of symmetry, and a hyperboloid of one sheet when rotated about the common axis between its two parts. The latter is often used for cooling towers, because it can be formed of straight lines (Fig. 15). Another doubly curved shape formed of straight lines is the conoid (Fig. 16), for which a straight line travels along

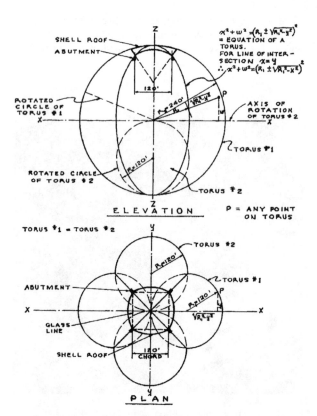

Fig. 14. Shell composed of intersection of two tori

Fig. 17. General doubly curved element

another straight line at one end and a curve at the other end. Shells of translation are generated by translating one curve along another. A circle-translated tangent to a straight line generates a cylinder, and if translated along another circle produces a torus.

Fig. 17 shows a doubly curved surface with two different curvatures. At any point on a surface, there are two principal radii of curvature that uniquely define the surface. Of all the curves on the surface that can be drawn through the point, the two principal radii of curvature will be the maximum and minimum that can exist at the point. The maximum radius of curvature for a cylinder is infinity, while the minimum is the radius of the circle (Fig. 18). Figs. 19(a and b) shows two pieces taken from the outside and inside of a torus, respectively. In the former, both radii of curvature lie on the same side of the surface, and the curvature is considered positive, while in the latter they are on opposite sides of the surface, and the curvature is negative.

All shells have either positive (bowl-shape) or negative (saddle-shape) curvature. The behavior of these two types of

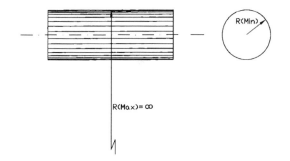

Fig. 18. Cylinder

Fig. 21. Another view of HP and structure built from it (Richard Bradshaw)

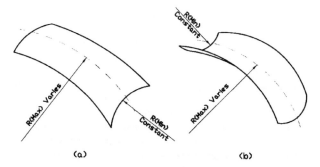

Fig. 19. (a) Cut from outside torus; (b) Cut from inside torus

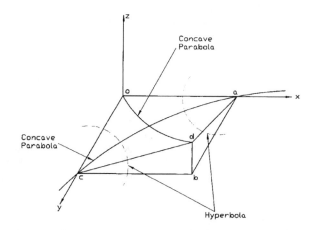

Fig. 20. Generating hyperbolic paraboloid by lifting one corner of plane

tive curvature (Fig. 22). The figures show that a simple change in the geometric parameters can result in a very different shape with greatly different structural behavior.

Analysis of Shells

The structural analysis of shells has had a long and difficult history. Shells were developed and reached their peak popularity just before the ready availability of computers and the FE method. This was unfortunate for the designers of these complex structures because in lieu of rigorous methods they went to considerable effort to verify their designs. Model analysis was one such technique, where plexiglass models, or rarely cementitious models, were strain gauged and loaded with weights. However, model analysis was laborious, expensive, and impractical for testing various trial shapes as easily as in the FE method.

Many cylindrical shells were analyzed using approximate methods, in that when extended in the long direction they approach beams in behavior, and when shortened in the same direction approach arches in behavior. Hence, they fall between the limiting cases of beams and arches. Corrugated iron, which is a collection of cylindrical shells side-by-side, may be analyzed as a beam of corrugated cross section. For short shells such as aircraft hangars, where spacing of the arches is small compared to their span, loads are mostly carried by the arches, not the shell itself.

Another method was to get the funicular or nonbending shape of the shell using hanging weights from a mesh. The Swiss engineer, Isler, froze suspended wet cloth to get the funicular. The dimensions of the prototype were then taken from measurements made on the model. A certain amount of error was thus introduced in the prototype. Also, the funicular shape for dead load is not the same as for partial span loads, which can occur with wind and

shells is very different. Positive curvature shells are subject to buckling, as the entire shell is subject to compression forces. In contrast, material failure is more common in negative curvature shells with brittle materials such as concrete.

Hyperbolic paraboloids (HP) are doubly curved surfaces with negative curvature. An HP can be generated by lifting one corner of a square shape as shown in Fig. 20. Lines parallel to the x- and y-axes remain straight lines. This is very important because the surfaces can be formed with straight forms, which are much more economical than curved forms. An HP can also be generated by translating a convex parabola along a concave one as shown in Fig. 20. Fig. 21 shows an HP in its more usual orientation and a structure built from it. If the convex parabola of Fig. 20 had been translated along another convex parabola instead of the concave parabola, it would have produced an elliptic parabola with a posi-

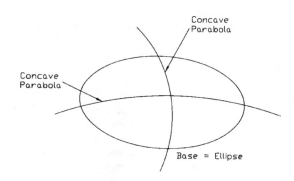

Fig. 22. Elliptic paraboloid shell

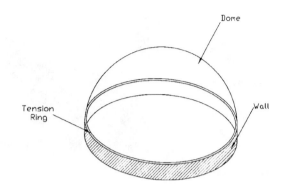

Fig. 23. Spherical shell on top of circular wall

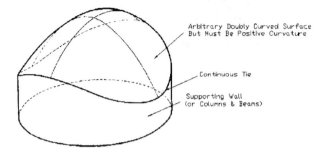

Fig. 24. Arbitrary doubly curved surface

snow loads, where partial span loading frequently governs. Other approximate methods used were to figuratively cut out pieces of the shell and analyze them for static equilibrium. One advantage of these approximate methods was that they forced the designer to develop an intuitive feel for the structural behavior of the shell, which is sometimes missing with the uncritical use of computers.

Of the more rigorous analytical methods, one can refer to the FE and the finite difference methods. In the FE method, the shell is cut into small pieces or finite elements and then "reassembled" using equilibrium and compatibility. In contrast, the finite difference method breaks down the governing equations of shells and solves them to an approximate solution. Prior to the availability of FE, some shells were analyzed by finite difference methods, which required a tedious convergence process without the use of computers. Today FE methods prevail for shell analysis. The shell element must consider anisotropy, creep, and the plastic flow of concrete. There have been failures of concrete shells after erection and initial loading, where creep and plastic flow have played a major role (Beles and Soare 1966).

Shells are usually modeled using triangular or quadrilateral plane elements. The former could be used to approximate any singly or doubly curved surface with positive or negative curvature, as the three corners of the element could always be made to fall on the shell. Although there are certain advantages to the use of quadrilateral elements, it is not always possible to approximate surfaces with them. Fig. 23 represents a doubly curved shell with no axis of symmetry. This surface could be approximated to any degree of accuracy with triangular elements. However, the same may not be true for a quadrangular element, as they are given a twist that is not included in their derivation. The amount of twist depends on the element size and the surface curvature. One characteristic of the hyperbolic paraboloid is that its twist is constant over the entire surface. Therefore, a flat quadrilateral cannot exactly fit its surface. The designer must estimate the consequences of this effect and verify the permissible twist in the FE program.

The analysis of shells requires that the three conditions of equilibrium, compatibility, and constitutive laws be satisfied simultaneously. The latter are the stress-strain properties for the materials of the shell. Most shells are designed with isotropic materials. With the development of advanced composites, their orthotropic and anisotropic behavior must be considered. However, composites have not been used for architectural shells to date. Roof structures are seldom designed for dynamic loads. Earthquake and wind loads may be treated as equivalent static loads.

The above three conditions result in three partial differential equations, two of the 2nd order and one of the 4th order, for the

most general case with two different radii of curvature and with combined bending and membrane actions. An early representation of these equations for cylindrical shells may be found in Donnell (1933). Bradshaw (1961) extended those equations to the general case of double curvature, which can describe any 1D member (beam), 2D member (plate), or 3D singly or doubly curved member (shell). If the 4th order bending-related terms are left out, the equations will represent only the membrane action, which is usually sufficient for part of the shell away from the abutments because flexural resistance of thin shells contributes little in this region. Equal radii of curvature result in equations for a sphere. Equations of a cylindrical shell are derived when one radius of curvature is set to infinity; when both are set to infinity, it will result in bending of a flat plate. Finally, the ordinary differential equation for bending of a beam is derived when plate width is set to unity.

Stress analysis of complete shells, such as pressure vessels, is much simpler than for architectural roofs because of the boundary conditions. When the shell is a portion of the sphere, it tends to spread outward at the discontinuous edge. To counteract this a ring is added, but the ring and the shell distort by different amounts, which results in bending stresses in the shell. These incompatible strains must be reconciled analytically, which is not too difficult a task for simple spherical shells. However, when the shell has isolated supports and few (if any) planes of symmetry, it is a severe problem at the discontinuous edges.

If we imagine the architectural shell to be cut from the complete shape, profound perturbations are introduced at the discontinuous edges. The resulting disturbance at the edge may be thought of as causing stress redistributions to flow across the entire shell with diminished effect as they move away from the edge. In many cases of shell analysis, the stresses resulting from the discontinuous edges will dominate the design. Physically, the shell boundaries are treated in various ways. It is sometimes possible to simply leave them as free discontinuous edges. Ribs are frequently added at the edges, though visually disruptive. One of the graceful aspects of Candela's shells is their lack of ribs. It is also possible to design the shell with the rib integrated within the shell itself. Compare, for example, Figs. 4(b and c) and 13 with Fig. 5.

There is a remarkable property of shells supported vertically at their edges. Fig. 23 shows a spherical dome supported on a wall. A tension tie is required around the perimeter at the intersection of the dome and the wall. This tie will be funicular, i.e., it will only carry axial tension forces. This principle has been known since antiquity for circular domes and ties. However, it is important to note that the tie will be funicular for *any shape of either the plan or elevation* (Csonka 1962) provided that the shell has positive curvature and continuous vertical support (Fig. 24). The support may be a continuous wall or stiff beams between ad-

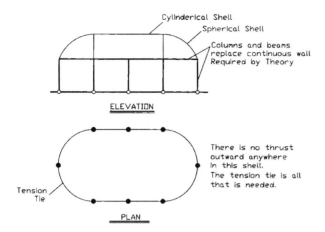

Fig. 25. Cylindrical shell combined with spherical shell

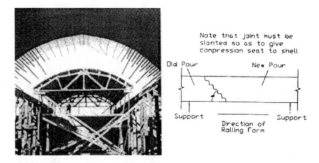

Fig. 26. Showing reuse of forms by rolling scaffolding (Richard Bradshaw)

equately spaced columns. It is interesting that the straight parts of the tie in Fig. 25 do not require ties across the building. The thrusts are taken by shear forces through the width of the shell, and only tension forces exist in the tie.

Buckling of Shells

Beles and Soare (1966) have reported buckling failure of shells. Unlike shells of positive curvature that are subject to buckling, in shells of negative curvature, such as hyperbolic paraboloids, buckling is prevented through the tension curvature in the other direction. Virtually all studies on shell buckling have focused on cylindrical, conical, and spherical shells made of metals, and usually on full 360° models rather than the much more complex architectural shells. Buckling of spherical shells has been studied for radial compressive loads. Cylinders and cones have been studied for radial compressive loads, axial, and shear (twisting) loads. Shallow spherical shells have been extensively tested for snap-through or "oil-canning" buckling. Applicability of these tests to large-scale concrete shells, however, is questionable. Initial imperfections in shells can result in their buckling at loads far below their theoretical capacity. Once a shell buckles, its collapse tends to be complete, contrary to plates, which have high post-buckling capacity.

Construction of Shells

Formwork has always been a major expense in shell construction. Several methods such as precasting or shotcreting over balloons and over reinforcing steel cages have been utilized to minimize this drawback. Double layer shells, spread apart, have been used to reduce weight. Full or partial use of straight forms for hyperbolic paraboloids, hyperboloids, cylinders, cones, and conoids also makes the formwork less costly. Rolling forms can be used for cylindrical shells, but the designer must pay attention to the cold joints (Fig. 26). Circumferentially moving forms may be used for shells of revolution, where pie-shaped pieces can act as temporary arches until the entire shell is in place. Joints must be made in places of low stress (Fig. 27).

Cylindrical shells have been cast on top of each other and then assembled with a crane. However, the pieces must be of the geometry shown in Fig. 28 to prevent dimensional creep. This makes the pieces slightly thicker in the middle than at the edges, but the difference is undetectable. Fig. 28 shows a doubly curved

toroidal shell, which uses the system of casting one piece on top of another. In this case, it is necessary to allow for the dimensional creep in both directions. The bottom shell was built over an earth form with the other shells cast on top of the first one. For precast shells, the cost of the cranes must be compared with the cost of forms.

Shells shotcreted over balloons have been used, particularly where high precision of dimensions is not important. When shotcrete is placed on a balloon, the weight distorts the balloon. This means the shell will not be exactly the initial shape of the balloon. This is not important for a small span shell. For a long span shell, however, this deviation from the spherical shape could be serious as shells are sensitive to buckling due to the initial roughness effect.

Future of Shells

At present, shells have lost their popularity compared to their heyday in the 1950s and 1960s, when architects eagerly adopted them as a new means for artistic expression. They were perhaps so eagerly adopted that they became a fad, and when a backlash inevitably set in, they were abandoned as quickly as they were first embraced. Shells were seldom the most economical way of covering a large space, especially when compared to lightweight tension membranes. Also, their formwork has always been a major cost factor.

There are signs, however, that shells are attracting interest among the new generation of architects and engineers. They will never become the vogue they once were, but they will regain some of their former popularity when used appropriately. There are also new materials such as fibercrete concrete and fiber reinforced polymer (FRP) composites that may be used in shells. At present, they may be too expensive for use in architectural shells, but with time that may change. Composite shells will require ortho- or anisotropic modeling, as well as careful buckling analysis, because they tend to be much thinner than concrete shells.

The future shells will take their place alongside other forms of architecture in structural engineering. When designed properly,

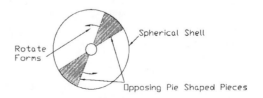

Fig. 27. Use of pie-shaped opposing forms

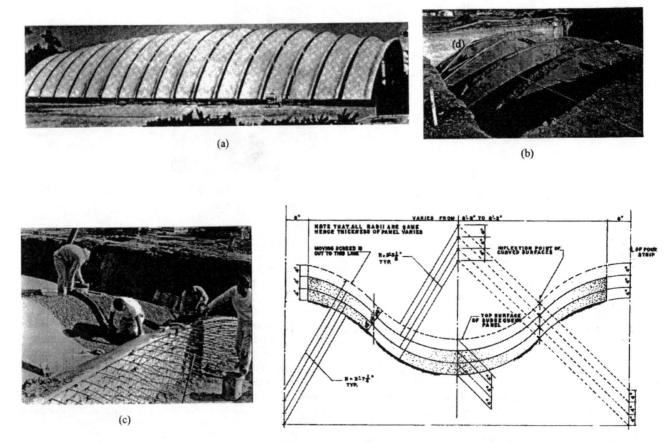

Fig. 28. Corrugated torus shell geometry: (a) Overall view of shell; (b) Earth form for bottom shell; (c) shells cast on each other; (d) geometry of precasting showing the elimination of dimensional creep (Richard Bradshaw)

they are among the most beautiful and efficient of architectural structures. Even those who are not professionals can sense the flow of forces through them. They will present both problems to be solved and opportunities to create for those who take the time to understand them.

Tension Membrane Structures

History of Tension Structures

Tension structures include a wide variety of systems that are distinguished by their reliance upon tensile only members to support load. They have been employed throughout recorded history as in rope bridges and tents. However, large permanent tension structures were generally a 19th century development in bridges and a 20th century development in buildings. The design of large tension membranes has been fully dependent upon the use of computers. Many of the developments in membranes have occurred in the last 30 years, precisely because of the accessibility of powerful computers. The pioneering work of Frei Otto was accomplished using physical models, which, while they well illustrate the desired form of a membrane, are not conducive to the precise determination of the membrane's structural characteristics in a manner necessary for the construction of large complex systems.

Large deployable membrane structures were used to cover touring public assembly events such as circuses and religious revival meetings in the 19th century. These were constructed of canvas and ropes with wood poles, as were contemporary tents. It was the fate of the "Big Top," the Barnum and Bailey Circus' main auditorium tent that burned (Martin and Wilmeth 1988), which created the most significant hurdle for membrane architecture in the United States: the issue of noncombustibility. Prior to the introduction of noncombustible structural fabrics, membrane structures were nomadic, and subsequent to the *"Big Top"* fire, they were relatively small. There were exceptions: Frank Lloyd Wright employed a tension membrane roof of canvas on his school and home, Taliesin West in Scottsdale, Ariz. in 1938. However, permanent tension membrane architecture began in North America in the 1970s.

Structural economy rather than aesthetics or architectural expression initially drove the modern use of membrane structures in North America. Thus, it is not surprising that the development of modern membrane structures was, with some exceptions (most notably John Shaver, the first American architect to develop permanent membrane architecture) primarily the work of engineers. However, these structures, like all spatial structures, are by their nature uniquely expressive, creating architecture that was in some instances a result rather than a goal.

Development of modern membrane structures began in the later half of the 20th century, and communications in the field were such that worldwide experience was quickly disseminated. The work of Frei Otto in Germany was particularly influential. As has been the case with other building technologies, World Expositions, particularly EXPO '70 in Osaka, Japan, were of great

Fig. 29. Walter Birdair atop a Birdair radome in 1956 (Birdair)

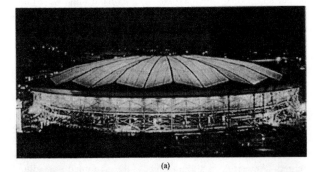

(a)

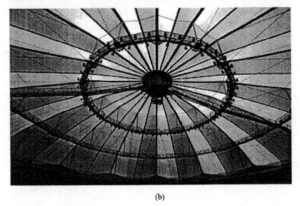

(b)

Fig. 30. Taoyuan Sports Arena, Taiwan, (a) outside view; (b) interior of Geiger Cabledome (Geiger Engineers)

significance in the development of membrane structures around the globe.

Simply considered, membrane structures were initially pursued in the United States for their cost-effectiveness. The first permanent membrane structures were the air-supported radar enclosures designed and built by Walter Bird (Fig. 29), as early as 1946. Walter Bird's successes with these pneumatic structures led to his founding of Birdair Structures in 1956.

Architectural membrane systems were a natural extension of tension structures of the 1950s and 1960s for long-span buildings. Fred Severud demonstrated the long-span potential of tension structures in benchmark projects such as the North Carolina State Fair Live Stock Pavilion in Raleigh, N.C., the Yale Univ. hockey rink, New Haven, Conn., and Madison Square Garden, New York. Severud's engineering practice was the incubator of membrane structure design in the United States. Designers of subsequent tension membranes included David Geiger, Horst Berger, Paul Gossen, and for brief periods, Edmund Happold and Frei Otto. Another engineering pioneer of prestressed tension-based structures was Lev Zetlin, who designed the Utica Memorial Auditorium roof and the New York State and Travelers Insurance Pavilions at the 1964 New York World Fair. Architects such as Eero Saarenen successfully exploited the unique architecture of these structures to create new forms for buildings.

While tension structures of Severud and others demonstrated the potential of such architecture in the United States, architectural membrane structures really began with EXPO '70 in Osaka, Japan, when David Geiger was commissioned to engineer the enclosure for the United States Pavilion. The initial pavilion design was abandoned when the project was unable to secure a sufficient appropriation from Congress. The pavilion program was maintained, but the project had to be realized for one-tenth of the original budget. In response to this challenge, Geiger invented the low profile cable-restrained, air-supported roof employing a superelliptical perimeter compression ring. This proved to be an exceedingly economical means of covering large clear span spaces, and quite interestingly, within 15 years of its completion, this structural system was employed to cover more than half the domed stadia in the world (Fig. 30).

Following the success of the United States Pavilion project, David Geiger considered applying it to permanent structures. However, such applications required a strong, noncombustible, durable material. An existing coated fiberglass fabric product, primarily used for conveyor belts in commercial ovens, seemed to have the desirable characteristics. David Geiger brought DuPont De Nemours Company, Owens-Corning Fiberglass Corporation and Chemical Fabrics Corporation together to develop this material for architectural applications. Marking a new era of permanent tension membrane architecture, the resulting product, *teflon coated fiberglass*, has since been employed around the world, as early as 1973 and 1974 for the Student Center at La Verne College, Calif. and the Steve Lacey Field House, Milligan College, Johnson City, Tenn. Architect John Shaver designed both of these buildings.

Horst Berger and David Geiger worked together between 1969 and 1984. While Geiger's interest in membrane structures was for their structural efficiency and economy, Berger did much to demonstrate the aesthetic potential of tension membrane forms in architecture. Together, they developed analysis and design tools and techniques indispensable in the design, documentation, and construction of complex tension structures.

Architects such as Paul Kennon of Caudill Rawlett and Scott, later known as CRS Sirrine, were quick to embrace membrane architecture in the early 1970s and designed a number of landmark projects. They explored the forms and the spaces created by the unique translucent envelope of tension membranes. Raul de Armas and his colleagues at Skidmore, Owings & Merrill were instrumental in bringing the new membrane architecture to the attention of the world with the Haj Terminal at the Jeddah International Airport, Saudi Arabia. His design solution to the unprecedented challenge of sheltering the Haj pilgrims while changing transport modes to Mecca was simple and brilliant. Drawn to tension membrane by the desire to create a translucent canopy, the modular tent forms, which evolved in the final design, were modern and at the same time reminiscent of the tents used for centuries by Haj pilgrims.

Canadian architect Eberhard Zeidler designed two landmark tension structures for EXPO '86 in Vancouver, B.C.: the Canada Pavilion, now the convention hall at Canada Harbour Place, and the Ontario Pavilion. Beginning with the Canada Place project, Eberhard Zeidler continued to explore the sculptural potential of tension membrane architecture beyond its structural origins.

The enclosure of large clear span space created great architectural opportunity for membrane structures. Early successes were a result of the application of this building technology to the recently emerged American building type, the "domed" stadium. Uniquely North American until the late 1980s, covered stadia require roof spans without precedence. Membrane structures have been employed more often in covering sports stadia than any other structural system. There is no other building type for which this is the case, due primarily to the economy of membrane structures. Initially, the Geiger low profile, air-supported roof system was used. Later, shortcomings of air-supported roofs led Geiger to invent a new system, the Cable Dome to cover a baseball stadium in St. Petersburg, FL, combining his experience in membrane structures with Buckminster Fuller's ideas of "tensegrity" and "aspension." These systems, their variants, and other tension membrane structures continue to be significant in covering long span spaces.

Architecture

Not long after its development, the light translucency of coated fiberglass became an obvious virtue, availing it as a cost-effective substitute to glazing in many commercial projects across the continent. Recently, Zeidler's work has successfully combined translucent tension membrane with glazing to enhance the day lighting as well as architectural composition. With noted exceptions, however, the architecture of membrane structures has been little explored in the United States beyond that driven by the economics.

Most architectural forms have developed from the nature of traditional building materials. Building forms that developed from a material, say masonry, are quite often built of other materials. These forms create an architectural vocabulary well understood by the general public. However, an architectural vernacular of membrane structures has yet to be established due to their recent development. As familiarity with tension membrane architecture increases, they will be employed more often for their architectural forms.

Applications

As with all spatial systems, tension membrane structures exploit 3D forms to support load. However, tension structures are unique in that they noticeably change form in response to loading. This attribute, which anyone who has walked on a trampoline has experienced, is why tension membrane systems are almost exclusively employed as building enclosures, particularly roofs, rather than platforms or floors.

Innovative and structurally efficient tension membrane systems have been mostly employed for roofs. In some applications, the roof is employed as the entire building envelope, but applications of tension membranes as enclosure material for walls alone remain a rare exception. This is not to say that the tension membranes are not suited to this application, only that such possibilities have not been explored. A list of notable tension membrane structures is provided in Table 1.

Table 1. Notable Tension Membrane Structures

Structure	Location
Raleigh Livestock Pavilion	Raleigh, N.C.
Memorial Auditorium	Utica, N.Y.
Museum of Automobiles	Win-Rock Farms, Ark.
Yale Hockey Rink	New Haven, Conn.
Madison Square Garden	New York
Pavilion of the Federal Republic of Germany at EXPO '67	Montréal
U.S. Pavilion at EXPO '70	Osaka, Japan
Munich Olympic Stadium	Munich, Germany
La Verne College Campus Center	La Verne, Calif.
Pontiac Silverdome	Pontiac, Mich.
Haj Terminal	Jeddah International Airport, Saudi Arabia
Stephen O'Connell Center	Gainesville, Fla.
Lindsay Park Sports Center	Calgary, Alberta
Olympic gymnastics and fencing arenas	Seoul, Korea
Schlumberger Cambridge Research Center	Cambridge, England
Redbird Arena at Illinois State University	Normal, Ill.
Martha Mitchell Pavilion	Woodlands, Tex.
Canada Harbour Place	Vancouver, BC
Tropicana Field	St. Petersburg, Fla.
San Diego Convention Center	San Diego
Stadio Olympico	Rome
Georgia Dome	Atlanta
Inland Revenue Center Amenity Building	Nottingham, U.K.
Hong Kong Stadium	Hong Kong
Denver International Airport	Denver
Akita Skydome	Akita, Japan
Gottlieb Daimler Stadium	Stuttgart, Germany
Millennium Dome	Greenwich, U.K.
Oita World Cup Stadium	Oita, Japan
Seoul World Cup Stadium	Seoul, Korea

Fabric Membranes

Almost all existing tension membrane structures use textiles in lieu of a true membrane material. Essentially, practical combinations of tensile strength, tear resistance, ductility, dimensional stability, and flexibility are currently only available in coated fabrics. In almost all cases, the curved surface of a membrane structure is fabricated from flat pieces of coated woven fabric cut from rolls. Seaming is most commonly accomplished by lapped heat seals but can also be done by mechanical means such as sewing or intermittent fasteners.

Fabrics are quite different from membranes as engineering materials. Commonly used coated fabrics are composite materials whose strength is primarily provided by the woven textile and yarn: weather protection, finish, and the jointing ability are provided by the coating. This results in materials that have very low shear stiffness in relation to their tensile stiffness and are also highly nonlinear. The orthotropic behavior of coated fabrics is complex, dependent on its stress history, and is dictated by micromechanics of the weave. Fortunately, most tension structures are not particularly sensitive to the stiffness of the fabric membrane. It is because the "geometric stiffness" of the membrane arising from change in geometry and membrane prestress is more significant than the extensional stiffness of the material. However, as a consequence, tension membrane structures exhibit first-order nonlinear behavior, complicating the analysis.

The primary "stress" directions of a fabric membrane are the weave directions of the textile: the warp and fill. The warp is the direction of the yarns, which are spooled out lengthwise in the loom or weaving machine. The fill or weft is the yarn in the cross machine direction, which "fills" in the weave. As a curved membrane surface is fabricated from flat pieces of fabric, the warp is generally parallel to the seams. Different weaves have different mechanical qualities, which are primarily governed by the convolutions of the yarn and the initial state of crimp in the weave. Generally, the initial crimp in the warp is different from that of the fill. This results in different initial elongation properties in the warp and fill direction. In general, elongation at service level stresses is dominated by weave crimp, rather than strain of the yarn fibers. Consequently, elongation behavior in service has more to do with the weave than yarn characteristics. The coatings employed in most structural fabric tend to attenuate this behavior, especially for transient changes in strain. This results in response to transient loads similar to membranes. However, as almost all currently employed coatings are polymeric in nature, the effects of the coating diminish with load duration as creep of the coating allows the yarn with its weave-dominated behavior to resist the loads.

All textiles have common attributes that are significant in structural applications. Tensile strength of fabrics is greater in uniaxial than in biaxial loading, and failure is almost always a result of tear propagation rather than tensile rupture. This belies the fact that current design practice establishes membrane resistance solely on uniaxial strength. Tear propagation in textiles can be roughly analogous to crack propagation in metals in direct tension. Tears are initiated at cuts, abrasions, or other discontinuities and propagate when the force at the head of the tear reaches a critical value. Tear resistance is dependent on both yarn and weave properties.

Structural Forms

The surface forms of tension membrane structures are architecturally unique. Because the load applied to the surface must be resisted by tensile stresses in the membrane, local curvature of the surface is required. In order to ensure that stresses remain within acceptable limits, it is usually desirable to establish initial curvature in the membrane. This requires that the membrane be placed in state of internal stress or "prestress." The problem of finding prestress equilibrium forms for membrane structures is of great interest, making it important to understand the nature of the forms that can be readily employed in tension structure architecture. At the risk of being somewhat simplistic, we may categorize some of the most common forms as follows:

- Conical "tent" shape, such as in the Haj Terminal, is a prestressed anticlastic surface;
- Ridge "tent" shape is an anticlastic form characterized by a catenary ridge line supporting the membranes between two point supports (masts) nominally at the edge of the structure. The same concept can be developed in a circular configuration;
- Pleated surface shape, where the membrane surface appears folded or pleated, to form an undulating surface of ridges and valleys. This differs from the previous category in that the surface is only slightly, if at all, anticlastic, and becomes synclastic when subjected to loads. Load is carried in one direction;
- Saddle form is characterized by a single anticlastic surface;

- Vault form is anticlastic, and is usually supported by parallel or crossed arches; and
- Pneumatic forms are all synclastic, and the prestress is established by internal pressure on the membrane.

These basic forms or their combinations have been used to create a myriad of large structures. Tension membrane structures have been successfully combined with tensile net, truss, and dome systems to create lightweight long-span structures. The use of membranes for these cable structures has the advantage over more conventional building materials in that the membrane can well accommodate the relatively soft structures without a need for special jointing or releases.

Some of these systems such as the Tensegrity domes were first realized as tension membranes. These "domes" combine Buckminster Fuller's ideas of tensegrity and aspension and are comprised of a network of continuous cables and "flying" struts prestressed within the confines of a perimeter compression ring. Similar in some respects to a spatial lenticular cable truss, except that these structures rely upon nested tension rings or hoops rather than continuous bottom chords. The first tensegrity-type dome of any scale was Geiger's Cabledome for the Olympic Gymnastics Arena in Seoul, Korea. He developed the Cabledome system in order to achieve the virtues of his air-supported roof structures without the disadvantages of mechanical support. Two variants of tensegrity type domes have been realized to date, Geiger Cabledome structures and the spatially triangulated dome variant proposed by Fuller and realized by Levy. Both of these have been employed in dome stadia with spans in excess of 200 m (656 ft). These roof structures are unprecedented in their low mass. The Cabledome covering Tropicana Field in St. Petersburg, FL spans 210 m (688 ft) with a unit weight of only 0.24 kN/m^2 (5 lb/ft^2), while it has allowable capacity to carry applied gravity loads of 0.67 kN/m^2 (14 lb/ft^2) or 2.8 times its unit weight.

Design

Design of large and complex tension membrane structures is more reliant upon computers than most structural systems, as they defy classical analysis. The nonlinear behavior of these structures coupled with the need to determine prestressed forms to meet specific design and boundary conditions as well as the loading analysis, necessitates a true "computer-aided" design and modeling technique. The procedures for prestressing the system are determined in similar fashion. Finally, the templates used to cut and fabricate the fabric membrane surface are typically computer generated. As with most design methodologies the process is iterative, such that anticipation of the results in the conception of a structure will reduce the general effort involved in the design and engineering of the system.

Tension membrane structures exhibit both geometric and material nonlinearities. The nature of tension membrane structures is such that much of their stiffness is achieved by virtue of initial prestress in the membrane and its supporting components. This prestress is an internal stress condition usually prescribed by the designer to achieve the desired performance of the structure and must be induced into the system in its construction.

Fabric membranes are selected for a given structure based upon their strength, durability, fire performance, optical properties, and finish. Standard practice is to establish the minimum required strength in the warp and fill based upon the uniaxial dry strip tensile strength of the material in the warp and fill. Minimum strip tensile strengths during the expected life of the membrane are established as 5 times the maximum service stress due to the worst service combination of prestress, dead load, live load, and snow load or 4 times the maximum stress of the worst combination of prestress, dead load, and wind load. Suitable tear resistance relies upon careful detailing, installation and inspection to eliminate stress concentrations and discontinuities, as well as to identify and repair minor cuts and damage from installation and handling. Cables are commonly employed in tension membrane structures.

Form Finding

In the simple case of air-supported structures, the prestress is achieved by loading a synclastic shaped membrane with a differential air pressure. The simplest form of air-supported structure for which the prestress can be easily determined is a spherical dome. Assuming that the unit weight of the membrane is small with respect to the internal operating pressure, the membrane stress at a given pressure is proportional to the radius of curvature of the sphere. While analysis of such a structure under wind loads is nontrivial, both membrane patterning and determination of prestress are easily accomplished without the aid of computing. Hence, it is not surprising that the first widely used air-supported membranes were the spherical air-domes.

Prestressed anticlastic tensile structures present a more difficult problem. A wide variety of complex forms can be determined from physical models. As demonstrated by Frei Otto, minimal surfaces can be created using soap films. However, none of these techniques can precisely communicate to the fabricator the prestress and surface geometry information required to fabricate and stress the membrane shape. This became a pressing issue as desirable materials suitable for permanent structures, such as teflon-coated fiberglass fabric became available. Coated fiberglass fabrics have desirable attributes such as their noncombustibility. However, they are significantly stiffer than other materials commonly used in tension membrane structures and consequently require greater precision in patterning. The development of algorithms for defining the surface form or shape of a general class of prestressed networks was the key to the general exploitation of tension membranes in structures of significant scale. There is a number of form-finding algorithms currently in use. All are iterative procedures, as follows:

- The force density approach is a matrix method that solves directly for the geometry of a general network of prestressed tensile components. Iterative techniques allow the designer to prescribe desired prestress conditions for cable and membrane elements,
- Alternatively, a matrix analysis algorithm can be employed for form finding. Basically, elements are assigned very low mechanical stiffness and a prescribed prestress. Equilibrium geometry is determined in an iterative analysis of the structure, and
- Another method of form finding in common use is the method of dynamic relaxation with kinetic damping.

While physical models can be utilized to study membrane forms, the geometric and stress conditions of the membrane surface are almost exclusively determined by using computers. Data from form finding, typically comprised of connectivity, nodal geometry, and element prestress represent a complete model description of the membrane structure and the element properties. Consequently, shape results with the addition of element properties can be employed directly in the analysis. Often, additional elements, such as struts or beams, are added to create an analysis model of a complete structural system.

In order to fabricate the surface established in the form-finding process, it is necessary to establish cutting patterns. The problem is to determine the pattern of flat strips of fabric, which when seamed together will approximate the form's surface. As the shape geometry is determined for a prestressed condition, patterns must be compensated for strain in the fabric. Compensated strip patterns are then used for cutting.

Structural Analysis

General analysis of all tension-based, specifically tension membrane structures requires geometric nonlinear techniques. It is necessary to account for the change in the geometry of the structural network. It has been demonstrated that deflection terms are of first-order significance in structural networks with initial prestress.

Typical matrix methods employ an iterative procedure using the Newton-Raphson method or a variant, often with a damped solution strategy. Common tensile structural systems initially go through strain softening but then exhibit strain hardening once sufficient load is applied. Consequently, nonlinear solution strategies that anticipate strain hardening have been used with success to speed convergence in most common problems. The dynamic relaxation method is also used with success for the general analysis of geometrically nonlinear problems. Most importantly, the principles of superposition do not apply to nonlinear systems. Therefore, all critical load combinations must be analyzed individually.

Material nonlinearity is rarely modeled, although it is inherent to most fabric materials. This is just as well, because material properties are often affected by stress history. While fabric material nonlinearity is typically not modeled, it will likely prove to be useful when the mechanics of fabric failures are better understood and utilized quantitatively in a limit states design approach. The fabric is commonly modeled utilizing linear strain or constant strain triangle membrane finite elements or a network of string elements. These approaches have been widely used with success; each has attendant limitations that the analyst must consider.

Construction and Stressing Analysis

The ability to visualize, analyze, design, and fabricate complex membrane forms can create difficult construction problems. Prestress is as much a property of tensile structures as element properties and geometry. A prestressed state for a structural system can be created in a computer model without regard for the manner in which the prestress would be developed in the structure. Consequently, with redundant structures, techniques are required to establish the sequence of stressing to ensure that the structure will in fact realize the design prestressed state. Moreover, in many complex tensile systems, analysis of the stressing sequence is necessary to assure that various components of the system are not over stressed during stressing.

A technique now commonly employed is the analytical disassembly of a prestressed structural system in reverse order of stressing. The erection and stressing sequence of many complex prestressed structural systems can be determined in this manner. Generally, the accurate construction of many complex structural systems is only possible using appropriate software and suitable techniques for the determination of the stressing sequence.

Space Grid Structures

A space grid structure (SGS) is a 3D system assembled of linear elements (Engel 1968), so arranged that forces are transferred in a 3D manner. The system is also called vector-active, which is made up of two-force members whose primary internal forces are axial tension or compression. A force applied on the space grid system, typically at a node, is distributed among the axial members. When SGSs have depth or thickness, they are commonly referred to as space frames, double-layer grids, or space trusses. Single layer semi-spherical space grids are commonly known as geodesic domes.

The characteristics that make SGSs popular include: the ability to create multipurpose column-free large architectural spaces; light weight reduces their susceptibility to seismic forces; use of small elements facilitates their mass production, transportation, and handling; ease of assembly without highly skilled labor and with limited access; aesthetic appeal, visual elegance, and interesting geometric patterns; and an open form that allows easy installation of mechanical and electrical services. Since the 1940s, SGSs have been developed for the construction market, and have been used for exhibition halls, gymnasia, auditoria, swimming pools, aircraft hangars, world's fair pavilions, and mostly anywhere that a large unobstructed space is required.

History of Space Grid Structures

Space frames or grids originated with railroad truss bridges in the 19th century (Condit 1961), although the truss system dates back much earlier. Railroad expansion not only brought the development of many common truss shapes, but also led to the development of modern truss analysis. Truss development led to an understanding of how vector-based structures functioned, and to an understanding of the importance of the nodes.

Even though Alexander Graham Bell is recognized for the invention of the space frame structures in the early 1900s (Wachsmann 1961), it was August Föppl who published the first treatise, *Theorie des Fachwerks*, on space frame structures in 1880 (Schueller 1983). This treatise aided Gustave Eiffel with the analysis of his tower in 1889. Bell's obsession with the development of the first flying machines led him to investigate light structural systems. He developed a series of kites that used a tetrahedral structure, and then built architectural objects such as a windbreak wall and an observation tower using the tetrahedral structure (Mainstone 1975).

The next step in the evolution of space frame structures was the development of the lamella structural system, invented in 1908 by Zollinger in Germany and refined by Keiwitt in the United States (Schueller 1983). The roof system is distinctive for its diamond-patterned vaulting, with the sides made of short members of equal length referred to as lamellas. The nodal principles learned from joining large numbers of lamellas particularly benefited the nodal development of space frame structures. One of the most notable lamella buildings was Nervi's precast concrete airplane hangar, which was constructed in 1938.

The first major commercial development of space frame structures began in the late 1930s. In 1939 Attwood received a patent for his space frame system (Condit 1961), which later became known as the Unistrut system. In 1940 Mengeringhausen developed a space frame system in Berlin (Schueller 1983), which later developed into the MERO system. In 1945 Wachsmann and Weidlinger received a patent for their Mobilar system, which differed significantly from the MERO and Unistrut systems in that

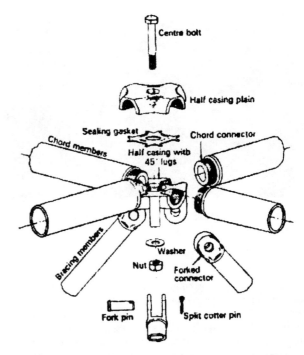

Fig. 31. Connection of struts to the node on a Nodus space frame system (Chilton 2000)

the nodes were not separated from the strut, and the geometry of the connection mechanism was not as rigid as in earlier systems. Throughout the 1940s and 1950s, these systems continued to be refined as others were being introduced, including the Triodetic system in Canada (Schueller 1983).

Geodesic domes were developed in the 1940s and 1950s by Fuller. The term *geodesic* refers to the shortest arc on a surface joining two points and was first studied by Bernoulli in 1697. Fuller studied the surfaces of a sphere or semisphere divided into large circles. Fuller's motivation in pursuing these structures was to develop an economical shape that could be used in all parts of the world. Geodesic domes have been developed from many materials including wood, steel, aluminum, concrete, and bamboo. The geodesic domes that are considered a part of space grid structures, such as the U.S. Pavilion at the 1967 Montreal World's Fair, are those whose structure is along the arc joining two points. Geodesic domes whose structure is along the surface of the polygons defined by the arcs, such as the Kaiser Dome, are considered shell structures.

The next major development for SGS came about with high-speed computers simplifying the FE analysis of complex structures and computer aided manufacturing.

Systems

A SGS acts as a network of struts and nodes. The connection methodology of the node determines all possible polyhedra within the system. The joint module determines the position of every point off direct connection from the chosen system (Wachsmann 1961). Each node must be connected with at least three noncoplanar struts to maintain stability and to prevent translation. The more axial members that can be accommodated at any given node the greater the number of morphological possibilities for the system (Gerrits 1994).

Fig. 32. Skew-chord Takenaka space truss for a school auditorium, Ferndale, Wash., Geiger Engineers

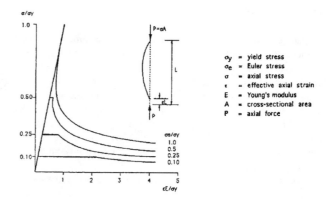

Fig. 33. Mean axial stress versus effective axial strain for axially loaded struts (Gargari 1993)

The vast majority of buildings with space grids are designed using one of many proprietary systems, such as A-Deck, Mero, Moduspan, Nodus, Ocatube, and Unistrut. This list by no means covers all of them. What makes each system unique (Fig. 31) is the geometry of the node, how the struts are connected to the nodes, the method of manufacturing the nodes and struts, and the polyhedral units possible with each system.

Very large space grids are commonly made from nonproprietary systems because of the economics of manufacturing. When designing such a system, the engineer needs to pay particular attention to the connection of the members. The system must be able to handle the rotation of the node caused by nonconcentric axial loads. It must also be able to handle the lack of fit of the members that can lead to residual stresses within the system.

Materials

Most buildings with SGS are made of high strength or mild steel tubes with circular or square shapes as well as channels and special forms, either hot rolled or cold formed. Aluminum, wood, and composites have also been used in different cross sections. The nodes for steel and aluminum grids have been designed in several shapes and forms based on their strength and aesthetics, as discussed earlier. Timber is used in the form of round poles, sawn square sections, and glued laminated elements. Members of timber grids are connected by metal pieces at their ends to each other or to metal nodes. Fig. 32 shows the interior of a skew-chord Takenaka space truss for the roof of an auditorium. The members are "pealer cores," a by-product of the plywood industry, and the nodes are cast steel. Large reinforced concrete space truss struc-

Table 2. Notable Space Grid Structures

Structure	Location	Designer	Year
Biosphere 2	Tucson, Ariz.	Margaret Augustin, Phil Hawes, John Allen with Pearce Systems International	1990
British Air 747 hanger at Heathrow Airport	London	Z.S. Makowski	1974
Climatron	St Louis	R. Buckminster Fuller	1960
Crystal Cathedral	Garden Grove, Calif.	Johnson/Burgee Architects and Severud, Peronne. Szegezdy and Strum	1980
Exhibition hall for PORTOPIA '81	Port Island, Japan	Masao Saitoh and Nikken Sekkei, Ltd	1980
Ford Rotunda Building	Dearborn, Mich.	R. Buckminster Fuller	1953
Grandstand Roof	Split, Croatia	Mero Systems	1978
Javits Center	New York	James Freed of Pei Cobb Freed and Matthys Levy	1988
Kansai Airport	Osaka, Japan	Renzo Piano Building Workshop with Ove Arup and Partners	1996
Louvre Pyramid	Paris	I.M. Pei of Pei Cobb Freed with Peter Rice	1989
McCormick Place Convention Center	Chicago	C. F. Murphy Associates	1970
Meishusama Hall	Shiga, Japan	Minoru Yamasaki and Associates and Yoshikatsu Tsuboi	1983
Palafolls Sports Hall	Barcelona, Spain	Arata Isozaki, J. Marínez-Calzón	1991
Sainsbury Visual Arts Center	Univ. of East Anglia, U.K.	Norman Foster Associates and Anthony Hunt Associates	1978
Sant Jordi Sports Palace	Barcelona, Spain	Arata Isozaki and Mamoru Kawaguchi	1990
Skydome	Toronto	Roderick Robbie and Adjeleian, Allen, Rubeli Limited	1989
Union Tank Car Company	Baton Rouge, La.	R. Buckminster Fuller	1958
United States Pavilion	Montréal	R. Buckminster Fuller	1967
World Expo Building	Osaka, Japan	Kenzo Tange, Tomoo Fukuda and Koji Kamiya, and Yoshikatsu Tsuboi	1969
World Memorial Hall	Kobe, Japan	Mamoru Kawaguchi	1984

Fig. 34. Guangdong Olympic Stadium in Guangzhou, China

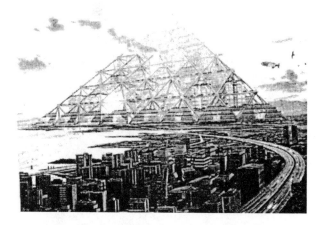

Fig. 36. Futuristic City-in-the-Air (Chilton 2000)

tures, although heavy, have been built, for example, for pavilions at a trade fair site in New Delhi, India.

Analysis and Design

SGSs are currently analyzed using linear elastic theory. The load-carrying capacity of a SGS is usually limited by the first member or set of members to fail. Connections are either made of readily available standard shapes or proprietary prefabricated pieces. Connection pieces are designed for structural efficiency or appearance. It is assumed that the connections will be strong enough so that any failure will take place in the struts or ties.

The struts and ties are treated as straight, axially loaded pin-ended members, for which the load-deformation relationship is linear up to buckling in compression or yielding in tension. Tension members ideally would yield, but may rupture in a brittle manner at the net section or at the connection. For slender compression members, there is a plateau at the maximum load in the load-deformation curve. However, when buckling stress is greater than one-half of the yield stress, failure of such members is sudden. Fig. 33 is a theoretical load-deflection graph based on the assumption that yielding in the extreme compression fiber in a straight pin-ended column limits its capacity. Although idealistic, the graph serves the purpose of showing how the plateau changes

with slenderness. In the practical range of slenderness ratios, behavior of the strut is brittle, and the collapse of the system can be initiated by the buckling of a few members (Schmidt et al. 1982).

There is a serious misconception in the behavior of a SGS. Although the members may be over-designed because of redundancy and sizing constraints, the structure may not be able to reach its full capacity predicted by elastic analysis. This occurs when compression members with practical slenderness ratios buckle before their postbuckling reserve capacity could be developed (Fig. 33). Many attempts have been made to modify the brittle behavior of a SGS, including (1) over-design of compression members to ensure that tension members yield first; (2) relying on nonlinear behavior of eccentrically loaded diagonals to redistribute forces in the chords; and (3) stress redistribution by means of force-limiting devices.

Future Direction of Space Grid Structures

Table 2 shows a list of notable SGSs in existence. The SGS will continue to develop with the extensive use of computers in both manufacturing and design. Computer aided manufacturing allows the cutting and drilling of elements with great precision, while computer aided design can help explore unprecedented complex configurations and geometries. Recent computer design programs have allowed the design of nonplanar forms. Once the form is set,

Fig. 35. Louvre Pyramid by Pei and Rice (Patrick Tripeny)

Fig. 37. Futuristic transportation system (Chilton 2000)

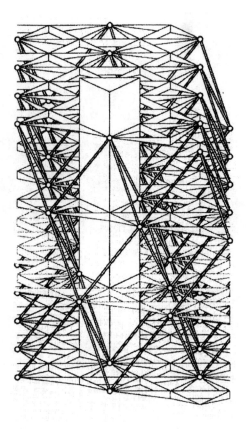

Fig. 38. Multilayer three-way grid hexmod cells (Chilton 2000)

computer programs translate the design into a space frame structure. A good example of this methodology is the roof for the Guangdong Olympic stadium in Guangzhou, China by Ellerbe Becket (Fig. 34).

Advanced work is already taking place in the area of prestressed space-frame structures, where certain members only carry tension forces in different loading conditions. The advantage of these systems is that tension members can be made relatively small in cross section, hence making the structure more transparent. These systems are commonly used in glass wall assemblies, such as the Louvre pyramid by Pei and Rice (Fig. 35).

Another advantage of using tension and compression members is that when assembled correctly they can become deployable structures that can go from a folded state to an expanded structure. While Piñero pioneered these types of structures in Spain in the 1960s, modern examples of deployable structures can be found in the work of Escrig and Hoberman (Chilton 2000).

The Shimizu Corporation in Japan has proposed the building of a pyramidal "city-in-the-air," as a 2,000-m (6,562 ft) high multi-layer grid (Fig. 36), to accommodate over 1 million people during working hours (Chilton 2000). The concept, titled "TRY2004," consists of a square-based, pyramidal, multilayer, space truss mega structure. The concept demonstrates the eminent suitability of multilayered space trusses for the construction of such large-scale projects, using tubular elements where the internal void may be used for transportation. Although this is still a dream, it represents one possible future for the use of space grids (Fig. 37).

The use of polyhedra in the design and construction of buildings of all sizes has been studied by Francois Gabriel, in particu-

lar, the architecture of high-rise buildings constructed using six-directional, multilayer, space-filling, space grids, composed of tetrahedral and octahedral (Fig. 38). With this type of space-filling lattice, it is possible to generate continuous horizontal plane grids by orienting the octahedral in two ways: with their long axis set vertically; and with one triangular face in the horizontal plane.

Concluding Remarks

Special structures are what our profession is most proud of analyzing, designing, and constructing. They are the symbols of our civilizations that brighten the horizons of our neighborhoods, and magnify the skylines of our downtowns and uptowns. History, state-of-practice, and potential future of three types of special structures were discussed: shells, tension membranes, and space grids. Considering the lack of standards and codes and direct training for special structures in most engineering curriculums, their design and construction would have not been feasible without the bravery of a few maverick engineers who used their fundamental engineering knowledge to create such landmarks around the globe. It is necessary, however, for academic programs to recognize the unique features of special structures and devote parts of the curriculum to discuss their analysis, design, and construction. If nothing else, these landmarks may serve as great motivation for the next generation of young structural engineers.

Acknowledgments

The writers would like to thank all members of the ASCE Special Structures Committee for their valuable comments, especially Professor Ronald Shaeffer of Florida A&M University and Professor George Blandford of University of Kentucky. The Committee has been very active over the last 10 years, producing a number of publications including those referenced earlier on lattice towers and guyed masts, tension fabric structures, and double-layer grids.

References

Beles, A. A., and Soare, M. V. (1966). *Space structures*, R. M. Davies, ed., Univ. of Surrey, Guilford, U.K.

Bradshaw, R. R. (1961). "Application of the general theory of shells." *J. Am. Concr. Inst.*, 58(2), 129–147.

Chilton, J. (2000). *Space grid structures*, Architectural Press, Boston.

Condit, C. (1961). *American building art: The twentieth century*, Oxford University Press, New York.

Csonka, P. (1962). *Simplified calculation methods of shell structures*, North Holland, Amsterdam, 219–234.

Cuoco, D. A., ed. (1997). "Guidelines for the design of double-layer grids." *Special Structures Committee Rep.*, ASCE, New York.

Donnell, L. H. (1933). "Stability of thin walled tubes under torsion." *Rep. No. 479*, National Advisory Committee for Aeronautics, Washington, D.C. (out of print).

Engel, H. (1968). *Structure systems*, Fredrick A. Praeger, Ind., New York.

Faber, C. (1963). *Candela: The shell builder*, Van Nostrand Reinhold, New York.

Gargari, M. (1993). "Behavior modification of space trusses." PhD thesis, Concordia Univ., Montréal.

Gerrits, J. M. (1994). "Morphology of structural connections of space frames." *Proc., 2nd Int. Seminar on Structural Morphology*, International Association for Shell and Spatial Structures, Institute for Lightweight Structures, Stuttgart, Germany, 47–56.

Gould, P. L. (1988). *Analysis of shells and plates*, Springer, New York.

Joedicke, J. (1963). *Shell architecture*, Van Nostrand Reinhold, New York.

Madugula, M. S. ed. (2002). "Dynamic response of lattice towers and guyed masts." *Special Structures Committee Rep.*, ASCE, Reston, Va.

Mainstone, R. (1975). *Developments in structural form*, MIT Press, Cambridge, Mass.

Martin, E., and Wilmeth, D. B. (1988). *Mud show: American tent circus life*, Univ. of New Mexico Press, Albuquerque, N.M.

Progressive Architecture. (1955). New York.

Schmidt, L. C., Morgan, P. R., and Hanaor, A. (1982). "Ultimate load testing of space trusses." *J. Struct. Div.*, 108(6), 1324–1335.

Schueller, W. (1983). *Horizontal-span building structures*, Wiley, New York.

Wachsmann, K. (1961). *The turning point of building; Structure and design*, T. E. Burton, translator, Van Nostrand Reinhold, New York.

American Society of Civil Engineers 1852 – 2002 *150th Anniversary Paper*
Building a Better World

Surveying and Mapping: History, Current Status, and Future Projections

Paul R. Wolf, M.ASCE[1]

Abstract: Since the founding of ASCE in 1852, remarkable changes have occurred in all areas of civil engineering practice, including surveying and mapping. Instruments employed in surveying and mapping in the United States have evolved from compass and chain, through a period of transits and tapes, into another era of optical-reading theodolites, electronic distance measuring equipment, aerial photogrammetry, and finally into the current stage of high-speed computers, the global positioning system, robotic total station instruments, digital photogrammetry, and satellite remote sensing systems. This paper includes three parts: (1) a discussion of the history of surveying and mapping; (2) a description of the current state of the art of surveying and mapping; and (3) some projections on how surveying and mapping may evolve in the future.

DOI: 10.1061/(ASCE)0733-9453(2002)128:3(79)

CE Database keywords: Mapping; Surveys; History.

Introduction

Since the founding of ASCE in 1852, remarkable changes have occurred in all areas of civil engineering practice. Certainly, surveying and mapping are no exception. In fact, during the most recent decades, changes have been occurring so rapidly and advancements have been so great even those closely associated with the profession have had difficulty keeping up with them. During the past 150 years, through the combined efforts of industry, government, and private practice, instruments employed in surveying and mapping in the United States have evolved from compass and chain, through a period of transit and tapes, into another era of optical-reading theodolites, electronic distance measuring equip-

[1]Deceased March 6, 2002; formerly, Professor Emeritus, Dept. of Civil and Environmental Engineering, Univ. of Wisconsin, Madison, WI 53706.

Note. Discussion open until January 1, 2003. Separate discussions must be submitted for individual papers. To extend the closing date by one month, a written request must be filed with the ASCE Managing Editor. The manuscript for this paper was submitted for review and possible publication on March 6, 2002; approved on March 6, 2002. This paper is part of the *Journal of Surveying Engineering*, Vol. 128, No. 3, August 1, 2002. ©ASCE, ISSN 0733-9453/2002/3-79–107/$8.00+$.50 per page.

ment, aerial photogrammetry, and finally into the current stage of high-speed computers, the global positioning system, robotic total station instruments, digital photogrammetry, and satellite remote sensing systems.

Surveying and mapping has had a long and important history throughout the world. It has always been recognized as a sign of progress. In the United States, surveying has been practiced by many distinguished individuals. To name just a few, George Washington was first a surveyor and later General of the Continental Army and President of the United States. He continued his surveying as opportunity permitted until the time of his death. Thomas Jefferson, Abraham Lincoln, and Herbert Hoover were other presidents who, for a time, were surveyors. Surveying and mapping has also always been recognized as an important area of practice within civil engineering. This was especially noticeable during the earlier years of ASCE. *The Transactions of ASCE*, which date from 1867, were the first publications of our society. They were heavily laden with articles on topographic surveying and mapping to support civil engineering planning and design. They also contained many construction surveying articles that described new methods for providing line and grade for building railroads, highways, harbors, and structures. Nearly 100 years ago, the *Cyclopedia of Civil Engineering* was published. Major portions of three of its eight volumes were devoted to surveying related matters. In 1926, the Surveying and Mapping Division was formed within ASCE, and beginning in 1956, it began publishing its own *Journal of the Surveying and Mapping Division*.

For this sesquicentennial publication, surveying and mapping is presented in three parts: (1) a discussion of its history; (2) a description of the current state of the art; and (3) some projections on how it may evolve in the future.

Part 1—History of Surveying

Surveying by the Ancients

Evidence clearly suggests that surveying was practiced at least as early as 2900 B.C., which dates the construction of the Great Pyramid of Gizeh. Modern surveying techniques have shown that the lengths of the pyramid's four sides (which each measure about 230 m) agree within ± 8 cm with each other, the elevations of all corners differ by no more than 14 mm, and the sides are aligned to within about 3 minutes of the cardinal directions. According to available records, the earliest boundary surveys were conducted in Egypt about 1400 B.C., although there is evidence that similar surveys had also been practiced in Babylonia, China, and India at about the same time, or even earlier. The earliest Egyptian boundary surveys marked and recorded plots of land ownership along the Nile River for taxation purposes. Each year the Nile flooded its banks sweeping away portions of these markers, and surveyors, who used graduated ropes for distance measurements, were appointed to replace them.

In about 120 B.C., a prominent Greek thinker named Heron authored several important publications that explained methods for surveying a field, drawing a

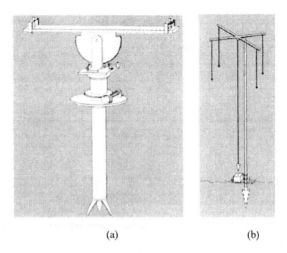

(a)   (b)

Fig. 1. Historical surveying instruments: (a) the diopter and (b) the groma

geometric plan, and making associated computations. He also described one of
the first surveying instruments, the diopter, shown in Fig. 1(a). This ingenious
instrument had a sighting mechanism that could be rotated by means of gears in
both the horizontal and vertical planes, thus enabling horizontal and vertical
angles to be measured. Not only was the instrument useful for property surveys,
but it also was employed for astronomical observations.

Surveying during Time of the Roman Empire

Surveying contributions made by the Romans date from about the 1st century
A.D., and relate primarily to construction projects throughout the empire such as
roads, aqueducts, and fortifications. They developed several useful surveying instruments including the *groma* [Fig. 1(b)], which employed a plumb bob for
setting up over a point, and was used for sighting. This instrument was particularly useful for laying out rectangular plots of land. Another instrument of Roman
origin, the *chrobates*, was used for leveling. It consisted of a straightedge approximately 20 ft long with legs rigidly fastened by braces at right angles to each
end. It could be leveled in either of two ways: (1) a groove that was cut along the
top of the straightedge could be filled with water and then adjusted until level; or
(2) by means of plumb bobs suspended from each end, the orientation of the
straightedge was adjusted until the plumb bob strings coincided with precise 90°
marks made on the legs. Once leveled, horizontal sightings could be made along
the top of the straightedge. Fortifications still in place today give testimony to the
high level of advancement achieved by Roman surveyors.

Continued Development of Surveying in Europe

After the fall of the Roman empire, there was little progress made in surveying until the 16th century. About 1570, an English surveyor, Thomas Diggs, published a book in which he described an instrument he called a *theodolitus* that was developed and used for topographic surveying. It consisted of a sighting mechanism that was incorporated with a horizontal disk graduated into 360°. The theodolite evolved from this instrument. By 1615, European surveyors used instruments of this type for *triangulation*, i.e., measuring the horizontal angles and one or more baselines in geometric figures and then trigonometrically computing the relative positions of each of the vertices.

Important developments of the 17th century that improved the accuracy and utility of surveying instruments include the invention of the *telescope* by the Italian, Galileo Galilei, in 1609 and the development of the *vernier* by the Frenchman, Pierre Vernier, in 1631. The telescope greatly facilitated making sightings with theodolites, extending their range and improving their accuracy. A French surveyor, Jean Picard, is credited with refining the telescope by installing *cross hairs* and then being the first to employ the device in surveying. The vernier made it possible for accurate readings to be made of fractions of degrees and minutes. Another important invention that furthered the development of surveying instruments occurred in 1666 when Melchisedech Thevenot introduced his *level vial*. This device, which consisted of a curved glass tube with fluid sealed inside to create a bubble, began to replace the use of plumb bobs and/or water for precise leveling of astronomical instruments and theodolites. Finally, in 1770, the *mechanical circle-dividing machine* was developed by an Englishman, Jesse Ramsden. It enabled the circles of surveying instruments to be precisely graduated, thereby significantly improving their accuracy. These inventions collectively provided the components of many precise surveying instruments to follow, including spirit levels used for determining elevations, theodolites, and others.

Surveying during the Colonial Period of the United States

The earliest surveys in the United States during colonial times were done to mark and record parcels of land ownership claimed by arriving settlers or granted by foreign crowns. Initially, plots of land were identified only by corners and/or sides that were marked with monuments such as trees, rocks, streams, fences, and roads. To record these parcels of ownership, *deeds* were prepared. This system was not based upon surveys, however, and as could be expected, it soon led to disputes arising from ambiguous descriptions and lost or destroyed monuments that could not be reliably replaced. Overlaps and gaps created by these substandard descriptions also resulted in problems. To improve the system, distances and directions between the corner markers were surveyed and made a part of the record.

Although rather sophisticated instruments had been developed in Europe, many were bulky and awkward in use. Thus, the majority of the early land surveys in the United States were done by using the relatively simple *Gunter's*

Fig. 2. Gunter's chain

chain and *surveyor's compass*. Gunter's chain, shown in Fig. 2, was developed by Edmund Gunter in England about 1620. It was made of heavy tempered wire, is 66 ft long, and consists of 100 links. This length was convenient for layout of land because of its relationship to the mile and acre, i.e., 80 chains = 1 mi and 10 square chains = 1 acre. As can be seen in Fig. 2, wire loops used to connect adjacent links resulted in hundreds of wearing surfaces, and this caused the chain to be unstable in length. Another version of Gunter's chain, the so-called *engineer's chain*, was of the same basic design and also had 100 links but was 100 ft long. Engineer's chains were more convenient than Gunter's chains for construction layout, particularly where the 100-ft station was involved.

The compass had its origins both in Europe and the Far East about 1200 A.D. The earliest use of this instrument was for navigation, but it was later employed in surveying for determining directions of lines and measuring angles in horizontal planes. As shown in Fig. 3, the surveyor's compass consists of a metal base plate (*A*) with two sight vanes (*B*) at the ends. The compass box (*C*) and two small level vials (*D*) for orienting the instrument in a horizontal plane are mounted on the base plate. Early compasses were supported by a single leg called a *Jacob staff*, but later versions like that shown in Fig. 3 were mounted on a tripod. Compasses gave magnetic directions of lines, but by applying the magnetic declination of the place and time, these could be corrected to directions based upon true north.

Westward Expansion and Growth of Surveying

After the Revolutionary War, the Colonies ceded their western lands to the central government. Some of this land was intended to provide rewards for soldiers; whereas, the balance was to be sold to raise money for the treasury of the new

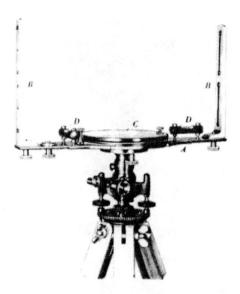

Fig. 3. Surveyor's compass

republic. To provide for the orderly distribution of land, the Continental Congress established The Land Ordinance of May 20, 1785. Later revisions enacted by the federal government altered the first ordinance somewhat, but its fundamentals have not changed. The law, which has become known as the United States *Public Lands Survey System* (PLSS), introduced a mathematically devised plan for the systematic subdivision of land and provided for the unique description of each parcel surveyed. It created a standard land unit, the *section*, having nominal dimensions of 1 mil^2, sides oriented in cardinal directions, and corners physically marked on the ground. Later, over a billion acres were added to the republic through acts such as the Louisiana Purchase and the cessions of foreign countries. Surveyors in large numbers were deputized and sent to the frontier, with the intent of subdividing all lands prior to their sale. Instruments used in the earlier years of this work were largely the compass and Gunter's chain. Overall, the PLSS, although not perfect, worked well and contributed enormously to the political and economic advancement of our country.

By the early to mid-1800s, the population of the United States was growing rapidly, due in part to the arrival of large numbers of immigrants. Westward expansion was reaching a fever pitch, maps of the western territories were in great demand, and huge quantities of new lands for settlers to purchase needed to be marked off according to the specifications of the PLSS. Also, roads, railroads, harbors, canals, and other projects had to be constructed, not only to accommodate the westward movement of settlers but also to support the development of American civilization. Preparation of the needed maps, subdivision of the public

Fig. 4. American engineer's transit

lands, and the collection of topographic information needed for planning and designing these projects and then constructing them, created huge demands for surveyors.

Surveying Instruments used in the United States from 1850 to 1950

Beginning in the mid-1800s, or at about the time of the birth of ASCE, *surveyor's tapes* began succeeding Gunter's chains and engineer's chains for measuring distances, and the *American engineer's transit* began replacing the compass for measuring angles and establishing directions of lines. Surveyor's tapes were generally 100 ft long and were made of steel. They were graduated at each foot, with hundredths of foot graduations for the first foot on one or both ends. They could be wound on reels, or done up in loops, and were far more convenient and efficient than chains.

Among the first American engineer's transits was one invented by William Young in Philadelphia in 1831. His instrument and other early versions were small and could only resolve angles to about 3 min, but by increasing the sizes of their circles, improving the quality of their circle graduations, and employing verniers to read the circles, their accuracy was improved to about 30 s for a single angle measurement. By taking the mean of repeated measurements, values accurate to within a few seconds were possible with transits. The important feature

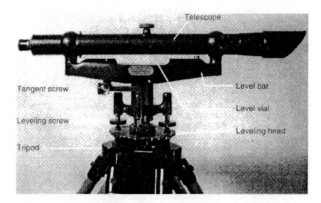

Fig. 5. Dumpy level

that distinguished American engineer's transits from the theodolites of European origin described earlier was their shorter telescopes mounted on higher standards (Fig. 4). This design enabled the telescope to be completely revolved around an axle through the standards, and it made the instrument more accurate because it could be used in both a *direct* and *inverted* mode, which compensated for errors. It also made the instrument more convenient for such tasks as prolonging straight lines and laying out curved alignments by deflection angles. American engineer's transits could be used to (1) measure angles in horizontal and vertical planes; (2) determine horizontal and vertical distances using the principles of stadia; (3) perform ordinary differential leveling (by creating a horizontal line of sight using the bubble mounted beneath the telescope); and (4) measure magnetic directions (using the compass box mounted above the base plate). They were rugged instruments, comparatively easy to operate, and could be conveniently adjusted in the field. They were arguably the most important instrument in the American surveyor's tool box for over 100 years.

Following the invention of the telescope and level vials, a new type of leveling instrument called the "spirit level" soon evolved. The early versions in use in the United States were of two basic types, the *dumpy level* and the *wye level*. As shown in Fig. 5, the dumpy level consists of a telescope rigidly mounted on a level bar. A level vial is set in the level bar. The bar is centered on an accurately machined vertical spindle that sits in the conical socket of a leveling head, an arrangement that allows the instrument to be revolved in azimuth. The leveling head screws onto a tripod. When the instrument is leveled by means of the level vial, a horizontal line of sight is created, which enables rod readings to be taken on points whose elevations are desired. The wye level is similar in most respects to the dumpy level, except that its telescope rests in supports called "wyes" on the level bar. Its advantage was that the telescope could be removed, a procedure that facilitated adjusting the instrument.

The *planetable and alidade* are among the oldest of surveying instruments employed for topographic mapping, and they were used extensively in the United

Fig. 6. Topographic mapping with planetable and alidade

States well into the 20th century. There is evidence that planetable mapping was done in Europe at least as early as about 1600. The earliest alidades did not have a telescope for sighting, but rather they employed open vanes like those of the compass in Fig. 3. In the planetable method, a map was complied directly in the field on a drawing board, which was mounted on a tripod (Fig. 6). The alidade, consisting of a telescope (later versions) mounted on a pedestal and rigidly attached to a straightedge at its base, was placed on the board. Using the telescope, a rod held at an object to be located was sighted, and a stadia interval and vertical angle were read. From these data, the horizontal distance to the object and its elevation were determined. A line drawn along the straightedge provided the direction to the object, the distance was scaled, and the point immediately located on the map. Features and contours were plotted directly in the field in planetable mapping as the work progressed. The procedure had the advantage that the map could be checked in progress for omissions and reliability. Planetable mapping eventually gave way to aerial photogrammetry.

Surveying in the United States from 1950 to 1980

As noted previously, the theodolite originated in Europe and had reached a fairly sophisticated state of design by the 18th century. However, these instruments were developed primarily for geodetic work such as extensive triangulation, and, as such, they were generally large and cumbersome. Early theodolite telescopes were long, and they could not be revolved about the axis of their standards like those of the American engineer's transit. Some of these instruments found their way to the United States, but they were used primarily for geodetic surveys rather than engineering surveys. Gradually, as machining techniques improved and op-

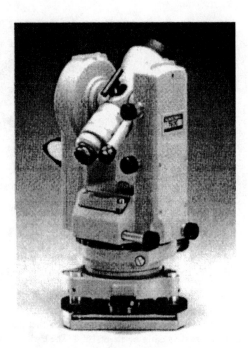

Fig. 7. Optical-reading theodolite

tical systems were perfected, European theodolites became smaller, their circles were etched on glass, and internal microscopic reading systems replaced verniers. These and other improvements made the instruments more accurate than American engineer's transits. Also, Japanese optical companies began producing theodolites of excellent quality, and the competition lowered prices. For these reasons, beginning in about the 1950s, these smaller optical-reading theodolites began replacing transits for general surveying and engineering work in the United States. Their use had become widespread by about the 1980s when total station instruments were introduced. Thereafter, their popularity gradually decreased to the point that few are in use today. An optical-reading theodolite is shown in Fig. 7.

About the middle of the 19th century, an automatic or self-leveling feature was introduced that led to today's so-called *automatic level* (Fig. 8). One version of a self-leveling device uses prisms suspended by thin wires to create a pendulum. The nature of the prisms, wire lengths, and support locations within the telescope is such that only horizontal rays are directed through the cross hairs. This produces a horizontal line of sight even though the telescope may be slightly tilted. Automatic levels are quickly set up by roughly centering a bull's-eye bubble, thereby greatly increasing the efficiency of leveling operations. They were

Fig. 8. Automatic level

quickly adopted for general surveying and engineering use and later also for precise geodetic leveling. Now automatic levels have almost totally replaced all other types.

Photogrammetry

Photography was invented in 1839 by a Frenchman, Louis Daguerre, and a year later *terrestrial photos*, those taken from a ground-based platform, were first used for surveying. It was recognized early that photos taken from high above the ground would give an advantageous view for planimetric mapping, and thus, some experiments were conducted in kite and balloon photography. However, these generally yielded unsatisfactory photos for mapping. The invention of the airplane in 1902 by the Wright Brothers provided the needed platform for taking *aerial photos*. The first mapping project using aerial photos taken from an airplane occurred in 1913. Aerial photos were used extensively during World War I, primarily for reconnaissance.

Between the two world wars, *photogrammetry*, as this developing science became called, emerged as a world-wide procedure for topographic mapping and other applications. Aided by the development of *stereoplotting* instruments, government agencies as well as private companies became engaged in the practice. One type of early stereoplotter, the *optical projection* type, was first used in Europe about 1920 for topographic mapping. These instruments were composed of optical and mechanical elements. When transpariencies of overlapping photos were properly oriented in their projectors, corresponding light rays from the two photos were projected through objective lenses to create an accurate model of the overlapping terrain, and a map of the model could be drawn. Coordinatographs were incorporated with later versions of these types of stereoplotters so that X, Y, and Z coordinates of points could be read and recorded. Fig. 9 shows the multiplex optical projection stereoplotter manufactured by Bausch and Lomb. This instrument was used in the United States beginning in the late 1930s and was widely employed by the U.S. Geological Survey in their topographic mapping program for more than two decades.

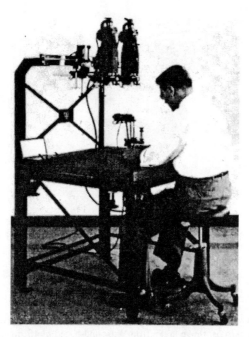

Fig. 9. Bausch and Lomb multiplex optical projection stereoplotter

Also during the 1920s, developments were ongoing in Europe on another type of stereoplotter, the *mechanical projection* instrument. Whereas objective lenses occupied the perspective centers of projectors in optical projection instruments, *gimball* joints were centered in the perspective centers in the mechanical projection type of instrument. Precisely manufactured *space rods* mechanically connected to the gimball joints simulated projected light rays. Transparencies of overlapping photos were viewed stereoscopically through internal optical systems by means of binoculars. Depending upon the design, an operator used either a hand- or foot-impelled device to impart movements to the space rods, which in turn shifted the viewing system as necessary to place a reference mark on points of interest. This enabled a map to be traced of the stereomodel, or if interfaced with a coordinatograph, the X, Y, and Z coordinates of points could be recorded. Fig. 10 illustrates a mechanical projection stereoplotter with attached coordinatograph. Both the optical and mechanical projection types of stereoplotters have now been replaced by more modern instruments.

With the advent of computers in the 1950s, *analytical photogrammetry* came into being. In this procedure, the object space positions of points are computed numerically. First, images of points whose spatial locations are desired must be selected and marked on one photo, their corresponding image locations transferred stereoscopically and marked on all overlapping photos upon which they appear, and their coordinates measured in the reference coordinate system defined

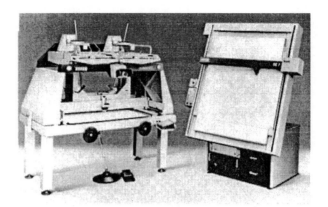

Fig. 10. Zeiss Planicart mechanical projection stereoplotter with coordinatograph

by the camera's fiducial marks. To enable computing the spatial positions of the selected points, a suitable number of ground control points that appear in the photos must also be selected. Their coordinates must be determined in the desired object space coordinate system, and their images included within the identification, stereoscopic transfer, and measuring mentioned above.

To complete the analytical determination, equations are developed that incorporate the ground control coordinates, photo coordinates, and camera parameters. The unknown ground coordinates of desired points are also included and calculated by solving the equations numerically. Highly redundant systems of equations are generally formed in analytical solutions, and the method enables rigorous accommodation for most systematic errors. This provides the highest accuracy possible in photogrammetry today. Analytical photogrammetry is widely used for densifying ground control networks in a process called *aerotriangulation*.

Analytical plotters represent the next generation of stereoplotter evolution. The first such instrument was commercially marketed in 1964, and improved versions are still widely used today. Analytical plotters combine a precise *stereocomparator* (a device for measuring photo coordinates), servomotors, and encoders, all interfaced with a computer. The servomotors and encoders are important parts of the system, because they enable the computer to drive the various components of the equipment so that a wide variety of photogrammetric operations can be performed in a highly automated fashion. In using the instrument, transparencies of a pair of overlapping photos are placed on the stages of the stereocomparator, and photo coordinates are measured with assistance from the computer. From externally input camera parameters and coordinates of ground control points, together with these measured photo coordinates, the computer determines the parameters of the instrument's orientation. Once this is completed, object space coordinates of points of interest can be computed in real time as an operator measures their image positions. Analytical plotters are very versatile instruments

Fig. 11. Electronic distance measuring instrument mounted on a theodolite

and are regularly used for aerotriangulation, topographic mapping, measuring elevations along preselected profile lines, cross sectioning, generating digital elevation models, and compiling layers of information in digital form for geographic information systems.

Electronic Surveying

An instrument for measuring distances electronically was first introduced by the Swedish physicist Erik Bergstrand in 1948. His *electronic distance measuring* (EDM) device, called the *geodimeter*, resulted from research to improve methods for measuring the velocity of light. The instrument transmitted visible light, measured its time of travel to a reflector and back, and converted the result to distance. This first EDM instrument was followed by a host of others, the early versions of which all were expensive, bulky, and inconvenient to use. Nevertheless, these instruments ushered in the age of electronic surveying, which practically eliminated the painstaking process of taping, and dramatically changed the practice of surveying.

During the decades from the 1950s through the 1970s, EDM instruments were improved significantly, making them smaller, more accurate, and easier to use. Second generation versions employed infrared or laser light, which greatly reduced power requirements. Soon they were small enough to be mounted on theodolites (Fig. 11), which increased surveying efficiency by making it possible to measure distances and angles at a point from a single instrument setup. During this same period of time, advancements were being made with so-called *electronic theodolites*, i.e., instruments that could automatically read and record angular values. These developments paved the way for further advancements in surveying technology.

Evolution of Federal and State Surveying Organizations

Through the years, the federal government, faced with the enormous tasks of surveying and managing the lands within the national domain, created several key agencies to manage these tasks. *The Survey of the Coast*, our nation's first civilian scientific agency, was established within the Commerce Department by President Thomas Jefferson in 1807. Its name was changed to the *Coast and Geodetic Survey* in 1878 and to the *National Geodetic Survey* in 1970. As implied by its original name, its initial tasks were to provide coastal charts and navigational aids. Its mission was soon broadened to include performing the geodetic control surveys needed to develop and maintain the *National Spatial Reference System*. This system consists of many thousands of horizontal control monuments and bench marks of elevation located throughout the country for use in originating local surveys.

In 1812, Congress established the *General Land Office* within the Treasury Department. In 1946, this agency was merged into the *Bureau of Land Management* and transferred to the Interior Department. The original purpose of this agency was to oversee the survey and disposition of federal lands. It is currently responsible for managing about one-eighth of the area within the United States and Alaska. The *Geological Survey* was established in 1879, also within the Department of the Interior. It was originally charged with examining our nation's geological structure and surveying its mineral resources. Its mission was later expanded to include national mapping. In this capacity, it has compiled nearly 70,000 different maps and distributes approximately 10 million copies annually.

Besides the federal agencies noted above, many others have performed extensive surveying and mapping for specialized purposes. They include the Corps of Engineers, Forest Service, National Park Service, International Boundary Commission, Bureau of Reclamation, Tennessee Valley Authority, Mississippi River Commission, Department of Transportation, and others.

Agencies have also evolved in all 50 states that perform specialized surveying and mapping. Departments of Transportation, for example, exist in all states to provide the surveying and mapping needed for highway planning and design. Departments of Natural Resources are another example. All states have enacted laws that require licensing of surveyors who perform boundary surveys. The individual states set their own education and experience requirements for granting licenses.

Professional surveying and mapping societies also have evolved through the years at the international, national, and state levels. In general, their goals are to advance knowledge in the field, encourage communication among members, and upgrade standards of practice and ethics. The *International Federation of Surveyors*, founded in 1878, holds international meetings and symposia where delegates from around the world gather for the exchange of information. On the national level, in addition to *ASCE*, the *American Congress on Surveying and Mapping*, founded in 1941, and the *American Society for Photogrammetry and Remote Sensing*, founded in 1934, are two leading surveying and mapping professional organizations in the United States. Most states also have professional surveyor societies with full membership open only to licensed surveyors. They are generally affiliated with the American Congress on Surveying and Mapping.

Part 2—Current State of the Art in Surveying

Computer Era and its Impact upon Surveying

There can be little doubt that the single most important technological development in recent years has been the computer. It has affected almost every facet of what most of us do in our daily lives, regardless of our field of endeavor. But computers have especially impacted the technical areas of practice. In surveying and mapping, they have completely changed the way data are collected, computations are made, maps and other products are developed and then disseminated to users.

Among the newest computer-driven technologies available to assist surveying engineers in their work of developing and processing spatial data are (1) total station instruments, including robotic systems; (2) the global positioning system (GPS); (3) digital photogrammetry and light detection and ranging (LIDAR) systems; (4) satellite remote sensing; and (5) geographic information systems (GIS). These new systems are enabling surveying engineers to provide new and better types of information at lower cost and in fractions of the time previously required. These systems are briefly described in the sections that follow.

Total Station Instruments

Beginning in about 1980, an EDM component, which also had been improved to enable automatic readout, was combined with an electronic theodolite to create a single instrument called the *total station*. The functions of the distance and angle measuring components were controlled by an interfaced computer. Modern total station instruments can now make slope distance measurements, automatically display the results, and also store the data in computer memory. They can also measure angles both in horizontal planes and vertical planes, and again, the results can be automatically displayed and stored. The on-board computer can use these measured data in real time to resolve horizontal and vertical distances, to calculate the positions and elevations of points, or to set points for construction projects. Total station instruments are probably the most commonly used and important instruments in modern surveying today, having practically replaced all transits, theodolites, and stand-alone EDM instruments. Fig. 12 illustrates a total station instrument.

Today's most modern types of total station instruments are equipped with servo-drive mechanisms and a built-in radio communication system. These de-

Fig. 12. Total station instrument

vices, called *robotic total stations*, are able to automatically follow a special type of prism, known as the *remote unit*, that is also equipped with a radio telemetry link. Robotic total station instruments can automatically and repeatedly measure distances and angles to a moving reflector and store the values in memory without the need of an operator. With its built-in computer it can also calculate instantaneous positions and relay them by radio to users. This has many potential applications, including real-time tracking and positioning for hydrographic surveys, dredging operations, and of particular interest, "stakeless" construction (described later).

Global Positioning System

The global positioning system is based upon observations made of signals transmitted from a constellation of satellites. Each satellite operates in a precisely known orbit, and each transmits a unique signal. The signals are picked up at ground stations by *receivers*. Highly exacting distances (ranges) from the satellites to the receivers are determined from the signal information, enabling receiver positions to be computed. Two different techniques are used in computing positions by GPS, *code matching* and *carrier-phase measurements*. Because the latter is more accurate, it is the method most often employed in surveys for engineering applications.

In the carrier-phase method of locating positions, *phase shifts* that occur in the carrier signal while it travels from satellites to receivers are recorded. The indi-

Fig. 13. Global positioning system receiver being used in the kinematic mode for construction staking

vidual phase shifts depend upon the travel distances of the signals. This is fundamentally the same principle that has been used for years in electronic distance measurement. A problem associated with this procedure is that only the indicated phase shift within the last cycle of the signal is measured, and the number of full wavelengths contained in the travel path is unknown. This is the so-called *cycle ambiguity*. A procedure called *differencing* (taking the difference between simultaneous measurements made at two or more stations to multiple satellites) can be used to eliminate most of the errors in the system and resolve the cycle ambiguities.

Field work with GPS using the phase-shift method usually employs *relative positioning* techniques, that is, two (or more) receivers occupying different stations simultaneously make observations to several satellites, usually four or more. The vector between receivers is called the *baseline,* and its dX, dY, and dZ coordinate difference components are computed as a result of the measurements by employing the differencing methods noted above. Relative positioning with GPS is capable of yielding locations of points to accuracies within a few centimeters or less.

Procedures have been developed so that GPS measurements can be made in either a *static* or *kinematic mode*. In the latter method, one receiver is permanently placed on a control station (point of known position), and the other, called the *rover*, is moved about from point to point. The rover can be hand carried or placed in a ground vehicle or aircraft, and readings are taken on the move. This makes GPS suitable for many specialized dynamic tasks such as hydrographic surveys, dredging, marine pipeline laying, stakeless grading, and others. But GPS is also equally adaptable for traditional applications such as control surveying, topographic mapping, boundary surveying, and construction surveying. A GPS receiver being used in the kinematic mode for construction surveying is shown in Fig. 13.

To ensure the highest order of accuracy, GPS measurement sessions must be carefully planned and designed. The geometric configuration of visible satellites must be taken into account as it can significantly affect results. Surveying engineers must also carefully analyze all data at the termination of each session to verify that it is all of a consistent nature and that no blunders exist. The coordinate system in which satellite and receiver positions are computed is three-dimensional rectangular. It is called a *geocentric* coordinate system because its origin is at the earth's mass center. Before measured GPS data can be used, it must be adjusted, and finally transformed into a local gravity and meridian-oriented reference coordinate system, either geodetic or plane, depending upon user requirements.

GPS has been in use now for several years, and the systems are continually improving and their costs decreasing. Surveying by GPS affords many advantages over other traditional methods, including speed, accuracy, and operational capability by day or night and in any weather. Also, intervisibility between stations is no longer required as it was in traditional methods of surveying. Of course a practical minimum of four satellites must be visible, so GPS has limited applications in mining, tunneling, and surveys in forests or high-rise urban areas. But it has rapidly gained acceptance in all other types of surveys.

Digital Photogrammetry

Surveying engineers have used photogrammetry for many years, but recently the manner by which it can be accomplished has dramatically changed. As noted earlier, in traditional photogrammetric procedures, the spatial locations of points were determined using either analog instruments known as *stereoplotters* or by employing analytical methods. In digital photogrammetry, the contents of overlapping photographs are obtained as a raster of *pixels* (picture elements), each of which has its row and column location within the raster and its unique density stored digitally. This can be done by scanning photos taken by traditional film cameras, but recently, digital cameras have been developed that produce images directly in a raster format, thereby eliminating the scanning step.

Selected pixels replace marked image points in digital photogrammetry. A pixel for each of the points whose spatial locations are desired is specified within the image. The selected pixels are identified by their row and column locations within the image rasters. Their corresponding locations on the overlapping photos are located by a computerized process known as *image correlation*, also called *image matching*. This automatic process, which replaces manual stereoscopic point transfer in the traditional analytical procedure, involves designating a small subraster of pixels called the *target area* that surrounds the selected pixel on one photo. A larger subraster of pixels, called the *search area*, is selected in the corresponding area of the overlapping photo such that it will encompass the target area.

In the image correlation process, the computer automatically selects the specific subraster of pixels within the search area that corresponds to the target area. This is done by systematically moving about the search area and comparing densities of the pixels of the target area to those of target-sized subrasters within

Fig. 14. Intergraph Image Station Z, digital photogrammetric instrument

the search area. Various algorithms have been developed for making these density comparisons. When the target-sized subraster is found within the search area that yields the smallest difference among compared pixels, correlation is completed, and the center pixel of that subraster is the desired corresponding point. It is designated by its row and column location. (Refined processing enables points to be located to subpixel levels.) This image correlation process is continued until all required corresponding points have been identified on all overlapping photos.

Again, as with traditional analytical photogrammetry, ground control points are required, and their images must be identified on the photos. This image identification can also be accomplished by image correlation. Finally, a set of equations is formulated that incorporates the ground coordinates of control points, the known row and column locations of the image points, and the camera parameters. The equations also contain the unknown ground coordinates of the desired points, and they are determined by solving the equations. A digital photogrammetric instrument is shown in Fig. 14.

Digital photogrammetry can be performed at a savings in labor and in cost. Its accuracy is dependent upon many variables, but one important one is pixel size, with smaller pixels yielding better results. Research has shown that the process is generally accurate to a half pixel or less. For photos at 1:5000 scale and a pixel size of 25 μm, a half pixel represents 12 cm on the ground.

Light Detection and Ranging Mapping Systems

LIDAR is a relatively new technology that appears to have the potential of significantly impacting surveying, particularly in aerial mapping. Airborne LIDAR mapping systems consist of a laser imaging device, an inertial navigation system, a GPS receiver, and a computer, all of which are carried onboard an aircraft. In operation, laser pulses are transmitted toward the terrain below. They are reflected, returned, and their time of travel accurately measured. The GPS receiver continually provides precise information on the aircraft's position in its flight path, and at the same time, the inertial navigation system records the aircraft's attitude. The computer processes all of this information to determine vector displacements (distances and directions) from known positions in the air to unknown positions on the ground, and as a result, it is able to determine the X, Y, Z locations of a dense network of ground points. Not only are positions computed, but an image of the ground is also generated. Airborne LIDAR systems are currently achieving accuracies in the range of 10–15 cm. These systems are also being employed in close-range applications such as industrial and architectural work. Its use for topographic mapping is also being investigated. Because they are capable of producing digital mapping products in real time, these systems hold great promise for the future.

Satellite Remote Sensing

The launch of the Landsat 1 satellite in 1972 by the United States ushered in a new era of earth-resource monitoring and data collection. This satellite system made possible the collection of image data for the first time on a global and repetitive basis. The *multispectral scanner* of this system collected rather coarse data by today's standards, as its resolution was only 80 m, with data collected in four spectral bands and in a range of 64 radiance levels. Subsequent Landsat satellites improved the resolution, first to 30 m, and finally to 15 m, with scanning occurring over seven spectral bands, and in a range of 256 radiance levels. Fig. 15 illustrates an image obtained by the second generation Landsat Thematic Mapper spacecraft over Madison, Wisconsin. Note the clarity with which roads, urban areas, and agricultural crops are shown.

The French government launched their first *Systeme Pour d'Observation de la Terre* (SPOT) satellite in 1986, and subsequent versions have followed. These systems employ *charge couple devices* (CCD) arranged in a linear array along a line perpendicular to the satellite's track and provide imagery with a resolution of 10 m. The imaging procedure requires no moving parts, and thus the geometric integrity and image quality of the data has been improved—so much so that the images look very much like small-scale aerial photos. SPOT has the capability of pointing its data collection optics by means of mirrors that are rotated according to commands from the ground. This enables specific areas to be covered stereoscopically from two different orbit tracks, a feature that enables the imagery to be used for small-scale topographic mapping.

In September 1999, Space Imaging launched the first commercial imaging satellite, IKONOS. Its imaging system has a resolution of 1 m, and hence, objects as small as individual trees and automobiles can be identified in the images. The imaging optics can be aimed off-nadir, either from side to side, or fore and aft—a feature that enables stereoscopic coverage to be obtained. Thus, the imagery is also suitable for small-scale topographic mapping.

Satellite imagery is in a digital format and is thus directly amenable to processing using digital photogrammetric techniques.

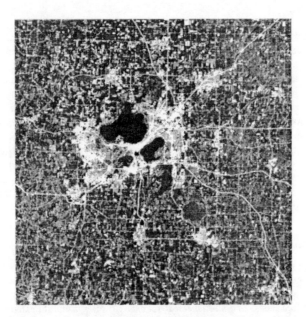

Fig. 15. Landsat Thematic Mapper image taken near Madison, Wisconsin

Digital Elevation Models and Digital Orthophotos

Digital elevation models (DEM) and *digital orthophotos* are two relatively new products that are being developed as a result of new aerial mapping technologies. Digital elevation models are a systematic array of points on the ground whose X, Y, and Z coordinates are known. Using digital photogrammetric techniques they can be generated by performing image correlation on a grid of points within overlapping aerial photos, and then solving for their ground coordinates analytically. DEM can also be produced using airborne LIDAR.

Orthophotos are perspective photos that have been modified so that they have a constant scale throughout. Thus, they are equivalent to maps, and distances and angles can be accurately scaled from them. Digital orthophotos are produced in a digital photogrammetric process that consists of shifting pixel locations to account for scale variations due to relief in the topography and tilt in the aerial photos. The pixels are then reprinted electronically to obtain an image having constant scale throughout. Special hardware packages called *softcopy workstations* and sophisticated software have been developed that facilitate the performance of the above described digital photogrammetric processes.

DEM and digital orthophotos are extremely useful for many applications. They provide the topographic information needed to plan grade lines for construction projects, determine earthwork quantities, and investigate drainage. Of course, contour maps, crosssection, and ground profiles can also be generated from them. DEM also provide the basic data needed to support a host of analytical functions within geographic information systems that depend upon terrain information.

As noted above, digital orthophotos are equivalent to maps, but they are more useful than maps in many instances because features are shown by their images rather than by lines and symbols. This makes them more easily interpreted. Because they are in a digital format and their pixels are all geographically referenced, they are ideal as base maps for engineering planning and design, as well as for use in geographic information systems.

Because satellite images are collected in digital format and their pixels are also geographically referenced, they too are convenient for use in geographic information systems, especially for certain small-scale applications. Satellite images are particularly suited for a variety of land and resource management applications. Also, with their improved resolution and image qualities, they are being used to compile small-scale maps that have a multitude of uses, and the images are also employed in a variety of image interpretation applications, both automated and manual. Examples of applications include exploration for engineering materials and minerals, performing water-quality analyses, generating forest-type maps and detecting stressed trees, preparing land-cover maps for improving hydrologic modeling, identifying crops in support of agricultural programs and erosion control activities, analyzing flooding, assessing wildlife habitat, and many others.

Merger of Global Positioning System and Photogrammetry

Time and costs associated with establishing the ground control required for photogrammetry can be very significant if done by conventional surveying methods, or even if GPS is used to locate the ground control points. Furthermore, in certain situations, due to vegetation or other constraints, it is nearly impossible to establish the needed ground control. By mounting a GPS unit in the aircraft, operating it in the kinematic mode, and correlating the timing of camera exposures with the operation of the GPS unit, the amount of ground control that is needed can be greatly reduced. This procedure provides ground coordinates of the exposure stations and fixes the scale of the photogrammetric solution. The technique is now being applied by many commercial firms worldwide, and is being used by some state governments as well as the federal government of the United States. The merger of these two technologies is making digital photogrammetric products more affordable.

Geographic Information Systems

The term *Geographic Information System* (GIS) is a relatively new one in surveying and engineering. A GIS is a computer-based system that enables storing, integrating, manipulating, analyzing, and displaying virtually any type of spatially related data about our environment. GIS have rapidly assumed a position of prominence in our modern information-oriented society and are being used worldwide at all levels of government, and by businesses, private industry, and public utilities to assist in planning, design, and management. Besides engineering, specific applications have occurred in many diverse areas and include natural resource management; facilities planning and management; land records modernization; demographic and market analysis; infrastructure management; emergency vehicle response and fleet operations; regional, national, and global environmental monitoring; and many others.

Spatial data that is timely and of good integrity is the key to the successful operation of any GIS. Many types of information are included within GIS databases, and the exact selection of what to include in any specific one will depend upon its purpose. Data may be derived from surveys (aerial or ground) dedicated to a specific GIS and may consist of digital orthophotos, digital elevation models, and digitized layers of various types of natural and cultural features. Data may sometimes also be obtained from existing maps, charts, aerial photos, satellite images, construction plans, statistics, tabular data, and other documents. Examples of specific types of frequently used data or "layers" include (1) boundaries of land ownership; (2) land cover; (3) soil types; (4) floodplains; (5) wetlands; (6) zoning; (7) infrastructure including roads, railroads, electrical transmission lines, water utilities, sewers, and many others.

In order for a GIS to function, all information entered into its database must be spatially related, that is, it must all be located in a common geographic reference framework. Only then can the different layers of information be physically related so they can be analyzed using computers to support decision making. This geographical position requirement will place a heavy demand upon surveying engineers in the future who will play major roles in GIS design, implementation, and management. Their work will include establishing basic control frameworks; conducting boundary surveys and preparing legal descriptions of property ownership; performing topographic and hydrographic surveys using ground, aerial, and satellite methods; and compiling and digitizing maps and assembling a variety of other digital data files. Quality control will be an important aspect in all of these activities.

Uses of Geographic Information Systems

It has been estimated that in the United States, on average, surveying and engineering account for approximately 40% of the total costs of civil projects. It has also been determined that if geographic information systems had been employed on applicable projects for planning and design up to 50% of those costs for surveying and engineering could have been saved. That represents a huge potential savings!

A key to maximizing the potential savings of geographic information systems is the sharing of digital information among different organizations. This can be particularly effective among cooperating governmental agencies. In the United States for example, the Geological Survey, the National Oceanic and Atmospheric Administration, the Bureau of Land Management, the Environmental Protection Agency, the Forest Service, and others are each generating large quantities of digital data. But experience shows that certain data developed for use by one agency is also suitable for others. This led to the creation of the *National Spatial*

Data Infrastructure (NSDI), which serves as a clearinghouse or pathway for finding information about available spatially referenced data. Goals of the NSDI are to promulgate policies, standards, and procedures for organizations to cooperatively produce and share geographic information. An important element among the standards deals with *metadata*, or information about the source, data of acquisition, accuracy, and other characteristics that become important in selecting available data.

The utility of geographic information systems is perhaps best described by example. An application in civil engineering involves performing flood analyses on a regional basis. Critical spatial data entered into this type of GIS for the region covered would be topography; soil types; land cover; number, sizes, locations, and configuration of drainage basins; existing stream networks with stream-gaging records; locations and sizes of existing bridges, culverts, and other drainage structures; data on existing dams and the water impoundment capacities of their associated reservoirs; and records of past rainfall intensity and duration. Given these and other data sets, together with a model to estimate runoff, a computer analysis can be performed to predict locations of potential floods and estimate their severity under varying conditions. Conditions that could be varied include rainfall intensity and duration and the area of the region where it occurs. A 100-year storm, for example, could be assumed over the entire region, enabling an assessment to be made of the magnitude of the resulting flood. Analyses could then be made of ways to mitigate the disaster by modeling the affects of adding dams and other flood-detention structures of differing sizes at varying locations.

The successful implementation of geographic information systems relies on people with backgrounds and skills in many different disciplines, but none are more important than surveying engineers. This is true because the very function of these systems relies on spatially related data that is accurate, timely, and complete.

Part 3—Future of Surveying and Mapping

Introduction

It is always risky to make predictions about the future. At this point in time, however, based upon successes already realized using new technologies now available in surveying and mapping, some reasonable predictions may be made about the near future. It is predictable, for example, that in the near future surveying-grade GPS units will become more affordable and that all persons engaged in surveying data collection will own and use them. These units will be applied extensively for all types of surveys including control, boundary, construction, topographic, hydrographic, and others. It is also predictable that GPS units and other guidance equipment carried in aircraft will completely eliminate the need for ground-control surveying in aerial mapping, whether done by photogrammetry, LIDAR, or other equipment yet to be perfected. Carrying this a step further, it also appears reasonable to expect that aerial mapping technologies will be modified and improved so that orthophotos and digital elevation models can be prepared in flight and instantaneously transmitted directly to engineering offices for immediate use in planning and design or for GIS applications.

In the future it can be expected that the technologies of GPS, GIS, and digital spatial information will be integrated in solving a host of problems, particularly dynamic applications. The following section suggests some of these applications.

Integration of New Technologies and Digital Products

A number of projects are now being visualized or experimentally pursued that capitalize on the integration of new surveying technologies and digital spatial information. Three examples are cited here. The first involves the use of GPS and digital elevation models in aircraft guidance. With the GPS unit mounted in an aircraft, real-time direction of flight, speed, and position is made available in three dimensions. This flight information would be superimposed onto a DEM, which also includes information about tall structures such as towers, buildings, overhead power lines, etc. A computer on board would analyze the digital database along the projected flight path in real time and give a timely warning anytime the aircraft was approaching a height less than a minimum threshold above the terrain or other feature. The system would be most beneficial in poor visibility and as aircraft approach airports, especially those situated near mountains or near the downtown areas of large cities.

A second example that illustrates the value of using integrated surveying technologies relates to *intelligent transportation systems* (ITS). A lower level of ITS that is operational now employs a GPS operating in real time in a vehicle while at the same time a digital planimetric map is visible to the driver on a screen. A cursor on the map indicates the vehicle's real-time location. The driver uses this information to select the route based upon current position and intended destination. The system can be enhanced by adding other information and allowing the computer to select the best or most direct route. Additional information can include data on traffic congestion, accidents, construction activities, detours, or any other activity that could possibly affect travel time. The computer performs dynamic routing, taking into account all of this data, but the driver still manually operates the vehicle. This system is currently in experimental stages.

A more advanced ITS visualized for the future is a totally automated guidance system. It is currently being tested on unsignalized roadways such as freeways and interstate highways. The system uses all of the information of the lower level ITS systems, but, in addition, it contains digital information that defines the shape and size of the road and precise information on its geometrical configuration, both horizontally and vertically. A guidance system would automatically drive the vehicle with no operator intervention. Its operation would be based on the continuous receipt of GPS data giving the vehicle's position and the simultaneous comparison of those positions to information about the roadway. In an even more advanced system visualized for unsignalized roadways, an additional database would include information on locations of traffic signals, their nature, function, and their sequences. Automated signal timing and operation will be an important aspect of the system. Obviously, at this level, information in the database must be timely and highly accurate and reliable—requirements that will challenge even the most astute surveyor!

The third example involves "stakeless" grading during construction. For this example, assume that a new expressway is to be constructed. Assume further that a DEM has been generated for the corridor through which the roadway will run and that the design has been prepared based on a DEM and a digital orthophoto. In design process, 3D coordinates (in the same reference system as the DEM and orthophoto) have been developed for expressway alignment, gradeline, and design template. Under these conditions, either of two different procedures could be followed that would enable the grading to be performed without setting out grade stakes; one uses GPS and the other employs robotic total-station instruments.

In employing GPS, one or more antennas would be mounted, for example, on the blades of bulldozers and graders. By operating in the kinematic GPS mode, the operators will have continuous 3D positions for their blades while moving about the construction area. At the same time, the computer can superimpose and compare the 3D plan information to blade locations, continually guide operators in position, and provide the adjustments necessary to achieve the desired cut or fill.

Alternatively, a robotic total-station instrument could be employed to accomplish stakeless grading. For this application, the remote unit is equipped with multiple reflectors mounted around its circumference. This guarantees that at least one prism will face the robotic total station regardless of the location or orientation of the remote unit. The remote unit would be mounted on the blade of a bulldozer or grader, the robotic total station would be set up on a strategic nearby ground control station, oriented in 3D position and azimuth, and then locked onto the remote unit. From continuously repeated angle and distance measurements made to the remote unit, the on-board computer can determine the 3D positions of the blade in real time as the operator moves about the construction area. Again, the computer can superimpose the design plan onto the locations of the blade so that appropriate cut and fill adjustments can be made.

In either of these approaches, grading can be accomplished without the need for construction stakes and with reduced assistance from grade foremen. Furthermore, when the work is completed, any deviations from plan will have been recorded, and therefore, a precise "as-built" plan will have been produced. Experimental projects are now being conducted using these procedures, and early results indicate that grading in this manner can be done faster and with significant cost savings.

New Birth in Surveying and Mapping: Geomatics

Surveying and mapping is in the midst of perhaps the most exciting period of its existence. Never before have so many modern high-speed technologies been available for performing the work of collecting and processing data about planet earth. Never before has civilization demanded so much information for making decisions in planning and designing proposed projects, managing available natural resources, and preserving the environment. Concurrent with these developments, geographic information systems have emerged and matured. These systems enable vast quantities of environmental data to be integrated and analyzed to render the information needed for decision making. All of these factors collectively have created a great opportunity for a rebirth in surveying and mapping.

In the future, large numbers of well educated and dedicated professionals will be needed in the surveying and mapping area. The challenges will be great. The long range and high accuracy potential of our new technologies will demand that geodetic procedures be followed in collecting and processing data. Those involved in working with GPS data must understand the fundamental differences between the earth-centered geocentric coordinate system employed in satellite surveying, and the local gravity and meridian oriented coordinate systems used in engineering design, and they must be capable of making transformations back and forth between these coordinate systems. Those in modern practice must be able to make adjustments of observed data by employing the method of least squares in order to comply with new standards of accuracy that are already in place.

As indicated in the preceding, the manner, breadth, and scope of surveying and mapping have changed dramatically in recent years. As a result, many in the field have concluded that the name "surveying" no longer adequately reflects the expanded and changing role of their profession, and new names have been proposed. Among the suggested names, *geomatics* appears to be preferred. During the past decade, this name has gained widespread acceptance in the United States, as well as in other English-speaking countries of the world, especially Canada, the United Kingdom, and Australia. In 1986, the name of the ASCE Surveying and Mapping Division was changed to the Surveying Engineering Division, but in 1997, the name was again changed, this time to the Geomatics Division. Many college and university programs formerly identified as surveying or surveying engineering are now called geomatics or geomatics engineering. Many believe that, although the name may not necessarily be perfect, it does have a modern connotation that begs the question, "What do geomatics engineers do?" This immediately provides an opportunity to explain the role of modern surveying engineers, and dispel the stereotype of a person wearing muddy boots and peering through a transit. Adoption of the new name can have positive results for surveying and mapping in the same way that geotechnical engineering did for soil mechanics, and environmental engineering did for water supply and sanitary engineering.

Concluding Remarks

It should be noted that many of the new technologies discussed above provide topographic and environmental data in 3D coordinate systems, and, in order to make the most effective use of the data and technologies, 3D designs are necessary. These circumstances will require that changes be made in our traditional civil engineering curriculums. Students in the future should be familiar with the basic operations of these new technologies, and they must understand the 3D reference coordinate systems that they employ. They must be capable of making transformations between geocentric coordinate systems used in data collection and the local gravity and meridian oriented design coordinate systems. They must also have training and experience in developing 3D designs.

Bibliography

American Society of Photogrammetry (ASP). (1980). *Manual of photogrammetry*, 4th Ed., C. C. Slama, ed., American Society of Photogrammetry, Falls Church, Va.

Antennucci, J., et al., (1991). *Geographic information systems*, Van Nostrand Reinhold, New York.

ASCE. (1976). *ASCE surveying and mapping division golden jubilee*, ASCE, New York.

Berry, R. M. (1976). "History of geodetic leveling in the United States," *Journal of Surveying and Mapping*, American Congress on Surveying and Mapping, Gaithersburg, Md.

Burnside, R. S., Jr. (1958). "The evolution of surveying instruments." *Journal of Surveying and Mapping*, American Congress on Surveying and Mapping, Gaithersburg, Md.

Colcord, J. E. (1976). "Surveying—Before 1776," *Proc., 36th Annual Meeting of the American Congress on Surveying and Mapping*, American Congress on Surveying and Mapping, Gaithersburg, Md.

Johnson, J. B. (1902). *The theory and practice of surveying*, 16th Ed., Wiley, New York.

Leick, A. (1995). *GPS satellite surveying*, 2nd Ed., Wiley, New York.

Lillesand, T. M., and Kiefer, R. W. (1999). *Remote sensing and image interpretation*, 4th Ed., Wiley, New York.

Multhauf, R. P. (1958). *"Early instruments in the history of surveying: Their use and their invention,"* American Congress on Surveying and Mapping, Gaithersburg, Md.

Peyton, H. J., Jr. (1951). "Early development of horizontal angle-measuring survey instruments." *Journal of Surveying and Mapping*, American Congress on Surveying and Mapping, Gaithersburg, Md.

Turneaure, F. E., ed. (1909). *Cyclopedia of civil engineering*, 8 volumes, American School of Correspondence, Chicago.

U.S. Department of Interior, Bureau of Land Management. (1973). *Manual of surveying instructions 1973*, U.S. Government Printing Office, Washington, D.C.

Whitten, C. A. (1976). "Surveying—1776 to 1976," *Proc., Fall Convention of the American Congress on Surveying and Mapping*, American Congress on Surveying and Mapping, Gaithersburg, Md.

Wolf, P. R., and Brinker, R. C. (1989). *Elementary surveying*, 8th Ed., Harper and Row, New York.

Wolf, P. R., and Dewitt, B. A. (2000). *Elements of photogrammetry: With applications in GIS*, 3rd Ed., McGraw-Hill, New York.

Wolf, P. R., and Ghilani, C. G. (2002). *Elementary surveying: An introduction to geomatics*, 10th Ed., Prentice-Hall, Upper Saddle River, N.J.

American Society of Civil Engineers 150th Anniversary Paper

1852 – 2002
Building a Better World

Development of Transportation Engineering Research, Education, and Practice in a Changing Civil Engineering World

Kumares C. Sinha[1]; Darcy Bullock[2]; Chris T. Hendrickson[3]; Herbert S. Levinson[4]; Richard W. Lyles[5]; A. Essam Radwan[6]; and Zongzhi Li[7]

Abstract: Transportation has been one of the essential areas within civil engineering since its early days. In commemoration of the 150th anniversary of ASCE, this paper presents a review of developments in research, education, and practice in transportation engineering. The review is based primarily on the issues of the *Journal of Transportation Engineering* over the past several decades. Main topics include transportation engineering practice, airport and highway pavements and materials, design and safety, planning and operations, pipelines, technology, and education. Historical appraisals and the current state-of-the-art for these topics are discussed. In conclusion, future directions in transportation engineering as a result of advances in technology and the attendant changing need of the transportation engineering profession in the 21st century are addressed.

DOI: 10.1061/(ASCE)0733-947X(2002)128:4(301)

Introduction

Transportation has been one of the essential components of the civil engineering profession since its early days. From time immemorial, the building of roads, bridges, pipelines, tunnels, canals, railroads, ports, and harbors has shaped the profession and defined much of its public image. As cities grew, civil engineers became involved in developing, building, and operating transit facilities, including street railways and elevated and underground systems. The role of civil engineers in providing transportation infrastructure to accommodate a growing population and economy was never more prominent than in the United States around the late 19th century and the early part of the 20th century. Transcontinental railroads, national highways, canals, petroleum and natural gas pipelines, as well as major urban transit systems,

are testimonials to the achievement of civil engineers. And, in the latter part of the last century, these achievements played a major role in developing the Interstate System, new rail transit lines, and major airports.

In the last 150 years, railroads, transit lines, ports, and airports have helped to increase the range of cities and reduce the isolation of rural areas. They have brought the nation closer together. Such major bridges as the Eads Bridge in St. Louis, the Brooklyn and George Washington Bridges in New York City, the Golden Gate and Bay Bridges in San Francisco, and the Mackinac Straights Bridge in Michigan have not only spanned major water crossings, but have also become a dramatic part of the national landscape (Billington 1985; Gies 1996; Petroski and Kastenmeier 1996). The great railroad terminals—sometimes electrified with tunneled approaches, as are Penn Station and Grand Central Terminal in New York, and the Union Stations in Chicago and Washington, D.C.—continue to serve as major gateways and urban monuments. The subway, elevated, and commuter rail lines built over the last century have made the centers of cities such as New York, Washington, Chicago, and Boston possible (Meeks 1995; Parissien 2001). Also, the Interstate Highway System has changed the national landscape. These are a few of the many transportation contributions to the nation by members of the civil engineering profession (Fig. 1).

To commemorate the 150th anniversary of ASCE, this paper has been prepared to document the development of the transportation engineering field within civil engineering. The paper provides a historical appraisal of how the field evolved to its current state-of-the-practice as well as a preview of future directions, and it draws heavily on volumes of the *Journal of Transportation Engineering* (JTE) and its predecessor journals.

The *Journal's* roots go back to the early years of ASCE, the first issue appearing in 1874. It currently contains the technical and professional articles of the Air Transportation, Highway, Pipeline, and Urban Transportation divisions of the Society. The

[1]Olson Distinguished Professor of Civil Engineering, Purdue Univ., West Lafayette, IN 47907-1284.

[2]Associate Professor of Civil Engineering, Purdue Univ., West Lafayette, IN 47907-1284.

[3]Duquesne Light Company Professor of Engineering, Carnegie Mellon Univ., Pittsburgh, PA 15213-3890.

[4]Transportation Consultant, New Haven, CT 06515.

[5]Professor of Civil and Environmental Engineering, Michigan State Univ., East Lansing, MI 48824-1226.

[6]Professor of Civil and Environmental Engineering, Univ. of Central Florida, Orlando, FL 32816-2450.

[7]Graduate Research Assistant in Civil Engineering, Purdue Univ., West Lafayette, IN 47907-1284.

Note. Discussion open until December 1, 2002. Separate discussions must be submitted for individual papers. To extend the closing date by one month, a written request must be filed with the ASCE Managing Editor. The manuscript for this paper was submitted for review and possible publication on December 5, 2001; approved on April 12, 2002 . This paper is part of the *Journal of Transportation Engineering*, Vol. 128, No. 4, July 1, 2002. ©ASCE, ISSN 0733-947X/2002/4-301–313/$8.00+$.50 per page.

Fig. 1. 1935 reproduction of the Oregon Trail, Illinois, as it looked in the mid-1800s

Fig. 2. I-10 and College Drive in Baton Rouge, Louisiana

Highway Division was established in 1922, Air Transportation in 1945, Pipelines in 1956, and Urban Transportation in 1971. In 1969, the *Journal of the Highway Division* was renamed the ASCE *Transportation Engineering Journal* and was a forum for publications in the fields of air transportation, aerospace, highway, pipeline, and urban transportation. In 1983 the journal was renamed *Journal of Transportation Engineering*, and the Aerospace Division started its own journal in 1988. In 1999 the divisional organization of the editorial function was replaced by functional grouping. At present the editorial functions are divided into Pavements and Materials, Planning and Operations, Design and Safety, and Pipelines. In addition, there are two new features, "Practitioner's Forum" and "Book Review."

Transportation as a Civil Engineering Profession

Rapid urbanization of the United States challenged civil engineers with the task of meeting the mobility needs in and around cities. Civil engineers served not only as developers and builders of transit facilities, but also as planners and operators of such facilities. For example, the Chicago Transit and Subway Commission, under the chairmanship of William Barclay Parsons, prepared in 1916 one of the most extensive studies ever made in the operation of public transportation systems, including the interrelations of these operations with population, employment and residential distribution, and commercial and industrial regions of the city (Condit 1982).

An early intercity transportation engineering challenge was the rapid development of canal systems and railroads during the 19th century. This development represented major challenges for new and stronger bridges and for geometric design. The importance of economic and environmental impacts of transportation facilities became evident in these early developments (Wellington 1887).

As the nation became motorized in the 20th century, civil engineers played a growing role in road construction. This involvement has continued to the present day. Such major roads as the Columbia River and Lincoln Highways, New York City's parkways, and Los Angeles' freeways are among their achievements. In the 1920s engineers played a role in preparing street and transit plans for many communities. They were among the pioneers in applying traffic engineering and management methods to streets and highways. During the 19th century civil engineers involved in transportation mainly worked for public agencies and transportation providers such as railroads. However, during the 20th century, consulting firms emerged. Their numbers and importance have grown steadily, except for during the two world wars and the Great Depression. In recent years, this growth has accelerated as governmental agencies have downsized.

After World War II, civil engineers' involvement in transportation projects increased dramatically as the Interstate System was developed and as transportation planning became a requirement for federal funds. However, while civil engineers were increasingly involved in transportation activities, the transportation engineer *per se* did not really emerge until the United States Department of Transportation (USDOT) was established in 1967 and state highway agencies became multimodal (Fig. 2).

At the beginning of the 21st century, transportation engineering has evolved into a mature subdiscipline within civil engineering, with clear functions of planning, design, construction, operation, and maintenance of multimodal systems for the transportation of people and goods. This subdiscipline has greatly expanded civil engineering into such areas as economics and financing, operations research, and management. With the rapid development of Intelligent Transportation Systems (ITS) in recent years, the transportation engineering profession has also started to make increasing use of information and communication technologies.

Transportation engineering, as it is practiced today, has three major components. One component involves design, construction, and maintenance of facilities, including roads, bridges, tunnels, railroads, airports, transit systems, and ports and harbors. The second component encompasses planning, project development, and financing and management. The third component covers operations and logistics, including traffic engineering and operations of transit, trucking, and other facilities, as well as business logistics. Specialties have emerged in each as the profession has continued to mature. The current focus is on intermodal transportation systems that emphasize the connectivity of modes over the entire portal-to-portal trip length.

A specific indication of the subcomponent of the transportation engineering field with current importance to civil engineers can be obtained by examining the topics of the technical committees of the transportation-related divisions in ASCE. There are 44 technical committees, and most involve the physical infrastructure of surface transportation modes. A review of descriptions of the scope of various committees indicates that, while facility

Fig. 3. Freight train leaving urban area

planning and design continue to be the core of the transportation engineering field, such areas as operations, logistics, network analysis, and financing and policy analysis are also important to civil engineers involved in the transportation field.

Another source of information, to gauge the relative emphasis placed by civil engineers in recent years on the need for new knowledge, can be attained by reviewing the papers published in the past several decades in the JTE. As the papers represent scholarly interests, which in turn should respond to the needs of the profession over a long period of time, an assessment of the state-of-the-art is presented in the next few sections (Fig. 3).

Highway and Airport Pavements and Materials

The Role in the Field

One hundred and fifty years ago, most roads between cities were unpaved except for a few plank roads. The power stone crusher (1858) and the steamroller (1859) made the use of crushed stone feasible for rural roads. Cobblestones and untreated blocks were used in cities. The first brick pavement was built in Charleston, West Virginia, in 1871, and the first sheet asphalt was placed on Pennsylvania Avenue, Washington, D.C., in 1879. In the first decade of the 20th century, portland cement concrete (PCC) was introduced in Bellefontaine, Ohio, and Wayne County, Michigan. The first theory of rigid pavement was developed at this time, and has progressively evolved since.

Papers published in the area of pavements and materials in the *Journal of Transportation Engineering* over the past 30 years can be grouped into following categories: design, construction, materials and testing, performance analysis, and system management. Discussions that follow represent not only highway and airport pavements but also different types, such as flexible, rigid, and composite pavements.

State of the Practice

Pavement Design

The stresses and deflections of flexible and rigid pavements were analyzed using Boussinesq theory and Westergaard theory as early as 1926. Among others, Yoder (1959) pioneered the exploration of pavement design principles in the 1950s. However, a great surge of research activities and subsequent practical applications took place in pavement design as a result of the American Association of State Highway Officials (AASHO) Road Test, which was conducted in Illinois from 1958 to 1960. Using data from that test, a set of widely accepted pavement design procedures for new construction or reconstruction, overlay, and rehabilitation of pavements was developed and first published in 1972, with the latest revision in 1993 (AASHTO 1993). For flexible pavements, selecting optimal thickness of various pavement components to achieve minimum total pavement costs was an important topic of investigation among researchers (Hegal et al. 1993; Garcia-Diaz and Liebman 1978). With regard to rigid pavement design, appropriate joint design and design of concrete block pavements were extensively explored (Fordyce and Yrjanson 1969; Rada et al. 1990). Recognizing that increasing pavement construction and rehabilitation costs make it imperative to have a quick and rational method of designing the overlay thickness, several papers dealt with the topic of overlay design (Bandyopadhyay 1982; Fwa 1991).

Currently, there are more than 2.6 million miles of low-volume roads that typically carry less than 500 vehicles per day. Pavement design for low-volume roads is especially challenging because cost is always a major factor and alternative designs and materials can be used (Kestler and Nam 1999). Design procedures were also developed for airfield pavements in terms of magnitude of applied loads, tire pressures, geometric section of pavements, and number of load repetitions applied to pavements during their design lives (Murphree et al. 1971; Ahlvin et al. 1974; Seiler et al. 1991).

The Strategic Highway Research Program (SHRP), in the 1980s, launched a major research activity in the area of pavements and materials. As a part of this program, a comprehensive 20-year study of in-service highway pavements (long-term pavement performance, LTPP) was undertaken.

Pavement Construction

A number of papers covered the issue of compaction of graded aggregate bases and subbases (Marek 1977; Halim et al. 1993). Benefits of the use of hot-mix asphalt were investigated by several researchers (Colony et al. 1982). The technique of non-fines concrete with single-sized coarse aggregates held together by a binder consisting of a paste of hydraulic cement and water was also discussed (Ghafoori and Dutta 1995).

Materials and Testing

Along with pavement design procedures, investigations were made on properties of new construction materials as well as on recycled materials. Examples include engineering properties of soil-lime mixes for stabilization (Sauer and Weimer 1978), tensile fracture and fatigue of cement-stabilized soil (Crockford and Little 1987), low-temperature fracture parameters of conventional asphalt concrete and asphalt-rubber mixture (Mobasher et al. 1997), field studies on polymer-impregnated concrete (Mehta et al. 1975), and service lives of pavement joint sealants (Biel and Lee 1997).

Examples of testing procedures for pavements and materials include variably confined triaxial testing, fatigue response of asphalt concrete mixtures, rut susceptibility of large stone mixtures, viscoelastic behavior of asphalt concrete, field impregnation techniques for highway concrete, moisture content in PCC pavements, and back calculation of moduli of pavement layers (Allen and Thompson 1974; Chen et al. 1977; Uzan 1994). The LTPP program also addressed key questions about the revised resilient modulus laboratory tests and procedures. They are geared to highway engineers, laboratory managers, and technicians.

Performance Analysis

The pavement serviceability-performance concept was first introduced by Carey and Irick (1960). To study the performance characteristics of flexible pavements, Hertz's theory of the deflection of an elastic plate on a fluid subgrade was used (Wiseman 1973). The relationship between the cumulative peak pavement deflections and condition of that system, the stress/strain response of asphalt concrete under cyclic loading, threshold values for friction index, and crack propagation between beam specimens and layered pavements were also investigated by a number of researchers (Highter and Harr 1975; Ramsamooj et al. 1998; Fulop et al. 2000; Castell et al. 2000).

The breaking load for rigid pavements was studied in early 1970s (Ghosh and Dinakaran 1970). Later, the use of finite-element analysis of pile-reinforced pavement systems was also introduced (Tabatabaie and Barenberg 1980). In the mid-1990s, the issue of probability that a continuously reinforced concrete (CRC) pavement section with a certain amount of distress manifestations would last at least a certain number of equivalent single axle load (ESAL) applications was addressed (Weissmann et al. 1994). By modeling a pavement structure as a beam resting on a viscoelastic foundation, a physical picture associating vehicle dynamics, road profile, and pavement response in a theoretical framework was constructed recently (Liu and Gazis 1999).

Lack of strain characteristics of rigid pavement overlays, the susceptibility of overlays to abrasion wear, fuel spillage, and stripping led to the research on this topic (Al-Qadi et al. 1994). The fracture behavior of interface between interlayer and asphalt overlay as well as the entire overlay pavement system was studied in recent years (Tschegg et al. 1998).

System Management

An increasing interest could be seen in the area of pavement system management over the past two decades. A framework for pavement management systems was the topic of several papers (Findakly et al. 1974; Kilareski and Churilla 1983). Utility theory was introduced in pavement rehabilitation decisions in the mid-1980s (Mohan and Bushnak 1985). An integrated project-level pavement management model, consisting of life-cycle cost analysis and cost-effectiveness method, was developed in the same period (Rada et al. 1986). In later years, a framework for evalu-

ating the effects of pavement age and traffic loading on pavement routine maintenance effectiveness was introduced (Al-Suleiman et al. 1991). Integrating pavement and bridge programs started to appear in mid-1990s (Ravirala and Grivas 1995). Project-level optimization and multiobjective optimization for pavement maintenance programming began to be implemented in recent years (Mamlouk et al. 2000; Fwa et al. 2000).

Transportation Design and Safety

Historical Appraisal of the Field

During the 20th century, the private automobile in the United States went from being rarely sighted to a ubiquitous presence as the supporting road system was methodically expanded and improved, making automobile use safe and convenient. As engineers were successful in these undertakings, the dependence of the public on other modes of transportation generally eroded. While other modes remain important, the domination of the automobile is, nonetheless, fairly complete. To be sure, the emergence of the automobile meant unprecedented freedom of movement for the population and is closely tied to the continued growth of the American economy. However, problems associated with its use are also widespread, such as urban sprawl and air pollution (Altshuler et al. 1993).

The transportation engineers have been engaged in a constant struggle to make the system safe and to overcome congestion, whether of horse-drawn vehicles in New York City long ago or on Los Angeles' freeways. Although the rates of occurrence of crashes and fatalities are lower than ever, the absolute numbers are still high (BTS 2000). For congestion, the engineer's response has largely been to increase system capacity. This has proven to be a short-term solution in the larger urban areas. Congestion, once restricted to the downtown areas of our older cities, now occurs daily in suburban areas as freeways are extended, arterials are widened, and local roads are designed to ever higher standards. The lack of adequate and continuous streets in many suburban settings, coupled with the lack of effective land-use controls, overloads many major arterials. And public transport service remains inadequate to provide congestion relief in suburban corridors, especially for circumferential trips. At the outset of the 21st century, many components of our transportation system are fast approaching or have already exceeded their design lives, and much of the vaunted roadway system is suffering from wear and tear due to higher than expected use. Since the era of building large systems or substantially enlarging existing ones seems to be at an end (except, perhaps, for new transit lines), the major challenges are now to rebuild the transportation system in place and to use it more efficiently and responsibly (Fig. 4).

State of the Practice

As in many fields, there appears to be a gap between the practitioner in the field and the cutting edge of research in system design and safety. Researchers dealing with issues in these areas are largely "tinkering at the margin" or with high-tech applications in order to develop suggestions for making the system still safer and operationally more efficient. Unfortunately, practitioners are often put in a position of taking more and more "on faith." The information explosion in transportation-oriented journals has put day-to-day exploration of new findings beyond the attention span of end users. For example, the widely used *Highway Capacity Manual* (HCM) has gone from a primer on high-

Fig. 4. Transit mall in Denver

way operations to a complex treatise that requires a solid background in traffic flow theory to even begin to understand it. Calculations of volume-to-capacity ratios and levels of service that were done with an adding machine and slide rule now require special-purpose software with routines that, for many practitioners, defy understanding or are simply viewed as a black box that may or may not work. While large agencies can and do employ transportation engineers who are specialized enough to understand these evolving approaches, there are many transportation engineers whose duties are so broad as preclude "keeping up" across the board.

Concurrent with evolving technology and burgeoning information, the safe use of the highway system has moved from being solely the user's responsibility to a model of shared responsibility that includes the users, the system providers, and the vehicle manufacturers. Notwithstanding this shared model, tort liability still looms large as a driving force in transportation safety-related endeavors. Indeed, in some instances it can be argued that there are institutional constraints to both the identification of problems on the highway system and the implementation of reasonable solutions due to the exposure that a state might incur if differences in opinion are publicly aired or problem sites identified.

Despite everything that transportation engineers and those in related fields know about the relative safety of the system, there is still a gap between what is really "known" and what is "done" in the field. Safety audits that find problems before crashes take place are a step in the right direction, as there remains great variation among agencies in a given state regarding when and how horizontal curves might be signed and the appropriate advisory speed determined, from paper and pencil analysis to use of sophisticated in-vehicle equipment to a guess in the field. Providing safer and more consistent designs remains a challenge. Given that the treatment of highway curves goes back virtually to the first road, it is alarming that there is still a significant divergence in deciding on treatments within a given state, let alone among states. The litany of problems being dealt with by transportation engineers as the 21st century is embarked upon is not that different from a similar list compiled 100 years ago—progress has been made, but much remains to be done.

Transportation Planning and Operations

From Wellington's classic study of railroad location (1887) to contemporary multimodal corridor studies, the planning of trans-

portation facilities has been an integral part of civil engineering profession. While the scope and focus of these activities has evolved and broadened, planning continues to be a major effort. Civil engineers developed the first systematic comparison of urban land use, formulated the methods for forecasting future travel demands (trip generation, trip distribution, mode split, and traffic assignment), and analyzed spacing requirements for urban freeways (Peterson 1960).

The Role in the Field

The role of planning in transportation is vital for strategic or tactical evaluations and predictions of travel demands, land use patterns, and air quality issues for various transportation modes for both passenger and freight movements. The body of knowledge about planning, traffic operations, control, and management has witnessed drastic growth over the past several decades, and papers published in the *Journal of Transportation Engineering* have contributed to this growth. The information published in the *Journal* mirrors the need to share experiences about problems facing engineers and planners, new technologies being deployed to help remedy the situation, what worked and what did not work, and the academic contribution to improve the understanding of why and how to use basic and applied research to solve these problems.

Planning and operations appear most prominently in the field of traffic engineering for surface transportation. However, transportation engineers also confront planning and operations issues in facility management, particularly for large complexes such as airports. Transportation engineers also often become involved with urban-planning processes in developing long-term transportation plans.

State of the Practice

Transportation Planning
Before the 1960s, transportation planning was primarily an exercise in physical planning. In the past several decades, however, such plans have been perceived as inadequate for meeting social needs. Important political, economic, and social trends that have affected the evolution of transportation planning include fiscal austerity as a theme of government policy, changing roles of automobiles, environmental concerns, and changing household characteristics, among others.

Since 1945, air transportation has seen tremendous growth, with ever-increasing levels of enplaning passengers and airfreight tonnage. In recent years, air transportation planning has attracted much attention as the design and operation of this mode of transportation is associated with long lead times of large investments. The cornerstone of airport planning over the past 20 years has been the Individual Airport Master Plan for commercial as well as general airports, as approved by the Federal Aviation Administration (FAA) (Fig. 5).

A large number of research studies have been conducted in transportation planning since the 1970s. Papers published in the *Journal* focused on a vast array of planning issues, ranging from political processes of planning (Ellis 1973), examination of modal split in recreational transport planning (Theologitis and Powell 1984), longitudinal assessment of transportation planning forecasts (Miller and Demetsky 2000), airport environmental planning practices (Orlick 1978), and present and future characteristics of air-transport infrastructure planning (Coussios 1991).

Fig. 5. Chicago O'Hare Airport Terminal

Traffic Operations

The construction of the freeway system started in the mid-1950s and spanned a decade. By the late 1960s the interstate system was completed except for short sections located in urban areas. Traffic increased at a much faster rate than forecasted, especially in urban areas. This resulted in congestion, excessive delays, and pavement deterioration.

Papers published in the 1970s addressed topics like better design of traffic control devices (Payne 1973), developing and improved traffic operations programs (La Baugh 1971), freeway congestion and roadway and street capacities (Priestas and Mulinazzi 1975), and improving on traffic data collection and weigh-in-motion (WIM) (Machemehl et al. 1975). Papers related to freeways in Chicago and Los Angeles in terms of congestion characteristics and traffic management strategies were published during this time period (McDermott 1980). The strategies of traffic surveillance ramp metering and Variable Message Sign (VMS) were discussed and tested at a few locations. At the city-street level, testing of new signal controllers and timing schemes were investigated and new traffic management strategies like signal priority strategies to special vehicles, parking controls, and signal coordination and optimization were conceived (Michalopoulos et al. 1978). Because of the complexity of road construction and maintenance and the potential risk to motorists and workers, negligence-suit and tort-liability issues were raised and studied.

In the 1980s the transportation profession witnessed major changes from the introduction of personal computers. Computing power became more affordable, and advancements in microelectronics and integrated circuits have resulted in better and smaller traffic controllers, video cameras, and loop detectors. Research related to incident detection on freeways was expanded and produced some better understanding of how traffic bottlenecks happen on freeways (Ahmed and Cook 1980). More sophisticated mathematical models were developed to better detect freeway-capacity-reducing incidents and estimate their impacts on traffic delays. More traffic surveillance strategies were demonstrated and tested on urban freeways. The microscopic simulation models that were developed in the 1970s to simulate street and freeway traffic like NETSIM and FREESIM became more attractive to research (Rathi and Santiago 1990; Cheu et al. 1998). The use of signal coordination software like PASSER and TRANSYT became a reality for some state agencies, and papers were published to educate the average traffic engineers about what the software can do for them (Jovanis and Gregor 1986; Chang et al. 1987). The ASCE's Urban Transportation Division sponsored three specialty

conferences on microcomputer applications, and a large number of papers presented at those conferences were refereed and published (Skabardonis 1986; Matthias et al. 1987). Computers also enabled extensive analysis of complicated systems such as airport operations or alternative development-transportation plans.

More and more of our streets and highways are deteriorating, and work zone operation and safety emerged to be a concern to the profession. The capacity of a freeway section or an arterial passing through a work zone was researched, and the results were well documented (Nemeth and Rouphail 1983). Knowledge-based expert systems were introduced as research and education tools to build on knowledge accumulated over the last few decades. The idea of tapping the expertise of an expert and mimicking his decision process using computer software caught on fairly fast. Some applications to this concept included traffic signal timing, budget allocation, traffic-crashes management and inventory, and pavement cracks recognition (Zozaya-Gorostiza and Hendrickson 1987; Lin 1991). The introduction of an adaptive signal control took place in the late 1980s. The idea here is to take advantage of improvements in signal-controller enhancements and better detection systems to make the controller more responsive to changes in traffic patterns at intersections. Initially, the work was fairly limited to short-term future flow patterns (Lin and Vijayakumar 1988; Young 1989), but it is now expanding.

The enhancement of software and hardware continued successfully through the 1990s. The price of personal computers continued to decline until about 1993. The industry held the price fairly constant and offered more power in terms of speed, memory, and storage capabilities. Graphical representation and three-dimensional visualization became possible and affordable in the late 90s. Traffic-simulation software became versatile and available to most local and state agencies to use for conducting technical studies as well as demonstrations in public hearing and decision making. Geographic information systems (GIS) is an offspring of this computing revolution (Abkowitz et al. 1990; Quiroga and Bullock 1996). The ability to integrate different databases into one system and then overlay information in different colors and layers has helped us to better understand how the aspects of the system interact, and it assisted us in making better decisions based on a sound approach.

Neural networks appeared on the research scene as a new and viable tool of simulating the human brain. Applications of neural networks included incident detection techniques, crash-type identification and recognition, data warehousing, and data mining (Nam and Drew 1998; Wolshon and Taylor 1999).

The evolution of the ITS discipline has given the transportation profession great opportunities and challenges. There were opportunities that allowed us to take advantages of advanced technological innovations that were created in the 1980s under defense contracts to be used in traffic operations and control. Examples of these innovations include image processing and machine vision, wireless communications, global positioning systems (GPS), high-speed networks, parallel processing, detection systems using laser, and others (Bullock and Hendrickson 1992). The challenges created by ITS are mostly political and social. The transition occurred rapidly without educating and training a cadre of professionals about what technology can do for them and about the complexity of incorporating the diverse population of motorists on transportation systems. Furthermore, transportation corridors cross several jurisdictions and counties; coordinating the communication, command, and control among those entities is a great barrier. The fact that we can install transponders and telematics on our vehicles permits us to receive real-time information around-the-clock about the status of the network, signal distress signals in case of emergencies, use toll roads without having to stop, and avoid collisions with other vehicles and fixed objects. The major hurdle of doing all that is the privacy issue and the acceptability of such systems.

Because of ITS deployment programs and the substantial funding invested in this discipline, large number of research activities and technical papers were reviewed and published. Some topics include dynamic traffic assignments, data mining, electronic toll collection, collision detection and avoidance using GPS, vehicle detection using laser and image processing technologies, and real-time information given to motorists. The use of wireless communication allowed researchers to collect more accurate data about behavior of traffic under different conditions. This data has been used to revisit the theoretical models that were developed two decades ago and fine-tune them.

Pipelines

The first pipeline was developed in 1865 to transport oil from northwestern Pennsylvania to a railroad station six miles away. The first long distance pipeline was built in 1878; the six-inch, 100-mile pipeline connected Corryville and Williamsport, in Pennsylvania. Today more than 225,000 miles of pipeline account for one-sixth of the total intercity ton-miles (BTS 2000).

Role in the Field

Pipelines represent a unique mode of freight transportation. Although the role of pipelines in the entire transportation sector is not large, it serves very specialized needs, and the *Journal of Transportation Engineering* has the distinction of serving as a forum for publishing new knowledge in this area. A review of the papers on pipelines published in the journal over the past three decades indicates the following major areas of interest: pipeline infrastructure engineering (consisting of design, testing, and construction); safety; economics, and implementation policy issues—all discussed below.

State of the Practice

Pipeline Infrastructure Engineering
This area involves such topics as design and testing, stress analysis in pipe junctions, corrosion, settlement, and soil/pipe interactions. There were several papers on pipeline design, including partially embedded pipes (Olander and Robertson 1970; Olander and Davidson 1980), undersea aqueduct design (Armstrong 1972), testing of large diameter pipes under combined loading (Bouwkamp and Stephen 1973), and surcharge loads on buried pipes (Shmulevich and Galili 1986; Potter 1985). To study the problem of stress at pipe junctions, finite-element methods were used in early 1970s (Godden 1973). In the same period, an extensive review of pipeline corrosion and corrosion control was carried out (Kinsey 1973). In the mid 1990s, the issue of corrosion was revisited by using probabilistic modeling (Ahammed and Melchers 1994).

The Soil Pipe Interaction Design and Analysis (SPIDA) program, developed by the American Concrete Pipe Association (ACPA), was a culmination of over 20 years of research and testing for improved methods of estimating load and pressure distribution on buried concrete pipes. The differences between the traditional indirect design method and the direct design using

Fig. 6. Port of Tacoma in state of Washington

SPIDA program by analyzing an actual installation was researched in early 1990s (Kurdziel and McGrath 1991).

Some major pipeline construction projects implemented over the last three decades included the Trans-Alaska pipeline project and the Similkameen pipeline suspension bridge at Princeton, British Columbia (Patton 1973; Chen and McMullan 1974). In addition, the construction of the Dallas water distribution system, the Schuylkill River major crossing project at Philadelphia, and the Water System Improvement Program at Eugene, Oregon, also took place in the same period (Hudson 1974; Missimer 1978; Brown 1980) (Fig. 6).

Safety, Economics, and Implementation Policy Issues
The Federal Natural Gas Safety Act of 1968, a major milestone in pipeline safety, required enforcement of minimum safety standards, record-keeping, compliance with maintenance standards, reporting of safety activities, and the development of reasonable and proper safety standards. The practice of Illinois Office of Pipeline Safety was discussed by Shutt (1972). A procedure for seismic risk probability analysis of buried pipelines was developed by Mashaly and Datta (1989). Accidental detonation of explosives near pressured gas pipelines may have severe consequences on pipelines. By taking into consideration the characteristics of explosion source, soil, and the pipelines, a prediction model for the safe distance of a pipeline from an explosion source with known explosive quantity was developed by Rigas and Sebos (1998).

In the early 1970s, economic evaluation of slurry pipelines was conducted by Wasp et al. (1971) in terms of reliability, immunity to escalation, minimal maintenance and operation personnel, and esthetics. Osborne and James (1973) applied the concept of marginal economics to pipeline design, obviating the need for a considerable amount of calculation normally required. Based on an annualized cost approach, comparison of pipeline with a rail and truck economy was made by Zandi et al. (1979), who indicated that pipeline was reasonably attractive in terms of per ton-mile of transport.

Several papers dealt with implementation policy issues. For instance, right-of-way aspects were addressed by Stastny (1972). Issues of acquisition of subsurface information for pipeline construction, cost, and, applicability as well as the contractor's use of and reliance upon such information were discussed by Carter (1978). In the same period, studies were conducted on criteria and procedures required in the preparation of design, construction methods, and maintenance plans to ensure safe and dependable pipeline river crossings (O'Donnell 1978). Trends in environmental regulations and their implications were also analyzed (Barningham and Ott 1980).

Technology in Transportation Engineering

Role in the Field

The major technology concepts of current transportation engineering were generally in place by the midpoint of the 20th century. Among others, notable technologies include the internal combustion engine with petroleum-based fuel for land and water transport, construction of all-weather roads, standardized signals and signs (longitudinal center stripes on roadways appeared in Wayne County, Michigan, as early as 1911), and intercity train speeds attained in the early 20th century.

Despite the stability of such fundamental concepts, there have been major improvements in the underlying technologies for airplanes, trains, motor vehicles, concrete mixes, system operations, and others. At the same time, information technology applications have been widely developed and applied within transportation engineering. New concepts have had profound effects, such as containerization for freight transport. Regulatory requirements have spawned new technologies, such as the requirement for handicapped access to public transit.

State of the Practice

The *Journal of Transportation Engineering* provides an interesting perspective on the introduction of new technologies into transportation engineering. The various special issues devoted to a single subject or application are good starting points. Since the creation of the *Journal* in its current form in 1983, there have been eight special issues with related sets of topical papers: expert systems in transportation engineering (1987); high-speed rail (1989); robotics and automatic imaging (1990); Intelligent Vehicle and Highway Systems (IVHS) (1990); real-time traffic control systems (1990); hazardous materials transportation (1993); advanced traffic-management systems (1993); and high-speed ground transportation (1997). Each special issue had a component of new technology and motivated continuing work.

Expert Systems in Transportation Engineering

The systems reflect a widespread interest in artificial intelligence, incorporating greater reasoning capabilities in computer software. Subsequent developments have implemented fuzzy logic and neural networks technologies for transportation engineering. The motivation for this work was improved decision making by such automated systems as traffic controllers. These techniques have now become common state-of-the-practice for software engineering.

High-Speed Rail and High-Speed Ground Transportation

These special issues represent the continuing effort to redesign the U.S. intercity passenger system to become more like those in Europe or Japan. These countries have a heavy emphasis and investment in high-speed rail transportation. This redesign has faced substantial market and organizational obstacles, reflected in the paper by Harrison (1995).

Robotics and Automatic Imaging

These topics continue to be an active area of research, and commercial products are appearing in many areas. Equipment trains

for asphalt roadway repaving are good examples of semiautomated procedures. Video traffic detectors are available commercially.

Intelligent Vehicle and Highway Systems

These special issues predate the expansion of this topic to the general area of ITS. Indeed, ITS technologies were featured in the *Journal* as early as 1988 (Sinha et al. 1988). Along with the special issues of real-time traffic control and advanced traffic management systems (ATMS), ITS development has represented a major infusion of new technologies into transportation engineering, for applications such as electronic toll collection (Al-Deek et al. 1997) and traffic control (Jovanis and Gregor 1986). Despite these advances, ITS remains a source of frustration because of slow implementation. For example, Chase and Hensen (1990) noted a 10-year lag in the use of new technology for traffic control applications. With the stagnation in signal control hardware, the lag has certainly increased substantially since they wrote in 1990.

Hazardous Materials Transportation

This special issue reflects new regulatory concerns and a new risk-management approach to this topic. Such changes often require new technologies. Requirements for handicapped access changed the vehicle designs and operating procedures for transit enterprises (Smith 1983). Environmental concerns profoundly influence designs (Johnston and Rodier 1999) and materials choices (Baybay and Demirel 1983). Supply chain effects also become concerns, as with the indirect energy required for transportation construction (Levinson et al. 1984).

The introduction of new technologies into transportation engineering is certainly not limited to these special issue topics. Some other major areas represented in the *Journal* include information technology for planning, design, and operations, and the invention of new control devices. The transportation engineering community has embraced new hardware and software for a variety of applications, including roadway systems management (Kilareski and Churilla 1983), traffic control design (Skabardonis 1986), and noise modeling (Cohn et al. 1983). New sensors provide data either more cheaply or in ways not previously available. Satellites provide a new vantage point for transportation engineering (Graham 1980) and much-improved location information. Field sensors are used routinely in roadway and facilities management decision making (Rollings and Pittman 1992; Uzarski and McNeil 1994).

Transportation Engineering Education

Early Professional Activities

Excluding Roman times, the 19th century railroads, and canals, one could argue that transportation education (at least in the United States) has its roots at the turn of the century, when Ford introduced the Model T and the federal government passed the Federal Aid Road Act of 1916 establishing a federal aid highway program. Those developments for the first time created an urgent need to apply scientific principles to the building of roads suitable for automobile travel. With the exception of traditional civil engineering surveying and earthwork practices, there was little reference or education material available on the engineering principles involved in building highways (Hickerson 1926). For the next 40 years, the legislative process dominated and transportation-related education primarily occurred on an *ad hoc* basis via existing organizations such as ASCE (1852) or emerging technical organizations such as AASHO (1914), the Highway Research Board (HRB 1921), and the Institute of Traffic Engineers (ITE 1930).

The first major efforts to develop reference material for educating professionals on the technology of building roads began occurring at the time of the Federal Aid Highway Act of 1938. That act directed the Bureau of Public Roads to study the feasibility of national toll roads and probably served as a motivating force to distill existing knowledge into recommended practice. This process of educating professionals on the construction of roads was initiated by AASHO in 1937 by organizing a committee to study the planning and design polices used for highways. The AASHO effort was completed for rural highways in 1954 and supplemented in 1957 for urban highways (AASHO 1957). Parallel to the effort of developing uniform design procedures, other committees developed procedures for uniform signs, striping, and control devices (PRA 1948) and for estimating the capacity of highways (USDOC 1950). These three efforts provided much of the educational material available to professionals involved in designing the Interstate System authorized in the Federal Aid Highway Act of 1956. Educational material related to pavement design and materials emerged slightly later, from the AASHO road test that occurred during 1955–1961. Several papers previously published in the JTE have provided improvements on these initial methods and have been incorporated into the recent editions of these manuals used today (AASHTO 1994; TRB 2001; USDOT 2000).

Engaging Universities in Transportation Education

During the first part of the century, state highway departments begin to team up with local universities. For example, the Texas Highway Department partnered with Texas A&M University in 1919, and the Indiana State Highway Commission partnered with Purdue University and formed the Joint Highway Research Project (JHRP) in 1937. Although these ventures were primarily intended to address fundamental road-building issues, locating them at universities provided a mechanism for quickly injecting research findings into curricula. In addition to these joint research ventures, Harvard University developed a nine-month nondegree Certificate of Highway Traffic in 1925. That program moved to Yale University in 1938 and remained there until 1968, when it moved to Pennsylvania State University and then dissolved in 1982 (Rankin 1997). In contrast to modern funding practices for graduate education, from 1925 to 1972 the Bureau received most of its funding from foundations and other private-sector sources.

In the 1970s, the education in transportation was greatly influenced by the funding made available through the university research and education programs of the USDOT and the Urban Mass Transportation Administration (UMTA). Similar programs were later initiated by the FAA and other modal agencies. In the 1980s, university transportation centers were established by the USDOT. These federal programs provided an impetus for multimodal transportation engineering education, mainly at graduate level. Currently, with the exception of a few short courses targeting very narrow topics, the dominant delivery mechanism today for transportation education is one or two undergraduate courses in a civil engineering curriculum or several courses in a civil engineering graduate degree curriculum specializing in transportation.

Emerging Textbooks and Curricula

In parallel to the ongoing research to develop these professional codes and standards, principles and practices from these efforts begin to be distilled into reference and textbook material adaptable to formal courses in the field of highway engineering. This material quickly broadened to cover other modes, system analysis, and planning (Woods 1960; Hickerson 1964; Wohl and Martin 1967). By the early 1970s, textbooks and university curricula were following two distinct paths: some undergraduate curriculums adopted more broad-based transportation references that provided introductory material to the variety of transportation modes (Wright and Ashford 1997); other undergraduate curricula adopted tightly focused references targeted specifically at highway design and construction (Mannering and Kilareski 1997). This philosophical debate of breadth-versus-depth transportation education remained unresolved and continues today with universities struggling to address additional issues related to policy, energy, environment, and technology.

Future Directions

Planning, Operations, and Technology

During the past several decades, the transportation planning area has matured and many of the procedures are now standardized. While early efforts on developing a transportation planning framework in the late 1950s and 1960s were made by civil engineers, the transportation planning area gained much from the disciplines of economics, urban geography, and planning and regional science. In fact, transportation planning is another instance where the sphere of the civil engineering profession has expanded to let other disciplines enter into the transportation engineering curriculum. Travel demand estimation, data mapping using GIS, and land-use transportation modeling are some examples that bear the results of collaborations of civil engineers with economists, geographers, and planners, respectively. This trend can be expected to continue with increased use of technology and real-time planning of operational actions.

Another development in transportation education and research that has gained popularity in recent years is the area of transportation logistics. While traditional transportation engineering is concerned with the construction and maintenance of roads, bridges, and other facilities, transportation logistics involves decisions that are made by transportation carriers, as well as shippers and receivers. Long the subject of interest to management and business professionals, transportation logistics has found its way into civil engineering transportation programs over the years in the United States, indicating yet another area where the collaboration with another discipline has expanded the boundary of the civil engineering. Potential exists for much growth in this area.

Traffic engineering as a discipline will continue to take advantage of the information technology revolution, and alternate delivery systems of travel information and services will be pursued. Traffic operation centers (TOC) of the future will all be centralized into one physical location, with the ability to have various agencies—like freeway management, emergency response, toll collection services, and signal systems—operate in a coordinated fashion yet run independently.

Indeed, transportation engineering technology has some exciting changes underway. The development of ubiquitous computing and high-band communications networks will enable vehicles and infrastructure elements to sense their environment and to communicate to operating agencies. Transportation engineers must determine how this flood of information can be best used. Environmental concerns will also drive new technologies. Hybrid and fuel-cell-powered vehicles are appearing. Compressed natural gas is well established as a fuel for fleet vehicles. Organizational innovation may be as critical to the introduction of these changes as technology development itself.

Pavements and Pipelines

As the LTPP data are analyzed, many of the pavement design procedures will be modified and emphasis will continue to be placed on maintenance and rehabilitation decisions. Emerging technologies will allow early detection of deterioration and non-destructive evaluation of pavements. With new high-quality materials and improved testing procedures, much progress can be expected in the provision of pavements and structures with longer lives and better service. Improved methods of data collection and monitoring will enable more reliable determination of various life-cycle costs and repair actions and associated trade-offs for maintenance and rehabilitation of both highway and airport pavements.

Similar trends can also be expected in the area of pipelines, including the use of technology in monitoring conditions. In addition, engineering and economic aspects of pipeline infrastructure maintenance and rehabilitation will remain critical, along with environmental, land use, and safety impacts.

Safety and Security

According to the World Health Organization (WHO), in 1998 road crashes were the largest cause of illness or early death for males between the ages of 15 and 44 worldwide, and the second-largest cause of illness or early death for males in all age groups in developing countries (BTS 2000). Although in the United States the fatality rate per 100 million vehicle miles of travel is lower than that of many Western European countries, more than 44,000 lives are lost each year with an estimated annual cost of more that $137 billion. Traffic safety remains a serious public health concern in both developed and developing countries, and the sustainability of automotive transportation cannot be achieved until the safety issue is resolved.

The trajic events of September 11, 2001, dramatically underscored the critical need for security of transportation systems. The spectre of terrorism worldwide has introduced a new dimension in transportation planning, design, and operation. The threat exists not only for air transportation, but also for the entire physical infrastructure and control for the transportation of people and goods, including pipelines, rails, waterways, mass transit, highways, bridges, and combinations of various modes and the interfaces among them. To face this challenge, transportation engineers must consider the risk and vulnerability associated with transportation facilities and services so that threats can be detected in a timely manner and measures can be taken to preserve security. Information and communication technologies will play a key part in the development of techniques and approaches to ensure the safety and security of our transportation systems.

Transportation Engineering Education and Practice

The skills required by a practicing transportation engineer no longer appear to be coincident with a general civil engineer. Basic education must be more flexible to accommodate more specialization early in academic training. Universities and other institutions must become more actively involved in providing continuing education programs for practitioners. While some professionals will keep up with changes because of their constant use, others will need assistance, through continuing education, to stay current. The involvement of line agencies in research and development is important regardless of whether that work is done in-house or by consultants and universities. As it stands, much is done in the name of safety and better design, but both pre- and postimplementation evaluations, especially the latter, are still relatively scarce. The link between transportation systems, land-use policy, and increasing or decreasing travel demand has long been reasonably well understood. As the transportation system becomes more congested in certain areas, it is clear that even better use of new technology will not solve congestion. Nor is it always feasible to build out of congestion. What we need are long-range efforts on land-use planning and "smart growth" policies, along with the use of pricing and other economic investments so that other modes, such as public transportation, can be viable alternatives. Policies including freight transportation also require a renewed look so that all modes—trucking, rail, water, air, and pipeline—operate on a level playing field.

In the last three decades transportation engineering education has broadened to include a vast spectrum of subdisciplines. Nevertheless, the pace of research has quickened and transportation-related publications are being published at a rapid rate. In 1997, the *Journal of Transportation Engineering* launched a Book Review section with publications generally covering three broad areas—transit studies, introductory transportation engineering, and professional handbooks and references. Notably lacking from this list are comprehensive textbooks on the design and management of ITS systems. With recent legislation aimed at mainstreaming information technology into transportation systems, the next two decades are likely to see a mainstreaming of ITS-related education material in college curriculum. From that, we will probably see the field of transportation education broaden even further

Conclusions

The early engineers, by their training, experience, and inclination, were often generalists. Many engineers worked pragmatically with a strong sense of physical and political reality. As the field became more diverse and complex, this was no longer possible. Growing federal, state, and local requirements called for a broad range of skills and capabilities, and new analysis tools. These new tools have in many ways transformed approaches to transportation engineering. But they have created many technical specialists who are often unfamiliar with or insensitive to many other aspects of transportation engineering. This growing dichotomy between the generalist and specialist has been aided by contemporary transportation education. Although universities have often broadened their curriculum programs, they are increasingly theoretical, sometimes at the expense of practicality; part of this dilemma stems from a growing emphasis on training for research rather than practice. It also stems from a growing number of faculty with little experience with and interest in practical matters. A related concern is how best to attract new talent. In today's society transportation engineering, despite its promise and importance, remains far less on the cutting edge than do fields like biomedical engineering and computer science.

Advances in technology and computer capabilities in just the last decade have brought an increasing amount of information to the engineer's fingertips for many applications. Many states have extensive files of road inventory, traffic and cost data, and access to GIS, to say nothing of ever more powerful programs that can be used for system design, analysis, and management. The systems exist but are not easily made compatible. Moreover, the extensive files often have data rife with errors. While the data were required, their subsequent use was given little importance. They were collected because they were required, not because they were actively used by anyone other than in an annual year-end report. This serves to illustrate that practicing engineers have increasing information with which to work but that few are using the information in proactive decision-making processes.

The role and need for transportation engineering will grow in the 21st century. A growing and more affluent population will increase demands for travel and improved transportation facilities and services. There will be a need for environmentally sensitive and creative designs, and ingenious management and operating strategies. There will be a need to achieve community consensus in making these a reality. Transportation engineers should be well positioned to meet these challenges. Transportation engineering in particular must provide an integrated approach that includes planning, statistics, economics, finance, public policy, operations, and management. It must provide a sense of physical, environmental, and political reality.

The transportation engineering profession in the 21st century must be able to adapt to changes in community needs and values. It must be able to plan, manage, and operate as well as build. Sound judgments and better integration of theory and practice are now essential. Flexibility and a strong sense of physical, political, financial, and environmental reality will help us in planning, designing, building, operating, and managing transport facilities in the 21st century.

Abbreviations

AASHO: American Association of State Highway Officials
AASHTO: American Association of State Highway and Transportation Officials
ACPA: American Concrete Pipe Association
ASCE: American Society of Civil Engineers
ATMS: Advanced Traffic Management Systems
BTS: Bureau of Transportation Statistics
CRC: Continuously Reinforced Concrete
ESAL: Equivalent Single-Axle Load
FAA: Federal Aviation Administration
GIS: Geographical Information System
GPS: Global Positioning System
HCM: Highway Capacity Manual
HRB: Highway Research Board
IVHS: Intelligent Vehicle and Highway Systems
ITE: Institute of Transportation Engineers
ITS: Intelligent Transportation Systems
JHRP: Joint Highway Research Project
LTPP: Long-Term Pavement Performance
MUTCD: Manual on Uniform Traffic Control Devices
NRC: National Research Council
PCC: Portland Cement Concrete
PRA: Public Roads Administration
SHRP: Strategic Highway Research Program
SPIDA: Soil Pipe Interaction Design and Analysis

TOC: Traffic Operation Center
TRB: Transportation Research Board
USDOC: United States Department of Commerce
USDOT: United States Department of Transportation
UMTA: Urban Mass Transportation Administration
VMS: Variable Message Sign
WHO: World Health Organization
WIM: Weigh-in-Motion

References

AASHO. (1957). *A policy on arterial highways in urban areas*, Washington, D.C.

AASHTO. (1993). *AASHTO guide for design of pavement structures*, Washington, D.C.

AASHTO. (1994). *A policy on geometric design of highways and streets*, Washington, D.C.

Abd El Halim, A. O., Phang, W. A., and Haas, R. C. (1993). "Unwanted legacy of asphalt pavement compaction." *J. Transp. Eng.*, 119(6), 914–932.

Abkowitz, M., Walsh, S., Hauser, E., and Minor, L. (1990). "Adaptation of geographic information systems to highway management." *J. Transp. Eng.*, 116(3), 310–327.

Ahammed, M., and Melchers, R. E. (1994). "Reliability of underground pipelines subject to corrosion." *J. Transp. Eng.*, 120(6), 989–1002.

Ahlvin, R. G., Chou, Y. T., and Hutchison, R. L. (1974). "Structural analysis of flexible airfield pavements." *Transp. Eng. J. ASCE*, 100(3), 625–641.

Ahmed, S. A., and Cook, A. R. (1980). "Time series models for freeway incident detection." *Transp. Eng. J. ASCE*, 106(6), 731–745.

Al-Deek, H. M., Mohammed, A. A., and Radwan, A. E. (1997). "Operational benefits of electronic toll collection: Case study." *J. Transp. Eng.*, 123(6), 467–477.

Allen, J. J., and Thompson, M. R. (1974). "Significance of variably confined triaxial testing." *Transp. Eng. J. ASCE*, 100(4), 827–843.

Al-Qadi, I. L., Gouru, H., and Weyers, R. E. (1994). "Asphalt portland cement concrete composite: Laboratory evaluation." *J. Transp. Eng.*, 120(1), 94–108.

Al-Suleiman, T. I., Sinha, K. C., and Riverson, J. D. (1991). "Effects of pavement age and traffic on maintenance effectiveness." *J. Transp. Eng.*, 117(6), 644–659.

Altshuler, A. A., Gomes-Ibanez, J. A., and Howitt, A. M. (1993). *Regulation for revenue: The political economy of land use exactions*, The Brookings Institution, Washington, D.C.; and The Lincoln Institute of Land Policy, Cambridge, Mass.

Armstrong, E. L. (1972). "The undersea aqueduct—A new concept in transportation." *Transp. Eng. J. ASCE*, 98(2), 303–310.

Bandyopadhyay, S. S. (1982). "Flexible pavement evaluation and overlay design." *Transp. Eng. J. ASCE*, 108(6), 523–539.

Barningham, M. L., and Ott, S. A. (1980). "Working within the environmental regulatory process." *Transp. Eng. J. ASCE*, 106(3), 309–315.

Biel, T. D., and Lee, H. (1997). "Performance study of portland cement concrete pavement joint sealants." *J. Transp. Eng.*, 123(5), 398–404.

Billington, D. P. (1985). *The tower and the bridge: The new art of structural engineering*, Princeton University Press, Princeton, N.J.

Bouwkamp, J. G., and Stephen, R. M. (1973). "Large diameter pipe under combined loading." *Transp. Eng. J. ASCE*, 99(3), 521–536.

Boybay, M., and Demirel, T. (1983). "Hydraulic cement from fly ash and phosphoric acid." *J. Transp. Eng.*, 109(4), 506–518.

Brown, J. H. (1980). "Water system improvement program in Eugene, Oregon." *Transp. Eng. J. ASCE*, 106(6), 685–695.

Bullock, D., and Hendrickson, C. (1992). "Advanced software design and standards for traffic signal control." *J. Transp. Eng.*, 118(3), 430–438.

Bureau of Transportation Statistics (BTS). (2000). *National transportation statistics 2000*, Washington, D.C.

Carey, W. N., and Irick, P. E. (1960). "The present serviceability performance concept." *Highway Research Board, Bulletin 250*, National

Research Council, Washington, D.C.

Carter, L. R. (1978). "Legal and contractual areas of subsurface investigations for pipelines." *Transp. Eng. J. ASCE,* 104(3), 379–383.

Castell, M. A., Ingraffea, A. R., and Irwin, L. H. (2000). "Fatigue crack growth in pavements." *J. Transp. Eng.,* 126(4), 283–290.

Chang, E. C. P., Marsden, B. G., and Derr, B. R. (1987). "PASSER II-84 microcomputer environment system—Practical signal-timing tool." *J. Transp. Eng.,* 113(6), 625–641.

Chase, M. J., and Hensen, P. J. (1990). "Traffic control systems—Past, present, and future." *J. Transp. Eng.,* 116(6), 703–717.

Chen, S. C., and McMullan, J. G. (1974). "Similkameen pipeline suspension bridge." *Transp. Eng. J. ASCE,* 100(1), 207–219.

Chen, W. F., Vanderhoff, J. W., Manson, J. A., and Mehta, H. C. (1977). "Field impregnation techniques for highway concrete." *Transp. Eng. J. ASCE,* 103(3), 355–368.

Cheu, R. L., Jin, X., Ng, K. C., Ng, Y. L., and Srinivasan, D. (1998). "Calibration of FRESIM for Singapore expressway using genetic algorithm." *J. Transp. Eng.,* 124(6), 526–535.

Cohn, L. F., Bowlby, W., and Casson, L. W. (1983). "Interactive graphics use in traffic noise modeling." *J. Transp. Eng.,* 109(2), 179–195.

Colony, D. C., Wolfe, R. K., and Heath, G. L. (1982). "Hot-mix paving in marginal weather." *Transp. Eng. J. ASCE,* 108(6), 632–649.

Condit, C. W. (1982). *American building,* University of Chicage Press, Chicago.

Coussios, D. C. (1991). "European air transport infrastructure planning: Present and future characteristics." *J. Transp. Eng.,* 117(5), 486–496.

Crockford, W. W., and Little, D. N. (1987). "Tensile fracture and fatigue of cement-stabilized soil." *J. Transp. Eng.,* 113(5), 520–537.

Ellis, R. H. (1973). "Reexamination of transportation planning." *Transp. Eng. J. ASCE,* 99(2), 371–382.

Findakly, H., Moavenzadeh, F., and Soussou, J. (1974). "Stochastic model for analysis of pavement systems." *Transp. Eng. J. ASCE,* 100(1), 57–70.

Fordyce, P., and Yrjanson, W. A. (1969). "Modern design of concrete pavements." *Transp. Eng. J. ASCE,* 95(3).

Fwa, T. F. (1991). "Remaining-life consideration in pavement overlay design." *J. Transp. Eng.,* 117(6), 585–601.

Fwa, T. F., Chan, W. T., and Hoque, K. Z. (2000). "Multiobjective optimization for pavement maintenance programming." *J. Transp. Eng.,* 126(5), 367–374.

Fülöp, I. A., Bogárdi, I., Gulyás, A., and Csicsely-Tarpay, M. (2000). "Use of friction and texture in pavement performance modeling." *J. Transp. Eng.,* 126(3), 243–248.

Garcia-Diaz, A., and Liebman, J. S. (1978). "Optimal flexible pavement design." *Transp. Eng. J. ASCE,* 104(3), 325–334.

Ghafoori, N., and Dutta, S. (1995). "Development of no-fines concrete pavement applications." *J. Transp. Eng.,* 121(3), 283–288.

Ghosh, R. K., and Dinakaran, M. (1970). "Breaking load for rigid pavement." *Transp. Eng. J. ASCE,* 96(1).

Gies, J. (1996). *Bridges and men,* Grosset & Dunlap, New York.

Godden, W. G. (1973). "Structural model studies of large pipe bifurcation." *Transp. Eng. J. ASCE,* 99(1), 83–109.

Graham, D. S. (1980). "Pipeline river crossings: A design method." *Transp. Eng. J. ASCE,* 106(2), 141–153.

Harrison, J. A. (1995). "High-speed ground transportation is coming to America—Slowly." *J. Transp. Eng.,* 121(2), 117–123.

Hegal, S. S., Buick, T. R., and Oppenlander, J. C. (1993). "Optimal selection of flexible pavement components." *J. Transp. Eng.,* 119(6).

Hickerson, T. F. (1926). *Highway curves and earthwork,* McGraw-Hill, New York.

Hickerson, T. F. (1964). *Route surveys and design,* McGraw-Hill, New York.

Highter, W. H., and Harr, M. E. (1975). "Cumulative deflection and pavement performance." *Transp. Eng. J. ASCE,* 101(3), 537–551.

Hudson, W. D. (1974). "A modern metropolis looks ahead." *Transp. Eng. J. ASCE,* 100(4), 801–814.

Johnston, R. A., and Rodier, C. J. (1999). "Automated highways: Effects on travel, emissions, and traveler welfare." *J. Transp. Eng.,* 125(3), 186–192.

Jovanis, P. P., and Gregor, J. A. (1986). "Coordination of actuated arterial traffic control systems." *J. Transp. Eng.,* 112(4), 416–432.

Kestler, M. A., and Nam, S. I. (1999). "Reducing damage to low volume asphalt-surfaced roads and improving local economies: Update on variable fire pressure project." *Cold regions engineering: Putting research into practice,* ASCE, Reston, Va., 461–471.

Kilareski, W. P., and Churilla, C. J. (1983). "Pavement management for large highway networks." *J. Transp. Eng.,* 109(1), 33–45.

Kinsey, W. R. (1973). "Underground pipeline corrosion." *Transp. Eng. J. ASCE,* 99(1), 167–182.

Kurdziel, J. M., and McGrath, T. J. (1991). "SPIDA method for reinforced concrete pipe design." *J. Transp. Eng.,* 117(4), 371–381.

La Baugh, W. (1971). "Optimizing traffic flow on existing street networks." *Transp. Eng. J. ASCE,* 97(2).

Levinson, H. S., Strate, H. E., Edwards, S. R., and Dickson, W. (1984). "Indirect transportation energy." *J. Transp. Eng.,* 110(2), 159–174.

Lin, F. B. (1991). "Knowledge base on semi-actuated traffic-signal control." *J. Transp. Eng.,* 117(4), 398–417.

Lin, F. B., and Vijayakumar, S. (1988). "Adaptive signal control at isolated intersections." *J. Transp. Eng.,* 114(5), 555–573.

Liu, C., and Gazis, D. (1999). "Surface roughness effect on dynamic response of pavements." *J. Transp. Eng.,* 125(4), 332–337.

Machemehl, R. B., Walton, C. M., and Lee, C. E. (1975). "Acquiring traffic data by in-motion weighing." *Transp. Eng. J. ASCE,* 101(4), 681–689.

Mamlouk, M. S., Zaniewski, J. P., and He, W. (2000). "Analysis and design optimization of flexible pavement." *J. Transp. Eng.,* 126(2), 161–167.

Mannering, F. L., and Kilareski, W. P. (1997). *Principles of highway engineering and traffic analysis,* J Wiley, New York.

Marek, C. R. (1977). "Compaction of graded aggregate bases and subbases." *Transp. Eng. J. ASCE,* 103(1), 103–113.

Mashaly, E. A., and Datta, T. K. (1989). "Seismic risk analysis of buried pipelines." *J. Transp. Eng.,* 115(3), 232–252.

Matthias, J. S., Radwan, A. E., and Sadegh, A. (1987). "Comparison of computer programs for isolated signalized intersection." *J. Transp. Eng.,* 113(4), 370–380.

Mehta, H. C., Manson, J. A., Chen, W. F., and Vanderhoff, J. W. (1975). "Polymer-impregnated concrete: Field studies." *Transp. Eng. J. ASCE,* 101(1), 1–27.

McDermott, J. M. (1980). "Freeway surveillance and control in Chicago area." *Transp. Eng. J. ASCE,* 106(3), 333–348.

Meeks, C. L. V. (1995). *The railroad station: An architectural history,* Dover, Mineola, N.Y.

Michalopoulos, P. G., Binseel, E. B., and Papapanou, B. (1978). "Performance evaluation of traffic actuated signals." *Transp. Eng. J. ASCE,* 104(5), 621–636.

Miller, J. S., and Demetsky, M. J. (2000). "Longitudinal assessment of transportation planning forecasts." *J. Transp. Eng.,* 126(2), 97–106.

Missimer, H. C. (1978). "Schuylkill River—Major crossing." *Transp. Eng. J. ASCE,* 104(1), 43–54.

Mobasher, B., Mamlouk, M. S., and Lin, H. M. (1997). "Evaluation of crack propagation properties of asphalt mixtures." *J. Transp. Eng.,* 123(5), 405–413.

Mohan, S., and Bushnak, A. (1985). "Multi-attribute utility pavement rehabilitation decisions." *J. Transp. Eng.,* 111(4), 426–440.

Murphree, E. L., Woodhead, R. W., and Wortman, R. H. (1971). "Airfield pavement systems." *Transp. Eng. J. ASCE,* 97(3).

Nam, D. H., and Drew, D. R. (1998). "Analyzing freeway traffic under congestion: Traffic dynamics approach." *J. Transp. Eng.,* 124(3), 208–212.

Nemeth, Z. A., and Rouphail, N. (1983). "Traffic control at freeway work sites." *J. Transp. Eng.,* 109(1), 1–15.

O'Donnell, H. W. (1978). "Considerations for pipeline crossings of rivers." *Transp. Eng. J. ASCE,* 104(4), 509–524.

Olander, H. C., and Davidson, J. W. (1980). "Study of pipe with concrete bedding—Part III." *Transp. Eng. J. ASCE,* 106(6), 631–646.

Olander, H. C., and Robertson, G. G. (1970). "Study of pipe with concrete bedding—part II." *Transp. Eng. J. ASCE,* 96(1).

Orlick, S. C. (1978). "Airport/community environmental planning." *Transp. Eng. J. ASCE,* 104(2), 187–199.

Osborne, J. M., and James, L. D. (1973). "Marginal economics applied to pipeline design." *Transp. Eng. J. ASCE,* 99(3), 637–653.

Parissien, S. (2001). *Station to station,* Phaidon, New York.

Patton, E. L. (1973). "The trans-alaska pipeline story." *Transp. Eng. J. ASCE,* 99(1), 139–144.

Payne, H. J. (1973). "Freeway traffic control and surveillance model." *Transp. Eng. J. ASCE,* 99(4), 767–783.

Peterson, J. J. (1960). "Freeway spacing in an urban freeway system." *J. Highway Div.* 86(3).

Petroski, H., and Kastenmeier, E. (1996). *Engineers of dreams: Great bridge builders and the spanning of america,* Knopt, New York.

Potter, J. C. (1985). "Effects of vehicles on buried, high pressure pipe." *J. Transp. Eng.,* 111(3), 224–236.

Priestas, E. L., and Mulinazzi, T. E. (1975). "Traffic volume counting recorders." *Transp. Eng. J. ASCE,* 101(2), 211–223.

Public Roads Administration (PRA). (1948). *Manual on uniform traffic control devices for streets and highways,* Joint Committee on Uniform Traffic Control Devices, Public Roads Administration, Federal Work Agency, Washington, D.C.

Quiroga, C. A., and Bullock, D. (1996). "Geographic database for traffic operations data." *J. Transp. Eng.,* 122(3), 226–234.

Rada, G. R., Perl, J., and Witczak, M. W. (1986). "Integrated model for project-level management of flexible pavements." *J. Transp. Eng.,* 112(4), 381–399.

Rada, G. R., Smith, D. R., Miller, J. S., and Witczak, M. W. (1990). "Structural design of concrete block pavements." *J. Transp. Eng.,* 116(5), 615–635.

Ramsamooj, D. V., Ramadan, J., and Lin, G. S. (1998). "Model prediction of rutting in asphalt concrete." *J. Transp. Eng.,* 124(5), 448–456.

Rankin, W. W. (1997). *Bureau of highway traffic: A history,* Bureau of Highway Traffic Alumni Association, Chesire, Conn.

Rathi, A. K., and Santiago, A. J. (1990). "Urban network traffic simulation: TRAF-NETSIM program." *J. Transp. Eng.,* 116(6), 734–743.

Ravirala, V., and Grivas, D. A. (1995). "Goal-programming methodology for integrating pavement an bridge programs." *J. Transp. Eng.,* 121(4), 345–351.

Rigas, F., and Sebos, I. (1998). "Shortcut estimation of safety distances of pipelines from explosives." *J. Transp. Eng.,* 124(2), 200–204.

Rollings, R. S., and Pittman, D. W. (1992). "Field instrumentation and performance monitoring of rigid pavements." *J. Transp. Eng.,* 118(3), 361–370.

Sauer, E. K., and Weimer, N. F. (1978). "Deformation of lime modified clay after freeze-thaw." *Transp. Eng. J. ASCE,* 104(2), 201–212.

Seiler, W. J., Darter, M. I., and Garrett, J. H., Jr. (1991). "An airfield pavement consultant system (AIRPACS) for rehabilitation of concrete pavements." *Proc. ASCE Specialty Conf. on Airfield/Pavement Interaction* ASCE, New York.

Shmulevich, I., and Galili, N. (1986). "Deflections and bending moments in buried pipes." *J. Transp. Eng.,* 112(4), 345–357.

Shutt, H. E. (1972). "The state's role in gas safety program." *Transp. Eng. J. ASCE,* 98(1), 45–50.

Sinha, K. C., Cohn, L. F., Hendrickson, C. T., and Stephanedes, Y. (1988). "Role of advanced technologies in transportation engineering." *J. Transp. Eng.,* 114(4), 383–392.

Skabardonis, A. (1986). "Microcomputer applications in traffic engineering." *J. Transp. Eng.,* 112(1), 1–14.

Smith, R. (1983). "Accessible public transit alternatives." *J. Transp. Eng.,* 109(1), 16–32.

Stastny, F. J. (1972). "Joint utilization of right-of-way." *Transp. Eng. J. ASCE,* 98(2), 299–302.

Tabatabaie, A. M., and Barenberg, E. J. (1980). "Structural analysis of concrete pavement systems." *Transp. Eng. J. ASCE,* 106(5), 493–506.

Theologitis, J., and Powell, D. (1984). "Modal split in recreational transport planning." *J. Transp. Eng.,* 110(1), 15–33.

Transportation Research Board (TRB). (2001). "Highway capacity manual." *Special Rep. 209, 4th Ed.,* Transportation Research Board, Washington, D.C.

Tschegg, E. K., Ehart, R. J. A., and Ingruber, M. M. (1998). "Fracture behavior of geosynthetic interlayers in road pavements." *J. Transp. Eng.,* 124(5), 457–464.

U.S. Department of Commerce (USDOC). (1950). "Highway capacity manual, practical applications of research." Committee on Highway Capacity, Dept. of Traffic and Operations, National Research Council, Bureau of Public Roads, Washington, D.C.

U.S. Department of Transportation (USDOT). (2000). *Manual on uniform traffic control devices,* Washington, D.C.

Uzan, J. (1994). "Dynamic linear back calculation of pavement material parameters." *J. Transp. Eng.,* 120(1), 109–126.

Uzarski, D., and McNeil, S. (1994). "Technologies for planning railroad track maintenance and renewal." *J. Transp. Eng.,* 120(5), 807–820.

Wasp, E. J., Thompson, T. L., and Aude, T. C. (1971). "Initial economic evaluation of slurry pipeline systems." *Transp. Eng. J. ASCE,* 97(2).

Wellington A. M. (1887). *The economic theory of railroad location,* Wiley, New York.

Weissmann, A. J., McCullough, B. F., and Hudson, W. R. (1994). "Reliability assessment of continuously reinforced concrete pavements." *J. Transp. Eng.,* 120(2), 178–192.

Wiseman, G. (1973). "Flexible pavement evaluation using Hertz Theory." *Transp. Eng. J. ASCE,* 99(3), 449–466.

Wohl, M., and Martin, B. (1967). *Traffic system analysis for engineers and planners,* McGraw-Hill, New York.

Wolshon, B., and Taylor, W. C. (1999). "Impact of adaptive signal control on major and minor approach delay." *J. Transp. Eng.,* 125(1), 30–38.

Woods, K. B. (1960). *Highway engineering handbook,* McGraw-Hill, New York.

Wright, P. H., and Ashford, N. J. (1997). *Transportation engineering: Planning and deign.* Wiley, New York.

Yoder, E. J. (1959). *Principles of pavement design,* 1st Ed., Wiley, New York.

Young, W. (1989). "Application of VADAS to complex traffic environments." *J. Transp. Eng.,* 115(5), 521–536.

Zandi, I., Mersky, R., and Warner, J. A. (1979). "Freight pipeline, rail, and truck cost comparison." *Transp. Eng. J. ASCE,* 105(4), 411–425.

Zozaya-Gorostiza, C., and Hendrickson, C. (1987). "Expert system for traffic signal setting assistance." *J. Transp. Eng.,* 113(2), 108–126.

American Society of Civil Engineers *150th Anniversary Paper*

Building a Better World

Forensics and Case Studies in Civil Engineering Education: State of the Art

Norbert J. Delatte, M.ASCE,[1] and Kevin L. Rens, M.ASCE[2]

Abstract: This paper reviews the state of the art in the use of forensic engineering and failure case studies in civil engineering education. The study of engineering failures can offer students valuable insights into associated technical, ethical, and professional issues. Lessons learned from failures have substantially affected civil engineering practice. For the student, study of these cases can help place design and analysis procedures into historical context and reinforce the necessity of lifelong learning. Three approaches for bringing forensics and failure case studies into the civil engineering curriculum are discussed in this paper. These are stand-alone forensic engineering or failure case study courses, capstone design projects, and integration of case studies into the curriculum. Some of the cases have been developed and used in courses at the United States Military Academy and the Univ. of Alabama at Birmingham, as well as at other institutions. Finally, the writers have tried to assemble many of the known sources of material, including books, technical papers, and magazine articles, videos, Web sites, prepared PowerPoint presentations, and television programs.

DOI: 10.1061/(ASCE)0887-3828(2002)16:3(98)

CE Database keywords: Engineering education; Case reports; Forensic engineering.

Introduction

In response to several well-publicized engineering failures, the ASCE Technical Council on Forensic Engineering (TCFE) was established in 1982. The purpose of the TCFE is to

- Develop practices and procedures to reduce failures;
- Disseminate information on failures and their causes, providing guidelines for conducting failure investigations;
- Encourage research and education in forensic engineering; and
- Encourage ethical conduct in forensic engineering practice.

In addition to ASCE's board of direction's executive committee, the TCFE consists of six units:

- Forensic engineering practice;
- Dissemination of failure information;
- Technology implementation;
- Practice to reduce failures;
- Publication; and
- Education.

The TCFE's Committee on Education encourages universities to include forensic engineering and failure case studies in civil engineering education (Rendon-Herrero 1993a,b). The mission of

[1]Assistant Professor, Dept. of Civil and Environmental Engineering, 1075 13th St. South, Suite 120, Univ. of Alabama at Birmingham, Birmingham, AL 35294-4440.

[2]Associate Professor, Dept. of Civil Engineering, 1200 Larimer St., Univ. of Colorado at Denver, Denver, CO 80217-3364.

Note. Discussion open until January 1, 2003. Separate discussions must be submitted for individual papers. To extend the closing date by one month, a written request must be filed with the ASCE Managing Editor. The manuscript for this paper was submitted for review and possible publication on November 9, 2001; approved on January 30, 2002. This paper is part of the *Journal of Performance of Constructed Facilities*, Vol. 16, No. 3, August 1, 2002. ©ASCE, ISSN 0887-3828/2002/3-98–109/$8.00+$.50 per page.

the Committee on Education is to develop resources to meet educational needs and to implement education programs. Although forensic engineering is a growing field, a void remains in forensic engineering education. This committee encourages the inclusion of forensic engineering topics and failure case studies in civil engineering education at the graduate and undergraduate levels, as well as continuing education programs. The committee also recommends activities to promote and advance the educational objectives of colleges and universities. Finally, the committee acts as a source of referral for educational material with a forensic engineering emphasis. It is this last point at which overlap exists with the TCFE's Dissemination of Failure Information Committee. In fact, if one looks at the deliverables of each TCFE committee, it could be argued that a large percentage of the council's activity is indeed geared toward education.

The study of engineering failures can offer students valuable insights into associated technical, ethical, and professional issues. Many writers have pointed out the need to integrate lessons learned from failure case studies in civil engineering education (Bosela 1993; Rendon-Herrero 1993a,b; Baer 1996; Delatte 1997; Rens and Knott 1997; Pietroforte 1998; Carper 2000; Delatte 2000; Jennings and Mackinnon 2000; Rens et al. 2000d).

The editor of ASCE's *Journal of Performance of Constructed Facilities* addressed this need in an editor's note in the February 1998 issue. Some educators have developed upper-level courses in failure analysis and forensic engineering as electives or graduate offerings, but few civil engineering undergraduates are able to take advantage of them. It is not practical to add another mandatory course to already crowded civil engineering undergraduate curricula. In some cases, failure investigations have been used as problems for capstone design courses. Another approach is to integrate case studies into existing courses.

This paper reviews the state of the art for using forensics and failure case studies in civil engineering education and discusses all three of the approaches introduced above.

Many of the key technical principles that civil engineering students should learn can be illustrated through case studies. For example, the first writer has discussed the Hyatt Regency walkway collapse in Kansas City, Missouri, the Tacoma Narrows Bridge failure in Washington, and other well-known cases with students in statics, mechanics of materials, and other courses. As another example, the second writer has discussed temporary steel bracing failures on masonry walls and temporary timber bracing on wood trusses in steel design and capstone design courses (Rens et al. 2000e). These cases help students to

- Grasp difficult technical concepts and begin to acquire an "intuitive feel" for the behavior of structures and the importance of load paths and construction sequences;
- Understand how engineering science changes over time as structural performance is observed and lessons are learned;
- Analyze the impacts of engineering decisions on society; and
- Appreciate the importance of ethical considerations in the engineering decision-making process.

In a survey conducted by the ASCE TCFE's education committee in December 1989, reported by Rendon-Herrero (1993a, b) and Bosela (1993), about a third of the 87 civil engineering schools responding indicated a need for detailed, well-documented case studies. The response from the Univ. of Arizona said "ASCE should provide such materials for educational purposes," and the response from Swarthmore College suggested "ASCE should provide funds for creating monographs on failures that have occurred in the past" (Rendon-Herrero 1993b).

The ASCE TCFE conducted a second survey in 1998 that was sent to all Accreditation Board for Engineering and Technology (ABET)-accredited engineering schools throughout the United States (Rens et al. 2000d). As with the 1989 survey, the lack of instructional materials was cited as a reason that failure analysis topics were not being taught. One of the unprompted written comments in that survey was "A selected bibliography is needed on the topic, which could be accessed via the Internet." Another comment went on to say "The best things TCFE can do are (1) provide instructional materials to make it easy for a teacher to incorporate failures in their courses, and (2) provide Internet materials so instructors can give self-guided homework assignments." Still another responder to the 1998 survey indicated "Need published case studies such as project designs, failures, evaluations, etc."

This paper provides some resources, but this is by no means a complete bibliography. A fairly comprehensive bibliography of references available through 1996 was published by Nicastro (1996). Puri (1998) and Carper (2000) list several additional references that could be used as well. The combination of the references in these three papers with the references in this state-of-the-art paper provides a fair representation of the published failure analysis work to date.

Forensic Engineering and Failure Case Study Courses

Several papers have been written describing courses on forensic engineering or failure case studies at various universities (Bosela 1993; Rens and Knott 1997; Pietroforte 1998). In the previously mentioned 1989 TCFE survey (Bosela 1993; Rendon-Herrero 1993), 8 of the 87 university respondents offered a course on failures of structures. Although not specifically asked for in the 1998 TCFE survey, 5 of the 112 university respondents specifically mentioned a stand-alone failure analysis course. One comment indicated, "We have two MS level courses (in which undergraduates can enroll) that are dedicated to failure analysis and performances in structural and geotechnical engineering." Another comment went on to say, "The department offers a graduate level course on infrastructure surety, which is devoted almost entirely to what may go wrong in construction situations." And finally, one institution indicated, "We will introduce a new elective undergraduate course on forensic engineering and failure investigations next spring." The two TCFE surveys indeed illustrate that stand-alone failure courses are rare.

Some universities offer forensic engineering courses as electives, although these courses mainly appear at the graduate level (Carper 2000). Forensic engineering courses typically have the educational objective of teaching students to become forensic engineers. Since forensic engineering is generally practiced by expert engineers, these courses are usually only available to graduate students and upper-division undergraduates. Furthermore, these types of courses may require faculty who are practicing forensic engineers. Baer (1996) pointed out that faculty with the necessary background to teach these types of course are rare.

These courses have been taught by Fowler at the Univ. of Texas (UT, taken by the first writer in 1994), Rens and Knott at the Univ. of Colorado, Denver (UCD) (Rens and Knott 1997), and Rendon-Herrero at Mississippi State Univ. (MSU). Table 1 shows some topics typically addressed in forensic engineering course syllabi at these three universities.

There are similarities and differences. All three courses rely heavily on case studies, written and oral student projects, and presentations. The UCD course combines forensic engineering for civil and mechanical engineers with nondestructive evaluation (NDE) methods. NDE is a tool used in many failure analysis investigations that allows students hands-on laboratory training with relatively inexpensive equipment. In the NDE portions of the courses, students are required to use standard samples in addition to making their own case studies.

In one example, students acquired flawed steel that contained the rolling/milling flaw of lamellar tearing. The student team was able to prove that the steel contained these flaws by using high-frequency ultrasound, nondestructive testing equipment. In another example, students illustrated how voids in large-grained building materials such as concrete and masonry can be detected by low-frequency ultrasound. Still other examples illustrated how dye penetrant and magnetic particle testing can be used to highlight cracks and flaws in metals that are not visible to the naked eye. Students are required to produce a professional report and to defend it orally.

Although the course is cotaught with a practicing failure analysis engineer (Dr. Albert Knott), it does involve guest speakers such as NDE engineers, other practicing forensic engineers, attorneys, and technicians. All of the assigned student projects for the UCD course are taken from actual case studies led by Knott. Because actual case studies are used, the reports are first sanitized to remove any ties to the engineers, clients, and other associated parties. Students are given photographs, actual field notes, actual police evidence, videotapes, and other information about expired projects. In some cases, students are allowed to interview the actual assigned engineer. As in the NDE portion of the course, student-written presentations and oral defenses are required.

The UT course concentrates on civil engineering, with emphasis on structural evaluations. The UT and MSU courses rely heavily on guest speakers, including an attorney, a petrographer, and a retired judge.

Table 1. Typical Forensic Engineering Course Syllabus

Topic	Hours
(a) CE 5806, Forensic analysis and condition assessment of civil and mechanical infrastructure—Univ. of Colorado, Denver (Rens and Knott 1997)	
Infrastructure inspections	1.5
Failures due to expansive soils	3.0
Nondestructive evaluation (NDE)—dye penetrant, ultrasound, magnetic particle methods	12.0
Hyatt Regency and other cases	1.5
Vehicular accident reconstruction	4.5
Ethical issues in vehicular accident reconstruction	1.5
Product failure investigation	1.5
NDE of timber structures	1.5
NDE of steel structures	1.5
NDE of masonry structures	1.5
NDE of concrete structures	1.5
Depositions and court testimony	3.0
Construction and product law	3.0
Ethics in engineering practice	1.5
(b) ARE 383/CE 397 Forensic engineering—Univ. of Texas	
Introduction—what is forensic engineering?	—
Qualifications, role, history of failure, failure statistics	
Causes of failures—definition, classification, causes, specific causes	—
Investigation—planning, client interface/schedule/ budget, team, site observations/testing/analysis, document search, historical information/visual documentation, literature search, synthesis, development of conclusions	—
Case studies—residential structure, concrete structure, concrete materials, masonry walls, building envelope	—
Engineer in dispute resolution—civil litigation process, pretrial responsibilities, trial responsibilities, alternate dispute resolution	—
Issues in forensic engineering—ethics, professionalism, liability	—
(c) CE 4003 Forensic engineering—Mississippi State Univ.	
Introduction, definitions, and discussion of legal process	4.5
Failure case studies and causes	1.5
Natural hazards and unusual loads	1.5
Engineer as expert witness	4.5
Learning from failures	3.0
Investigation	6.0
Fire, industrial, product liability	4.5
Traffic accident and transportation	3.0
Environmental systems failures	1.5
Case studies	4.5

As with the UCD course, an important feature of the UT course is that each student must independently carry out a building investigation, write a report, and present the results. Campus buildings are used in order to avoid issues of liability and confidentiality. The university also benefits from free consulting services, although the staff is advised to consider the reports carefully before taking any action.

The MSU course has evolved considerably over the years, and the topics shown in Table 1 reflect the Fall 2001 course offering.

A review of course syllabi from fall 1991 onward shows that the broad themes addressed have remained the same, with topics such as dispute resolution techniques, construction safety, nonstructural failures, and dams and bridges being occasionally covered. Three to five guest lecturers are used, and some offerings include in-class mock trials or debates.

Pietroforte (1998) described a failure case study course recently developed at Worcester Polytechnic Institute (WPI) in Massachusetts. This course was developed in conjunction with a new Master Builder program—a five-year program leading to the award of a combined BSCE and MSCE degree. The course title is "Construction Failures: Analysis and Lessons." Five case studies are presented and discussed in detail in class, and students write term projects on cases from the literature and present the results. In contrast to the courses at UCD and UT, the WPI course focuses on failures and lessons learned, rather than the practice of forensic engineering. However, it should be noted that lessons learned from failures are an implied intrinsic by-product of the discussion of a failed engineering project—especially in an educational setting.

Both Pietroforte and the first writer (Delatte) took a course at the Massachusetts Institute of Technology (MIT) in fall 1984, taught by Irwig and Becker, that incorporated two Boston failure case studies—Hotel Vendome and 2000 Commonwealth Avenue. This MIT course provided a useful stimulus toward the inclusion of case studies in engineering education. The first writer has taught a course at the Univ. of Alabama at Birmingham (UAB), "Engineering the Environment," that introduced a number of basic civil engineering concepts through failure case studies and other historical examples.

At this point it is useful to consider separately courses that teach forensic engineering and courses that teach engineering practice through failure case studies because these different categories of courses have different objectives. The courses in the former category (e.g., UCD, UT, and MSU) are intended to introduce graduate students with sound technical backgrounds to investigate methods, depositions, testimony, and other details of forensic engineering practice such as writing and speaking, although some time is spent on technical topics. A proper definition of forensic engineering, after all, is that it is engineering practiced in a public forum, such as within the legal system.

In the second category, courses on failure case studies can be of considerable value. In addition to technical material, valuable lessons in engineering practice, ethics, and professionalism can be taught. Another option is to use such a course to teach research and technical communication skills. However, it should be noted that no matter which type of failure analysis course one is considering, each should incorporate lectures involving case studies as these examples essentially make up the textbook.

Outside the United States, students participating in a "learning from disasters" exercise at Queen's Univ., Belfast, investigated a number of failures and disasters. The exercise was carried out in the first two years of the bachelor's and master's degree curricula. These researchers (Jennings and Mackinnon 2000) noted, "It has become clear how little students generally know about key disasters which are common knowledge to their elders and which have had a profound effect on the profession." The same can be said of the majority of U.S. students as well. Even though well-published cases such as the Hyatt Regency walkway collapse, Willow Island cooling tower, (West Virginia), or Hartford Civil Center (Connecticut) failure are common knowledge to moderately seasoned engineers, they tend to be new issues to 20-year-old engineering students.

Design Capstone Courses and Forensic Engineering

Under the urging of ABET, many universities have developed a capstone or senior design course that gives students actual engineering problems to solve. These courses are becoming increasingly important in many engineering programs (Dutson et al. 1997). Students often work in teams on projects gathered from the surrounding community. Among the problems to solve may be failures or rehabilitation of existing facilities. One respondent to the 1998 TCFE survey did indicate that failure analysis topics were introduced in the design capstone courses.

An excellent example of how to incorporate failure analysis in the capstone design course can be found at MSU, where Professor Oswald Rendon-Herrero has been incorporating failure analysis topics into the civil engineering curriculum since 1973. A recent paper discusses how a term project in the foundation design capstone course involves failure analysis (Rendon-Herrero 1998). This course is offered on an annual basis each spring and regularly has an enrollment of around 30 students.

As a background to the course, it should be noted that the MSU region is known for its highly expansive soils, which are native to the area. Many homes suffer from distresses that range from either minor cosmetic problems to general foundation problems. The cosmetic problems range from dry wall cracking, inoperable doors and windows, elevation problems, or other functional types of distresses, while the general foundation problems can involve heaved piers or foundations, concrete cracking, or other safety issues. Before the semester begins, Rendon-Herrero has a meeting with the local newspaper and runs a short story about the upcoming semester term project. The article is designed to solicit owners of homes containing functional or structural safety distresses to submit an application to be a case study for the semester-long course. MSU is located in the town of Starksville, Miss., which has a population of around 35,000 people. According to Rendon-Herrero, in a typical semester, around 35 homeowners respond to the solicitation.

At the beginning of the semester, each of the homeowners is interviewed by the students and each residence is ranked based on a variety of parameters such as magnitude of problem, availability of and access to the home, availability of design information (that is, design plans or specifications), age of home, and friendliness of the homeowner. After the initial interview process, the top 10 or so homes are chosen based on a three-student-per-house ratio.

The term project involves six or seven tasks with milestone benchmarks at certain intervals during the semester. As an example, the first benchmark, set a short period after the semester begins, involves a second detailed interview of the homeowner, a detailed inspection of the residence involving visual inspections and photography, and a literature review of the published USDA soil survey report. At each benchmark, Rendon-Herrero reviews reports and offers advice to the student teams. Another benchmark involves obtaining actual soil borings with a hand auger to produce a core in order to develop boring logs. Other soil laboratory tests are performed and are compared to the published USDA soil report, usually with a favorable correlation.

Three weeks before the semester is over, students are required to put the entire puzzle together and come up with a theory of the residence behavior, source of problems (usually drainage or other water source or faulty construction), and a tentative rehabilitation scheme. Each student team defends a written and oral report. The final deliverable is a wrap-up inspection report that is given to the owner. Rendon-Herrero then offers his own failure theory and rehabilitation advice to the owner in the form of a marked-up student report with a typical disclaimer.

Although the term project is a tremendous amount of work, given that normal foundation lectures and homework such as footing and retaining wall analysis and design are also required, Rendon-Herrero is anxious each spring to continue the tradition.

A similar project at North Dakota State Univ., describing forensic analysis of slope stability problems around a lake, is discussed in a paper by Padmanabhan and Katti (2002). Eleven groups of six students each investigated the problem and proposed mitigation solutions. Although a failure analysis project was discussed in the paper, most capstone projects at that university do not involve failures or forensic engineering.

Integration Into Curriculum

The 1998 TCFE survey revealed that perhaps the best way to introduce undergraduates to failure analysis and lessons learned from failures was to allow case studies to permeate undergraduate courses. Nearly all the respondents felt that a few lectures were sufficient to reinforce basic engineering concepts. Few respondents indicated the need for four or more lectures. Several unprompted written comments were received, each with a theme similar to the following comment: "(design) courses should include failure information as a part of lecture and laboratory efforts. This information should be provided at the undergraduate level on an ongoing basis as opposed to stand-alone courses." The respondent went on to say that "Stand-alone courses should be reserved for the Master's degree level courses." This final statement echoes many comments provided.

Therefore, one solution is to integrate case studies and lessons learned into existing courses. The main obstacle to this is that many faculty do not have time to research and prepare case studies. The review below builds on previous work (Delatte 2000) to suggest some courses, lesson topics, and cases. Available sources of well-developed case studies and teaching materials are also identified.

Use of Case Studies

The first writer has used case studies such as these at the United States Military Academy (USMA) and at the UAB in such courses as statics, mechanics of solids, and reinforced concrete. Some of the ways to integrate failure case studies and a suggested format were reviewed in Delatte (1997, 2000). These include

- Introductions to topics: Use the case to illustrate why a particular failure method is important. Often the importance of a particular mode of failure only became widely known after a failure; examples include the wind-induced oscillations of the Tacoma Narrows Bridge and the Air Force warehouses (Ohio and Georgia) without reinforcement for shear and dimensional changes, discussed below;
- Class discussions: Link technical issues to ethical and professional considerations;
- Example problems: Calculate the forces acting on structural members and compare them to design criteria and accepted practice; and
- Group and individual projects: Have students research the cases in depth and report back on them. This will also help build a database of cases for use in future classes.

In the writers' opinion the best way to incorporate cases into classroom discussions is to link them to specific topics, as sug-

Table 2. Courses, Topics, and Case Studies (based on Delatte 2000)

Course	Topic	Case study
Statics	Free-body diagram	Hyatt Regency walkway collapse
		T. W. Love Dam cantilever form failure
Dynamics	Mass moment of inertia and stiffness	Tacoma Narrows Bridge collapse
Mechanics of materials (solids)	Kinetics: dynamic forces	Bomber crash into Empire State Building
	Stress and strain	Shrinkage of concrete masonry units
		and swelling of brick masonary
	Structural deformation as warning	Hartford Civic Center
	of impending collapse	
	Elastic buckling	Stepped roof structure, Elwood, N.Y.
Structural analysis	Loads on structures	Bomber crash into Empire State Building
	Load paths	L'Ambiance Plaza collapse
	Structural deformation	Quebec Bridge
		Hartford Civic Center
	Checking computer results	Hartford Civic Center
Reinforced concrete design	Structural integrity of formwork	New York Coliseum
	Strength development of concrete	Willow Island cooling tower
		2000 Commonwealth Avenue, Boston
		Bailey's Crossroads, Virginia
	Punching shear in concrete slabs	Harbor Cay condominium, Florida
	Reinforcement development length	Pittsburgh Midfield Terminal precast beam collapse
Steel design	Connections	Hyatt Regency walkway collapse
		Steel frame connections in Northridge earthquake
	Buckling	Stepped roof structure, Elwood, N.Y.
		Quebec Bridge
Introduction to engineering or capstone	Professional ethics	Citicorp Tower

gested in this section. Student response to the cases developed so far has been enthusiastic. The most successful case studies are those that inspire students to go out, do their own research, and learn more about their chosen profession.

Courses and Lesson Topics

Lesson topics should be identified for required courses in a civil engineering curriculum. Once these have been identified, it is possible to suggest case studies to support the topics. Some courses, topics, and case studies are suggested in Table 2.

Engineering Mechanics

For the purposes of this discussion, engineering mechanics refers to courses in statics, dynamics, and mechanics of materials (also called strength of materials or mechanics of solids). At some schools, such as USMA or UAB, the civil engineering faculty often teach these courses and can easily incorporate appropriate case studies. Where these courses are taught by other departments, it may be necessary to address these topics in a later course, such as structural analysis. An earlier paper discussed seven case studies developed for engineering mechanics courses (Delatte 1997). A discussion by Puri (1998) provided comments as well as additional exercises and examples.

The free body diagram is the basic equilibrium analysis tool used to determine forces acting on a body. If the diagram is not drawn correctly, the forces cannot be calculated accurately and the design may be unsafe. The importance of a correct free body diagram may be shown through analysis of the Kansas City Hyatt Regency walkway collapse. A free body diagram of the original detail, on the left in Fig. 1, shows that the nut-to-beam connection supports the weight of a single deck, or P. In contrast, the free

body diagram of the as-built detail now requires the nut-to-beam connection to transfer $2P$. The load on the connection was doubled, and it failed. An excellent discussion of this case, with emphasis on ethical issues, is provided by Roddis (1993).

This case was revisited, with considerable new information and analysis, in four papers published in a special issue of ASCE's *Journal of Performance of Constructed Facilities* (Gillum 2000; Luth 2000; Moncarz and Taylor 2000; Pfatteicher 2000). In addition, all four writers published and presented abbreviated versions of ASCE's 2nd forensic congress (Rens et al. 2000c).

Structural Analysis

At institutions where engineering mechanics courses are taught outside the civil engineering department, it may be desirable to address some of the cases in engineering mechanics in later courses. Case studies appropriate for inclusion in a structural analysis course are discussed as follows.

Loads Acting on Structures

Accurate prediction of loads acting on structures is difficult, but extremely important. This topic is addressed in Chapter 2 of Feld and Carper (1997). The bomber crash into the Empire State Building (New York) (Delatte 1997) as well as the Oklahoma City Federal Building (Oklahoma) bomb are examples of extraordinary loads acting on buildings. In addition, introduction of basic wind, snow, or other live loads tends to be a topic overlooked in many engineering courses—that is, usually the loads are a given parameter. It is important for instructors to discuss the nature and variability of both ordinary and extreme loads in lectures.

Original   As-Built

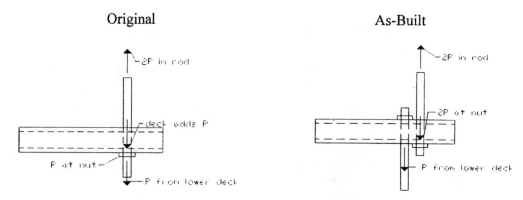

Fig. 1. Original and as-built hanger details

Load Paths

It is important that the designer provide a continuous load path at all times to transfer all loads safely to the foundation. There are many competing theories as to why the L'Ambiance Plaza (Connecticut) towers collapsed while under construction, but each theory is based on a break in the load path. The state of construction just before collapse is illustrated in Fig. 2, which shows an elevation of the building with the packages of slabs being jacked up together. Six theories are discussed in Martin and Delatte (2000). Several of the theories focus on the lift heads (Fig. 3) used to lift the slabs in position.

Calculating Structural Deformations

Accurate calculation of structural deformation is important for two reasons. The first is that excessive deformations may lead to serviceability problems, causing nonstructural damage or making continued use of the facility difficult or impossible. This is why codes limit these deformations. The second is that if deformations during construction or while the building is in service greatly exceed predictions, this is a warning that the structure may be in danger of collapse.

In two cases, higher-than-expected deformations were ignored until it was too late. Before the Quebec River Bridge (Canada) collapsed in 1907, killing 82 workers, compression members were observed to be distorted by up to 57 mm (2 $\frac{1}{4}$ in.), indicating incipient buckling. An excellent account of this tragedy is provided by Roddis (1993). The collapse of the Hartford Civic Center in 1978 also occurred after excessive structural deformations observed during construction had been ignored. Bracing for the

compression members of the roof space truss (Fig. 4) proved to be inadequate, and several members failed by buckling. This case is reviewed in detail in Martin and Delatte (2001).

Checking Computer Results

Petroski (1985) suggests that design of the Hartford Civic Center roof would never have been attempted without computers because the space truss would be very difficult to analyze by hand methods. He also suggests that uncritical acceptance of the computer solution played an important role in the catastrophe. Computer results can never substitute for understanding structural behavior. Several errors in application of finite-element analysis have been discussed (Bell and Liepins 1997). The engineer should know the approximate answer before sitting down in front of the computer and must be able to distinguish an accurate solution from one that is absurd but appears precise.

An excellent example of known failures as a result of computer misuse can be found in Puri (1997), where a 50-page paper was presented by the panel at ASCE's 1st forensic congress (Rens 1997). This report outlines 52 cases on computer misuse, failure types, error sources, and lessons learned.

Reinforced Concrete Design and Concrete Materials

A large number of case studies relate to concrete design and construction. Chapters 7 and 8 of Feld and Carper (1997) address this topic. Topics such as formwork, shoring, and other temporary structures are covered in the section titled "Structural Integrity during Construction."

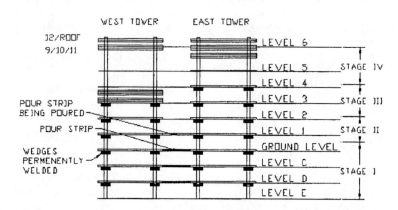

Fig. 2. L'Ambiance Plaza construction status just before collapse

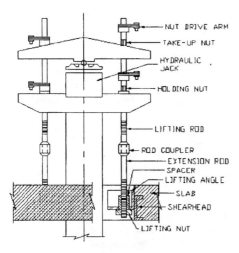

Fig. 3. L'Ambiance Plaza lift heads

Strength Development of Concrete

Twenty-eight day cylinder strength tells one how strong a concrete cylinder cured in a laboratory is at 28 days, but doesn't necessarily tell you much about the strength of the as-built structure. The strength gain of concrete is highly dependent on ambient temperature. In cold weather, the concrete in a structure will be much weaker at an early age than laboratory cylinder strengths cured at a different temperature.

The Willow Island, West Virginia, cooling tower collapsed while under construction on April 27, 1978, killing 51 workers in the worst construction disaster in U.S. history (Lew et al. 1979; Ross 1984; Kaminetzky 1991; Feld and Carper 1997; LaCome et al. 2000). A jump-form system was being used, with the forms secured by bolts in one-day and three-day-old concrete; the forms were designed to be progressively moved up the tower as it was built. The temperature had been in the mid-thirties at night. The National Bureau of Standards found that the concrete had not attained enough strength to support the forms. The report concluded that "the most probable cause of the collapse was the imposition of construction loads on the shell before the concrete of lift 28 had gained adequate strength to support these loads"

(Lew et al. 1979). LaCome et al. (2000) further show how the concrete maturity method can be used with the actual published information to determine the strength of the concrete at failure.

An investigation into the collapse of a 17-story concrete high-rise under construction at 2000 Commonwealth Avenue, Boston, disclosed a number of irregularities and deficiencies (Kaminetzky 1991), including the following, among others:
1. Lack of proper building permit;
2. Insufficient concrete strength;
3. Insufficient length of rebars;
4. Lack of proper field inspection;
5. Various structural design deficiencies;
6. Improper formwork;
7. Premature removal of formwork;
8. Inadequate placement of rebars; and
9. Lack of construction control.

Four workers were killed and 20 injured. Fortunately, the collapse occurred slowly enough for many of the workers to escape. The collapse occurred on January 25, and low temperatures had certainly retarded strength gain. Cores showed concrete compressive strengths as low as 4.83 MPa (700 psi) (Kaminetzky 1991; Feld and Carper 1997).

Yet another formwork collapse blamed on inadequately cured concrete occurred at Bailey's Crossroads, Virginia, in March 1973. Fourteen workers were killed and 30 were injured. Shores were removed between the 22nd and 23rd floors of the building while concrete was being placed on the 24th. The collapse tore an 18 m (60-ft) wide gap through the building all the way to the ground. The concrete, when tested, turned out to be well below its expected strength. The floor slab failed in punching shear at the columns (Ross 1984; Kaminetzky 1991; Feld and Carper 1997).

Dimensional Changes in Concrete and Shear Reinforcement

Two warehouse roofs at U.S. Air Force bases in Ohio and Georgia cracked and collapse under combined load, shrinkage, and thermal effects in 1955 and 1956. Continuous 122 m (400 ft) lengths of reinforced concrete roof girders functioned as single units because of defective expansion joints. Other warehouses, built to the same plans, survived because separation was maintained between adjacent 61 m (200 ft) bays. These failures led to more stringent shear steel requirements in subsequent editions of the ACI Build-

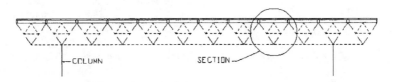

Space Frame Roof

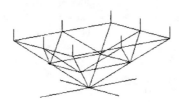

Section of the Space Frame Roof

Fig. 4. Hartford Civic Center roof

ing Code. In these structures, the concrete alone, with no stirrups, was expected to carry the shear forces, and the members had no shear capacity once they cracked (McKaig 1962; Feld and Carper 1997).

Reinforcing Steel Placement and Punching Shear

Numerous design errors were uncovered when the five-story Harbor Cay Condominium, Cocoa Beach, Florida, collapsed under construction in 1981, killing 11 workers and injuring 23. Incredibly, no punching shear calculation had been made for the concrete floor slabs. Furthermore, the slabs were only 200 mm (8 in.) thick and should have been 280 mm (11 in.) thick to satisfy the ACI Building Code minimum. The chairs used to support the slab steel were 108 mm (4 $\frac{1}{4}$ in.) high, which, coupled with the thin slabs, led to a very small effective depth (Lew et al. 1982; Kaminetzky 1991).

Development Length of Reinforcing Steel

Unless a sufficient development length of steel is embedded in concrete, the bar will pull out before it yields. In 1990 a portion of the Pittsburgh Midfield Terminal (Pennsylvania) failed during construction. In a precast concrete beam the bottom reinforcing bar was embedded only 185 mm (7 $\frac{1}{4}$ in.), which is much shorter than necessary (Thornton and DeScenza 1997).

Structural Steel Design

A number of case studies involving steel structures are reviewed in Chapter 6 of Feld and Carper (1997). Others are discussed in Kaminetzky (1991). Critical issues for steel structures are connection detailing and buckling. The cases of the Hyatt Regency for connections and the Hartford Civic Center for buckling can be used if they have not been addressed earlier in the curriculum.

Connection Details

Many cases in structural engineering have illustrated the importance of careful attention to connections. The Northridge Earthquake of 1994 showed that the special moment-resisting frame welded connection behaved much worse than anticipated during the event [four papers in Rens (1997), pp. 219–257]. This is an excellent example of the dangers of extrapolating behavior from small test specimens to full-scale structures and of why codes and standards continue to evolve.

Soil Mechanics and Foundation Engineering

Courses in soil mechanics and foundation engineering often discuss the Leaning Tower of Pisa, but many other case studies are available. Chapter 3 of Feld and Carper (1997) addresses earthwork, soil, and foundation problems, and Shepherd and Frost (1995) list foundation and geoenvironmental case studies. Papers published in ASCE's *Journal of Performance of Constructed Facilities* address several topics:

- Problem fill materials, soil shrinkage, and expansive soils (Gnaedinger et al. 1987; Richardson et al. 1987; Raghu and Hsieh 1989; Day 1992, 1994a, b, 1995; Meehan and Karp 1994).
- Lateral earth pressure and retaining walls (Lin and Hadipriono 1990; Day 1993; Diaz et al. 1994; Leonards et al. 1994).
- Soil bearing capacity (Whitlock and Mossa 1996; Amini and Khalilian 1997).

Two books (Handy 1995; Shallow 1995) provide a thorough but understandable discussion of engineering geology and soil mechanics and their influence on the performance of facilities.

Structural Integrity during Construction

Although a specific course is not available that teaches all the basics of temporary design construction issues such as formwork, shoring, masonry wall bracing, and timber truss bracing, it is a topic that can permeate other design courses, such as concrete design, timber design, steel design, and foundation design. Perhaps the best course to address this topic would be a capstone senior design course. Chapter 11 of Feld and Carper (1997) is specifically dedicated to bracing issues and failures.

Temporary Bracing

As temporary bracing is usually a construction-sequencing event, many times the bracing and construction process is left for the field crew to decide. According to Feld and Carper (1997), in the United States alone, an average of 2,200 fatalities occur yearly in the construction industry. The economic loss is also staggering—around $9 billion lost annually to accidents. According to Kaminetzky (1991), in steel construction alone insufficient bracing accounts for 25% of the failures during construction. Temporary bracing is also a common failure issue in concrete construction. It is common for contractors to remove shoring and formwork as quickly as possible. Again, sequencing is always an issue as pressure is always mounting to move on to the next phase of construction in order to meet certain milestone deadlines. Improperly braced timber trusses and masonry walls are also common failure events. The second writer has been involved in several related case studies (Rens et al. 2000 e).

But who is responsible? Contractors in charge of construction routinely have the attitude, "if we work fast enough, we won't have to brace it, and nothing is likely to happen" (Feld and Carper 1997). However, it is generally the contractor's responsibility to provide the necessary bracing and make decisions about the most efficient construction sequence (Goldstein 1999). By common practice and custom, engineers provide designs for completed structures, including permanent bracing but not temporary bracing.

In any event, engineering education can make temporary bracing and construction sequencing a topic in capstone and structural design courses. The second writer has presented a paper involving two case studies of bracing failures and has suggested ways that simple application of related examples could integrate design and construction (Rens et al. 2000e). For example, the first case study involved an inadequately braced 6.1 m (20 ft) tall masonry wall. A typical windy Colorado evening caused a midsized commercial building to collapse. The case incorporated the determination of wind loads, also an often-overlooked civil engineering topic, and possible failure modes, including bracing buckling (steel design); connection design (steel design); punching failure (concrete/masonry design); and soil anchorage failure (foundation design). The second case also involved wind interacting with the construction of a large school gymnasium. Again, a typical wind event caused 11–24.4 m (36–80 ft) long timber trusses to collapse. To make matters worse, the failure occurred twice during the setting of the trusses, each occurring over the weekend—no lessons were apparently learned. This case also involved the determination of wind loading and forces on connections (timber design).

Structural Integrity of Formwork

There is an economic incentive for a builder to keep formwork as inexpensive as possible and to remove it as quickly as possible so that it can be reused. Formwork is a structure, and like any structure can only stand if it is stable and load paths are maintained.

Table 3. Available Case Study and Disaster Videotapes

Title	Description and source
"To Engineer is Human: The Role of Failure in Successful Design"	1997, written and presented by Henry Petroski, producer Alec Nesbitt, 51 min, published by Princeton, JH: Films for the Humanities and Sciences: FFH 7378, distributed under license from BBC Worldwide Americas, Inc., also available through Films Incorporated Education, 5547 N. Ravenswood, Chicago, IL 60640-1199
"The Day the Earth Shook"	1996, NOVA, written and produced by Simon Campbell-Jones and Suzanne Campbell-Jones, Executive producer Paula S. Apsell, 60 min, WGBH Educational Foundation ⟨www.wghb.org⟩
"Earthquake"	1990, NOVA, written and produced by Carl Charlson, Executive producer Paula Apsell, 60 min, WGBH Educational Foundation ⟨www.wgbh.org⟩
"Earthquakes: The Terrifying Truth"	1994, written and produced by Alex Gregory, executive producer, Dennis B. Kane, ABC Video Publishing, 50 min
"Chernobyl Nuclear Disaster"	1990, MPI home video presentation of an ABC news production, 30 min
"Disaster Proof—Architectural Failures"	1996, written and produced by John Borst, SCI-TREK—The Discovery Channel, 30 min ⟨www.discovery.com⟩
"The New Detective—Case Studies in Forensic Science"	1997, produced by Tom Naughton and Nicolas Valcon, written by Stephen Zorn, 30 min, SCI-TREK—The Discovery Channel ⟨www.discovery.com⟩
"Fatal Flaw—A Skyscraper's Nightmare"	1996, A&E Investigative Reports, executive producer Andrea Miller, directed by Roger Parsons, Kurtis Production Inc. ⟨www.aande.com⟩—also discussed in Morganstern (1997)
"Academic Integrity: The Bridge to Professional Ethics"	1995, Center for Applied Ethics, Duke Univ., written by P. Aarne Vesilind, directed by Jody McAuliff, 35 min
"Engineering Disasters"	From the leaning tower of Pisa to Soyuz 11, here are the fascinating—and sometimes tragic—tales of engineering gone wrong ⟨www.aande.com⟩

At the New York Coliseum in 1955 (McKaig 1962; Kaminetzky 1991), about 929 m^2 (10,000 sq ft) of main exhibition hall collapsed during construction, killing one worker and injuring 50 others. The forms were two-stories high, supported on 3.35 m long, 89 mm square (11 ft 4×4) timbers linked together by a cross beam at midheight. The crossbeams did not provide bracing against lateral instability. Buggies were used to transport the concrete for the slab being poured, and eight buggies were on the formwork at the time of collapse. According to the district attorney's office, the cause of failure was "inadequate provisions in the formwork to resist lateral forces" [McKaig (1962), p. 16]. Without proper bracing, the structure became unstable under the dynamic loading of the buggies. Formwork designs that had been safe before the use of buggies proved unsafe under the heavier loads. McKaig (1962) also discusses 14 other formwork failures.

Sources for Case Materials

There are many sources for case studies. These include books, technical papers, and magazine articles, videos, Web sites, prepared Microsoft PowerPoint presentations, and television programs.

Books

Three excellent texts are Feld and Carper (1997), Kaminetzky (1991), and Levy and Salvadori (1992). McKaig (1962) is also

good. Ross (1984) contains cases reprinted from *Engineering News Record*, a weekly publication covering the construction industry that often contains examples of recent failures. Shepherd and Frost (1995) contains short summaries of a wide variety of cases. Two excellent recent sources of case studies are the proceedings of the 1st and 2nd ASCE congresses on forensic engineering (Rens 1997; Rens et al. 2000 c).

Some books, such as Levy and Salvadori (1992) and Petroski (1985), do an excellent job of explaining fundamental structural behavior without relying on complex theories or mathematics and are particularly appropriate for lower-division undergraduate students.

Papers and Articles

Engineering News Record, addressed briefly above, is a good source of news on recent cases. Another excellent source is the quarterly *Journal of Performance of Constructed Facilities*, published by ASCE. Other ASCE journals, such as the *Journal of Structural Engineering* and *Journal of Professional Issues in Engineering Education and Practice*, often feature useful case studies. A useful bibliography on failures was assembled in a paper by Nicastro (1996), as noted earlier.

Video

An excellent video illustrating case studies is "When Engineering Fails," written and presented by Henry Petroski. This videotape

Table 4. Available Case Study Web Sites (Based on Delatte 2000)

Topic or case	URL
UAB REU site case studies	⟨http://www.eng.uab.edu/cee/REU_NSF99/rachelwork.htm⟩
	⟨http://www.eng.uab.edu/cee/REU_NSF99/reu_nsf00/carlos›ebpage2.htm⟩
	⟨http://www.eng.uab.edu/cee/REU_NSF99/reu2001/King.htm⟩
Assorted case studies	⟨http://carbon.cudenver.edu/~mclark/⟩
	⟨http://www.eos.uoguelph.ca/webfiles/james/FamousEngrgDisasters.htm⟩
Investigations—SGH	⟨http://www.sgh.com/investig.htm⟩
Tacoma Narrows	⟨http://www.bergen.org/AAST/Projects/Timeline/Transportation20/tacoma/index.htm⟩
	⟨http://www.nwwf.com/wa003a.htm⟩
	⟨http://www.math.uconn.edu/~kmoore/tacoma.html⟩
	⟨http://www.stkate.edu/physics/phys111/curric/tacomabr.html⟩
	⟨http://www.me.utexas.due/~uer/papers/paper_jk.html⟩
Hyatt Regency	⟨http://www.uoguelph.ca/~ajenney/webpage.htm⟩
	⟨http://lowery.tamu.edu/ethics/ethics/hyatt/hyatt1.htm⟩

very closely parallels the book "To Engineer is Human" (Petroski 1985) and provides dramatic footage of the Hyatt Regency walkway collapse, the Tacoma Narrows Bridge collapse, and other cases.

Another quality video that illustrates a specific failure analysis case study, professional ethics, and rehabilitation can be found at the Arts and Entertainment Channel (A&E)'s investigative reports. This story, titled "Fatal Flaw—A Skyscrapers' Nightmare," describes when the engineer of record William LeMessurier went public about the inadequate design of the Citibank skyscraper in New York City. A paper chronicling the events can be found in Morgenstern (1997). The video has several clips of LeMessurier documenting his trouble with peace of mind when he discovered, after the skyscraper was already constructed and inhabited by office staff, that the lateral bracing was inadequately designed. The video illustrates how local officials worked with emergency response individuals to develop a plan to evacuate a several-block region should winds reach a critical magnitude. In the end, working in the evenings, a major rehabilitation was accomplished and failure was avoided without public knowledge (and ensuing

panic). LeMessurier presented a talk on this issue at the 1 ASCE forensic congress (Rens 1997). Table 3 describes these and several other educational videos dealing with engineering failures, natural disasters, and ethics.

Internet

Several Web sites also provide case studies, or images to go with case studies, as shown in Table 4. Rachel Martin's Web site, the first listed, provides links to many of the others. The UCD also has a Web site with cases and is continuing to collect more (Rens et al. 2000a, b).

Presentations

The TCFE's Committee on the Dissemination of Failure Information has prepared a set of presentations, "Failure Vignettes," which are targeted at architects but may also be of value to engineering educators (Zickel 2000).

Table 5. Case Study Television Channels and Programs

Channel	Web site	Typical program title and description
The Discovery Channel	⟨www.discoverychannel.com⟩	"New Detectives: Case Studies in Forensic Science"—Some of the best clues come from the least likely places. Baffling crimes have been solved and criminals betrayed through evidence provided by insects, beer bottles, and other seemingly meaningless objects
The Learning Channel	⟨www.tlc.com⟩	"Without Warning: Bridge Collapse"—In 1995, the old Songsu Bridge in Seoul, Korea, collapsed. News reports show the gaping hole in the popular commuter bridge and the death and destruction below
The Arts and Entertainment (A&E) Channel	⟨www.aande.com⟩	"Greatest Blunders of the 20th Century"—The 20th century witnessed a long parade of stunning achievements, from the popularization of the automobile to the landing on the moon. But there is a flipside to these stories, for the path to progress has not always been smooth

Television

Several excellent, educational regular television programs deal with forensic science. Some of these programs are centered on the subject of pathology, but many of the procedures are well suited to engineering failure analysis. In particular, the gathering of evidence, interviewing witnesses, photography, document preservation, and interpretation of results are common topics. A look at the television listings via the Internet revealed the programs listed in Table 5. In addition, many of the videotapes shown in Table 3 are occasionally broadcast.

Summary, Recommendations, and Conclusions

The formation of the TCFE and its representative committees in the early 1980s marked an important benchmark for failure analysis and education. Since that time, numerous articles and a wealth of other sources have been published on the subject. The cases and sources identified in this paper represent a starting point in the development of case studies in civil engineering education. This paper also reflects 20 years worth of evolution of failure analysis education. It is hoped that the interest will grow exponentially over the next 20-year period as well.

Acknowledgments

Rachel Martin and Suzanne King gathered much of the background material for this paper; Martin prepared Figs. 1–4. Their work was supported by the National Science Foundation Research Experiences for Undergraduates site at the University of Alabama at Birmingham under Grant No. EEC-9820484. Professor Oswald Rendon-Herrero of Mississippi State University provided copies of course syllabi for a number of past offerings of his CE 4003 forensic engineering course.

References

Amini, F., and Khalilian, A. (1997). "Old post office foundation failure investigation." *J. Perform. Constr. Facil.*, 11(1), 13–17.

Baer, R. J. (1996). "Guest editorial (Are civil engineering graduates adequately informed on failure? A practitioner's view)." *J. Perform. Constr. Facil.*, 10(2), 46.

Bell, G. R., and Liepins, A. A. (1997). "More misapplications of the finite element method." *Forensic engineering*, K. L. Rens, ed., ASCE, New York, 258–267.

Bosela, P. A. (1993). "Failure of engineered facilities: Academia responds to the challenge." *J. Perform. Constr. Facil.*, 7(2), 140–144.

Carper, (2000). "Lessons from failures: Case studies as an integral component of the civil engineering curriculum." *Civil and structural engineering education in the 21st century*, Southampton, U.K.

Day, R. W. (1992). "Damage to two apartment buildings due to moisture variation of expansive soil." *J. Perform. Constr. Facil.*, 6(3), 169–176.

Day, R. W. (1993). "Performance of utility-trench shoring: Case study." *J. Perform. Constr. Facil.*, 7(1), 20–26.

Day, R. W. (1994a). "Performance of fill that contains organic matter." *J. Perform. Constr. Facil.*, 8(4), 264–273.

Day, R. W. (1994b). "Performance of slab-on-grade foundations on expansive soil." *J. Perform. Constr. Facil.*, 8(2), 128–138.

Day, R. W. (1995). "Case study of the settlement of gravelly sand backfill." *J. Perform. Constr. Facil.*, 9(3), 184–193.

Delatte, Jr., N. J. (1997). "Integrating failure case studies and engineering ethics in fundamental engineering mechanics courses." *J. Prof. Issues Eng. Educ. Pract.*, 123(3), 111–116.

Delatte, N. J. (2000). "Using failure case studies in civil engineering education." *Forensic engineering*, K. L. Rens, O. Rendon-Herrero, and P. A. Bosela, eds., ASCE, Reston, Va., 430–440.

Diaz, C. F., Hadipriono, F. C., and Pasternack, S. (1994). "Failures of residential building basements in Ohio." *J. Perform. Constr. Facil.*, 8(1), 65–80.

Dutson, A. J., Todd, R. H., Magleby, S. P., and Sorensen, C. D. (1997). "A Review of Literature on Teaching Engineering Design Through Project-Oriented Capstone Courses," *J. Eng. Educ.*, 86(1).

Feld, J., and Carper, K. (1997). *Construction failure*, 2nd Ed., Wiley, New York.

Gillum, J. D. (2000). "The engineer of record and design responsibility." *J. Perform. Constr. Facil.*, 14(2), 67–70.

Gnaedinger, J. P. (1987). "Open hearth slag—A problem waiting to happen." *J. Perform. Constr. Facil.*, 1(2), 78–83.

Goldstein, E. W. (1999). *Timber construction for architects and builders*, McGraw-Hill, New York.

Handy, R. L. (1995). *The day the house fell*, ASCE, New York.

Jennings, A., and Mackinnon, P. (2000). "Case for undergraduate study of disasters." *J. Perform. Constr. Facil.*, 14(1), 38–41.

Kaminetzky, D. (1991). *Design and construction failures: Lessons from forensic investigations*, McGraw-Hill, New York.

LaCome, M. L., Blankespoor, A., and Rens, K. L. (2000). "Concrete maturity: A valid nondestructive evaluation and forensic tool." *Proc., 2nd Forensic Congress*, ASCE, Reston, Va. 172–182.

Leonards, G. A., Frost, J. D., and Bray, J. D. (1994). "Collapse of geogrid-reinforced retaining structure." *J. Perform. Constr. Facil.*, 8(4), 274–292.

Levy, M., and Salvadori, M. (1992). *Why buildings fall down: How structures fail*, Norton, New York.

Lew, H., Fattel, S., Shaver, J., Reinhold, T., and Hunt, B. (1979). *Investigation of construction failure of reinforced concrete cooling tower at Willow Island, W.V.*, U.S. Dept. of Labor, OSHA/National Bureau of Standards, Washington, D.C.

Lew, H. S., et al. (1982). "Investigation of construction failure of Harbour Cay Condominium in Cocoa Beach, Florida." *Rep. S/N 003-003-0245-8*, U.S. Dept. of Commerce, National Bureau of Standards, Washington, D.C.

Lin, H.-M., and Hadipriono, F. C. (1990). "Problems in deep foundation construction in Taiwan." *J. Perform. Constr. Facil.*, 4(4), 259–270.

Luth, G. P. (2000). "Chronology and context of the Hyatt Regency collapse." *J. Perform. Constr. Facil.*, 14(2), 51–61.

Martin, R., and Delatte, N. (2000). "Another look at L'Ambiance Plaza collapse." *J. Perform. Constr. Facil.*, 14(4), 160–165.

Martin, R., and Delatte, N. (2001). "Another look at Hartford Civic Center Coliseum collapse." *J. Perform. Constr. Facil.*, 15(1), 31–36.

McKaig, T. (1962). *Building failures: Case studies in construction and design*, McGraw-Hill, New York.

Meehan, R. L., and Karp, L. B. (1994). "California housing damage related to expansive soils." *J. Perform. Constr. Facil.*, 8(2), 139–157.

Moncarz, P. D., and Taylor, R. K. (2000). "Engineering process failure—Hyatt Walkway collapse." *J. Perform. Constr. Facil.* 14(2), 46–50.

Morgenstern, J. (1997). "The fifty-nine story crisis." *J. Prof. Issues Eng. Educ. Pract.*, 123(1), 23–29.

Nicastro, D. H. (1996). "Annotated bibliography of forensic engineering." *J. Perform. Constr. Facil.*, 10(1), 2–4.

Padmanabhan, G., and Katti, D. (2002). "Using community-based projects in civil engineering capstone courses." *J. Prof. Issues Eng. Educ. Pract.*, 128(1), 12–18.

Petroski, H. (1985). *To engineer is human*, St. Martins, New York.

Pfatteicher, S. K. A. (2000). "The Hyatt horror: Failure and responsibility in American engineering." *J. Perform. Constr. Facil.*, 14(2), 62–66.

Pietroforte, R. (1998). "Civil engineering education through case studies of failures." *J. Perform. Constr. Facil.*, 12(2), 51–55.

Puri, S. P. S. (1997). "Computer misuse—Are we dealing with a time bomb? Who is to blame and what are we doing about it?" *Forensic engineering*, K. L. Rens, ed., ASCE, New York, 285–336.

Puri, S. P. S. (1998). "Discussion of 'Integrating Failure case studies and engineering ethics in fundamental engineering mechanics courses,' by N. J. Delatte." *J. Prof. Issues Eng. Educ. Pract.,* 124(4), 123–124.

Raghu, D., and Hsieh, H.-N. (1989). "Performance of some structures constructed on chromium ore fills." *J. Perform. Constr. Facil.,* 3(2), 113–120.

Rendon-Herrero, O. (1993a). "Including failure case studies in civil engineering courses." *J. Perform. Constr. Facil.,* 7(3), 181–185.

Rendon-Herrero, O. (1993b). "Too many failures: What can education do?" *J. Perform. Constr. Facil.,* 7(2), 133–139.

Rendon-Herrero, O. (1998). "Experience teaching civil engineering failures 1973–1997." *Structural Engineering World Congress,* San Francisco.

Rens, K. L., ed. (1997). *Forensic engineering,* ASCE, New York.

Rens, K. L., and Knott, A. W. (1997). "Teaching experiences: A graduate course in condition assessment and forensic engineering." *Forensic engineering,* K. L. Rens, ed., ASCE, New York, 178–185.

Rens, K. L., Clark, M. J., and Knott, A. W. (2000a). "A failure analysis case study information disseminator." *J. Perform. Constr. Facil.,* 4(3), 127–31.

Rens, K. L., Clark, M. J., and Knott, A. W. (2000b). "Development of an Internet failure information disseminator for professors." *Forensic Engineering,* K. L. Rens, O. Rendon-Herrero, and P. A. Bosela, eds., ASCE, Reston, Va.

Rens, K. L., Rendon-Herrero, O., and Bosela, P. A. eds. (2000c). *Forensic Engineering,* ASCE, Reston, Va.

Rens, K. L., Rendon-Herrero, O., and Clark, M. J. (2000d). "Failure of constructed facilities in the civil engineering curricula." *J. Perform. Constr. Facil.,* 4(1), 27–37.

Rens, K. L., Royston, H. J., and Lacome, M. L. (2000e). "Temporary bracing during construction (fact or fiction): Case studies." *Proc., Forensic Engineering: 2nd Congress,* San Juan, Puerto Rico 652–661.

Richardson, D. N., Stephenson, R. W., and Molloy, D. (1987). "Soil-shrinkage induced structural failure." *J. Perform. Constr. Facil.,* 1(4), 219–228.

Roddis, W. M. K. (1993). "Structural failures and engineering ethics." *J. Struct. Eng.,* 119(5), 1539–1555.

Ross, S. (1984), *Construction disasters: Design failures, causes, and prevention,* McGraw-Hill, New York.

Shepherd, R., and Frost, J. D. (1995). *Failures in civil engineering: Structural, foundation, and geoenvironmental case studies,* ASCE, New York.

Thornton, C. H., and DeScenza, R. P. (1997). "Construction collapse of precast concrete framing at Pittsburgh's Midfield Terminal," *Forensic engineering:* K. L. Rens, ed., ASCE, New York, 85–93.

Whitlock, A. R., and Mossa, S. S. (1996). "Foundation design considerations for construction on marshlands." *J. Perform. Constr. Facil.,* 10(1), 15–22.

Shallow Foundations Committee of the Geotechnical Division. (1995). *So your home is built on expansive soils,* W. K. Wray, chairman, ASCE, New York.

Zickel, L. L. (2000). "Failure vignettes for teachers." *Proc., 2nd Forensic engineering,* K. L. Rens, O. Rendon-Herrero, and P. A. Bosela, eds., ASCE, Reston, Va., 421–429.

American Society of
Civil Engineers
1852 – 2002
Building a Better World

150th Anniversary
Paper

ROLE OF ASCE IN THE ADVANCEMENT OF COMPUTING IN CIVIL ENGINEERING

By Steven J. Fenves,[1] Honorary Member, ASCE, and
William J. Rasdorf,[2] P.E., Fellow, ASCE

ABSTRACT: Computing has emerged as a major focus area in civil engineering, just as it has in other disciplines. This paper examines the role of the American Society of Civil Engineers (ASCE) in the advancement and development of this focus on computing in civil engineering. The paper documents the technical activities of ASCE that contributed to this evolution, particularly the committees of the Structural Division (now the Structural Engineering Institute) and the Technical Council on Computer Practices (now the Technical Council on Computing and Information Technology). Emphasis is placed on the initial activities and the current status of each group. A broad survey of ASCE activities that contribute to the dissemination of information on computing in civil engineering is presented. The role of ASCE publications in this effort is examined. The ASCE conferences and congresses on computing are documented and evaluated. Finally, observations are made about the society's overall impact on computing in civil engineering.

INTRODUCTION

Computers and computer-based tools have been an integral part of civil engineering for nearly 50 years. Much of today's civil engineering practice, particularly its reliance on rapid communication and interaction, analytical modeling and simulation, and geometric modeling and visualization, would be inconceivable without the aid of computing. A great deal of the computing underlying civil engineering practice and research has become so routine that it is simply not reported. Innovations, research results, and new developments based on computing capabilities tend to be reported in the conferences and journals of the appropriate subdiscipline within civil engineering most directly affected—structures, water resources, transportation, construction, and the like.

Paralleling and, to an extent, integrating the discipline-specified advances, there has emerged a broad, discipline-wide emphasis on the rapid evolution of computing, its supporting basic sciences, and its enabling information technologies. This emphasis is coupled with the exploration of potentials for improving civil engineering research, education, and practice by the exploitation of the emerging developments. In this paper, computing is used in its broadest sense to include computer-based communication; sharing and display of graphical, textual, and symbolic information; and calculation of numeric data. Computing thus defined affects every sphere of human endeavor. This paper specifically addresses computing in civil engineering.

The paper provides a compilation and review of a portion of the historical record of computing in civil engineering. This historic review is not meant to be complete and comprehensive. Rather, it is intended to sketch the evolution of those aspects of computing in civil engineering that have shaped today's roles and perceptions. The review is highly idiosyncratic, reflecting the writers' experiences, or, rather, their memory of those experiences, supplemented with material from the writers' files and contributions from colleagues. Furthermore, the review concentrates largely on the role of the American Society of Civil Engineers (ASCE) and its technical activities, conferences, and publications as a window on the evolution of the profession's practical, educational, and research activities in computing, rather than on the activities themselves.

The paper includes some new compilations of data on computing in civil engineering. It also summarizes and refers to other resources where more complete compilations of information already exist. Thus, it is intended to serve as a starting point for understanding the evolution of computing in civil engineering.

The activities and events described were the result of the work of many ASCE members. The writers wish that they had the space to acknowledge the contributions of all of these dedicated volunteers, but due to space limitations, they had to restrict themselves to recognizing only the most salient participants. The writers apologize in advance to those whose work could not be individually recognized.

ORGANIZATION OF PAPER

This section provides a brief introduction to the organization of the paper. The "Early History of Computing in Civil Engineering" section discusses the first steps in the introduction of computing into the civil engineering profession. The section describes the earliest civil engineering computing applications and identifies early difficulties and inhibitions to computing. The emergence and subsequent disappearance of user groups are described.

ASCE, the leading professional society of civil engineers, has played a significant role in the evolution of computing in

[1]PhD, NIST, 100 Bureau Dr., Build. 304, Stop 8262, Gaithersburg, MD 20899-8262.

[2]PhD, Prof., North Carolina State Univ., Civ. Engrg. Dept., Box 7908, Raleigh, NC 27695-7908.

Note. Discussion open until March 1, 2002. To extend the closing date one month, a written request must be filed with the ASCE Manager of Journals. The manuscript for this paper was submitted for review and possible publication on October 19, 2000; revised February 20, 2001. This paper is part of the *Journal of Computing in Civil Engineering*, Vol. 15, No. 4, October, 2001. ©ASCE, ISSN 0887-3801/01/0004-0239–0247/$8.00 + $.50 per page. Paper No. 22170.

civil engineering. This section examines what that role has been and what it is today. It identifies those ASCE activities that have had the most significant role in shaping computing and its use in civil engineering.

The ASCE *Journal of Technical Topics* (*JTT*) was one of the earlier journals to focus on computing in civil engineering. It was followed by the ASCE *Journal of Computing in Civil Engineering* (*JCCE*). The "Publications" section examines what has been published in these and other related journals. Some trends are identified and the more recent shifts in computing in civil engineering directions are identified. Other society publications are examined as well. The "Summary and Observations" section provides a synthesis of ideas and frames a window on the profession's use of computing and ASCE's overall impact on it.

EARLY HISTORY OF COMPUTING IN CIVIL ENGINEERING

Civil engineers began to use computers quite early. The two major impetuses for the introduction of computing in the 1950s and 1960s were, first, the defense and space programs and, second, the interstate highway system. Defense and space efforts required a whole new range of structural response modeling capabilities, while the vast interstate construction program required extensive design calculations at an enormous scale. Without a doubt, the single most successful civil engineering application of the early decades was the earthwork quantity ("cut-and-fill") calculation program, churning out daily earthwork volumes for miles and miles of highways without having to draw and measure by planimeter the cross sections at every station. Optimization, in the sense of balanced cut and fill, became a practical reality by rerunning the calculations with changed grade elevations. The program further increased its usefulness, thus establishing the principle of capturing data at the source, when it began to be used not just for design ("estimated quantities"), but also for computing payments to contractors ("final quantities") using the original ground data in conjunction with the final survey results.

The first computers acquired by engineering organizations, whether public agencies or private consulting and construction firms, came with some rudimentary programming tools—either native machine or assembly languages or not much more effective interpreted systems—and essentially no software. Thus, civil engineering organizations faced the daunting task of programming all the applications they needed.

Not surprisingly, user groups were quickly formed in order to share development efforts as well as to provide a united voice to the hardware vendors, particularly in clamoring for better software development tools. User groups tended to cluster around particular brands of computers, such as the Bendix G-15, the LGP30, and the IBM 650 and 1620 (all long ago consigned to their well-deserved trash piles). User groups sharing application interests began exchanging programs, tentatively and haphazardly at first, becoming better organized as time progressed, with coordinated planning, explicit development responsibilities, and elaborate accounting schemes for credits earned by submitting programs.

The IBM 1130, introduced around 1960 with the first practical FORTRAN compiler for small computers, became popular with civil engineering firms, and eventually the 1130 Users' Group, renamed Civil Engineering Program Applications (CEPA) (formed in 1965; the acronym was retained after the group was incorporated as the Society for Computer Applications in Engineering, Planning, and Architecture), became the preeminent civil engineering user organization (Civil 1966). In response to CEPA demands for a structural analysis tool, IBM provided a version of STRESS (Fenves et al. 1964, 1965) for the 1130. A second user group, the Highway Engi-

neering Exchange Program, attracted the state transportation departments as well as their consultants. Many firms belonged to both groups. Somewhat later, the Integrated Civil Engineering System (ICES) (Roos 1966, 1967) User Group (IUG) became another user organization, notable in that it was concerned not with program development, but with the maintenance and use of ICES and its constituent programs, particularly STRUDL (Logcher et al. 1965), an outgrowth of STRESS, and COGO (Miller 1961; Roos and Miller 1964).

Eventually, professional issues related to the cost of computer program development, changing policies for computer use, and legal responsibilities began to dominate discussions at user group meetings. CEPA and similar groups felt that they needed representation at a higher professional level than that provided by their separate organizations. As discussed below, CEPA members were subsequently influential in directing ASCE into activities dealing with professional practice.

As computing in civil engineering practice increasingly moved away from in-house software development toward acquired third-party software, the role of user groups diminished substantially. In retrospect, it is difficult to determine how effective the actual program exchange function of these organizations was. Essentially all programs acquired from external sources had to be modified to some degree to reflect local practices. Suites of interrelated programs developed by a number of organizations did not mesh as well as planned. The primary intangible benefit of membership in the user groups turned out to be the networking the groups provided, binding together a collection of civil engineering computing professionals who were aware of the pioneering work they were doing.

ASCE

The divisions within ASCE that have been the most active in computing in civil engineering are the Structural Division and the Technical Council on Computing and Information Technology (TCCIT). These are discussed in detail in the following sections. Additionally, the current activities of all institutes, divisions, and councils active in computing are discussed.

Structural Division

Initial Committee Activities

The Structural Division's Committee on Electronic Computation was formed in the fall of 1957. Its first chairman was Dr. Nathan M. Newmark, head of the Department of Civil Engineering at the University of Illinois, a pioneering researcher and educator in structural engineering. The contact member from the Executive Committee of the division was Dr. Elmer Timby, a distinguished educator and consultant. It undoubtedly required the prestige of these two gentlemen to convince the division to establish a committee devoted to a device that was then still considered largely a laboratory curiosity. In the initial round of correspondence about potential committee activities, one member suggested the task of "keeping a record of civil engineers in this country who are actively doing work in this field of electronic computation." Had such a record been implemented, it would have contained a few hundred entries, at most. Fenves, then a graduate student and instructor at the University of Illinois, was made secretary of the committee.

The initial subcommittees established by the Committee on Electronic Computation were as follows:

- Programming and Coding
- Mathematical Methods

TABLE 1. Conferences on Electronic Computation and Analysis and Computation

Number	Date	Location	Document	Editor(s)	Library number	International Standard Book Number
1	11/20–11/21/58	Kansas City, Mo.	—	—	—	—
2	9/8–9/9/60	Pittsburgh	—	—	—	—
3	6/19–6/21/63	Boulder, Colo.	(89)ST4,8/63	—	—	—
4	9/7–9/9/66	Los Angeles	(92)ST6,12/66	—	—	—
5	8/31–9/2/70	Lafayette, La.	(97)ST1,1/71	—	—	—
6	8/7–8/9/74	Atlanta	(101)ST4,4/74	—	—	—
7	8/6–8/8/79	St. Louis	—	—	—	—
8	2/21–2/23/83	Houston	—	James K. Nelson Jr.	59-65010	0-87262-351-3
9	2/23–2/26/86	Birmingham, Ala.	—	Kenneth M. Will	85-73831	0-87262-512-5
10	4/29–5/1/91	Indianapolis	—	Oktay Ural and Ton-Lo Wang	91-12684	0-87262-802-7
11	4/24–4/28/94	Atlanta	—	Franklin Y. Cheng	94-7104	0-87262-974-0
12	4/15–4/18/96	Chicago	—	Franklin Y. Cheng	96-11789	0-7844-0163-2
13	7/19–7/23/98	San Francisco	—	N. K. Srivastava	—	0-08-042845-2
14	2000	Philadelphia	—	—	—	—

Note: Dashes indicated unavailable information. All entries in the Document column represent conference proceedings except for four, which were published in the journals noted. Library number refers to the Library of Congress card catalog number. The first 10 conferences are referred to as Conferences on Electric Computation. Presently, and including 11, 12, 13, and 14, the conferences are referred to as Conferences on Analysis and Computation.

- Data Processing
- Current Progress and Planning
- Publications and Technical Programs

The first subcommittee eventually produced a series of manuals (with the first manual starting with machine language programming, of course, before proceeding to index registers), and the second, a number of annotated bibliographies, comparative book reviews, and program feature comparisons. In very short order, however, the Publications and Technical Programs Subcommittee became the primary focus of the committee, starting with the organization of the first session on computing at the February 1958 ASCE convention in Chicago. Dr. Elvind Hognestad, the Technical Program Committee chairman for the convention, asked the committee to provide "a broad, simple, and informative subject presentation designed to be a 'first meeting' of the civil engineers of the audience with a new tool, the digital computer, which has the future potentiality of becoming as important to him as the slide rule." Fenves recalls the subcommittee rounding up the first three speakers, all Doctors of Philosophy (PhDs), and then working hard to find a fourth speaker who did not have an advanced degree, reflecting the subcommittee's strong belief that eventually it would not be necessary to have a PhD to be able to write a civil engineering application program.

Starting with Kansas City, Mo. in 1958, the committee organized a series of Conferences on Electronic Computation that have continued to this day (Table 1). Papers presented at the early conferences dealt with a great variety of analysis and design methods and programs; others addressed important issues of the day, such as analyzing very large structures on computers with 1–2 kilobyte words of memory and setting up early users' groups. Many other papers had impacts that are still felt today—the first matrix and finite-element analysis papers by Ray Clough (Clough 1958, 1960), the first consistent mass matrix paper by John Archer (Archer 1963), and others. The successive conference technical program committees exercised stringent quality control on the papers accepted. As an example, following C. L. Miller's paper on "Man-Machine Communication in Civil Engineering" at the Third Conference on Electronic Computation (Miller 1963) (apologies for the gender insensitivity of the time), papers describing programs that depended on users providing numeric codes for selecting program options were summarily rejected.

As time progressed, the committee attempted to address a range of issues broader than structural analysis and design.

A subcommittee on professional problems was formed in 1967, and one on interdisciplinary activities was formed in 1970. Such activities were reviewed as exceeding the scope of the Structural Division within ASCE, and some had to be curtailed or eliminated at the request of the division's Executive Committee. As a result, some committee members began to look for alternate means of action within the society's framework.

In retrospect, from the mid-1950s to the mid-1970s, the Structural Division's Committee on Electronic Computation, chaired in succession by Nathan M. Newmark, Sidney Shore, Robert E. Fulton, and Donald McDonald, reflected quite accurately the developments of computing as they related to the technical aspects of structural engineering. Its conference proceedings and other publications showed the profession's response to the rapid, if not revolutionary, changes in support technologies such as programming languages, hardware and communication technologies, new application areas such as computer-aided design and computer graphics, and in research leading to new modeling and analysis capabilities. The committee's position within the ASCE organizational structure made it unsuited to address professional issues transcending structural engineering and affecting a larger proportion of the society's members.

Current Committee Activities

The Structural Division has since evolved into the Structural Engineering Institute. The Committee on Electronic Computation has continued to function to this day. Following the 10th Conference on Electronic Computation in 1991, the committee was renamed the Committee on Analysis and Computation in 1992. Its activities are summarized in a later section.

Over a span of 33 years, from 1958 to 1991, the Structural Division sponsored 10 Conferences on Electronic Computation (Table 1), averaging one approximately every three years. The third, fourth, fifth, and sixth conferences were republished as special issues of the *Journal of the Structural Division*. The 11th–14th Conferences on Analysis and Computation (sponsored by the Structural Engineering Institute) have been offered on a two-year cycle. The primary focus of these conferences is on computing in structural engineering.

Other Divisions

Initial Committee Activities

Progressions very similar to that of the Structural Division took place in other divisions. Specifically, by the late 1960s the Geotechnical and Hydraulics Divisions both had technical committees similar in scope to the Committee on Electronic Computation of the Structural Division, although with less aggressive schedules of publications and specialty conferences. Contact among these groups across divisions was either through personal contacts or through organizations external to ASCE.

Current Committee Activities

Today, ASCE members are involved in computing through committee activities in a number of divisions. The following is a list of the ASCE committees specifically involved in computing (ASCE 2000), and their purposes.

1. Urban Transport Division, Computing in Transportation Committee. Its purpose is to facilitate the dissemination of information on new developments in computing for planning, design, operations, and management of urban transportation systems. The committee will encourage education and research necessary to advance computing applications in academia and practice.
2. Geo Institute, Computer Applications and Numerical Methods Committee. Its purpose is to serve the geo-profession areas of general computing use, including (1) software availability; (2) information technology; (3) numerical methods; and (4) general computing issues in professional practice. The committee provides support in these areas, both in leading roles and in cooperation with other technical committees and organizations.
3. Construction Division, Computing in Construction Committee. Its purpose is to simulate and influence the effective application of computer technology in construction.
4. Geomatics Division, Geographic Information Systems Committee. Its purpose is to serve as a resource to the profession in providing recommended guidelines for the development and selection of processes, procedures, techniques, and technologies associated with the collection, management, dissemination, and utilization of spatially related data; to identify and formulate the responsibility and contribution of civil engineering to integrated geographic information systems; to foster studies, research, and development; to develop and maintain appropriate technical standards and procedures; and to provide a focal point for interdivisional and interdisciplinary cooperation.
5. Structural Engineering Institute, Analysis and Computation Committee. Its purpose is to report, encourage the development of, and evaluate innovative methods of structural analysis; to foster the use of digital computer and other modern computing devices to obtain more effective analyses and improved designs; to encourage the development and reporting of innovations in the use of computers that may have practical significance; and to report on professional problems involving the multidisciplinary use of computers.
6. Structural Engineering Institute, Emerging Computing Technology Committee. Its purpose is to provide a mechanism by which various emerging computing technologies that may be applicable to some part of the structural engineering process can be identified and brought to the attention of the structural engineering community. These technologies include parallel and distributed computing, databases and information systems, Web-based and collaboration technologies, artificial intelligence, field based computing, and automation.

Research Council on Computer Practices

In 1970, a group of CEPA members who were also members of ASCE petitioned the society "to appoint a coordinating council to deal with the legal, professional, and educational aspects of computer use, cutting across the technical divisions of ASCE." The initial petition brought the CEPA group in contact with like-minded people in the Structural and Geotechnical Divisions, and eventually led to the establishment of the Technical Council on Computer Practices (TCCP) via the short-lived Research Council on Computer Practices.

At the time the CEPA petition was made, technical councils were viewed within ASCE as proving grounds for fledgling divisions. The society had never before received a petition for a council that clearly stated that the proposed council had no intention of every becoming a division. As a result, the CEPA petition was denied. However, since research was one potential activity of such an eventual council, and since the Committee on Research was willing to act as a sponsor, the petition was changed to that for a Research Council on Computer Practices. The Task Committee on Computer Application Research of the Committee on Research was charged with organizing the research council and clarifying its relationship to an eventual council with broader goals. At its first meeting, the task committee resolved that "this group intends to be the focal point in studying and pursuing the establishment of a Computer Practices Council."

The Research Council on Computer Practices was authorized in 1972, with Elias C. Tonias, a consulting engineer—later a software developer—and CEPA member, as chair and Fenves as vice-chair. The Research Council proceeded to define the scope of the council-to-be in the following areas:

- Coordination within ASCE among the computing-related activities of the divisions
- Coordination outside ASCE with professional and users groups
- Education
- Professional practice
- Research

The aspects of professional practice included

- Computer pricing policies
- Computer facility organization and management
- Credibility and legal aspects of computer-aided design
- Software development costs and recovery procedures
- Job description and salary structure
- Legal and professional aspects of program sharing

This latter list makes clear the ambitious scope of professional practice issues that awaited the attention of the proposed council-to-be, quite distinct from the technical issues addressed by committees of the technical divisions. The Research Council on Computer Practices was abolished in 1973 as soon as the Technical Council on Computer Practices was authorized.

Technical Council on Computer Practices

Initial Committee Activities

As stated above, the first petition for a Technical Council on Computer Practices was denied because it was too broad in scope and cut across the traditional boundaries of the technical divisions. Coincidentally, while the second petition was being circulated in draft form, the ASCE Technical Activities Council (TAC), which oversees the divisions, was performing its own major restructuring study. The study group, chaired by Dr. Albert D. M. Lewis, professor of structural engineering at

Purdue University, and considering a "matrix management" structure, with divisions representing the "depth" dimension and some other organizational form representing the "breadth" dimension. The draft petition was embraced by the restructuring study group, which recommended "technical council" as the designation of the groups dealing with the breadth dimension.

The Technical Council of Computer Practices was authorized on January 30, 1973, as the first council under the new TAC organization. Subsequently, four other technical councils were formed—Codes and Standards, Lifeline Earthquake Engineering, Research, and Cold Regions. TCCP's objectives were "to establish the means by which the civil engineering profession will be able to properly utilize the impact of the electronic computer and its related software in civil engineering practice, research, and education." The first 12 TCCP chairs were Robert L. Schiffman, Elias C. Tonias, Robert C. Y. Young, Augustine J. Fredrich, Norman R. Grieve, Steven J. Fenves, Vincent J. Vitagliano, Charles V. Smith, J. Crozier Brown, Charles Hodge, Morton B. Lipetz, and Richard L. Bland.

Five sets of early activities of TCCP are worth singling out. First, the Computer Practices Committee (CPC) produced an ASCE manual on computer pricing (ASCE 1973; Grieve 1976) and a number of reports on professional issues. Building on activities started in 1971, CPC cooperated with CEPA and later with IUG on the following three projects:

- Sponsoring a National Science Foundation (NSF)-funded Special Workshop on Engineering Software Coordination (1971) (Schiffman 1972a,b,c)
- Conducting an NSF-funded study entitled "Definition of a National Effort to Promote Effective Application of Computer Software in the Practice of Civil Engineering and Building Construction" (1973–76) (Civil 1975a,b; McGrory et al. 1975)
- Incorporating the National Institute for Computers in Engineering "to provide an information service which will assist in promoting the effective use of computers and software as tools of the practicing engineer" (1980–82) (Beck (1981)

The latter effort, important as it was for its time, came to naught with the advent of the personal computer and the subsequent blossoming of a competitive free-enterprise market in engineering software.

Second, starting with the First Conference on Computing in Civil Engineering at Atlanta in June 1978 (organized by Lipetz, Vitagliano, and Leroy Z. Emkin), the Committee on Coordination Outside ASCE (COCOA) became the organizer of national conferences and later congresses on computing in civil engineering. The first three conferences were organized in conjunction and cooperation with CEPA and the ICES User Group.

Third, COCOA organized the first international conference, while the following three were organized by other groups. Eventually, COCOA took over the organization of the fifth international conference and became instrumental in forming an organization that would oversee subsequent international conferences. That organization, the International Society for Computing in Civil and Structural Engineering (ISCCSE), came into being in 1993 and was formally ratified in 1995. All international conferences since that date have been conducted with the auspices of ISCCSE.

Over a span of 22 years, from 1978 to 2000, a total of 21 national and international conferences and congresses were sponsored. These conferences covered the entire range of computing topics throughout civil engineering. At a number of conferences, entire tracks of sessions were devoted to special topics and were separately identified as symposia. Admittedly, at later conferences and congresses, the coverage of the computing practices track has often not been as extensive as that of the academically oriented research track, and the two tracks have not always been well integrated. Nevertheless, these conferences and congresses have become one of the preeminent media for the exchange of information on the rapidly evolving aspects of computing in civil engineering. Details regarding the national and international conferences and congresses are presented in Tables 2–4.

Fourth, the Publications Committee undertook an aggressive publication schedule from the start, initially publishing in the *Journal of the Technical Councils*, then in the *Journal of Technical Topics*, and eventually in the *Journal of Computing in*

TABLE 2. National Conferences on Computing in Civil Engineering

Number	Date	Location	Editor(s)	Library number	International Standard Book Number
1	6/26–6/29/78	Atlanta	Morton B. Lipetz and Leroy Z. Emkin	—	—
2	6/9–6/13/80	Baltimore	David R. Schelling	80-66141	0-87262-246-0
3	4/2–4/6/84	San Diego	Charles S. Hodge	80-66141	0-87262-396-3
4	10/27–10/31/86	Boston	W. Tracy Lenocker	86-25911	0-87262-569-9
5	3/29–3/31/88	Alexandria, Va.	Kenneth M. Will	88-3771	0-87262-635-0
6	9/11–9/13/89	Atlanta	Thomas O. Barnwell Jr.	89-37697	0-87262-722-5
7	5/6–5/8/91	Washington, D.C.	Louis F. Cohn and William J. Rasdorf	91-12813	0-87262-803-5
8	6/7–6/9/92	Dallas	Barry J. Goodno and Jeff R. Wright	92-13050	0-87262-869-8

Note: Dashes indicate unavailable information. Library number refers to the Library of Congress card catalog number.

TABLE 3. National Congresses on Computing in Civil Engineering

Number	Date	Location	Editor(s)	Library number	International Standard Book Number
1	6/20–6/22/94	Washington, D.C.	Khalil Khozemich	94-19972	0-7844-0026-1
2	6/5–6/8/95	Atlanta	J. P. Mohsen	95-15012	0-7844-0088-1
3	6/17–6/19/96	Anaheim, Calif.	Jorge Venegas and Paul Chinowsky	96-19480	0-7844-0182-9
4	6/16–6/18/97	Philadelphia	Teresa M. Adams	—	0-784-0250-7
5	10/18–10/21/98	Boston	Kevin C. P. Wang	—	0-784-0381-3

Note: Dashes indicate unavailable information. Library number refers to the Library of Congress card catalog number.

TABLE 4. International Conferences on Computing in Civil Engineering

Number	Date	Location	Country	Library number	International Standard Book Number
1	5/12–5/14/81	New York	United States	81-66346	0-87262-270-3
2	1985	Hangzhou, Zhejiang	China	4950-61[a]	0-444-99560-9
3	8/10–8/12/88	Vancouver	Canada	—	—
4	1991	Tokyo	Japan	—	—
5	6/1–6/9/93	Anaheim, Calif.	United States	93-17524	0-87262-915-5
6	1995	Berlin	Germany	—	90 5410 556 9
7	1997	Seoul	Korea	—	—
8	8/14–8/16/00	Stanford, Calif.	United States	—	0-7844-0513-1

Note: Dashes indicate unavailable information. Library number refers to the Library of Congress card catalog number.
[a]Science Press book number.

Civil Engineering. These activities are detailed in forthcoming sections. Fifth, the State of Computer Practices (SOCP) Committee responded to the perceived lack of information about practice-oriented information on computing available to ASCE members by launching the *CE Computing Review* newsletter, described in a later section.

Current Committee Activities

In 1997, the TCCP officially became the TCCIT. Its committees, and their purpose, as stated in the official ASCE register, are as follows (ASCE 2000).

1. Education—to study and promote educational uses of computers in the field of civil engineering; to promote the exchange of information regarding computer methods; to recommend those computer-related activities that promote educational objectives of colleges and universities
2. Publications—to administer the solicitation, review, and editing of manuscripts submitted for publication in the *JCCE* (The publications section discusses journals in greater detail.)
3. Imaging Technologies—to coordinate and promote the exchange of technical information about current and emerging imaging technologies and their application to civil engineering problems and projects
4. Intelligent Computing—to gather, maintain, and disseminate information on the application of expert systems and artificial intelligence to civil engineering, and to keep the society membership aware of developments in this rapidly growing field
5. State of Computer Practices—to gather, maintain, and disseminate information on existing and new developments in computer technology as it relates to civil engineering, and to promote an understanding within the engineering community of the benefits and limitations of that technology
6. Database and Information Management—to increase the appropriate use of databases and information management technologies in civil engineering teaching and practice (Included in the committee scope are the representation, management, storage, and retrieval of civil engineering information, product and process modeling, data/object/knowledge repositories and interoperability standards, and information infrastructure issues.)
7. Coordination of Computing Activities—to promote computer-related activities through coordination with groups internal and external to ASCE, including the presentation of specialty conferences and joint meetings with other organizations

Activities undertaken by these committees vary widely, and include committee meetings, international meetings, and inter-organizational meetings; surveys; workshops; special issues and sections of the journal; conferences and conference sessions, tracks, and panels; as well as special publications and manuals.

Education Surveys

Adequate computing resources, expertise in the teaching of computing, and computing requirements in the curriculum have been engineering education concerns of both the academic community and the professional community. To a large extent, the first two of these have been ameliorated over time. The third, however, still presents a dilemma to engineering education. To better understand these concerns and the steps to be taken to overcome them, the TCCP undertook a series of computing surveys in 1986, 1989, and 1995. Consideration is currently being given to a follow-up survey in 2001.

In 1986, the Education Committee of the TCCP conducted a survey to determine the availability of computing resources in civil engineering departments and to determine the attitude of faculty (and professionals) toward education in computing in civil engineering. The following specific areas that needed exposure were identified by the survey:

1. Technology of computers
2. Computers as problem-solving tools
3. Computers as engineering simulators to assist in design

In 1987, TCCP formed a task committee to conduct a second survey to assess the status of computing in the civil engineering curriculum and to determine how to address these key areas. The task committee conducted the survey, analyzed and presented the results, suggested three different scenarios for incorporating problem-solving computing tools and concepts into the curriculum, and enumerated lists of pros and cons for each scenario. The results of the work were published in Law et al. (1989, 1990a,b), and were distributed widely to engineering faculty. Today they serve as a benchmark from which we can measure progress, reassess needs, and plan for the continued future roles of ASCE in computing in civil engineering (O'Neill et al. 1996a,b).

In 1995, the TCCP Education Committee again conducted a survey, aimed at both educators and practitioners, to determine their perspectives on the then-current role of computing in civil engineering. Because of the steady improvements and advances in computer technology, the committee made the case that surveys must be repeated on a regular basis to ensure that the computing needs of the profession are being properly defined and met, and that curriculum changes can be made when appropriate (O'Neill et al. 1996a,b).

The results of the 1995 practitioners' survey were surprisingly close to those of the 1989 survey, with the seven highest ranked items from 1989 (computer-aided design and drafting, spreadsheets, databases, programming, graphics, word processing, and expert systems) appearing in the list of nine ranked items in 1995. The new survey also identified the growing importance of the Internet by including communications in its top nine items, whereas this item was ranked 11th in 1989 (O'Neill et al. 1996b). The results of the 1995 educators' survey were also close to those of the 1989 survey, with six of the eight highest ranked items from 1989 (spreadsheets, word processing, computer-aided design, programming, databases, and expert systems) appearing in the list of nine ranked items in 1995. There is also close agreement between the practitioner and academic surveys, indicating broad agreement on computing skill needs.

However, the 1995 survey clearly indicated that academicians had still not reached a consensus on what to teach with respect to a programming language course (O'Neill et al. 1996a). It was also observed that there was no consensus definition for computer fluency, a constantly moving target. Finally, the results of the survey indicated a strong appreciation for advances in information technology (communications and the Internet), as well as for the use of computing to assist students in communicating their ideas (equation solvers and presentation software) (O'Neill et al. 1996a).

PUBLICATIONS

A number of ASCE publications have contributed to the advancement of computing in civil engineering. These include its archival journals, special journal issues, conference proceedings, special publications, and textbooks. Some of its conventional publications, such as *ASCE Magazine* and *ASCE News*, publish feature articles on computing in civil engineering. These and other publications are described in this section.

Journal of the Technical Councils (JTC) and Journal of Technical Topics

When the Technical Council on Computer Practices was established in 1973, it had no immediate outlet for information dissemination through publications. For the 14 years between 1973 and 1987, the Publications Committee of the TCCP worked first through the *JTC* and then through the *JTT* to publish research findings and results.

In 1977, ASCE first published the *JCT* for the purpose of presenting the research activities of the Technical Councils. The *JTC* was published aperiodically from 1977 through 1982 (volume 103, number 1, December 1977 through volume 108, number 2, November 1982).

In 1983, the *JTC* of ASCE became the *JTT in Civil Engineering*. It was then published aperiodically through December of 1985 (volume 109, number 1, April 1983 through volume 111, number 1, December 1985). At that time, ASCE split the content of the *JTT*, establishing individual journals for each of the councils. Thus, the first issue of the *JCCE* appeared in 1987 (volume 1, number 1, January 1987), establishing the one primary archival publication of the TCCP.

Journal of Computing in Civil Engineering

The goal of the *JCCE* is to serve as an archival resource on computing in civil engineering for the professional and academic communities alike. It is published four times per year. The journal is abstracted in *ASCE Publications Information*, *Transactions of ASCE*, and in the Civil Engineering Database (online). It is indexed in the *Science Citation Index*. The journal covers new programming paradigms, information management systems, computer-aided engineering, intelligent computing, robotics and automation, and implementation strategies, as well as organizational impacts of and for computing resources (Rasdorf 1997).

Five-Year Review

All journals of ASCE are periodically evaluated for quality assurance. The *JCCE* underwent such a five-year review for the period of 1990–94 (Lakmazaheri and Rasdorf 1995, 1996). Although the review contained some of the elements of "what the journals contain," its primary focus was on "what people think of the journal." Insights into this question were gained through surveys of reviewers and authors, readers, and previous subscribers.

One of the findings of the five-year review was that the quality and scope of the journal were satisfactory. Another key finding was that the majority of those responding to the survey indicated that the *JCCE* did not influence their professional practice; these readers viewed the papers as too theoretical and suggested that the balance between research and practice was unsatisfactory. On the other hand, the needs of the academic community for a society publishing outlet in computing in civil engineering that would provide for a forum of information exchange on research and education topics, act as a resource and graduate teaching and research, and act as a filter to evaluate the degree of success, applicability, or utility of new computer science concepts and methodologies applied to civil engineering were perceived as being successfully met.

10-Year Analysis

A 1997 study provided a window on computing in civil engineering through the eyes of the *JCCE* (Lakmazaheri and Rasdorf 1997, 1998). That study conducted a detailed analysis of the journal over the 10-year period from 1987 to 1996, the first decade of publication.

The analysis was made to obtain a general perspective on the contributions and technical content of the journal. Its purpose was to help the computing in civil engineering community gain a better understanding of the nature and dynamics of (1) the research community and research areas and topics; and (2) the literature forming the knowledge infrastructure of the published work. The study tabulated authors and their contribution patterns; computing topics and civil engineering subdisciplines covered; and references cited, categorized (journals, conferences, reports, and theses), and their sources tabulated.

This study resulted in a valuable compilation of information. The study did not draw conclusions, but it provided a wealth of information regarding the publication of knowledge about computing in civil engineering. The report highlighted the significant role of ASCE in the advancement of computing in civil engineering through the *JCCE*.

CE Computing Review

By 1989, the TCCIT's SOCP Committee was increasingly concerned about how to get information about practice-related computing in civil engineering to ASCE's membership. At about the same time, the society's Publications Department was pursuing several options for creating new breeds of periodicals, including discipline-specific newsletters. SOCP agreed that a newsletter would fit its goals, and in June of 1989, the first monthly issue of *CE Computing Review* was published. *CE Computing Review* was edited by a professional editor employed by the society's Publications Department. SOCP provided the majority of the newsletter's editorial board. Articles were submitted to the editor and, although often reviewed by the editorial board, the final decision on publication was made by the editor.

CE Computing Review's content explored several avenues, with emphasis on computer use by practicing civil engineers. There were several reviews of popular civil engineering software products written by engineers who used the products. The newsletter also provided basic personal computer user topics; at the time, this was the only place that many ASCE members were exposed to such information. During the life of *CE Computing Review*, SOCP continued to provide technical assistance and aid to the editors in creating a popular periodical. Many of the articles written for the newsletter were authored by SOCP committee members. Two SOCP members, Tracy W. Lenocker and Charles S. Hodge, authored monthly columns for several years for the newsletter.

During its first three years, *CE Computing Review* was financially successful, in that its subscription fees offset the cost of the editor, publishing, and mailing. In the following two years, the newsletter did not break even and became a financial burden on ASCE. Its last issue was published in June of 1994.

Other ASCE Publications

ASCE journals have sponsored a number of special issues or sections directly or indirectly focused on computing in civil engineering. Table 5 provides a compilation of these, and identifies the theme, subject, or title of the special issue, as well as the journal sponsoring it. These special issues and sections provide insight into computing areas deemed to be timely and of interest to the readers of various civil engineering subdisciplines. There is emphasis on topics broadly grouped into artificial intelligence [expert systems, robotics, neural networks, and even intelligent vehicle highway systems (IVHS)] and information technology (databases, object-oriented systems, geographic information systems, and data, product, and process modeling).

Civil Engineering, ASCE's monthly magazine, devotes a full issue of feature articles to computing each year, usually in June. Individual articles in other issues, such as that by Dewberry (2000), focus on the application of computing or new automation technologies in the practice of civil engineering. Each issue of the magazine features a Software Reviews section and a Computing section that focuses on the Web, software, and hardware. As a service to ASCE members, each issue also contains a significant amount of advertisements related to computing in civil engineering.

ASCE sponsors a Web page. This site assists civil engineers by providing information of interest to the profession. Additionally, the site supports activities of the Publications Division through a database of publications information.

The Civil Engineering DataBase, a comprehensive electronic index that presently contains all articles published in all

29 ASCE journals from the present back to 1973, provides a variety of indexes and information access formats. Conference proceedings papers, ASCE Press books, standards, committee reports, manuals of practice, *Civil Engineering*, and *ASCE News* have also been added to the database. Finally, an active forward link to the actual text of the article has been added for all on-line journals.

Beginning in January of 1999, full text versions of all ASCE journals appeared on-line, providing direct access to the text of the articles. Furthermore, the ASCE references cited in an article point to their citation in the Civil Engineering DataBase. Thus, each article points to all of its ASCE references, and any ASCE journal reference (since January 1995) points to the actual text of the article it cites. ASCE also offers a Personal Journal, whereby a reader subscribes to a fixed number of articles per year, rather than to one or more entire journals, and may search the entire journals database and extract or download up to the specified number of articles from any of the journals.

On the horizon is the "virtual journal." The virtual journal focuses on a timely topic of interest in a specialized area. An ASCE editor selects articles from all 29 journals that relate to the topic and synthesizes them into a virtual journal, which points to each article in its home journal.

SUMMARY AND OBSERVATIONS

In the first two decades of the "computer era," emphasis on computing in civil engineering was dispersed between organizations outside ASCE, notably computer user groups, and committees within ASCE divisions that addressed computing as a component of the subdisciplines within each division's scope. Of these committees, the Committee on Electronic Computation of the Structural Division had the most ambitious program of publications and conferences. However, as modern methods of structural analysis became more general, much of the literature in the field moved out of ASCE publications into journals with a broader scope. Thus, ASCE publications did not, and could not, become the premier archival repository on, say, the finite-element method. Furthermore, the segmentation of ASCE into divisions precluded any organizational entity that could address computing issues across the breadth of the profession.

At the end of this formative period, the Technical Council on Computer Practices emerged in response to the need for addressing within ASCE a range of broad professional issues arising out of the increased use of computing. The TCCP also

TABLE 5. Journal Special Issues and Sections

Journal	Volume (issue)	Date	Category	Subject
JCEM	119(2)	June 1993	Issue	"Applications of Microcomputers and Workstations in Construction"
JTE	113(2)	March 1987	Section	"Use of Expert Systems in Transportation Engineering"
JTE	113(4)	July 1987	Issue	"Microcomputers in Transportation"
JTE	116(3)	May/June 1990	Section	"Robotics and Automatic Imaging in Transportation Engineering"
JTE	116(4)	July/August 1990	Section	"Intelligent Vehicle/Highway Systems"
JCCE	1(4)	October 1987	Issue	"Expert Systems in Civil Engineering"
JCCE	4(3)	July 1990	Section	"Computational Geometry"
JCCE	5(1)	January 1991	Section	"Expert Systems in Planning and Design"
JCCE	6(1)	January 1992	Issue	"Databases"
JCCE	6(3)	July 1992	Issue	"Object Oriented Systems"
JCCE	7(3)	July 1993	Issue	"Geographic Information Analysis"
JCCE	8(2)	April 1994	Issue	"Neural Networks"
JCCE	8(4)	October 1994	Issue	"European Computing"
JCCE	10(3)	July 1996	Section	"Data, Product, and Process Modeling"
JCCE	13(1)	January 1999	Issue	"Computing and Information Technology in AEC Education"
JCCE	13(2)	April 1999	Issue	"Imaging Technologies in Civil and Environmental Engineering"
JCCE	15(1)	January 2001	Issue	"Information Technology for Life Cycle Infrastructure Management"

Note: *JCEM* = *Journal of Construction Engineering and Management*, *JTE* = *Journal of Transportation Engineering*, and *JCCE* = *Journal of Computing in Civil Engineering*.

pointed the way for ASCE to form other councils dealing with cross-disciplinary issues. The TCCP, together with its successor, TCCIT, have satisfied the objectives of the initial petitioners, in at least two respects.

First, the TCCP provided the ASCE "seal of approval" on publications dealing with professional issues such as computer pricing policies at a time when such a seal was needed. Most of these issues have by now become so much a part of civil engineering business practices that the ASCE seal is no longer needed.

Second, the TCCP publications and proceedings of conferences and congresses have provided a forum for the discussion and dissemination of innovations in computing in civil engineering. The record is by no means perfect. As the surveys indicate, the practice-oriented segment of ASCE feels poorly served by the *JCCE*, but this is a perennial issue across all ASCE publications. The conference and congress proceedings have a larger contingent of practice-oriented material, but here too research themes predominate. This is not a bad thing by itself; after all, today's research is the seed of tomorrow's practice. What makes the situation in computing in civil engineering more precarious is the extremely rapid changes in computing and information technology (IT). Much of computing in civil engineering research is motivated by the perceived shortcomings in practice. However, with very few exceptions, the vast effort by the civil engineering industry organizations and software vendors has taken place with essentially no direct input from the research community. The best that can be said is that practice has benefited from some of the research explorations that have produced awareness and illustrations of new approaches.

In parallel with the TCCP, the institutes, divisions, and councils of ASCE have continued a deep involvement with computing in civil engineering, which now permeates every aspect of the profession. This frequently raises the question on where a new paper is to be published or the search for a prior paper initiated: Is the content more clearly identified by the kind of computing involved or with the subdiscipline specific phenomenon addressed? This overlap, and potential abuses of it, requires the vigilance of all journal editors and conference program chairpersons. Finally, ASCE as a whole has continued to adopt aspects of computing in civil engineering through its magazine and other IT initiatives.

REFERENCES

Archer, J. S. (1963). "Consistent mass matrix for distributed mass systems." *J. Struct. Div.*, ASCE, 89(4), 161–178.

ASCE. (2000). "Official Register." Reston. Va.

ASCE Computer Practices Committee, Technical Council on Computer Practices. (1973). "Computer pricing practices." *Rep. No. 59*, New York.

Beck, C. F. (1981). "Business plan for the establishment of National Business Institute for Computers in Engineering (NICE)." *J. Tech. Councils of ASCE*, ASCE, 107(1), 169–189.

Civil Engineering Program Applications (CEPA). (1996). *Newsletter*, Oct.

Civil Engineering Program Applications (CEPA). (1975a). *Newsletter*, Aug.

Civil Engineering Program Applications (CEPA). (1975b). "A proposal for a national institute for computers in engineering." *Newsletter*, Oct.

Clough, R. W. (1958). "Structural analysis by means of a matrix algebra program." *Proc., 1st Conf. on Electronic Computation*, ASCE, New York, 109–132.

Clough, R. W. (1960). "The finite element method in plane stress analysis." *Proc., 2nd Conf. on Electric Computation*, ASCE, New York, 345–378.

Dewberry, S. O. (2000). "Easing the way for E-permitting." *Civ. Engrg.*, ASCE, 70(9), 54–57.

Fenves, S. J., et al. (1964). *STRESS: A user's manual*, MIT Press, Cambridge, Mass.

Fenves, S. J., et al. (1965). *STRESS: A reference manual*, MIT Press, Cambridge, Mass.

Grieve, N. R. (1976). "Computer pricing policy and methods." *Engrg. Issues—J. Profl. Act.*, 102(4), 437–446.

Lakmazaheri, S., and Rasdorf, W. (1996). "A review and assessment of the *Journal of Computing in Civil Engineering*." *J. Comp. in Civ. Engrg.*, ASCE, 10(2), 95–96.

Lakmazaheri, S., and Rasdorf, W. (1998). "Foundation for research in computing in civil engineering." *J. Comp. in Civ. Engrg.*, ASCE, 12(1), 9–18.

Lakmazaheri, S., and Rasdorf, W. J. (1995). "Contents review of the *Journal of Computing in Civil Engineering*." *Tech. Rep.*, ASCE, New York.

Lakmazaheri, S., and Rasdorf, W. J. (1997). "The first 10 years: A foundation for research in computing in civil engineering." *Tech. Rep.*, ASCE, New York.

Law, K. H., Rasdorf, W., Karamouz, M., and Abudayyeh, O. Y. (1990a). "Computing in the civil engineering curriculum: Needs and issues." *J. Profl. Issues in Engrg.*, ASCE, 116(2), 128–141.

Law, K. H., Rasdorf, W. J., Karamouz, M., and Abudayyeh, O. Y. (1990b). "The role of computing in the civil engineering curriculum." *Proc., ASCE 1990 Nat. Forum on Educ. and Continuing Profl. Devel. for the Civ. Engr.*, ASCE, New York, 337–343.

Law, K. H., Rasdorf, W. J., Karamouz, M., and Abudayyeh, O. Y. (1989). "The role of computing in civil engineering education." *Proc., 6th Conf. on Computing in Civ. Engrg.*, ASCE, New York, 442–450.

Logcher, R. D., et al. (1965). "A user's manual for on-line use of the structural design language." *Proj. MAC Memo. M-234*, Massachusetts Institute of Technology, Cambridge.

McGrory, H. M., et al. (1975). "The National Software Center." *ASCE Nat. Convention Meeting, Preprint 2626*, ASCE, New York.

Miller, C. L. (1961). "COGO—A computer programming system for civil engineering problems." *Tech. Rep.*, Department of Civil Engineering, Massachusetts Institute of Technology, Cambridge.

Miller, C. L. (1963). "Man-machine communication in civil engineering." *J. Struct. Div.*, ASCE, 89(4), 5–30.

O'Neill, R. J., Henry, R. M., and Lenox, T. A. (1996a). "Role of computing: Educators' perspective." *Proc., ASEE Annu. Conf.*, American Society for Engineering Education, Washington, D.C., Session 3215.

O'Neill, R. J., Henry, R. M., and Lenox, T. A. (1996b). "Role of computing: Practitioners' perspective." *Proc., 3rd Congr. on Computing in Civ. Engrg.*, ASCE, New York, 670–676.

Rasdorf, W. J. (1997). "*Journal of Computing in Civil Engineering: Editorials.*" *Tech. Rep.*, Department of Civil Engineering, North Carolina State University, Raleigh.

Roos, D. (1967). *ICES system design*, 2nd Ed., MIT Press, Cambridge, Mass.

Roos, D., and Miller, C. L. (1964). "COGO-90: Engineering user's manual." *Tech. Rep. R64-12*, Department of Civ. Engineering, Massachusetts Institute of Technology, Cambridge.

Schiffman, R. L. (1972a). "Papers prepared for the Special Workshop on Engineering Software Coordination." *Computing Ctr. Rep. 72-4*, University of Colorado.

Schiffman, R. L. (1972b). "Special Workshop on Engineering Software Coordination." *Comp. Ctr. Rep. 72-2*, University of Colorado, Boulder.

Schiffman, R. L. (1972c). "Transcript of the Special Workshop on Engineering Software Coordination." *Computing Ctr. Rep. 72-17*, University of Colorado, Boulder.

Engineering, Design and Construction of Lunar Bases

Haym Benaroya[1]; Leonhard Bernold, M.ASCE[2]; and Koon Meng Chua, F.ASCE[3]

Abstract: How do we begin to expand our civilization to the Moon? What are the technical issues that infrastructural engineers, in particular, must address? This paper has the goal of introducing this fascinating area of structural mechanics, design, and construction. Published work of the past several decades about lunar bases is summarized. Additional emphasis is placed on issues related to regolith mechanics and robotic construction. Although many hundreds of papers have been written on these subjects, and only a few tens of these have been referred to here, it is believed that a representative view has been created. This summary includes environmental issues, a classification of structural types being considered for the Moon, and some possible usage of in situ resources for lunar construction. An appendix provides, in tabular form, an overview of structural types and their lunar applications and technology drivers.

DOI: 10.1061/(ASCE)0893-1321(2002)15:2(33)

CE Database keywords: Design; Construction; Moon.

Introduction

Concepts for lunar base structures have been proposed since long before the dawn of the space age. We will abstract suggestions generated during the past quarter century, as these are likely to form the pool from which eventual lunar base designs will evolve. Also, one concept will be suggested that has particularly attractive qualifications for the surface lunar base. Significant studies have been made since the days of the Apollo program, when it appeared likely that the Moon would become a second home to humans. For an early example of the gearing up of R&D efforts, see the Army Corps of Engineers study (Department of the Army 1963) (note the date of this report!). During the decade between the late 1980s to mid-1990s, these studies intensified, both within NASA and outside the government in industry and academe. The following references are representative: Benaroya and Ettouney (1989, 1990), Benaroya (1993a, 1995), Duke and Benaroya (1993), Ettouney and Benaroya (1992), Galloway and Lokaj (1994, 1998), Johnson and Wetzel (1998, 1990a), and Johnson (1996), Mendell (1985), Sadeh et al. (1992). Numerous other references discuss science on the Moon, the economics of lunar development, and human physiology in space and on planetary bodies. An equally large literature exists about related policy issues. These topics are outside the scope of this paper.

[1]Professor, Dept. of Mechanical and Aerospace Engineering, Rutgers Univ., Piscataway, NJ 08854 (corresponding author). E-mail: benaroya@rci.rutgers.edu

[2]Professor, Dept. of Civil Engineering, North Carolina State Univ., Raleigh, NC 27695.

[3]Professor, Dept. of Civil Engineering, Univ. of New Mexico, Albuquerque, NM 87131.

Note. Discussion open until September 1, 2002. Separate discussions must be submitted for individual papers. To extend the closing date by one month, a written request must be filed with the ASCE Managing Editor. The manuscript for this paper was submitted for review and possible publication on September 19, 2001; approved on October 23, 2001. This paper is part of the *Journal of Aerospace Engineering*, Vol. 15, No. 2, April 1, 2002. ©ASCE, ISSN 0893-1321/2002/2-33–45/$8.00+$.50 per page.

Unfortunately, by the mid-1990s, the political climate turned against a return to the Moon to stay and began to look at Mars as the "appropriate" destination, essentially skipping the Moon. The debate between "Moon First" and "Mars Direct" continues, although it is clear that the latter will do no more for the expansion of civilization into the solar system than did the Apollo program. It is also clear that we do not have the technology and experience to send people to Mars for an extended stay. Physiological and reliability issues are yet unresolved for a trip to Mars; the Moon is our best first goal.

The emphasis below is on structures for human habitation, a technically challenging fraction of the total number of structures likely to comprise the lunar facility. The test for any proposed lunar base structure is how it meets certain basic as well as special requirements. On the lunar surface, numerous constraints, different from those for terrestrial structures, must be satisfied by all designs. A number of structural types have been proposed for lunar base structures. These include concrete structures, metal frame structures, pneumatic construction, and hybrid structures. In addition, options exist for subsurface architectures and the use of natural features such as lava tubes. Each of these approaches can in principle satisfy the various and numerous constraints, but differently.

A post-Apollo evaluation of the need for a lunar base has been made (Lowman 1985) with the following reasons given for such a base:

- Advancing lunar science and astronomy;
- Stimulus to space technology and test bed for technologies required to place humans on Mars and beyond;
- Utilization of lunar resources;
- Establishment of U.S. presence;
- Stimulation of interest of young Americans in science and engineering; and
- Beginning of long-range program to ensure survival of species.

The potential for an astronomical observatory on the Moon is very great, and it could be serviced periodically in a reasonable fashion from a lunar base. Several bold proposals for astronomy from the Moon have been made (Burns et al. 1990). Nearly all of these proposals involve use of advanced materials and structural concepts to erect large long-life astronomy facilities on the Moon. These facilities will challenge structural designers, constructors, and logistics planners in the 21st century (Johnson 1989; Johnson et al. 1990). One example is a 16 m diameter reflector with its supporting structure and foundation currently being investigated by NASA and several consortia.

Selection of the proper site for a lunar astronomical facility, for example, involves many difficult decisions. Scientific advantages of a polar location for a lunar base (Burke 1985) are that half the sky is continuously visible for astronomy from each pole and that cryogenic instruments can readily be operated there due to the fact that there are shaded regions in perpetual darkness. Disadvantages arise from the fact that the sun will essentially trace the horizon, leaving the outside workspace in extreme contrast, and will pose practical problems regarding solar power and communications with Earth; relays will be required.

Environment

The problem of designing a structure to build on the lunar surface is a difficult one, discussed here in a necessarily cursory way. Many issues are not discussed, but will need to be tackled eventually. Some important topics not discussed here, but necessary in a detailed study, include the following:

- Relationships between severe lunar temperature cycles and structural and material fatigue, a problem for exposed structures;
- Structural sensitivity to temperature differentials between different sections of the same component;
- Very-low-temperature effects and the possibility of brittle fractures;
- Outgassing for exposed steels and other effects of high vacuum on steel, alloys, and advanced materials;
- Factors of safety, originally developed to account for uncertainties in the Earth design and construction process, undoubtedly need adjustment for the lunar environment, either up or down, depending on one's perspective and tolerance for risk;
- Reliability (and risk) must be major components of lunar structures, just as they are of significant Earth structures (Benaroya 1994);
- Dead/live loads under lunar gravity;
- Buckling, stiffening, and bracing requirements for lunar structures, which will be internally pressurized; and
- Consideration of new failure modes such as those due to high-velocity micrometeorite impacts.

In a light, flexible structural system in low gravity, light structural members (for example, composite cylinders that have a wall thickness of only a few $1/1,000^{th}$ of an inch) are sometimes designed to limit their load-carrying capacity by buckling when that limit is met. In turn, the load would have to be redistributed to other, less-loaded structural members. Such an approach offers possibilities for inflatable and other lunar surface structures where it would be simpler and less costly to include limit-state and sacrificial structural elements. Some of these discussions are under way (Benaroya and Ettouney 1992a,b), in particular regarding the design process for an extraterrestrial structure.

Our purpose in this paper is to discuss the technical issues and provide some historical context. Important issues such as financing the return to the Moon, enhancing human physiological understanding, and many others are beyond the scope here. The focus for us, again, is to provide the reader with a brief glimpse of the structural and structural-related engineering issues for human habitation on the Moon.

Important components in a design process are the creation of a detailed design and prototyping. For a structure in the lunar environment, such building and realistic testing cannot be performed on the Earth or even in orbit. It is not currently possible, for example, to experimentally assess the effect of suspended (due to $1/6$ g) lunar regolith lunar soil fines on lunar machinery. Apollo experience may be extrapolated, but only to a boundary beyond which new information is necessary.

Another crucial aspect of a lunar structural design involves an evaluation of the total life cycle that is, taking a system from conception through retirement and disposition, or the recycling of the system and its components. Many factors affecting system life cannot be predicted due to the nature of the lunar environment and the inability to realistically assess the system before it is built and utilized.

Finally, it appears that concurrent engineering will be a byword for lunar structural analysis, design, and erection. Concurrent engineering simultaneously considers system design, manufacturing, and construction, moving major items in the cycle to as early a stage as possible in order to anticipate potential problems. Here, another dimension is added to this definition. Given the extreme nature of the environment contemplated for the structure, concurrency must imply flexibility of design and construction. Parallelism in the design space must be maintained so that at each juncture alternate solutions exist that will permit continuation of construction, even in the face of completely unanticipated difficulties. This factor needs to be further addressed and its implications clearly explored. A discussion of lunar design codes has already started (Benaroya and Ettouney 1992a,b).

Loading, Environment, and Regolith Mechanics

Any lunar structure will be designed for and built with the following prime considerations:

- *Safety and reliability:* Human safety and the minimization of risk to "acceptable" levels are always at the top of the list of considerations for any engineering project. The Moon offers new challenges to the engineering designer. Minimization of risk implies in particular structural redundancy and, when all else fails, easy escape for the inhabitants. The key word is "acceptable," a subjective consideration deeply rooted in economic considerations. What is an acceptable level of safety and reliability for a lunar site, one that must be considered highly hazardous? Such questions go beyond engineering considerations and must include policy considerations: Can we afford to fail?

- *1/6 g gravity:* A structure will have, in gross terms, six times the weight-bearing capacity on the Moon as on the Earth; or, to support a certain loading condition, one-sixth the load-bearing strength is required on the Moon as on the Earth. In order to maximize the utility of concepts developed for lunar structural design, mass rather than weight-based criteria should be the approach of lunar structural engineers. All of NASA's calculations have been done in kg_{force} rather than newtons. Calculations are always without the gravity component; use kilogram feet per square centimeter as pressure, for example.

In the area of foundation design, most classical analytical approaches are based on the limit-state condition, in which the design is based on the limit of loading on a wall or footing at the point when a total collapse occurs—that is, the plastic limit. Since many of the structures on the Moon require accurate pointing capabilities for astronomy, communication, and so on, a settlement-based design method would be more useful. Chua et al. (1990) propose a nonlinear hyperbolic stress-strain model that can be used for the lunar regolith in a finite-element analysis. The paper also shows how the finite-element method can be used to predict settlement of the railway under a support point of a large telescope. Chua et al. (1992) show how a large deformation-capable finite-element program can be used to predict the load-displacement characteristics of a circular spud-can footing, which was designed to support a large lunar optical telescope.

A note against assuming that less gravity means a footing can support more load: if soil can be assumed to be linearly elastic, then the elastic modulus is not affected by gravity. However, the load-bearing capacity of a real soil depends on the confining stress around it. If the soil surrounding the point of interest were heavier because of larger gravity, the confining stress would be higher and the soil at the point of interest could support a higher load without collapsing.

The area of lunar soil (regolith) mechanics was exhaustively explored in the 1970s. Much of the work was approached from interpretation based on classical soil mechanics. Newer work and development of nonlinear stress-strain models to describe the mechanics of the lunar regolith can be found in Johnson et al. (1995b) and Johnson and Chua (1993). Chua et al. (1994) show how structure-regolith simulations can be done using the finite-element approach.

- *Internal air pressurization:* The lunar structure is in fact a life-supporting closed environment. It will be a pressurized enclosed volume with an internal pressure of nearly 15 psi. The enclosure structure must contain this pressure and must be designed to be "fail-safe" against catastrophic and other decompression caused by accidental and natural impacts.

- *Shielding:* A prime consideration in the design is that the structure be able to shield against the types of hazards found on the lunar surface: continuous solar/cosmic radiation, meteorite impacts, and extreme variations in temperature and radiation. In the likely situation that a layer of regolith (lunar soil) is placed atop the structure for shielding, the added weight would partially (in the range of 10–20%) balance the forces on the structure caused by internal pressurization mentioned above.

Shielding against micrometeorite impacts is done by providing dense and heavy materials, in this case compacted regolith, to absorb the kinetic energy. Lunar rocks would be more effective than regolith because the rocks have fracture toughness, but may be more difficult to obtain and much more difficult to place atop surface structures.

Much effort in this country has been devoted to determining the damage effects on human beings and electronics resulting from nuclear weapon detonation, and little is being done to determine long-term, sustained low-level radiation effects such as those that would be encountered on the Moon. According to Silberberg et al. (1985), during the times of low solar activity, the annual dose-equivalent for humans on the exposed lunar surface may be about 30 rem (radiation equivalent man), and the dose-equivalent over an 11 year solar cycle is about 1,000 rem, with most of the particles arriving in one or two gigantic flares lasting 1 to 2 days. It appears that at least 2.5 m of regolith cover would be required to keep the annual dose of radiation at 5 rem, which is the allowable level for radiation workers (0.5 rem for the general public). A shallower cover may be inadequate to protect against the primary radiation, and a thicker cover may cause the second-ary radiation (which consists of electrons and other radiation as a result of the primary radiation hitting atoms along its path).

In recent years, there has been a move away from silicon-and germanium-based electronic components toward the use of gallium arsenide. Lower current and voltage demand and miniaturization of electronic components and machines would make devices more radiation hardened.

Radiation transport codes can be used to simulate cosmic radiation effects, which is not possible in the laboratory. One such code that has been found to be effective is LAHET (Prael et al. 1990), developed at the Los Alamos National Laboratory.

- *Vacuum:* A hard vacuum surrounds the Moon that will preclude the use of certain materials that might not be chemically or molecularly stable under such conditions. This is an issue for research.

Construction in a vacuum has several problems. One would be the possibility of outgassing of oil, vapors, and lubricants from pneumatic systems. Hydraulic systems are not used in space for this reason. The outgassing is detrimental to astronomical mirrors, solar panels, and any other moving machine parts because these structures tend to cause dust particles to form pods. For more discussion of construction challenges in the extraterrestrial environment, see Chua and Johnson (1991). Another problem is that surface-to-surface contact becomes much more abrasive in the absence of an air layer. The increase in dynamic friction would cause fusion at the interfaces, for example, a drill bit fusing with the lunar rock. This is of course aggravated by the fact that the vacuum is a bad conductor of heat. The increase in abrasiveness at interfaces also increases wear and tear on any moving parts, such as railways and wheels.

Blasting in a vacuum is another interesting problem to consider. When the explosive in a blast hole is fired, it is transformed into a gas, the pressure of which may sometimes exceed 100,000 terrestrial atmospheres. How this would affect the area around the blast on the Moon and the impact of ejecta resulting from the blast is difficult to predict. Keeping in mind that a particle set in motion by the firing of a rocket from a lander could theoretically travel halfway around the Moon, the effects of surface blasting on the Moon would be something to be concerned about. Discussion of the tests involving explosives that were performed on the Moon can be found in Watson (1988). Joachim (1988) discussed different candidate explosives for extraterrestrial use, and the Air Force Institute of Technology (Johnson et al. 1969) studied cratering at various gravities and/or in vacuum. Bernold (1991) presented experimental evidence from a study of blasting to loosen regolith for excavation.

- *Dust:* The lunar surface has a layer of fine particles that are easily disturbed and placed into suspension. These particles cling to all surfaces and pose serious challenges for the utility of construction equipment, air locks, and all exposed surfaces (Slane 1994).

Lunar dust consists of pulverized regolith and appears to be charged. The charge may be from the fractured crystalline structure of the material or may be of a surficial nature, for example, charged particles from the solar wind attaching themselves to the dust particles. Criswell (1972) reported that the dust particles levitated at the lunar terminator (line between lunar day and lunar night) and that this may be due to a change in polarity of the surficial materials. Johnson et al. (1995a) discuss the issue of lunar dust and its effects on operations on the Moon. Haljian (1964) and Seiheimer and Johnson (1969) studied the adhesive characteristics of regolith dust.

- *Ease of construction:* The remoteness of the lunar site, in conjunction with the high costs associated with launches from Earth, suggests that lunar structures be designed for ease of construction so that the extravehicular activity of the astronaut construction team is minimized. Construction components must be practical and, in a sense, modular in order to minimize local fabrication for initial structural outposts.

Chua et al. (1993) discuss guidelines and the developmental process for lunar-based structures. They present the governing criteria and also general misconceptions in designing space structures. For example, a device that is simple and conventional looking and has no moving parts is preferred over one that involves multiple degrees of freedom in an exotic configuration involving a yet-to-develop artificial intelligence control, if the former meets the functional requirements. Other misconceptions are that constructing on the Moon is simply a scaling of the effects of similar operations on Earth, and that theoretical predictive tools, especially those performed with computers, can accurately predict events. It is also a misconception that astronauts would have to work around the structure, rather than that, the structure would be designed as to make construction easy for the astronauts.

- *Use of local materials:* This is to be viewed as extremely important in the long-term view of extraterrestrial habitation, but feasibility will have to wait until a minimal presence has been established on the Moon. Initial lunar structures will be transported for the most part in components from the Earth (Fig. 1).

The use of local resources, normally referred to as ISRU (in situ resource utilization), is a topic that has been studied, more intensely now than ever, because of the possibility of actually establishing a human presence on the Moon, near-earth orbit, and Mars. Discussions are found in Johnson and Chua (1992) and Casanova and Aulesa (2000).

Possible Structural Concepts

Various concepts have been proposed for lunar structures. In order to assess the overall efficiency of individual concepts, decision science and operations research tools have been proposed, used (Benaroya and Ettouney 1989), and demonstrated (Benaroya and Ettouney 1990). Along these lines, various concepts are compared (Richter and Drake 1990) using a points system for an extraterrestrial building system, including pneumatic, framed/rigid foam, prefabricated, and hybrid (inflatable/rigid) concepts.

In a very early lunar structural design study, Johnson (1964) presented the then-available information with the goal of furthering the development of criteria for the design of permanent lunar structures. In this work, the lunar environment is detailed, lunar soil from the perspective of foundation design is discussed, and excavation concepts are reviewed. An excellent review of the evolution of concepts for lunar bases up through the mid-1980s is available (Johnson and Leonard 1985), as is a review of more recent work on lunar bases (Johnson and Wetzel 1990b). Surface and subsurface concepts for lunar bases are surveyed (Hypes and Wright 1990) with a recommendation that preliminary designs be considered that focus on specific applications. America's future on the Moon is outlined as supporting scientific research, exploiting lunar resources for use in building a space infrastructure, and attaining self-sufficiency in the lunar environment as a first step in planetary settlement. The complexities and costs of building such a base will depend on the mission or missions for which such a base is to be built.

Fig. 1. Two versions of LESA modules emplaced on Moon by Boeing in 1963 [reprinted with permission from Lowman (1985, p. 37)]

A complete Earth-Moon infrastructure (Griffin 1990) uses proven technologies and the National Space Transportation System for early development of a lunar outpost (Fig. 2). Transfer vehicles and surface systems are developed so that the payload bay of the Space Shuttle can be utilized in transport. The lunar outpost structural scheme separates radiation protection from module support, allowing easy access, installation, and removal of elements attached to the shuttle trusses.

Several types of structures have been proposed for lunar outposts. A preliminary design of a permanently manned lunar surface research base has been briefly studied by Hoffman and Niehoff (1985), with criteria for the base design to include scientific objectives as well as the transportation requirements to establish and support its continued operations.

Inflatables

A pillow-shaped structure proposed by Vanderbilt et al. (1988) as a possible concept for a permanent lunar base (Fig. 3) consists of quilted inflatable pressurized tensile structures using fiber composites. Shielding is provided by an overburden of regolith, with accommodation for sunlight ingress. These studies of the inflatable concept are continued by Nowak et al. (1990) with consideration of the foundation problem and additional reliability concerns and analysis (Nowak et al. 1992). This concept is a significant departure from numerous other inflatable concepts in that it shows an alternative to spheroidal inflatables and optimizes volume for habitation. Inflatable structural concepts for a lunar base are proposed (Broad 1989) as a means to simplify and speed

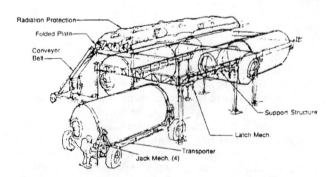

Fig. 2. Modules for lunar base [reprinted with permission by ASCE from Griffin (1990, p. 397)]

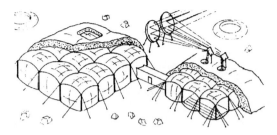

Fig. 3. Inflatable structure [reprinted with permission by ASCE from Vanderbilt et al. (1988, p. 353)]

up the process while lessening the costs. The inflatable structure is suggested as a generic test-bed structure for a variety of application needs for the Moon (Sadeh and Criswell 1994). Design criteria are also put forward (Criswell et al. 1996).

Another pressurized membrane structure, proposed by Chow and Lin (1988, 1989) for a permanent lunar base, is constructed of a double-skin membrane filled with structural foam. A pressurized torus-shaped substructure provides edge support, and shielding is provided by an overburden of regolith. Briefly, the construction procedure requires shaping the ground and spreading the uninflated structure upon it, after which the torus-shaped substructure is pressurized. Structural foam is then injected into the inflatable component, and the internal compartment is pressurized. The bottoms of both inflated structures are filled with compacted soil to provide stability and a flat interior floor surface. Backfilling is a difficult operation to carry out through an airlock. It will, of course, be crucial to ensure that the interior is dust-free (Fig. 4).

A detailed architectural master plan is also proposed for a horizontal inflatable habitat (Kennedy 1992). Finite-element simulations of inflatable structures are needed because it is very difficult to reproduce a hard vacuum and low-gravity condition on Earth. The finite-element modeling would have to be large-deformation capable and have membrane elements (which are essentially beam elements that are without bending stiffness) and axial tensile stiffness, but not axial compression stiffness. The program should also ideally be able to model regolith-structure interaction. GEOT2D (Chua et al. 1994) is a program that has the capabilities needed to simulate inflatable structure-regolith interaction.

Erectables

An expandable platform suggested by Mangan (1988) as a structure on the Moon consists of various geometrically configured 3D trussed octet or space frame elements used both as building blocks and as a platform for expansion of the structure. Examples of the shapes to be used include tetrahedral, hexahedral, octahedral, and so on. This effort is primarily qualitative.

A concept proposed by King et al. (1989) would use the liquid oxygen tank portions of the Space Shuttle external tank assembly for a basic lunar habitat. The modifications of the tank, to take place in low Earth orbit, will include the installation of living quarters, instrumentation, air locks, life-support systems, and environmental control systems. The habitat is then transported to the Moon for a soft landing. This idea, if proven economically feasible, may provide the most politically palatable path to the lunar surface, with the added advantage that many of the necessary technologies already exist and only need resurrection (similar to Fig. 2).

A semiquantitative approach to lunar base structures provided by Kelso et al. (1988) gives some attention to economic consid-

erations, and the structural concepts included could be developed in the future. A modular approach to lunar base design and construction is suggested by Schroeder et al. (1994b) as a flexible approach to developing a variety of structures for the lunar surface. In a related vein, a membrane structure is suggested for an open structure that may be used for assembly on the lunar surface by Schroeder et al. (1994a). A tensile-integrity structure has been suggested as a possible concept for larger surface structures by Benaroya (1993b) (Fig. 4).

Concrete and Lunar Materials

A structural analysis and preliminary design of a precast, prestressed concrete lunar base is reported by Lin et al. (1989). In order to maintain structural integrity, and thus air tightness, when differential settlement is possible, a floating foundation is proposed. All materials for such a lunar concrete structure, except possibly hydrogen for the making of water, may be derivable from lunar resources. Horiguchi et al. (1998) study simulated lunar cement.

The use of unprocessed or minimally processed lunar materials for base structures, as well as for shielding, may be made possible (Khalili 1989) by adopting and extending terrestrial techniques developed in antiquity for harsh environments. A variety of materials and techniques discussed are candidates for unpressurized applications.

The use of indigenous materials is considered by Happel (1992a,b) for the design of a tied-arch structure. The study is extensive and detailed and also includes an exposition on lunar materials.

Construction of layered embankments using regolith and filmy materials (geotextiles) is viewed as an option using robotic construction (Okumura et al. 1994), as are fabric-confined soil structures (Harrison 1992).

In order to avoid the difficulties of mixing concrete on the lunar surface due to lack of water, Gracia and Casanova (1998) have suggested examining use of sulfur concrete because sulfur is readily available on the Moon.

Lava Tubes

Ideas regarding the utility of constructing the first outposts under the lunar surface have been proposed. A preliminary assessment is provided by Daga et al. (1990) of a lunar outpost situated in a lava tube. They conclude that an architectural solution is needed to the problems surrounding the development of a lunar outpost, but that lunar surface structures are not the best approach. Rather subselene development offers real evolutionary potential for settlement.

In another structural approach, fused regolith structures are suggested by Cliffton (1990) and Crockett et al. (1994). In this case, the structures are small and many and reside on the surface. A prime advantage offered for planning numerous smaller structures is safety and reliability. The premise of this work is to use the sun's energy to fuse regolith into components.

Construction in New Environment

Site plans (Sherwood 1990) and surface system architectures (Pieniazek and Toups 1990) are forcefully presented as being fundamental to any development of structural concepts. One of the challenges to the extraterrestrial structures community is that of construction. Lunar construction techniques have differences

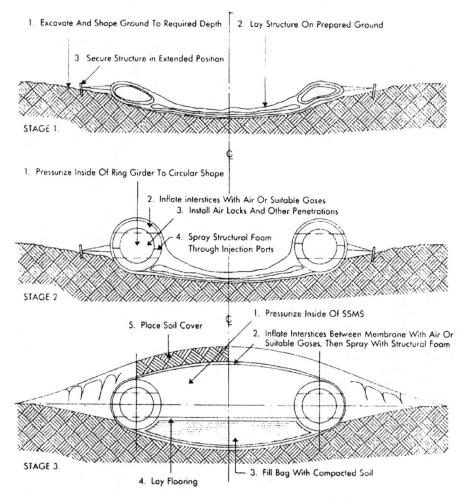

Fig. 4. Inflatable membrane structure—construction sequence [reprinted with permission by ASCE from Chow and Lin (1988, p. 372)]

from those on Earth; for example, the construction team will likely operate in pressure suits, motion is dominated by 1/6 *g*, solar and cosmic radiation are not shielded by an Earth-type atmosphere, and suspended dust exists in the construction site. Toups (1990) assesses various construction techniques for the classes of structures and their respective materials.

Structural and architectural designs along with manufacturing plants and construction methods are discussed by Namba et al. (1988b) for a habitable structure on the Moon using concrete modules. The module can be disassembled into frame and panels.

A qualitative study by Drake and Richter (1990) is made of the design and construction of a lunar outpost assembly facility. Such a facility would be used to construct structures too large for transport to the Moon in one piece. The assembly facility would also be used to support operation and maintenance operations during the functional life of the lunar outpost. A series of trade studies is suggested on the construction of such an assembly facility.

Construction of a lunar base will at least partially rest on the capabilities of the Army Corps of Engineers. Preparations that are now under way are outlined by Simmerer (1988) and challenges discussed by Sargent and Hampson (1996).

All the above are contingent on the "practical" aspects of building structures on the Moon. These aspects include the sort of machinery needed to move equipment and astronauts about the surface; the methods needed to construct in 1/6 *g* with an ex-

tremely fine regolith dust working its way into every interface and opening; and the determination of the appropriate layout of structures considering human safety and operations needs. Using harsh Earth environments such as the Antarctic as test beds for extraterrestrial operations is advocated by Bell and Neubek (1990).

The performance of materials and equipment used for lunar construction needs to be examined in terms of the many constraints discussed so far. Structures that are unsuitable for Earth construction may be adequate for the reduced-gravity lunar environment (Chow and Lin 1989). Several research efforts have been directed to producing construction materials, such as cement, concrete, and sulfur-based materials, from the elements available on the Moon (Lin 1987; Agosto et al. 1988; Leonard and Johnson 1988; Namba et al. 1988a; Yong and Berger 1988; Strenski et al. 1990).

The appendix to this paper provides a long list of structures that require a study not only of the materials that could be used for construction, but also of the necessary tools/equipment, methods of operation/control, and most importantly, how to construct structures with and within the lunar environment (that is regolith, vacuum, 1/6 *g*). Because most of the construction methods developed since the beginning of mankind are adapted to fit and take advantage of terrestrial environments (that is, soil characteristics, atmosphere with oxygen, and 1 *g* gravity), technologies that are common on Earth either will not work on the Moon or are too

a) Backhoe   b) Cableway   c) Clamshell

Fig. 5. Common "Earth" excavators that depend on gravity

costly or inefficient. The following sections will address some of the unique problems and circumstances that we face.

Creating Base Infrastructure

The availability of an adequate infrastructure is key to the survival and growth of any society. "In all human societies, the quality of life depends first on the physical infrastructure that provides the basic necessities such as shelter, water, waste disposal, and transportation," wrote Grigg (1988.) Today, and especially for the lunar base, we have to add communication and power as part of the physical infrastructure. All of these constructed facilities have one issue in common, namely the interaction with lunar surface materials: (1) rocks; (2) regolith; and (3) breccias. Lunar soil, referred to as regolith, differs from soil on Earth in several respects that are significant for construction. While the soil that establishes the top layers (10–20 cm) is loose and "powdery," easily observable in Apollo movies, the regolith reaches the relative density of 90–100% below 30 cm. The grain size distribution of a common regolith, as well as its high density below the top layers, is hardly found in the terrestrial environment. This creates unique problems for excavating, trenching, backfilling, and compacting the soil (Goodings et al. 1992). These operations, however, are needed to create (1) building foundations; (2) roadbeds; (3) launch pads; (4) buried utilities (power, communication); (5) shelters and covers; (6) open-pit mining; and (7) underground storage facilities.

Excavating "Hard" Lunar Soil

Bernold (1991) reported about efforts to study the unique problems related to digging and trenching on the Moon. All the common excavation technologies used on Earth depend on the effect of gravitational acceleration that turns mass into forces that are needed to cut, scoop, and move soil (Fig. 5).

Because of the drastically reduced gravity, transporting the masses and material to the lunar surface would be prohibitively expensive. Dick et al. (1992) presented the result of experimental work to study an alternative to traditional excavation of soil, namely the use of explosives to loosen the dense soil so it can be excavated with a limited amount of force. Fig. 6 presents images of the effect of a small amount of explosives on lunar simulants.

Fig. 6(a) shows the direction and position of ejecta clumps 43.16 and 52.24 ms after detonation, while Fig. 6(b) presents an overview and 6(c) a cross section of the crater created by a small amount of explosives. Although the ejection of regolith would not be acceptable on the lunar surface, since the resulting dust would travel far, research showed that explosives buried deep enough would not create craters but loosen the soil very effectively. In fact, Fig. 5(b) demonstrates how a lightweight bucket pulled by a cable was slicing into the lunar soil simulant that had been loosened in this manner. Furthermore, the sensor-equipped lightweight backhoe excavator required a drastic reduction of energy to dig the loosened soil (Lin et al. 1994).

Building Transportation Infrastructure

The creation of durable roads without using asphalt or concrete as a top requires the planning/cutting of the existing surface and the compaction of fill material. The main objectives of the road base and road surface are to distribute the point loads under the wheels/tracks to the maximum allowable bearing capacity, to provide stable traction resistance for the needed rim pulls (force at the rim of the wheel to allow motion) and breaking forces, and for abrasion resistance. Earthbound equipment that achieves these objectives depends on a large mass, gravitational force, and a

Surface
Before
Detonation

Fig. 6. Effect of buried explosive charges on lunar soil simulants: (a) ejecta milliseconds after detonation; (b) effect of 1/8 g PETN buried at 7.1 cm; and (c) detonation chamber, chimney, and crater after removal of loose soil

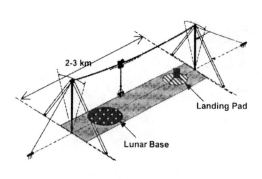

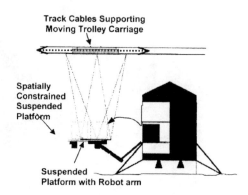

a) Robotic Cable Way System Over Base and Landing Pad   b) Cable Way with Cargo Handling Robot Attachment

Fig. 7. Robotic transportation system for lunar base

sufficient power source. It is obvious that the size of each would make it cost-prohibitive to deploy on the Moon, even if the power source were switched from diesel engines to electric. In addition, Bernold (1994b) showed that the compaction of lunar soil necessary for creating a stable roadbase would create unique problems. Preliminary research data indicated that the normal size distribution of soil particles would make it impossible to achieve needed density and strength using common methods of static or vibratory compaction.

However, reducing the percentage of fines present in the regolith can increase the compacted strength of the dry and mainly cohesionless lunar soil. In addition, the surface has to be covered with larger-size stones that have to be crushed from rock, requiring additional equipment such as rock-drill and pyrotechnic equipment, loaders, rock crushers, vibrating screens, and conveyor feeders. It is apparent that the construction of trafficable and stable roads and/or pads on the Moon will require many different machines capable of pushing, loading, cutting, sizing, and compacting regolith as well as the crushing, transportation, spreading, and compaction of rock. The use of multipurpose equipment will certainly be desirable but, on the other hand, slow the operation. If one wants to rely on "Earth-proven" technologies, significant disadvantages will have to be overcome. The most significant handicap is the large reduction of gravitational acceleration that is the basis for the efficient operation of terrestrial roadbuilding equipment.

As an alternative to wheel-or water-based transportation, Bernold (1994a) proposed a cable-based transportation system: "Lunar tramway systems can take advantage of the reduced gravity, which permits building wider spans and/or using smaller cable diameter for lifting and transporting heavy loads. The use of luffing masts and a unique semistable rigging platform provide many opportunities for reaching wide areas on the lunar surface and performing various tasks needed for handling material, construction, servicing and maintaining facilities needed on a lunar base." Fig. 7 presents the basic elements of a cable-based transportation system that could cover large areas.

As depicted in Fig. 7, the track cables are attached to two (or more) masts, thus being able to span long distances (for example, 3,000 m) because the lower gravity reduces not only the weight of the load to 1/6, but also the weight of the cable itself, leading to much smaller cable deflections than on Earth. Two electrowinches and cables at each mast provide the mechanism for luffing (sideways rotation around its base socket) the track cables. As indicated in Fig. 7(a), the luffing mechanism adds a significant capa-

bility in that it allows the transportation system to cover a rectangular area. In addition, the mast and cables can be lowered all the way to the ground during the final landing approach of a cargo ship. Fig. 7(b) presents the concept of a trolley-carriage that is being moved along the track by a haul cable. Attached to the carriage are either three or six lift cables that can be individually operated with winches. By combining these cable-based mechanisms, a spatially controlled platform can be established (Richter et al. 1998). While the system can be used to unload a lunar lander, it can also support construction and mining operations. Fig. 8 portrays a platform attachment capable of (1) excavating trenches to bury cables and pipes; (2) removing rock boulders; (3) collecting rocks for the crusher; (4) deploying soil or rock drills; and (5) mining open pits for processing.

As shown in Fig. 8, the same spatially constrained platform supporting a cargo-handling robot can be reconfigured to carry a shovel excavator capable of loading regolith and rock boulders into a pan mounted on top of the platform. For other operations, such as trenching, the same robot arm could reconfigure itself to work as a backhoe, drill boom, or other desirable end-effectors.

A major problem that needs to be considered in the design of robotic systems for construction is the question of control. The complexities of working in a totally new environment will make it impossible to have the lunar construction equipment operate autonomously. Kemurdjian and Khakhanov (2000) discuss several specific aspects of the problem. These include the effects of low gravity on traction, the amount of power to be consumed, and

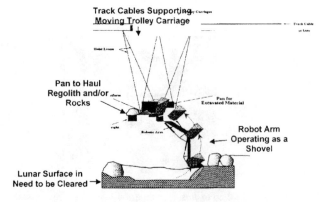

Fig. 8. Cable-based robot shovel excavator at work

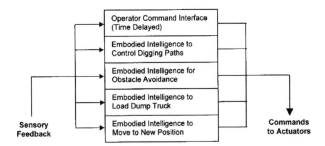

Fig. 9. Layered control architecture for Earth-based teleoperation of lunar equipment

Fig. 10. Lunar habitat assembled out of components delivered by automated cargo flights. Pressurized rovers, logistics modules, and spacesuit maintenance and storage module combine to provide living and working quarters for crew (NASA/JSC image #S93-45585).

most importantly, the dynamics of such vehicles. At the same time, the problems associated with exposing astronauts to long prolonged extravehicular activities (EVA) and the cost of deploying human operators on the lunar surface make it unlikely that each piece of equipment will be steered by an operator who rides on it.

Robotic Control of Construction Equipment

One of the main problems in robotic control of equipment is the time that signals need to travel through vacuum, atmosphere, or fiber-optic or other communication lines. The time it takes a signal to travel through a network is commonly referred to as latency. Nelson et al. (1998) report about their work on the issue of latency: "Teleoperation is commonly used in the remote control of terrestrial mining equipment. Teleoperating mining equipment on the Moon from the Earth is attractive but involves a transmission loop time delay of 4 to 10 s. A human operator can handle time delays of about 1 second in simple teleoperation applications." A variety of control schemes that help alleviate the problems caused by excessive signal delays have been developed. They range from teleautonomous, to predictive teleoperation, to semiautonomous operation (Conway et al. 1990). In this context, one effort by researchers is to equip mobile computerized equipment with the robustness and intelligence to react to the dynamics of the environment. To do this, it is critical "to build complete agents which operate in dynamic environments using real-time sensors. Internal world models which are complete representations of the external environment, besides being impossible to obtain, are not at all necessary for agents to act in a competent manner" (Brooks 1986, 1990). Since work in construction always requires moving within, and interacting with, a complex environment while handling messy materials that have to be joined, layered on top, inserted, and so on, a distributed intelligence embodied in the site equipment and sensors and communicating via networks may serve as a uniquely qualified approach to creating a semiautonomous fleet of equipment.

Fig. 9 presents a partial model of layered control architecture for a teleoperated backhoe operation that integrates human control with intelligent control modules that work in parallel rather than in sequence. The layered control architecture for robotic excavation was first proposed by Huang and Bernold (1993, 1994). The key feature of this approach is the distribution of the control task to the most efficient module.

Issue of Water on Moon

In a recent development, it appears that there may be water-ice in some craters near the poles of the Moon. It was suggested that water-laden comets and asteroids may have deposited the water. If water does exist in those craters, it was conjectured by Chua and Johnson (1998) that the moisture distribution may consist of water-ice mixing with the regolith to saturation or near saturation, and reducing outward according to the matric suction pressure (which is influenced by the particle size distribution). Since the gravitation potential is relatively small compared to the matric suction potential, the water would have been drawn laterally or even upwards over some distance. (Note: Since the regolith has no clays, unlike Earth, there would not be an osmotic suction component to influence moisture migration.) The extent of this unsaturated zone is primarily influenced by how fast the water vapor condensed at the bottom of the crater, which have temperatures as low as $-230°C$. The Lunar Prospector Mission team indicated that the moisture content in the regolith at the bottom of the crater might be between 0.3 and 1%.

Issue of using Geosynthetics in Extraterrestrial Environment

Some recent papers suggested using geosynthetics as soil reinforcement to construct earth structures such as berms, walls, and slopes. Several problems have to be considered in order for this to be a reality.

- Plastic materials are susceptible to degradation when subjected to radiation;
- The glass transition temperature of many if not all of the geosynthetics used on Earth is well above the cold temperatures encountered on candidate sites, including that on the Moon, which would make the plastics brittle, thus rendering them useless as reinforcing elements; and
- There is little experience on how geosynthetics fare in a hard vacuum and respond to the relatively more abrasive regolith.

Conclusion

We have presented a summary of current thinking regarding some of the issues surrounding the engineering and construction of structures for long-term lunar human habitation. We close here with a NASA vision of how a lunar base may look (Fig. 10).

Acknowledgments

Professor Manohar Kamat of the Georgia Institute of Technology, the past editor of this journal, is thanked for his invitation to submit this paper for this special issue. This paper is dedicated to the vision and those who are willing to make it happen.

Appendix: Building Systems

Types of Applications

Habitats
- People (living and working)
- Agriculture
- Airlocks: ingress/egress
- Temporary storm shelters for emergencies and radiation
- Open volumes

Storage Facilities/Shelters
- Cryogenic (fuels and science)
- Hazardous materials
- General supplies
- Surface equipment storage
- Servicing and maintenance
- Temporary protective structures

Supporting Infrastructure
- Foundations/roadbeds/launchpads
- Communication towers and antennas
- Waste management/life support
- Power generation, conditioning, and distribution
- Mobile systems
- Industrial processing facilities
- Conduits/pipes

Application Requirements

Habitats
- Pressure containment
- Atmosphere composition/control
- Thermal control (active/passive)
- Acoustic control
- Radiation protection
- Meteoroid protection
- Integrated/natural lighting
- Local waste management/recycling
- Airlocks with scrub areas
- Emergency systems
- Psychological/social factors

Storage Facilities/Shelters
- Refrigeration/insulation/cryogenic systems
- Pressurization/atmospheric control
- Thermal control (active/passive)
- Radiation protection
- Meteoroid protection
- Hazardous material containment
- Maintenance equipment/tools

Supporting Infrastructure
- All of the above
- Regenerative life support (physical/chemical and biological)
- Industrial waste management

Types of Structures

Habitats
- Landed self-contained structures
- Rigid modules (prefabricated/in situ)
- Inflatable modules/membranes (prefabricated/in situ)
- Tunneling/coring
- Exploited caverns

Storage Facilities/Shelters
- Open tensile (tents/awning)
- "Tinker toy"
- Modules (rigid/inflatable)
- Trenches/underground
- Ceramic/masonry (arches/tubes)
- Mobile
- Shells

Supporting Infrastructure
- Slabs (melts/compaction/additives)
- Trusses/frames
- All of the above

Material Considerations

Habitats
- Shelf life/life cycle
- Resistance to space environment (UV/thermal/radiation/abrasion/vacuum)
- Resistance to fatigue (acoustic and machine vibration/pressurization/thermal)
- Resistance to acute stresses (launch loads/pressurization/impact)
- Resistance to penetration (meteoroids/mechanical impacts)
- Biological/chemical inertness
- Reparability (process/materials)

Operational Suitability/Economy
- Availability (lunar/planetary sources)
- Ease of production and use (labor/equipment/power/automation and robotics)
- Versatility (materials and related processes/equipment)
- Radiation/thermal shielding characteristics
- Meteoroid/debris shielding characteristics
- Acoustic properties
- Launch weight/compactability (Earth sources)
- Transmission of visible light
- Pressurization leak resistance (permeability/bonding)
- Thermal and electrical properties (conductivity/specific heat)

Safety
- Process operations (chemical/heat)
- Flammability/smoke/explosive potential
- Outgassing
- Toxicity

Structures Technology Drivers

Mission/Application Influences
- Mission objectives and size
- Specific site—related conditions (resources/terrain features)
- Site preparation requirements (excavation/infrastructure)
- Available equipment/tools (construction/maintenance)
- Surface transportation/infrastructure
- Crew size/specialization
- Available power
- Priority given to use of lunar material and material processing
- Evolutionary growth/reconfiguration requirements
- Resupply versus reuse strategies

General Planning/Design Considerations
- Automation and robotics
- EVA time for assembly
- Ease and safety of assembly (handling/connections)
- Optimization of teleoperated/automated systems
- Influences of reduced gravity (anchorage/excavation/traction)
- Quality control and validation
- Reliability/risk analysis
- Optimization of in situ materials utilization
- Maintenance procedures/requirements
- Cost/availability of materials
- Flexibility for reconfiguration/expansion
- Utility interfaces (lines/structures)
- Emergency procedures/equipment
- Logistics (delivery of equipment/materials)
- Evolutionary system upgrades/changeouts
- Tribology

Requirement Definition/Evaluation

Requirement/Option Studies
- Identify site implications (lunar soil/geologic models)
- Identify mission-driven requirements (function and purpose/staging of structures)
- Identify conceptual options (site preparation/construction)
- Identify evaluation criteria (costs/equipment/labor)
- Identify architectural program (human environmental needs)

Evaluation Studies
- Technology development requirements
- Cost/benefit models (early/long-term)
- System design optimization/analysis

References

Agosto, W. N., Wickman, J. H., and James, E. (1988). "Lunar cements/concretes for orbital structures." *Engineering, construction, and operations in space*, S. W. Johnson and J. P. Wetzel, eds. ASCE, New York, 157–168.

Bell, L., and Neubek, D. J. (1990). "Antarctic testbed for extraterrestrial operations." *Engineering, Construction, and Operations in Space II*, S. W. Johnson and J. P. Wetzel, eds. ASCE, New York, 1188–1197.

Benaroya, H., ed. (1993a). "Applied mechanics of a lunar base." *Appl. Mech. Rev.*, 46(6), 265–358.

Benaroya, H. (1993b). "Tensile-integrity structures for the Moon." *Applied Mechanics of a Lunar Base, Appl. Mech. Rev.*, 46(6), 326–335.

Benaroya, H. (1994). "Reliability of structures for the Moon." *Struct. Safety*, 15, 67–84.

Benaroya, H., ed. (1995). "Lunar structures." *Special Issue, J. British Interplanetary Society*, 48(1).

Benaroya, H., and Ettouney, M. (1989). "Framework for the evaluation of lunar base structural concepts." *9th Biennial SSI/Princeton Conf., Space Manufacturing*, Princeton, 297–302.

Benaroya, H., and Ettouney, M. (1990). "A preliminary framework for the comparison of two lunar base structural concepts." *Engineering, Construction, and Operations in Space II*, S. W. Johnson and J. P. Wetzel, eds., ASCE, New York, 490–499.

Benaroya, H., and Ettouney, M. (1992a). "Design and construction considerations for lunar outpost." *ASCE, J. Aerosp. Eng.*, 5(3), 261–273.

Benaroya, H., and Ettouney, M. (1992b). "Design codes for lunar structures." *Engineering, Construction, and Operations in Space III*, W. Z. Sadeh, S. Sture, and R. J. Miller, eds., ASCE, New York, 1–12.

Bernold, L. E. (1991). "Experimental studies on mechanics of lunar excavation." *ASCE, J. Aerosp. Eng.*, 4(1), 9–22.

Bernold, L. E. (1994a). "Cable-based lunar transportation system." *ASCE, J. Aerosp. Eng.*, 7(1), 1–16.

Bernold, L. E. (1994b). "Compaction of lunar-type soil." *ASCE, J. Aerosp. Eng.*, 7(2), 175–187.

Broad, W. J. (1989). "Lab offers to develop an inflatable space base." *New York Times*, Nov. 14.

Brooks, R. A. (1986). "A robust layered control system for a mobile robot." *IEEE Journal of Robotics and Automation*, RA-2, 14–23.

Brooks, R. A. (1990). "Elephants don't play chess." P. Mae, ed., *Designing autonomous agents: Theory and practice from biology to engineering and back*, MIT Press, Cambridge, Mass., 3–15.

Burke, J. D. (1985). "Merits of a lunar polar base location." *Lunar bases and space activities of the 21st century, Proc., Lunar and Planetary Institute*, Houston, 77–84.

Burns, J. O., Duric, N., Taylor, G. J., and Johnson, S. W. (1990). "Astronomy on the Moon." *Sci. Am.* 262(3), 42–49.

Casanova, I., and Aulesa, V. (2000). "Construction materials from in-situ resources on the Moon and Mars." *Proc., Space 2000: 7th Int. Conf.*, S. W. Johnson, K. M. Chua, R. G. Galloway and P. I. Richter, eds., ASCE, Reston, Va., 638–644.

Chow, P. Y., and Lin, T. Y. (1988). "Structures for the Moon." *Engineering, construction, and operations in space*, S. W. Johnson and J. P. Wetzel, eds., ASCE, New York, 362–374.

Chow, P. Y., and Lin, T. Y. (1989). "Structural engineering's concept of lunar structures." *ASCE, J. Aerosp. Eng.*, 2(1), 1–9.

Chua, K. M., and Johnson, S. W. (1991). "Foundation, excavation and radiation shielding emplacement concepts for a 16 meter large lunar telescope." *SPIE, Intl. Soc. Opt. Engrg., Proc.*, 1494.

Chua, K. M., and Johnson, S. W. (1998). "Martian and lunar cold region soil mechanics considerations." *ASCE, J. Aerosp. Eng.*, 11(4), 138–147.

Chua, K. M., Johnson, S. W., and Nein, M. E. (1993). "Structural concepts for lunar-based astronomy." *Appl. Mech. Rev.*, 46(6), 336–357.

Chua, K. M., Johnson, S. W., and Sahu, R. (1992). "Design of a support and foundation for a large lunar optical telescope." *Engineering, construction and operations in space III*, W. Z. Sadeh, S. Sture, and R. J. Miller, eds., ASCE, New York, 1952–1963.

Chua, K. M., Xu, L., and Johnson, S. W. (1994). "Numerical simulations of structure-regolith interactions." *Computer Methods and Advances in Geomechanics*, H. J. Siriwardane and M. M. Zaman, eds., Balkema, Rotterdam, The Netherlands.

Chu, K. M., Yuan, Z., and Johnson, S. W. (1990). "Foundation design of a large diameter radio telescope on the Moon." *Engineering, construction and operations in space II*, S. W. Johnson and J. P. Wetzel, eds., 707–716.

Cliffton, E. W. (1990). "A fused regolith structure." *Engineering, construction, and operations in space II*, S. W. Johnson and J. P. Wetzel, eds., ASCE, New York, 541–550.

Conway, L., Volz, R. A., and Walker, M. W. (1990). "Teleautonomous systems: Projecting and coordinating intelligent actions at a distance." *IEEE Trans. Rob. Autom.*, 6(2), 146–158.

Criswell, D. (1972). "Lunar dust motion." *Proc., 3rd Lunar Science Conf.*, NASA, Washington, D.C., 2671–2680.

Criswell, M. E., Sadeh, W. Z., and Abarbanel, J. (1996). "Design and performance criteria for inflatable structures in space." *Engineering, construction, and operations in space*, S. W. Johnson, ed., ASCE, New York, 1045–1051.

Crockett, R. S., Fabes, B. D., Nakamura, T., and Senior, C. L. (1994). "Construction of large lunar structures by fusion welding of sintered regolith." *Engineering, construction, and operations in space IV*, R. K. Galloway and S. Lokaj, eds., ASCE, New York, 1116–1127.

Daga, A. W., Daga, M. A., and Wendell, W. R. (1990). "A preliminary assessment of the potential of lava tube-situated lunar base architecture." *Engineering, construction, and operations in space II*, S. W. Johnson and J. P. Wetzel, eds., ASCE, New York, 568–577.

Department of the Army. (1963). *Special study of the research and development effort required to provide a U.S. Lunar Construction Capability*, Office of the Chief of Engineers.

Dick, R. D., Fourney, W. L., Goodings, D. J., Lin, C.-P., and Bernold, L. E. (1992). "Use of explosives on the Moon." *ASCE, J. Aerosp. Eng.*, 5(1), 59–65.

Drake, R. M., and Richter, P. J. (1990). "Design and construction of a lunar outpost assembly facility." *Engineering, construction, and operations in space II*, S. W. Johnson and J. P. Wetzel, eds., ASCE, New York, 449–457.

Duke, M., and Benaroya, H. (1993). "Applied mechanics of lunar exploration and development." *Applied mechanics of a lunar base, Appl. Mech. Rev.*, 46(6), 272–277.

Ettouney, M., and Benaroya, H. (1992). "Regolith mechanics, dynamics, and foundations." *ASCE, J. Aerosp. Eng.*, 5(2), 214–229.

Galloway, R. G., and Lokaj, S., eds. (1994). *Engineering, construction, and operations in space IV*, ASCE, New York.

Galloway, R. G., and Lokaj, S., eds. (1998). *Space 98*, ASCE, Reston, Va.

Goodings, D. J., Lin, C.-P., Dick, R. D., Fourney, W. L., and Bernold, L. E. (1992). "Modeling the effects of chemical explosives for excavation on the Moon." *ASCE, J. Aerosp. Eng.*, 5(1), 44–58.

Gracia, V., and Casanova, I. (1998). "Sulfur concrete: A viable alternative for lunar construction." *SPACE 98*, R. G. Galloway and S. Lokaj, eds., ASCE, Reston, Va., 585–591.

Griffin, B. N. (1990). An infrastructure for early lunar development." *SPACE 90*, S. W. Johnson and J. P. Wetzel, eds., ASCE, New York, 389–398.

Grigg, N. S. (1988). *Infrastructure engineering and management*, Wiley, New York.

Halajian, J. D. (1964). "Soil behavior in a low and ultrahigh vacuum." *Contribution 64-WA/AV-14*, ASME, New York.

Happel, J. A. (1992a). "The design of lunar structures using indigenous construction materials." MS thesis, Univ. of Colorado, Boulder, Colo.

Happel, J. A., William, K., and Shing, B. (1992b). "Prototype lunar base construction using indigenous materials." *Engineering, construction, and operations in space III*, W. Z. Sadeh, S. Sture, and R. J. Miller, eds., ASCE, New York, 112–122.

Harrison, R. A. (1992). "Cylindrical fabric-confined soil structures." *Engineering, Construction, and Operations in Space III*, W. Z. Sadeh, S. Sture, and R. J. Miller, eds., ASCE, New York, 123–134.

Hoffman, S. J., and Niehoff, J. C. (1985). "Preliminary design of a permanently manned lunar surface research base." *Lunar bases and space activities of the 21st century, Proc., Lunar and Planetary Institute*, Houston, 69–76.

Horiguchi, T., Saeki, N., Yoneda, T., Hoshi, T., and Lin, T. D. (1998). "Behavior of simulated lunar cement mortar in vacuum environment." *Space 98*, R. G. Galloway and S. L. Lokaj, eds., ASCE, Reston, Va., 571–576.

Huang, X., and Bernold, L. E. (1993) "Towards an adaptive control model for robotic backhoe excavation." *Transportation Research Record 1406*, Transportation Research Board, Washington, D.C., 20–24.

Huang, X., and Bernold, L. E. (1994). "Control model for robotic backhoe excavation and obstacle handling." *Proc., Robotics for challenging environments*, L. A. Demsetz and P. R. Klarev, eds., ASCE, New York, 123–130.

Hypes, W. D., and Wright, R. L. (1990). "A survey of surface structures and subsurface developments for lunar bases." *Engineering, construc-*

tion, and operations in space II, S. W. Johnson and J. P. Wetzel, eds., ASCE, New York, 468–479.

Joachim, C. E. (1988). "Extraterrestrial excavation and mining with explosives." *Engineering, construction and operations in space*, S. W. Johnson and J. P. Wetzel, eds. ASCE, New York, 332–343.

Johnson, S. W. (1964). "Criteria for the design of structures for a permanent lunar base." PhD dissertation, Univ. of Illinois, Urbana, Ill.

Johnson, S. W. (1989). *Extraterrestrial facilities engineering*, 1989 Yearbook, Encyclopedia of Physical Science and Technology, Academic Press, San Diego.

Johnson, S. W., ed. (1996). *Engineering, construction, and operations in Space: Space '96*, ASCE, New York.

Johnson, S. W., and Chua, K. M. (1992). "Assessment of the lunar surface and in situ materials to sustain construction-related applications." Joint Workshop (DOE/LANL, NASA/JSC and LPI) on New Technologies for Lunar Resource Assessment.

Johnson, S. W., and Chua, K. M. (1993). "Properties and mechanics of the lunar regolith." *Appl. Mech. Rev.*, 46(6), 285–300.

Johnson, S. W., Chua, K. M., and Burns, J. O. (1995a). "Lunar dust, lunar observatories, and other operations on the Moon." *J. British Interplanetary Society*, 48, 87–92.

Johnson, S. W., Chua, K. M., and Carrier, III, W. D. (1995b). "Lunar Soil Mechanics." *J. British Interplanetary Society*, 48(1), 43–48.

Johnson, S. W., and Leonard, R. S. (1985). "Evolution of concepts for lunar bases." *Lunar Bases and Space Activities of the 21st Century, Proc., Lunar and Planetary Institute*, Houston, 47–56.

Johnson, S. W., Smith, J. A., Franklin, E. G., Moraski, L. K., and Teal, D. J. (1969). "Gravity and atmosphere pressure effects on crater formation in sand." *J. Geophys. Res.*, 74(20), 4838–4850.

Johnson, S. W., and Wetzel, J. P., eds. (1988). *Engineering, construction, and operations in space*, ASCE, New York.

Johnson, S. W., and Wetzel, J. P., eds. (1990a). *Engineering, construction, and operations in space II*, ASCE, New York.

Johnson, S. W., Burns, J. O., Chua, K. M., Duric, N., and Taylor, G. J. (1990). "Lunar astronomical observatories: Design studies." *ASCE, J. Aerosp. Eng.*, 3(4), 211–222.

Johnson, S. W., and Wetzel, J. P. (1990b). "Science and engineering for space: Technologies from SPACE 88." *ASCE, J. Aerosp. Eng.*, 3(2), 91–107.

Kelso, H. M., Hopkins, J., Morris, R., and Thomas, M. (1988). "Design of a second generation lunar base." *Engineering, construction, and operations in space*, S. W. Johnson and J. P. Wetzel, eds., ASCE, New York, 389–399.

Kemurdjian, A., and Khakhanov, U. A. (2000). "Development of simulation means for gravity forces." *Robotics 2000*, W. C. Stone, ed., ASCE, Reston, Va., 220–225.

Kennedy, K. J. (1992). "A horizontal inflatable habitat for SEI." *Engineering, construction, and operations in space III*, W. Z. Sadeh, S. Sture, and R. J. Miller, eds., ASCE, New York, 135–146.

Khalili, E. N. (1989). "Lunar structures generated and shielded with on-site materials." *ASCE, J. Aerosp. Eng.*, 2(3), 119–129.

King, C. B., Butterfield, A. J., Hyper, W. D., and Nealy, J. E. (1989). "A concept for using the external tank from a NSTS for a lunar habitat." *Proc., 9th Biennial SSI/Princeton Conf. on Space Manufacturing*, Princeton, AIAA, Washington, D.C., 47–56.

Leonard, R. S., and Johnson, S. W. (1988). "Sulfur-based construction materials for lunar construction." *Engineering, construction, and operations in space*, S. W. Johnson and J. P. Wetzel, eds., ASCE, New York, 1295–1307.

Lin, T. D. (1987). "Concrete for lunar base construction." *Concr. Int.*, 9(7).

Lin, C. P., Goodings, D. J., Bernold, L. E., Dick, R. D., and Fourney, W. L. (1994). "Model studies of effects on lunar soil of chemical explosions." *J. Geotech. Eng.*, 120(10), 1684–1703.

Lin, T. D., Senseney, J. A., Arp, L. D., and Lindbergh, C. (1989). "Concrete lunar base investigation." *ASCE, J. Aerosp. Eng.*, 2(1), 10–19.

Lowman, P. D. (1985). "Lunar bases: A post-Apollo evaluation." *Lunar Bases and Space Activities of the 21st Century, Proc., Lunar and Planetary Institute*, Houston, 35–46.

Mangan, J. J. (1988). "The expandable platform as a structure on the Moon." *Engineering, construction, and operations in space*, S. W. Johnson and J. P. Wetzel, eds., ASCE, New York, 375–388.

Mendell, W., ed. (1985). "Lunar bases and space activities of the 21st century." *Proc., Lunar and Planetary Institute*, Houston.

Namba, H., Ishikawa, N., Kanamori, H., and Okada, T. (1988a). "Concrete production method for construction of lunar bases." *Engineering, construction, and operations in space*, S. W. Johnson and J. P. Wetzel, eds., ASCE, New York, 169–177.

Namba, H., Yoshida, T., Matsumoto, S., Sugihara, K., and Kai, Y. (1988b). "Concrete habitable structure on the Moon." *Engineering, construction, and operations in space*, S. W. Johnson and J. P. Wetzel, eds., ASCE, New York, 178–189.

Nelson, T. J., Olson, M. R., and Wood, H. C. (1998). "Long delay tele-control of lunar equipment." R. G. Galloway and S. Lokaj, eds., *Space 98*, ASCE, Reston, Va., 477–484.

Nowak, P. S., Criswell, M. E., and Sadeh, W. Z. (1990). "Inflatable structures for a lunar base." *Engineering, construction, and operations in space II*, S. W. Johnson and J. P. Wetzel, eds., ASCE, New York, 510–519.

Nowak, P. S., Sadeh, W. Z., and Criswell, M. E. (1992). "An analysis of an inflatable module for planetary surfaces." *Engineering, construction, and operations in space III*, W. Z. Sadeh, S. Sture, and R. J. Miller, eds., ASCE, New York, 78–88.

Okumura, M., Ohashi, Y., Ueno, T., Motoyui, S., and Murakawa, K. (1994). "Lunar base construction using the reinforced earth method with geotextiles." *Engineering, construction, and operations in space IV*, R. G. Galloway and S. Lokaj, ASCE, New York, 1106–1115.

Pieniazek, L. A., and Toups, L. (1990). "A lunar outpost surface systems architecture." *Engineering, construction, and operations in space II*, S. W. Johnson and J. P. Wetzel, eds., ASCE, New York, 480–489.

Prael, R. E., Strottman, D. D., Strniste, G. F., and Feldman, W. C. (1990). "Radiation exposure and protection for Moon and Mars missions." *Rep. LA-UR-90-1297*, Los Alamos National Laboratory, Los Alamos, N.M.

Richter, P. J., and Drake, R. M. (1990). "A preliminary evaluation of extraterrestrial building systems." *Engineering, construction, and operations in space II*, S. W. Johson, and J. P. Wetzel, eds., ASCE, New York, 409–418.

Richter, T., Lorenc, S. J., and Bernold, L. E. (1998). "Cable based robotic work platform for construction." *15th Int. Symp. on Automation and Robotics in Construction*, 137–144.

Sadeh, W. Z., and Criswell, M. E. (1994). "A generic inflatable structure for a Lunar/Martian base." *Engineering, Construction, and Operations in Space IV*, R. G. Galloway and S. Lokaj, eds., ASCE, New York, 1146–1156.

Sadeh, W. Z., Sture, S., and Miller, R. J., eds. (1992). *Engineering, construction, and operations in space III*, ASCE, New York.

Sargent, R., and Hampson, K. (1996). "Challenges in the construction of a lunar base." *Engineering, construction, and operations in space, S. W. Johnson*, ASCE, New York, 881–888.

Schroeder, M. E., and Richter, P. J. (1994a). "A membrane structure for a lunar assembly building." *Engineering, construction, and operations in space IV*, R. G. Galloway and S. Lokaj, eds., ASCE, New York, 186–195.

Schroeder, M. E., Richter, P. J., and Day, J. (1994b). "Design techniques for rectangular lunar modules." *Engineering, construction, and operations in space IV*, R. G. Galloway and S. Lokaj, eds., ASCE, New York, 176–185.

Seiheimer, H. E., and Johnson, S. W. (1969). "Adhesion of comminuted basalt rock to metal alloys in ultrahigh vacuum." *J. Geophys. Res.*, 74(22), 5321–5330.

Sherwood, B. (1990). "Site constraints for a lunar base." *Engineering, construction, and operations in space II*, S. W. Johnson and J. P. Wetzel, eds., ASCE, New York, 984–993.

Silberberg, R., Tsao, C. H., Adams Jr., J. H. and Letaw, J. R., (1985). "Radiation transport of cosmic ray nuclei in lunar material and radiation doses." *Lunar bases and space activities of the 21st century*, W. W. Mendell, ed., Lunar and Planetary Institute, Houston, Tex.

Simmerer, S. J. (1988). "Preparing to bridge the lunar gap." *ASCE, J. Aerosp. Eng.*, 1(2), 117–128.

Slane, F. A. (1994). "Engineering implications of levitating lunar dust." *Engineering, construction, and operations in space IV*, R. G. Galloway and S. Lokaj, eds., ASCE, New York, 1097–1105.

Strenski, D., Yankee, S., Holasek, R., Pletka, B., and Hellawell, A. (1990). "Brick design for the lunar surface." *Engineering, construction, and operations in space II*, S. W. Johnson and J. P. Wetzel, eds., ASCE, New York, 458–467.

Toups, L. (1990). "A survey of lunar construction techniques." *Engineering, construction, and operations in space II*, S. W. Johnson and J. P. Wetzel, eds., ASCE, New York, 399–408.

Vanderbilt, M. D., Criswell, M. E., and Sadeh, W. Z. (1988). "Structures for a lunar base." *Engineering, construction, and operations in space*, S. W. Johnson and J. P. Wetzel, eds., ASCE, New York, 352–361.

Watson, P. M. (1988). "Explosives research for lunar applications: A review." *Engineering, construction and operations in space*, S. W. Johnson and J. P. Wetzel, eds., ASCE, New York, 322–331.

Young, J. F., and Berger, R. L. (1988). "Cement-based materials for planetary materials." *Engineering, construction, and operations in space*, S. W. Johnson and J. P. Wetzel, eds., ASCE, New York, 134–145.

American Society of Civil Engineers 1852–2002 Building a Better World

150th Anniversary Paper

Cementitious Materials—Nine Millennia and A New Century: Past, Present, and Future

Arnon Bentur[1]

Abstract: Cementitious materials have been used in construction for several millennia and through several civilizations. Their use has been revitalized in the recent century, and they have become the major construction material for housing and infrastructure. The amalgamation of the science and engineering of these materials in recent decades has advanced the state of the art in their understanding and development. In a new century where sustainable construction is becoming an overriding consideration, the new scientific concepts can have direct impact on the way we use and formulate cementitious materials. On top of this, the scientific advances have provided the means for the potential use of the current portland cement as a platform to develop composite formulations of a high level of performance equivalent to and in some respects exceeding those of steel and ceramics. This may provide the basis for advanced and efficient construction techniques. The present article reviews the past, present, and future of cementitious materials and highlights the need for a comprehensive approach to maximize the advantages of the newly emerging cementitious materials and concretes.

DOI: 10.1061/(ASCE)0899-1561(2002)14:1(2)

CE Database keywords: Cements; History; Concrete.

Introduction

Construction has been an important element since civilization began and continues its vital role as a key industry in the rapidly changing modern society. It supplies the basic human need for housing and provides the infrastructure for industry and commercial activities, which are the driving forces for the advancement and well-being of society. Materials are at the heart of the construction industry. They determine the quality of the end product and the technology by which it is manufactured. Thus, advancement of the performance of a variety of types of structures and their efficient and industrialized production are directly related to the characteristics of the materials involved. Innovation in construction is highly linked with development of advanced construction materials.

Cementitious materials are a major class of construction materials that have been with us for more than nine millennia. To a nonexpert, they seem quite simple—a powder that is mixed with water, fillers, and aggregates to form a fluid mass that can be easily shaped and molded and thereafter hardens spontaneously in normal environmental conditions. This apparent simplicity is not matched by any other existing material and makes this class of materials unique.

Most cementitious materials can be considered calcareous, and their overall composition is defined within the ternary diagram $CaO\text{-}SiO_2\text{-}Al_2O_3$ shown in Fig. 1. The ancient cementitious materials were lime alone or lime in combination with natural pozzolans, as well as gypsum (the latter is outside the ternary diagram, as it is composed of calcium sulfates), while the modern ones are largely portland cement. The trend at present and in the foreseeable future is to increase the use of combinations of portland cement and large contents of mineral additives such as slag and fly ash.

The development of cementitious materials has a long historical record. In the early ages, their advent was to a large extent based on tradition and experience that crossed a range of civilizations. In the last century a major leap in the technology of cementitious materials has occurred, based on a unique bridging between the science and engineering of these systems.

The simplicity in production of components, engrained in the fresh fluid nature of the system, can be drastically improved using methods based on sophisticated concepts of surface chemistry and rheology. The properties in the mature state can be enhanced, approaching the level of "higher-grade" materials through control of microstructure and chemical reactions. The concepts of materials science are routinely applied in cementitious materials, providing the infrastructure for the development of systems with the superior performance and environmental friendliness required in this new 21st century. Among the various classes of cements, portland cement and composite cements based on portland cement are and will continue in the foreseeable future to be the main binder used in construction. The viability of this binder is demonstrated in the consumption trends for the last 50 years presented in Fig. 2. This article provides a review of the past and present and a projection of some future trends.

[1]National Building Research Institute-Faculty of Civil Engineering Technion, Israel Institute of Technology.

Note. Associate Editor: Antonio Nanni. Discussion open until July 1, 2002. Separate discussions must be submitted for individual papers. To extend the closing date by one month, a written request must be filed with the ASCE Managing Editor. The manuscript for this paper was submitted for review and possible publication on June 20, 2001; approved on August 29, 2001. This paper is part of the *Journal of Materials in Civil Engineering*, Vol. 14, No. 1, February 1, 2002. ©ASCE, ISSN 0899-1561/2002/1-2–22/$8.00+$.50 per page.

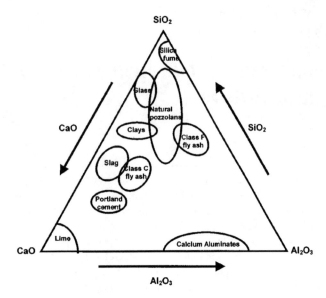

Fig. 1. Ternary diagram representing overall composition of calcareous cementitious materials

Historical Notes

The early beginning of cementitious materials dates back to 7000 B.C., to the Galilee in Israel, where a mortar floor was discovered in 1985 in Yiftah El (Malinowski and Garfinkel 1991; Ronen et al. 1991). It consists of several layers of $CaCO_3$ (upper finish of 5 mm and core of 50 mm), which was apparently manufactured from hydrated lime produced by burning of limestone and slaking thereafter. The microstructure of the binder is quite dense (Fig. 3), and the strength of the material evaluated in recent studies exceeded 30 MPa (Ronen et al. 1991). There is evidence for the use of similar cementitious floor construction on the Danube River, at Lepenski Vir in Yugoslavia, which dates to 5600 B.C. (British Cement Association 1999). The use of lime implies that ancient society had the technology required to burn limestone, and this is confirmed by several archeological findings.

A more intensive use of cementitious materials dates back to the ancient Egyptian civilization, at about 3000 B.C. It was suggested that the stone blocks used in the construction of the pyramids were bonded by cementitious materials, including gypsum (burned gypsum) and lime (burned limestone) [e.g., Snell and Snell (2000)]. It is believed that this technology was diffused around the Mediterranean countries, to be adopted by the Greeks and later by the Romans.

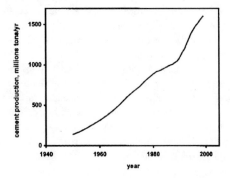

Fig. 2. Consumption of portland cement in recent 50 years

Fig. 3. Microstructure of 7000 B.C. lime revealed in Yiftah El, in the Galilee, Israel. Right-hand side is top layer and left is bottom layer (after Ronen et al. 1991)

The Romans, during the period between 300 B.C. and 200 A.D., improved this technology (Vitruvius 13 B.C.; Malinowski 1979; British Cement Association 1999). They used slaked lime in a mixture with volcanic ash (called pozzolana) found near Pozzuoli at Naples Bay. It is likely that the first use of this mixture resulted from observation that when ash was added as a fine filler (sand replacement), the hardened material performed better. The experience that led to this discovery resulted in what is known as pozzolanic cement, where improved properties in the hardened state are based on chemical reactions between hydrated lime (slaked lime) and the amorphous aluminosilicates. In his handbook for Roman builders, Vitruvius (13 B.C.) described this material as one that hardens both in air and underwater.

During the Roman Empire, this pozzolanic cement served for the construction of such large-scale structures as the theater in Pompeii in 75 BC, as well as infrastructure components such as aqueducts and sewers (Malinowski 1979). The Romans also developed the concept of lightweight concrete by casting jars into wall arches as well as the use of pumice aggregate, which was obtained by crushing a porous volcanic rock (British Cement Association 1999). The arches of the Colosseum and the Pantheon dome were made with such materials.

The art of using cementitious materials and making concretes was essentially lost after the fall of the Roman Empire. The reemergence of cementitious materials dates largely to the 17th and 18th centuries (Klemm 1989; Idorn 1997; Blezard 1998; British Cement Association 1999). A significant step toward the development of modern portland cement was made by John Smeaton in England in 1756, when he was involved in the reconstruction of the Eddystone lighthouse. He was driven by the need to develop a masonry construction that would be durable in a marine environment. His concept was to carry out experiments in order to produce a binding lime mortar that would not dissolve in seawater. For that purpose he burned different types of limes and mixed them later with a trass ("tarras") that possessed pozzolanic characteristics. He discovered that argillaceous limestone, namely limestone that contains some clay, provided better performance than "pure white chalk." This class of materials may be defined as hydraulic limes and can be considered as midway between slaked lime and modern portland cement. Additional improvements of this class of binders by Bay Higgins in 1779 and James Parker in 1796 followed thereafter (Klemm 1989).

Evolution of Modern Portland Cement

A significant step toward the development of modern portland cement was made by Louis-Joseph Vicat in France in his publication, "Recherches experimentales sur les chaux de construction," in 1818. He discovered that hydraulic lime can be made not only from argillaceous limestone, but also by burning an artificial mixture of clay with a high-grade limestone or quicklime (i.e., burned limestone before slaking—CaO). The latter mode of production was preferred and was named "artificial lime twice kilned." Various patents and products that may be defined as hydraulic limes or perhaps low-grade hydraulic cements were issued in that time period by Maurice St. Leger in 1818, John Tickel in 1820, Abraham Chambers in 1821, and James Frost in 1822 (Klemm 1989).

A three-stage process for producing hydraulic cement was first developed by Joseph Aspdin in 1824: calcining the limestone, burning it with clay, and recalcining the mixture to obtain the final product. This process has some similarities to Vicat's "twice-calcined" product. The Aspdin process is more adaptable to a variety of raw materials and assumes more intimate and efficient reaction between the raw ingredients during the firing stages. This process was patented as "an improvement in the modes of producing artificial stone," which was claimed to be similar to portland stone, considered one of the best quarried building stones in England. Aspdin called this product "portland cement," and he was quite successful in marketing it to the construction industry, convincing them that it performed better than the variety of hydraulic limes.

Portland cement was introduced in the United States around 1870, and the first plant was established by David O. Saylor, in Coplay, Pa. It is of interest to note that prior to that time, a binder named "natural cement" was used in the United States that was produced by burning a natural rock, which was argillaceous limestone, with a composition almost ideal for cement production. These cements competed for some time with portland cement imported from Europe. The natural cement was weaker, but its price was lower (Klemm 1989).

The development of portland cement in the 19th and 20th centuries was classified by Blezard (1998) into three stages: "Proto-Portland," "Meso-Portland," and "Normal-Portland." The difference between them is in the improved proportioning of the raw materials and in the burning process, reaching an adequate clinkering temperature. Thus, the Proto-Portland cement was essentially a somewhat improved hydraulic lime, not much different from the others of its generation, at the first half of the 19th century; the Meso-Portland cement, with better clinkering, consisted largely of dicalcium silicate and only a little tricalcium silicate; and the Normal-Portland had a significant content of tricalcium silicate and little free lime.

The advancement from the Meso-Portland to Normal-Portland stage was the result of a better choice of the raw mix and also a much better control of the burning process, in particular in the range of the clinkering temperatures, where the calcium silicates are being formed. The development of the rotary kiln to replace the shaft kiln had a considerable impact on the quality of the product and the efficiency of the production process, which became continuous rather than the batch type operation in the shaft kiln (Klemm 1989; Blezard 1998; British Cement Association 1999).

The development in concrete technology might be quantified in terms of the strength of the portland cement, evaluated by the standard test of 1:3 cement:sand mortars cured continuously in water (Fig. 4). The improved quality reflects the mineralogical composition of cements, with a higher C_3S to C_2S ratio, and a better control of the grinding of the clinker.

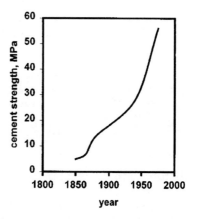

Fig. 4. Development over time of quality of portland cement as characterized by strength of 1:3 mortar cubes (adapted from Blezard 1998)

Evolution of Scientific Foundations

The foundations for the chemistry of cement were laid by Henry Le Chatelier from France in his doctoral thesis of 1887 (Le Chatelier 1887). He established that the main mineralogical phase in portland cement responsible for its cementing properties had the formula $3CaO \cdot SiO_2$ (C_3S using the then current cement chemistry notation). Le Chatelier based his investigation on microscopic petrographic observations that revealed that C_3S was one of four phases. Using chemical analysis, he established guidelines for the composition of raw material required for producing good cements, expressed in terms of ratios named "moduli":

$$\frac{CaO + MgO}{SiO_2 + Al_2O_3} < 3 \tag{1}$$

$$\frac{CaO + MgO}{SiO_2 - (Al_2O_3 + Fe_2O_3)} > 3 \tag{2}$$

He suggested that optimal composition is obtained when ratio (1) is about 2.5 to 2.7 and ratio (2) is 3.5 to 4.0.

The four-phase composition was confirmed by Torrebohm, who named them alite, belite, celite, and felite. Alite was shown to be essentially C_3S, and belite was largely β-C_2S. In 1915, G. A. Rankin and F. E. Wright (Klemm 1989) presented a fundamental study of the ternary system CaO-SiO_2-Al_2O_3 within which portland cement is placed. They provided the cement chemists' notations of C for CaO, S for SiO_2, A for Al_2O_3, F for Fe_2O_3, and $\bar{S}$ for SO_3 (Klemm 1989).

Studies of the mineralogical composition of the cement that served as a foundation for control of the production process were accompanied by investigations of the hydration reactions, to establish the mechanisms of setting and hardening. The crystalline concept was suggested by Le Chatelier in 1881, while W. Michaelis in 1893 resorted to a concept of formation of colloidal-gelatinous material (Lea and Desch 1956). Controversy was struck at that time and continued into the early part of the 20th century regarding which of the two hypotheses better describes the hydration products. A controversy of similar intensity was sparked years later in the 1960s and 1970s regarding the internal structure of the hydrated C-S-H and the significance of surface area measurement with respect to characterization of its structure: H_2O adsorption (Powers and Brownyard 1948, Powers 1960) versus N_2 adsorption (Feldman and Sereda 1968).

Another important characteristic of portland cement chemistry, highlighted already at the end of the 19th century and the beginning of the 20th century, was regulation of setting time by the use of gypsum. From 1876 to 1878, Dyckerhoff of Germany studied the effect of up to 2% additions of gypsum and reported set retardation from 20 to 840 min. The reactions with gypsum were studied by Candolt, and Le Chatelier commented in 1905 on the formation of $Al_2O_3 \cdot 3CaO \cdot 3(CaO \cdot SO_3) \cdot 30H_2O$ (known as ettringite), which may cause expansion deterioration in seawater.

Early 20th Century Technological Developments

The development of portland cement was followed by advancement of technologies for efficient use of this new binder (British Cement Association 1999). A reinforcing system based on the use of twisted steel rods was patented by Earnest L. Ransom in 1884, and the first reinforced concrete skyscraper, the Ingalls Building, was constructed in 1904. In 1927 Eugene Freyssinet developed the prestressing concept, and in 1933 Eduardo Torroja designed the first thin-shelled roof at Algeciras (Torroja (1958)) and Pier Luigi Nervi developed a thin shell construction for airplane hangars. The Hoover Dam, the first massive concrete dam, was built in the United States in 1936.

Architectural concepts to use exposed concrete, particularly precast elements, were demonstrated first in 1923 in France in the Cathedrale Notre-Dame du Haut in Le Raincy. Technologies for surface finishing of precast elements were developed during the 1930s. The precast concrete industry took a sudden rise after World War II, in view of the massive need for quick housing. This development was the result of an integrative effort dealing with the materials and production technology, curing practices, and mechanization of construction sites by such equipment as cranes.

High-Temperature Chemistry and Cement Production

The technology of production of high-grade cement is based on high-temperature chemistry. The high-temperature reactions for clinker formation can be divided into five steps (Taylor 1997; Jackson 1998; Macphee and Lachowski 1998):

1. Decomposition of clay materials ($\sim$500 to 800°C);
2. Decomposition of calcite, calcining ($\sim$700 to 900°C);
3. Reactions of calcite (or lime formed from it), quartz, and decomposed clays to form C_2S ($\sim$1000° to 1300°C);
4. Clinkering reactions at 1,300 to 1,450°C to form C_3S. A melt of aluminate and ferrite is formed that acts as a flux to facilitate the formation of C_3S through the reaction of C_2S and lime (CaO); and
5. Processes occurring during cooling where liquid crystallizes, forming the aluminate and ferrite phases.

Bogue developed a calculation for estimating the phase composition of portland cement clinker assuming a composition of C_3S for alite, C_2S for belite, C_3A for aluminates, and C_4AF for ferrites.

It should be noted that the portland cement particles, which are obtained by grinding the clinker in the presence of gypsum, are not separated into the individual phases, but rather one particle may contain several phases (Scrivener 1989).

The high-temperature chemistry studies resulted in the establishment of the ternary phase diagram CaO-Al_2O_3-SiO_2, which serves as the basic guideline for characterizing the high-

temperature reactions and stable-phase compositions. The raw materials for producing such compositions are limestone, which is the major source of CaO, and suitable clays and shales containing 55 to 60% SiO_2, 15 to 25% Al_2O_3, and 5 to 10% Fe_2O_3, with mineralogical composition consisting of clay minerals, quartz, and sometimes iron oxide.

Much attention was given to the optimization of the clinkering process and the grinding thereafter with gypsum. The considerations of the clinkering process were the composition of the cooled clinker, the burnability of the mix (i.e., the ease with which free lime can be reduced to an acceptable value in the kiln), and the role of minor components, energy consumption, and environmental effects. In recent years, much of the change in the production processes of portland cement has been driven by energy and environmental constraints and requirements (European Commission 1999). Modern clinker is produced in a dry process using suspension preheaters and precalciners with shorter rotary kilns. The fuels commonly used are pulverized coal, oil, natural gas, and lignite. There is an increasing tendency to use wastes as partial replacement for the fuels and raw feed (Odler and Skalny 1995).

The overall composition and burning process are affected to a large extent by the ratio between the four major oxides in the raw feed (CaO, SiO_2, Al_2O_3, and Fe_2O_3). This is demonstrated in conventional calculation of the composition using the Bogue equation and the guidelines for optimal choice of the three moduli: LSF (lime saturation factor)—$CaO/(2.80SiO_2 + 1.18Al_2O_3 + 0.65Fe_2O_3)$; SR (silica ratio)—$SiO_2/(Al_2O_3 + Fe_2O_3)$; and AR (alumina ratio)—Al_2O_3/Fe_2O_3. Increase in SR lowers the content of the fluxing liquid and thus reduces burnability. The AR modulus controls the ratio between the aluminate and ferrite phases, which may have considerable influence on the properties of hydrated cements.

Special attention has been given in recent years to the influence of minor components. The role of some of them has changed due to environmental and health considerations. A special case is that of the alkalis, sodium and potassium (both the content and the ratio between the two). The sources for alkalis are usually in the fuel, mainly coal and shale. A cycle of gas flow is established in the kiln in which gas is flowing in a direction opposite to the movement of solids, in order to save energy by the use of the hot gas to heat the raw material. As a result, alkalis in the gas react with sulfate to form alkali-sulfates, which may condense and be removed from the gas phase to be combined in a variety of ways with the silicate melts. As a result of such processes, the contents of alkalis in cement can become higher. This may have a variety of influences, some positive and some negative, on the performance of portland cement (Taylor 1997).

Materials Science of Cement and Concrete

The scientific foundations for the behavior of hydrating portland cement and the resulting concrete end product were set first on the basis of the chemistry of hydration reactions. These were followed by studies and characterization of the microstructure and internal bonding using techniques such as electron microscopy, magnetic resonance, atomic force microscopy, adsorption isotherms, and mercury porosimetry.

The understanding of the relations between internal structure and performance in the fresh and mature state serves as a basis for developing advanced systems and solving a variety of technological limitations in existing systems. Numerous books and confer-

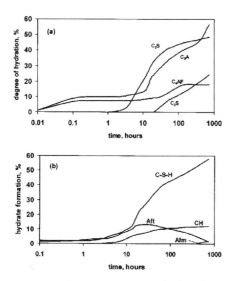

Fig. 5. Portland cement hydration reactions: kinetics of hydration and nature of products formed (adapted from Odler 1998)

ences have been published addressing these issues, e.g., Soroka (1979); Skalny (1989, 1991, 1992); Skalny and Mindess (1995, 1998); Taylor (1997); Hewlett (1998); and Ramachandran (2001).

Chemistry of Cement

Early studies of hydration reactions revealed the differences in the kinetics and nature of hydration reactions and products of the various pure individual minerals. Although the fundamentals established were found applicable to many portland cement systems, some significant differences had to be considered due to a variety of influences (Odler 2000; Taylor 1997), such as the following:

- Mineral phases in portland cement are not "pure," but are doped with various foreign ions;
- Alkalis and gypsum present in portland cement affect modes and rates of hydration reactions; and
- Minerals in portland cement are not in form of separate grains.

These differences exert considerable influence on the various processes involved in the hydration reactions, due to several factors:

- Nature of dissolution of mineral phases in contact with water,
- Nucleation and growth of hydration products, and
- Rate of diffusion of dissolved ions through hydrated products engulfing unreacted cement grains.

Thus a relevant study of cement hydration requires evaluation of the cement system itself. The nature of the hydration reactions in terms of the rates of reactions of the individual minerals and the hydration products formed are shown in Fig. 5. The process can be divided into several stages: preinduction period (first few minutes, where some dissolution of the minerals occurs and all the alkali-sulfate dissolves, contributing K^+, Na^+, and SO_4^{2-} ions to the solution); dormant period (few hours); acceleration stage (3 to 12 h, characterized by nucleation of hydration products); and postacceleration period, when the reactions become diffusion controlled. Details of the mechanisms involved are reported in several publications [e.g., Taylor (1997); Odler (1998)] and are beyond the scope of this review.

An important element in the processes at this stage is the composition of the liquid phase, which is highlighted here in view of the significance of the composition of the aqueous solution on other characteristics, particularly the durability of the mature system. The alkali-sulfates are highly soluble and enter the liquid phase immediately, as K^+, Na^+, and SO_4^{2-} ions. The Ca^{2+} ions dissolve into the liquid at a somewhat slower rate, to become saturated with respect to $Ca(OH)_2$ after a few minutes. The OH^- ion concentration in the liquid phase is controlled by dissolution of $Ca(OH)_2$ in the hydration of C_3S and by the dissolution of alkalis present in the clinker materials. Thus the OH^- concentration is greater than expected on the basis of $Ca(OH)_2$ dissolution only, and therefore the pH of the pore solution eventually attains values greater than about 12.5, which is characteristic of a saturated $Ca(OH)_2$ solution. Values as high as 13.5 have been reported, depending largely on the soluble alkali contents.

Composition and Microstructure of Hydrated Cement

It is well known that the major phases of the hydration products are C-S-H, CH, and a variety of sulfoaluminate hydrates, such as ettringite ($C_3A \cdot 3C\bar{S} \cdot 31H$) and monosulfate ($C_3A \cdot C\bar{S} \cdot 12H$). Although there is a rough agreement on their overall composition and structure, the detailed composition and microstructural characteristics are still debated and explored. This is largely because some of the phases, in particular the C-S-H, are largely amorphous and show variability in their structure, which may depend on a variety of factors related to the cement composition, water/cement ratio, and curing conditions [e.g., Taylor (1997); Odler (1998), Richardson (1999)]. Also, it should be borne in mind that the cement paste, even after many years, is still a "live" system: residues of unhydrated grains always exist, and the hydrated phases are not in equilibrium, i.e., some of the pahses are in a metastable state and therefore are prone to changes under a variety of environmental conditions, such as carbonation.

Several models of the structure of hydrated cement pastes have been developed over the years, based on a variety of concepts, some of them complementary and some in contradiction to each other. Three of the more "traditional" models are presented in Fig. 6. The Powers model is based on a monumental series of studies carried out at the Portland Cement Association (PCA) in the United States during the 1940s, 1950s, and 1960s (Powers and Brownyard 1948; Powers 1968). Applying the physical chemistry concepts available at that time, the PCA group developed many of the fundamental concepts of the structure of the pores and the C-S-H phase, which are still valid: its high surface area and colloidal nature, with the implication of these characteristics to strength generation and volume changes. The concept of two types of pores—capillary and gel pores—can provide a working hypothesis for understanding engineering characteristics of the material. Yet studies with more advanced tools indicate that the pore-size distribution is more gradual in nature, and a large range of pore sizes—over several orders of magnitude—need to be considered (Diamond 1999), (Table 1).

Considerable controversy developed regarding the microstructure of the C-S-H, as reflected to some extent in the models presented in Fig. 6. In part, the differing views were related to the concept of interlayer water [metacrystals in Ishai's (1968) model]: whether they are part of the C-S-H or separate gel pores. Their role was assumed to be greater in the Feldman-Sereda model, with respect to explaining volume changes induced in drying and wetting conditions. The differing views were also reflected in interpretations of surface area measurements of the hydrated material; they are typically about 150 m^2/g obtained by BET measurement using water adsorption, about 15 m^2/g for BET measurement obtained by N_2 absorption, and about 15 m^2/g

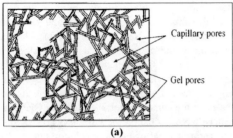

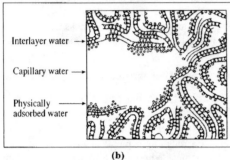

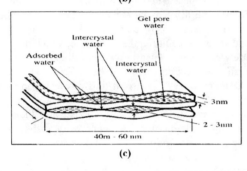

Fig. 6. Three traditional models of structure of hydrated portland cement paste: (a) Powers (1960); (b) Feldman-Sereda (1968); (c) Ishai (1968)

calculated from the pore-size distribution obtained by mercury intrusion porosimetry. The technique of low-angle X-ray scattering, which allows surface area measurement at different moisture states, revealed areas of about 800 m^2/g for saturated samples (Winslow and Diamond 1974), which are reduced at lower relative humidity. The differences in surface area measurements reflect the accessibility of different spaces to the various adsorbing or intruding materials.

A common element in all the models is the implication that the basic unit is some kind of a C-S-H sheet, and that these units are organized in a variety of modes and varying spaces (gradient of spaces). This is clearly described in the model of Ishai (1968). The nature of interactions involved in removal or adsorption of

Table 1. Characteristic Size Range of Porosities in Hydrated Cement Paste

Type of pore	Size of pore (μm)
Interparticle spacing between C-S-H sheets	0.001–0.004
Capillary pores	0.01–1.6
Entrained air bubbles	70–1,200
Entrapped air voids	1,200–4,000

Table 2. Typical Characteristic-Average Compositional Parameters in Hydrated Portland Cement Paste

Characteristic compositional parameters	C-S-H	CH	AFm
C/S ratio	1.67	25	25
C/(A+F) ratio	9.1	25	1.8
C/$\overline{S}$ ratio	33	250	4.2

water depends on the space: very low physical interactions for bigger spaces, and stronger ones, approaching those of structural interlayers, for smaller ones.

These structural characteristics have considerable implications for understanding, modeling, and predicting the behavior of the cementitious systems in changing moisture conditions.

Results of modern research on the microstructure and microcomposition of hydrated cements suggest that many of the features of the structure should be characterized in terms of average parameters and gradients. Microchemical analysis using various methods of testing shows variabilities in composition in space that can be interpreted in terms of three main compositional regions:

1. Regions high in Ca and Si, low in Al and Fe—characteristic of spots high in C-S-H;
2. Regions very high in Ca, low in Si, Al, and Fe—characteristic of spots high in CH; and
3. Regions high in Ca, Al, and Fe, low in Si—characteristic of spots high in monosulfate (AFm) and ettringite (AFt).

Average compositions of these phases are provided in Table 2, showing that none of them can be considered "pure." However, one should address with caution such average values, since they represent in fact a much more complex structure (Richardson 1999). For example, distinctions should be made between the inner and the outer C-S-H and the C-S-H formed in the presence of pozzolans. The ratio of Ca/Si (C/S ratio) is often used to characterize the C-S-H, but the average value may range from 1.5 to 2.0, with spot values spreading over a wider range. Lower C/S ratio is characteristic of C-S-H formed in the presence of pozzolans.

When considering the structure of the C-S-H, due attention should also be given to the nature of the Si-O bonding in the SiO_4 hydrated units. At first the structure is that of hydrated monomeric units (SiO_4), and the progress of hydration results not only in the formation of more hydrates, but also in the formation of dimeric units and sometimes even larger units. This is essentially a process in which some polymerization may occur, i.e., silicate polymerization, when silica tetrahedra become linked together in –Si–O–Si–bonds. Various techniques have been used to assess the degree of polymerization, and they indicate that polymerization may occur as hydration progresses (Young and Sun 2001). In a mature paste, polymerization may be driven by conditions such as carbonation and drying. A process of this kind has been suggested as one of the causes of irreversible shrinkage under carbonation and drying conditions. It may be simply perceived as bonds that are formed as the C-S-H foils approach each other on drying, when the interlayer water is removed.

Composition and Microstructure of Concrete

The structure of concrete has often been considered in simplistic terms as a portland cement paste with diluting inclusions of aggregates. This implies that the microstructure of the paste matrix

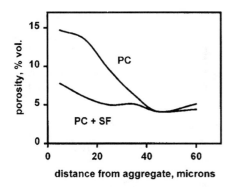

Fig. 7. Gradients of porosities at ITZ zone measured by backscattered electron imaging (after Scrivener et al. 1988)

in the concrete is identical to that of portland cement paste. However, this is an oversimplified approach, as several levels of heterogeneity may be induced in the concrete, revealed first by French researchers studying a composite system of cement paste cast against aggregate surface (Farran 1956) and later in studies of the concrete itself [e.g., Ollivier et al. (1995); Diamond (1999)], that do not occur in the paste. One such level is associated with the rheological characteristic of the concrete, affecting the distribution of the cement and filler particles in the mass, as well as effects such as bleeding and a second level of heterogeneity that may be due to aggregate-paste chemical interactions.

The processing of the concrete in its fluid state is intended to provide a uniform dispersion of all of the constituents. However, ideal dispersion is difficult to achieve. There is an inherent "wall effect" that leads to a gradient in the dispersion of the fine particles (cement and fillers) in the vicinity of the aggregate surface: the coarser cement and filler particles cannot pack readily near the aggregate surface. Since the cement particle size distribution is typically in the range of 1 to 100 μm, a zone of about 10 to 50 μm is formed around the aggregate in which the effective water/cement ratio is higher (i.e., less cement in this region, with the cement particles being of the smaller size range) (Goldman and Bentur 1994; Ollivier et al. 1995; Alexander et al. 1999; Bentz and Garboczi 1999).

Obviously, the hydrated structure in this zone will tend to be more porous, although the rate of hydration will be higher since the zone consists of smaller particles. The greater porosity will also result in a greater tendency for the deposition of large CH crystals. Gradients of microstructure that develop in this zone, called the interfacial transition zone (ITZ), are shown in Fig. 7. The introduction of fine fillers, such as silica fume, with particles smaller than that of cement (~0.1 μm) results in densification of this zone (Fig. 7).

The more porous nature of the ITZ is considered a factor affecting strength and transport properties. The extent of its detrimental influence is still in debate. Probably its influence on strength and modulus of elasticity is moderate or small (~20%), and its influence on transport properties could either be direct (the result of a more porous path near the aggregates) or indirect by creating a weaker zone that is more prone to cracking (Bentur 1998a).

An additional influence on concrete microstructure—caused by the properties of the fresh concrete—is bleeding, which creates a different microstructure beneath aggregate and steel bars, in addition to zones of varying porosity throughout the concrete

mass. These microstructural characteristics are affected by the rheological nature of the concrete.

Additional influence due to chemical reactions between the aggregate and the cementing paste matrix are rather limited. The exception is the deleterious alkali-aggregate reaction, whereby aggregates containing amorphous silica phases react more readily with the alkaline matrix.

Modeling of Concrete Behavior

Modern materials science concepts have been used to develop models to describe the behavior of concrete, addressing a variety of issues such as mechanical performance and transport characteristics. The development of these models is based on a large number of fundamental studies resolving many characteristics, such as (1) cement hydration and development of microstructure in pastes and concretes; (2) transport characteristics considering the microstructure and processes, such as diffusion, permeation, electromigration, and convection by capillary suction; (3) time-dependent volume changes and their dependence on microstructure, considering a variety of physical processes, such as surface tension, capillary stresses, and hindered adsorption; and (4) mechanical behavior based on such concepts as composite materials and fracture mechanics.

Microstructural Modeling

Computer modeling of cement hydration and microstructure enables us to simulate the microstructure obtained in electron microscope analysis and has the potential of becoming a tool for optimizing the composition of the material by means such as control of particle size of cement and fillers, [e.g, Bentz et al. (1995); Van Breugel (1996); Bentz (1997); Bentz and Garboczi (1999)]. The advent of computer modeling is linked with the improved understanding of the processes by which microstructure evolves, as is also the increase in computational power. Computer modeling allows us to consider features differing by several orders of magnitude in size. Such models have been used to quantify the microstructure and phase composition and their variability in the concrete, showing in particular effects such as the ITZ, which is characterized by initial higher porosity. The computer models can be used to predict the effect of combinations of various mineral fillers on the transport properties, considering the variability in microstructure.

Modeling of Transport Properties

Models to predict transport properties have been developed based on the basic laws describing transport of substances through porous media, superimposed on the special microstructural characteristics and composition of pastes and concretes. A variety of transport processes that depend on the transported substance and driving force for the transport were resolved by analytical methods. These were quantified in terms of tests developed to characterize relevant parameters such as diffusion and permeability coefficients [e.g., Kropp and Hilsdorf (1995)]. The processes modeled were diffusion of ions (usually chloride ions, the driving force being concentration differences); electromigration (similar to diffusion but having the electric field as the driving force—this is less common in practice but useful for accelerated tests to determine diffusion coefficients); permeation of gases and liquids (the driving force being pressure difference); and convection mass flow (the flow of a substance transported in the fluid in which it is contained) (Bentz et al. 1995, 1998; Marchand et al. 1998; Nilsson and Ollivier 1996). The convection occurs in the case of permeation of fluids and is of acute significance in convection, which is the result of capillary suction of water into dry concrete pores in contact with water.

The modeling of the transport phenomena frequently considers combined flow mechanisms of diffusion and convection, diffusion and permeation, and diffusion and electromigration. In modeling of this kind, mass balance equations are used to predict the processes, and due consideration is given to some complicating processes such as chemical binding of some of the migration substances, which is time-dependent-process, and physical binding, which is the result of adsorption on the colloidal high-surface-area particles of hydrated cement (Marchand et al. 1998). On the basis of these concepts, tests to evaluate transport characteristics were developed that tend to be used to set performance criteria for concrete durability (Kropp and Hilsdorf 1995).

Modeling of Mechanical Properties

Mechanical modeling of cementitious systems has a particularly long history. The earlier concepts were based on developing relations between strength and porosity based on prevailing concepts in porous materials in general, e.g., ceramics (Rice 1984), and in cement systems in particular (Beaudoin et al. 1994; Roy 1988; Brown et al. 1991). This concept is behind the early relations developed between strength and water/cement ratio by Feret in 1896 [Eq. (3)] and Abrams in 1919 [Eq. (4)]:

$$f_c = K \left(\frac{c}{c+w+a} \right)^2 \tag{3}$$

$$f_c = K_1 / K_2^{(w/c)} \tag{4}$$

where K, K_1, K_2 = constants; c = cement content; w = water content; and a = air content.

This empirical approach is adequate for many engineering calculations and is applied even today. However, it is not sufficient to develop advanced systems in which the role of aggregates, particulate fillers, and fibers is becoming increasingly important. For this purpose, concepts of composite materials (assessing the bonding between different phases) and fracture mechanics have been applied. Models describing these systems on the meso and micro level have been advanced [e.g., Van Mier (1997); Van Mier and Vervurt (1999)].

On the meso level, the cement paste is considered a continuous and uniform phase that interacts with inclusions such as aggregates and fibers. The matrix-inclusion interaction is characterized in terms of a bonding function that in the most simplistic form is modeled by the single parameter of bond strength. A more realistic approach requires addressing the nature of bonding in terms of distribution of properties as well as the rigidity of the ITZ, reflecting its special structure.

Micromechanical models have been used to predict the rigidity of concretes considering the properties of the aggregates and the nature of bonding, as well as to predict the strength of cementitious composites. However, when the overall response of the material is to be considered, to characterize whether its behavior is ductile or brittle, fracture mechanics concepts had to be introduced, combined with characterization of interfacial properties (Shah and Ouyang 1992). Finite-element analysis with inputs of characteristic properties of paste matrix, aggregates, and interfaces can predict the overall behavior of the system. Such modeling is of particular interest in high-strength concrete systems, in which aggregates play a major role in the mechanics of the system and cannot be considered simply as a diluting component in a paste matrix.

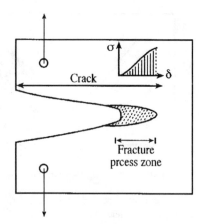

Fig. 8. Fracture process zone in front of advancing crack

Micromechanical models have also become common tools in predicting the properties of fiber-cement composites, such as strength and toughness (Li et al. 1995). The concepts applied within these models include fiber-matrix interactions and fracture processes [e.g., Bentur and Mindess (1990); Balaguru and Shah (1992); Bentur et al. (1995)].

With such models it is possible to predict the properties required to achieve high-strength and pseudo-plastic composites. In consideration of the fracture behavior of cementitious materials, the concepts of linear and nonlinear fracture mechanics have been applied (Shah et al. 1995). The nonlinear concept is modeled in terms of a fracture process zone in front of the advancing crack, which consumes energy by a variety of mechanisms induced under the stress field in front of the crack (Fig. 8).

Modeling of Time-Dependent Deformation

Time-dependent volume changes is an area where much effort has been expended for decades to model and develop rules in codes and standards. This pertains to shrinkage during drying and creep under load in combination with drying. Many of the earlier models were empirical ones describing the effect of cement content, water/cement ratio, aggregate content, size of component, environmental conditions, and level of loading.

With a better understanding of the microstructure of pastes and concretes, models based on fundamental physical and chemical processes were developed taking into account a variety of interaction of water molecules with the hydrated material (Hansen and Young 1991). The concepts used for modeling include capillary stresses and disjoining pressures at higher levels of relative humidity and surface tension and interlayer water movements at lower relative humidity [e.g., Bazant (1995)]. Irreversible volume changes were considered in terms of collapse of pore structure and chemical interactions, in particular silicate polymerization (Hansen and Young 1999).

Fresh Concrete: Rheology and Admixtures

The properties of fresh concrete have usually been addressed from a technological point of view to proportion a mix that can be readily transported, placed, and consolidated. The use of traditional water-reducing admixtures has generally been considered in this context.

The advent of high-performance concrete materials is to a large extent based on developing mixes of drastically improved

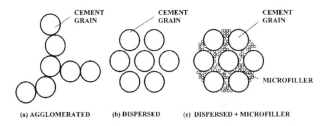

Fig. 9. Schematic presentation of cement particles: (a) in water-flocculated state; (b) in dispersed state; and (c) dispersed in combination with microfillers

flow properties by controlling particle size distribution in combination with chemicals engineered for the purpose of controlling the interparticle attractions in the fresh material (i.e., high-range water-reducing admixtures). For this purpose a much more comprehensive understanding of the flow properties—and their characterization in terms of tests that are more meaningful than the slump test—are required.

The early foundations for such treatment were laid down at the Portland Cement Association, where the principles of rheology were applied systematically to pastes and concretes (Powers 1968). In the past two decades the area was advanced considerably due to the needs outlined above [e.g., Beaupre and Mindess (1998); Struble et al. (1998)]. Rheological measurements were interpreted in terms of the various modes of suspension of particles in water, in flocculated or dispersed states [Figs. 9(a–b)]. The results of rheological measurements of cementitious suspensions, characterizing stress-strain rate behavior of cement pastes and concretes, can be approximated and modeled as a Bingham fluid, with a yield stress and apparent viscosity (Fig. 10). This behavior can be interpreted in terms of a pseudo-yield stress, which causes the breakdown of the flocculated network, and the plastic viscosity thereafter, reflecting the behavior of a highly concentrated dispersed suspension. A more-detailed characterization can reveal additional aspects related to the thixotropic behavior and irreversible breakdown of the structure at increasing strain rates.

Standard rheological testing techniques based on testing the response of a thin layer (of a few millimeters) of fluid to a strain rate effect, producing a stress, can be applied more readily to cement paste of sufficient fluidity. Such tests can be clearly interpreted in terms of stress and strain rates. However, such instru-

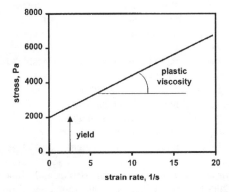

Fig. 10. Stress-strain rate curve of fresh cementitious systems that can be modeled by Bingham-type behavior (adapted from Struble et al. 1998)

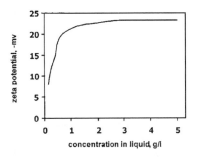

Fig. 11. Buildup of surface charge (zeta potential) with increase of superplasticizers that adsorb on cement grains (adapted from Mollah et al. 2000)

mentation is not valid for concrete where aggregate particles are present, and special rheometers had to be developed for this purpose (Wallevik and Gjorv 1990; de Larrard and Chang 1996; Wallevik 1998).

Addition of water to the mix improves the fluidity without breaking the flocculated structure of the cement particles. However, in the presence of larger amounts of water the particles comprising the concrete mix tend to segregate more, once agitation is stopped, and bleeding occurs. Careful grading of the mix to increase its cohesiveness may compensate for some of the extra segregation and bleeding. Thus, water content and the content and composition of the fine aggregate fraction in the mix were traditionally optimized to obtain a required flow with minimum segregation.

The constraints on improvement in flow properties while minimizing segregation are largely overcome in modern concrete technology by the use of high-range water-reducing chemical admixtures (superplasticizers). These admixtures were first developed and applied in Japan (Mighty admixture) and Germany (Melment admixture), and they represented a breakthrough in concrete technology that opened the door to the new generation of concretes such as high-strength/high-performance and self-compacting concretes. These admixtures are synthetic high-molecular-weight water-soluble polymers [e.g., Sakai and Daimon (1995)]. That are effective in dispersing the flocculated cement grains. Solubility is achieved by the presence of adequate hydroxyl, sulfonate, or carboxylate groups attached to the main organic unit, which is usually anionic.

These polymers adsorb on the surface of the grains and facilitate dispersion by three mechanisms: (1) buildup of a negative charge on the surface (i.e., increase of zeta potential) sufficiently large to cause repulsion (-20 to -40 mV) (Fig. 11); (2) increase in solid-liquid affinity; and (3) steric hindrance when oriented adsorption of nonionic polymers weakens the attraction between solid particles (Fig. 12) Haneharo and Yamada 1999; Mollah et al. 2000. These admixtures do not cause set retardation and air

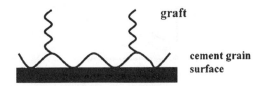

Fig. 12. Schematic description of adsorbed graft polymer superplasticizer causing steric hindrance (adapted from Haneharo and Yamada 1999)

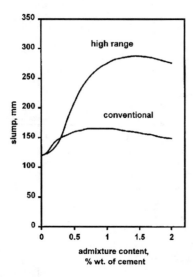

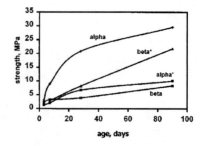

Fig. 15. Hydraulic activity of various forms of belite-rich cements estimated from strength development of 0.60 w/c ratio pastes (adapted from Odler 2000). Main belite phases are alpha (cooling rate of 50,000°C/min), alpha′ (cooling rate of 3,000°C/min), beta* (cooling rate of 50,000°C/min), and beta (cooling rate of 2°C/min)

Fig. 13. Influence of traditional and modern high-range water reducers on flow of concrete (adapted from Edmeades and Hewlett 1998)

entrainment, as may occur with conventional water reducers, and therefore they can be added at much higher dosages of about 2% by weight of cement. The difference in the performance of traditional water reducers and modern high-range water reducers is demonstrated in Fig. 13. Development of more-efficient high-range water reducers by better control and optimization of their molecular structure is currently at the forefront of concrete technology.

Another advanced concept of improving workability in low water/binder ratio mixes, which is based on uniform dispersion of cement grains with very fine fillers between them, is shown in Fig. 9(c). Here, the microfiller particles serve to improve flow properties by acting as small rollers between the bigger and rougher cement grains. In a system of this kind, excellent flow can be obtained at a very low water content, thus achieving a mix with hardly any segregation at a low water/binder ratio. The filler is considered part of the binder, alongside the portland cement. This is the fundamental concept underlying the new generation of high-strength and self-compacting concretes using silica fume. The characterization of such concretes by rheological measurements is shown schematically in Fig. 14.

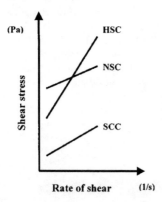

Fig. 14. Schematic description of characteristic rheological behavior of normal strength concretes (NSC), high-strength concretes (HSC), and self-compacting concretes (SCC) (adapted from Wallevik 1998)

Sustainability and Cementitious Materials

The awareness of the ecological impact of technology and the need to address it in our modern society is becoming a factor of increasing importance in the field of cement and concrete. The influence here is at several different levels: (1) producing binders that consume less energy and emit less greenhouse gases, in particular CO_2; (2) incorporating industrial by-products and recycled materials in the cementitious binder as well as in the concretes; and (3) producing structures that would function more efficiently over time, in terms of their durability performance.

Binders

Two main ecological issues are currently of concern in portland cement production: energy consumption, and release of CO_2 during the decalcination stage of clinker production. In modern kilns with preheaters and precalciners using the dry process, the energy required is about 3,000 kJ/kg of cement. About two-thirds of this energy is consumed in the decalcination and clinkering reactions, and one-third in energy losses. This energy range is significantly lower than typical values of a few decades ago, which exceeded 5,000 kJ/kg of cement. The theoretical energy required is dependent on the LSF, being in the range of 1,570 to 1,800 kJ/kg for LSF in the range of 80 to 100% (Lawrence 1998). There is a rough correlation between the CO_2 emission and energy consumption, since decalcination is an energy intensive step.

Several strategies are currently being considered or applied to reduce energy and CO_2 emission:
- Improving current technology to produce a cement of similar composition with less energy. This however is not expected to lead to drastic reduction in CO_2 emission (European Commission 1999).
- Improving the reactivity of the clinker to enable the use of leaner mixes and reducing the clinkering temperature by developing efficient fluxes. It should be borne in mind that reduction in the clinkering temperature by about 100°C will provide only modest energy savings of about 5%.
- Producing cements in which the major reactive phase is belite (C_2S), thus reducing the CaO content and CO_2 emissions. Strategies to improve the reactivity of C_2S need to be developed for this approach to be viable (Ishida and Mitsuda 1998; Odler 2000). Rapid quenching or incorporation of minor components (e.g., alkalis, chromates) can lead to the formation of the α or α′ polymorph of belite, which is more reactive (Fig. 15). It should be noted that the potential savings by this strat-

Table 3. Range of Composition of Common Mineral Additives (adapted from Odler 2000)

| Oxide | Content of Oxide (% wt) | | | |
	Granulated Blast Furnace Slag	Fly Ash (Low Lime)	Fly Ash (High Lime)	Silica Fume
SiO_2	27–40	34–60	25–40	85–98
Al_2O_3	5–33	17–30	8–17	0–3
Fe_2O_3	<1	2–25	5–10	0–8
CaO	30–50	1–10	10–38	<1
MgO	1–21	1–3	1–3	0–3
SO_3	—	1–3	1–5	—
Na_2O+K_2O	<1	<1	0–3	1–5
C	—	<5	<5	0–4

egy are significant but not drastic, in the range of 20%.

- Cements based on combinations of sulfoaluminates and calcium silicates. In these compositions the clinkering temperature can be reduced significantly, and the hydraulic properties are generated not only by reactive C_2S but also by various types of sulfoaluminates. Cements produced in this way can also be adjusted to achieve early strength and expansion for self-stressing (Odler 2000).
- Activation of slags by alkalis, an approach that has been used in Eastern Europe.
- Blends of gypsum with calcium silicate phases (Bentur et al. 1994).
- Blended cements consisting of portland cement and a variety of natural pozzolans and supplementary materials such as fly ash. Currently, and for at least the next decade, this is probably the most-effective practical means to deal with environmental considerations, to reduce energy and CO_2 emission, and to use industrial by-products, (Brandstetr et al. 1997). This is dealt with in greater detail in the next section on composite cements.

Composite Cements

Environmental considerations and economic incentives have resulted in considerable use of a variety of minerals—either natural or industrial by-products—as supplements for portland cement.

They are added as a component of the cement or can be incorporated directly into the concrete mix. Their content can be quite high, replacing in extreme cases up to 70% of the portland cement. Typical ranges of composition of the most common mineral additives are given in Table 3. The interest in the use of such mineral additives resulted in more detailed specifications in the United States and Europe for blended cements. The range of compositions as specified by the U.S. and European standards are given in Table 4. These minerals additives can be classified in terms of the nature of their reactivity:

- Latent hydraulic minerals that have some self-cementing properties, where a small amount of activator such as portland cement is needed. They consist of glassy (amorphous) calcium aluminosilicates, with the most notable example being granulated blast furnace slag. Metakaolin, which is made of a clay heated to ~700°C, is another example that is receiving attention as a component for producing high-quality concretes.
- Pozzolanic minerals that have no self-cementing properties and require activation by calcium hydroxide. They are made of an amorphous material consisting mainly of SiO_2 and Al_2O_3 with a relatively low content of CaO. Fly ash and silica fume are examples of industrial by-products that are classified in this group.
- Nonreactive minerals that may affect the grindability of the clinker and modify the rheological characteristics and some properties of the mature concrete. Limestone filler, which is added at levels of less than 10%, is included in this category.

There are also trends to develop composite cements consisting of portland cement and more than one mineral additive.

It should be noted that composite cements may be adjusted to have properties superior to those of portland cement, and this has become one of the driving forces for their increased use. Notable examples are improved resistance to sulfate attack and alkali-aggregate reactions for blended cements containing fly ash.

Recycling

Recycling is receiving growing attention in the construction industry in general, and in cement and concrete technology in particular. The major area where recycling is of significant impact is

Table 4. Blended and Composite Cements according to U.S. Standard ASTM C595 and European Specifications EN 197

	Name	Portland cement content	Blended minerals
ASTM C595	Portland blast furnace slag	30–75%	Granulated blast furnace slag
	Slag-modified portland cement	>75%	Granulated blast furnace slag
	Portland pozzolan cement	60–85%	Pozzolan
	Pozzolan modified portland cement	>85%	Pozzolan
	Slag cement	<30%	Granulated blast furnace slag
EN 197	CEM I portland cement	95–100	Minor additional constituents
	CEM II[a]	65–94	Blast furnace slag, silica fume, pozzolans (natural or calcined), fly ash, burnt shale, limestone
	CEM III[b] blast furnace cement	5–64	Blast furnace slag
	CEM IV[c] pozzolanic cement	45–89	Silica fume, pozzolans, fly ash
	CEM V[d] composite cement	20–64	Blast furnace slag, pozzolans, fly ash

[a]Includes subclassification depending on type of blended mineral.

[b]Includes subclassification depending on content of slag: 36–65, 66–80, 81–95%.

[c]Includes subclassification depending on content of pozzolans (silica fume+pozzolans+fly ash): 11–35, 36–55%

[d]Includes subclassification depending on content of blended minerals (blast furnace slag+pozzolans+fly ash): 36–60, 62–80%.

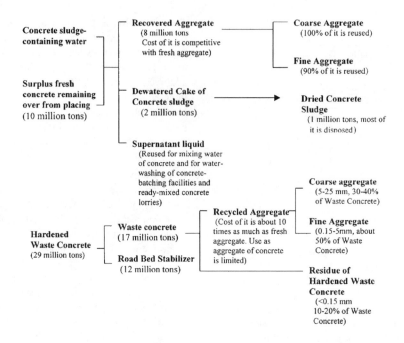

Fig. 16. Modes of recycling concrete wastes and their contents in Japan in 1992 (after Uchikawa 2000)

obviously in the use of industrial by-products such as fly ash and blast furnace slag as components in blended and composite cements and as direct additions to the concrete.

A second developing area of recycling is that of the concrete itself, where two direct sources can be identified: concrete wastes discharged in the manufacturing process, and dismantling of old concrete structures (Tomasawa and Noguchi 1996; Chandra 1997; Hendriks et al. 1998; Hendriks 2000; Uchikawa 2000). A schematic presentation of the wastes and their use in Japan is presented in Fig. 16. At this stage, technologies have been developed to use the recycled concretes as a source of aggregate, but this is quite limited in view of the technological difficulties and cost limitations (Dhir et al. 1998).

Special attention is given to the recycling of water in concrete plants, and this is now covered by standards, such as ASTM C94 and PrEN 1008 (Sandrolini and Franzoni 2001).

Long-Term Performance

Issues of long-term performance of concrete and reinforced concrete are receiving greater attention than ever before, due to a variety of causes:

1. The ecological and economic constraints result in greater use of marginal aggregates that are more prone to alkali-aggregate attack. This issue is further aggravated by higher contents of alkalis due to environmental restrictions on cement production, leading to the incorporation of a greater content of alkalis in the cement.

2. Concrete structural design was driven for many years primarily by considerations of optimization of the structural stability. This led also to the development of more strength-efficient cements, with which higher strength is achieved at the earlier age of 28 days by increasing the alite content and grinding fineness. The result is the use of leaner mixes and loss of reserve for hydration of the cements at later ages. Thus, the more "economic" structures and more "efficient" cements resulted in many instances in design where long-

term performance was unintentionally compromised. In reinforced structures, this is related to such parameters as concrete cover depth and water/cement ratio.

3. Societal needs in cold countries resulted in greater use of $CaCl_2$ for deicing of concrete transportation structures (e.g., highways, bridges), leading to accelerated rates of steel corrosion in concrete.

4. Workmanship quality has declined in many instances, resulting in poor finishing of concrete.

This state of affairs resulted in intensive research followed by code developments to address long-term performance as an important component in the design of concrete materials and structures (ACI 1992; Kropp and Hilsdorf 1995; Pigeon and Pleau 1995; Bentur et al. 1997; Hobbs 1998; Schiessel 1998).

A major emphasis has been given to the transport properties of concrete, quantifying mechanisms of diffusion of ions, permeation of fluids, and capillary absorption. On this basis models to predict service life have been developed, most notably for steel corrosion in concrete, as demonstrated in Fig. 17. The models for corrosion of steel in concrete are based on a comprehensive understanding of the electrochemical nature of steel corrosion. Concrete properties required for durability performance and design

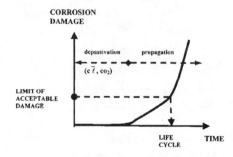

Fig. 17. Schematic description of stages and mechanisms involved in corrosion of steel in concrete and estimation of life cycle

Table 5. Classification of Exposure Conditions and Recommended Limiting Values for Concrete Composition and Properties according to EN 206

Type and level of risk			Maximum w/c	Minimum[a] strength class	Minimum cement content	Minimum air content
No risk	X0		—			
Carbonation-induced corrosion	XC1	Dry/permanently wet	0.65	C20/25	260	—
	XC2	Wet/rarely dry	0.60	C25/30	280	—
	XC3	Moderate humidity	0.55	C30/37	280	—
	XC4	Cyclic wet/dry	0.50	C30/37	300	—
Corrosion induced by chlorides other than sea water	XD1	Moderate humidity	0.55	C30/37	300	—
	XD2	Wet/rarely dry	0.55	C30/37	300	—
	XD3	Cyclic wet/dry	0.45	C35/45	320	—
Corrosion induced by chlorides from sea water	XS1	Airborne salt, no direct contact	0.50	C30/37	300	—
	XS2	Permanently submerged	0.45	C35/45	320	—
	XS3	Tidal splash and spray zone	0.45	C35/45	340	—
Freeze-thaw attack	XF1	Moderate water saturation, no de-icing salts	0.55	C30/37	300	—
	XF2	Moderate water saturation, with de-icing salts	0.55	C25/30	300	4.0
	XF3	High water saturation, no de-icing salts	0.50	C30/37	320	4.0
	XF4	High water saturation, with de-icing salts	0.45	C30/37	340	4.0
Chemical attack	XA1	Slightly aggressive chemical	0.55	C30/37	300	—
	XA2	Moderately aggressive chemical	0.50	C30/37	320[b]	—
	XA3	Highly aggressive chemical	0.45	C35/45	360[b]	—

[a]Characteristic strength grade, cylinder/cube.

[b]Sulfate resistant cement.

can be found in the literature and have been used to develop codes that are prescriptive in nature. The more modern codes, such as the European one, provide guidelines based on the nature of the aggressive mechanism (Table 5).

Advanced means to provide extensive service life in severe conditions have been developed and are used in practice, such as (1) use of fly ash and other pozzolans to improve chemical resistance; (2) use of low water/binder ratio concretes with extreme resistance to the transport of aggressive species; (3) development of specialty admixtures to deal with specific durability problems, such as inhibitors to stabilize the depassivation film over the steel bars embedded in concrete; and (4) sealing of the concrete by treatments with polymeric and inorganic materials.

In the drive to develop sophisticated means to deal with durability issues, there is sometimes a tendency to overlook the impact of the more conventional influences and of "good practice" in relation to concrete "cover" and concrete "skin," as defined and shown schematically in Fig. 18(a) (Schonlin and Hilsdorf 1987). In the structure itself, deficient curing has a markedly adverse effect on the permeability of the exposed surface of the concrete, while the "core" of the concrete, ~20 mm away from the surface, will remain relatively wet for prolonged periods [Fig. 18(b)]. This highlights the need to address separately the "cover" concrete, which responds to curing quite differently from the "core" concrete. Practices of achieving durable performance cannot rely only on mix design to achieve impermeable concrete, and curing practices to achieve impenetrable concrete cover are required [Fig. 18(c)].

Tools of life cycle cost analysis are being developed to provide a rational approach for the design for durability. The code requirements such as EN 206 are based to a large extent on a life cycle of 50 years. The life cycle cost analysis should provide the de-

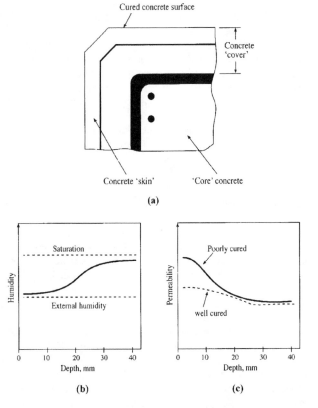

Fig. 18. Schematic description of (a) concepts of concrete cover and concrete skin; (b) its moisture state; and (c) influence of curing on permeability gradients

signer with tools to extend beyond the codes and explore the advantage and economy of more stringent requirements than those stated in the codes.

Four major durability issues have been of particular concern and thus have been studied extensively: steel corrosion in concrete (Bentur et al. 1997), freeze-thaw deterioration [e.g., Pigeon and Pleau (1995)], sulfate attack [e.g., Marchand and Skalny (1999)], and alkali-aggregate attack [e.g., Helmuth and Stark (1992)].

The use of high-performance concretes and concretes with a variety of mineral admixtures to meet the durability needs for most of these issues has been advanced considerably. Special additional means and technologies have been developed to deal with steel corrosion and resistance to cold climate.

The mechanisms of corrosion of steel in concrete have been studied extensively. The role of passivation of the steel in the high-pH environment was established in conjunction with the electrochemical nature of the process in the concrete matrix. Models have been developed to account for the conditions and rates of depassivation and propagation (Fig. 17) and quantify the life cycle and its dependence on concrete cover composition and thickness as well as environmental conditions (chlorides, CO_2, temperature, and humidity conditions). This fundamental know-how serves as a basis for durability design and development of standards and codes, as well as the advent of sophisticated means to combat steel corrosion in concrete [e.g., Bentur et al. (1991)] such as corrosion-inhibiting admixtures (e.g., calcium nitrite), treatments of the steel surface (e.g., epoxy coating, galvanizing), electrochemical means (e.g., cathodic protection), and special high-performance concretes.

The mechanisms of deterioration of concretes in cold climate, in particular with respect to freeze-thaw effects, have been resolved and quantified in terms of models that serve to predict life cycle and develop means to eliminate the damage associated with such freezing effects as internal microcracking and surface scaling [e.g., Pigeon and Pleau (1995)]. In this context, development of air-entraining admixtures and the accompanying air void spacing factor concept can be considered a breakthrough with an enormous impact on the successful construction of concrete structures in cold zones. Current codes, design practices, and accelerated testing techniques are based on an in-depth understanding of the physical mechanisms associated with freezing of water in the special structure of the cement matrix.

Advanced Cementitious Systems

In recent decades, cementitious systems of advanced performance that can compete with steel and sometimes with ceramics have been developed, and some are even used. A notable example is the use of high-strength concretes in high-rise buildings as the structural material rather than steel. The trends in these systems and the methodologies involved will certainly be guiding us in the next decade and beyond.

The underlying concept in many of the advanced systems is that the basic binder, portland cement, is a commodity and economical material, and advanced performance should be achieved by using it almost as is. This reflects the limitations in producing specialty cements and making them economically available, since economy of scale is an overriding factor in the construction industry. This state of affairs is clearly demonstrated in the decline in availability of special cements such as sulfate resistant (ASTM Type V) and low heat (ASTM Type IV) and the means taken to

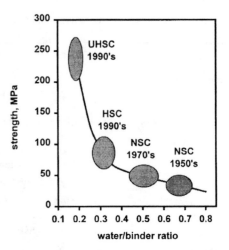

Fig. 19. Concept of strength-water/binder relationship and development of concrete strength over time

achieve sulfate resistance and low heat by using normal portland cement (ASTM Type-I) in combination with mineral admixtures.

In view of this state of affairs, the major thrust in development of superior cementitious systems is based on the combination of portland cement with other components tailored specifically for this purpose. This category would include specialty chemical admixtures, fillers to achieve optimal grading, and incorporation of fibers and polymers in cement composites. Often some of these approaches are applied simultaneously to achieve synergistic effects.

High-Strength/High-Performance Systems

The concept of achieving high strength and improved durability by making the concrete highly impermeable has been known for decades and is shown schematically in Fig. 19. This is the w/c concept laid down by Feret and Abrams years ago. To obtain such concretes of low water/cement ratio, there was a need to address the rheology of the fresh concrete to enable the production of flowing concrete having a low water content with rheological properties that would be adaptable to conventional technologies of production (mixing, pumping, vibrating, and so on).

The key to the solution that enabled the breakthrough into the era of high-strength/high-performance, low water/binder ratio concretes was the combination of effective high-range water reducers and dense packing of the binder, as achieved by incorporation of silica fume [Fig. 9(c)]. Concretes with strength grades of 100 to 150 MPa can be achieved with extremely low diffusivity, orders of magnitude smaller than normal concretes, using normal mixing procedures. They are competing with steel as the structural material in high-rise buildings and in offshore structures where a combination of strength and enhanced durability in a cold marine environment is essential (Nawy 1996; Aitcin 1998; Breitenbucher 1998). The tallest buildings in the world and the deepest offshore structures were built with such concretes.

The conventional high-strength/high-performance concretes are characterized by a water/binder ratio that usually does not go below 0.30. At a lower water/binder ratio, the aggregates become the weakest link in terms of strength, and therefore specialty aggregates such as bauxite need to be used to mobilize the extra strength of the very low water/binder ratio matrix. This is the concept underlying ultra-high-strength systems known as DSP

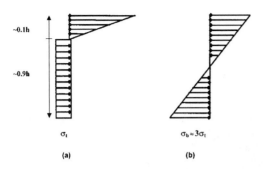

Fig. 20. Concepts for calculation of flexural strength of fiber-reinforced concretes: (a) actual stress distribution of elastic-plastic behavior; and (b) equivalent elastic stress distribution

systems (densified cement/ultrafine particle-based material), where the grading of the cement and microfillers is better controlled (usually with higher contents of silica fume) (Bache 1981).

The issue of grading has thus received a new attention and more fundamental treatment, as well as sophisticated design procedures, which include also the coarser aggregates (Dewar 1999; de Larrard 1999).

The emergence of high-strength concretes of low water/binder ratio matrix resulted in renewed interest in the sensitivity to early age cracking, which is more acute in such concretes. This is associated with combined effects of autogenous shrinkage of the low water/binder ratio matrix and heat liberation induced by the higher binder content characteristic of such concretes. These influences have been studied extensively to predict and model their influence as well as develop innovative means to eliminate them, such as internal curing by lightweight aggregates (e.g., Persson and Fagerlund 1999; Baroghel-Bouny and Aitcin 2000; Kovler and Bentur 2001).

These high-strength concretes possess special workability characteristics associated with their optimal grading, which shows up as improved flow that is not accompanied by segregation. These special characteristics led to the development of self-compacting concrete in which the slump in a conventional test is over 200 mm (Skarendahl 2001).

Fiber-Reinforced Concretes

Fiber-reinforcement of cementitious materials has its roots in ancient history when straw was used for reinforcement. In modern times the first extensive application of fiber-cement composite was in the asbestos-cement industry, which is now in decline due to health risks associated with asbestos production and use.

In recent decades, fundamental studies have led to the development of specialty fibers that are compatible with the concrete matrix. This compatibility includes composites that are more durable in the highly alkaline cement matrix (such as alkali-resistant glass fibers), fibers that can be more readily mixed uniformly in the concrete (special hydrophilic treatment), and special geometries that allow for enhanced bond by generating anchoring mechanisms (Bentur and Mindess 1990; Balaguru and Shah 1992; Bentur 1998b).

Low contents (0.1% volume) of low-modulus polymeric fibers are used commercially for control of plastic shrinkage cracking (Bentur and Mindess 1990; Balaguru and Shah 1992). Higher contents of steel fibers (0.5 to 1.5% by volume) are used for reinforcement of hardened concrete to replace conventional steel mesh in applications such as slabs on grade and shotcrete (ACI

1999). In mining shotcrete applications, new synthetic fibers have largely replaced steel, at a fiber content of 7 to 9% by volume. The influence of the fibers at this content range is on the post-cracking load-bearing capacity, inducing a pseudo-plastic behavior. The flexural resistance of such composites can be calculated based on the concept shown in Fig. 20, and comparative design of fiber-reinforced and steel mesh-reinforced members can be advanced based on this concept.

One of the attractive advantages of the fiber reinforcement concept is the simplification of the construction process by eliminating the need to place a steel mesh. This has led fiber-reinforced concrete to be the dominant technology in shotcreting, replacing mesh-reinforced shotcrete. The fibers used and the mix design are such that conventional concrete production technology and equipment can be used with the fiber-concrete mixes (ACI 1999).

Attention is currently being given to development of the use of high-performance fiber-reinforced cements based on advanced materials and fibers as well as design methods based on micromechanical concepts [e.g., Naaman and Reinhardt (1995); Li et al. (1995)]. Special attention is given to synthetic fibers characterized by a geometry of very small diameter filaments ($\sim$10 to 40 μm and less) grouped in a bundled form. The extremely high surface area provides reinforcing efficiency due to inherent high bond, but at the same time they induce difficulties in mixing. These fibers are classified as microfibers (size similar to that of cement grains) in contrast to steel fibers, which are usually of a macro size of 0.1 mm diameter or more. Polymeric macrofibers with complex geometrical shape, similar to steel fibers, are currently being introduced to compete with steel for the reinforcement of hardened concrete.

The current main application of fibers is largely for the purpose of crack control, either in the fresh or hardened state. The control is for cracks induced by environmental loads (humidity and temperature changes) and low-level structural loads, that exist in pavements and concrete linings. Thus, fibers are effective in improving the serviceability of such structures.

Cement-Polymer Composites

Considerable research and development efforts have been directed toward advancing cement-polymer composites using two main concepts: polymer impregnation of the concrete pores, and incorporation of polymer latex dispersions to produce interpenetrating networks of polymer film and hydrated cement (Ohama and Ramachandran 1995).

Impregnation has been considered mainly for production of concrete components of extreme durability performance. The technique can be applied for surface impregnation only (ACI 1997). However, in view of the complexity of the technology (drying of pores, efficient impregnation of monomer, and in situ polymerization) it has not been applied extensively, and the interest in it has somewhat subsided.

The use of cement-polymer latex has gained greater applicability, especially for a variety of mortars intended for coatings, resurfacing, and a variety of repair and adhesive formulations (ACI 1995; 1997). The formation of a polymer film within the mortar improves bond, ductility, flexural strength, and impermeability; it is thus suitable for specialty mortar formulations. A range of polymer latex compositions has been developed that are compatible with the cementitious matrix, giving special attention to the durability of the polymer film in the highly alkaline matrix and elimination of entrainment of air in the fresh mix. The recent trend is for the polymer to be in a powder form, thus enabling the

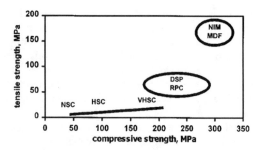

Fig. 21. Relations between tensile and compressive strength of conventional concretes and newly developed"high-tech"systems VHSC (very-high-strength concretes), DSP&RPC, and NIM&MDF

use of prebagged mixes formulated for different types of applications. This is the basis for many sophisticated products, with some containing in addition to the polymers other additives such as fibers (for coatings and repair) and lightweight aggregates (mortar rendering for thermal insulation).

Cementitious Materials of High Tensile Strength

Cementitious materials, like any typical brittle ceramic, are characterized by a low tensile strength, about one-tenth of the compressive value. This is also true for the high-strength–low water/binder compositions. This characteristic has a direct impact on construction technology and design, requiring concrete to be used with reinforcing steel. The potential advantages to be gained by drastically changing the ratio between tensile and compressive strength are enormous, and a variety of approaches have been advanced to achieve this purpose (Young 1985).

The simplest approach is to reinforce the cementitious matrix with a relatively large content of high-modulus fibers such as steel: reinforcement at a level of ~1.5% is sufficient to change the behavior from elastic-brittle to elastic-plastic; 10% volume reinforcement will increase tensile strength significantly, several-fold, resulting in a strain-hardening composite. Since such a content cannot be added by simple mixing, special production techniques have been developed such as slurry infiltration into a bed of fibers laid in a mold (SIFCON). Flexural strength values of 50 MPa have been reported (Lankard 1985; Naaman and Homrich 1989). However, the tedious production process and high cost of 10% fibers limit the use of this material to special applications.

A completely different technology was developed in the past two decades, leading to a class of materials known as macro-defect-free (MDF) and new inorganic materials (NIM). They are made of a water soluble polymer [usually PVA (polyvinyl alcohol), poly (acrylamide), or methyl cellulose] and a cement that is either calcium aluminate cement or portland cement (Young 1991). The material requires special processing involving high-shear mixing and pressure, resulting in a dough that can be extruded or shaped in similar ways. Compressive and flexural strength values of 300 and 150 MPa, respectively, were reported. Earlier theories attributed these properties to the formation of a material with reduced flaw sizes. Later studies attributed its characteristics to the polymer phase, with the cement component hardly reacting, but serving as a source of Ca ions to cross link the polymers and as a high-quality chemically bonded filler. The material made with calcium aluminate cement is stronger but is sensitive to moisture, which leads to strength reduction and swelling. The complexity of production and the high cost prohibit extensive use of this material in conventional civil engineering applications.

A recent development is an approach based essentially on the combination of a low water/binder matrix of well-graded portland cement and fillers with short fibers, at a content of about 6% by volume. This concept was first advanced by Bache (1981), using a DSP system (cement+silicafume+superplasticizers) with 6% of short steel fibers. The concept was further improved in France, combining in the graded fine material a fraction of graded quartz with particle sizes smaller than 600 μm. The term reactive powder concrete (RPC) was used for this material (Cheyerzy et al. 1995; Richard and Cheyerzy 1995; Bonneau et al. 1997). A modified concept was suggested by Rossi (1997) based on a strategy of using different types of fibers (short and long) with this densified matrix. This material was named HPMFRCC—high-performance multimodal fiber-reinforced cement composite. The water/binder ratio in these materials is in the range of 0.10 to 0.20, with flexural and compressive strengths being 30 to 60 and 170 to 230 MPa, respectively. The components in these systems are less exotic than in the MDF/NIM materials, and the workability is such that they can be produced by conventional technological means. Obviously the material's cost is high ($\sim$\$700/m^3), and it lends itself for application of thin layer components. A recent study by Collepardi et al. (1997) suggests that there is a potential for combining coarse aggregates with RPC matrix while still maintaining many of the superior properties. See Fig. 21 for a comparison of these systems.

The incorporation of special fine aggregate in these systems, such as bauxite and metallic powder, can result in a hardened material with compressive strength as high as 800 MPa. This class of emerging materials is at the cutting edge of concrete technology, and they have been labeled by Aitcin (2000) as high-tech concretes.

Construction with Concrete and Concrete Technology

Much of the review in this paper is centered around concrete as a material and the binder composition. It should be emphasized that the marked advances in the materials aspects are followed by and inspire technological innovations of construction methods and design. Some of these aspects are briefly highlighted here.

The reciprocal relations between the materials and construction technologies are most vividly demonstrated in the case of offshore structures, primarily those intended for oil exploration in the deep seas. The structural and durability needs for offshore structures and the inherent potential advantages of concrete for these purposes were the driving force behind the development of high-strength concrete with improved workability (Nawy 1996).

To facilitate efficient construction, mechanization and automation techniques have been advanced. The improved understanding and control of concrete rheology in the fresh state is frequently a key element in such developments. Slipforming, shotcreting, and roller compaction are examples of techniques allowing rapid and continuous construction in mechanized ways. However, for such methods to be satisfactory there is a need for a special and careful control of concrete rheology. Methods of this kind lend themselves to automation. Robotization, which may be the way of the future, can be more readily adapted for construction with technologies of this kind (Nawy 1997).

A high level of industrialization of the concrete mix production, transportation, and placing has been achieved and is currently the conventional state of the art, using ready-mixed plants, concrete delivery trucks, and on-site pumping. Here too the combination of equipment and control of concrete rheology is a substantial element. Special technologies have been developed for automated finishing, such as screeders and trowlers, which can be readily mechanized and automated.

Technologies for production of concrete members with special properties, in particular high strength and impermeability, were developed to a limited extent. They include preplaced aggregate concrete and vacuum processing. These are of less interest at present because the developments of high-strength–low water/binder ratio mixes allows the achievement of many of these characteristics using conventional production and placing techniques. Special technologies are applied for underwater concreting, combining methods of operation and control of fresh concrete properties. Antiwashout agents are important elements.

When considering technologies and methods of construction, one should keep in mind that with current concrete properties (and in the foreseeable immediate future), the construction material needed is essentially reinforced concrete. The innovations in prestressing and posttensioning (bonded and unbonded) systems are key achievements. They include the development of new design and application procedures to achieve performance, such as larger spans, technologies for rehabilitation and strengthening of buildings, development of improved tensioning and anchoring technologies, and advanced methods and materials to improve long-term performance (CEB 1992; CEB-FIP 1995; Nawy 2000).

Future Trends

It is very difficult to predict the future, yet some trends can be established, and, perhaps more important, some ideas can be put forward to suggest how we can influence the future. In discussing such trends, a variety of points of view should be addressed:

1. Changes and trends that might be predicted by extrapolation of recent developments in cement and concrete technology, as outlined in the sections on sustainability and advanced cementitious systems.
2. General trends of development related to economy and society, such as
- Expected increase in world population and urbanization, mainly in developing countries;
- Growing needs for pollution and waste management; and
- Technological innovation of a generic nature such as information technology and biotechnology. Some of these trends may have direct or indirect effects on cement and concrete technology.

Reference to some of these considerations and their implications to the future of cementitious materials have been the subject of several recent publications, which the present section will draw upon (Mather 1994; Mehta 1994; Neville 1994; ACI 2001). In the following discussion, a classification is made between near-future trends for the next 5 to 10 years and long-term trends beyond that time.

Near-Future Trends

The pressure of ecological constraints and environmental regulations is bound to increase in the coming years. This will lead to continuation of the trend for greater use of supplementary cementitious materials that are in many instances industrial by-products, and seems to be a more viable alternative in the near future than producing special low-energy cements. A complementary alternative that may develop is low-temperature activation of locally available aluminosilicates to be used in combination with portland cement.

This trend may change the concept of the cement industry, which may turn from a "portland cement clinker industry" to a "hydraulic binder industry" (Aitcin 2000). The cement will become only one component in a range of fine powders produced and handled by this industry, to be blended and marketed as a variety of products to meet different needs. The fine powders may be industrial by-products as well as such artificial supplementary materials as metakaolin and calcined clay. The end products may serve for production of "normal" concrete as well as speciality "high-tech" binders with controlled composition and particle-size distribution. This versatility might result in binders with a lower content of portland cement as well as portland cement clinker of lower alite (C_3S) and grinding fineness that will be more compatible with the mineral admixtures and high-range water reducers.

The increasing demand to consider durability as a performance criterion as important as strength and structural stability will result in greater use of concretes of low water/binder ratio, which can readily meet both criteria at the same time and may be more efficient than the use of "strong" cements. Cements for low water/binder ratio mixes will have to be adjusted in composition to achieve adequate rheological behavior, in particular compatibility with high-range water-reducing admixtures rather than high-strength (early age and beyond). For this purpose and for the production of high-tech binder, it may be sufficient to relax the requirements for high C_3S content and grinding fines and provide more attention to optimizing the content of sulfate and alkalis (Aitcin 2000).

In order to deal with the need for sustainable construction with cementitious material, a more comprehensive approach will have to be advanced (Hoff 1996; Rostam 1996; Sakai 1996). Concrete technologists, structural designers, and experts in construction management and economics will have to work together more closely. Looking at sustainability from each of the individual points of view, represented by each of these three disciplines individually, is not going to be effective and may even create conflicts. An example pointing toward integration of approaches is the recent establishment of activities to develop life-cycle cost methodologies to address strategies to deal with steel corrosion in concrete. The advent of superior concrete and admixtures to deal with this issue, which has been the thrust in the admixture and concrete industry in the last two decades, will not result in their application unless their advantage in terms of life-cycle cost is established. This is true for prototype evaluation of innovative technologies as well as the use of advanced materials in specific projects.

Aitcin (2000) reflected on this issue by predicting that the trend for BOT (build, own, operate, and transfer) projects may lead to greater use of sophisticated concrete compositions, such as RPC, where the high price of such materials ($\sim\$700/m^3$ for RPC) may not be a deterrent, and if the total cost and life-cycle cost are considered, it may even lead to savings. An example of this kind was cited by Aitcin (2000), demonstrating a cost saving of about 8% for bridge construction with RPC.

Advancing new materials and concepts will require the development of innovative testing techniques. Special attention will have to be given to site tests based on a variety of principles (electrical, electromagnetic, wave propagation, and so on) to assess the internal structure of the concrete. Based on such data, predictions of the transport properties and long-term durability could be made. This is a necessity if alternatives to testing and specification of concrete based on strength only are to be advanced. Such tests and specifications, which are performance related, could pave the way for new and versatile cementitious systems and combinations that will be tailored for a variety of end uses. The ASTM standard for cements, ASTM C1157, which is a performance standard, is an example.

Long-Term Trends

The impact of environmental considerations has already affected this industry and is bound to continue in a variety of ways by increasing the use of industrial by-products and recycling. However, if drastic changes are to be achieved, one should consider the option of revolutionizing the way in which we use cementitious materials to build structures. This implies developing materials and modes of construction that will drastically reduce the quantities of materials without affecting the performance of the structure. The aim should be for a reduction of the quantities consumed per unit of structure by an order of magnitude. This might be achieved with the development of a new generation of cementitious materials, which will be accompanied by an integrative approach, as outlined below.

Ultra-high-performance cementitious materials with superior tensile strength and ductility can be developed based on the current portland cement composition. Yet the bottleneck for application of these ultra-high-performance materials will be the establishment of innovative construction technologies and design procedures to take advantage of their enhanced properties and consider limitations such as materials cost and processing. The approach that will need to be taken should include structural characteristics as well as constructability, which is becoming an increasingly important issue in view of the rising cost of labor and the scarce availability of skilled professionals on site. This may require the development of concretes with markedly improved rheological properties and self-curing characteristics, as well as new innovative lightweight systems based on cementitious composites with high tensile strength and ductility.

Information technology has its impact on design and production control. However, it should be extended and combined with sensors to develop interactive materials, following ideas and concepts being developed and implemented in other engineering disciplines. Smart buildings are already around, and they may be complemented by smart cementitious materials.

A trend that is currently gaining momentum is the mathematical modeling of the material. This will be enhanced to compile together comprehensive models that will address simultaneously the whole structure on different scales, starting from the material microstructure to the overall behavior under mechanical and environmental loads [e.g., the HYMOSTRUC model—Van Breugel (1995, 1996); (Van Breugel and Lokhorst 2001); CEMY 3D model—Bentz (1997, 1999); Bentz et al. (1995, 1998); DuCON Model—Maekawa et al. (1999)]. The impact from the materials end will be based on the quantitative concepts of hydration and physics of the cementitious systems and their interactions with the mechanics of the system, as expressed in terms of micromechanical concepts and modeling [e.g., Ulm and Coussy (1998)].

Summary and Conclusions

Cementitious materials have been used for thousands of years in a variety of compositions and for a wide range of applications. The level of sophistication in ancient times was quite remarkable and can be attributed to ingenuity based on experience and the learning-by-doing process.

In modern times the development and use of cementitious materials is to a large extent driven by scientific concepts mobilized to push the engineering frontiers. Materials science concepts are routinely applied to develop sophisticated compositions and formulations that can potentially provide performance approaching that of metals and ceramics. A comprehensive approach is consistently being advanced to deal systematically with the cementitious materials and structures in an integrated mode, taking into consideration a variety of aspects such as safety, long-term performance, life-cycle cost. With an integrated approach of this kind, cementitious materials seem to have a competitive advantage for remaining the main construction material well into the new century.

Acknowledgment

The assistance of Ms. Yael Armane in compiling the literature and the bibliographical survey is gratefully appreciated.

References

Aitcin, P.-C. (1998). *High performance concrete*. E & FN Spon, London.

Aitcin, P.-C. (2000). " Cements of yesterday and today, concrete of tomorrow." *Cem. Concr. Res.,* 30, 1349–1359.

Alexander, M. G., Arliguie, G., Ballivy, G., Bentur, A., and Marchand, J. (1999). "Engineering and transport properties of the interfacial transition zone in cementitious composites," Report 20, RILEM, Essen, Germany.

American Concrete Institute (ACI). (1992). "Guide to durable concrete." *ACI 201.2R-92*, Detroit.

American Concrete Institute (ACI). (1995). "State of the art report on polymer modified concrete." *Report 584, 3R-95, ACI Manual of Concrete Practice*, Part 5, 1999, Detroit.

American Concrete Institute (ACI). (1997) ."Guide for the use of polymers in concrete," *Report 548, 1R-97, ACI Manual of Concrete Practice*, Part 5, 1999, Detroit.

American Concrete Institute (ACI). (1999). "Fiber reinforced shotcrete." *Report 506, 1R-98, ACI Manual of Concrete Practice*, 1999, Detroit.

American Concrete Institute (2001). "Vision 2030: A vision for the U.S. concrete industry," *Concr. Int.,* 23, 25–34.

Bache, H. H. (1981). "Densified cement/ultra-fine particle-based materials," *2nd International Conference on Superplasticizers in Concrete*, Ottawa.

Balaguru, P. N., and Shah, S. P. (1992). *Fiber reinforced cement composites*. McGraw-Hill, New York.

Baroghel-Bouny, V., and Aitcin, P.-C. (2000). "Shrinkage of concrete: Shrinkage 2000," *Proceedings, PRO 17*, RILEM, Essen, Germany.

Bazant, Z. P. (1995). "Creep and damage in concrete." 355–390, *Materials Science and Concrete—IV*, J. Skalny and S. Mindess, eds., American Ceramic Society.

Beaudoin, J. J., Feldman, R. F., and Tumidajski, P. J. (1994). "Pore structure of hardened portland cement pastes and its influence on properties." *Adv. Cem. Res.,* 8, 163–166.

Beaupre, D., and Mindess, S. (1998). "Rheology and fresh concrete: Principles, measurement and applications." *Materials science of concrete—V*, American Ceramic Society, J. Skalny and S. Mindess, eds., 149–190.

Bentur, A. (1998a) "ITZ structure and its influence on engineering and transport properties: Concluding remarks to the conference." A. Katz, A. Bentur, M. Alexander, and G. Arliguie, eds., *The interfacial transition zone in cementitious composites*, E & FN Spon, London, 335–338.

Bentur, A. (1998b). "Long-term performance of fiber-reinforced cementitious composites." *Materials science of concrete—V*, American Ceramic Society, 513–536.

Bentur, A., Diamond S., and Berke, N. S. (1997). *Steel corrosion in concrete*. E & FN Spon, London.

Bentur, A., Kovler, K., and Goldman, A. (1994). "Gypsum of improved performance using blends with portland cement and silica fume." *Adv. Cem. Res.,* 6, 109–116.

Bentur, A., and Mindess, S. (1990). *Fibre reinforced cementitious composites.* Elsevier Applied Science, U.K.

Bentur, A., et al. (1995). "Fiber-matrix interfaces." in *High performance fiber reinforced composites,* E & FN Spon, London, A. E. Naaman and H. W. Reinhardt, eds. 149–192.

Bentz, D. P. (1997). "Three-dimensional computer simulation of cement hydration and micro-structure development." *J. Am. Ceram. Soc.,* 80, 3–21.

Bentz, D. P. (1999). "Modeling cement microstructure: Pixels, particles and property prediction." *Mater. Struct.,* 32, 187–195.

Bentz, D. P., and Garboczi, E. J. (1999). "Computer modeling of interfacial zone: Microstructure and properties." *Engineering and transport properties of the interfacial zone in cementitious composites,* RILEM, Essen, Germany. M. G. Alexander et al., 349–385.

Bentz, D. P., Garboczi, E. J., and Lagerren, E. S. (1998). "Multi-scale microstructural modeling of concrete diffusivity: Identification of significant variables." *J. Cem., Concr., Aggregates (ASTM),* 20, 129–139.

Bentz, D. P., Schlangen, E., and Garboczi, E. J. (1995). "Computer simulation of interfacial zone microstructure and its effect on properties of cement based composites." Materials science of concrete—V, American Ceramic Society, J. Skalny and S. Mindess, 155–200.

Blezard, G. (1998). "The history of calcareous cement." *Lea's chemistry of cement and conceret,* P. C. Hewlett, ed., Arnold, London, 1–18.

Bonneau, M., Lachemi, M., Dallaire, E., Dugat, J., and Aitchin, P.-C. (1997). "Mechanical properties and durability of two industrial reactive powder concretes." *ACI Mater. J.,* 94, 286–290.

Brandstetr, J., Havlica, J., and Odler, I. (1997). "Properties and use of solid residue from fluidized bed coal combustion," *Waste materials used in concrete manufacturing,* Noyes, S. Chandra, ed., 1–47.

Breitenbucher, R. (1998). "Development of applications of high-performance concrete." *Mater. Struct.,* 31, 209–215.

British Cement Association (1999). *Concrete through the ages: From 7000 BC to AD 2000.*

Brown, P. W., Shi, D., and Skalny, J. (1991). "Porosity/permeability relationships." *Materials science of concrete—I,* J. Skalny, ed., American Ceramic Society, 83–110.

(CEB). (1992). *Durable concrete structures: Design guide.* Thomas Telford, London.

(CEB-FIP). (1995). *"High performance concrete"* CEB, Switzerland.

Chandra, S., ed. (1997). *Waste materials used in concrete manufacturing.* Noyes.

Cheyerzy, M., Maret, V., and Frouin, L. (1995). "Microstructural analysis of RPC (reactive powder concrete)." *Cem. Concr. Res.,* 25, 1491–1500.

Collepardi, S., Coppola, L., Troli, R., and Collepardi, M. (1997). "Mechanical properties of modified reactive powder concrete." *Superplasticizers and other chemical admixtures in concrete,* ACI SP-173, V. M. Malhotra, ed., *Admixtures in concrete,* ACI SP-173, American Concrete Institutes, Detroit, 2–21.

de Larrard, F. (1999). *Concrete mixture proportioning.* E & FN Spon, London.

de Larrard, F., and Chang, H. (1996). "The rheology of fresh high-performance concrete." *Cem. Concr. Res.,* 26, 283–294.

Dewar, J. D. (1999). *Computer modeling of concrete mixtures.* E & FN Spon, London.

Dhir, R. K., Henderson, N. A., and Limbachiya, M. C., eds. (1998). "Sustainable construction: Use of recycled concrete aggregate." *Proc., Int. Conf.,* Thomas Telford, London.

Diamond, S. (1999). "Aspects of concrete porosity revisited." *Cem. Concr. Res.,* 29, 1181–1188.

Edmeades, R. M., and Hewlett, P. C. (1998). "Cement admixtures." P. C. Hewlett, ed., *Lea's Chemistry of cement and concrete,* Arnold, London, 837–901.

European Commission (1999). *Integrated pollution prevention control: Reference document on best available techniques in the cement and lime manufacturing industries.* Institute for Prospective Technological Studies, Joint Research Center,

Farran, J. (1956). "Contribution mineralogique a l'etude de l'adherence entre les constituants hydrates des ciments et les materiaux enrobes." *Rev. Mat. Constr.,* 490–491, 155–72.

Feldman, R. F., and Sereda, P. J. (1968). "A model for hydrated portland cement paste as deduced from sorption length change and mechanical properties." *Mater. Struct.,* 1, 509–520.

Goldman, A., and Bentur, A. (1994). "Properties of cementitious systems containing silica fume or non-reactive microfillers." *Adv. Cem. Based Mater.,* 1, 209–215.

Haneharo, S., and Yamada, K. (1999). "Interaction between cement and chemical admixture from the point of cement hydration, adsorption behavior of admixture and paste rheology." *Cem. Concr. Res.,* 29, 1159–1166.

Hansen, W., and Young, J. F. (1991). "Creep mechanisms in concrete." *Materials science of concrete—I,* J. Skalny, ed., American Ceramic Society, 185–200.

Helmuth, R., and Stark, D. (1992). "Alkali-silica reactivity mechanisms." *Materials science of concrete—II,* J. Skalny, ed., American Ceramic Society, 131–208.

Hendriks, C. (2000). *Durable and sustainable construction materials.* Aeneas, The Netherlands.

Hendriks, C., Pieterson, H. S., and Fraay, A. F. A. (1998). "Recycling of building and demolition waste—An integrated approach." *Sustainable construction: Use of recycled concrete aggregate,* R. K. Dhir et al., eds. 419–431.

Hewlett, P. C., ed. (1998). *Lea's chemistry of cement and concrete.* Arnold, U.K.

Hobbs, D. W., ed. (1998). *Minimum requirements for durable concrete.* British Cement Association, U.K.

Hoff, G. C. (1996). "Toward rational design of concrete structures—Integration of structural design and durability design." *Integrated design and environmental issues in concrete technology,* K. Sakai, ed., E & FN Spon, London, 15–30.

Idorn, M. G. (1997). *Concrete progress from antiquity to the third millennium.* Thomas Telford, London.

Ishai, O. (1968). "The time-dependent deformation behavior of cement paste, mortar and concrete." *Proc. Conf. The Structure of Concrete and Its Behavior under Load,* Cement and Concrete Association, London. 345–364.

Ishida, H., and Mitsuda, T. (1998). "Science and technology of highly reactive β-dicalcium silicate: Preparation, hydration and gels." *Materials science of concrete—V,* J. Skalny and S. Mindess, eds., American Ceramic Society, 1–44.

Jackson, P. J. (1998). "Portland cement: Classification and manufacture." *Lea's chemistry of cement and concrete,* P. C. Hewlett. eds., Arnolds, 25–94.

Klemm, W. A. (1989). "Cementitious materials: Historical notes." *Materials science of cement and concrete,* J. Skalny, ed., 1–26.

Kovler, A., and Bentur, A. (2001). "Early age cracking in cementitious systems." *Proc. RILEM Int. Conf.,* RILEM, Essen, Germany.

Kropp, J., and Hilsdorf, H. K., ed. (1995). "Performance criteria for concrete durability." *RILEM Rep. 12:*E & FN Spon, London.

Lankard, D. R. (1985). "Slurry infiltrated fiber concrete (SIFCON): Properties and applications." "Very high strength cement based materials," *Mat. Res. Soc. Proc. 42,* J. F. Young, Materials Research Society, 277–286.

Lawrence, C. D. (1998). "The production of low energy cements." *Lea's chemistry of cement and concrete,* P. C. Hewlett, ed., Arnold, London, 421–470.

Lea, F. M., and Desch, C. H. (1956). *Chemistry of cement.* Edward Arnold, London.

Le Chatelier, (1987). *Experimental researches on the constitution of hydraulic mortars.* English translation—1905, McGraw-Hill, New York.

Li, V. C., et al. (1995). "Micromechanical models of mechanical response of HPFRCC." *High performance fiber reinforced cement composites 2,* A. E. Naaman and H. W. Reinhardt, eds. E & FN Spon, London, 43–100.

Macphee, D. E., and Lachowski, E. E. (1998). "Cement components and their phase relations." *Lea's chemistry of cement and concrete*, P. C. Hewlett, ed., Arnold, London, 95–130.

Maekawa, K., Chaube, R., and Kishi, T. (1999). *Modeling of concrete performance*. E & FN Spon, London.

Malinowski, R. (1979). "Concrete and mortars in ancient aqueducts." *Concr. Int.,* 1, 66–79.

Malinowski, R., and Garfinkel, Y. (1991). "Prehistory of concrete." *Concr. Int.,* 13, 62–68.

Marchand, J., Gerard, B., and Delagrave, A. (1998). "Ion transport mechanisms in cement-based materials." *Materials science of cement and concrete*, J. Skalny and S. Mindess eds., American Ceramic Society, 307–400.

Marchand, J., and Skalny, J. P. (1999). *Sulfate attack mechanisms*. American Ceramic Society,

Mather, B. (1994). "Concrete—Year 2000 revisited." in *Concrete Technology: Past, Present and Future*, P. K. Mehta, ed., *ACI SP-144*, American Concrete Institute, 31–40.

Mehta, P. K. (1994). "Concrete technology at the crossroads—Problems and opportunities." *Concrete technology: Past, present and future*, P. K. Mehta, ed., *ACI SP-144*, American Concrete Institute, 1–30.

Mollah, M. Y. A., Adams, W. J., Schennach, R., and Cocke, D. L. (2000). "A review of cement—Superplasticizer interactions and their models." *Adv. Cem. Res.,* 12, 153–161.

Naaman, A. E., and Homrich, J. R. (1989). "Tensile stress-strain properties of SIFCON." *J. Am. Concr. Inst.,* 86, 244–251.

Naaman, A. E., and Reinhardt, H. W. (1995). *High performance fiber reinforced cement composites 2 (HPFRCC 2)*. E & FN Spon, London.

Nawy, E. G. (1996). *Fundamentals of high strength high performance concrete*. Longman Group, U.K.

Nawy, E. G., ed. (1997). *Concrete construction engineering handbook*. CRC Press, Boca Raton, Fla.

Nawy, E. G. (2000). *Prestressed concrete: A fundamental approach*. Prentice-Hall, Upper Saddle River, N.J.

Neville, A. (1994). "Concrete in the year 2000." *Advances in concrete technology*, V. M. Malhotra, ed., CANMET, Canada. 1–18.

Nilsson, L.-O. and Ollivier, J.-P. (1999). "Fundamentals of transport properties of cement-based materials and general methods to study transport properties." M. G. Alexander et al., *Engineering and transport properties of the interfacial zone in cementitious composites*, RILEM, Essen, Germany, 131–148.

Odler, I. (1998). "Hydration, setting and hardening of Portland cement." *Lea's chemistry of cement and concrete*, P. Hewlett, ed., Arnold, London, 241–148.

Odler, I. (2000). *Special inorganic cements*. E & FN Spon, London.

Odler, I., and Skalny, J. (1992). "Potential for the use of fossil fuel combustion by the construction industry." *Materials science of concrete—II*, J. Skalny, ed., American Concrete Society, 319–336.

Ohama, Y., and Ramachandran, V. S. (1995). "Polymer modified mortar and concrete." *Concrete admixtures handbook*, V. S. Ramachandran ed. Noyes, Park Ridge, N.J., 558–656.

Ollivier, J. P., Maso, J. C., and Bourdette, B. (1995). "Interfacial transition zone in concrete." *Adv. Cem. Based Mater.,* 2, 20–38.

Persson, B., and Fagerlund, G., eds., (1999). "Self-desiccation and its importance in concrete technology." *Proc. Int. Seminar*, Lund, Sweden.

Pigeon, M., and Pleau, R. (1995). *Durability of concrete in cold climates*. E & FN Spon, London.

Powers, T. C. (1960). "Physical properties of cement pastes." 577–613. *Proc. Symp. Chemistry of Cement*, Vol. II, Washington.

Powers, T. C. (1968), *The properties of fresh concrete*. Wiley, New York.

Powers, T. C., and Brownyard, T. L. (1948). "Studies of physical properties of hardened portland cement paste." *Research Department Bulletin No. 22*, Portland Cement Association, Chicago.

Ramachandran, V. S. (2001). *Handbook of analytical techniques in concrete science and technology*. Noyes, Park Ridge, N.J.

Rice, R. W. (1984). "Pores as fracture origins in ceramics." *J. Mater. Sci.,* 19, 895–914.

Richard, P., and Cheyerzy, M. (1995). "Composition of reactive powder concrete." *Cem. Concr. Res.,* 25, 1501–1511.

Richardson, I. G. (1999). "The nature of C-S-H in hardened cements." *Cem. Concr. Res.,* 29, 1131–1148.

Ronen, A., Bentur, A., and Soroka, I. (1991). "A plastered floor from the neolithic village Yiftahel (Israel)." *Paleorient,* 17, 149–155.

Rossi, R. (1997). "High performance multimodal fiber reinforced cement composites (HPMFRCC): The LCPC experience." *J. Am. Concr. Inst.,* 94, 478–483.

Rostam, S. (1996). "Service life design for the next century." *Integrated design and environmental issues in concrete technology*, K. Sakai, ed., E & FN Spon, London, 51–67.

Roy, D. M. (1988). "Relationships between permeability, porosity, diffusion and microstructure of cement pastes, mortar and concrete at different temperatures." *Proc. Symp. on Pore Structure and Permeability of Cementitious Materials*, Materials Research Society, 179–189.

Sakai, E., and Daimon, M. (1995). "Mechanisms of superplastification." *Materials Science of Concrete—IV*, J. Skalny and S. Mindess, eds. American Ceramic Society, 91–112.

Sakai, K., ed. (1996). *Integrated design and environmental issues in concrete technology*. E & FN Spon, London.

Sandrolini, F., and Franzoni, E. (2001). "Waste wash water recycling in ready-mixed concrete plants." *Cem. Concr. Res.,* 31, 485–489.

Schiessl, P. (1998). "Durability design strategies for concrete structures under severe conditions." *Concrete under severe conditions*, O. E. Gjorv, K. Sakai, and N. Banthia, eds., E & FN Spon, London, 17–26.

Schonlin, K., and Hilsdorff, H. (1987). "Evaluation of the effectiveness of concrete in structures." Concrete durability, J. M. Scanlon ed. *ACI SP-100*, American Concrete Institute, Detroit, 207–226.

Scrivener, K. L. (1989). "The microstructure of concrete." *Materials science of concrete—I*, J. Skalny, ed., American Ceramic Society, 127–162.

Scrivener, K. L., Bentur, A., and Pratt, P. L. (1988). "Quantitative characterization of the transition zone in high strength concrete." *Adv. Cem. Res.,* 1, 230–237.

Shah, S. P., et al. (1995). "Toughness characterization and toughening mechanisms." *High performance fiber reinforced cement composites 2(HPFRC-2)*, A. E. Naaman and H. W. Reinhardt, eds. E & FN Spon, London, 193–228.

Shah, S. P., and Ouyang, C. (1992). "Measurement and modeling of fracture processes in concrete," *Materials science in concrete—III*, J. Skalny ed. American Ceramic Society, , 243–270.

Skalny, J., ed. (1989). *Materials science of concrete—I*. American Ceramic Society.

Skalny, J., ed. (1991). *Materials science of concrete—II*. American Ceramic Society,

Skalny, J., ed. (1992). *Materials science of concrete—III*. American Ceramic Society,

Skalny, J., and Mindess S., eds. (1995). *Materials science of concrete—IV*. American Ceramic Society.

Skalny, J., and Mindess, S., eds. (1998). *Materials Science of Concrete—V*. American Ceramic Society.

Skarendahl, A. (2001). "Self-compacting concrete: State of the art report." *RILEM Rep. 23, RILEM*, Essen, Germany.

Snell, L. M., and Snell, B. G. (2000). "The early roots of cement." *Concr. Int.,* 22, 83–84.

Soroka, I. (1979). *Portland cement paste and concrete*. MacMillan, New York.

Struble, L., Xihuang, Ji, and Salenas, G. (1998). "Control of concrete rheology." *Materials science of concrete: The Sidney Diamond symposium*, M. Cohen, S. Mindess, and J. Skalny, eds., American Ceramic Society, 231–245.

Taylor, H. F. W. (1997). *Cement chemistry*. 2nd Ed., Thomas Telford, London.

Tomosawa, F., and Noguchi, T. (1996). "Towards completely recyclable concrete." *Integrated design and environmental issues in concrete technology*, K. Sakai, ed., E & FN Spon, London, 263–272.

Torroja, E. (1958). *The structures of Eduardo Torroja*. F. W. Dodge Corporation,

Uchikawa, H. (2000). "Approaches to the ecologically benign system in cement and concrete industry," *J. Mater. Civ. Eng.,* 12(1), 320–329.

Ulm, F. J., and Coussy, O. (1998). "Coupling in early-age concrete: From material modeling to structural design." *Int. J. Solids Struct.,* 35, 4295–4311.

Van Breugel, K. (1995). "Numerical simulation of hydration and microstructural development in hardening cement-based materials. II: Applications." *Cem. Concr. Res.,* 25, 522–530.

Van Breugel, K. (1996). "Models for prediction of microstructural development in cement-based materials." *Proc. Workshop,* July 1994, NATO ASI series E: Applied Sciences, 304, 91–106.

Van Breugel, L., and Lokhorst, S. J. (2001). "The role of microstructure development in creep and relaxation of hardening concrete." 245–257 A. Kovler and A. Bentur, "Early age cracking in cementations systems."

Van Mier, J. G. M. (1997). *Fracture processes of concrete—Assessment of material parameters for fracture models.* CRC Press, Boca Raton, Fla.

Van Mier, J. G. M., and Vervurt, A. (1999). "Test methods and modeling for determining the mechanical properties of the ITZ in concrete." "Engineering and transport properties of the interfacial transition zone in cementitious composites." *Report 20,* M. G. Alexander et al., RILEM, Essen, Germany, 19–52.

Vitruvius, P. (13 BC). *De Architectura Libra Decem, (Ten Books on Architecture),* English translation by F. Granger, London (1931).

Wallevik, O. (1998). *Rheological measurements on fresh cement paste, mortar and concrete by use of a coaxial cylinder viscometer. Icelandic Building Research Institute,* Iceland.

Wallevik, O., and Gjorv, O. E. (1990). "Development of a coaxial cylinder viscometer for fresh concrete." *Proc. RILEM Colloquium, Properties of Fresh Concrete,* Chapman and Hall, U.K., 213–224.

Winslow, D., and Diamond, S. (1974). "Specific surface area of hardened portland cement as determined by SAXS." *J. Am. Ceram. Soc.,* 57, 193–197.

Young, J. F., ed. (1985). "Very high strength cement based materials." *Mater. Res. Soc. Symp. Proc.* 42, Materials Research Society,

Young, J. F. (1991). "Macro-defect-free cement: A review," pp. 101–121 in *Mater. Res. Soc. Symp. Proc.,* 179, Materials Research Society,

Young, J. F., and Sun, G. (2001). *Silicate polymerization analysis. Handbook of analytical techniques in concrete science and technology,* V. S. Ramachandran, ed. Noyes, Park Ridge, N.J. 629–657.

American Society of Civil Engineers

150th Anniversary Paper

1852 – 2002

Building a Better World

Wood and Wood-Based Materials: Current Status and Future of a Structural Material

Kenneth J. Fridley, M.ASCE[1]

Abstract: Wood is one of the earliest construction materials, and the structural use of wood and wood-based materials continues to steadily increase. In fact, new wood-based materials continue to be developed and successfully introduced into the engineering and construction marketplace. Supporting the increase in use and development of new materials has been an evolution of our understanding of wood as a structural material. The primary aim of this paper is to provide a summary of the status and future of engineered wood and wood-based materials. This paper is not intended to be a comprehensive review of the literature or an all-inclusive state-of-the-art report on all types of wood and wood-based materials. Rather, the focus is on the past, present, and future of engineered wood and wood-based materials for use in civil engineering applications. It also provides a vision for addressing anticipated future research needs. At the conclusion of this paper, the reader may review and consider the recommended actions for advancing the state-of-the-art use and utilization of wood and wood-based materials.

DOI: 10.1061/(ASCE)0899-1561(2002)14:2(91)

CE Database keywords: Wood; Structural materials; Composite materials.

Introduction

Wood is one of the earliest building materials used by humankind. As such, its use has often times been based more on tradition and past experience than engineering principles. Additionally, and perhaps as a result of this, the engineering image of wood and wood-based materials has been one of a "low-tech" or "second-tier" material. In opposition to this misperception, however, is the fact that the use of wood and wood-based materials in construction each year, by weight, exceeds that of steel or concrete (LeVan 1998). The role of wood and wood-based materials is equally important from an economic standpoint. The wood and wood-fiber sector of the forest products industry employs nearly 14 million people. Additionally, in 1994, the value of primary and secondary forest products accounted for approximately 2% of the Gross Domestic Product (GDP) for goods and services and approximately 1% of the total GDP (LeVan 1998). Finally, one-tenth of the global economy is directly linked to the construction of homes and commercial office complexes, with a significant portion of that sector of construction being dominantly wood and wood-based buildings. Worldwide, housing alone accounts for approximately 25% of the demand for solid wood products (LeVan

1998). From this brief overview, the importance and role of wood in both today's and tomorrow's engineering community, as well as society in general, is easily recognized.

While wood is one of the earliest construction materials, the structural use of wood and wood-based materials also has increased steadily in recent times. The terms "wood" and "wood-based" are traditionally used to draw a distinction between materials that are produced directly from a log (e.g., a log itself or a sawn piece of lumber) and materials that are produced by processing or modifying the raw wood material into a composite (e.g., plywood, oriented strand board, I-joists, or wood-plastic composites). In fact, new wood-based materials continue to be developed and successfully introduced into the engineering and construction marketplace. The primary driving force behind the increased use of wood and wood-based material, as alluded to above, is the ever-increasing need to provide economical housing for the world's population. Supporting this demand, however, has been an evolution of our understanding of wood as a structural material and our ability to analyze and design safe and functionally efficient wood and wood-based structures.

The primary aim of this paper is to provide a summary of the status and future of engineered wood and wood-based materials. This paper is not intended to be a comprehensive review of the literature or an all-inclusive state-of-the-art report on all types of wood and wood-based materials. Such a task would be too demanding for this forum. Rather, the focus is on the past, present, and future of engineered wood and wood-based materials for use in civil engineering applications, and providing a vision for addressing anticipated future research needs. This includes, but is not limited to wood and wood-based materials used in residential and light-commercial buildings, transportation structures, utility support and distribution structures, and waterfront structures. While many and varied reference sources were consulted during the development of this paper components of the report "Wood

[1]Professor of Civil Engineering and Associate Dean of Research and Information Technology, Howard R. Hughes College of Engineering, Univ. of Nevada, Las Vegas, 4505 Maryland Pkwy., Box 454005, Las Vegas, NV 89154-4005.

Note. Associate Editor: Antonio Nanni. Discussion open until September 1, 2002. Separate discussions must be submitted for individual papers. To extend the closing date by one month, a written request must be filed with the ASCE Managing Editor. The manuscript for this paper was submitted for review and possible publication on October 17, 2001; approved on October 26, 2001. This paper is part of the ***Journal of Materials in Civil Engineering***, Vol. 14, No. 2, April 1, 2002. ©ASCE, ISSN 0899-1561/2002/2-91–96/$8.00+$.50 per page.

Table 1. General Size and Use Categories of Solid Lumber

Category	Name	Example of Sizes[a]
Boards	Stress-rated boards	1×2, 1×4, 1×6
Dimension lumber	Structural light framing	2×2, 2×4, 4×4
	Light framing	2×2, 2×4, 4×4
	Studs	2×4, 2×6, 4×4, 4×6
	Structural joists and planks	2×6, 2×8, 2×10, 4×6, 4×8, 4×10
	Decking	4×2, 6×2, 6×4
Timbers	Beams and stringers	6×10, 6×12, 8×12, 10×14, 10×16
	Posts and timbers	6×6, 6×8, 8×8, 8×10, 10×10, 10×12

[a] "Sizes" should be considered names that reflect actual size, similar to W-sections in steel. For example, a 2×4 is 38 mm (1.5 in.) by 89 mm (3.5 in.), and a 4×10 is 89 mm (3.5 in.) by 235 mm (9.25 in.). For additional information on sizes, see AF&PA (1996, 1997).

Engineering in the 21st Century: Research Needs and Goals" (Fridley 1998) are referenced extensively. This is because the aim of the workshop that resulted in the report was, in part, synonymous with the aim of this paper. At the conclusion of this paper, the recommended actions for advancing the state-of-the-art use and utilization of wood and wood-based materials, as concluded by the workshop, are provided for review and consideration by the reader.

Review of Existing Materials and Recent Trends

To provide a basis from which to project the future, a general review of existing wood and wood-based materials, along with recent associated trends, is provided herein. The discussion is organized with respect to the following material types: solid wood products, wood-based composites, and hybrid composites.

Solid Wood Products

When wood elements are sawn or formed directly from a log, it is considered a "solid wood product." If any further processing occurs where single elements are glued or "built-up" into a separate element, it would be considered a wood-based composite. The traditional forms of solid wood products include dimension lumber (from 2×2 up to 4×16), timbers (6×6 and larger), and poles (round timbers). Table 1 provides a general listing of the size and use categories for lumber. Within each size and use category are a variety of grades that delineate the structural design properties for each combination of size and grade. While these traditional solid sawn materials have been available and used successfully for an extended period of time (i.e., centuries), research and technical advances continue. For example, continued improvements in sawing and drying technologies, grading techniques and machinery, and chemical treatments for biological and fire resistance have advanced the functional use and utilization of solid wood products.

From an engineering perspective, some of the key recent advances and trends involve improving our ability to assess the raw material resource through the grading and evaluation processes. Following the grading process, improvements in assigning appropriate design values for use by both design codes and design engineers has also been a targeted effort, particularly with the recent introduction of the *Load and Resistance Design Specification (LRFD) for Engineered Wood Construction* (American 1996; ASCE 1995).

Grading of wood products was traditionally one that relied heavily on visual inspection and empirical relations between the visual appearance and structural properties. More recently though, the use of nondestructive evaluation, or NDE, of the material has improved the grading process significantly. Two basic NDE classifications are made for grading lumber: machine-stress-rated (MSR) lumber; and machine-evaluated lumber (MEL). MSR lumber is run through a series of rollers in which a bending load is applied to the minor axis of the cross section and the modulus of elasticity of each piece is determined. Assuming an empirical relation between the weak-axis modulus of elasticity and strong-axis strength properties, a grade is then assigned. Because of the testing procedure, MSR is limited to thin material, typically 2-by or smaller. MEL is a more recent development in the grading of lumber. MEL employs radiographic inspection to measure density, which is used to improve the accuracy of strength value predictions over simply using the modulus of elasticity in the empirical relation. Regardless, visual inspection, however, remains a vital component of the grading process.

In addition to the grading process, improvements in the assignment of design values have been made possible with the recent in-grade test program. Prior to the 1991 edition of the *National Design Specification* (AF&PA 1991), design values for structural size members were determined based on empirical relations between the characteristic properties of small, clear specimens (2×2 members with straight, clear grain) and that of full-size members, including various strength reducing features such as knots, slope of grain, etc. The in-grade test program (Green 1989; Green and Evans 1987) was a large-scale effort undertaken jointly by the lumber industry and the U.S. Forest Products Laboratory (FPL). The purpose of the in-grade program was to test full-size dimension lumber that had been graded following existing current practice. Direct empirical relations were then established between the grading criteria and actual mechanical properties and thus design values. The design values published for the 1991 and 1997 editions of the NDS, as well as the new wood LRFD (American, 1996), are based in part on clear-wood data and in part on the full-size in-grade data.

Preservative and Fire-Retardant Treatments

Generally, if it is protected (not exposed to the weather or in direct contact with the ground) and is used in relatively low humidity environments (typically a covered structure), wood performs well without need for chemical treatment. When required, chemicals can be impregnated into wood through a pressure-treating process. The purpose of the treatment is not to abate moisture, but rather to protect wood from the instruments that can destroy it namely decay, termites, marine borers, and fire. Many species readily accept treatment, while others do not accept the pressure treatment. The three basic types of pressure preservatives are creosote and creosote solutions, oilborne treatments (pentachlorophenal and others dissolved in one of four hydrocarbon solvents), and waterborne oxides. There are a number of variations in each of these categories and the specific choice of treatment and required retention (concentration) depends on the application. The reader is referred to the American Wood Preservers Institute (AWPI) for additional information. Methods and technologies to protect wood from decay, biological attack, and fire is an area of continuing research and development.

Wood-Based Composites

Almost without exception, wood-based composite materials have been developed as substitute materials that can be used to replace existing construction materials, typically wood and wood-based construction materials. One of the earliest examples of this was the substitution of plywood sheathing for board sheathing used in floor, roof, and wall assemblies. The sheathing application is again changing with the continued increase in use of oriented strand board, or OSB, as a sheathing material. Numerous other examples of this substitution-based evolution of wood-based composites exist.

The driving force behind the development of many wood-based composites is two-fold. First is the ever-changing wood resource. Today's use of small diameter, plantation grown trees limits both the size and quality of solid wood products. Second is the need to meet increased performance demands for the material. By engineering and producing a composite, larger elements can be produced and undesirable natural characteristics (also referred to as defects) can be randomly distributed resulting in a minimized effect on the element. For example, knots may significantly reduce the capacity of a solid piece of dimension lumber. If, however, that same log is cut into veneers and the veneers are used to produce a piece of laminated veneer lumber (LVL), that now may be distributed through the length of the element, thus minimizing its local effect. This process also has the effect of reducing overall variability, thereby increasing the confidence and reliability of the material.

Early structural wood-based composites were dominantly produced from large wood elements, such as lumber and veneer. Examples of these composites, which still receive considerable attention and use, include glued laminated timber, or glulam, and plywood. More recently, LVL has emerged as a common structural composite. Even more recently, the trend has been toward strand-based composites as both a lumber substitute and for panel applications. Examples of strand-based composites include OSB, and oriented strand lumber (OSL) or composite strand lumber (CSL). Particle- and fiber-based composites have been commercially available for some time, though these are used predominately in nonstructural applications such as interior furniture. Examples of particle- and fiber-based composites include particleboard, hardboard, medium-density fiberboard (MDF), and insulation board. Finally, fiber- and particle-based composites, as well as strand- and veneer-based composites, to a lesser extent, are commonly used to form nonstructural molded wood-based products, such as automobile interior components, furniture, and building finishings (doors, baseboards, door trim, etc.).

Table 2. General Classification and Examples of Wood-Based Composites

Material application	Constituent wood element				
	Lumber	Veneer	Strand	Particle	Fiber
Beam/column	Glulam	LVL	OSL/CSL	N/A	N/A
Structural panel/plate	Glulam	Plywood	OSB	N/A	N/A
Nonstructural panel/plate	N/A	Plywood	OSB	Particleboard	Hardboard and MDF

Note: LVL=laminated veneer lumber; OSL=oriented strand lumber; CSL=composite strand lumber; MDF=medium density fiberboard; N/A = not available.

Commercial wood-based composite materials may best be classified by the constituent elements used in production of the material as well as the materials' intended application. Table 2 provides a simple classification of existing composites. As noted, these composites have been developed and are used primarily as substitution materials for solid wood equivalents.

In addition to the material-based composites discussed previously and overviewed in Table 2, a separate class of composite *wood structural components* should be recognized. One example would be the wood composite I-joist, which is a substitute component for lumber joists and headers. The flanges of wood composite I-joists could be made from solid lumber, or any of the lumber substitutes such as LVL or OSL. The web of the I-joists likewise could be made using any of the composite panel/plate products such as plywood or OSB. Structural insulated panels, or SIPs, are another example of a wood composite structural component. SIPs are prefabricated wall, roof, and floor panels in which two outer sheets, or faces, of plywood or OSB are separated by an insulated (e.g., expanded polystyrene) core, which also acts to transfer shear forces between the faces. When structural demands increase, lumber or wood-composite framing is also used to transfer the shear forces. Even preengineered wood trusses could be considered a composite wood structural component.

While from an industry perspective, wood composite materials may appear as self-competing, often times the substitution is a needed result of the changing natural resource. A decreasing quantity of large solid structural timbers and decreasing quality of dimension lumber, particularly with respect to use characteristics such as straightness, consistency, and visual appearance, has promoted the development and use of composites. In addition, the introduction of wood-based composites has simultaneously increased building efficiency. Improved use characteristics, high strength-to-weight ratios, and increased availability of both size and configuration have resulted in increased open spans for floor and roof assemblies.

Hybrid Wood Composites

The latest trend in wood-based composite materials is toward the use of synthetic materials in combination with wood or wood-based materials. The use of synthetic materials in combination with wood has been and continues to be used to repair and retrofit existing wood elements (e.g., Avent 1993). Various repair techniques have been explored for wood elements.

While the use of synthetic materials in combination with wood is not a new concept, significant advances have been made recently with this class of materials. Additionally, a dramatic increase in the commercialization and consumer acceptance of the

Table 3. Hybrid Wood and Synthetic Materials

Technology	Example of hybrid materials
Synthetic reinforcement	Glulam timbers
Inorganic bonding	Siding, roofing, wallboard, tile-backer
Thermoplastics	Decking, floor and door molding, window casements, countertops
Polymer impregnation	Flooring

materials has been noted recently as well. Table 3 provides a list of synthetic material technologies along with some examples of their application with wood. The use of synthetic reinforcement (e.g., glass or carbon fiber) in glulam timber has the aim of increasing strength and stiffness, thus allowing even greater free spans. Additionally, reinforcing glulams with synthetic fiber composites reduces the coefficient of variation in strength, thus increasing the relative design values for the material as well. Other hybrid composites tend to target improved durability; that is, with the introduction of synthetic materials into a wood-based composite, improved resistance to moisture, biological degradation, and general wear is targeted.

Existing Technical Resources

Given the diversity of wood and wood-based materials available for use in civil engineering applications, as well as the diversity in the material property and characteristics within a given product, engineers and design professionals must be able to readily access specific and detailed technical information. In addition to ASCE, several professional societies concern themselves with wood and wood-based materials, including the Society of Wood Science and Technology (SWST) and the Forest Products Society (FPS), both headquartered in Madison, Wis. SWST publishes *Wood and Fiber Science*, a technical journal that often includes papers focused on wood and wood-based materials. FPS publishes the *Forest Products Journal*, a more applied journal that covers the broad spectrum of forest products disciplines, including wood engineering and materials, and *Wood Design Focus*, an applied publication dedicated to design issues relative to wood structures. Additionally, FPS publishes conference proceedings and other technical literature in the topical area of wood materials.

Several series of technical conferences are focused on wood materials and engineering, including the World Conference on Timber Engineering (WCTE) series and the International Wood Composite Materials Symposium (IWCMS) (formerly the Particle/Composite Materials Symposium). The WCTE was most recently held in Vancouver, British Columbia, and will be held in Malaysia in 2002, and in Finland in 2004. The IWCMS is held annually at Washington State University.

The Forest Products Laboratory (FPL) in Madison, Wis. maintained by the U.S. Department of Agriculture Forest Service, serves the public as the nation's laboratory for wood-related research. The FPL produces and maintains numerous publications relevant to wood and wood-based materials. Of specific interest is the publication *Wood Handbook: Wood as an Engineering Material* (Forest 1999). This single publication offers extensive information on the manufacture and utilization of wood and wood products, including detailed information on the physical and mechanical properties of wood and wood products.

From a design standpoint, the American Wood Council (AWC) of the American Forest and Paper Association (AF&PA) is a key organization offering technical assistance to engineers and design professionals. The AWC/AF&PA maintains both the allowable stress formatted National Design Specification (AF&PA 1997) and the LRFD design specification (American 1996) in addition to many other technical resources. The APA—The Engineered Wood Association provides technical support for a broad spectrum of wood products, as does the American Institute for Timber Construction (AITC). Several regional associations, including the Southern Forest Products Association (SFPA) and the Western Woods Products Association (WWPA) also offer technical assistance. Each of these associations can, in turn, direct designers to other, more focused groups specializing in, for example, chemical treatments, fire retardants, and uses.

Future Trends and Directions

With the preceding general review of the status of and current trends in wood and wood-based materials used as background, the following future trends are envisioned. The discussion is again organized with respect to material type: solid wood products, wood-based composites, and hybrid composites. As a general theme, the continued impact of a changing resource is seen as a driving force in coming years. The wood materials and products industry will see increasing amounts of wood and wood-fiber derived from small diameter and/or fast-grown, plantation trees. Wood-based materials will contain more juvenile wood, which is generally less dense than mature wood and exhibits increased variability in both mechanical and physical properties and more significant dimensional changes under moisture cycling.

Solid Wood Products

Solid wood products, as a material, will likely see little attention in the coming years beyond that currently afforded to it. As mentioned, the changing resource will drive changes in the fundamental physical and mechanical properties of solid wood products, and such changes must be monitored and reported on through programs like the in-grade test program (Green 1989; Green and Evans 1987). Related to this, however, are areas where active research and development investments will be made in the future such as processing and evaluation. Enhancements of existing nondestructive evaluation techniques for grading solid wood materials and products, both during production and in situ, are needed for better utilization of the raw material and recycling of previously used material. However, the greatest opportunity for growth and expanded use of solid wood materials may not be as a product in itself, but as a component used in combination with other wood and/or nonwood materials as a composite element (e.g., glulam or fiber-reinforced glulam).

Wood-Based Composites

In addition to the changing virgin resource as discussed previously, new resource supplies, namely that of recycled wood products, will find an increased role in wood-based composites. The interest in recycling post-consumer wood materials that was noted during the 1990's (Falk 1997) will continue in the coming years. Both consumer interest and traditional virgin resource limitations will drive further utilization of recycled materials. Currently, several examples exist, namely particleboard and MDF mills, wherein the raw material streams are supplemented with recycled wood-fiber. The primary limiting factor for further use and utilization of recycled materials for wood-based composites is one of sorting and handling. Research will need to focus on methods and systems that can produce a consistent and clean supply of recycled materials for reprocessing as new wood-based composites.

Durability is perhaps the one over-riding issue needing focused research attention for wood-based composites. For use and acceptance to continue and grow, improved understanding and performance is needed with respect to biodegradation, moisture resistance, dimensional stability, creep, creep-rupture, and fatigue. Related to this, accelerated test methods for assessing the durability of wood-based composites are truly needed. Considerations should include accurate representation of end-use conditions and the relationship between test results and engineering design. Accelerated tests and the resulting data must be usable for comparing product performance as well as allow engineers to account for the performance in design.

One path to improve durability and possibly counter the effect of the ever-changing resource stream is the use of chemical additives as stabilizing agents. The downside of this is the potential for contamination of the material and its effect on both subsequent reuse and consumer perceptions. Another potentially viable and productive alternative is the combination of wood and wood-fiber with other materials, particularly synthetic materials.

Hybrid Wood Composites

The recent trend to combine wood and wood-fiber with other materials, particularly synthetic materials, will continue. It will lead to new markets and expanded opportunities for engineered wood-based products. Improvements in fiber-reinforcing technologies, such as that currently used with glulam beams, will continue and expand to other product lines. The result will be structural properties formerly unattainable with wood-based products. Advances in this area, though, will depend on improvements in our understanding of bonding and durability issues and enhanced manufacturing processes. The opportunities for reinforcing schemes are not, however, limited to aramid, carbon, and glass fibers. Polymer-cellulose and other cellulose technologies may develop into viable alternatives in the future. Additionally, the viability and effectiveness of simple steel reinforcing will continue to be explored.

Beyond the simple reinforcement of wood and wood-based materials with other materials, direct combinations of wood and other materials will continue to see significant attention in the future. Wood-plastic composites have already made significant headway into commercial markets. Future efforts will be focused on optimizing the combined use of wood fiber, in some form, with synthetic materials to produce an efficient and cost-effective structural material. New technologies and materials will be sought that improve dimensional stability and resistance to moisture and biological degradation with reduced use of chemical treatments to ward off biodegradation.

While the preceding discussion of hybrid composites is general in application, as engineers we typically direct our attention toward structural material applications. Nonstructural wood-based composites, particularly hybrid composites, that are used in window casings, doors, siding, trim, and roofing, significantly contribute to the total cost of a building. Additionally, since these materials are often the primary point of contact between the occupant and the building structure, they are high profile from a consumer prospective. Furthermore, these same materials frequently have a significant role in preserving the durability of the building through their function as a part of the building envelope.

Court records alone are evidence of the need for further research and development with respect to these nonstructural materials.

Finally, beyond direct replacement and substitution as described above, the potential exists to positively alter current building practice and convention with innovative, preengineering of materials in concert with novel approaches to the entire building system. For example, with society's increasing interest in and demand for recyclability, imagine designing buildings, building systems, and materials within these systems to be functionally recyclable from the onset. The flexibility and adaptability offered by hybrid wood-based composites, considering both formulation and processing, may lead to the realization of custom designed materials, elements, and systems for buildings that are not tied to past conventions, but based on future projected needs.

Critical Research Needs

Part of the aim of the workshop "Wood Engineering in the 21st Century: Research Needs and Goals" (Fridley 1998) was to establish critical research needs that would significantly advance wood engineering, including wood and wood-based materials, in the first decade of the 21st century. Three broad-based critical research needs were identified. These were "Behavior and Performance of New and Existing Wood Structures," "Integrated Analysis, Design, and Construction Methodologies," and "Development of Wood-Based Composite Materials." Within these broad themes, the following specific needs related to wood and wood-based materials were identified:

- Develop an understanding of the behavior and performance of wood and wood-based structures and materials. This understanding must be translated into design and performance criteria that can be easily integrated into the design process;
- Develop performance and assessment criteria for serviceability issues, including durability and creep;
- Develop hybrid systems that capitalize on the economy and flexibility of wood and the unique characteristics of other materials to create economy and efficiency in the final product;
- Develop technologies to evaluate and monitor performance over time using economical and accurate nondestructive and nonintrusive methods;
- Develop and promote new means to prevent deterioration of new and existing products that are low cost, reliable, and easily implemented;
- Develop technologies and methods to assess residual strength and performance of damaged, deteriorated, or distressed materials. Included in this is the development of repair practices that are robust and verifiable;
- Assess environmental effects including durability on wood and wood-based materials and include appropriate design criteria for such factors in the design process;
- Develop a comprehensive understanding of fundamental properties of material used in wood-based composites, including both wood and non-wood components;
- Understand how changes in the wood resource, considering both virgin and recycled sources, will influence the performance and economic future of wood-based materials;
- Aggressively support the development and understanding of hybrid wood composites, functional wood elements, and structural systems;
- Improve the durability of wood-based materials, specifically addressing long-term performance as affected by biodegradation, moisture resistance, dimensional stability, creep, creep-rupture, and fatigue; and

- Produce accelerated test methods and procedures for assessing durability. Such methods and procedures must reflect actual potential field use and be usable for ultimately being adopted into design codes and specifications.

References

American Forest and Paper Association (AF&PA). (1996). *Load and resistance design specification (LRFD) for engineered wood construction*, Washington, D.C.

AF&PA. (1991). *National design specification for wood construction*, Washington, D.C.

AF&PA. (1997). *National design specification for wood construction*, Washington, D.C.

ASCE. (1995). "Standard for load and resistance design specification (LRFD) for engineered wood construction." *ASCE 16–95*, New York.

Avent, R. (1993). "Structural design for epoxy repair of timber." *Wood Des. Focus,* 3(3), 4–6.

Falk, R. H. (1997). "Wood recycling: Opportunities for the woodwaste resource." *Forest Products. J.,* 47(6), 17–22.

Forest Products Society. (1999). "Wood handbook: Wood as an engineering material." Reprint Forest Products Laboratory General, *Tech. Rep. FPL-GTR-113*, Madison, Wis.

Fridley, K. J., ed. (1998). "Wood engineering in the 21st century: Research needs and goals." *Proc., ASCE and the Structural Engineering Inst. Com. on Wood Workshop*, ASCE, Reston, Va, 1–157.

Green, D. W., ed. (1989). *In-grade testing of structural lumber*, Forest Products Society, Madison, Wis.

Green, D. W., and Evans, J. W. (1987). "Mechanical properties of visually graded lumber: Volume I, A summary." U.S. Dept. of Agriculture, Forest Service, Forest Products Laboratory, Madison, Wis.

LeVan, S. L. (1998). "Benefits from wood engineering research." *Wood engineering in the 21st century—Research needs and goals*, ASCE, Reston, Va, 1–4.

Use of Fiber Reinforced Polymer Composites as Reinforcing Material for Concrete

Taketo Uomoto[1]; Hiroshi Mutsuyoshi[2]; Futoshi Katsuki[3]; and Sudhir Misra[4]

Abstract: Although the potential of continuous fiber reinforced polymers (FRPs) was recognized more than 50 years ago, bottlenecks such as vulnerability to static fatigue, ultraviolet radiation, alkaline environment, etc., continued to restrict their use as a construction material. Extensive research across the world during approximately the last 25 years has led to a better understanding of properties and behavior of the FRPs under different conditions. Given their lightweight, noncorrosive, and nonmagnetic nature and their high tensile strength, FRP composites are being used as reinforcement in concrete, especially in prestressed concrete construction, anchors for slope stabilization, and use in special structures such as high-speed linear motor railway tracks, MRI units of hospitals, and repair and rehabilitation works. An attempt has been made in this paper to comprehensively put together the various aspects of FRPs, including their structural and durability characteristics, the ongoing research in the areas related to their utilization as a reinforcing material for concrete structures, and the application of FRP sheets in rehabilitation of deteriorated concrete structures. Though issues related to the manufacture of FRP composites and design procedures have not been considered in detail, they have been included for information and completeness. The paper also presents the current thinking on the design and construction procedures using FRP material in conjunction with concrete and provides a brief discussion of some of the documented applications of FRP composites—rods and sheets—as reinforcement in concrete construction. It may also be mentioned that the paper is largely based on the case studies and research work carried out in Japan; appropriate references to related work in other parts of the world are also included, though in principle there is very little difference in the approaches being adopted across the world.

DOI: 10.1061/(ASCE)0899-1561(2002)14:3(191)

CE Database keywords: Fiber reinforced materials; Concrete, reinforced; Polymers; Construction materials.

Introduction

Continuous fiber reinforced polymer (FRP) composites have been used in automobile, electronics, and aerospace engineering for several decades, but their application in concrete engineering as a reinforcing material is relatively recent in origin. Besides the advances in the field of development of new fibers, etc., several developments in the construction industry have accelerated the efforts to apply FRPs as a reinforcing material in concrete. One of

[1]Professor of Civil Engineering and Director, International Centre for Urban Safety Engineering, Institute of Industrial Science, Univ. of Tokyo, Meguro Ku 4-6-1, Tokyo. E-mail: uomoto@iis.u-tokyo.ac.jp

[2]Professor of Civil and Environmental Engineering, Faculty of Engineering, Saitama Univ., Saitama-shi, Shimo-ohkubo 255, Saitama. E-mail: mutuyosi@mtr.civil.saitama-u.ac.jp

[3]Assistant Professor of Civil Engineering, Shibaura Institute of Technology, Minato-ku Shibaura 3-9-14, Tokyo. E-mail: katuki@sic.shibaura-it.ac.jp

[4]Associate Professor, Dept. of Civil Engineering, Indian Institute of Technology Kanpur, Kanpur-208 016, India. E-mail: sud@iitk.ac.in

Note. Associate Editor: Antonio Nanni. Discussion open until November 1, 2002. Separate discussions must be submitted for individual papers. To extend the closing date by one month, a written request must be filed with the ASCE Managing Editor. The manuscript for this paper was submitted for review and possible publication on December 26, 2001; approved on January 17, 2002. This paper is part of the *Journal of Materials in Civil Engineering*, Vol. 14, No. 3, June 1, 2002. ©ASCE, ISSN 0899-1561/2002/3-191–209/$8.00+$.50 per page.

the chief causes is the realization that the reinforcing steel, both in reinforced and prestressed concrete construction, is corrosion prone, though the malady may take more than 20–30 years to show symptoms! This realization obviously prompts the desire to use noncorrosive materials such as FRP, especially in environments where steel has been shown to be vulnerable.

Fig. 1 is a representation of how the predominant construction material has changed in the last about 7,000 years. It can be seen the FRPs are the youngest in the family of civil engineering construction materials. Extensive research across the world during the last 25 years or so has led to a better understanding of the properties and behavior of the FRPs under different conditions, and more extensive use of FRPs is likely to be seen in the coming years (e.g., Taerwe et al. 2001). It may be pointed out that the application of FRPs in other disciplines is mostly in the form of essentially 2D thin-shell or plate-like elements, where buckling of the elements is the main design criteria. On the other hand, the use of FRPs as a reinforcing material in concrete construction largely focuses on utilization of FRPs arranged in a single direction and subjected to high levels of tensile load. In other words, concrete engineering applications of FRPs need a more careful consideration of such properties as creep, fatigue, and modulus of fracture.

Besides being noncorrosive, the FRPs have a much better strength-to-weight ratio than steel and this makes them an ideal material for applications in repair, rehabilitation, and strengthening works. Also, in certain special applications (e.g., high-speed

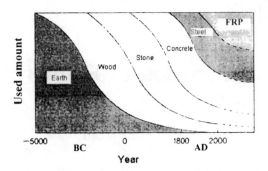

Fig. 1. Dominant construction materials in different ages

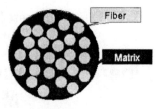

Fig. 2. Schematic representation of cross section of continuous fiber reinforced polymer composite

linear motor railway tracks in Japan) use of a nonmagnetic material such as FRPs has an added advantage. Though research in several areas pertaining to the utilization of the FRPs as a reinforcing material is still ongoing, engineers have used a blend of caution and judgment to proceed with the development of application-related technology, and several cases of actual application have been reported (e.g., Jesse and Curbach 2001; ACC 1995, 1998).

An effort has been made in this paper to provide an overview to the properties and behavior of commercially available FRPs and to briefly discuss the application of FRPs as a reinforcing material in reinforced and prestressed concrete construction. Some details of the application of continuous FRP sheets, especially in the area of retrofitting and strengthening of concrete structures, have also been presented. This study presents the current thinking on the design and construction procedures using FRP material in conjunction with concrete. Although most of the examples cited in the paper are drawn from the developments in Japan, the basic trends and overall approach to the use of FRP as a reinforcing material is the same in other parts of the world.

Historical Notes

Given the high strength and lightweight nature of the FRPs, their potential as a reinforcing material in the prestressed concrete construction was realized early; it was the subject of research in the U.S., the erstwhile Soviet Union, and the U.K. as early as the 50s and 60s (Kobayashi et al. 1987). In the tests carried out at that time, it was found that though a tensile strength of as much as 3,620 kgf/mm^2 was observed using a 3μm monofilament, the strength of the rods or other composite materials was much lower (about 700 kgf/mm^2). Besides this reduction in the load-carrying capacity when made into a composite material, it was found that the FRPs were susceptible to premature failure under sustained load. It was estimated that if the load was applied over a long period of time, the rods could fail at load levels as low as 60–70% of the tensile strength observed in the laboratory. Thus, assuming that a maximum load of about 70% of the tensile capacity is applied at the time of prestressing, the actual prestressing load is only about 45–50% of the tensile strength observed in the laboratory. Even so, it must be borne in mind that a permissible load level of even 400 kgf/mm^2 was higher than the levels of 160 kgf/mm^2 permissible in steel at that time.

Premature failures in the FRPs were attributed to the phenomenon of static fatigue, which resulted from the irregular distribution of tensile stresses among the fibers in an FRP composite.

Such a distribution of stresses resulted from the fact that all the fibers were not properly aligned and thus did not fully participate in carrying the applied tensile load. The problem persisted till the process of pultrusion for the manufacture of FRP products was developed in the 70s (Kobayashi et al. 1987).

German scientists used FRP products using glass fiber and polyester resin as a matrix to build an experimental bridge with a span of only 6.5 m in the early 80s (Waaser and Wolf 1986). The tensile strength and the modulus of elasticity of the FRP used were 150 kgf/mm^2 and 5,100 kgf/mm^2, respectively. Over the last 20 years, substantial progress has been made in all areas, including manufacture, quality control, and applications. Some of the more important features are discussed in this paper.

Besides overcoming the problem of static fatigue, persistent research efforts were required to overcome such problems as the durability of glass fibers in the highly alkaline environment of concrete (Uomoto 2001) and the rapid deterioration of epoxy resins upon exposure to ultraviolet radiation (Yamaguchi 1998). Development of appropriate test methods and standards for the FRP composites also placed a major challenge to the professional institutions and researchers. It may be noted that comprehensive specifications for the design and construction of concrete structures using FRPs have been published only recently (JCI 1998; ACI 2001; JSCE 1997, 2001).

Continuous Fiber Reinforced Polymer Materials

General

As against short fibers of steel that are sometimes used to improve the tensile strength and postcracking behavior of the brittle concrete matrix, continuous long fibers can be used as a replacement of the steel as a reinforcing material for reinforced and prestressed concrete construction.

From the schematic representation of the cross section of continuous FRP composites shown in Fig. 2, it is clear that they can be seen as made up of extremely fine fibers embedded in a matrix. The volume fraction of the fibers in commonly available FRPs ranges from about 50 to 65%. As discussed in the following paragraphs, the properties of FRP composite products are related to (1) the properties of the fibers used, (2) the properties of the matrix used, and (3) the volume fraction of the fibers.

Constituent Fibers

Fibers made from carbon, aramid, glass, and polyvinyl alcohol are commonly used for the manufacture of FRP. Table 1 gives a brief overview of the properties of the commonly used fibers in FRPs used in the construction industry. As can be seen, the specific gravity of the fibers ranges between 1.4 for aramid fibers to

Table 1. Properties of Fibers used in Fiber Reinforced Polymer Composites

Item	Carbon fiber				Aramid fiber		Glass fiber		Polyvinyl alcohol fiber (high strength)
	Polyacrylic Nitril Carbon		Pitch Carbon		Kevlar 49 Twaron	Tech-nora	E-Glass	Alkali-Resistant Glass	
	HIGH STRENGTH	HIGH YOUNG'S MODULUS	ORDINARY	HIGH YOUNG'S MODULUS					
Tensile strength (MPa)	3430	2,450–3,920	764–980	2,940–3,430	2,744	3,430	343 0–3 528	1,764–3 430	2254
Young's modulus (GPa)	196–235	343–637	37–39	392–784	127	72.5	72.5–73.5	68.6–7 4.5	59.8
Elongatio n (%)	1.3–1.8	0.4–0.8	2.1–2.5	0.4–1.5	2.3	4.6	4.8	2–3	5.0
Density (gm/cm³)	1.7–1.8	1.8–2.0	1.6–1.7	1.9–2.1	1.45	1.39	2.6	2.27	1.30
Diameter (μm)	5–8		9–18		12		8–12		14

about 2.6 for glass fibers. The diameters of the fibers range from about 6 microns for carbon fibers to about 15 microns in the case of aramid and glass fibers.

Fig. 3 gives a diagrammatic representation of the stress-strain curves for some commonly used FRP composites. Curves for steel and PC tendon have also been included in the figure only for reference. It can be seen that though all the fibers exhibit a higher tensile strength than steel, the elongation before the material finally fails is very small for the fibers. In fact, it is this lack of extensibility before failure that often makes the FRP composites unusable as normal reinforcement in RC structures, where large deformations before collapse are a design requirement. The carbon fibers have a higher modulus of elasticity and a lower elongation as compared with others. This reduces the toughness of the carbon FRPs and also makes carbon fiber products vulnerable to impact loads. The polyvinyl alcohol fiber is light and has better extensibility, but it rather belongs with other fibers with a relatively low tensile strength and Young's modulus. The following paragraphs give a brief description of the properties of the individual fibers (CCC 1998).

Carbon Fiber

These are made from either petroleum or coal pitch and polyacrylic nitril (PAN). Each fiber is an aggregate of imperfect fine

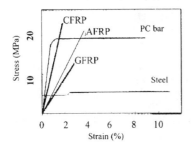

Fig. 3. Typical stress-strain curves for some fiber reinforced polymer (FRP) composites

graphite crystals, and its characteristics vary depending on the composition and the orientation of the crystals. Commercially available carbon fibers are usually classified into pitch carbon and PAN carbon fibers, the pitch carbon being ordinary or high Young's modulus fiber; the modulus of elasticity for the ordinary fiber of the pitch carbon family is about 20% of those of the PAN family. High Young's modulus fibers, on the other hand, provide an identical or higher Young's modulus than the fibers of PAN family.

Aramid Fiber

This is a highly oriented organic fiber, and three types of these fibers are available—Kevlar, Twaron, and Technora—which are essentially trade names. Kevlar and Twaron are aromatic polyamide fibers that have their benzene nuclei bonded linearly by aramid. Technora, on the other hand, is an aromatic polyetheramide fiber whose nuclei are bonded by ether.

Glass Fiber

Two types of glass fibers are commercially available for use—E-glass fiber and alkali-resistant glass fiber. The former contains large amounts of boric acid and aluminate, while the latter contains a considerable amount of zirconia, which serves to prevent corrosion by alkali attacks from cement matrix.

Polyvinyl Alcohol Fiber

As the name suggests, these are manufactured using polyvinyl alcohol of a high degree of polymerization. The fiber is rolled during the manufacturing process to provide added strength and elasticity.

Mixed Fibers

Recent years have also seen the introduction of products, where fibers of different origins are combined to produce a final product having desired properties. Nishimura reported the development of a combined fiber "hybrid" with glass fibers in the middle, surrounded by aramid fibers (Nishimura et al. 1996). Fig. 4 shows a schematic representation of the hybrid FRP composite. Such a structure provides the glass fibers extra protection in the alkaline

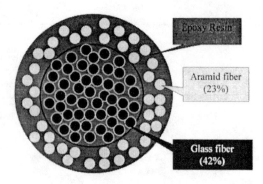

Fig. 4. Schematic representation of hybrid fiber reinforced polymer composite

environment of the concrete. NEFMAC, which stands for "new fiber composite material for concrete," is another example of such a material that uses continuous carbon and glass fibers impregnated with resin and formed into a lattice, as schematically given in Fig. 5.

Resins

Though there is limited evidence of use of special cement or silica fume, most of the FRPs use various kinds of resins, including epoxies, vinyl esters, and polyesters (Fig. 2). The specific gravity of the resins used varies between 1.1 and 1.4. Although the resin does not carry any tensile load directly, it acts as a filler material and holds the fibers together. The resins also play a crucial role in transferring the load and protecting the fibers in much the same manner as the surrounding concrete protects the embedded reinforcing bars. For example, the durability of glass-fiber-reinforced composites can be substantially increased if individual fibers are coated with a very thin layer of resins.

It is important that the fracture strain of the resins and the compatibility with the fibers themselves are taken into consideration in an appropriate manner. For example, epoxy resins have a low strain capacity and work well in conjunction with carbon fibers, which also have low extensibility, but the results are not so encouraging when epoxy resins are used with aramid or glass fibers, which have larger strains at failure. Vinyl ester formulations serve as better "filler" products in such an instance.

Substantial lowering in the tensile strength of the FRP rods has been reported in cases when the impregnation with the resin is improper. This lowering of the tensile strength can be attributed to stress concentrations in certain fibers as a result of the uneven

distribution of applied stress arising from the improper impregnation of the matrix (Hodhod 1992).

In the case of application of continuous FRP sheets, the resins are usually applied in situ while wrapping or attaching the sheet, and care needs to be taken to ensure that hardening and strength development proceed at a satisfactory rate.

Manufacture of Fiber Reinforced Polymers

As mentioned above, earlier attempts to utilize FRPs failed primarily on account of manufacturing difficulties, which led to static-fatigue-induced failure in the rods. Most of the present-day manufacture of FRP products used for reinforcement is by "pultrusion," a continuous process that combines pulling and extrusion for manufacturing composite sections that have the same cross section and shape. The process consists of pulling a fiber material through a resin bath and then through a heated shaping die, where the resin is cured. Important steps of the process and other methods of manufacture have been described in the literature (e.g., Mallick 1988; Agarwal and Broutman 1990).

The applied tension on the fibers during the manufacture eliminates twists and straightens them. When such products are actually stressed while testing or in a real application, a more uniform distribution of stresses among all the fibers is achieved, thus preventing static fatigue (at low load levels). It may be pointed out that the tension applied at time of manufacture is small, so as not to impair the performance of the composite as a construction material.

As far as the production of 2D and 3D FRP products is concerned, the grids are manufactured by impregnating bundles of fibers with resins and alternately laying them in X and Y directions under tension and pressure before shaping them into grids. The three-dimensional textiles are manufactured either by alternately laying bundles of fibers in X, Y, and Z directions in several layers while impregnating them with resins or by impregnating fibers already shaped into textiles with resins. Fibers woven into textiles using special cement or silica fume are produced as flat or curved plates.

FRP sheets are another variety of FRP materials available for use, especially in jobs related to the reinforcement and strengthening of existing structures (Iwahashi et al. 1996; Toritani et al. 1997; Yagishita et al. 1997). These sheets consist essentially of fibers laid out in one or two directions and held in place by a light adhesive. They can be classified as (1) one-way textile, (2) two-way textile, or (3) one-way pre-impregnated textiles. The resin and hardeners are applied to the sheets at the time of actual application, as discussed later in this paper (JCI 1998).

Properties of Fiber Reinforced Polymer Products

General

Mechanical properties such as the strength and modulus of elasticity of a product are related to the properties and proportions of the constituent materials. A laboratory study has demonstrated the possible use of the rule of mixtures, to estimate the properties of the FRP rods from the properties and the volume fraction of the parent fibers (Uomoto and Nishimura 1994). The study used the following equations:

$$E_{\text{FRP}} = E_1 V_f + E_2 (1 - V_f) \qquad (1)$$

$$\sigma_{\text{FRP}} = \sigma_1 V_f + \sigma_2 (1 - V_f) \qquad (2)$$

Fig. 5. Schematic representation of NEFMAC

Table 2. Comparison of Calculated and Experimental Values of Tensile Strength using Rule of Mixtures

V_f	Item	AFRP	GFRP	CFRP
45	Experimental (1)	134.5	140.1	124.4
	Calculated (2)	175.1	113.0	150.8
	(1)/(2)	0.77	1.24	0.83
55	Experimental (1)	168.9	168.9	133.5
	Calculated (2)	214.0	138.1	184.3
	(1)/(2)	0.79	1.22	0.72
66	Experimental (1)	204.2	176.9	148.4
	Calculated (2)	256.7	165.7	221.1
	(1)/(2)	0.80	1.07	0.67* (0.81)

As an approximation, if the contributions of the matrix phase are neglected, the above equations can be written as follows:

$$E_{FRP} = E_1 V_f \qquad (3)$$

$$\sigma_{FRP} = \sigma_1 V_f \qquad (4)$$

Table 2 shows the variation in the values of the estimated and observed values of the tensile strength and modulus of elasticity, assuming that all the tensile load is carried only by the fibers. It can be seen that the two are in reasonable agreement. The following paragraphs discuss briefly some of the important aspects of properties of the FRP, which are relevant to their application as a construction material.

Tensile Strength and Modulus of Elasticity

Given the fact that FRPs are used as tensile reinforcement in concrete construction, it is important that the behavior of the FRPs under tension is accurately understood. An extensive study was undertaken by Uomoto to better understand the variation in these properties with the type and content of the fibers and the statistical variation therein (Uomoto and Nishimura 1994). 100 specimens were tested at each test condition, and Table 3 gives the representation of the results obtained, including the statistical

Table 3. Tensile Strength and Modulus of Elasticity of Fiber Reinforced Polymer Rods

Parameter	V_f	Fiber type	AFRP	GFRP	CFRP
Strength (kgf/mm²)	45	Mean Value	135	140	124
		SD	9.49	5.83	11.8
		COV	0.0703	0.0416	0.0952
	55	Mean Value	169	169	134
		SD	14.8	8.53	11.2
		COV	0.0876	0.0505	0.0836
	66	Mean Value	204	177	148
		SD	5.73	12.2	27.0
		COV	0.0281	0.0689	0.1824
Elastic modulus (kgf/mm²)	45	Mean Value	3748	4274	11202
		SD	216.1	52.2	259.5
		COV	0.0576	0.0122	0.0232
	55	Mean Value	4570	5211	13530
		SD	194.5	47.6	236.3
		COV	0.0426	0.0091	0.0175
	66	Mean Value	5460	6024	15714
		SD	241.2	101.2	379.4
		COV	0.0442	0.0168	0.0241

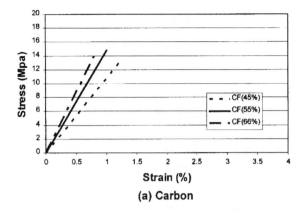

(a) Carbon

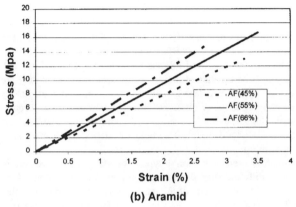

(b) Aramid

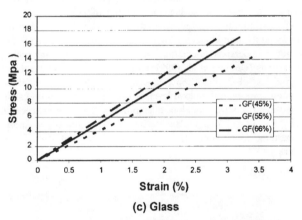

(c) Glass

Fig. 6. Typical stress-strain curves obtained using fiber reinforced polymer rods

parameters, such as mean and standard deviation. Typical stress-strain curves obtained using rods having different fibers and fiber volumes are given in Fig. 6. Given the fact that the properties depend on various factors such as the type and content of the fibers and the type of the resin, it is difficult to make generalized conclusions about the properties of the bars. However, the following are generally known principles that guide the present thinking on application of FRP as a construction material.

- The stress-strain behavior of all FRP materials is essentially linear till the point of failure, which is sudden;
- The fracture strains are much smaller when compared with steel and hence the material (FRP) should be used with great caution for applications when large deformations or ductility in the member is required (for example, in flexural failure);

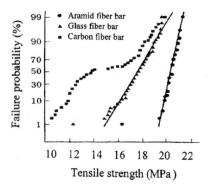

Fig. 7. Representation of tensile strength of fiber reinforced polymer rods in terms of cumulative probability versus stress

- The modulus of elasticity is about half that of steel (except in some cases of carbon FRPs); and
- The tensile strength of FRPs is much higher than that of steel. It may also be mentioned that the properties of the FRP may not necessarily follow a "normal distribution," research effort has been directed to apply different statistical distributions and represent the failure strength in terms of a cumulative probability distribution (Hodhod 1992). A sample of such a representation is shown in Fig. 7.

Fig. 8 is a representation of typical failure patterns observed in the FRP rods and strip, depending upon the volume content of the fibers and the properties of the matrix. It can be clearly seen that FRP composites show several modes of failure besides the rupture of the fibers—these include cracking of the matrix, partial interfacial adhesion failure, debonding between the fiber and the matrix, and longitudinal cracking. Studies using acoustic emission to better understand the failure of FRP also corroborate such a hypothesis (Uomoto et al. 1992). It has also been pointed out that aramid and glass fibers show consistency in the mode of failure regardless of the failure load level. However, the carbon FRPs fail in different modes, including fiber pull-out from the matrix and simultaneous failure at the anchorage, fan-type failure, and failure containing components of both the above (Uomoto and Nishimura 1994).

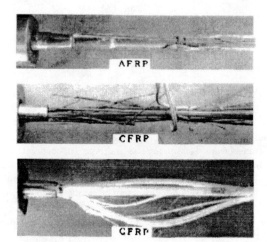

Fig. 8. Representation of typical patterns of tensile failure of fiber reinforced polymer rods

Fatigue Strength, Creep and Relaxation

As mentioned above, static fatigue was one of the major impediments in the application of the FRPs in concrete structures. Thus a substantial amount of research work has been carried out to better understand the increase in the strain at constant load (creep) and the decrease in the load at constant strain (relaxation) behavior of the FRPs (e.g., Uomoto 1995, 1998; Saadatmanesh and Tannous 1999). It has been pointed out that the extent of creep deformation observed in FRP composites is almost twice that observed in steel. However, the lower value of the modulus of elasticity can be used to an advantage when losses are to be accounted for (as in the case of prestressed concrete), as the loss of a certain amount of deformation leads to a lesser extent of force reduction in case FRP reinforcement is used.

Compressive Strength

Because of the inherent difficulty in testing FRP specimens in compression because of their tendency to buckle, there is only limited information available on the subject (Mallick 1988; Piggott and Harris 1980). It has been reported that the compressive strength and the modulus of elasticity of Kevlar fibers are significantly lower in compression than the corresponding values in tension. In the case of carbon and glass fibers, the difference is much less, and there is virtually no difference in the values obtained for boron-reinforced FRP composites. Given the scarcity of the information, it remains a challenge to adapt published data to the analysis and design procedures, as far as the application of FRPs to a reinforcing material in concrete is concerned.

Bond with Concrete

Whereas excellent bond between the prestressing reinforcement and the surrounding concrete is a requirement in normal reinforced and pretensioned prestressed concrete construction, the property is only of marginal significance in the case of posttensioned construction. Since there is an apprehension of inadequate bond between the concrete and the FRP reinforcement, most of the application of FRPs has been reported in posttensioned prestressed concrete construction. However, if special measures like braiding are taken, it has been shown that adequate bond can indeed be ensured (JSCE 1997). It has also been found that immersion in water tends to increase the bond between the surrounding concrete and the FRP reinforcement due to expansion of the polymers on account of reaction with water (Katz et al. 2001). Eq. (5) has been proposed to estimate the development length, which may be taken to be a measure of the bond strength (JSCE 1997):

$$f_{\text{bond}} = \frac{0.28 \alpha f_{ck}^{2/3}}{\gamma_c} \tag{5}$$

In the above expression, α is a modification factor in the estimation of the bond strength for continuous FRPs. The JSCE recommends a value of 1.0 in cases where the bond strength is equal to or greater than that of the deformed bars. In other cases, a lower value is to be determined on the basis of experimental results. Fig. 9 shows a comparison between the actual and estimated values of the bond strength. It can be seen that the equation used estimates the bond strength reasonably well.

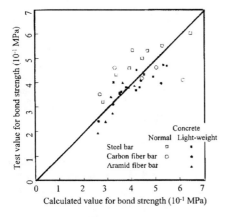

Fig. 9. Comparison of estimated and actual bond strength between fiber reinforced polymer rods and concrete

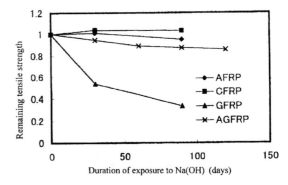

Fig. 10. Variation of observed tensile strength with continued exposure to sodium hydroxide

Durability

In cases when the FRPs are used within normal concrete, they need to show high durability in an alkaline environment, whereas if they are used for external prestressing, durability against UV radiation could be an additional requirement. Results of experimental research in the area of durability characteristics of FRP materials have been published by several researchers (Katsuki 1997; Yamaguchi 1998; Uomoto 2001). Tests were carried out using fibers *per se* and FRP rods to study the static fatigue fracture, fatigue fracture, alkali resistance, ultraviolet ray resistance, freeze-thawing resistance, high temperature and fire resistance. The results summarized in Table 4 indicate that

• Carbon fibers and FRP rods have good durability characteristics.
• Aramid fibers and FRP rods show good properties except under static fatigue, UV radiation, and acidic environment.
• Glass fibers may show poor durability characteristics as far as their alkali resistance is concerned, though they were found to be satisfactory in the acidic and freeze-thaw environment.
• FRP materials in general show poor performance at high temperatures, and therefore their use should be avoided when fire resistance is a requirement.
• Sheets made with carbon fibers need to checked for durability against UV radiation.

• There is also a need to limit the tensile load depending on the duration of the load, in cases when the FRPs are used as an internal reinforcement.

Fig. 10 shows the variation of the tensile strength observed for the different FRP rods under the continued action of sodium hydroxide. It can be seen that the glass fibers are expectedly most vulnerable. Also, the performance of the hybrid fiber shown in Fig. 4 clearly shows the effectiveness of the aramid fibers in protecting the interior core of glass fibers.

It may be mentioned here, however, that the long-term durability of the FRP products is related to various factors, such as the properties of the resin, fiber, interfacial characteristics, fabrication process, and the environment. Thus, the results of accelerated and other tests carried out in the laboratory are only representative in nature, and an appropriate test program taking into account the actual usage should be developed, more or less on a case-to-case basis. Some of the standards that have been developed in Japan for testing of FRP rods and sheets have been listed in the appendixes of this paper.

Commercially Available Fiber Reinforced Polymer Composites

Though it is, in principle, possible to produce FRP products (cable, rods, flat bars, tendons, grids, and fabrics) with a great

Table 4. Durability of Fibers and Fiber Reinforced Polymer Rods

Particulars	Fiber			Notes
	Carbon	Aramid	Glass	
Alkali resistance	95%	92%	15%	NaOH, 40°C, 1,000 h
Acidic resistance	100%	60–85%	100%	HCl, 40°C, 120 days
Ultraviolet ray resistance	100%	45%	81%	0.2 MJ/m^2/h, 1,000 h

Particulars	FRP rods			Notes
	CFRP	AFRP	GFRP	
Static fatigue	91%	46%	30%	20°C, 100 years (calculated)
Cyclic fatigue	85%	70%	23%	100 MPa Amp., 2 million cycles
Alkali resistance	100%	98%	29%	NaOH, 40°C, 120 days
Ultra-violet ray resistance	77%	69%	90%	3 years exposure
Freeze-thaw resistance	100%	100%	100%	−20 to +15°C, 300 cycles
High temperature resistance	80%	75%	80%	−10 to 60°
Fire resistance	75%	65%	75%	350°C

Table 5. Commonly Available Fiber Reinforced Polymer Composites for Application as Reinforcement

Material	Type of FRP	Diameter (mm)	Tensile strength (MPa)	Elongation at rupture (%)	Young's modulus (GPa)	Prestressing method	Introduced prestress Tensile strength
Aramid	Round bar	6.1	1,744	3.66	4.8	Compression friction type	Stress of 4.9 MPa at
	Round bar	6	1,911	3.2	5.7	Compression friction type	0.31, 0.4
	Braided bar with and without sand	8	1,421	2.2	6.5	Pretension	0.2, 0.4, 0.6
	Braided bar with sand	9, 11, 13.5, 15.5, 17.5	1,215–1,382	1.92–2.3	5.9–6.5	Pretension Posttension	0.29, 0.32, 0.47, 0.50
	Eight round bars	6	1,921	4.0	4.8	Posttension	0.5
	Rope	8.3	1,921	2.5	7.8	Pretension	0.3
	Deformed bar	6	1,862	3.6	5.3	Posttension	0.49
Carbon	Deformed bar round bar	8	1,431	1.0	14.3	Pretension Block beam	Pretension 0.2 Block 0.6
	Twisted seven FRP bars	9.8	2,068	1.5	13.7	Pretension	0.13–0.19
	Bars put in sand	9.5	1,617		14.8	Pretension	0.17–0.50
	Twisted seven simple	5, 6	1,862	1.3	13.7	Pretension	0.4, 0.55, 0.7
	Twisted seven FRP bars	12.1	1,803	1.32	13.7	Pretension	0.55
	Twisted seven FRP bars	6	1,548	1.3	11.9	Pretension	0.4
	Eight round bars	8×8 dia.	1,803	1.21	14.7	Posttension	0.5, 0.65
	Eight round bars	8×8 dia.	1,803	1.2	13.9	Posttension	0.47
	Twisted 7 or 19 CFRP bars	12.5, 17.5, 25.5	2,068, 1,931, 1,911	1.5	14.2, 13.6, 12.9	Posttension	0.27, 0.44, 0.51, 0.54, 0.75
Glass	Grid		657	2.18	3.1	Pretension	0.37, 0.41, 0.5
	Deformed bars	6, 8	1,205, 1,313	3.1–3.3	4.0–4.3	Pretension	0.27

deal of flexibility depending upbon the requirements, some of the more commonly available products are listed in Table 5 along with their basis properties. In addition to the above products, FRP sheets are also available, which are being increasingly used for the repair and reinforcement of concrete structures.

FRP materials for use as reinforcing materials in concrete construction are available in different shapes. Fig. 11 shows a sample of some of the common shapes of FRP materials used in the construction industry—grid, spiral rope, braided rope, deformed bar, round bar, and flat bar. It can be seen from the figure that special efforts are made at times to improve the bond between the FRP reinforcement and the surrounding concrete or the grout. Such efforts include braiding of fibers or giving some kind of special surface treatments. Fibers may also be wound around the surface in a twill or spiral pattern or may be coated with sand. It may also be pointed out that the 3D FRP products are rarely used, as they pose a major problem in their transportability.

Testing Fiber Reinforced Polymer Composites

Given the diversity of the products available and the relatively short history of the use of FRP materials in the construction industry, it is only natural that the relevant test methods and specifications are still in their infancy. Some of the standards that have been developed in Japan for testing of FRP rods and sheets have been listed in the Appendixes 1 and 2 of this paper. Only one aspect of properly holding the composite at the ends has been taken up for a brief discussion here.

Fig. 12(a) shows an FRP being held in a specially designed V-shaped split chuck for tensile testing. It has been found that unless the restraining compressive load is applied uniformly along the entire surface held within the chuck, the FRP could rupture on account of split tension or stress concentrations, even before the fibers reach their full load-carrying capacity (Uomoto and Nishimura 1994). In fact, the existence of two clear peaks in the probable failure loads of carbon fibers has been clearly demonstrated (Uomoto and Nishimura 1994). Two possible approaches have emerged to achieve the desired objective in this regard:

1. Shape the interior of the split chuck appropriately in terms of the end angle and the cross section of the V; and
2. Embed the FRP in a threaded pipe using an appropriate expansive agent. The threads are then used to connect to the chuck and in applying the tensile load. Fig. 12(b) shows an example of a holding device based on this principle.

Use of Fiber Composites in Concrete Engineering

General

Some of the basic properties of the available FRP materials have been discussed above. In the last several years, considerable research effort has also been directed to study the possible application of FRP as a reinforcing material in concrete construction. There are several aspects to the research work; the following broad classification has been made to facilitate discussion in this paper:

1. Application of FRP rods to the construction of new reinforced concrete structures;
2. Application of FRP rods to the construction of new prestressed concrete structures; and

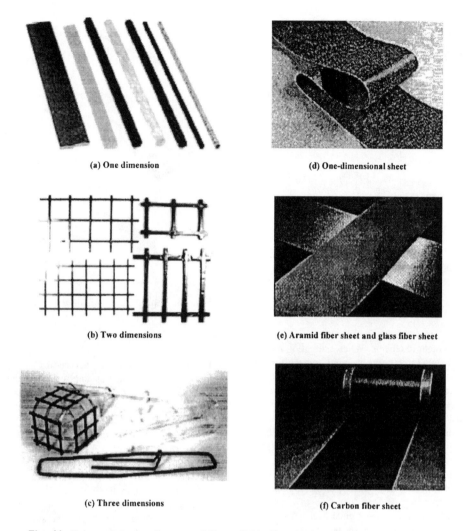

(a) One dimension

(d) One-dimensional sheet

(b) Two dimensions

(e) Aramid fiber sheet and glass fiber sheet

(c) Three dimensions

(f) Carbon fiber sheet

Fig. 11. Common shapes of commercially available fiber reinforced polymer composites

3. Application of FRP sheets in repair and rehabilitation of different kinds of concrete structures

Whereas the following paragraphs discuss the application of FRP rods in concrete construction, the issues relating to the use of FRP sheets are taken up in the next section of this paper.

Application of Fiber Reinforced Polymer Rods in Construction of New RC Structures

Given the almost linear stress-strain behavior of FRP reinforcement, RC beams reinforced with FRP rods generally behave as continuous elastic bodies until cracking starts in concrete, and the behavior can be modeled using conventional models used for the steel-reinforced beams. Once the tensile stress in the bottom (concrete) fiber reaches the tensile strength of the concrete, and flexural cracking is initiated, the behavior is quite different. Concepts like that of "a balanced section" are equally valid in the case of an RC beam reinforced with FRP rods. In cases when FRP is used as a reinforcing material, care needs to be taken to define the concept, taking into account the variation of the fiber volumes, etc., in the reinforcement used.

Initial Flexural Cracking

It is well known that although the occurrence of flexural cracks in concrete is related to the strength of concrete, the distribution of the cracks and the spacing is related to the bond between the reinforcing material and the surrounding concrete. It may be noted that if certain basic precautions are taken, the bond between the concrete and the FRP rods does not pose a major problem. In the case of grid-form FRP composites, flexural cracks occur in the vicinity of the intersection points of the FRP grids. It has also been reported that the maximum crack intervals is about two times the grid spacing, and that the width of the cracks is not as large as that observed in cases when steel bars are used and is nearly proportional to the elongation of the principal reinforcing materials.

Flexural Rigidity After Cracking

When flexural cracking starts, it is known that the deflections increase and the neutral axis in the cracked portions rises. Now, since the modulus of elasticity of FRP rods is considerably lower than that of steel (except for a few carbon-fiber-reinforced FRPs), the deflections and the extent of the rise of the neutral axis are considerably more than those observed for conventional steel bars.

Ultimate Strength

Basically speaking, the ultimate load-carrying capacity is determined by either the compression failure of the concrete or the

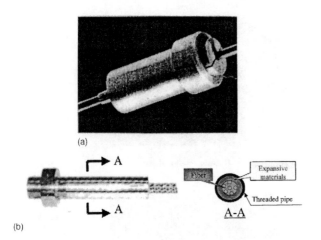

(a)

(b)

Fig. 12. (a) Typical V-shaped split chuck to hold fiber reinforced polymer composites during tensile testing; (b) special arrangement for holding fiber reinforced polymer composites during tensile testing

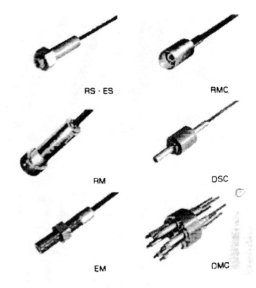

Fig. 13. Representation of some of the chucks used for prestressing fiber reinforced polymers tendons during construction

failure of the reinforcement in tension. When estimating the latter moment capacity, it needs to be borne in mind that the FRP materials do not have a very large fracture strain, and therefore the fracture is very brittle. In fact, if the reinforcement (FRP rods) is arranged in several layers, as it often happens for RC beams, it is likely that the capacity is decided as the extreme FRP fiber layer reaches fracture strain (even though the strands in others layers, i.e., those closer to the NA, are well below that strain level). In other words, in such cases, the full area of the FRP reinforcement is not effectively utilized. This is in contrast to the behavior seen in normal steel reinforcement, where because of the large deformability of the steel, the reinforcement continues to resist load, even after the extreme layer reaches "yield levels"—in other words, failure occurs only when the entire tensile reinforcement reaches yield level strains.

Application of Fiber Reinforced Polymer Rods in Construction of New PC Structures

In spite of lower modulus of elasticity, a relatively high strength, light weight, corrosion resistance, and low relaxation makes FRP a very suitable material for use as reinforcement in prestressed concrete construction. Continuous fibers have been effectively used in several prestressed concrete applications using both pretensioning and posttensioning methods. It may be noted that round bars are usually used only in posttensioned applications because of the lower bond that is achieved. In the absence of clear-cut design procedures and lack of understanding of the behavior of PC members, it is common to introduce prestressing up to the extent of about only 50% of the tensile strength of the FRP tendon. The following paragraphs briefly discuss the results from some of the research carried out in this direction.

Prestressing using Fiber Reinforced Polymers Tendons

Some of the issues involved in the tensile testing of the FRP composites discussed above in this paper are extremely relevant to the design of an appropriate system for prestressing the FRP composites. Fig. 13 shows a schematic representation of some of the holding chucks that are used at the time of prestressing FRP rods.

Load-Deformation Characteristics

Basically, the principles applicable to the prestressed concrete members using steel reinforcement are also applicable in case FRP tendons are used as reinforcement. It needs only to be borne in mind that, in the absence of a yield region in the stress-strain behavior of the FRP rods, the failure can only be either by the failure of the concrete (in compression) in the compression zone, or the breaking of the FRP rods.

Results from a study of prestressed concrete beams using FRP rods, and tested in flexure, have been reported. Several beams were tested varying the type of the fiber used, the level of the applied prestressing force, and the amount of fiber reinforcement (analogous to the different reinforcement ratio in conventional beams). It was found that the initial cracking load and the maximum strength increase as the introduced prestressing force and the volume of fibers in the aramid FRP are increased. A typical load-deflection diagram obtained using flexural members having FRP tendons is given in Fig. 14. It was further found from the tests that

1. The behavior before initiation of flexural cracking is linear and elastic. It may be pointed out that the values of deflection obtained match well with those estimated using conventional bending theory.

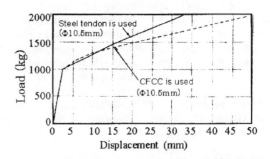

Fig. 14. Typical load deflection diagram for flexural member with prestressed fiber reinforced polymer tendons

Fig. 15. Application of fiber reinforced polymer rods in soil stabilization

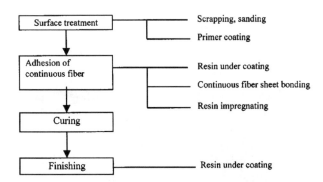

Fig. 16. Depiction of different steps in fiber reinforced polymer sheet application

2. The values of deflection at loads higher than cracking loads can be predicted using conventional theories. Further, it should be pointed out that because of the lower values of the modulus of elasticity, the deflections in the case of FRP reinforcement are significantly higher than those in PC members reinforced with steel;
3. The failure load observed is in close agreement with the calculated value, estimated assuming complete bond between the FRP rods and the surrounding concrete. The failure is governed by the load-carrying capacity of the concrete (in compression), and is therefore sudden and brittle; and
4. Failure by compression of concrete occurs if the initial prestressing force is $0.6\square_u$ or less ($\square_u$: tensile strength of FRP) and the FRP are likely to rupture if the value is $0.6\square_u$ or more.

It has also been pointed out that in the case of unbonded members, the maximum load-carrying capacity of the member is about 60% of that of the bonded members.

Soil Stabilization

Fig. 15 shows application of prestressed FRP tendons in slope stabilization. The presence of the prestressing force in the anchor bolts increases the internal friction along the failure surface; this stabilizes the slopes and allows more steep cuts than can be made without the use of such bolts. It can be seen that because of the noncorrosive nature of the FRP rods, there is hardly a need to use protective cement mortar, which is an essential component in the schemes using steel reinforcement. Thus the holes used can be much smaller and the exercise becomes more economic and efficient.

Use of Fiber Reinforced Polymer Sheets

General

Besides the use of FRP rods and strips, as described above, continuous FRP sheets weighing between 200 and 410 g/m^2 have also evolved as a major application of FRP materials in the construction industry, especially in the repair and rehabilitation of deteriorated concrete structures (JCI 1997; JSCE 2001; ACI 2001). Application of steel plates for the rehabilitation of deteriorated structures has the basic drawback that the plate tends to separate from the concrete surface and may not thus behave in a manner monolithic with the concrete member. Also, it is difficult to ensure proper adhesion because of the uneven nature of the concrete surface. The development of FRP sheets addresses both

of these issues effectively as the sheets adhere closely to the concrete surface and no special effort needs to be made to ensure a proper contact between the sheet and the concrete surface. Fig. 16 gives a brief description of the various steps in the operation, which can be summarized as follows:

1. Surface preparation, including impregnation of cracks and application of smoothing agents, if required, followed by primer coats;
2. Application of FRP sheets along with impregnating resins and hardeners. The number of layers of sheets is adjusted to suit the structural requirements;
3. Curing of FRP sheets applied; and
4. Finishing coat application on the FRP sheets

Some of the main features of the materials used and other aspects of using FRP sheets in repair and rehabilitation of structures are briefly described below.

Materials Used

Continuous Fiber Sheet

The unit weight of the sheet used is a direct reflection of the fiber content in the composite sheet, and indirectly the structural capacity of the sheet. It should be pointed out that resin and hardener penetration in very high density sheets is difficult, and therefore there is an optimum density of sheets for best structural performance, especially in cases when several layers of the sheets are used (CCC 1998).

Impregnation Resin

The principles that govern the choice of filler resins in FRP rods are also applicable for FRP sheets. Epoxy, vinyl ester, and acrylic resins are generally used for the purpose of impregnation. Structural properties of the hardened FRP sheet, such as strength and Young's modulus, rely upon the monolithic behavior of all the FRP layers. This can be ensured only if the resin completely impregnates the spaces between the fibers while it hardens. During application, the resin should have the required pot life and viscosity to hold the sheet in place. It may be pointed out that these properties are also related to the atmospheric temperature at the time of application, and therefore the resin should be carefully selected (CCC 1998).

Primer

Primer is applied to the concrete surface to ensure proper adhesion between the concrete surface and the FRP sheet and to pre-

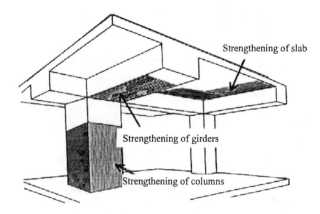

Fig. 17. Attachment of one-way reinforced fiber reinforced polymer sheets for increasing flexural capacity

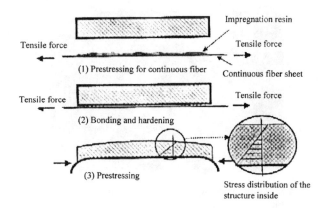

Fig. 19. Steps in application of prestressed fiber reinforced polymer sheets for repair and rehabilitation

vent formation of air bubbles at the interface, either during application of sheets or at a later time during curing.

Smoothing Agent

Unevenness in the concrete surface could become a source for accumulation of voids and wrinkles in the sheets. Such defects reduce the effectiveness of sheet application. Thus, in principle, all unevenness on the concrete surface should be removed before the FRP sheets are applied. Different methods, including application of a disc sander or a surface grinder, are used to suit the requirements. Sometimes, however, application of a "smoothing agent" may be sufficient for the purpose. The agent used should be selected keeping the required properties, temperature of application, etc., in mind. Epoxies have been reported to be used for this purpose.

Other Materials

Besides the materials outlined above, the application of FRP sheets is often preceded by impregnation of cracks in the concrete with an appropriate grout. Further, the FRP application may need to be provided with a finishing coat to improve its resistance to the environmental forces and prevent degradation on account of sunlight. Depending upon the environment of the structure, a fireproofing coat may also be required.

Construction Method

Surface Preparation

This step should include rounding of the corners of members, abrading uneven surfaces using disc sanders, impregnation of cracks, and application of smoothing agents.

Application of Nonprestressed Fiber Reinforced Polymer Sheets

The surface is primed with resins before the sheets are pasted, and entrapped air is removed using rollers while coating with resins.

In the case of multiple-layer application, the process is repeated after the application of each layer. Fig. 17 shows a schematic representation of application of FRP sheets for the reinforcement of different structural elements—columns, beams, and slabs. It may be noted that fibers in the sheets could be directed only in one direction or in two directions, again depending upon the desired results. For example, in the case of beams and columns, the fibers are aligned along the axes, whereas a two-way reinforcement is provided when using the sheets to reinforce a slab.

The different ways of attaching or wrapping the sheets (in a single- or multilayer application) are also shown in Fig. 18 for a beam-and-slab type of construction. It should be pointed out that it is often considered better to "wrap" the sheets, as shown in Fig. 18(d), to provide better structural behavior. However, it should be pointed out that FRP sheets can be more easily wrapped around such structures as chimneys or columns, where all the sides are usually accessible, rather than beams, where the compression face is usually not easily accessible. Another variation in applying the sheets is the use of preformed FRP plates around the structural members and making the joints of such plates monolithic subsequently. The space between the concrete and the FRP plates is filled or injected with cement or polymer mortar (Fujimura et al. 1998). In this case, one-way reinforced continuous FRP sheets are attached to the tensile face of a flexural member with fibers oriented along the member axis to provide increased flexural capacity (Kimura et al. 1997).

Application of Prestressed Continuous Fiber Reinforced Polymer Sheets

In the methods described above, the FRP sheets are simply bonded to the concrete surface using an impregnation resin. The effectiveness and reliability of such a sheet application has been found to be reduced on account of the facts that the FRP sheets do not participate in resisting dead load and that they play a part in resisting only live load. Further, such FRP sheets have a tendency to peel off even before the concrete member reached the ultimate state. To overcome the above problems, a method using pre-

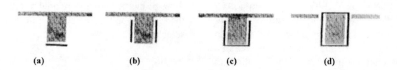

(a)   (b)   (c)   (d)

Fig. 18. Schematic representation of different wrapping methods to increase flexural and shear capacity of RC members

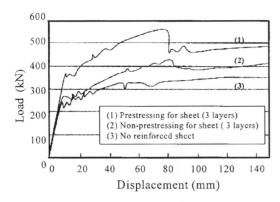

Fig. 20. Effectiveness of prestressed fiber reinforced polymer sheets in enhancing capacity of concrete members

stressed FRP sheets for bonding to the concrete surface has been developed and briefly described here (Wu et al. 2001). The principle of the method is illustrated in Fig. 19. The basic steps in the process are

1. Preparation of concrete surface in the usual manner (as for nonprestressed sheets);
2. Prestressing the continuous FRP sheet;
3. Applying the adhesive resin along the plane of adhesion and bonding to the concrete surface;
4. Curing and hardening of hardener with the tensile force held in place. At the ends, the reinforcement is carried out using a sheet fitted with steel plates, which are anchored to the concrete using bolts; and
5. Removing the externally applied load once sufficient hardening has been achieved, introducing a compressive "prestress" in the FRP sheets. Excess sheets at the ends are cut off.

Results of research work carried out to study the effectiveness of the application of prestressed FRP sheets in increasing the flexural capacity of RC members have been reported (Wu et al. 2001). Fig. 20 shows the results reported in the study. It was found that, besides toughness, the cracking load, yield load for the reinforcing bars, and the breaking loads are higher than those observed with simple (unprestressed) FRP sheet applications. It was also found that upon introduction of prestressing force, no peeling at the surface was observed, and that the failure was by rupture of the sheet.

Curing

During the hardening process of the hardeners (resins), especially in outdoor applications, it is important that the resins are protected from the action of rain, sand, dirt, etc. The temperature of the atmosphere also needs to be carefully monitored to ensure proper and adequate strength development in the resins (JRTRI 1996, 1997; JSCE 2001).

Finishing

After the resin has sufficiently hardened, the finishing coat may be applied with the object of ensuring long-term durability of the FRP sheet, fire protection, aesthetic appeal, etc. It may also be pointed out here that in order to ensure that the sheets do not debond from the concrete surface due to differential deformation during loading, the sheet is sometimes wrapped at the ends also, as shown in Fig. 18(d). This end protection provides an effect similar to that provided by the driving of anchors into the concrete in the case of use of steel plates for the strengthening, which

is far more cumbersome, especially in cases when the beams are heavily reinforced and it is difficult to find space for the anchors.

Design Methodology

General

The Japan Society of Civil Engineers (JSCE) and the American Concrete Institute (ACI) have published recommendations for the design and construction of concrete structures using FRP reinforcement (ACI 2001; JSCE 1997, 2001). These documents cover the requirements for materials used, their characteristic strength, and design recommendations for flexure and shear. The design for flexural members is based on the principles of equilibrium and compatibility and the constitutive laws of the materials. The recommended design methodology is based on the limit-state philosophy, and the design is checked for fatigue, creep, and serviceability criteria.

In addition to the application of FRP rods and tendons in the construction of new structures, FRP sheets have been extensively used in Japan in the retrofitting and upgrading of existing structures, especially after the Kobe earthquake in 1995. The FRP sheets have been used to increase the flexural and shear strength as well as ductility in many RC structures. Though methods such as jacketing with steel plates, etc., had been traditionally used for reinforcement, the development of this method has given the engineers a very useful option for rehabilitation of structures.

Design using Fiber Reinforced Polymer Rods or Tendons

Although only the details relating to design for flexure and shear have been briefly discussed in this paper, the JSCE also explains procedures for the design of prestressed concrete structures, design for fatigue, details for working out parameters like the flexural crack width, provisions for the cover thickness, and construction details—minimum concrete strength, handling, etc. (JSCE 1997, 2001).

The principles are similar to those used for normal RC design using steel reinforcement, and the flexural capacity of a section having FRP reinforcement can be calculated using the following assumptions:

- Strain in the concrete and the FRP reinforcement is proportional to the distance from the neutral axis (i.e., a plane section before loading remains plane after loading),
- The maximum usable compressive strain in concrete is assumed to be 0.003,
- The tensile strength of concrete is ignored,
- The tensile behavior of the FRP reinforcement is linearly elastic until failure, and
- A perfect bond exists between concrete and FRP reinforcement.

For flexural compressive failure, the compressive stress distribution in the concrete may be assumed to be identical to the rectangular compressive stress distribution (equivalent stress block). Concepts such as that of the reinforcement ratio (area of reinforcement in relation to the cross-sectional area of the member) can be used for FRP reinforced concrete members also. However, in determining the balanced reinforcement (when the tensile and compressive load-carrying capacities are the same), percentage, especially from the point of view of design, care needs to be taken to appropriately account for the brittle nature of the FRP composite failure.

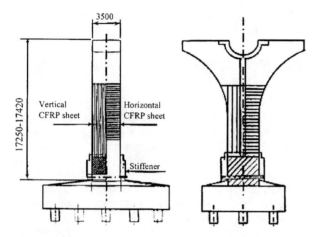

Fig. 21. Improvement in ductility of structure using fiber sheets in conjunction with steel plates

Now, the basic capacity of the section is determined either by the rupture of FRP due to tension or the crushing of concrete in compression. The implications of having the tensile reinforcement laid out in several layers have already been discussed in this paper.

Although the design procedure for the utilization of FRPs is still in its infancy, the following may be noted about the JSCE recommendations: (1) They ignore the presence of the FRP composites' presence in the compression zone; (2) modification factors are provided to account for the brittle failure of the FRPs; (3) issues like combined biaxial bending and axial forces are also addressed; and (4) the treatment of shear resistance is essentially an adaptation of the traditional method of adding individual contributions from concrete, shear reinforcement, and bent-up tendons.

Design Methodology for Upgrading using Fiber Reinforced Polymer Sheets

Fig. 21 shows a typical application of FRP sheets for the reinforcement of deteriorated concrete structures. It may be noted that the application is often carried out by using multiple layers of FRP sheets laid one after another, appropriately hardened together using a suitable hardener. Fig. 22 shows the variation of the strain and stress across the face of a beam reinforced with n layers of FRP sheets at the bottom. It can be seen that the resultant of all the layers is taken to be acting through one point and the thickness of the individual sheets has been neglected.

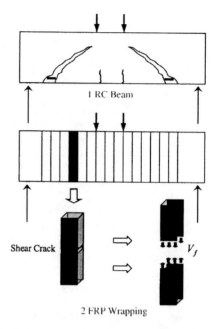

Fig. 23. Typical representation of truss model in study of contribution of continuous fiber reinforced polymer sheets to shear capacity of RC members

The sheets thus attached to concrete members contribute to (1) increasing the flexural and shear capacity of the section, and (2) increasing the ductility of the section (JSCE 2001; Katsuki et al. 2000; Nakai et al. 2000). As a result of the contributions to the flexural and shear capacities and the ductility of the members, members reinforced with FRP sheets also exhibit a larger deformation capacity compared with unreinforced members. Fig. 23 shows a schematic representation of the traditional truss model, which can be used to evaluate the contribution of FRP sheet(s) to the shear capacity of the RC member.

Examples of Application

In spite of their low modulus of elasticity and deformability at the ultimate load, FRP rods, grids, and sheets have been increasingly used as reinforcing material both in reinforced concrete, prestressed concrete, and repair and rehabilitation of concrete structures in the last decade. The nonmagnetic and noncorrosive nature of the FRPs also gives these materials an edge in specific appli-

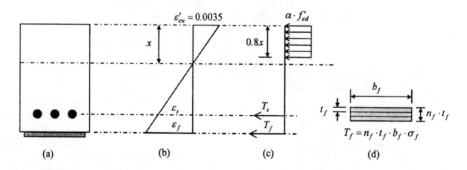

Fig. 22. Variation of strain and stress across beam face reinforced with n layers of CFRP sheets at bottom

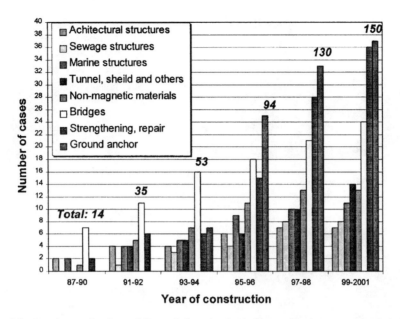

Fig. 24. Growing applications of fiber reinforced polymer composites in construction industry

cations. Fig. 24 gives a representation of some of the applications of the FRP reported in the literature. Though the list used is by no means exhaustive, it is clearly a pointer to the wide spectrum of possible applications of the FRP in the construction industry. In fact, the figure clearly shows that not only the number of applications show an increasing trend, but also that FRPs are being used in newer areas (ACC 1995, 1998).

General Application of Rods

Marine Concrete Structures

The noncorrosive nature of the FRPs makes them an excellent reinforcing material in the marine environment. One such application is shown in Fig. 25, where the Hexagonal Marine Structure constructed using FRP rods was used for tensioning of diagonals. Another application of FRP in marine structures is given in Fig. 26, where the rods were used in the marine slope stabilization. It may also be noted that FRP rods are also at times used in conjunction with other corrosion-resistant reinforcement, such as epoxy-coated bars.

Bridges (PC)

FRP composites have been used extensively in bridges, especially in prestressed concrete construction. The first actual structure to be constructed using continuous fibers as tendons was the Ulenberg Bridge in Germany, built in 1986. The bridge is a continuous slab bridge and its tendons are multiple strands made of 7.5-mm-diameter FRP. This was preceded by a small experimental bridge spanning only about 6 m, also in Germany, in the early 1980s.

A pedestrian bridge (Marienfede Bridge) in Berlin Park, the Ludwigshafen Bridge, and the Lunen'sche-Gasse Bridge are examples of FRP applications in Germany (ACI 2001). Notsch Bridge, constructed in 1991 in Austria, also used GFRP tendons as prestressing tendons. A precast posttensioned bridge of 9 m span and 5.2 m width was erected at a cement plant in Rapid City, South Dakota, in 1992. Cables of three different materials were used to prestress the slab. Thirty GFRP cables were used to pre-

Fig. 25. Hexagonal marine structure

Fig. 26. Application of fiber reinforced polymer in marine slope stabilization

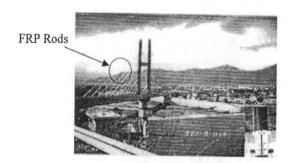

FRP Rods

Fig. 27. Use of fiber reinforced polymer composites as cables in cable-stayed bridge

Fig. 28. Use of fiber reinforced polymer rods as external cables in rehabilitation project

Fig. 29. Use of nonmagnetic fiber reinforced polymer reinforcement in compass alignment apron area of Haneda airport, Japan

stress one-third of the length of the slab, 30 FRP cables were used to prestress the next third of the length, and steel cable prestressing was adopted for the remaining length (ACI 2001). FRP bars were also used in the U.S. for the deck construction of Pierce Street Bridge in Lima, Ohio, in 1999 and the redecking of Dayton, Ohio's Salem Avenue Bridge in 1999 (ACI 2001).

The "Birdie Bridge" is one of the first applications of continuous FRP material in bridge construction in Japan. In this bridge, a total of 16 cables each consisting of 8 AFRP flats were used as tendons. Besides being used as reinforcing materials within concrete, the FRP rods have also been used as cables in suspension and cable-stayed bridges, such as is shown in Fig. 27.

Besides examples of use in new construction, as discussed above, FRP rods have also been used in the repair and rehabilitation works. Fig. 28 shows an example where such rods were used

Fig. 30. Use of fiber reinforced polymer in conjunction with conventional reinforcement in parking area

as external cables in the rehabilitation work.

Transport Facilities and Tunnel Linings
FRP rods have also found application as reinforcing materials in applications where the magnetic nature of the conventional steel reinforcement becomes a bottleneck. Such examples include the experimental facility for the high-speed linear motor-rails. Fig. 29 shows the application of FRP tendons in the construction of the compass alignment apron area of the Haneda airport in Tokyo, Japan, where the presence of magnetic materials in the slab interfere with the operation of devices. Fig. 30 shows an example where FRP rods are used in conjunction with conventional reinforcement in a parking area. The nonmagnetic FRP reinforcement is used only in those areas where cars are designated to park and magnets are used for operational purposes.

FRP grids are becoming increasingly popular as a replacement of steel meshes used in the shotcreting in tunnels. Fig. 31 shows an example of the application of NEFMAC fibers in an underground oil storage facility in Kagoshima, in Japan.

Building Structures
Glass FRP bars have also been used for the beams and walls in a University building and the construction of the MRI unit of a hospital building in San Antonio (ACI 2001). Fig. 32 shows the application of a 2D FRP composite in building construction. Several applications of FRP using carbon fibers in curtain walls of buildings have also been reported.

Fig. 31. Use of fiber reinforced polymer grids in shotcreting in underground oil storage facility in Kagoshima, Japan

Fig. 32. Use of two-dimensional fiber reinforced polymer reinforcement in building construction

Fig. 33. Use of fiber reinforced polymer sheets to strengthen RC columns

Fig. 34. Chimney repair using fiber reinforced polymer sheets

Applications of Continuous Sheets for Repair and Rehabilitation

Sheets made from carbon and aramid fibers have been found to be especially useful in retrofitting works to improve the seismic performance of buildings and bridges, rehabilitation of deterioration, and damaged structures. Projects have involved the use of FRP sheets for several hundred to several thousand square meters of area. Some of the examples are given in this section to show the diversity of the application.

Fig. 35. Strengthening of beams and slabs in school building

For example, a change in the design wheel loads, in Japan in 1993, necessitated retrofitting a large number of highway bridges. Carbon FRP sheets have been used to strengthen flexural and shear capacities of structures including highway piers. Reference may be made to Fig. 21 for the application of continuous FRP sheets in conjunction with steel plates to reinforce piers. Fig. 33 shows another example of the use of continuous FRP sheets for the reinforcement of columns. FRP sheets have also been applied to repair the deterioration of jetties and other structures on account of chloride-induced corrosion, tunnel linings to strengthen the concrete walls and help resisting the ground pressure and repair of cracks in tunnels. Fig. 34 shows the example of application of FRP sheets in the repair works of a chimney, damaged on account of reinforcement corrosion, exhaust fumes, and temperature-related distress.

Fig. 35 shows an example of the use of FRP sheets for the strengthening of the floor slabs and beams in a school slab. It can be seen that FRP sheets are used on both the top and bottom surfaces of the slab to improve the structural integrity and prevent further propagation of cracks. It may be noted that use of FRP sheets as discussed here allows the designer the flexibility to use different number of layers in different directions in cases when there is a differential requirement for strengthening in the two directions.

Concluding Remarks

An attempt has been made in this paper to briefly discuss the different aspects related to the utilization of continuous FRP composites as a reinforcing material in concrete construction. Given the diversity involved in the issues, this paper has largely focused on providing information only on the core engineering aspects—the structural and durability characteristics of FRP composite, research work related to the utilization of FRPs as a reinforcing material, and applications of FRP composites, including sheets. Several examples of utilization in the field have been included to demonstrate the range of possible applications. Since issues related to structural design were not the primary focus of this paper, the issue has been only briefly discussed. Other issues, such as the manufacture of FRPs, have not been discussed in any length.

As can be clearly seen from the information presented in this paper, there is clearly a growing need and acceptability of FRP composites in the field of concrete engineering, and FRP composites are being used in a growing number of applications. Thus, the coming years are likely to see a rapid increase in the development of appropriate test methods and specifications, design methods in the field of reinforced and prestressed concrete construction, and increased field applications. More research effort is called for in areas such as making the FRPs cost-effective and developing appropriate construction technology to implement designs in the field.

Acknowledgments

The writers are grateful to the members of the American Concrete Institute Committee 440, Technical Committee of the Japan Concrete Institute, the Concrete Committee of the Japan Society of Civil Engineers, and the Architectural Institute of Japan for their cooperation and support. Thanks is also due to Advance Composite Cables Club for the permission to use some photographs from their publications.

Appendix I. Test Methods for Continuous Fiber Reinforcing Materials (JSCE 1997)

1. Test method for tensile properties of continuous fiber reinforcing materials (JSCE-E 531-1995)
2. Test method for flexural tensile properties of continuous fiber reinforcing materials (JSCE-E 532-1995)
3. Test method for creep failure of continuous fiber reinforcing materials (JSCE-E 533-1995)
4. Test method for long-term relaxation of continuous fiber reinforcing materials (JSCE-E 534-1995)
5. Test method for tensile fatigue of continuous fiber reinforcing materials (JSCE-E 535-1995)
6. Test method for coefficient of thermal expansion of continuous fiber reinforcing materials by thermal-mechanical analysis (JSCE-E 536-1995)
7. Test method for performance of anchorages and couplers in prestressed concrete using continuous fiber reinforcing materials (JSCE-E 537-1995)
8. Test method for alkali resistance of continuous fiber reinforcing materials (JSCE-E 538-1995)
9. Test method for bond strength of continuous fiber reinforcing materials (JSCE-E 539-1995)
10. Test method for shear properties of continuous fiber reinforcing materials (JSCE-E 540-1995)

Appendix II. Test Methods for Continuous Fiber Sheets (JSCE 2001)

1. Test method for tensile properties of continuous fiber sheets (JSCE-E 541-2000)
2. Test method for overlap splice strength of continuous fiber sheets (JSCE-E 542-2000)
3. Test method for bond properties of continuous fiber sheets to concrete (JSCE-E 543-2000)
4. Test method for bond properties of continuous fiber sheets to steel plate (JSCE-E 544-2000)
5. Test method for direct pull-off strength of continuous fiber sheets with concrete (JSCE-E 545-2000)
6. Test method for tensile fatigue strength of continuous fiber sheets (JSCE-E 546-2000)
7. Test method for accelerated artificial exposure of continuous fiber sheets (JSCE-E 547-2000)
8. Test method for freeze-thaw resistance of continuous fiber sheets (JSCE-E 548-2000)
9. Test method for water, acid, and alkali resistance of continuous fiber reinforcing materials (JSCE-E 549-1995)
10. Test method for flexural tensile strength of continuous fiber reinforcing materials (DRAFT)
11. Test method for surface incombustibility of protective materials for continuous fiber reinforcing materials (DRAFT)

Notation

The following symbols are used in this paper:

E_{FRP} = modulus of elasticity of FRP composite, e.g., rod, etc.;
E_1 = modulus of elasticity of fiber in composite;
E_2 = modulus of elasticity of matrix phase in composite;
f_{ck} = characteristic compressive strength of concrete in N/mm^2;
n = number of layers of FRP sheets used to reinforce member;
V_f = volume fraction of fibers in FRP composite;
α = modification factor for estimation of bond strength of FRP composite with concrete;
γ_c = constant $= 1.3$;
σ_{FRP} = stress in FRP composite, e.g., rod;
σ_1 = stress carried by fiber phase in composite;
σ_2 = stress carried by matrix phase in composite; and
$\square_u$ = tensile strength of FRP composite.

References

Advanced Composite Cables. (1995). *Applications and construction using new materials*, ACC Club, Tokyo (in Japanese).
Advanced Composite Cables. (1998). *Applications and construction using new materials*, Vol. 2, ACC Club, Tokyo (in Japanese).
Agarwal, B. D. and Broutman, L. J. (1990). *Analysis and performance of fiber composites*, Wiley, New York, 44–45.
American Concrete Institute. (2001). "Guide for the design and construction of externally bonded FRP systems for strengthening concrete structures." *Rep. by ACI Committee 440*.
CCC. (1998). *Fiber-reinforced composites for construction*, CMC Co., T. Nakatsuji, ed. (in Japanese).
Fujimura, M., et al. (1998). "Strength and behavior of RC column strengthened by FRP plates." *Proc. Japan Concrete Institute*, 20(3), 1201–1206 (in Japanese).
Hodhod, H. A. A. (1992). "Employment of constituent properties in evaluation and interpretation of FRP rods mechanical behavior." DEngrg. thesis, Univ. of Tokyo.
Iwahashi, T., et al. (1996). "Studies on the repairing and strengthening of existing RC column." *Proc. JCI*, 18(2), 1499–1504 (in Japanese).
Japan Concrete Institute. (1997). *Report of technical committee on continuous fiber-reinforced concrete* (in Japanese).
Japan Concrete Institute. (1998). "Technical report on continuous fiber-reinforced concrete." *JCI TC952*.
Japan Railway Technical Research Institute. (1996). *Recommendation in the design and practice for seismic strengthening the pier of railway bridges by carbon fiber sheet* (in Japanese).
Japan Railway Technical Research Institute. (1997). *Recommendation in the design and practice for seismic strengthening the pier of railway bridges by aramid fiber sheet* (in Japanese).
Japan Society of Civil Engineers. (1997). "Recommendations for design and construction of concrete structures using continuous fiber reinforcing materials." *Concrete Engineering Series, No. 23*.
Japan Society of Civil Engineers. (2001). "Recommendations for upgrading of concrete structures with use of continuous fiber sheets." *Concrete Engineering Series, No. 41*.

Jesse, F., and Curbach, M. (2001). "The present and the future of textile reinforced concrete." *Proc., 5th Int. Conf. on Fiber-Reinforced Plastics for Reinforced Concrete Structures*, C. J. Burgoyne, ed., Cambridge, U.K.

Katsuki, F. (1997). "Evaluation of alkali resistance of fiber reinforcement for concrete." Thesis submitted to the Univ. of Tokyo for D. Engrg. degree (in Japanese).

Katsuki, F., et al. (2000). "Ductility evaluation of RC member retrofitted with continuous fiber sheet." *Proc. Jpn. Concr. Inst.*, 22(3), 1537–1542 (in Japanese).

Katz, A., Bank, L. C., and Puterman, M. (2001). "Durability of FRP rebars after four years of exposure." *Proc., 5th Int. Conf. on Fiber-Reinforced Plastics for Reinforced Concrete Structures*, C. J. Burgoyne, ed., Cambridge, U.K.

Kimura, K., et al. (1997). "Flexural behaviors reinforced concrete members with FRP plates." *Proc. Jpn. Concr. Inst.*, 19(2), 267–272 (in Japanese).

Kobayashi, K., Uomoto, T., and Cho, R. (1987). "Structure of prestressed concrete using one-dimensional fiber composites as reinforcement." *J. Jpn. Soc. Compos. Mater.*, 13(5).

Mallick, P. K. (1988). *Fiber-reinforced composites—materials, manufacturing and design*, Marcel Decker, New York.

Nakai, H., et al. (2000). "Evaluation of shear capacity of linear members retrofitted by continuous fiber sheets." *Proc. Jpn. Concr. Inst.*, 22, 493–498 (in Japanese).

Nishimura, T., Uomoto, T., Katsuki, F., and Kamiyoshi, M. (1996). "Development of alkali resistant AGFRP rods." *Proc., 51st Annual Meeting of the JSCE* (Part V) (in Japanese).

Piggott, M. R., and Harris, B. (1980). "Compression strength of carbon, glass and Kevlar-49 fiber reinforced polyester resins." *J. Mater. Sci.*, 15, 2523.

Saadatmanesh, H., and Tannous, F. E. (1999). "Relaxation, creep and fatigue behavior of carbon fiber reinforced plastic tendons." *ACI Mater. J.*, March–April, 143–153.

Taerwe, L., Matthys, S., Pilakouts, K., and Guadagnini. (2001). "European activities on the use of FRP reinforcement, fib Task Group 9.3 and the ConFiberCrete network." *Proc., 5th Int. Conf. on Fiber-Reinforced Plastics for Reinforced Concrete Structures*, C. J. Burgoyne, ed., Cambridge, U.K.

Toritani, T., et al. (1997). "Experimental study on aseismic strengthening of existing reinforced concrete columns, Part 2, discussion on experimental results." *Summaries of Technical Papers of Annual Meeting*, AIJ, 23330, 659–660 (in Japanese).

Uomoto, T., Tajima, T., and Nishimura, T. (1992). "Relation between acoustic emission and absorbed energy of reinforced concrete and fiber-reinforced plastic." *Progress in Acoustic Emission VI*, Japanese Society for NDI.

Uomoto, T., and Nishimura, T. (1994). "Static strength and elastic modulus of FRP rods for concrete reinforcement." *Concrete Library of the Japan Society of Civil Engineers*, No. 23.

Uomoto, T., Ohga, H., Nishimura, T., and Yamaguchi, T. (1995). "Fatigue strength of FRP rods for concrete reinforcement." *Proc., EASEC*, Y. C. Loo, ed.

Uomoto, T., Nishimura, T., Ohga, H., Katsuki, F., Yamaguchi, T., and Kato, Y. (1998). "Strength and durability of FRP rods for prestressed concrete tendons." *Rep. of the Institute of Industrial Science*, Univ. of Tokyo, 39(2), (Serial 244) (in Japanese).

Uomoto, T. (2001). "Durability considerations for FRP reinforcements." *Proc., 5th Int. Conf. on Fiber-Reinforced Plastics for Reinforced Concrete Structures*, C. J. Burgoyne, ed., Cambridge, U.K.

Waaser, E., and Wolf, R. (1986). *Ein neuer werkstoff für spannbeton HLV Hochleistungs-Verbundstab aus glasfersern*, Beton H 7 (in German).

Wu, Z., et al. (2001). "Experimental study on strengthening method of PC flexural structures with externally bonded prestressed fiber sheet." *Proc. Jpn. Concr. Inst.*, 23(1), 1099–1104 (in Japanese).

Yagishita, K., et al. (1997). "Experimental study on aseismic strengthening of existing reinforced concrete columns, Part 1—outline of experiment." *Summaries of Technical Papers of Annual Meeting*, AIJ, 23329, 657–658 (in Japanese).

Yamaguchi, T. (1998). "Deterioration due to ultraviolet radiation and the creep failure of FRP rods." DEngrg. thesis, Univ. of Tokyo (in Japanese).

American Society of Civil Engineers 1852–2002 Building a Better World

150th Anniversary Paper

History of Hot Mix Asphalt Mixture Design in the United States

Freddy L. Roberts, M.ASCE[1]; Louay N. Mohammad, M.ASCE[2]; and L. B. Wang, M.ASCE[3]

Abstract: Asphalt has been used as a construction material from the earliest days of civilization, but though it has long been used as a waterproofing material in shipbuilding and hydraulics, its use in roadway construction is much more recent. A recent survey revealed a total of over 2.3 million miles of hard-surfaced (asphalt or concrete) roads in the United States, of which approximately 96% have asphalt surfaces. Asphalt mixture consists of asphalt, coarse and fine aggregate, and a number of additives occasionally used to improve its engineering properties. The purpose of mixture design is to select an optimum asphalt content for a desired aggregate structure to meet prescribed criteria. The demands and reliance upon America's roadways for mobility and commerce have increased substantially over the past three decades. The highway network is not only the economic backbone of the country, but it also provides the only transportation access to a growing number of communities. This paper presents a review of the past, present, and future trends in asphalt mixture design as the methods have evolved in an attempt to meet the ever-increasing demands of traffic.

DOI: 10.1061/(ASCE)0899-1561(2002)14:4(279)

CE Database keywords: History; Asphalt mixes; Design; Asphaltic concrete.

Introduction

Asphalt has been used as a construction material from the earliest days of civilization. The first reference to use of asphalt occurs in Genesis 6:14, where the Lord told Noah, "So make yourself an ark of cypress wood; make rooms in it and coat it with pitch inside and out." Pitch is asphalt that floats to the surface of bodies of water from fissures in the Earth's crust that leak crude oil. When crude oil is exposed to air the volatile compounds evaporate, leaving the heaviest molecules, which form a sticky brown material having about the same unit weight as water. The Egyptians used asphalt in the process of mummification. The Romans, being more practical-minded than the Egyptians, used asphalt to seal their baths and to make waterproof hydraulic connections for their water distribution systems. It has been rumored that Columbus would not have made the return trip from the Americas to tell the world of his discoveries had he not stumbled upon asphalt

[1]T. L. James Professor of Civil Engineering, Civil Engineering Dept., Louisiana Technical Univ., Ruston, LA 71272.

[2]Associate Professor of Civil Engineering, Civil Engineering Dept. and Louisiana Transportation Research Center, Louisiana State Univ., 4101 Gourrier Ave., Baton Rouge, LA 70808.

[3]Assistant Professor of Civil Engineering, Civil Engineering Dept., Louisiana State Univ. and Southern Univ., CEBA Building, Louisiana State Univ., Baton Rouge, LA 70803.

Note. Associate Editor: Antonio Nanni. Discussion open until January 1, 2003. Separate discussions must be submitted for individual papers. To extend the closing date by one month, a written request must be filed with the ASCE Managing Editor. The manuscript for this paper was submitted for review and possible publication on March 7, 2002; approved on April 1, 2002. This paper is part of the *Journal of Materials in Civil Engineering*, Vol. 14, No. 4, August 1, 2002. ©ASCE, ISSN 0899-1561/2002/4-279–293/$8.00+$.50 per page.

deposits that enabled his men to pitch their badly leaking ships. The use of asphalt in roadways is much more recent than its use in shipbuilding and hydraulics.

There has been a need for a method to design hot mix asphalt since the first mixes were placed in the late 19th century. In the 1880s asphalt road mixes in the United States were marketed as patented products made primarily from natural asphalt from the lake deposits of Trinidad near the village of La Brea, British West Indies. To construct a project, this asphalt was transported into the construction area in barrels, blended with local petroleum fluxes, and then mixed with local aggregates using proportions worked out on a trial-and-error basis by the patent holder. The first asphalt paving project in the United States may have been built in New York City by DeSmedt prior to 1875 (New York Section of the Society of the Chemical Industry 1907). There is evidence that S. H. Robertson of the District of Columbia may have built a hot mix asphalt pavement in 1874 (Robertson 1958). The largest early project was placed on Pennsylvania Avenue in Washington, D.C., in 1876. The asphalt binder on the Pennsylvania Avenue project was a blend of 50% Trinidad asphalt and 50% heavy petroleum oil mixed with local aggregate.

It is interesting to note that before 1900 crude oils produced east of the Rocky Mountains could not be reduced to a paving grade asphalt using the refining tools available at that time. As a result there was no competition for Trinidad asphalt, which had become the standard against which all asphalts were measured. However, on the West Coast, the California crude oils were easily reduced to asphalt using the shell stills available at that time. California-produced asphalt had been used in a job in Los Angeles in 1895 (Welty and Taylor 1958), by 1904, 38 California refineries were producing asphalt (Krchma and Gagle 1974). The combined output of these refineries exceeded the market requirements, so asphalt began to be transported by ship and arrived at East Coast ports at a cost 10% lower than for Trinidad asphalt.

Rail delivery brought California asphalt to the East Coast at a price only 8% higher than Trinidad asphalt. Besides, boasted the California producers, our asphalt is 100% asphalt, and the Trinidad asphalt is 35–50% filler. The Trinidad asphalt producers boasted that their asphalt was *natural* and that their filler was *unique*, not just any ordinary filler. This competition reduced the price of paving in New York from $3.36 per square yard in 1901 to $1.52 per square yard by about 1910 (Union Oil Co. 1912).

Production of asphalt from Texas crude oils entered the supply picture in 1903 with the use of air blowing. Texas crude could not be reduced to grade using the shell stills, but when air blowing was patented in 1893 by Byerly and in 1899 by Culmer, technology had provided a method for increasing the viscosity of the Texas crude sufficiently to produce a paving grade asphalt rivaling that of the easily reduced California crude (Krchma and Gagle 1974). It was not until 1912 that East Coast refiners began to receive Mexican crude, which could be easily reduced to asphalt in a shell still. However, it should be noted that most refiners did not produce paving-grade asphalt directly in the stills, but rather produced solid asphalt that had to be fluxed before being used in paving. The solid asphalt was shipped in barrels to the job site, and local flux materials were used to reduce the viscosity. However, by World War I refining directly to grade was a general practice, even after pumps had been developed that allowed lines and tanks to be installed for storing and blending various materials.

It was not the paving of city streets that produced the widespread use of asphalt paving in the United States, but rather the effects of the automobile on the existing highways that was the impetus to paving rural highways. The best of the rural highways, when automobiles came on the scene, was the water-bound macadam, which consisted of large rocks that were keyed together by compaction; the voids in the large rocks were then filled with fine material by washing the material into place with water. The automobile, traveling at high speed over these roads, produced clouds of dust during dry weather. This dust was the fine material from the water-bound macadam surface, which became airborne, eventually denuding the surface and leaving only a mass of loose large stones (Spencer 1979). The magnitude of this problem can be appreciated only when considering the rapid growth of ownership of automobiles in the United States. In 1895 four experimental automobiles were reported in the United States, but by 1904 the number of U.S. automobiles had grown to 55,290 and by 1910 to 468,500 (Labatut and Lane 1950). It is not surprising that the problem of wind erosion of fines produced by the automobile had reached crisis proportions.

Logan Waller Page, director of the Office of Public Roads, noted in the February 26, 1910, issue of *Engineering News Record* that while horse's feet caused the production of dust by abrading the surface rocks, automobiles denuded the surface of the road of the fine material needed for a stable road. This problem became so significant that the ASCE, the Highway Research Board, the American Association of State Highway Officials, the American Road Builders Association, and the American Society for Testing Materials jointly called for a series of formal meetings to discuss these problems and to seek a solution. The discussions brought into focus the divergent ideas of engineers from across the nation, providing a forum for describing experiments with asphalt materials and developments in road construction equipment.

The reader is reminded that the problems described above centered on rural highways, not city streets. Because rural highway mileage was so vast compared to the mileage of city streets, the problem was one of scale and economy. The existing rural roads had to be protected from the damaging effects of the automobile, which sucked the dust from the surface, thereby removing the dust binder, but the solution had to be of a low enough cost to be affordable. An early technique for dealing with this problem involved using heavy road oils as a dust palliative, implemented by James W. Abbott, special agent with the Office of Public Road Inquiry, Los Angeles, as reported in the 1902 Yearbook of the U.S. Department of Agriculture (Abbott 1902). In 1898 Abbott oiled 6 mil of road in Los Angeles County for the sole purpose of laying the dust. It should be noted that asphalt-base crude oils were readily available in southern California at that time, and the news of the beneficial effect of asphalt in retaining the dust in the macadam surfaces soon spread across the nation.

It was not until the early 20th century, when manufactured asphalt became more widely available and automobile ownership became more common, that public works engineers began to seriously consider developing specifications to describe the properties of low-cost road mixtures purchased by their organizations. Early methods of describing an acceptable oil mix paving material included the "pat test," which involved taking a sample of asphalt mixture, kneading it into a ball in one's fist, and patting it into a pancake shape. A piece of brown paper was then pressed into the surface of the pat. If the liquid asphalt made a heavy stain on the paper, the mix had too much asphalt, and if the stain was very light, there was not enough asphalt; but if the appearance of the mix was just right, there was enough asphalt and construction could begin. Obviously, calibrated eyeballs were needed to judge when the appearance was just right! Calibrated eyeballs could only be developed by experience, and all the inspectors had to achieve this calibration by correlating their eyeballs with eyeballs that had already been calibrated, a time-consuming and very imprecise process.

The first formal design method for asphalt mixtures was the Hubbard-Field method, which was originally developed to design sand-asphalt mixtures and later modified for aggregates (Asphalt Institute 1965). This design method included the compaction of 2 in. specimens at a range of asphalt contents. Each specimen was heated to 140°F in a water bath and placed in the testing mold, which was placed in the 140°F water bath, and a compressive load applied at a rate of 2.4 in./min. The 2 in. specimen was forced through a restricted orifice 1.75 in. in diameter. The maximum load sustained, in pounds, is the Hubbard-Field stability. A sketch of the test apparatus is shown in Fig. 1.

After testing all the prepared specimens, the average stability at each asphalt content was calculated and plotted to determine the optimum asphalt content. As indicated earlier, this test was designed for sand-asphalt mixtures and later adapted for aggregate mixtures. There are obviously serious disadvantages to using this procedure for mixes containing aggregate sizes larger than about 1/2 in. if the 4:1 ratio of mold diameter to maximum particle size is to be maintained, and many questions remained about what the test actually modeled relative to the performance of paving mixtures. Because of these limitations and concerns, other methods of mixture design moved forward on several fronts.

Such was the state of technology of asphalt mix design in 1927 when Francis Hveem was assigned to his first oil mix job as a project engineer for the California Division of Highways. Oil mixes were a new type of surface for California rural roads, and the primary design consideration was the determination of the proper oil content for the aggregate mixture. The oil content for oil mixes was determined by an experienced man who knew the right color of a mix containing the proper amount of oil (Stanton

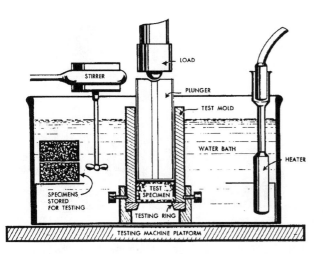

Fig. 1. Test apparatus for Hubbard-Field stability (Asphalt Institute 1965)

1927). Since Hveem had no experience with which to judge the color of a satisfactory mix, he decided to measure everything about the mix that he thought might relate to establishing the right oil content (Vallerga and Lovering 1985).

Development of Hveem Mix Design Method

After World War I three types of pavements were generally available for construction in California (Vallerga and Lovering 1985):

1. Hot mix asphalt layers of 6 in. or more were placed with header boards, and aggregates were mixed and spread by hand and roller compacted. The only element of design involved selecting aggregate proportions based on experience developed in the eastern United States. The design method typically used was the Hubbard-Field method, originally developed to design sand-asphalt mixtures and later modified to include aggregates. Construction experience in California demonstrated that minor aggregate variations often produced mixes that would bleed and rut.

2. Penetration surface treatments were widely used, primarily as a dust palliative. The asphalt was sprayed on the surface and allowed to penetrate into the surface of either the subgrade soil or natural gravel, if available. Only locally available materials were used. The road oil on the surface helped to minimize dust generated by passing vehicles and to minimize raveling of the surface. These mixes developed potholes very badly and developed ruts when much truck traffic was present.

3. Bituminous macadam surfaces are penetration treatments in which asphalt is sprayed on the surface of a prepared macadam base. Macadam bases consist of a layer of large rocks that are keyed together using compaction and then choked with smaller aggregate that is sprayed with asphalt until the aggregates are tightly stuck together. These mixes have a very rough surface and provide a very rough ride.

All three types of construction had serious disadvantages. Hot mix asphalt was very expensive to construct and consequently was used only in cities where heavy traffic occurred. The penetrative surface treatment, while fairly inexpensive to construct, was very expensive to maintain and, in many cases, did not have adequate strength to meet the traffic demands. The bituminous macadam surfaces were almost as expensive to build as the hot

mix asphalt, but provided a ride quality that was greatly inferior. There was a great need for an intermediate type of surface for use in rural areas that was stronger than the penetrative surface treatments, but not as strong or as expensive as the hot mix asphalt or the bituminous macadam surfaces. In 1924 California began to experiment with treating gravel roads using an "oil mat" process developed in Oregon in 1923 (Vallerga and Lovering 1985). These mixtures were produced by mixing a fairly good gravel and slow-curing cutback asphalt. In California cutback asphalt was called *fuel oil* in order to get a lower freight rate, and as a result these mixes were called "oil mixes" in departmental correspondence and in the technical literature. These oil mixes were mixed in place on the road using agricultural harrows and discs, spread smooth using motor graders, and compacted by traffic.

By 1928 oil mixes began to be cold mixed in central plants where more accurate proportioning and thorough mixing occurred at little additional cost. The mixes were still dumped on the road, bladed back and forth during curing of the slow-curing cutback, and compacted by traffic. Observations by engineers with the California highway department showed that the moisture content of the aggregate significantly affected the amount of blading required to cure the mixes. They subsequently set a maximum water content, which led to drying the aggregate before mixing with cutback. Dryers began to be widely used in the 1930s, and contractors found that dry aggregates allowed plant operations to speed up because mixing and coating occurred faster. Contractors also found that the mixes were more workable when heavy liquid asphalt was used with an 80 to 94% residue of an 80% penetration asphalt in a heavy oil.

Mechanical pavers were introduced about 1937, and it became possible to place mixes made with heavier asphalt without using header boards. With successful use of the paving machine, the trend was toward replacing the heavy liquid asphalt with paving grade asphalt. While construction equipment advances were occurring rapidly, such was not the case with mixture design, and as a result there was significant variation in the performance of mixtures because there was no standardized way of determining the proper amount of oil needed to ensure good performance.

In 1927 Hveem began a detailed study of oil mixes to identify the elements of these mixtures that affected the proper oil content. He made many more measurements than normal on aggregate gradation, moisture content, temperature, humidity, and oil content. From these measurements he concluded that there appeared to be a relationship between gradation and the amount of oil needed to produce the proper color of the mix, and that as the gradation varied, the amount of oil required to produce the same color and appearance had to be varied. He remembered reading an article in *Engineering News Record* that said that the only property of an aggregate grading that varied as the grading varied was the surface area of the particles. He concluded that perhaps the amount of oil required in a mixture was the amount that would coat the surface of all particles to some uniform thickness.

He searched the state archives and found an article by the Canadian engineer Captain L. N. Edwards, of Toronto, that described a method for calculating the surface area of aggregates used to design portland cement concrete mixes (Edwards 1918). Hveem used the surface area concept to develop a methodology for predicting the amount of asphalt needed for an oil mix. Hveem also noted that the moisture content of the aggregate had a potent effect on mix behavior. Mixes with excess moisture were more plastic and had a greasy, over-oiled appearance when compared to mixtures without excess moisture. If the oil content was reduced for those mixes with excess moisture to produce the de-

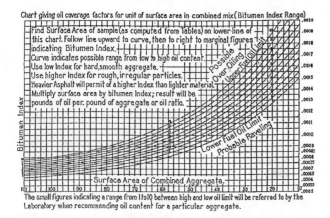

Fig. 2. Chart for determining weight ratio of oil to aggregates or oil ratio (Stanton and Hveem 1934)

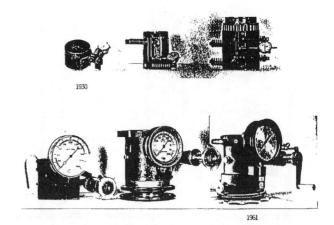

Fig. 3. Evolution of Hveem stabilometer from small expansible metal cylinder to hydraulic type with adjustable base for variable specimen height (Vallerga and Lovering 1985)

sired appearance, the mix had barely enough oil to hold it together once the excess moisture evaporated (Vallerga and Lovering 1985).

A few comments on Edwards's work are in order at this point because that work became the cornerstone of the surface area method of oil mixture design and later of the Hveem method of mixture design for hot mix asphalt. A review of Edwards's 1918 work showed that he physically counted particles ranging from material passing the 2 in. sieve to particles retained on the #100 sieve. He assumed that these particles were spheres and calculated the surface area of the particles retained on each sieve in terms of square feet of surface area per pound of material retained. Edwards also developed an equation to calculate the surface area of particles of various sizes based on the diameter of the particle.

A review of the literature has led the writers of this paper to conclude that Hveem used Edwards's equation to generate his surface area factors. However, a careful comparison of surface area factors generated using Edwards's equation for the sieve sizes used by Hveem does not produce the exact surface area factors proposed by Hveem, but a set of factors very close to those reported by Hveem. None of the published papers by Hveem provides a detailed description of the development of these surface area factors; rather, Hveem simply presents the values as being representative of aggregates used in California mixes. Hveem first presented the surface area factors and their use in mixture design to a western audience (Hveem 1932) and then to a national audience (Hveem 1942).

As a result of his stellar research on oil mixes, Hveem was invited to join the California Division of Highways' central laboratory staff in Sacramento. He was given the job of investigating the performance of oil mixes throughout the state and asked to prepare a report describing causes of failures and to identify the factors necessary for the successful design and performance of oil mixtures. It was during the period before 1932 that the surface area method of oil mix design was completed, based on the premises that

1. The proper amount of oil in a mix of different-size particles depended on the surface area of the gradation; and
2. The average film thickness for similar aggregates must decrease as the average particle diameter decreases, as illustrated in Fig. 2 (Stanton and Hveem 1934).

Later refinements to the surface area method included the addition of the centrifuge kerosene equivalent test to evaluate the

absorption and surface roughness effects of the aggregate on the amount of oil required for mixture design (Hveem 1942).

Once the surface area method was developed to estimate the optimum asphalt content for oil mixes, there was no assurance that the resulting oil mix, even under the most favorable conditions, would have sufficient shear strength or stability to resist deformations under truck traffic loads. By 1929 Hveem and others in the California highway department recognized that a testing device was needed that could evaluate the stability characteristics of oil mixes. Hveem set out to develop a device that could be used to determine whether an aggregate mix with an asphalt content determined using the surface area method would be stable enough to produce a hard, smooth road surface that would not deform under traffic. In developing the stability device, the following factors were thought to be important (Vallerga and Lovering 1985):

1. Failure in most road surfaces is not catastrophic, but rather is the result of an accumulation of a large number of rapid shoves/pushes;
2. The rate of flow for a particular mixture is dependent on such factors as the temperature, amount, and consistency of the oil, the aggregate gradation, and the amount of filler dust;
3. The testing device should test specimens that are typical of the loading conditions in the field, that is, a thickness of 2.5 in., which was typical of existing roads, and a diameter of 4 in., which was typical of tire-loaded areas at the time;
4. The specimen should be loaded vertically on its circular face, and the amount of load transmitted horizontally would indicate the degree of plasticity of the oil mix being tested; and
5. Testing other materials in the device should produce predictable behavior:
   • A perfectly rigid solid would transmit no lateral pressure; and
   • A perfect liquid would transmit a lateral pressure equal to the vertical pressure.

Using these requirements, Hveem began to develop a test device capable of measuring the stability of an oil mix. Stages in the development of the stabilometer are demonstrated in Fig. 3 (Vallerga and Lovering 1985). The various versions of the stabilometer shown in Fig. 3 are briefly described in Table 1.

Table 1. Stages of Development of Hveem Stabilometer

Version	Device description	Action of device and use
V-1: top left device in Fig. 3	Thin cylindrical sleeve split down one side and held together by spring	Apply load; sleeve expands against spring. Movement measured via dial gauge. Differences were measured for different mixes, but specimens could not be compacted and then placed in the sleeve for testing very easily. Many oil mixes fell apart in attempts to place them in split sleeve
V-2: top middle device in Fig. 3	Added second inner split sleeve for specimen preparation; outer split sleeve held together with springs	Remove inner split sleeve, insert into solid pipe, and prepare specimen. Remove specimen in split sleeve and insert into outer sleeve with splits on opposite sides. Apply load and measure lateral movement. Difficult to maintain constant friction between inner and outer sleeves. Device calibrated by filling inner split sleeve with 1/2 in. steel balls, adding oil between sleeves until constant results were achieved
V-3: top right device in Fig. 3	Refined V-2 by adding hydraulic cell for calibration instead of 1/2-in. steel balls	Hydraulic cell was inserted inside inner split sleeve and used to calibrate amount of lubricant required between sleeves to achieve constant results. Note: While demonstrating V-3, someone asked why hydraulic cell was not arranged to surround specimen and do away with split steel sleeves. This suggestion led to development of first hydraulic stabilometer
V-4: lower left device in Fig. 3	Steel outer shell replaced with interior cylindrical rubber diaphragm held in place with steel rings at top and bottom. Annular space filled with water. Displacement pump attached to outer shell so water could be added and so 5 psi pressure could be applied to bring diaphragm in contact with specimen. Glass gauge was added to measure volumetric strain	Vertical load applied to specimen and horizontal shear displacement produces pressure increase in fluid surrounding specimen as measured with pressure gauge. Variable results were traced to variable amounts of air in pressure cell. This recognition was partially due to rusting of steel inside stabilometer, which reduced quantity of trapped oxygen, producing changes in amount of air in cell and causing variable results
V-5: lower middle device in Fig. 3	Length of inside chamber of stabilometer was increased to allow for adjustable base, glass gauge removed, outer shell made from brass instead of steel, and liquid in stabilometer changed from water to oil	Variability in results decreased and adjustment of stabilometer position easier with adjustable base. Encountered difficulty in calibrating device by getting right amount of air into main body of stabilometer
V-6: lower right device in Fig. 3	Replaced balance of steel inside hydraulic chamber with brass, added a small reservoir for air outside main body of stabilometer, and modified displacement pump to make it easier to turn	By 1934 stabilometer design had been fully developed and has not changed in design since last modifications were implemented

During the same time that the stabilometer was being developed, each version was immediately pressed into service in the laboratory to evaluate mixes being produced on jobs being constructed by the California highway department. The result was a continual process of evaluation of the implementation potential of the stabilometer and comparisons of test results with the performance of field mixes. In addition to measurements of the strength of mixtures, issues of specimen preparation were being addressed since the mixtures produced in the laboratory needed to simulate mixes being constructed in the field. As a result, comparisons were made between the density produced in the laboratory, using impact energy from the 8 lb slide hammer dropped 5 in., and that measured from field cores.

The conclusion was that field densities were greater than those produced using the laboratory compaction process, and besides, the method of achieving density in the field was more of a knead-

ing action than an impact action. As a result, a decision was made to change the laboratory compaction to a kneading action, which produced particle orientation in laboratory specimens similar to that obtained under field rollers. Making that change to kneading compaction was expected to produce specimens having a stability similar to that observed in field-produced mixes. The kneading compactor was developed to the point that it was adopted as the standard method for compacting specimens in 1938 (Vallerga and Lovering 1985).

Determining a minimum acceptable stability required that field studies be conducted to measure stability of mixes that provided different levels of field performance, ranging from good to poor. Results from an early study were reported by Stanton and Hveem (1934) and are shown in Fig. 4 in terms of applied vertical load (pound) and stabilometer gauge reading (pounds per square inch). It was noted that all of the projects that had satisfactorily perform-

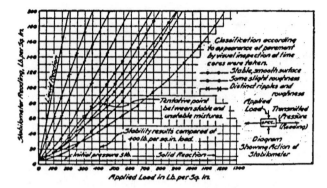

Fig. 4. Results of early correlation study to establish stabilometer reading that distinguished between stable and unstable mixtures (Stanton and Hveem 1934)

ing pavements sustained stabilometer gauge readings in excess of 80 psi. It was necessary to convert the stabilometer gauge readings to percent stability, which was accomplished using a conversion chart. By 1938, W. R. Lovering (unpublished memorandum to F. N. Hveem, March 5, 1938) had developed the following formula to calculate stability:

$$S = \frac{22.2}{\dfrac{P_h D_2}{400 - P_h} + 0.222}.$$

where P_h = horizontal pressure, that is, stabilometer gauge reading, in pounds per square inch (kilopascals) (taken the instant that the vertical pressure is 400 psi); and D_2 = displacement on specimen, in number of turns of the displacement pump handle.

Field data were used to develop limiting criteria on stability that were used in mixture design. Studies over several years involving several hundred projects showed the following relationship between Hveem stability and stable mix performance in the field: stabilometer value <30 (unstable under traffic); stabilometer value >30 but <35 (borderline stability under traffic); and stabilometer value >35 (stable under traffic).

During the time after 1938, several changes were made to the mixture design procedure when shortcomings were identified. The major changes represent additions to the procedure to evaluate the susceptibility of mixtures to field problems that surfaced after the basic Hveem method was developed. Changes included

1. The moisture susceptibility test was added in 1946 to evaluate the adverse effects of water vapor entering the pavement from the bottom. The method involves exposing a specimen to water vapor before testing in order to determine whether the mix has the potential to be damaged by the stripping effects of the water; and

2. The sand equivalent test was developed in the early 1950s to evaluate the detrimental effects of small amounts of clay on the disintegration of mixes under the influence of repeated heavy traffic loads. The sand equivalent test was added to the design procedure in 1954.

By 1959 the Hveem method of asphalt mix design had developed its final form. The complete procedure is described in the Asphalt Institute Manual Series 2 and in ASTM 1557. A survey of the states conducted by Kandhal and Koehler (1985) identified the states using both the Hveem and Marshall mixture design procedures. The Hveem method has been primarily used by the five western states and a smattering of five other states.

Marshall Mixture Design Method

Bruce G. Marshall developed the Marshall method in the period just before World War II in an attempt to produce a procedure that would use available laboratory equipment to determine the asphalt content of asphalt mixtures. There is not much information on the development and use of the method before the U. S. Army Corps of Engineers began to investigate the procedure in 1943 (War Department 1943). Prior to 1943, the method had been used on an experimental basis by several highway departments in the south for about 4 years.

The Marshall method initially involved preparing specimens using 25 blows of the standard Proctor soil compaction hammer, well spaced around the perimeter of the 4 in. mold, followed by application of a 5,000 lb static load for 2 min, to level the surface of the specimen. An earlier version of the procedure appears to have involved applying 15 blows with the standard Proctor hammer in two increments, including 10 blows in the first increment, after which the extension of the mold was removed and 5 additional blows were applied in the second increment. After compaction with the standard Proctor hammer a 5,000 lb static load was applied for 2 min. After releasing the load, the Marshall compaction mold containing the specimen was placed in a water bath for 1 min, after which the specimen was extracted from the mold. The specimen was then placed in the cool water bath for 5 min, removed from the water bath, and then air cooled at room temperature for a period of time ranging from 1 h to a maximum of 24 h.

After cooling and immediately before testing, each specimen was placed in a 140°F water bath for 1 h, removed from the water bath, immediately placed in the Marshall compression device, and loaded vertically at a deformation rate of 2 in. per minute until the maximum load occurred. Test measurements included the maximum load carried by the specimen and the vertical deformation of the specimen at the maximum load. Test results included the maximum load in pounds, called Marshall stability, and the deformation, called flow, measured in 1/32 in. Initially the deformation was measured by hand. The flow meter was developed later as an improvement in repeatability of test results. Specimens were tested at different asphalt contents, and a plot of stability versus asphalt content was prepared and used to select the optimum asphalt content. Criteria used in mixture design included a recommended minimum stability of 300 lb and a flow between 4/32 and 8/32 in. The recommended density target for the field was suggested to be 97% of the laboratory density at optimum asphalt content. It was thought that the flow measurement would be most useful in field control of plant mixes as it was observed that mixes with asphalt content exceeding the optimum experienced a rapid increase in flow.

An earlier study by the Tulsa District (War Department 1943) of the U.S. Army Corps of Engineers recommended that the Corps adopt the Hubbard-Field method for asphalt mixture design. The next year the Waterways Experiment Station (WES) at Vicksburg, Mississippi, initiated a study to compare the relative suitability of the Hubbard-Field and the Marshall methods for designing asphalt mixtures being trafficked by 100 psi tire pressure military aircraft of World War II. The findings from this study included

1. Specimen failure: The Marshall test stresses and fails the entire specimen by breaking it into four pieces, while the Hubbard-Field test produces failure only around the perimeter of the specimen at the base as the material is forced through the base orifice. Because only the small portion at

Table 2. Recommendations for 50-Blow Marshal Criteria to be used to Establish Laboratory Optimum Asphalt Content based on Results from Waterways Experiment Station Studies (White 1985)

	50-Blow Marshal Method (Surface Cource)			
	Select Asphalt		Limiting Criteria	
Test property	Asphalt concrete	Sand asphalt	Asphalt concrete	Sand asphalt
Stability (lb)	Peak	Peak	500 lb minimum	500 lb minimum
Flow (0.01 in.)	—	—	20 maximum	20 maximum
Unit weight total mix (pcf)	Peak	Peak	—	—
Percent voids total mix (%)	4	6	3–5	5–7
Percent voids filled with asphalt (%)	80	70	75–85	65–75

the specimen lower edge fails in the Hubbard-Field method, larger aggregates in that region of the specimen cause very erratic stability test results. As a result, a large number of tests are required to produce accurate, repeatable results.

2. Equipment: The Hubbard-Field equipment is heavy, bulky, and designed only for laboratory use; specimen failure loads are often in excess of 5,000 lb; the strain rate is 2 in. every 50 s, which requires a heavy turning wheel and special gears for manual load application. The Marshall equipment is lighter and more compact than the Hubbard-Field equipment. Marshall specimens fail at about 800 lb at a strain rate of 2 in./min. These loads and strain rates can easily be produced with an ordinary hand crank.

3. Testing time and effort: Tests can be conducted more rapidly and with less effort using the Marshall test apparatus than with the Hubbard-Field test apparatus.

4. Correspondence with good performing field materials: The optimum density achieved with the Marshall mix design procedure appears to match field densities better than those from the Hubbard-Field method. A direct result of this observation is that the optimum asphalt content from the Marshall method corresponds more closely to the asphalt content of good performing field mixes than does the Hubbard-Field, in which the laboratory optimum asphalt content is greater than that which produces good performance in the field.

As a result of these observations, the Waterways Experiment Station staff recommended that the Marshall method be selected for further evaluation and application to the design of airfield pavements. An extensive evaluation to adapt the Marshall method to design of asphalt mixes for airfield pavements was conducted by the Waterways Experiment Station (Department of the Army 1948). This evaluation, which occurred during the period from 1944 to 1948, included work on the following several elements of the Marshall mix design procedure:

1. Laboratory compaction methods including weight of hammer, size and shape of the foot of the compaction hammer, and number of compaction blows before applying the leveling load. The principal objective of the compaction study was to adopt a sample preparation procedure that would identify the optimum asphalt content while requiring a minimum amount of effort and time;

2. Mixture properties to be used in selecting the optimum asphalt content; and

3. Combined laboratory and field studies conducted to determine the relationship between properties of mixes designed in the laboratory, constructed in the field, and then subjected to accelerated aircraft loading until failure occurred. The accelerated testing provided the opportunity to compare the performance of mixes designed at different compaction levels and asphalt contents without having to wait several years for typical traffic to produce enough performance data to perform the analyses.

Optimum asphalt content for all field mixtures was determined from the 15 blow and 5,000 lb static load compaction version of the Marshall method. During mixture design, seven properties were either recorded or calculated for each test conducted in order to have a variety of mixture properties available from which to develop the criteria for selecting the optimum asphalt content. These seven properties were flow, stability, total specimen unit weight, aggregate unit weight, percent voids in the mineral aggregate (VMA), air voids in the compacted specimen, and percent voids filled with asphalt. In this laboratory study a number of material variations were considered to evaluate the responsiveness of the paving mix properties to changes in mixture constituents, that is, aggregate gradation, filler content, and asphalt content.

Observations from field construction showed that the original 15 blows with the 10 lb modified Proctor hammer and 5,000 lb load compaction effort produced a density in the laboratory that corresponded to the density achieved in the field under normal pavement construction. This same density was found to be consistently 2% lower than the density achieved with 50 blows of the 10 lb hammer with the 3 7/8 in. foot dropped 18 in. One advantage of using the hammer with the 3 7/8 in. foot was that a leveling load was no longer needed to produce a smooth face on the specimen. One conclusion from the field and compaction studies was that 98% of the density achieved in the laboratory with 50 blows of the 10 lb hammer corresponded to the average compacted density of field mixes that performed well under aircraft gear loading. A summary of the field study materials, loads, and results can most easily be found in a paper by White (1985), to which the reader is directed for additional details of that study.

During the design of the mixtures to be constructed, it was recognized that field mixes should be constructed over a range of asphalt contents. As a result field mixes for each aggregate combination were constructed at optimum, 10% lean of optimum, and 20% lean of optimum and subjected to aircraft loading to observe the performance at each asphalt content. The pavements were subjected to aircraft gear loads of 15 kips at 50 psi tire inflation pressure, 37 kips at 110 psi tire inflation pressure, and 60 kips at 110 psi tire inflation pressure. Observations of distress were noted as loads were applied up to 3,500 coverages of the gear loads. Distress measurements included rutting and shoving, area of cracking in the loaded areas, and settlement. Visual observations were made for tire printing in the surface, raveling or shelling, upheaval, and longitudinal movement. In addition to distress measurements, samples were cut from the pavement surfaces, and the

Table 3. Marshall Mixture Design Criteria for High-Pressure Tires (White 1985)

	75-Blow Marshall Method (Surface Course)	
Test property	Select asphalt content (for asphalt concrete)	Limiting criteria (for asphalt concrete)
Stability (lb/kg)	Peak	1,000 minimum
Flow (0.01 in./mm)	—	16 maximum
Unit weight aggregate only (pcf/kg/m^3)	Peak	—
Unit weight total mix (pcf/kg/m^3)	Peak	—
Percent voids in total mix (%)	5	4–6
Percent voids filled with asphalt (%)	78	75–82

seven mixture properties mentioned above were determined.

The results from the field study were used to identify the ranges of asphalt contents that produced mixtures that performed well under traffic. The laboratory mixture properties were used to establish the limiting criteria for mixture properties as part of the mixture design procedure. The criteria established in this part of the study are included in Table 2.

The optimum asphalt content was to be the average of the asphalt contents at each of the criteria listed in the second or third columns of Table 2, depending on the mix type. Once the average was determined, each test property had to be checked to ensure that each criterion listed in the second or third columns was met at the selected optimum asphalt content. White (1985) presents discussions describing the logic for eliminating three of the seven mix properties mentioned earlier from the calculation of optimum asphalt content.

It should be pointed out that the recommendations in Table 2 were not implemented by the U.S. Army Corps of Engineers. Originally, Charles R. Foster had proposed that the stability value be used only in determining the optimum asphalt content and that stability be rejected as a test to judge the quality of field mixes (Foster 1982). Foster's contention was that the field studies at the WES demonstrated that Marshall stability showed no relationship to field performance, and therefore its use should be limited to helping to determine the optimum asphalt content. Foster's recommendation was rejected by the chief engineer's office, and the WES engineers were told to include a minimum Marshall criterion. Table 2 criteria were developed and recommended, and the supporting logic for the minimum of 500 lb for stability was

> In the evaluation ... it was determined that stability alone was not a satisfactory tool for evaluating the ability of a pavement to resist displacement under repetitive wheel loads .... This analysis showed that the minimum stability values that were tested were satisfactory and there was no apparent advantage in obtaining higher stability values .... A minimum stability requirement of 500 lbs. was tentatively established. The selection of 500 as a minimum design requirement has merit since it can be easily and economically obtained ... will insure better gradation, closer control, and a mix that will lay and roll satisfactorily .... (Foster 1982)

Notice in the quotation from Foster that his contention is that the Marshall stability test result is not an indicator of the ability of an asphalt mixture to resist permanent deformation. Yet a review of the state's use of Marshall test results reported by Kandhal and Koehler (1985) shows that minimum stability values of 1,800 to 2,000 lb were commonly specified, allegedly to minimize permanent deformation being produced in asphalt pavements by heavy truck traffic.

It should be remembered that one of the basic criteria used to develop empirical laboratory compaction requirements was that the density produced in laboratory-prepared specimens should be the same as the densities achieved in the field after the mix has served traffic for a few years. Such a comparison was faithfully made in the WES studies. Unfortunately, few states made these comparisons before making changes in mixture design criteria. When permanent deformation continued to occur in their Marshall mixes, they increased the stability in an attempt to decrease rutting, following a path that Foster clearly pointed out as a flawed procedure.

During the late 1940s and early 1950s, military aircraft continued to increase in size and weight. Rather than increase the gear size to handle the additional weight, aircraft designers increased the tire inflation pressure to 200 psi, which produced a smaller pavement contact area. The result was higher stresses directly under the wheel load. It was observed that pavements subjected to the new loads continued to density even after they had reached a stable density under the lighter loads with lower tire pressures. In some cases the higher stresses caused failure in the binder courses that had satisfactorily performed under the 100 psi loads (White 1985). Concern for these problems caused the WES to initiate a study of these heavier loads and higher tire pressures. Some areas of the test lanes from the original study had not been trafficked, so these lanes were subjected to wheel loads of 30 kips with 200 psi inflation pressures. The following studies were conducted:

1. To identify the laboratory compaction effort needed to produce laboratory densities that would correspond to the field densities produced by these heavier loads and higher tire pressures; and

2. To modify the mixture design criteria for heavier traffic conditions (Department of the Army 1950a).

Before these studies were completed, an additional load was added to the study. A new Navy aircraft with gear loads of 15 kips and tire inflation pressure of 240 psi was added to the list. Loads simulating this aircraft were also applied to test lanes of the original experiment that had not been trafficked. Observations of performance of the Navy gear loads confirmed that pavements continued to density under higher inflation pressure tires, even at lower load magnitudes (Department of the Army 1950b). Laboratory compaction tests were conducted to determine the compaction effort required to produce densities in the laboratory that matched those achieved in the field under the higher tire pressures. After these studies were completed a new set of mix design criteria was developed for mixtures serving loads applied with 200 psi tires, as shown in Table 3.

During the 1950s WES provided technical support for construction and performance evaluations of airfields around the world. In spite of changes in mix design and quality control criteria, rutting was observed in a number of the projects, and subsequent analyses indicated that natural sands were thought to be a major contributing factor to the rutting. In an attempt to eliminate these sands, the stability was arbitrarily raised to 1,800 lb minimum. Subsequent field results showed that increasing the minimum stability did not solve the problem, so the Corps arbitrarily limited the amount of natural sand to 10%. Later this percent of natural sand was increased to 15%.

Application of the Marshall mix design method by the Corps of Engineers involves two distinct phases: the traditional laboratory mixture design to establish the gradation and asphalt content for the field mix, and adjustments to the mix once the mix began to be produced in the field. The first step in field verification involves compacting specimens using the field-produced mixture and testing the specimens to determine if the test results meet the Marshall field criteria. In most cases adjustments are needed in the field in order to bring the field-produced material in line with the field criteria. Most often, the asphalt content requires some adjustment to bring the air voids into line with the desired limits. After the plant is in production, samples of plant-produced materials are secured, Marshall specimens are compacted and tested and the air voids, Marshall stability, and flow values are monitored to ensure that the plant process is producing uniform materials. If the aggregates change, a new mixture design is developed in the laboratory and verified in the field.

The Marshall mixture design procedure was used successfully by 38 states and the Corps of Engineers for many years (Kandhal and Koehler 1984). One of the primary reasons justifying the use of the Marshall mixture design procedure by 38 states is that the equipment required to prepare and test laboratory specimens for the Hveem design procedure is too expensive for departments of transportation (DOTs) and contractors to purchase for all of their laboratories. As a result the Marshall method has been the preferred mix design method for a large majority of the hot mix asphalt industry in the United States. During the 1980s many of the interstate and other freeways with heavy truck traffic began to experience premature rutting or rapid development of rutting during some portion of their service life. In many cases, too much asphalt was blamed for the problem. Many engineers believed that the impact compaction used to produce specimens for the Marshall method could not satisfactorily produce the densities observed in these pavements. This issue was addressed in a national study on asphalt materials called the Strategic Highway Research Program (SHRP). One of the results from this very large research program was the development of a new mixture design method, Superpave, which is discussed in the next section.

Superpave Mixture Design Method

Superpave is part of SHRP, a program started in 1987 and completed in 1992 at a funding level of $150 million and dedicated to improving the performance of highway facilities. Funding for SHRP's asphalt research program was $50 million, and the program's objective was to identify and define properties of asphalt binders, aggregate, and hot mix asphalt that influence pavement performance and to develop test methods for performance-based specifications. The product was a new system for mix design and performance-testing of hot mix asphalt named Superpave, (superior performing asphalt pavements), which is gradually being implemented by state DOTs and provides a mix design system to deal with extreme temperatures and heavy loads. Superpave standardizes the hot mix design process among the 50 states with only minor variations from state to state and incorporates the elements of aggregate selection, asphalt binder selection based on climate, compaction based on traffic, and selection of an aggregate skeleton and asphalt binder content based on volumetric properties.

The Superpave system consists of three interrelated areas: (1) a performance-graded asphalt binder specification and tests that are based on the range of temperatures experienced by the pavement; (2) aggregate criteria and tests; and (3) a mixture design system using both a volumetric mixture design with a Superpave gyratory compactor and an analysis/performance prediction element. The mixture analysis portion of the system to predict pavement performance remains the one element of the system still under research and development. Each of the areas is described in the following paragraphs.

The asphalt binder grading system used in Superpave is called a performance-grading (PG) system. It is considered an improvement over the viscosity and penetration systems because testing is conducted at conditions that better simulate actual pavement conditions, it relies on engineering parameters related to the actual failure mechanisms leading to pavement deterioration, and it incorporates reliability concepts.

Climate, traffic, and age are known to be the most critical factors affecting asphalt performance in pavements. These three critical factors are simulated in the Superpave asphalt testing system to achieve the performance grading. To simulate climate conditions, testing is conducted at three design pavement temperatures that represent a maximum, intermediate, and low pavement design temperature. These temperatures are obtained from weather databases. Models are then used to convert air temperature to pavement temperature at selected depths within the asphalt layers. The loading mode and rate are selected to represent traffic speeds commonly observed on open highways to simulate traffic conditions. Pavement aging is simulated by testing the asphalt binder at two different aging conditions, representing the immediately-after-construction condition and long-term in-service condition of pavements. Short-term aging is simulated through the rolling thin film oven (RTFO) test, while long-term aging is simulate through the pressure-aging vessel (PAV) test (Bahia and Anderson 1994).

The engineering parameters considered in grading the asphalt binders are related to the following failure mechanisms: permanent deformation or rutting, fatigue cracking, and thermal cracking. At high temperatures typical of asphalt mixture production temperatures, asphalt binders are mainly fluid, and thus the PG includes a maximum limit on viscosity to ensure proper handling during pumping and mixing (Bahia and Anderson 1994).

The workability of asphalt binder is determined with a rotational viscometer, which is used to determine the viscosity of the asphalt binder at high construction temperature (135°C) to ensure that the asphalt binder is sufficiently fluid for pumping and mixing (McGennis et al. 1994).

The dynamic shear rheometer (DSR) is used to characterize the viscoelastic behavior of the asphalt binder and evaluate its rutting and fatigue cracking potential. The basic principle used for DSR testing is that asphalt behaves like elastic solids at low temperatures and as a viscous fluid at high temperatures. These behaviors can be defined by measuring the complex shear modulus (G^*) and phase angle (δ) of an asphalt binder under a specific temperature and frequency of loading. The G^* and δ parameters are measured by the DSR by applying a torque on the asphalt binder placed between a fixed and an oscillating plate. The resulting angle of twist is measured and is used to compute the shear strain. The complex shear modulus G^* is defined as the ratio of shear stress to shear strain, and delta is the time lag between the peak stress and the peak strain in the asphalt binder (Christensen and Anderson 1992; McGennis et al. 1994).

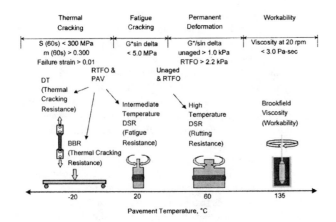

Fig. 5. Performance grading system tests

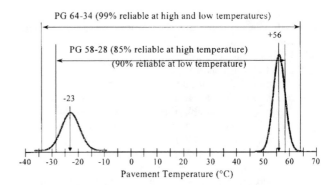

Fig. 6. Example of reliability in grade selection

The rutting factor is $G^*/\sin$ delta. Minimum values of $G^*/\sin$ delta are specified before and after short-term aging at the maximum pavement design temperature to verify that the asphalt can be used. These minimum values are intended to provide better resistance to rutting. The temperature that results in a $G^*/\sin$ delta of 1.0 kPa with no aging and a value of 2.2 kPa after RTFO aging is the minimum high PG temperature grade (Bahia and Anderson 1995).

The fatigue factor parameter $G^* \times \sin$ delta is used to characterize fatigue resistance and is also measured using the DSR. A maximum limit on the value of $G^* \times \sin$ delta of 5,000 kPa is required and is measured on the asphalt binder samples that are RTFO and PAV aged (Bahia and Anderson 1995).

The bending beam rheometer (BBR) is used to evaluate the thermal cracking potential at low temperature. The asphalt physical properties are controlled to minimize buildup of thermal stress and to avoid using asphalts that are too brittle. Small asphalt beams are made and tested in the bending beam rheometer. The creep stiffness (S) and the logarithmic creep rate (m) measured from the BBR are used in the specification. A maximum S of 300 MPa and a minimum m value of 0.300 are specified on asphalt binder after RTFO and PAV aging. In modified binders, measuring the stiffness is not sufficient to characterize the low-temperature cracking potential. This is because these binders may have a higher stiffness, but they may not be susceptible to cracking as they are more ductile and can stretch without breaking. The direct tension test is used to characterize such asphalt binders. It is used for testing binder that has a stiffness exceeding the specified stiffness at low temperatures. A minimum failure strain of 1.0% is specified on asphalt binder after RTFO and PAV aging (McGennis et al. 1994).

Fig. 5 presents the PG system in which each of the main pavement failure criteria is tested by the different testing systems. The maximum, intermediate, and minimum temperatures at which the specification limits are met define the PG for an asphalt binder. For example, a PG 64-28 represents an asphalt grade for a binder that has an adequate physical property to use in a climate in which the maximum consecutive 7 day average pavement design temperature will not exceed 64°C and the minimum pavement design temperature will not drop below −28°C. In this case, it is assumed that traffic conditions are typical of open highway speed and that these temperatures are typical of a hot summer and a cold winter. Summer and winter temperatures vary from one year to the next, and traffic speeds vary from one highway to the other. The Superpave system is therefore designed to allow asphalt

grade selection based on probability of temperatures from one year to the other and based on traffic speed (McGennis et al. 1994).

Climate is the main factor used in selecting the specific asphalt grade for a project. The Federal Highway Administration (FHWA), as part of the Long Term Pavement Performance Program (LTPP) has developed a computer program called LTPP-Bind. By entering a county or city name, the program searches through thousands of weather stations to select those closest to the project. It then automatically gives the data for the high and low air temperatures and the standard deviation for the weather station. The program also gives the corresponding pavement temperatures at selected depths. Using the standard deviations, the program gives a list of PGs that would fit the pavement temperatures along with the reliability factor for each grade. The maximum consecutive 7-day pavement temperature and single-event coldest temperature are used to determine the pavement grade. An example of reliability is shown in Fig. 6 for the local conditions of Topeka, Kansas. As shown in the figure, two grades could be chosen to achieve different reliability factors. Designers could specify grades based on availability, cost, and reliability required. Typical reliability for highways is 98% for both high and low temperature grades.

A number of standard grades have been selected and included in the AASHTO standard specification MP1. Table 4 lists these grades, which are spaced at 6°C intervals. As shown in the table, PGs range between a high temperature of 82°C and a low temperature of −46°C. It is well recognized that while pavement temperatures could reach −46°C in northern U.S. climates, no pavement has been reported to reach a temperature of 82°C. The purpose of providing these high temperature grades is to allow consideration of high traffic volumes and/or slow traffic speeds. In other words, the effects of changing loading rate or traffic speed could be accounted for by a temperature change. Using this

Table 4. Binder Performance (PG) Grades

HTG[a] (°C)	Low temperature grade (°C)						
PG 46	−34	−40	−46	—	—	—	—
PG 52	−10	−16	−22	−28	−34	−40	−46
PG 58	−16	−22	−28	−34	−40	—	—
PG 64	−10	−16	−22	−28	−34	−40	—
PG 70	−10	−16	−22	−28	−34	−40	—
PG 76	−10	−16	−22	−28	−34	—	—
PG 82	−10	−16	−22	−28	−34	—	—

[a]HTG = high temperature grade.

Table 5. PG Shifting to Account for Speed and Traffic Volume

Design ESALs[a] (million)	Traffic speed		
	Standing	Slow	Standard
<0.3	+1	None	None
0.3 to <3	+2	+1	None
3 to <10	+2	+1	None
10 to <30	+2	+1	+1
<30	+2	+1	+1

+1: increase high PG grade by six degrees (one grade); that is, PG 64-22 to PG 70-22.

[a]ESALs = equivalent single-axle loads used in AASHTO pavement design procedures.

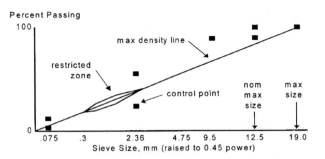

Fig. 7. 0.45 gradation chart

concept of rate-temperature equivalency, it is assumed that the effect of slower traffic speed on asphalt binders could be considered by selecting a shift in the climate grade high temperature. Using the same theme, the grading system recommends shifting the grade temperature. Table 5 shows the recommended grade shifting for different traffic speed and design traffic conditions. It is interesting to note that this is the first time that a binder selection procedure is tied to design traffic volume and speed.

In the case of mineral aggregates, Superpave has specified criteria for the use of certain test properties called consensus properties, such as coarse aggregate angularity, fine aggregate angularity, flat, elongated particles, clay content based on the location of mix, and traffic volumes. In addition to the consensus aggregate properties, certain other aggregate characteristics were believed critical and are termed source properties. Those properties are toughness, soundness, and deleterious materials. The most important recommendation in Superpave for aggregates is in the case of aggregate gradation. Superpave recommends certain con-

trol points and a restricted zone in the usual 0.45 power chart that is used for plotting gradation. Control points and the restricted zone are provided to avoid the use of undesirable gradations, which can result in tender mixes. Candidate aggregate gradations must pass between control points and avoid the restricted zone. Fig. 7 shows examples of control points and restricted zones for a 12.5 mm nominal maximum aggregate size (McGennis et al. 1994).

Four steps are involved in the Superpave mixture design: (1) materials selection; (2) design aggregate structure; (3) design asphalt content; and (4) moisture-susceptibility evaluation. The mixture design is based on compaction of mixes with the Superpave gyratory compactor and measurement and analysis of the resulting mixture's volumetric properties. The effort for compaction of the hot mix asphalt samples is specified in terms of the number of gyrations. Based on the intended use of the resulting mixture for application in the field, four levels of compaction are specified in the mix design (Table 6).

Table 6. Superpave Gyratory Compaction Effort

Design ESALs[a] (millions)	Compaction parameters			Typical roadway application[b]
	$N_{initial}$	N_{design}	N_{final}	
<0.3	6	50	75	Applications include roadways with very light traffic volumes, such as local roads, county roads, and city streets where truck traffic is prohibited or at a very minimal level. Traffic on these roadways would be considered local in nature, not regional, intrastate, or interstate. Special-purpose roadways serving recreational sites or areas may also be applicable to this level
0.3 to <3	7	75	115	Applications typically include many collector roads or access streets. Medium-trafficked city streets and majority of county roadways may be applicable to this level
3 to <30	8	100	160	Applications include many two-lane, multilane, divided, and partially or completely controlled access roadways. Among these are medium to highly trafficked city streets, many state routes, U.S. highways, and some rural interstates
<30	9	125	205	Applications include vast majority of U.S. interstate system, both rural and urban in nature. Special applications such as truck-weighing stations or truck-climbing lanes on two-lane roadways may also be applicable to this level

Note: When specified by agency and top of design layer ≥ 100 mm from pavement surface and estimated design traffic level ≥ 0.3 million ESALs, decrease estimated design traffic level by one, unless mixture will be exposed to significant mainline construction traffic prior to being overlaid. If less than 25% of construction lift is within 100 mm of surface, lift may be considered to be below 100 mm for mixture design purposes. When estimated design traffic level is between 3 and <10 million ESALs, agency may, at its discretion, specify $N_{initial}$ at 7, N_{design} at 75, and N_{final} at 115.

[a]Anticipated project traffic level expected on design lane over 20-year period. Regardless of actual design life of roadway, determine design ESALs for 20 years.

[b]As defined by *A Policy on Geometric Design of Highways and Streets* (1994), AASHTO.

Table 7. Superpave HMA Volumetric Design Requirements

Design ESALs (millions)	Required density (% of theoretical maximum specific gravity)			Voids in mineral aggregate (%), minimum Nominal Maximum Aggregate Size (mm)					Voids filled with asphalt (%) minimum	Dust-to-binder ratio
	$N_{initial}$	N_{design}	N_{max}	37.5	25.0	19.0	12.5	9.5		
<0.3	≤91.5								70–80	
0.3 to <3	≤90.5								65–78	
3 to <10										
10 to <30										
≥30	≤89.0	96.0	≤98.0	11.0	12.0	13.0	14.0	15.0	65–75	0.6–1.2

Within each of these levels of compaction are three points at which the mixture is evaluated:

- $N_{initial}$—Number of initial gyrations: This parameter indicates a tender mix during field compaction, caused by either an inappropriate gradation or excessive asphalt content. It is undesirable for the mix to achieve a high degree of compaction at a low number of gyrations.
- N_{design}—Design number of gyrations: The number of gyrations for which the mix is designed.
- N_{final}—Final number of gyrations: The number of gyrations the mixtures are compacted to. It is undesirable for the mix to obtain less than 2% air voids at this point as this would indicate long-term instability under traffic.

Table 7 shows the Superpave specification for the volumetric properties.

The final step in the Superpave mix design system is an evaluation of the moisture sensitivity of the design asphalt mixture. Design asphalt mixtures having tensile strength retained ratios, as defined in AASHTO T-283, less than 0.8 may be sensitive to moisture damage. While moisture damage is not a distress form by itself, it can accelerate the development and propagation of actual forms of distress.

Future Trend

It is very challenging to discuss future trends at present, especially long-term future trends, as technology development and implementation are much faster than a few decades ago. Nevertheless, some short-term and/or medium-term trends can be envisioned from the research achievements of the last few years and especially the recent National Cooperative Highway Research Program (NCHRP) research projects. Different views are possible of future trends in such areas as performance-related specifications, new test methods, and development of new and better performance models. We chose the performance models as the linking string for our discussions according to the following considerations: (1) the intended performance models in the Superpave mix design are still not available for adoption; (2) the performance models link mix design with pavement design; and (3) mix design and pavement design should be an integrated system.

Near-Term Future Trends

The near-term future trends are to enhance the Superpave mix design methods, especially the Superpave performance model and analysis system. Better distress models and advanced materials characterization models are anticipated for predicting the performance of asphalt concrete, and the models' corresponding laboratory characterization methods, especially the Simple Performance Tests (Goodman et al. 2002; Kaloush and Witczak 2002;

Pellinen and Witczak 2002), will be ready for adoption in the next 5 to 10 years. A significant feature of all these future trends is the integration of the mix design and pavement structural design into a consistent system, where common distress models of rutting and fatigue cracking and the advanced materials characterization models provide the links. This can be supported by the fact that the state-of-the-practice performance models and supporting test methods identified by NCHRP Project 1-37A, "Development of the 2002 Guide for the Design of New and Rehabilitated Pavement Structures: Phase II," might be adapted to the requirements of the Superpave mix design and analysis method, which is planned to be completed in 2005. In this way, materials engineers and pavement design engineers may work together to determine a final mix that is specific to the structural, traffic, and environmental conditions of a project. It is anticipated that SPTs will be the major test methods to calibrate the models and screen problem mixes. These trends are supported by the research activities and interim achievements of the following NCHRP projects: Project 9-19, "Superpave Support and Performance Models Management;" Project 9-29, "Simple Performance Tester for Superpave Mix Design;" and Project 9-30, "Experimental Plan for Calibration and Validation of HMA Performance Models for Mix and Structural Design." In addition, better test methods may be developed for evaluating moisture damage, as evidenced by the research for Project 9-34, "Improved Conditioning Procedure for Predicting the Moisture Susceptibility of HMA Pavements."

Some requirements on mixture volumetric characteristics such as VMA and VFA, gradation blending such as the Restricted Zone, and aggregate characteristics resulting from better imaging-based test methods may be changed as a result of the research efforts of Project 9-31, "Air Void Requirements for Superpave Mix Design"; Project 9-25, "Requirements for Voids in Mineral Aggregate for Superpave Mixtures"; Project 9-14, "Investigation of the Restricted Zone in the Superpave Aggregate Gradation Specification"; and Project 4-30, "Test Methods for Characterizing Aggregate Shape, Texture, and Angularity."

Considering the progress of these projects and the efforts of state DOTs, the new trends may be implemented in the next 5 to 10 years. However, it may take much longer to implement the advanced materials characterization models (rigorous constitutive models) to be selected by NCHRP Project 9-19. It should be noted that performance models in the near-term future trends are mainly empirical models, as no rigorous constitutive models are available to permit consistent stress-strain analysis. Complex boundary value problems to incorporate phenomena such as thermal-mechanical coupling, the effect of damage resulting from microcracking on rutting, and the effect of cumulative permanent deformation on the initiation of fatigue cracking cannot be solved. This situation favors the adoption of SPE, where the stress/strain conditions are simple.

Medium-Term Future Trends

It is anticipated that the complex properties of asphalt concrete will be better understood as phenomena, both qualitatively and quantitatively, in the near-term future, which will prepare the ground for the medium-term future trends. During the medium term, rigorous constitutive models—that is, the SHRP type of model (Monismith et al. 1994) will be developed that may include thermal mechanical coupling, damage evolution and healing (Kim et al. 2001), and the microstructure (as internal variables) of the material. In addition to the fundamental tests, simulative tests will be more widely used in conjunction with analytical tools, as the constitutive models and computational technology permit numerical simulation of complex boundary-value problems. During this period, the results from micromechanics characterization, simple performance tests, and simulative tests will drastically enhance the understanding of the performance (or the distress) mechanisms. However, at this stage the models are still mainly phenomena models, which may include microstructural quantities as internal variables. Numerical techniques, including the finite-element method (Kose et al. 2000; Masad et al. 2001) and the boundary element method (Bjorn et al. 2002), will provide tools to analyze complex boundary-value problems. It is therefore possible to simulate scaled tests to provide more accurate calibration of the models, predictions of the pavement performance (full-scale), or even field performance if environmental conditions can be better characterized as realistic thermal-mechanical coupled models may become available.

During this medium term, the advanced models are mainly in the continuum scheme [that is, Monismith et al. (1994); Baburamani (1999); Ghuzlan and Carpenter (2000); Seibi et al. (2000); Kim et al. (2001)], which may to a certain degree incorporate internal variables to represent the internal structure of the mixture. The continuum type of constitutive model is numerically easy to implement, doesn't have restrictive requirements on computational capability such as memory, and keeps pace with the overall computational capability anticipated in the next 10 to 15 years. There is no drastic change in mix design practice. This situation may last for 10 years or more. Experimentwise, both simple performance tests and simulative tests (Kenis and Wang 1997; Brown et al. 2001a, 2001b; Crince and Baladi 2002; Zhang et al. 2002) may be adopted. However, it is anticipated that simulative tests will gain more applications due to their wider availability and the confidence that practitioners have gained in them in the last two decades. Tomography imaging and digital image analysis methods, such as the pattern recognition method by Wang et al. and the digital image correlation method by Seo et al. (2002) and Wen and Kim (2002) make it possible to measure strain fields in a specimen rather than the strains at one or a few points. It is believed that the strain field measurements will present much more information for advanced model calibration. Of the medium-term future trends, development of analytical methods to obtain fundamental properties out of the simulative tests will be an important effort.

Long-Term Future Trends

The long-term trends are predicted based on current individual research efforts other than the NCHRP research efforts. It is hard to predict the trends timewise, as more research is needed to stage the long-term trends based on better understanding of the fundamental mechanism of the various types of distresses. Before proceeding to the discussion of the long-term trends, an understanding of the limitations of the current methods will be helpful.

In the WesTrack project, the coarse mix didn't perform as well as the fine mix. However, the coarse mix at the NCAT performed better than the fine mix. These contradictory discoveries indicate that something (that is, film thickness, void size distribution, relative stiffness of mastics to aggregate), or in general, some parameters about the internal structure of the material, was missing in the mix design procedure. In this regard, micromechanics (Chang and Meegoda 1997, 1999; Buttlar et al. 1999; Li et al. 1999; Krishnan and Rao 2000; Buttlar and You 2001; Guddati et al. 2002) and microstructure characterization (Kose et al. 2000; Zhai et al. 2000; Masad et al. 2001; Wang et al. 2001), may present a breakthrough in enhancing the development of the performance-related mix design method. Better internal structure quantities characterized from tomography imaging may provide an additional criterion (Wang et al. 2001, Wang et al. unpublished manuscript, 2002) to otherwise differentiate the same mixes.

As research on micromechanics and microstructure characterization is currently considered fundamental research, efforts in this area are still at the level of "free contributions" by university professors. However, the micromechanics and microstructure characterization approach is the method most suited to asphalt mixtures. As a mixture of asphalt binder, aggregates, and voids, the microstructure of asphalt concrete is complicated, which is related to the gradation of aggregates, the orientation and number of contacts of aggregate particles, the properties of the aggregate-binder interface, the void-size distribution, and the interconnectivity of voids. As a result, the properties of asphalt concrete are very complicated. However, the properties of its components are relatively much less complicated and easier to characterize.

For example, aggregates can be considered linear elastic; asphalt binder can be considered viscoelastic/viscoplastic; and the resistance from air voids is negligible. Therefore, if the microstructure of asphalt concrete can be obtained, its properties can be computed from the properties of its components and its microstructure. In recent years, X-ray computed tomography has become a reliable tool in obtaining the microstructure of asphalt concrete. This technique, in conjunction with numerical techniques such as FEM and/or the DEM, presents promising perspectives to directly use the microstructure of asphalt concrete and the properties of its components to compute the effective properties of asphalt concrete to simulate the SPT and simulative tests.

The long-term future trends can be predicted to feature the application of computational heterogeneous mechanics and microstructure characterization. In these long-term trends, mix design is computation-based for all the performance models. The pavement structure design and mix design are thoroughly integrated because (1) the phenomena models (resulted from medium-term development) are available for performing structural analysis and prediction of performance (distress) using the macroscopic models; and (2) the mix microstructure is available to allow computational simulation (for example, large-scale parallel computation) of the performance tests at the mix design level. It is anticipated that the technology for the new generation of mix design methods will become matured and widely available in about 10 to 15 years. However, it may take another 10 to 15 years for the adoption of this new technology.

Revolutionary Change

Modified Binder

The enhancement of binder properties through binder modifications (Chollar and Memon 1997; Isacsson and Lu 1999; Panda and Mazumdar 1999; Wegan and Brule 1999; Fee 2001; Murayama and Himeno 2002) will also be important, and even more important than any of the mix design methods. The reason is that most of the problems in asphalt concrete result from the complex and poor properties of asphalt binder. It is anticipated that the true revolution will come from this area rather than other areas of mix design. For example, if the molecular structure of a binder can be optimized so that the binder will be less sensitive to temperature changes, then we may not have the rutting and fatigue cracking distresses. With recent development in nanotechnology, it may become realistic that binders of required properties may be manufactured through the use of nanotechnology.

References

Abbott, J. W. (1902). *Year book*, U.S. Department of Agriculture, Washington, D.C.

Asphalt Institute. (1956). *The asphalt handbook*, Manual Series No. 4 (MS-4), College Park, Md., 46–48.

Baburamani, P. (1999). "Asphalt fatigue life prediction models—A literature review." *Research Rep. ARR 334*, ARRB Transport Research Ltd., Vermont South, Victoria.

Bahia, H. U., and Anderson, D. A. (1994). "The pressure aging vessel (PAV): A test to simulate rheological changes due to field aging." *Special technical publication 1241*, ASTM, Philadelphia.

Bahia, H. U., and Anderson, D. A. (1995). "The SHRP binder rheological parameters: Why are they required and how do they compare to conventional properties." *Preprint Paper Number 950793*, Transportation Research Board, Washington, D.C.

Bjorn, B., Boonchai, S., and Reynaldo, R. (2002). "Prediction of the viscoelastic response and crack growth in asphalt mixtures using the boundary element method." *2002 TRB CD*, Transportation Research Board, Washington, D.C.

Brown, E. R., Kandhal, P. S., and Zhang, J. N. (2001a). "Performance testing for hot mix asphalt (Executive Summary)." *Report No. 01-05A, NCAT.*

Brown, E. R., Kandhal, P. S., and Zhang, J. N. (2001b). "Performance testing for hot mix asphalt." *Rep. No. 01-05, NCAT.*

Buttlar, W. G., Bozkurt, D., Al-Khateeb, G. G., Waldhoff, A. S., Shashidhar, N., and Shenoy, A. (1999). "Understanding asphalt mastic behavior through micromechanics (with discussion and closure)." *Transportation Research Record 1681*, Transportation Research Board, Washington, D.C., 157–169.

Buttlar, W. G., and You, Z. (2001). "Discrete element modeling of asphalt concrete: Microfabric approach." *Transportation Research Record 1757*, Transportation Research Board, Washington, D.C., 111–118.

Chang, K.-G., and Meegoda, J. N. (1997). "Micromechanical simulation of hot mix asphalt." *J. Eng. Mech.*, 123(5), 495–503.

Chang, G. K., and Meegoda, J. N. (1999). "Micromechanical model for temperature effects of hot-mix asphalt concrete." *Transportation Research Record 1687*, Transportation Research Board, Washington, D.C., 95–103.

Chollar, B., and Memon, M. (1997). "The search for the optimal asphalt." *Public Roads*, 61(2), 52–53.

Christensen, D. W., and Anderson, D. A. (1992). "Interpretation of dynamic mechanical test data for paving grade asphalt cement." *Journal of AAPT*, 61.

Crince, J. E., and Baladi, G. Y. (2002). "Field and laboratory characterization of asphalt mixes for the design of flexible pavements." *2002 TRB CD*, Transportation Research Board, Washington, D.C.

Dept. of the Army. (1948). "Investigation and control of asphalt paving mixtures." *Tech. Memorandum No. 3-254*, Corps of Engineers, Mississippi River Commission, U.S. Army Engineers Waterways Experiment Station, Vicksburg, Miss.

Dept. of the Army. (1950a). "Effect of traffic with high pressure tires on asphalt pavements." *TM 3-312*, U.S. Army Engineers Waterways Experiment Station, Vicksburg, Miss.

Dept. of the Army. (1950b). "Effect of traffic with small high pressure tires on asphalt pavements." *TM 3-314*, U.S. Army Engineers Waterways Experiment Station, Vicksburg, Miss.

Edwards, L. N. (1918). "Proportioning the materials of mortars and concretes by surface area of aggregates." *Proc., American Society for Testing Materials*, Vol. XVIII, Part II, Technical Papers, American Society for Testing Materials, Philadelphia, 235–283.

Fee, F. (2001). "Extended-life asphalt pavement: New approaches to increase durability." *TR News*, 215, 12–14.

Foster, C. R. (1982). "Development of Marshall procedures for designing asphalt paving mixtures." *Information Series 84*, National Asphalt Pavement Association, Riverdale, Md.

Ghuzlan, K. A., and Carpenter, S. H. (2000). "An energy-derived/damage-based failure criterion for fatigue testing." *2000 TRB CD*, Transportation Research Board, Washington, D.C.

Goodman, S. N., Hassan, Y., and El Halim, A. (2002). Shear properties as viable measures for characterization of permanent deformation of asphalt concrete mixtures. *TRB CD*, Transportation Research Board, Washington, D.C.

Guddati, M. N., Feng, Z., and Kim, R. Y. (2002). "Towards a micromechanics-based procedure to characterize fatigue performance of asphalt concrete." *TRB CD*, Transportation Research Board, Washington, D.C.

Hveem, F. N. (1932). "Determination of oil content for oil road mixes by surface area analysis." *California Highways and Public Works*, 10(October–November), 8–11, 22, 23, and 28.

Hveem, F. N. (1942). "Use of the CKE as applied to determine the required oil content for dense graded bituminous mixtures." *Proceedings*, Vol. 13, Association of Asphalt Paving Technologists, St. Paul, Minn. 9–40.

Isacsson, U., and Lu, X. (1999). "Laboratory investigations of polymer modified bitumens." *J. Assoc. Asphalt Paving Technol.*, 68, 35–63.

Kaloush, K. E., and Witczak, M. W. (2002). "Simple performance test for permanent deformation of asphalt mixtures." *TRB CD*, Transportation Research Board, Washington, D.C.

Kandhal, P. S., and Koehler, W. S. (1985). "Marshall mix design practice: Current practices." *Asphalt paving technology*, Vol. 54, Association of Asphalt Paving Technologists, St. Paul Minn., 284–314.

Kenis, W., and Wang, W. J. (1997). "Calibrating mechanistic flexible pavement rutting models from full scale accelerated tests." *Proc., 8th Int. Conf. on Asphalt Pavement*, Seattle.

Kim, Y., Kim, Y. R., Little, D. N., Lytton, R. L., and Williams, D. (2001). "Microdamage healing in asphalt and asphalt concrete, volume II: Laboratory and field testing to assess and evaluate microdamage and microdamage healing." *FHWA-RD-98-142, Research Rep. 7229*, Washington, D.C.

Kose, S., Guler, M., Bahia, H. U., and Masad, E. (2000). "Distribution of strains within hot-mix asphalt binders: Applying imaging and finite-element techniques." *Transportation Research Record 1728*, Transportation Research Board, Washington, D.C., 21–27.

Krchma, L. C., and Gagle, D. W. (1974). "A U.S.A. history of asphalt refined from crude oil and its distribution." *Asphalt Paving Technol.*, 43A, 25–88.

Krishnan, J. M., and Rao, C. L. (2000). "Mechanics of air voids reduction of asphalt concrete using mixture theory." *Int. J. Eng. Sci.*, 38, 1331–1354.

Labatut, E. S., and Lane, W. J. (1950). *Highways in our national life: A symposium*, Princeton Univ. Press, Princeton, N.J., 95.

Li, G., Li, Y., Metcalf, J. B., and Pang, S.-S. (1999). "Elastic modulus prediction of asphalt concrete." *J. Mater. Civ. Eng.*, 11(3), 236–241.

Masad, E., Somadevan, N., Bahia, H. U., and Kose, S. (2001). "Modeling and experimental measurements of strain distribution in asphalt mixes." *J. Transp. Eng.*, 127(6), 477–485.

McGennis, R. B., Shuler, S., and Bahia, H. U. (1994). "Background of superpave asphalt binder test methods." FHWA, *Rep. No. FHWA-SA-94-069*, Federal Highway Administration, Washington, D.C.

Monismith, C. L., et al. (1994). "Permanent deformation response of asphalt aggregate mixes." *SHRP-A-415*, Strategic Highway Research

Program, National Research Council, Washington, D.C.

Murayama, M., and Himeno, K. (2002). "Effects of microstructures on the deformation characteristics of modified asphalt mixtures at high temperature." *TRB 2002 CD,*

New York Section of the Society of the Chemical Industry. (1907). *Recent progress in the asphalt paving industry*, May 17.

Panda, M., and Mazumdar, M. (1999). "Engineering properties of EVA modified bitumen binder for paving mixes." *J. Mater. Civ. Eng.,* 11(2), 131–137.

Pellinen, T. K., and Witczak, M. W. (2002). "Use of stiffness of hot-mix asphalt as a simple performance test." *TRB CD,* Transportation Research Board, Washington, D.C.

Robertson, S. H. (1958). "The pavement of presidents." *Asphalt Inst. Q.,* 10(2).

Seibi, A. C., Sharma, M. G., Ali, G. A., and Kenis, W. J. (2000). "Constitutive relations for asphalt concrete under high rates of loading." *TRB CD,* Transportation Research Board, Washington, D.C.

Seo, Y., Kim, R. Y., Witczak, M. W., and Bonaquist, R. (2002). "Application of the digital image correlation method to mechanical testing of asphalt-aggregate mixtures." *TRB CD,* Transportation Research Board, Washington, D.C.

Spencer, H. (1979). "History of asphalt paving." *Bituminous materials: Asphalts, tars, and pitches, Vol. II: Asphalts,* A. J. Hoiberg, ed., Krieger, New York, 27–58.

Stanton, T. E., (1927). "The 'oil mix' method," *California Highways and Public Works*, November.

Stanton, T. E., and Hveem, F.-N. (1934). "Role of the laboratory in the preliminary investigation and control of materials for low cost bituminous pavements." *Proc., Highway Research Board*, Part II, Washington, D.C.

Union Oil Co. (1912). *Paving and Road Construction Bulletin*, Union Oil Company.

Vallerga, B. A., and Lovering, W. R. (1985). "Evolution of the Hveem stabilometer method of designing asphalt paving mixtures." *Asphalt Paving Technol.,* 54, 243–265.

Wang, L. B., Frost, J. D., and Shashidhar, N., (2001). "Microstructure study of Westrack mixes from x-ray tomography images." *Transportation Research Record 1767,* Transportation Research Board, Washington, D.C. 85–94.

War Department. (1943). *Comparative laboratory tests on rock asphalts and hot mix asphaltic concrete surfacing materials,* Corps of Engineers, U.S. Army, U.S. Engineer Office, Tulsa District, Tulsa, Okla., October.

Wegan, V., and Brule, B. (1999). "The structure of polymer modified binders and corresponding asphalt mixtures." *J. Assoc. Asphalt Paving Technol.,* 68, 64–88.

Welty, E. M., and Taylor, F. J. (1958). *The black bonanza*, McGraw-Hill, New York.

Wen, H. F., and Kim, R. Y. (2002). "A simple performance test for fatigue cracking of asphalt concrete based on viscoelastic analysis of indirect tensile testing and its validation using Westrack asphalt mixtures." *TRB CD,* Transportation Research Board, Washington, D.C.

White, T. D. (1985). "Marshall procedures for design and quality control of asphalt mixtures." *Asphalt Paving Technol.,* 54, 265–284.

Zhai, H., Bahia, H. U., and Erickson, S. (2000). "Effect of film thickness on rheological behavior of asphalt binders." *Transportation Research Record 1728,* 7–14.

Zhang, J. N., Cooley, L. A., and Kandhal, P. S. (2002). "Comparison of fundamental and simulative test methods for evaluating permanent deformation of hot mix asphalt." *TRB CD,* Transportation Research Board, Washington, D.C.

American Society of
Civil Engineers **1852 – 2002**
Building a Better World

150th Anniversary
Paper

Review of Stabilization of Clays and Expansive Soils in Pavements and Lightly Loaded Structures—History, Practice, and Future

Thomas M. Petry, P.E., F.ASCE[1], and Dallas N. Little, P.E., F.ASCE[2]

Abstract: Expansive clay soils—those that change significantly in volume with changes in water content—are the cause of distortions to structures that cost taxpayers several billion dollars annually in the United States. Much has been learned about their behavior over the past 60 years, and relatively successful methods have been developed to modify and stabilize them. This paper reviews some of the key advances developed over the past 60 years in improving our understanding of the nature and methods of modifying and stabilizing expansive clay soils. The state of the practice in stabilization is presented, and practical and research needs to help improve the state of the practice are discussed.

DOI: 10.1061/(ASCE)0899-1561(2002)14:6(447)

CE Database keywords: Stabilization; Clays; Expansive soils; Earth structures; Distortion.

Introduction

For centuries humankind has wondered at the instability of earth materials, especially clays: one day they are dry and hard, and the next wet and soft. Clays have always presented problems for lightly loaded structures, including pavements, by consolidating under load and by changing volumetrically along with seasonal moisture variation. The results are usually excessive deflections and differential movements resulting in damage to foundation systems, structural elements, and architectural features. In a significant number of cases the structures become unusable or uninhabitable. Even when efforts are made to improve clay soils, the lack of appropriate improvements sometimes results in volumetric changes that are responsible for billions of dollars of damage each year, as discussed by Wiggins et al. (1978). Therefore, as a profession we need to do more to develop our knowledge of proven methods to deal with expansive soils, support research for further improvement of these methods, and work toward better quality control and quality assurance for their application.

Looking back at 150 years of civil engineering, we realize that early "civil engineers" learned to cope with clay soil behavior by trial and error. But how far have we come? Have we successfully used scientific principles to define the behavior of these problematic soils? Can we stabilize them, or at least successfully modify

their behavior? This paper identifies the approximate state of the knowledge and the directions toward which we should consider moving in order to properly stabilize clay soils.

This paper highlights some key advances in expansive clay stabilization during the past 60 years. The writers have attempted to include those individuals who have made major contributions, but also wish to acknowledge the efforts of many others who have been unintentionally overlooked.

Our Fundamental Knowledge of Clay Behavior

Historical Perspective

As we look back through approximately 60 years of geotechnical literature, we notice significant attention to clay behavior, probably because damage resulting from clay soils contributes to such a huge proportion of overall structural damage, as described by Wiggins et al. (1978). Arthur Casagrande (1932) wrote one of the earliest treatises, which discussed clay structure and its impact on foundation engineering. Simpson (1934) published a paper describing experiences in foundations on clay in Texas. These were followed by observations on moisture changes beneath Texas pavements and the effects of such changes. Grim (1953) published his landmark book on clay mineralogy, Skempton (1953) described the effect of colloidal activity of clays, and Palit (1953) described a method to determine swell pressure of black cotton soils. Also, Jennings and Knight (1957) described how heave can be predicted using oedometer test data, and Felt (1953) described the influence of vegetation on the moisture content of clays.

In the 1950s a series of international conferences on soil mechanics and foundation engineering provided opportunities for researchers and practitioners to describe their experiences with clay. Altmeyer (1955) discussed engineering properties of expansive clays, and Holtz and Gibbs (1956) provided further information on this subject. Near the end of that decade, McDowell (1959)

[1]Professor of Civil Engineering, Univ. of Missouri-Rolla, 1870 Miner Circle, Rolla, MO 65409-0030.

[2]E.B. Snead Chair Professor of Civil Engineering, Texas A&M Univ., Suit 603 CE/TTI Building, College Station, TX 77843-3135.

Note. Associate Editor: Antonio Nanni. Discussion open until May 1, 2003. Separate discussions must be submitted for individual papers. To extend the closing date by one month, a written request must be filed with the ASCE Managing Editor. The manuscript for this paper was submitted for review and possible publication on June 10, 2002; approved on June 19, 2002. This paper is part of the ***Journal of Materials in Civil Engineering***, Vol. 14, No. 6, December 1, 2002. ©ASCE, ISSN 0899-1561/ 2002/6-447–460/$8.00+$.50 per page.

published a paper on the relationships between laboratory testing and design of pavements and other structures on expansive soils. The Texas Method of pavement design was a result of McDowell's work. This was one of the first steps toward introducing mechanics techniques in pavement design. In essence, McDowell used Boussinesq's single-layer theory to calculate minor and major principal stresses at locations under a design wheel load in order to define a Mohr-Coulomb envelope that describes the state of stress within the subgrade layer. This stress state was compared to the Mohr-Coulomb shear failure envelope of the subgrade soil in question, which had been subjected to an extended period of capillary soak to simulate a design state in the pavement. The information from that analysis was used to determine the required pavement thickness. This approach was a forerunner of mechanistic design and is still used. The end of the 1950s established a basis for the design of lightly loaded structures and pavements on clay subgrades.

The decade of the 1960s saw a notable increase in significant research and field experiments in expansive clays. Lambe (1960) developed a method to identify swelling clays and to assess their swelling potential via the potential vertical rise (PVR) meter. Seed et al. (1962a) discussed swelling characteristics of compacted clays. The same year Jennings (1961) compared laboratory predictions and field observations of heave in desiccated soils. Seed et al. (1962b) published information on how to predict swell in compacted clays. The Colorado Department of Highways and the University of Colorado (1964) published a review of the literature on swelling soils.

An important milestone was reached in the report by the Building Research Advisory Board (BRAB 1968) establishing criteria for selection and design of residential slabs on grade. This publication contained the first widely accepted design criteria for lightly loaded foundations on expansive clays and is still an authoritative reference. Designs using the BRAB system were considered too conservative for most builders and owners to accept, and somewhat less conservative designs began to be used that were based to some extent on the BRAB designs. Efforts to characterize expansive clays and to design for expansive clay properties were the subject of Johnson's (1969) review of the literature on expansive clay soils. Significant strides were made in the 1960s, and our understanding of clay soil behavior and the steps necessary to mitigate its impacts in design began to become more clear.

The decade of the 1970s opened new doors in the design of lightly loaded foundations on clays, and a new concept of how to determine the potential moisture change and swell of desiccated clay emerged. This concept—soil suction—is considered to be a negative pore-water pressure that is a result of the clay's inherent need for moisture to balance the physicochemical energy levels in the soil. Relationships were established between measured soil suction in a clay subgrade and resulting swell in the presence of moisture, and the approach became part of some design protocols.

Methods to expediently measure soil suction were developed and validated as well. Many of those from North America who contributed significantly to this effort in the 1970s and 1980s are included in the following discussion. Some of their numerous publications are listed in the reference section, and their contributions have moved the state of the art forward substantially.

Chen (1973), Lytton (1994, 1995, 1997), and Wray (1978, 1984, 1987) established new approaches for slabs on grade built on expansive clays. Johnson (1973, 1977), Johnson and Snethen (1978), McKeen (1980, 1981), Snethen (1979, 1984), and Snethen et al. (1975) developed the concept of soil suction and

methods to measure it. Vijayvergiya and Ghazzaly (1973), among others, started the process of creating predictive models. Tucker and Poor (1978) studied the behavior of existing slabs and assessed the causes of damage in them. Mitchell (1976), among his other publications, provided the profession with a landmark and comprehensive text on soil behavior, containing a significant treatment of expansive clays. By the end of the 1980s, geotechnical engineers began to feel they had a better handle on expansive clays and how to design for them.

Seven international conferences on expansive clay soils have been convened since the mid 1960s. The first two were held at Texas A&M University in 1965 and 1969; the third at the Technion Israel Institute of Technology in 1973; the fourth in Denver in 1980; the fifth in Adelaide, Australia, in 1984; the sixth in New Delhi, India, in 1987; and the seventh, an international conference on expansive soils, in Dallas in 1992. Since that time attention has shifted to include understanding the behavior of partially saturated soils and expansive clays.

Early in the 1980s a widely attended national workshop, held at Colorado State University, was hosted by Nelson and Miller, who published the results of the workshop in their book on expansive soils (Nelson and Miller 1992). The Post-Tensioning Institute (PTI) published its first manual on the design of posttensioned slabs on expansive clays (PTI 1980), and Chen (1988) presented his views on foundations in expansive soils. Subsequent to these, the major written contributions have addressed the behavior of unsaturated soils. The most complete of these is the text by Fredlund and Rahardjo (1993). Most opportunities for interaction on the technical aspects of expansive soils have recently been sponsored by the geotechnical group of the ASCE and its modern counterpart, the Geoinstitute. ASCE's Geotechnical Special Publication No. 39, edited by Houston and Wray (1993); No. 68, edited by Houston and Fredlund (1997); and No. 115, edited by Vipulanandan et al. (2001), contain some of the most recent and widely published information on expansive soil behavior.

Dempsey et al. (1986) used soil suction to estimate moisture content in the subgrade soils in a moisture equilibrium model. Such a model can be used as part of a mechanistic empirical pavement design scheme. Since the resilient modulus of the subgrade is a key parameter in pavement design and is highly sensitive to moisture, the ability to predict moisture content at equilibrium is an important step. The first step is to equate soil suction with the differences between the negative pore pressure of a soil at a certain position above the water table and the product of soil compressibility and overburden pressure. Compressibility can be calculated as a function of soil plasticity; overburden is calculated in the traditional manner; and negative pore pressure due to the position gradient is the product of position head above the water table and the unit weight of water. Soil suction is directly calculated, and since a unique relationship between moisture and suction exists for a specific soil, the moisture content can be calculated. The resilient modulus is thus determined.

Lytton presented three landmark papers on the subjects of moisture movement in expansive clays (Lytton 1994), foundations and pavements on unsaturated soils (Lytton 1995), and design of engineering structures on expansive soils (Lytton 1997). Lytton (1994) described how simple laboratory tests can be used to determine important properties of expansive soils, including the compression indices due to matrix suction and mean principal stress, the slope of suction versus water content curve, and unsaturated permeability and diffusivity. The simple tests referenced in the paper are Atterberg limits, hydrometer, water content, dry density, and sieve analysis. Lytton further described the dependence of the prediction of moisture movement on the Thornthwaite moisture index and the unsaturated permeability of soil. Lytton's work also discussed the unusually destructive effect that trees have when they grow near the edge of a foundation and the effectiveness of vertical root and moisture barriers around the perimeter of the foundation in reducing the moisture variation distance and the differential movement.

Lytton (1997) described constitutive equations for volume change that can be used in design and explained how to use classification of profiles and suction patterns in design. He described ranges in Gibbs free energy on a logarithmic scale that define the depth of the active zone. Although Lytton explained a detailed protocol to assess the active zone, he also offered some simple and direct clues by which to predict the suction profile and indicate the depth of the moisture active zone. Lytton (1995) described the necessity of using characteristics of unsaturated soils in numerical computation procedures in order to properly design structures on expansive unsaturated clays. The material properties of these soils are stress and suction dependent and are variable. These soils undergo large strains under service conditions in the field. These characteristics of unsaturated soils make small-strain, elastic analyses generally inadequate. Realistic characterization of these soils is necessary for analysis.

McKeen and Johnson (1990) described simple, rational methods to calculate the active zone depth and slab-edge penetration distance in clay soils. This analysis was based on moisture diffusion assuming a periodic variation in surface soil function. Based on that approach, the depth of change of the active zone is a function of the maximum suction change considered significant, climate frequency, and the field diffusion coefficient. McKeen (1992) was also able to classify the heave potential of expansive soils on the basis of the ratio of suction change to moisture content change and the soil compression index. Petry et al. (1992) published a paper about their study on the effects of pretest stress environment on the swell measured. Snethen and Huang (1992) described soil suction-heave prediction methods. Houston et al. (1994) provided further information on the use of filter paper to measure suction. Also, Ridley and Wray (1996) published a review of the current theory and practices for suction measurement. By the start of the new millennium a considerable knowledge base was available for use by practicing geotechnical engineers.

Current Design Practice Accounting for Expansive Clay Soils

Although substantial progress has been made in defining the behavior of expansive clays and developing theories that can assist in the prediction of their behavior, the practice of geotechnical engineering as applied to soils has not advanced in the same way. Few practitioners are measuring soil suction and using it to determine potential volume change. Perhaps the reason for this is the lack of a widely accepted test method. The filter paper method, while very useful if applied correctly, is sensitive and somewhat inconsistent in results from laboratory to laboratory. Research is needed to improve the repeatability and efficiency of this important test.

The primary sources of data for determination of volume change potential are water content profiles, Atterberg limit profiles, and various forms of the oedometer swell tests. In addition, swell test protocols vary, and the time allowed for swell to occur is often limited. Generally, secondary or long-term swell is ignored for the sake of economy of testing. Unfortunately, this has led to numerous situations where long-term swell resulting in

significant structural damage has been overlooked. In states most affected by this phenomenon, professional technical groups are organizing to both ascertain the methods used in practice and develop appropriate standards of practice for their communities.

The types of foundation systems used on and in expansive soils have not changed markedly since the 1980s, when design of posttensioned slabs was standardized by the PTI. Conventionally reinforced slabs are used less and less because of their relative expense compared to posttensioned slabs. Deep pier-supported and structurally suspended foundation systems are used only for relatively expensive structures. In fact, the initial cost of foundation systems on expansive clay soils is the driving factor in choice of design, with long-term performance given less consideration. This is also why modification and/or stabilization of expansive soils are often not given sufficient consideration. The process of geotechnical engineering is often controlled by short-term, bottom-line economics instead of life-cycle performance or even long-term costs. It becomes imperative for today's geotechnical and materials engineers to arm themselves with a knowledge of the history of the development of methods to mitigate expansive clays and a clear understanding of the current standards of best practice. This emphasizes the importance of continuing education regarding the latest innovations and techniques and the continuation of research to improve them.

Stabilization of Expansive Clays

Historical Perspective

Engineers, architects, and contractors have tried many ways to reduce the damaging effects of expansive clays. Their actions were often based on trial-and-error approaches. They have used mechanical stabilization to the extent practical but have found it also necessary to alter the physicochemical properties of clay soils in order to permanently stabilize them. Geotechnical engineers have borrowed from the knowledge of soil scientists and geologists to find ways to improve clay behavior. Barshad (1950), a mineralogist, discussed the effects of interlayer cation types on expansion of clays. Soil scientists Allison, Kefaumer, and Roller (1953) discussed ammonium fixation in clay soils. Earlier descriptions of soil stabilization may be found listed in Huang's (1954) selected bibliography.

By the mid-1950s engineers described their successes in modifying clay behavior. Dubose (1955) described how to control heavy clays using compaction. Jones (1958) discussed stabilization of expansive clay using hydrated lime and portland cement. Taylor (1959) explained the process of ion exchange in clays. McDowell (1959) described how Texas soils were being stabilized with lime and lime-fly ash combinations. Most of the basic ideas in use today regarding improvement of clay behavior had been published by 1960. The means for improving properties would come from chemical modification of the soils and/or mechanical modification via compaction.

Hilt and Davidson (1960) discussed lime fixation in clays heralding a decade of advancing technology in lime and portland cement stabilization of clay soils. Mitchell and Hooper (1961) described the effect of time of mellowing on the properties of lime-modified clay soils and noted that this mellowing period affected the workability and compaction response. Eades and Grim (1963) developed their quick test to determine required lime content, which is based on the concept that when a caustic, calcium-based compound is added to clay soil, a reaction occurs that is based on soil-silica and soil-alumina solubility at a high pH. Thus pH evolved as a quick test to identify the optimum lime content. Since initial lime-soil reactions occur within about 1 h of mixing, the test is a 1-h test.

Eades et al. (1963) identified lime-soil reaction products and also presented compelling data showing how stabilization of soils of different minerologies may indeed require vastly different amounts of lime. In their study they compared a Georgia kaolinite that required 4% hydrated lime for stabilization with a Wyoming montmorillonite that was not reactive until 10% hydrated lime was added. These data further bolstered the need for mixture design and the value of a quick test to assess lime demand. Glenn and Handy (1963) added to the literature of understanding of basic lime-soil reactions. Anday (1961, 1963) provided further evidence of the importance of properly curing lime-stabilized soils to fully develop the pozzolanic reaction, which is the key to strength gain and long-term stability.

In the 1960s, Iowa State University became a leader in research into the stabilization of soils with lime and portland cement. Mateos (1964) described some of this early work. Davidson et al. (1965) described the effects of pulverization and lime migration in treated soils, while Townsend and Klym (1966) discussed the durability of lime-stabilized soils. Diamond and Kinter (1965) published their findings of the mechanism of soil stabilization with lime, which provided an in-depth and much needed understanding of the mechanism of how the calcium hydroxide molecules interact at the surface of the clay minerals to alter the surface and stabilize the clay. Diamond and Kinter's (1965) classic paper clarifies the need for the pozzolanic surface reaction to achieve strength gain as well as plasticity reduction in plastic clay soils, especially those that are calcium-dominated prior to stabilization.

In the late 1960s Thompson (1966, 1969) [and Thompson and Dempsey (1969)] at the University of Illinois became perhaps the most prolific writer, and certainly one of the most highly regarded leaders, in the application of stabilization to the transportation arena. Thompson's work continues today. Although Thompson's contributions are many (see citations in the reference section), one of the most important is the Thompson (1970) method of mixture design. This is the basis of the method used today by the National Lime Association and developed by Little (1999a,b).

Stabilization by high-pressure lime slurry injection began in the 1960s. Lundy and Greenfield (1968) were among the first to describe this process, and the process of deep in situ soil stabilization was described by Higgins (1965). By the end of the 1960s, lime stabilization methodologies had become some of the most widely used to tame expansive clay soils.

Arman and Munfakh (1970) described how lime can treat even highly organic soils provided that enough lime is used to overcome the organic masking effect. Other important citations in the 1970s include publications by Fohs and Kinter (1972) on the migration of lime in compacted soils, Marks and Haliburton (1972) on the acceleration of lime reactions with the addition of salt, and by Thompson (1972) on deep-plow lime stabilization for pavement construction. The National Lime Association (1991) published its first manual on lime stabilization construction in 1972. Strategies to provide deeper stabilization through lime slurry pressure injection were the subject of papers by Wright (1973). This methodology was used, along with deeper mixing of lime, on one of the largest clay improvement projects ever attempted, the construction of the Dallas/Fort Worth airport (Thompson 1972). Unfortunately, as time passed, inadequate lime contents were specified on many jobs, and as a result the process began to lose favor in the geotechnical engineering community.

The use of chemical agents other than lime to improve clay soils was introduced when Carroll and Starkey (1971), clay mineralogists, published their findings on reactivity of clay minerals with acids and alkalines. O'Bannon et al. (1976) described the use of electroosmosis and potassium to improve a clay highway subgrade. The Portland Cement Association, beginning in the 1970s, described how portland cement, heretofore relegated to low-or moderate-plasticity soils, could be used to modify and even stabilize higher-plasticity clays via the use of more efficient rotary mixing machines.

Prewatering expansive clay subgrades also received considerable attention during the 1970s. McDonald (1973) reported on the use of moisture barriers and waterproofing the surface for clays in South Dakota. At the Haifa conference on expansive soils, Lee and Kocherhans (1973) described the use of moisture barriers to stabilize clays. McKinney et al. (1974) described the effectiveness of ponding to reduce clay swell potential in a project near Waco, Texas. Steinberg (1977) described the success of ponding in a clay cut. A most interesting study (Poor 1978) combined the effects of prewetting by ponding and sprinkling with moisture barriers to stabilize the subgrade of a residential slab. Tucker and Poor (1978) reported on the factors that influence moisture stability under slabs on grade. Prewetting and water content stabilization were a proven performer by the end of the 1970s.

Terrell et al. (1979) directed the Federal Highway Administration-funded development of a national guide to stabilization. During the 1970s the U. S. Air Force developed the Soil Stabilization Index System (SSIS), which was pioneered at Texas A&M University by Epps et al. (1970). The heart of the SSIS approach was a method to select the appropriate soil stabilizer based on two well-established but simple soil index properties: plasticity index and percent passing the number 200 sieve. The SSIS was validated at the U.S. Air Force F. J. Seiler Research Lab (Currin et al. 1976), and later the Air Force developed the SSIS concept into a technical manual (Little et al. 1992).

Ford et al. (1982) reported on mix designs for lime treatment of soils from the southeastern United States. Kennedy (1988) and Kennedy and Tahmovessi (1987) provided summaries of the effective use of lime in clay soils. Little (1987) described the fundamentals of lime-soil reactions for the National Lime Association, and the Transportation Research Board (1987) published its *State of the Art Report 5* on lime stabilization of soils. Petry and Lee (1989) discussed the benefits of using quicklime slurries over hydrated lime slurries in the treatment of clays, and Petry and Wohlgemuth (1989) described the effect of degree of pulverization during mixing on the products of lime and portland cement treatment of heavy clays.

Mitchell (1986), in a Terzaghi lecture, reintroduced the profession to calamities associated with sulfate-induced heave in lime-stabilized clay soils. This landmark paper was followed by a comprehensive study by Hunter (1988), which provided a geochemical analysis of the phenomenon of sulfate-induced heave in clay soils stabilized with calcium-based, caustic stabilizers. Both papers focused on a case study of failure at Stewart Avenue in Las Vegas, Nevada.

The injection of lime to improve the behavior of clay subgrades was better defined in the 1980s. Petry et al. (1982) discussed short-term property changes that occur in soils injected with lime and fly ash. Boynton and Blacklock (1986) wrote a manual on lime slurry pressure injection published by the National Lime Association. The injection of chemicals to stabilize clay subgrades below foundations was discussed in a paper by

Blacklock and Pengelly (1988). Since then, a variety of chemicals have been tried in the injection process.

The use of deep vertical fabric moisture barriers was updated by Steinberg (1981, 1985). Summaries of how to treat expansive soil were the subject of an ASCE article by Jones and Jones (1987) and a Transportation Research Board paper by Petry and Armstrong (1989). At the beginning of the 1990s, stabilization of expansive clays was described in a book by Nelson and Miller (1992).

The phenomenon of sulfate-induced heave received much attention during the 1990s and continues to pose problems for those stabilizing soils where soluble sulfates are present, especially in the western United States (although similar problems have occurred in unlikely venues such as England). Better understanding of this phenomenon was provided by Mitchell and Dermatas (1992), Little and Petry (1992), Petry and Little (1993), Petry (1995), and Rollings et al. (1999). The result of their efforts is a set of guidelines on how to test for sulfates and how to deal with differing levels of sulfates (Little and Graves 1995) in clay soils that are to be treated with calcium-based agents such as lime, portland cement, and fly ash.

Holmquist and Little (1992) and others proposed a successful approach for stabilization with lime at the new Denver International Airport. The approach is based on the basic stoicheometrics of the expansive minerals, the time required for their formation, and the amount of water required for solubility of the components. The use of soil conductivity for location of sulfates was proposed by Bredenkamp and Lytton (1995). There is a need however to provide better field information to map the locations of sulfates so that proper treatment decisions can be made. Another problem noted in the late 1980s and early 1990s, when lime was used to treat expansive clays, was the possible leaching of beneficial calcium from the chemical environment of the clays. Studies reported by McCallister and Petry (1990) indicated that, in order to provide durability, soils must be stabilized with sufficient lime to develop pozzolanic reactions.

The utilization of slag, especially with portland cement, has taken on a much larger role in stabilization. Pozzolanic mixtures have been successfully used to stabilize soils in Australia, Europe, and the United States. Generally, such mixtures are not especially beneficial for clay soils since clay itself is a pozzolan, but since some clay soils are simply not reactive enough with lime to produce substantial strength gain, the use of a pozzolan additive provides a means of strength enhancement.

Johnson and Pengelly (1993) described the efficacy of using potassium and combinations of potassium and ammonium compounds for injection stabilization. A study by Addison and Petry (1998) discussed the process used to optimize a multiagent and multiinjected mix of potassium and ammonium to modify swell behavior. From the 1990s to the present several clay-modification agents have been introduced that are classified as hydrogen ion exchangers. Controversy remains within the engineering community over their efficacy in the modification of clay subgrades.

The subject of water content stabilization of expansive clay subgrades was a primary topic within the published literature in the 1990s. Petry (1993) reported on the short-term success of a project to moisture stabilize a clay subgrade under an existing building using an injected lime and fly ash grout curtain. Near the end of the decade, Steinberg (1998) published work on the use of geomembranes to control expansive soils in construction. The application of moisture stabilization by prewetting via injection of water, the placement of geomembrane barriers around the perimeters of slabs, or occasionally both has continued to increase to the present time.

In 2002, the National Cooperative Highway Research Program will introduce a revised version of its pavement design guide. When validated by the participating states, this will become the 2002 version of the AASHTO pavement design guide. In anticipation of this the National Lime Association has developed an approach to the design of lime-soil mixtures that will define not only the optimum stabilizer content, but also the engineering properties that can be expected from the lime-soil mixture as part of a mechanistic pavement design approach. Little (1999a,b, 2000) has established a comprehensive basis for determining the engineering properties of lime-soil mixtures based on both laboratory and field evaluations.

Current Practice of Stabilization of Expansive Clays

Stabilization of clay subgrades is a popular alternative for geotechnical engineers considering the economics of construction with expansive clay soils. They must choose among traditional stabilizers and new, nontraditional stabilizers that are not backed by as much field experience or independent research results. Geotechnical engineers do not always have the opportunity to use as complete a comparative test program as they would like. Because of uncertainties with new, nontraditional stabilizers, traditional stabilizers such as lime and portland cement are the most widely used of the chemical additives. However, the most widely used overall process is moisture stabilization in its many forms. It is imperative that geotechnical engineers stay current with knowledge of the agents and methods available to them and that the research community continue to provide independent results to validate the use of agents and methods.

Most prewetted clay subgrades are injected with water to a depth that will nearly eliminate subgrade swell potential. After this process, they are either covered with a polyethylene sheet to retain moisture, kept wet by sprinkling, or built on fairly quickly. Vertical or horizontal moisture-loss barriers have been installed in only a few cases in the past, but their use is becoming more frequent. The combined effect of a moistened subgrade and moisture barriers, when applied correctly, can be very successful. The application of moisture treatment coupled with proper degrees of compaction has been utilized extensively.

Two construction factors have strongly affected the quality of chemical stabilization: moisture and degree of pulverization. First, subgrades are rarely prewetted but are brought to the required water content during construction; often they are worked dry of optimum, resulting in incomplete chemical reactions due to lack of water. Second, the degree of pulverization is often substandard. Lime must react intimately with particles to enact the pozzolanic process; the larger the soil particles, the longer that process will take.

One of the most useful factors in overcoming sulfate-induced heave is prewetting the soil to a moisture content of three to five percentage points above optimum for the treated soils and keeping the moisture at this level until final compaction. The additional moisture maximizes the quantity of sulfates solubilized, making them available to react with calcium from lime and aluminum from clay to form the stabile mineral ettringite. That factor alone can in some cases significantly reduce the damage caused by sulfate-induced heave by forcing formation of the expansive calcium-aluminate-sulfate-hydrate compounds before compaction and by forcing a more stable (sulfate-rich) form of the expansive minerals from the outset. It is important to note that, in spite of all the concerns over sulfate-induced heave, the vast majority of lime and lime-fly ash-treated clay subgrades and lower plasticity portland cement treated subgrades never exhibit sulfate-induced heave.

State of Practice in Stabilization of Clay Subgrades and Subbases

A variety of stabilizers are available. These may be divided into three groups: traditional stabilizers (hydrated lime, portland cement, and fly ash); byproduct stabilizers (cement kiln dust, lime kiln dust, and other forms of byproduct lime); and nontraditional stabilizers (sulfonated oils, potassium compounds, ammonium chloride, enzymes, polymers, and so on). The traditional stabilizers generally rely on calcium exchange and pozzolanic reactions to effect stabilization. This is also the primary mechanism for stabilization with many of the byproducts. The nontraditional stabilizers are identified as such because they rely on a different stabilization mechanism. For example, sulfonated oils typically rely on hydrogen ion penetration into the clay lattice. Hydrogen alters the clay structure primarily by reducing its water-holding capability.

The formation of aluminum hydroxide interlayers in the low pH environment may also aid the stabilization process. Potassium ions are known to enter the clay lattice of smectite clays that exhibit high-volume changes with changes in water content and cause them to act more like Illite clays, which are less active. Ammonium compounds are also effective ion exchangers that can reduce clay activity. Enzyme stabilizers rely on organic catalysts, which rapidly carry a reaction to completion without becoming part of the reaction. Enzymes usually combine with organic molecules to form a reactant intermediary, which exchanges with the clay lattice (Scholen 1992) causing it to break down and causing a cover-up effect that prevents further sorption of water by the clay. Latex acrylic copolymers are emulsions with 40 to 60% solids, which stabilize through a nonchemical coating effect similar to stabilization with asphalt emulsions.

With the flood of nontraditional stabilizers on the market, the engineer is often perplexed as he or she must digest an array of chemical data, process explanations, and engineering data. Unfortunately, the flood of confusing data can penalize viable nontraditional stabilizers. In sifting through the maze of information, the engineer should seek a reasonable and technically satisfying explanation of the mechanism of stabilization. She or he should use engineering judgment to assess whether or not the concentration of active ingredients is reasonable. Some nontraditional stabilizers offer a valid stabilization mechanism at reasonable concentrations but are too diluted at recommended field applications to be effective. A well-designed laboratory-testing experiment at various stabilizer concentrations can be used to assess the level (additive rate) at which the stabilizer becomes effective—if it does at all.

The objective of clay stabilization may be to (1) stabilize volume change characteristics; (2) modify plasticity and improve workability; or (3) modify plasticity and volume change characteristics while substantially improving strength. Generally, the engineer seeks to achieve objective (3) because the subgrade layer must not only be volumetrically stable, but must also support traffic or building loads. The issue then becomes what protocol to use to achieve stabilization and how to validate stabilization in clay soils. The following sections briefly review the current mixture design and construction protocols advocated by the trade organizations representing the traditional stabilizers. Subsequent sections critique these protocols to assess shortcomings, information gaps, and the next step.

National Lime Association (NLA) Approach

Lime or calcium hydroxide is the most widely used chemical stabilizer for clay soil subgrades. Lime-soil reactions are complex (Diamond and Kinter 1965; Little 1995; Yusuf 2001); however, understanding of the chemistry involved and the results of field experience are sufficient to provide design guidelines for successful lime treatment of a range of soils. The sustained (and relatively slow) pozzolanic reaction between lime and soil silica and soil alumina (released in the high-pH environment) is key to effective and durable stabilization in lime-soil mixtures. Mixture design procedures that secure this reaction must be adopted.

Design of lime-stabilized mixtures is usually based on laboratory analysis of desired engineering properties. Several approaches to mix design currently exist. In addition to engineering design criteria, users must consider whether the laboratory procedures adequately simulate field conditions and long-term performance. In terms of roadway subgrades, aspects of these procedures are likely to be superseded as AASHTO shifts to a mechanistic-empirical approach.

Laboratory testing procedures include determining optimum lime requirements and moisture content, preparing samples, and curing the samples under simulated field conditions. Curing is important for chemically stabilized soils and aggregates—particularly lime-stabilized soils—because lime-soil reactions are time and temperature dependent and continue for long periods of time (even years). Pozzolanic reactions are slower than cement-hydration reactions and can result in construction and performance benefits, such as extended mixing times in heavy clays (more intimate mixing) and autogenous healing of moderately damaged layers, even after years of service. On the other hand, longer reactions may mean that traffic delays are associated with using the pavement. In addition, protocols for lime-soil mixture design must address the impact of moisture on performance.

Males states in Little et al. (2000) that lime-stabilization construction is relatively straightforward. In-place mixing (to the appropriate depth) is usually employed to add the proper amount of lime to a soil, mixed to an appropriate depth. Pulverization and mixing are used to combine the lime and soil thoroughly. For heavy clays, preliminary mixing may be followed by 24 to 48 h (or more) of moist curing prior to final mixing. This ability to mellow the soil for extended periods and then remix it is unique to lime. During this process, a more intimate mixing of the lime and the heavy clay occurs, resulting in more complete stabilization. For maximum development of strength and durability, proper compaction is necessary; proper curing is also important.

The performance of lime-stabilized subbases and bases has been somewhat difficult to assess in the current AASHTO design protocol because the measure of structural contribution—the structural layer coefficient—cannot be ascertained directly. Indirectly determined coefficients for lime-stabilized systems, however, have been found to be structurally significant. As AASHTO shifts to a mechanistic-empirical approach, measurable properties, such as resilient moduli, will be used to assess stress and strain distributions in pavement systems, including stabilized bases and subbases. These properties will be coupled with shear-strength properties in assessing resistance to accumulated deformation.

The lime industry (Little 1999a,b, 2000) has submitted a three- to four-step design and testing protocol to be considered for inclusion in the AASHTO design protocol:

Table 1. Summary of Properties from Clay Subgrades on Selected Mississippi Projects

Pavement	Structural layer	Lab UCS (kPa) (24 h capillary soak)	CBR (%) (from DCP)	Lab E_p (MPa) (24 h capillary soak)	FWD back-calculated modulus, MPa	Dielectric value (from GPR)
U.S. 45 (20 years service)	Natural clay soil	20	10	60	125	—
	Lime treated clay	1,900	133	370	1,462	—
U.S. 61 (20 years service)	Natural clay soil	50	15	0	97	—
	Lime treated clay	2,000	500	410	425	9–13
U.S. 82E (20 years service)	Natural clay soil	30	12	0	119	—
	Lime treated clay	1,950	150	200	2,466	6–8
U.S. 82W (20 years service)	Natural clay soil	10	4	0	123	—
	Lime treated clay	1,600	47	260	1,462	7–10

- Step 1: Determine optimum lime content using the Eades and Grim pH test (ASTM D-6274);
- Step 2: Simulate field conditions. Use AASHTO T-180 compaction and 7-day curing at 40°C to represent good-quality construction techniques. After curing, subject samples to 24 to 48 h of moisture conditioning via capillary rise (soak);
- Step 3: Verify compressive strength, stiffness, and moisture sensitivity, and measure unconfined compressive strengths using ASTM D-5102 methods. [For most applications, the above three steps are sufficient because design parameters such as flexural strength, deformation potential, and stiffness (resilient modulus) can be approximated from unconfined compressive strength. For more detailed (Level 2) designs, a direct measure of resilient modulus may be required]; and
- Step 4: Perform resilient modulus testing using AASHTO T-294-94 or expedited (and validated) alternatives, such as the rapid triaxial test. This protocol and its mechanistic-empirical basis provide a sound foundation for future lime-stabilization applications.

The National Lime Association (NLA) approach to mixture design was recently validated on four projects in the state of Mississippi (Yusuf et al. 2001). In these projects the performance of stabilized clay subgrade soils was evaluated via in situ testing: falling weight deflectometer (FWD) testing to determine in situ moduli; dynamic cone penetrometer (DCP) testing to determine in situ strengths of the stabilized and natural clay soils; and ground penetration radar (GPR) to assess in situ layer thicknesses (for FWD back-calculations) and in situ dielectric properties of the lime-stabilized layers.

These pavements were between 15 and 20 years old at the time of testing. Samples of the native clay were collected and subjected to the four-step NLA mixture design protocol. Engineering properties (compressive strengths and resilient moduli) from the laboratory mixture design protocol were then compared to field results. Following this comparison, the pavement structures were modeled using layered elastic theory, GPR-determined layer thicknesses, and back-calculated resilient moduli to assess the impact of the stabilized layer (typically 200 mm thick) on the pavement performance life. The impact was deemed to be substantial as it typically extended pavement life by several hundred percent. Table 1 summarizes the laboratory- and field-derived values.

An extended study of approximately 10 lime-treated subgrade soils in Mississippi and approximately 40 lime-treated subgrade soils in Texas was accomplished (Little et al. 1994). This study showed a substantial structural effect of lime stabilization after many years of service on in situ strength (derived from DCP measurements) and on resilient moduli (derived from back-calculated moduli from FWD measurements). Figs. 1 and 2 summarize the back-calculated California bearing ratio (CBR) values and resilient moduli from pavements in the Texas study.

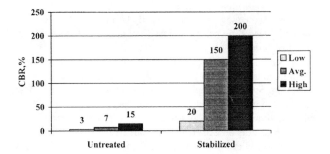

Fig. 1. CBR values (calculated from DCP data) on 40 Texas pavements

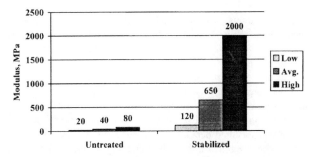

Fig. 2. Resilient moduli (from FWD back-calculations) on 40 Texas pavements

From these data it is apparent that a mixture design protocol that selects optimum lime content on the basis of optimal strength can produce strong, durable lime-stabilized pavement layers if the clay soil is lime reactive. It is also apparent that if substantial strengths and moduli can be achieved in a realistic laboratory protocol, they can be achieved in the field if good construction techniques are employed. Little (1999a,b) described structural benefits of lime stabilization in four categories: (1) strength and deformation properties, (2) resilient properties, (3) fracture and fatigue properties, and (4) durability properties. Within each category Little tabulated available data from key studies and described the current methods of measurement and the impact of lime on structural performance within each category. Little's (1999a,b) summary of structural benefits derived from lime stabilization bolster the description of benefits described on specific Mississippi and Texas pavement subgrades in this section.

Portland Cement Association Approach

Prusinski (Little et al. 2000) states that since 1915 more than 140,000 km of equivalent 7.5 m wide pavement bases have been constructed from cement-stabilized soils. Cement has been found effective in stabilizing a wide variety of soils, including granular materials, silts, and clays.

Soil-cement bases have more stringent requirements than cement-modified soil subgrades. For soil-cement bases, two types of testing have typically been used—durability tests and strength tests. The Portland Cement Association has developed requirements for AASHTO soils A-1 to A-7 that make it possible to determine the durability of cement on the basis of maximum weight losses under wet-dry (ASTM D559) and freeze-thaw (ASTM D560) tests. Many state departments of transportation (DOTs) currently require minimum unconfined compressive strength testing (ASTM D1633) in lieu of these durability tests (Little et al. 2000).

This requirement is often based on many years of experience with soil-cement. The advantage of using these strength tests is that they can be conducted more rapidly than the durability tests (7 days versus 1 month) and require less laboratory equipment and technician training. However, achievement of a specified strength does not always ensure durability (Little et al. 2000).

Typical minimum strength varies from 1,400 to 5,250 kPa. For cement-modified soils, the engineer selects an objective and defines the cement requirements accordingly. Objectives may include one or more of the following: reducing the plasticity index (Atterberg limits, ASTM D4318); increasing the shrinkage limit; reducing the volume change of the soil (AASHTO T116); reducing clay/silt-sized particles (hydrometer analysis); meeting strength values/indexes such as the California bearing ratio (ASTM D1883) or triaxial test (ASTM D2850); and improving resilient modulus (ASTM D2434). Cement has been incorporated successfully into soils in the field with plasticity indexes ranging as high as 50 (Little et al. 2000).

Construction of soil-cement and cement-modified soil is normally a fast, straightforward process. Cement can be incorporated into soil/aggregate in a number of ways. The most common method is to spread dry cement in measured amounts on a prepared soil/aggregate and blend it in with a transverse single-shaft mixer to a specified depth. Cement slurries—in which water and cement are combined in a 50/50 blend with a slurry-jet mixer or in a water truck with a recirculation pump—have been used successfully to reduce dusting and improve mixing with heavy clays (Little et al. 2000).

Compaction is normally a minimum of 95% of either standard or modified proctor density (ASTM D588 or ASTM D1557, respectively), with moisture content +2% of optimum. Prusinski (Little et al. 2000) explains that soil-cement shrinks as a result of hydration and moisture loss. Shrinkage cracks develop in the base and can reflect through narrow bituminous surfaces as thin (<3 mm) cracks at a spacing of 2 to 12 m. If proper construction procedures are followed, shrinkage cracks may not reflect through, and if they do, they generally pose no performance problem. However, cracks can compromise performance if they become wide and admit significant moisture. A number of techniques have been used to minimize this problem, including

- Compaction at a moisture content slightly drier than optimum;
- Precracking through inducement of weakened planes or early load applications; and
- Delayed placement of surface hot mix, reduced cement content, and use of interlayers to absorb crack energy and prevent further propagation.

Prusinski (Little et al. 2000) explains that many state DOTs now use compressive strength testing as the sole criterion for determining cement content in soil-cement; additional research is needed to ensure that durability is also achieved at the specified strength for a variety of soil types. He believes that new durability tests may need to be developed for this purpose. A rapid and reliable test for assessing the impact of wet-dry and freeze-thaw cycles on durability remains a key need as well.

Prusinski further explains that as cracking appears to be the greatest performance concern, further research is needed to confirm the applicability of the techniques discussed above for various soil types. Particularly promising is the use of precracking techniques, which can be as simple as applying a vibratory roller 24 to 48 h after final compaction.

Prusinski further explains (Little et al. 2000) that though no longer widely used, probably because of higher first costs, combination treatment with lime and portland cement of plastic clay soils offers an attractive alternative. The initial lime treatment offers the ability to reduce plasticity and improve mixing efficiency, and since the lime allows a mellowing period of up to three or four days, time is not a constraint. The addition of a relatively low amount of portland cement (approximately 2 to 3%) upon final mixing provides the required strength even in clays that have moderate or low reactivity with lime.

American Coal Ash Association Approach

Stabilization of soils and pavement bases with coal fly ash is an increasingly popular option for design engineers. Fly ash stabilization is used to modify the engineering properties of locally available materials and produce a structurally sound construction base. Both non-self-cementing and self-cementing coal ash can be used in stabilization applications. However, the use of fly ash to stabilize plastic clay soils must usually be in concert with lime or cement. Class F fly ash is useless as a stabilizer without the addition of lime as a source of calcium.

Stewart (Little et al. 2000) explains that lime-Class C fly ash combinations have been successfully added to clay soils for stabilization. Successful application is normally achieved by adding lime first to modify soil properties and then adding fly ash to react with the lime pozzolanically to achieve a specified strength gain. This approach is attractive when a specified strength level cannot be achieved with fly ash alone. Although a two-stabilizer, two-step process may be economically disadvantageous under certain circumstances, it offers certain advantages, including the ability and time to modify the soil with lime before adding the fly ash for stabilization. Lime allows an extended mellowing period, which is necessary with some expansive clays.

Stewart (Little et al. 2000) further explains that with the passage of the Clean Air Act in the 1970s, many utilities began burning low-sulfur subbituminous coals. An unexpected benefit of burning this lower-sulfur coal was the production of a new type of fly ash, designated by ASTM as Class C coal fly ash. This material is self-cementing because of the presence of calcium oxide (CaO) in concentrations typically ranging from 20 to 30%. Most of the CaO in Class C fly ash, however, is complexly combined with pozzolans, and only a small percentage is "free" lime. This characteristic may impact the suitability of the material for stabilization of plastic clay soils. Subbituminous coals are now shipped by rail to power plants throughout the United States, although the largest concentrations of subbituminous coal combustion are west of Ohio. Class C fly ashes are shipped by truck and rail into many construction markets throughout the United States. According to ASTM D 5239, "Standard Practice for Characterizing Fly Ash for Use in Soil Stabilization," the use of self-cementing fly ash can result in improved soil properties, including increased stiffness, strength, and freeze-thaw durability; reduced permeability, plasticity, and swelling; and increased control of soil compressibility and moisture.

Although these ashes have properties similar to those of portland cement, they also have unique characteristics that must be addressed by both the mix design and construction procedures. The primary design consideration is the rate at which the fly ash hydrates upon exposure to water. Recognizing and properly addressing the hydration characteristics of the ash can result in a significant enhancement of the potential benefits of its use.

Even self-cementing ashes can be enhanced with activators such as portland cement or hydrated lime. This is particularly true if the self-cementing ash does not have enough free lime to fully develop the pozzolanic reaction potential. In clay soil stabilization the activator (lime or cement) may play a dual role: modifying the clay and activating the fly ash. A key role for fly ash in lime-clay stabilization is to provide the required pozzolanic-based strength gain for "non-(lime) reactive" clays.

Stewart (Little et al. 2000) explains that to achieve optimum results, a thorough understanding of the influence of the compaction delay time and moisture content of the stabilized materials is essential. Ash hydration begins immediately upon exposure to water. Strict control of the time between incorporation of the fly ash and final compaction of the stabilized section is required. A maximum delay time of 2 h can be employed if contractors are not experienced in ash stabilization or if achieving maximum potential strength is not a primary consideration for the application. This limits the clay mellowing period; thus a two-step process is often required in lime-fly ash clay stabilization. The first step is lime modification followed by fly ash addition. A maximum compaction delay of 1 h is commonly specified for stabilization of pavement base or subbase sections when maximum potential strength is required. Achieving final compaction within the prescribed time frame generally requires working in small, discrete areas, an approach that differs from the methods used for lime stabilization.

Stewart (Little et al. 2000) explains that the second major design consideration is that there is an optimum moisture content at which maximum strength will be achieved. This optimum moisture content is typically below that for maximum density—often by as much as 7 to 8%. The strength of the stabilized material can be reduced by 50% or more if the moisture content exceeds the optimum for maximum strength by 4 to 6%. An understanding of both the influence of compaction delay and moisture control of the stabilized material is essential to achieving the optimum benefit from stabilization applications that use self-cementing fly ashes.

Priority Needs

Engineers Must Establish Direction

It is extremely important for geotechnical engineers practicing in areas where expansive clay soils that benefit from the application of stabilization techniques occur to continue to be active in the economic and political decisions affecting their practice. This means becoming active in statewide geotechnical engineering groups, such as those affiliated with the Geoinstitute of ASCE. It is imperative that engineers continue to set the standards for practice and not default to political interests.

Engineers Must Remain Technically Keen

Geotechnical engineers must become even more active in continuing education, whether their state professional engineer board legislation requires it or not. But it is equally important for continuing education activities to be carefully planned and directed toward valuable topics that will expand the capabilities of practicing engineers. The local engineering leadership should identify critical and timely topics and viable speakers.

Geotechnical and materials engineers may need to refresh their basic science and chemistry knowledge to appreciate some of the arguments set forth when candidate stabilizers are compared. Although some nontraditional stabilizers are attractive alternatives, some of their producers are overly aggressive in promoting their product. Some of these promotions attack traditional stabilizers in an attempt to discredit them and obtain market share. These attacks often demonstrate a poor level of understanding of the mechanisms involved. A recent, overzealous brochure uses quotations from the *Handbook for Lime Stabilization of Subbases and Subgrades* (Little 1995) to cite properties of lime-treated soils that, they say, make them undesirable and susceptible to leaching. The properties include lower density and higher permeability.

However, these changes, when taken in proper context, can also be used as an argument for the physicochemical effectiveness of lime. The brochure ignores the numerous field studies documented in the same handbook that describe the durability of properly lime-stabilized subgrades and subbases. In fact, the brochure misrepresents the handbook. The handbook attempts not only to identify the favorable structural properties that can be derived by lime treatment, but also to make it clear that proper lime concentrations must be used in order to achieve these properties, and that "under-treatment" can be ineffective. This requires that each lime-soil mixture must be properly designed and that the appropriate lime content is soil specific. The same brochure quotes a Texas Transportation Institute (TTI) laboratory study (Rajendran and Lytton 1997) in which their product performed well in sulfate-bearing clays, but ignores a subsequent TTI field study (Scullion et al. 2001) in which their product was deemed ineffective in sulfate-bearing clays on the basis of both laboratory (strength, swell, and moisture susceptibility testing) and field testing (dynamic cone penetrometer and falling weight deflectometer).

So how do we resolve such conflicts? The key lies in understanding the fundamentals behind the stabilization processes and making sure that valuable and revealing studies such as Rajendran and Lytton (1997) and Scullion et al. (2001) are quoted in context. Recently, Rauch et al. (2001) completed a comprehensive study of selected nontraditional stabilizers that demonstrated such an approach. They evaluated a representative "ionic" stabilizer (whose active ingredients were *d*-limonone and sulfuric acid) that was used to treat a montmorillonite clay and found that at sufficiently high application mass ratios the lattice was altered. Rauch et al. (2001) and Katz et al. (2001) concluded that ionic stabilizers applied at sufficiently high mass ratios might improve the properties of some soils.

Rauch et al. (2001) develop a 10-step protocol for soil-stabilizer mixing, compaction, and curing to allow objective comparisons among candidate stabilizers. Each treated and untreated soil was characterized in terms of Atterburg limits, compaction unit weight, 1D fines swell potential, and untrained triaxial shear strength. The test results did not show significant changes in properties of soils as a result of treatments with either of the three stabilizers evaluated (ionic, polymer, or enzyme). However, Rauch et al. (2001) specify that high application rates may yield more favorable results and that addition laboratory study is indeed warranted.

Research Needs in Expansive Soil Stabilization

Establish Accepted Protocol to Evaluate Candidate Stabilizers

Several areas of research are clearly important. First, it is necessary to establish a fundamental protocol to compare candidate stabilizers. The protocol must be based on the desired and expected use of the stabilizer: volume change control, strength development, and so on. A standardized approach for comparing candidate stabilizers based on their intended use will assist the geotechnical engineer in stabilizer selection, especially when the choice is between traditional and nontraditional stabilizers. The use of performance-based comparative tests that carefully consider local environmental effects is extremely important in making informed choices.

Establish Protocol to Address Sulfate Heave

Clay soils containing sulfate salts or the potential to develop sulfate salts—for example, through the oxidation of pyritic sulfur—continue to plague stabilization efforts when the traditional, calcium-based stabilizers are used. The writers have been moderately successful in identifying construction protocols to minimize sulfate heave potential and in empirically identifying threshold soluble sulfate contents beyond which stabilization with calcium-based stabilizer is impractical. The construction techniques are typically based on adding as much water as possible to solubilize sulfates, specifying an extended mellowing period to allow for expansive mineral formation prior to compaction and minimizing the future ingress of water into the stabilized soil.

The next step should be to collaborate with soil scientists and soil chemists to develop a protocol to screen for sulfate deposits without having to do time-consuming soluble sulfate testing (nonintrusive conductivity measurements); screen for other pertinent properties where sulfates are high (for example, soluble alumina, pH, pOH, $pSiO_2$, and so on); and use chemical stability diagrams to determine the potential for future development of expansive minerals. In fact this work has already begun and is being pioneered by the Chemical Lime Company in Texas and California. Finally, with the more and more widespread use of GIS and GPS, it is now possible to link soil survey data to potential construction sites.

Update Our Understanding of Mechanisms of Stabilization

It is always important and will continue to be important to clearly identify the mechanisms of modification and stabilization of clay soils. Only when these have been clearly identified can we address problematic reactions and improve the product. Recent research (Yusuf 2001) demonstrates that soil matrix suction and surface energy play key roles in defining the engineering properties of lime-stabilized clay soils and that these effects may rival pozzolanic effects in determining properties and performance. This means that physical processes in addition to or even in lieu of traditionally considered pozzolanic reaction and cation exchange may well be needed to explain part of the modification/stabilization process, even in the traditional stabilizers. One very important and promising area is to re-look at the effect of cations on altering the thickness of the double-diffused water layers that surround the basic clay sheets. These layers define the volume stability and strength of the soil.

Recent work by Wilding and Tessier (1990) takes a new and refreshing view of the clay structure, which they view as composed of quasi-crystals or tactoids. These quasi-crystals are made up of an array of clay sheets held together via cation linkage. The type and concentration of cation determines the stiffness or rigidity of the quasi-crystals by determining the number of clay plates that comprise the quasi-crystals and the water-filled spacing between clay particles. These quasi-crystals then act as structural members, and the members interact to form the structural array. The structural array defines the ability of the soil to resist shrink-swell phenomena (based on the rigidity and arrangement of quasi-crystals) and the overall strength of the soil mass. This approach provides a refreshing and promising way to view the effect of stabilization on soils. Furthermore, as Wilding and Tessier (1990) show, these structural effects can be readily validated via transmission electron microcopy. Such work as this must be encouraged and funding must be identified.

In Situ Pavement Properties

In the area of pavements, the long-standing issue of assessing field properties from laboratory measurements remains a key topic, as has been discussed in this paper. Other long-standing issues include shrinkage cracking problems in stabilized expansive clays, propagation of shrinkage cracks through thin asphalt pavement surface, durability of stabilized clays, and the quality control of field stabilization activities.

Moving toward Mechanistic Design

In order to utilize stabilized layers as structural pavement components, regardless of the type of traditional stabilizer used, it is necessary to be able to reliably predict the pertinent in situ properties that affect performance. The most important of these parameters are resilient modulus and plastic deformation potential. Little et al. (1995) recommended a process by which to assign structural significance to lime-stabilized subgrade layers. This protocol can be extended to cement and lime-fly ash or cement-fly ash stabilized clay soils as well; however, modifications in the curing protocol will have to be made as the pozzolanic reactions are faster with cement and fly ash than with lime. The first step is to assign a realistic approximate resilient modulus to the subgrade layer. This can be done by either laboratory-resilient modulus testing or from preexisting field data. If laboratory testing is selected, then the resilient modulus (K1) should be determined in accordance with AASHTO T-294-94 after curing for 5 days at 38°C. If laboratory facilities are not available for resilient modu-

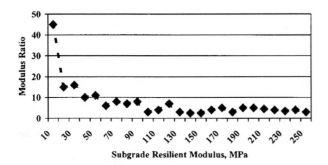

Fig. 3. Relationship between in situ modulus of natural subgrade soil as determined by FWD measurements and moduli ratio (lime-stabilized layer to natural subgrade layer) as determined by FWD measurements

lus testing following AASHTO T-294-94, then the resilient modulus properties can be approximated from compressive strengths determined following ASTM D-5102 after a curing period of 5 days at a temperature of 38°C.

In the majority of cases, the resilient modulus of the stabilized layer will have to be approximated based on a correlation between compressive strength and modulus. Although a number of relationships between compressive strength and modulus exist, Little et al. (1995) determined that a correlation developed by Suddath and Thompson (1975), further described by Thompson and Figurera (1989) and verified by Little et al. (1995), provides the most reliable approximation of moduli from compressive strength data when compressive strengths are between about 0.7 and 2.1 MPa. This relationship was further verified by CTL/Thompson (1998) for a variety of Denver soils. Based on this work a reasonable approximation of the resilient modulus of lime-stabilized soils from compressive strength data is $E_R = 0.125$ (UCS) + 10, where UCS is in psi and E_R is in ksi.

It is reasonable that the resilient modulus of the stabilized subgrade should also be affected by the level of support provided by the natural subgrade. Fig. 3 is a plot of subgrade resilient modulus versus the ratio of the modulus of the lime-stabilized subgrade (from FWD back-calculations) to the modulus of the natural subgrade (from FWD back-calculations). These data indicate that the level of modulus achieved by the lime-tested layer is limited by the subgrade modulus. If the material subgrade moduli are below about 50 MPa, the modulus ratio is typically around 10 or slightly above. For subgrade moduli between 50 and 200 MPa, the modulus ratio is between 5 and 10, and for subgrade moduli exceeding 200 MPa, the modulus ratio is less than about 5. This concept is reasonable in that one would expect that the in situ modulus ultimately provided by the stabilized layer is influenced by the supporting subgrade. For example, a very stiff stabilized layer over a soft subgrade will crack, leading to a reduction in the response modulus of the layer. It is also reasonable that the stiffness of the stabilized layer has an upper limit.

In a mechanistic analysis, the resistance of a subbase or base-course layer to plastic damage may be assessed by means of a number of approaches/models. Probably the most widely used and straightforward is the strain-hardening model relating permanent strain, p, to number of load cycles, N, in a repeated load triaxial test: $\log \varepsilon_p = a + b \log N$ where a and b are regression constants. Thompson and Smith (1990) found a good correlation between the ratio of deviator stress and shear strength as a means of differentiating between stable (low deformation potential) and unstable (high deformation potential) material. Little et al. (1995)

used a similar approach to evaluate deformation potential by suggesting that the compressive stress within a granular or cohesive layer should always be less than one-half of the compressive strength. He found that a study of hundreds of compressive strength tests on unbound granular bases and of lime-stabilized soils revealed a consistent pattern, that the stress-strain curve is approximately linear at stress levels below one-half the compressive strength but becomes nonlinear at stresses in excess of that point.

Summary

This paper serves three purposes: (1) highlight the historical development of the knowledge base in clay soils and their effective stabilization for light structures and pavements; (2) review the current state of the practice in clay stabilization for light structures and pavements; and (3) identify future research needs in these areas. The historical review reveals the valuable and varied input of an array of talented and dedicated engineers. It also reveals a difference and diversity of philosophies by which clay soils and stabilized clay soils are evaluated. These range from very practical and applied approaches based on simple test properties, which are empirically linked to performance, to more fundamental measurements of materials properties, which are mechanistically linked to performance.

The paper points out a history of success in stabilization of clay soils to reduce swell potential and improve strength for structural purposes using traditional, calcium-based stabilizers. The paper describes the introduction of by product and nontraditional stabilizers, which may offer a viable alternative. The paper warns that competent engineers who apply their professional knowledge judiciously must judge the effectiveness and appropriateness of all stabilizers. This knowledge is not cheap but must be gained through a dedicated effort to learn the fundamental concepts of stabilization and to keep abreast of the state of knowledge as it develops nationally and in a given region. This requires dedication to continuing education and a proactive mindset in attitude by the engineer.

The paper identifies future research needs, including establishing more effective ways to stabilize sulfate-bearing clays, reduce the potential for shrinkage cracking, improve the ability to predict in situ structural properties from laboratory measurements, and develop a satisfactory protocol by which to judge the efficacy of candidate stabilizers and continuously assess and update models that define the mechanisms of mechanical behavior of the natural and stabilized clay soil matrix.

References

Addison, M. B., and Petry, T. M. (1998). "Optimizing and multi-agent, multi-injected swell modifier." *Transp. Res. Rec. 1611*, Transportation Research Board, 38–45.

Allison, G. E., Kefauver, M., and Roller, E. M. (1953). "Ammonium fixation in soils." *Proc., Soil Science Society of America*, Vol. 17, 107–110.

Altmeyer, W. T. (1955). "Discussion of engineering properties of expansive clays." *Proc., ASCE*, 81 (separate no. 658), 17–19.

Anday, M. C. (1961). "Accelerated curing for lime-stabilized soil." *HRB Bulletin 304*, Highway Research Board, National Research Council, Washington, D.C., 1–13.

Anday, M. C. (1963). "Curing lime-stabilized soils." *HRB Record 29*, Highway Research Board, National Research Council, Washington, D.C., 13–26.

Arman, A., and Munfakh, G. A. (1970). "Lime stabilization of organic soils." *Engineering Research Bulletin No. 103*, Louisiana State Univ., Baton Rouge, La.

Barshad, I. (1950). "The effect of interlayer cations on the expansion of mica-type crystal lattice." *Am. Mineral.*, 35, 225–238.

Blacklock, J. R., and Pengelly, A. D. (1988). "Soil treatment for foundations on expansive clays." *Special topics in foundations*, B. M. Das, ed., ASCE, New York, 73–92.

Boynton, R. S., and Blacklock, J. R. (1986). "Lime slurry pressure injection bulletin." *Bulletin No. 331*, National Lime Association, Arlington, Va.

Bredenkamp, S., and Lytton, R. L. (1995). "Reduction of sulfate swell in expansive clay subgrades in the Dallas district." *Research Rep. 1994-5*, Texas Transportation Institute, Texas A&M Univ., College Station, Tex.

Building Research Advisory Board (BRAB). (1968). "Criteria for selection and design of residential slabs-on-ground." *National Academy of Sciences Rep. No. 33 to Federal Housing Administration, Publ. 1571, NTIS No. PB-261 551*, Washington, D.C.

Carroll, D., and Starkey, H. C. (1971). "Reactivity of clay minerals with acids and alkalines." *Clays Clay Miner.*, 19, 321–333.

Casagrande, A. (1932). "The structure of clay and its importance in foundation engineering." *J. Boston Soc. of Civil Engrs.*, 19(4), 168–209.

Chen, F. H. (1973). "The basic physical property of expansive soils." *Proc., 3rd Int. Conf. on Expansive Soils*, Vol. 1, 17–25.

Chen, F. H. (1988). *Foundations on expansive soils*, Elsevier Science, New York.

Colorado Department of Highways and University of Colorado. (1964). "A review of the literature on swelling soils." Colorado State Department of Highways, Denver.

CTL/Thompson, Inc. (1998). "Pavement design standards and construction specifications." *Prepared for the Metropolitan Government Pavement Engineers Council*, Denver.

Currin, D. D., Allen, J. J., and Little, D. N. (1976). "Validation of the soil stabilization index system with manual development." *FJSL-TR-76-0006*, Frank J. Seiler Research Laboratory, United States Air Force Academy, Colorado Springs, Colo.

Davidson, L. K., Demeril, T., and Handy, R. L. (1965). "Soil pulverization and lime migration in soil lime stabilization." *HRB Record 92*, Highway Research Board, National Research Council, Washington, D.C., 103–126.

Dempsey, B. J., Herlache, W. A., and Patel, A. J. (1986). "Climatic-materials-structural pavement analysis program." *Transp. Res. Rec. No. 1095*, Transportation Research Board, Washington, D.C., 111–123.

Diamond, S., and Kinter, E. B. (1965). "Mechanisms of soil-lime stabilization: An interpretive review." *HRB Record 92*, Highway Research Board, National Research Council, Washington, D.C., 83–101.

Dubose, L. A. (1955). "Compaction control solves heaving clay problems." *Civil Eng.*, 25, 232–233.

Eades, J. E., and Grim, R. E. (1963). "A quick rest to determine lime requirements for lime stabilization." *Highway Research Bulletin 139*, Highway Research Board, National Research Council, Washington, D.C., 61–72.

Eades, J. E., Nichols, F. P., and Grim, R. E. (1963). "Formation of new minerals with lime stabilization as proven by field experiments in Virginia." *Highway Research Bulletin 335*, National Research Council, Washington, D.C.

Epps, J. A., Dunlap, W. A., and Gallaway, B. M., (1970). "Basis for the development of a soil stabilization index system." *U.S. Air Force Contract No. F29601-90-C-008*, Air Force Weapons Laboratory, Texas A&M University, College Station, Tex.

Felt, E. J. (1953). "Influence of vegetation on soil moisture contents and resulting soil volume changes." *Proc., 3rd Int. Conf. Soil Mechanics and Foundation Engineering*, 1, 24–27.

Fohs, D. G., and Kinter, E. P. (1972). "Migration of lime in compacted soil." *Public Roads*, 37(1), 1–8.

Ford, C. M., Moore, R. K., and Hajek, B. F. (1982). "Reaction products of lime-treated Southeastern soils." *Transp. Res. Rec. 839*, Transportation Research Board, Washington, D.C., 38–40.

Fredlund, D. G., and Rahardjo, H. (1993). *Soil mechanics for unsaturated soils*, Wiley, New York.

Glenn, G. R., and Handy, R. L. (1963). "Lime-clay mineral reaction products." *Highway Research Record No. 29*, Highway Research Board, Washington, D.C., 70–78.

Grim, R. E. (1953). *Clay mineralogy*, McGraw-Hill, New York.

Higgins, C. M. (1965). "High-pressure lime injection." *Research Rep. 17, Interim Rep. 2*, Louisiana Department of Highways, Baton Rouge, La.

Hilt, G. H., and Davidson, D. T. (1960). "Lime fixation in clayey soils." *HRB Bulletin No. 262*, Highway Research Board, National Research Council, Washington, D.C., 20–32.

Holmquist, D. V., and Little, D. N. (1992). "Lime stabilization for airfield pavement design." *Prepared for the City and County of Denver, Job No. 17,350*, Denver International Airport, Denver.

Holtz, W. G., and Gibbs, H. J. (1956). "Engineering properties of expansive clays." *Transactions ASCE*, 121, 641–677.

Houston, S. L., and Fredlund, D. G. (1997). "Unsaturated soil engineering practice: Proceedings at Geo-Logan '97, Logan, Utah, July 15–19, 1997" *Geotechnical Special Publication No. 68*, ASCE, New York.

Houston, S. L., Houston, W. N., and Wagner, A.-M. (1994). "Laboratory filter paper suction measurements." *Geotech. Test. J.*, 17(2), 185–194.

Houston, S. L., and Wray, W. K. (1993). "Unsaturated soils." *Geotechnical Special Publication No. 39*, ASCE, New York.

Huang, E. Y. (1954). *Selected bibliography on soil stabilization*, Engineering Research Institute, Univ. of Michigan-Ann Arbor, Mich.

Hunter, D. (1988). "Lime-induced heave in sulfate-bearing clay soils." *J. Geotech. Eng.*, 114(2), 150–167.

Jennings, J. E. B. (1961). "A revised effective stress law for use in the prediction of the behavior of unsaturated soils." *Pore pressure and suction in soils*, Butterworth's, London, 26–30.

Jennings, J. E. B., and Knight, K. (1957). "The prediction of total heave from the double oedometer test." Transactions, South African Institute, *Civil Eng.*, 7, 285–291.

Johnson, L. D. (1969). "Review of literature on expansive clay soils." *Misc. Paper S-73-17*, U.S. Army Engineers Waterways Experiment Station, Vicksburg, Miss.

Johnson, L. D. (1973). "Influence of suction on heave of expansive soils." *Miscellaneous Paper S-73-17*, U.S. Army Engineers Waterways Experiment Station, Vicksburg, Miss.

Johnson, L. D. (1977). "Evaluation of laboratory suction tests for prediction of heave in foundation soils." *Technical Rep. S-77-7*, U.S. Army Engineers Waterways Experiment Station, Vicksburg, Miss.

Johnson, L. D., and Pengelly, A. D. (1993). "Chemical and lime stabilization of expansive clay." *3rd Int. Conf. on Case Histories in Geotechnical Engineering*.

Johnson, L. D., and Snethen, D. R. (1978). "Prediction of potential heave of swelling soils." *Geotech. Test. J.*, 1(3), 117–124.

Jones, C. W. (1958). "Stabilization of expansive clay with hydrated lime and with portland cement." *Highway Research Bulletin 193*, Highway Research Board, National Research Council, Washington, D.C., 40–47.

Jones, Jr., D. E., and Jones, K. A. (1987). "Treating expansive soils." *Civil Eng.*, 57(8), 62–65.

Katz, L. E., Rauch, A. F., Liljestrand, H. M., Harmon, J. S., Shaw, K. S., and Albers, H. (2001). "Mechanisms of soil stabilization with liquid ionic stabilizer." *Transp. Res. Rec. 1757*, Transportation Research Board, Washington, D.C., 50–57.

Kennedy, T. W. (1988). "Overview of soil-lime stabilization." *Presented at Effective Use of Lime for Soil Stabilization Conf.*, National Lime Association, Arlington, Va.

Kennedy, T. W., and Tahmovessi, M. (1987). "Lime and cement stabilization." *Lime notes, updates on lime applications in construction*, Issue No. 2, National Lime Association, Arlington, Va.

Lambe, T. W. (1960). "The character and identification of expansive soils: Soil PVC meter." *FHA 701*, Technical Studies Program, Washington, D.C., Federal Housing Administration.

Lee, J. J., and Kocherhans, J. G. (1973). "Soil stabilization by use of moisture barriers." *Proc., 3rd Int. Conf. on Expansive Soils*, 295–300.

Little, D. N. (1987). "Fundamentals of the stabilization of soil with lime." *Bulletin No. 332*, National Lime Association, Arlington, Va.

Little, D. N. (1995). *Stabilization of pavement subgrades and base courses with lime*, Kendall/Hunt Publishing Company, Dubuque, Iowa.

Little, D. N. (1999a). *Evaluation of structural properties of lime stabilized soils and aggregates. Vol. 1: Summary of finding*, National Lime Association, Arlington, Va.

Little D. N. (1999b). *Evaluation of structural properties of lime stabilized soils and aggregates. Vol. 2: Documentation of findings*, National Lime Association, Arlington, Va.

Little, D. N. (2000). *Evaluation of structural properties of lime stabilized soils and aggregates. Vol. 3: Mixture design and testing protocol for lime-stabilized soils*, National Lime Association, Arlington, Va.

Little, D. N., and Graves, R. (1995). "Guidelines for use of lime in sulfate-bearing soils." Chemical Lime Company.

Little, D. N., Males, E. H., Prusinski, J. R., and Stewart, B. (2000). "Cementitious stabilization." *79th Millenium Rep. Series*, Transportation Research Board, Washington, D.C.

Little, D. N., and Petry, T. M. (1992). "Recent developments in sulfate-induced heave in treated expansive clays." *Proc., 2nd Interagency Symp. on Stabilization of Soils and Other Materials*, COE, USBR, SCS, FHWA, EPA, and NAVFAC, 1-5-1-18.

Little, D. N., Scullion, T., Kota, P., and Bhuiyan, J. (1994). "Identification of the structural benefits of base and subgrade stabilization." *Rep. 1287-2*, Texas Transportation Institute, Texas A&M Univ., College Station, Tex.

Little, D. N., Scullion, T., Kota, P., and Bhuiyan, J. (1995). "Guidelines for mixture design and thickness design for stabilized bases and subgrades." *Rep. No. 1287-3F*, Texas Transportation Institute, Texas A&M Univ., College Station, Tex.

Lundy, L., and Greenfield, B. J. (1968). "Evaluation of deep in situ soil stabilization by high pressure lime slurry injection." *Highway Research Record*, Highway Research Board, National Research Council, Washington, D.C.

Lytton, R. L. (1994). "Prediction of movement in expansive clays." *Vertical and horizontal deformations of foundations and embankments*, A. T. Yeung and G. Y. Felio, eds., ASCE, New York, 1827–1845.

Lytton, R. L. (1995). "Foundations and pavements on unsaturated soils." *Proc., 1st Int. Conf. on Unsaturated Soils*, Vol. 3, International Society of Soil Mechanics and Foundation Engineering, Paris, 1201–1210.

Lytton, R. L. (1997). *Engineering structures in expansive soils*, Solos Nao Saturados, Rio de Janeiro, Brazil.

Marks, B. D., and Haliburton, T. A. (1972). "Acceleration of lime-clay reactions with salt." *J. Soil Mech. Found. Div.*, 98(4), 327–339.

Mateos, M. (1964). "Soil lime research at Iowa State University." *J. Soil Mech. Found. Div.*, 90(2), 127–153.

McCallister, L. D., and Petry, T. M. (1990). "Property changes in lime treated expansive clays under continuous leaching." *Technical Rep. GL-90-17*, U.S. Army Corps of Engineers, Vicksburg, Miss.

McDonald, E. B. (1973). "Experimental moisture barrier and waterproof surface." South Dakota Department of Highways.

McDowell, C. (1959). "The relation of laboratory testing to design for pavements and structures on expansive soils." *Quart. Colorado School of Mines*, 54(4), 127–153.

McKeen, R. G. (1980). "Field studies of airport pavements on expansive clays." *Proc., 4th Int. Conf. on Expansive Soils*, Vol. 1, 242–261.

McKeen, R. G. (1981). "Design of airport pavements for expansive soils." Federal Aviation Agency, U.S. Department of Transportation, Washington, D.C.

McKeen, R. G. (1992). "A Model for predicting expansive soil behavior." *Proc., 7th Int. Conf. on Expansive Soils*, 1–6.

McKeen, R. G., and Johnson, L. D. (1990). "Climate-controlled soil de-

sign parameters for mat foundations." *J. Geotech. Eng.*, 116(7), 1073–1094.

McKinney, R. L., Kelly, J. E., and McDowell, C. (1974). "The Waco ponding project." *Research Rep. 118-7*, Center for Highway Research, University of Texas at Austin, Tex.

Mitchell, J. K. (1976). *Fundamentals of soil behavior*, 1st Ed., Wiley, New York.

Mitchell, J. K. (1986). "Practical problems from surprising soil behavior." *J. Geotech. Eng.*, 112(3), 255–289.

Mitchell, J. K., and Dermatas, D. (1992). "Clay soil heave caused by lime-sulfate reactions." *Special Technical Publication No. 1135*, ASTM, West Conshohocken, Pa.

Mitchell, J. K., and Hooper, D. R. (1961). "Influence of time between mixing and compaction on the properties of a lime-stabilized expansive clay." *HRB Bulletin 304*, Highway Research Board, National Research Council, Washington, D.C., 14–31.

National Lime Association. (1991). "Lime stabilization construction manual." *Bulletin 226*, Arlington, Va.

Nelson, J. D., and Miller, D. J. (1992). *Expansive soils, problems and practice in foundation and pavement engineering*, Wiley, New York.

O'Bannon, D. E.r, Morris, G. R., and Mancini, F. P. (1976). "Electrochemical hardening of expansive clays." *Transp. Res. Rec. 593*, Transportation Research Board, Washington, D.C., 46–50.

Palit, R. M. (1953). "Determination of swelling pressure of black cotton soil—A method." *Proc., 3rd Int. Conf. on Soil Mechanics and Foundation Engineering*, Vol. 1.

Petry, T. M. (1993). "Remedial treatment of a slab on highly active clay utilizing a lime-fly ash grout curtain." *Transp. Res. Rec. 1362*, Transportation Research Board, Washington, D.C., 126–130.

Petry, T. M. (1995). "Studies of factors causing and influencing localized heave of lime treated clay soils (sulfate induced heave)." *Final Rep. to U.S. Army Engineers Waterways Experiment Station*, Vicksburg, Miss.

Petry, T. M., and Armstrong, J. C. (1989). "Stabilization of expansive clay soils." *Transp. Res. Rec., 1219*, Transportation Research Board, Washington, D.C., 103–112.

Petry, T. M., Armstrong, J. C., and Chang, D. (1982). "Short-term active soil property changes caused by injection of lime and fly-ash." *Transp. Res. Rec. No. 830*, Transportation Research Board, Washington, D.C., 25–32.

Petry, T. M., and Lee, T. W.r (1989). "Comparison of quicklime and hydrated lime slurries for stabilization of highly active clay soils." *Transp. Res. Rec. 1190*, Transportation Research Board, Washington, D.C., 31–37.

Petry, T. M., and Little, D. N. (1993). "Update on sulfate-induced heave in treated clays: Problematic sulfate levels." *Transp. Res. Rec. 1362*, Transportation Research Board, Washington, D.C., 51–55.

Petry, T. M., Sheen, J.-S., and Armstrong, J. C. (1992). "Effects of pretest stress environments on swell." *Proc., 7th Int. Conf. on Expansive Soils*, 39–44.

Petry, T. M., and Wohlgemuth, S. K. (1989). "The effects of pulverization on the strength and durability of highly active clay soils stabilized with lime and portland cement." *Transp. Res. Rec. 1190*, Transportation Research Board, Washington, D.C., 38–45.

Poor, A. R. (1978). "Experimental residential foundation design on expansive clay soils." *Rep. No. TR-3-78, Final Rep.*, Construction Research Center, Univ. of Texas at Arlington, Tex.

Post-Tensioning Institute (PTI). (1980). *Design and Construction of Post-Tensioned Slabs-On-Ground*, Phoenix, Ariz.

Rajendran, D., and Lytton, R. L. (1997). "Reduction of sulfate swell in expansive clay subgrades in the Dallas district." *Rep. TX-98/3929-1*, Texas Transportation Institute, Tex.

Rauch, A. F., Harmon, J. S., Katz, L. E., and Liljestrand, H. M. (2001). "Liquid soil stabilizers: Measured effects on engineering properties of clay." *81st Annual Transportation Board Meeting*.

Ridley, A. M., and Wray, W. K. (1996). "Suction measurement—A review of current theory and practices." *Proc., 1st Int. Conf. on Unsaturated Soils*, Vol. 3, 1295–1322.

Rollings, R. S., Burkes, J. P., and Rollings, M. P. (1999). "Sulfate attack on cement-stabilized sand." *J. Geotech. Geoenviron. Eng.*, 125(5), 364–372.

Scullion, T., Sebesta, S., and Harris, J., (2001). "Identifying the benefits of nonstandard stabilizers in high sulfate clay soils: A status report." *Research Rep. No. 408330-1*, Texas Transportation Institute, Texas A&M University, College Station, Tex.

Seed, H. B., Mitchell, J. K., and Chan, C. K. (1962a). "Studies of swell and swell pressure characteristics of compacted clays." *Highway Research Board Bulletin 313*, National Research Council, Washington, D.C., 12–39.

Seed, H. B., Woodward, R. J., and Lundgren, R. (1962b). "Prediction of swelling potential for compacted clay." *J. Soil Mech. Found. Div.*, 88(3), 53–87.

Scholen, D. E. (1992). "Non-standard stabilizers." *FHWA-FLP-92-011*, Federal Highway Administration, Washington, D.C.

Simpson, W. E. (1934). "Foundation experiences with clay in Texas." *Civil Eng.*, ASCE, New York.

Skempton, A. W. (1953). "The colloidal activity of clays." *Proc., 3rd Int. Conf. on Soil Mechanics*, Foundation Engineering, Vol. 1, 57–61.

Snethen, D. R. (1979). "An evaluation of methodology for predicting and minimization of detrimental volume change of expansive soils in highway subgrades." *Rep. No. FHWA-RD-79-49, Final Rep.*, Washington, D.C.

Snethen, D. R. (1984). "Evaluation of expedient methods for identification and classification of potentially expansive soils." *Proc., 5th Int. Conf. on Expansive Soils*, 22–26.

Snethen, D. R., and Huang, G. (1992). "Evaluation of soil suction-heave prediction methods." *Proc., 7th Int. Conf. on Expansive Soils*, 12–17.

Snethen, D. R., Townsend, F. C., Johnson, L. D., Patrick, D. M., and Vedros, P. J. (1975). "A review of the engineering experiences with expansive soils in highway subgrades." *Rep. No. FHWA-RD-75-48, Interim Report to FHWA*, Washington, D.C.

Steinberg, M. L. (1977). "Ponding on expansive clay cut." *Transportation Research Record No. 661*, Transportation Research Board, Washington, D.C., 61–66.

Steinberg, M. L. (1981). "Deep vertical fabric moisture barriers under swelling soils." *Transp. Res. Rec. No. 790*, Transportation Research Board, Washington, D.C., 87–94.

Steinberg, M. L. (1985). "Monitoring the use of impervious fabrics: Geomembranes in the control of expansive soils." *Dept. Research Rep. No. 187-12*, State Department of Highways and Public Transportation, Austin, Tex.

Steinberg, M. L. (1998). *Geomembranes and the control of expansive soils in construction*, McGraw-Hill, New York.

Suddath, L. P., and Thompson, M.R. (1975). "Load deflection behavior of lime stabilized layers." *Technical Rep. M-11B*, Construction Research Laboratory, Champaign, Ill.

Taylor, A. W. (1959). "Physico-chemical properties of soils: Ion exchange phenomena." *J. Soil Mech. Found. Div.*, 85(2), 19–30.

Terrel, R. L., Epps, J. A., Barenberg, E. J., Mitchell, J. K., and Thompson, M. R. (1979). "Soil stabilization in pavement structures: A user's manual." FHWA-IP-80-2, Vol. I, Department of Transportation, Federal Highway Administration, Washington, D.C.

Thompson, M. R. (1966). "Lime reactivity of Illinois soils." *J. Soil Mech. Found. Div.*, 92.

Thompson, M. R. (1969). "Engineering properties of lime-soil mixtures." *J. Mater.*, ASTM, Vol. 4.

Thompson, M. R. (1970). "Soil stabilization of pavement systems—State of the art." *Technical Rep.*, Department of the Army, Construction Engineering Research Laboratory, Champaign, Ill.

Thompson, M. R. (1972). "Deep-plow lime stabilization for pavement construction." *J. Transp. Eng. Div.*, 98(2), 311–323.

Thompson, M. R., and Dempsey, B. J. (1969). "Autogenous healing of lime-soil mixtures." *Highway Research Record 263*, National Research Council, Washington, D.C.

Thompson, M. R., and Figureora, J. L. (1989). "Mechanistic thickness design procedure for soil-lime layers." *Transp. Res. Rec. No. 754*, Transportation Research Board, Washington, D.C.

Thompson, M. R., and Smith. (1990). "Repeated Triaxial Characterization of Granular Bases." *Transp. Res. Rec. No. 1278*.

Townsend, D. L. and Klym, T. W. (1966). "Durability of lime stabilized soils." *HRB Record 139*, Highway Research Board, Transportation Research Board, Washington, D.C., 25–41.

Transportation Research Board (TRB). (1987). "Lime stabilization—Reactions, properties, design and construction." *State of the Art Report 5*, Washington, D.C.

Tucker, R. L., and Poor, A. R. (1978). "Field study of moisture effects on slab movements." *J. Geotech. Eng. Div.*, 104(4), 403–414.

Vijayvergiya, V. N., and Ghazzaly, O. I. (1973). "Prediction of swelling potential for natural clays." *Proc., 3rd Int. Conf. on Expansive Soils*, 227–236.

Vipulanandan, C., Addison, M. B., and Hansen, M., eds. (2001). *Expansive clay soils and vegetative influence on shallow foundations*, ASCE, Reston, Va.

Wiggins, J. H., Slossan, J. E., and Krohn, J. P. (1978). *Natural hazards: Earthquake, landslide, expansive soil, Rep. for National Science Foundation under grants ERP-75-09998 and AEN-74-23993*, J. H. Wiggins Co.

Wilding, L. P., and Tessier, D. (1990). "Genesis of vertisols: Shrink-swell phenomena." Texas A&M University Press, College Station, Tx.

Wray, W. K. (1978). "The effect of climate on expansive soils supporting on-grade structures in a dry climate." *Transp. Res. Rec. 1137*, Transportation Research Board, Washington, D.C., 12–23.

Wray, W. K. (1984). "The principle of soil suction and its geotechnical engineering applications." *Proc., 5th Int. Conf. on Expansive Soils*, 114–118.

Wray, W. K. (1987). "Evaluation of static equilibrium soil suction envelopes for predicting climate-induced soil suction changes occurring beneath covered surfaces." *Proc., 6th Int. Conf. on Expansive Soils*, Vol. 1, 235–240.

Wright, P. J. (October 1973, July 1974). "Lime slurry pressure injection tames expansive clays." *Civil Eng.*, ASCE, New York.

Yusuf, S. (2001). "Validation of the new mixture design and testing protocol for lime stabilization." MSci thesis, Texas A&M University, College Station, Tex.

Yusuf, S., Little, D. N., and Sarkar, S. L. (2001). "Evaluation of structural contribution of lime stabilization of subgrade soils in Mississippi." *Transp. Res. Rec. 1757*, Transportation Research Board, Washington, D.C., 22–31.

American Society of Civil Engineers — 150th Anniversary Paper — 1852–2002 — Building a Better World

Atmospheric Corrosion Resistance of Structural Steels

Pedro Albrecht, M.ASCE,[1] and Terry T. Hall Jr.[2]

Abstract: The main objective of this paper is to provide engineers with data on thickness loss of structural steel members resulting from corrosion. To this end, the writers have collected atmospheric exposure data from many research reports and journal papers. The data are presented in graphs of thickness loss versus exposure that demonstrate the effects of types of environment and steel. The environments include rural, industrial, and marine, becoming increasingly severe in that order. The steels under consideration are A242, A588, copper, and carbon steels, with their corrosion resistance decreasing in that order. Comparisons of the data with the medium corrosivity bands for weathering and carbon steels help to determine the severity of environments and the corrosion resistance of steel compositions. For bare exposed structures, a corrosion allowance should be added to all member thicknesses arrived at by stress calculations can be estimated by extrapolating an applicable thickness loss curve to the end of the service life of the structure. The corrosion allowance thickness loss data obtained from simple test specimens, mostly of the 150×100 mm size of an index card, must be applied carefully to complex structures such as a bridge, which has a variety of details with different exposure conditions. When portions of a bare, exposed structure remain damp or wet for long periods of time, thickness losses can be much higher than those reported for test specimens exposed at 30° facing south. Good structural detailing that prevents moisture accumulation is of utmost importance.

DOI: 10.1061/(ASCE)0899-1561(2003)15:1(2)

CE Database keywords: Corrosion; Weathering; Steel; Structures; Bridges.

Introduction

Scope

The original intent of this paper was to appraise how structural steels have evolved, to describe the state of the art, and to project future trends. The most important properties of structural steels are tensile and yield strengths, fatigue strength, fracture toughness, and corrosion resistance. Each of these properties could easily be addressed as a single topic, but the authors chose to write about only the corrosion of steels, an often-neglected topic. Readers and the authors need only to look back to their professional education to realize how little time was spent on corrosion control relative to other topics on steel design. Yet, failure to adequately protect steel against excessive corrosion carries a huge cost to the public at large.

[1]Professor, Dept. of Civil Engineering, Univ. of Maryland, College Park, MD 20742. E-mail: pedroalb@eng.umd.edu

[2]Structural Engineer, Whitman, Requardt, and Associates, 801 South Caroline Street, Baltimore, MD 21231;

formerly Graduate Student, Dept. of Civil Engineering, University of Maryland, College Park, MD 20742.

Note. Associate Editor: Antonio Nanni. Discussion open until July 1, 2003. Separate discussions must be submitted for individual papers. To extend the closing date by one month, a written request must be filed with the ASCE Managing Editor. The manuscript for this paper was submitted for review and possible publication on August 23, 2002; approved on September 23, 2002. This paper is part of the *Journal of Materials in Civil Engineering*, Vol. 15, No. 1, February 1, 2003. ©ASCE, ISSN 0899-1561/2003/1-2-24/$18.00.

There are three general categories of corrosion protection: surface coatings, cathodic systems, and reliance on the corrosion resistance of bare, exposed, so-called weathering steel. The type of structure and the environment will determine which of the three approaches is most suitable and economical. As Buck (1915) wrote in an early paper on the effect of copper in steel as a retarding influence on its deterioration: "The ideal, of course, is not corrosion retardation only, but corrosion prevention."

This paper focuses on the corrosion resistance of bare carbon, copper-bearing, and weathering steels. The latter steel holds the hope of building bare steel structures that have sufficient corrosion resistance to outlast the useful service lives of the structures.

Cost of Corrosion

A new study on the cost and preventive strategies mandate by the U.S. Congress estimated the total direct cost of metal corrosion in 26 industrial sectors to be $276 billion per year (Koch et al. 2002). Of that amount, highway bridges made of steel and concrete account for $3.79 billion per year. This includes the following.

(1) "Cost to replace structurally deficient bridges; and (2) corrosion associated life-cycle cost for remaining (non-deficient) bridges, including the cost of construction, routine maintenance, patching, and rehabilitation. Life-cycle analysis estimates indirect costs to the user due to traffic delays and lost productivity at more than 10 times the direct cost of corrosion" (Koch et al. 2002, pp. 24, D29).

No stronger argument could be made for steel producers to develop even better corrosion-resistant steels and for engineers to design structures with corrosion prevention in mind.

Structural Steels

With respect to corrosion resistance, structural steels are classified in the literature as being of the carbon, copper-bearing, and weathering types. Carbon steels have by definition less than 0.02% copper (Cu) and, hence, little atmospheric corrosion resistance. They are classified in the ASTM specifications as follows:

- *A7 steel for bridges and buildings*, originally adopted in 1936 and discontinued in 1965
- *A36 carbon structural steel*, originally adopted in 1960.

Other steels such as A572 high-strength low-alloy columbium-vanadium structural steel could be included in this category because they are not alloyed with Cu and, thus, have atmospheric corrosion resistance equivalent to that of low-carbon steel with <0.02% Cu.

Copper-bearing steels, hereafter simply called copper steels, contain at least 0.20% Cu and, thus, have moderate atmospheric corrosion resistance. They are classified in the ASTM specifications as follows:

- A7 steel for bridges and buildings, when copper steel is specified, originally adopted in 1936 and discontinued in 1965
- A36 carbon structural steel, when copper steel is specified, originally adopted 1960
- *A440 high-strength structural steel*, originally adopted in 1959 and discontinued in 1979
- *A441 high-strength low-alloy structural manganese vanadium steel*, originally adopted in 1960 and discontinued in 1975.

Weathering steels have substantial atmospheric corrosion resistance that is gained by alloying it with various amounts of phosphorus (P), silicon (Si), nickel (Ni), chromium (Cr), and Cu. They are classified in the ASTM specifications as follows:

- *A242 high-strength low-alloy structural steel*, originally adopted in 1941
- *A588 high-strength low-alloy structural steel with 345 MPa minimum yield point to 100 mm thick*. originally adopted in 1968.

Moreover, the ASTM specification A709 for structural steel for bridges lists weathering steels of grades 345W and 690W, which correspond to those of the A588 and 514 specifications, as well as a newly developed high-performance weathering steel of grade HPS 485W. Producers are currently developing additional weathering steels. The thickness loss data for carbon, copper, and weathering steels presented in this paper are applicable to the steels listed above.

Objective

Epstein and Horton (1949), Copson (1960), and Larrabee and Coburn (1962) performed the atmospheric corrosion tests of steels alloyed with P, Si, Ni, Cr, and Cu that, by far, contributed most toward enhancing the corrosion resistance of steels. In contrast, recent studies focused on testing mainly carbon, copper, A242, and A588 steels. The carbon and copper steels have served as references against which modified compositions of the A242 and A588 weathering steels are compared. Many results of atmospheric exposure tests have been published in various journals and research reports, but others remain unpublished.

Still lacking to date is a comprehensive database of thickness loss of structural steels caused by corrosion. To satisfy this need, the writers have collected all available data and are presenting the information in plots useful to engineers and designers.

Readers are cautioned that the thickness loss data presented in this paper were compiled mostly from tests of thin plate speci-

mens 100 mm long by 50 mm wide by 0.25 to 5 mm thick. The specimens were mounted on racks facing south at an angle of 30° from the horizontal, an ideal exposure condition. The specimens are sun-dried, the skyward side is rain-washed, and water runs off freely. That is not the case in most structures.

Larrabee (1953), Albrecht and Naeemi (1984), and Albrecht et al. (1989) described in-service conditions under which bare, exposed structural members made of weathering steel corrode more than the same material exposed as a specimen. The in-service conditions include shelter, orientation, angle of exposure, debris accumulation, surrounding vegetation, atmospheric pollutants, airborne marine salt, contamination by deicing salt, distance from bodies of salt water, clearance from vegetation and ground, structural details that trap moisture, in short, any condition that prolongs the wetness of a structural component. Although corrosion test data obtained with standard specimens are very useful, engineers should carefully evaluate whether the environment and the site are suitable for a bare, exposed steel structure.

Historical Development

This brief history traces the development in the United States of what are familiarly known as copper steel and weathering steel, for which the most promising current application is highway bridges. They were part of a family of steels termed "high-strength low-alloy" (HSLA). The products evolved from an iron and steel base.

Copper Steels

The most authoritative reviews of the development of copper steel were written by Buck (1920) in the United States and Daeves (1926) in Germany

Early Studies
According to Daeves (1926), the first observation of the effect of copper on corrosion resistance is found in a patent granted to Vazie in 1822. It was found that adding one part brass—an alloy consisting mainly of copper and zinc—to 100 parts of cast iron greatly increased the corrosion resistance of equipment exposed to acidic water in mines. The corresponding copper content was about 0.70%.

In Germany, Karsten (1827) reported on the effect of copper on the tensile properties of foundry products. He noted that adding 0.29% copper increased six fold the time needed to dissolve metal in sulfuric acid. As a result of this observation, people began to evaluate the corrosion performance of different steel compositions by the rate at which they dissolved in 20% sulfuric acid.

Stengel (1836) observed the effect of copper content on the formation of the oxide coat on steel. He found it annoying that in quenched steel rails containing 0.20 to 0.40% Cu, the mill scale did not come off easily. Instead, a dense, adherent, smooth, dark bluish coating formed on the steel surface. Apparently, these early studies were forgotten.

Encouraged by Hadfield's tests on nickel steel, Williams (1900) conducted a simple experiment. He selected a series of copper-free and copper-bearing (0.08 to 0.20% Cu) wrought irons and Bessemer steels and wetted them with water several times daily for 1 month while exposed outdoors. Only the copper-bearing products significantly retarded corrosion. Williams clearly understood the importance of his findings and suggested that iron ores containing copper should be used. Stead and Wigham (1901) immersed samples in Middlesborough, England, town water for a period of several weeks. Breuil (1907) alluded to the remarkable properties of copper steel to resist the attack of sulfuric acid.

The subject of copper in steel and iron was frequently discussed during the first decade of the existence of the newly formed American Society for Testing and Materials (1900–1910) as a result of initial studies conducted in England and the United States. Initially, the samples were tested in water and sulphuric acid, but in time, the focus shifted to atmospheric exposure.

Burgess and Aston (1913) added single elements to pure electrolytic iron and exposed the samples in Madison, Wisc. They found that "adding 0.089 to 7.05% copper consistently enhanced the atmospheric corrosion performance of the iron, with markedly beneficial results being obtained even with small amounts of added copper." It was suggested that Sudbury (Ontario, Canada) ores contained nickel and copper, and that such natural ores could provide steels with improved properties without any further alloy additions.

Buck's Contribution

In 1910, having observed the excellent performance exhibited by a 0.07% copper steel sheet, Buck (1913a,b) of the American Sheet and Tin Plate Company, a Division of the US Steel Corporation, conducted the first large-scale atmospheric exposure test of copper steel. The study involved open hearth and Bessemer steels; 0.0%, 0.06 to 0.07%, and 0.16 to 0.34% Cu; 610-mm-wide ×2,440-mm-long×1.52 and 0.42-mm-thick sheets corrugated into panels in the usual way; and 50-mm-wide×100-mm-long specimens cut from the same sheets as the corrugated specimens. The specimens were exposed in industrial coke regions of Pennsylvania, "where the air contains notable amounts of sulfurous and sulfuric acids and other fumes from the coke ovens;" in marine Atlantic City, N.J. "where the air carries sodium chlorides;" and along the rural Allegheny River above rural Kittanning, Pa., "where the air is quite pure and free from added corrosive agents." Visible perforation of a panel was considered to be the sign of failure.

The 50×100 mm specimens were carefully weighed, then mounted in wooden racks at the three sites, with free access to the weather. After 8.8, 9.3, and 11.5 months of exposure, respectively, "they were taken down and reweighed, after removing all rust by a solution of ammonium citrate, which takes off the oxide without attacking the underlying iron." The findings for the 50 ×100 mm specimens were as follows (Buck 1913a,b, p. 450).

- "In every case the steels with copper additions have shown a marked resistance to corrosion as compared to the non-copper steels, having on the average nearly twice the life."
- On the basis of mass loss, "there appears to be very little difference between the grades containing 0.15% Cu and those with 0.24 to 0.34%, while the material with 0.06 to 0.07% Cu takes an intermediate position."

The 0.42-mm-thick corrugated panels were installed at each location on a skeleton wooden building, 12×24 m, at a slope of 18%, as on a roof. The buildings were entirely open and free to the passage of air on all four sides, and the roofs were uncovered until the sheets were put on, the purlins being 2.34 m apart, thus allowing for a 50 mm hold at the end of the sheets. Open spaces were left between each course. The findings for the 0.42-mm-thick panels were as follows:

- The oxide was bright red and loosely adherent on the noncopper steels, a little darker on the 0.06% Cu steel, and dark brown and closely adherent on the 0.16 to 0.34% Cu steels.
- After 7.2 months of exposure in the industrial coke region of

Pennsylvania, the three noncopper panels "were failing and falling off, having rusted entirely through" and "had entirely disappeared after 12.6 months."

- The 0.06% Cu panel was "failing to the extent of dropping off" after 12.6 months and "had entirely disappeared" after 15.2 months.
- The copper steel panels, containing 0.16 to 0.34% Cu, "were still in excellent condition" after 12.6 months and "still in place" after 15.3 months.
- The 0.42-mm-thick panels exposed in marine Atlantic City, N.J. "were in about the same condition as those in the coke region were some months earlier."
- Of the 0.42-mm-thick panels exposed 14.6 months in rural Kittanning, Pa., those containing 0.06% Cu were "in very poor condition and will fail very soon, much before the other copper-bearing steels."

The 1.52-mm-thick panels exposed at the same three sites, being four times as thick, lasted longer and their results were reported later as follows (Buck 1915):

- After 3.2 years of exposure in the industrial coke region of Pennsylvania, two of the copper-free steels had "absolutely failed" while the third was in "poor condition."
- The copper-bearing panels were "nearly as good as when first installed." So far as could be determined by observation, "they have not changed in any way in the past two years."
- Similar results were obtained from the marine Atlantic City, N.J. and rural Kittanning, Pa., sites.

Buck (1913a,b, p. 451) concluded that "the copper-bearing steels resist the atmosphere from 1.5 to 2 times as well as normal steels without copper."

Buck (1915) and Buck and Handy (1916) performed a more detailed test in which they exposed, in Scottsdale and McKeesport (both sites being near Pittsburgh, Pa.), 700 full-size roofing sheets that varied in copper levels from 0.004 to 2.12%. In addition, they obtained a number of "pure" irons that contained from 0.014 to 0.018% copper. Specimens 50 mm wide by 100 mm long were cut from the sheets to measure mass loss, which can be converted to thickness loss per exposed surface. At the end of 10.9 months, Buck (1915, p. 1234) found that

- "A copper content of 0.25% materially increases the life of steel and iron. More copper, up to 2%, gives little, if any additional benefit. Lesser amounts than 0.25% have great influence in lowering the corrosion rate. A content of 0.15% is in most cases quite as good as 0.25%. Tests now underway prove that much lower amounts, down to as little to 0.04% and 0.6%, while not giving the best results, are remarkably superior to normal steels carrying only the usual traces of copper." Thus, they were able to locate the point for effective copper levels in carbon steel. Later studies confirmed these results (Fig. 1).
- "It appears that the rate of corrosion of copper-bearing sheets lessens as the time of exposure increases."
- "The rust formed on copper-bearing steel is dark brown, closely adherent, and smooth. The rust on a normal steel or iron, which is a limonite red in color, is rough, loose, and spongy in character. That the latter holds water to a greater extent is strikingly demonstrated by examining sheets that have been exposed for several months, both with and without copper, immediately after a rain. Copper-bearing sheets dry very quickly, while the loose, spongy rust on the sheets without copper holds the moisture a considerable time, allowing corrosion to proceed for several hours after it has ceased on the copper alloy sheets."

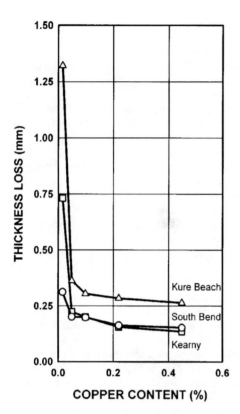

Fig. 1. Effect of copper on 15.5-year thickness loss (Larrabee and Coburn 1962)

ASTM Committee A-5

Allerton Cushman, a member of ASTM Corrosion Committee A-5, objected to the facts reported by Buck (1915). He insisted that a carefully supervised test be repeated under the sponsorship of Committee A-5 of the ASTM in cooperation with the National Bureau of Standards and that the specimens be exposed at government sites. Two major long-term exposure studies were then carried out with bare, corrugated, non-copper-bearing and copper-bearing sheets of dimensions 762×660×0.76 mm. During the preparation of the specimens, it was pointed out (ASTM Committee A-5 report in the 1916 proceedings) that the iron specimens have a high copper content because of the use of eastern Pennsylvania ores and that an effort would be made to find ores substantially free of copper such as typified by the newly available Mesabi Range ores. The specimens were inspected for perforations twice yearly. Committee members included Buck and Cushman or their representative.

In the first major exposure study conducted by members of Committee A-5, the sites and exposure times were as follows: Naval Academy at Annapolis, Md., 1916–1952; Fort Sheridan, Ill., 1917–1928; and Fort Pitt in Pittsburgh, Pa., 1916–1923.

The ASTM Proceedings contain Buck's (1919) interim report on the steels with copper levels varying from 0.01% up to 0.25% that were exposed at the aforementioned test sites. The findings, following 18 months of exposure, indicated that beginning with 0.04% copper the atmospheric corrosion rate slowed down considerably, and that a copper level of 0.15% was ample.

Buck (1920) summarized and presented again the interim results of the exposure tests that were started in 1916. He concluded that based on the overwhelming evidence the time to perforation of copper-bearing steel sheets was twice that of non-copper-bearing steels. In that article, Buck referred to the earliest use he could find for copper, in 1627, wherein the physical properties of iron were improved. He related that Richardson and Richardson (1916) reported that copper additions to steel resulted in greater improvement than equal copper additions to pure iron.

In the second major exposure study conducted by ASTM Corrosion Committee A-5, the sites and exposure times were as follows: Altoona, Pa., 1926–1939; State College, Pa., 1926–1954; Sandy Hook, N.J., 1926–1938; and Key West, Fla., 1926–1932. Larrabee (1954, p. 122) summarized the results of the two major studies and concluded on behalf of Committee A-5 that, "The ratios [of time to perforation] vary from about 1-to-1 to 4-to-1, depending upon the type of steel and location. The average life of all sheets with 0.20% minimum copper is about *twice* that of sheets with residual copper."

Forty years since Buck's pioneering studies, any lingering arguments about the beneficial effect of copper in raising the atmospheric corrosion resistance of carbon steel have been settled. As part of a very large study involving 270 steel compositions, Larrabee and Coburn (1962) reported the effect of copper on thickness losses of 150×100 mm specimens exposed in rural South Bend, Pa., industrial Kearny, N.J., and moderate marine Kure Beach, N.C. (Fig. 1). Clearly, adding 0.05% Cu contributes most to corrosion resistance, while amounts greater than 0.20% Cu are only marginally beneficial.

Carbon steel was defined later as having 0.02% Cu maximum, and copper steel was defined as having 0.20% Cu minimum. Both were included as references in most future tests leading to the development of weathering steels. As stated above, in the past copper structural steel met the requirements of the ASTM specifications A7, A440, and A441; it now meets the requirements of the ASTM specification A36, when copper steel is specified.

Weathering Steel

Early Studies

Porter and Repas (1982) attribute the origin of high-strength low-alloy steel, which encompasses weathering steels, to Williams (1900) who called attention to the benefit of the increased corrosion resistance of copper steel. US Steel, through its American Sheet and Tin Plate Division, began marketing copper-bearing steel sheets in 1911 (Buck 1920). At about this time, Byramji D. Saklatwalla came from England to Pittsburgh, Pa., to investigate ways of incorporating vanadium into carbon steel. He was aware of the activities of Buck and began some experiments in this field. By 1926, in cooperation with one of US Steel's mills (Vandergrift, Pa.), he obtained a patent covering the addition of several elements other than copper to carbon steels to both enhance atmospheric corrosion resistance and increase the yield strength beyond that achieved by copper, which was 220 to 250 MPa. Meanwhile, from 1920 to 1930, US Steel was developing a family of high-strength low-alloy steels intended primarily for the railroad industry.

At the same time in Pittsburgh, Pa., Jerome Straus, paralleling Saklatwalla's work, also developed a high-strength low-alloy steel composition for which a patent was granted in 1935. US Steel acquired the patent rights of both of these inventors to round out their product line of HSLA steels.

An article by Cone (1934) reviewed new proprietary HSLA steels then available, including Cor-Ten steel. Tables 1 and 2 compare their mechanical properties and chemical compositions.

USS Cor-Ten steel, the first and best known of the weathering type high-strength low-alloy steels, was first marketed in 1933. Its

Table 1. Mechanical Properties of 6–20 mm Thick Proprietary High-Strength Low-Alloy Steels (Cone 1934)

Steel type	Yield strength (MPa)	Tensile strength (MPa)	Elongation (%)
Cromansil	450	635	25
Manten	445	645	19
Granite City	380	555	19
Centralloy	380	490	25
R.D.S.	445	510	31
Yolloy	410	610	27
Inland-Hi-Steel	410	555	22
Armco HT 50	325	480	28
Lorig & Krause	445	555	27
Cor-Ten A	350–410	450–520	27–22

acronym, Cor-Ten, was derived from the two properties that distinguish it from mild carbon steel and copper-bearing steel. These are improved atmospheric corrosion resistance and higher yield and tensile strengths.

The most prominent competing steel was trademarked Mayari R (Bethlehem Steel Corporation), based on the fiery appearance of the natural nickel-chromium ore obtained from the Mayari District in northeastern Cuba. The "R" stood for enhanced atmospheric corrosion resistance. Youngstown Steel distributed its Yolloy brand of atmospheric corrosion-resistant steel. Madison (1966) of Bethlehem Steel began referring to its Mayari R steel as weathering steel. The use by the construction industry of the term weathering steel now makes it recognizable by the general public.

The major patents covering Cor-Ten steel expired in 1940 and 1950, respectively. Thereafter, many steel firms developed their own composition of weathering steel for applications where atmospheric corrosion resistance was desired.

Major Studies

Following the commercialization of HSLA steels during the 1930s, several research reports and journal articles described the enhanced atmospheric corrosion resistance of weathering steel. Cited below are the three major studies. Epstein and Horton (1949) reported the results of a comprehensive program in which the effects of 15 different alloying elements on three base compositions were determined by means of atmospheric testing. The base compositions were carbon steel, copper steel, and Mayari R steel. The 15 alloying elements were carbon, manganese, phosphorus, sulphur, silicon, nickel, chromium, copper, aluminum, ar-

senic, cobalt, molybdenum, tin, tungsten, and vanadium. About 18,000 specimens of nearly 300 different compositions were exposed in the industrial atmospheres of Bethlehem, Pa., Pittsburgh, Pa., and Columbus, Ohio, for from 7 to 16 years. Fifty years later, Townsend (2000) used the data to estimate the thickness loss of weathering steels of different compositions. With the closure of many steel plants and the enforcement of clean air policies by the Environmental Protection Agency, air pollution at the test sites is much lower now than it was 50 years ago.

Copson (1960) performed atmospheric exposure tests of steels with 76 different compositions including various amounts of P, Si, Ni, Cr, and Cu. The test sites were located in industrial Bayonne, N.J., marine Block Island, R.I., and moderate marine Kure Beach, N.C. The exposure times at the three sites were 18.1, 17.1, and 15.5 years, respectively.

Larrabee and Coburn (1962) studied the contribution to corrosion resistance made by the elements copper, phosphorus, silicon, nickel, and chromium. They demonstrated the interplay of the various combinations by testing 270 steels with three variations of chromium, five of copper, three of phosphorus, three of silicon, and two of nickel. Duplicate specimens of 270 compositions were exposed in semirural South Bend, Pa., industrial Kearny, N.J., and moderate marine Kure Beach, N.C. Exposure times were 0.5, 1.5, 3.5, 7.5, and 15.5 years, beginning in 1942 and ending in 1958. In total, 8,000 specimens were tested. Only the 15.5-year data were released. The corrosion index in the G101 *Guide for Estimating the Atmospheric Corrosion Resistance of Low-Alloy Steels* is based on the 15.5-year data from Kearny, N.J. Gallagher (1976) reported a 40% drop in corrosivity at the Kearny site between 1942, when the test was started, and 1974. An even larger drop can be expected at the time of this writing.

These three outstanding and comprehensive studies mark the contemporary history of the HSLA steels and the efforts to place on a firm basis the necessary alloy content, as well as the possible operating mechanism and the influence of widely differing environments.

The steels designated as ASTM A588 in 1968 were part of a broader effort to develop HSLA steels capable of being welded in thick sections. The development effort began after World War II, when concern about the brittle long-running fractures experienced in ships, pressure vessels, and some bridges focused attention on the fracture toughness and the ductile-to-brittle transition temperature of plate steels. Corrosion-resistant HSLA steels could be produced economically by hot rolling, but weldability generally

Table 2. Chemical Composition of 6–20 mm Thick Rolled Plates of Proprietary High-Strength Low-Alloy Steels (Cone 1934)

Steel type	Year rolled	Chemical Composition (%)						
		C	Mn	P	Si	Ni	Cr	Cu
Cromansil	1937	0.20	1.24	—	0.76	—	0.47	0.50 maximum[a]
Manten	1937	0.23	1.40	—	0.23	—	—	0.30
Granite City	1937	0.14	0.90	—	0.20	—	0.12 maximum	0.30
Centralloy	1937	0.15	0.60	—	—	0.25 maximum	0.25 maximum	0.60
R.D.S.	1937	0.08	0.70	—	—	0.75	—	1.40
Yolloy	1937	0.20	0.67	(a)	0.25	2.00	—	1.00
Inland-Hi-Steel	1937	0.10	0.50	0.12	0.15	0.50	—	1.00
Armco HT 50	1937	0.12	0.60	0.10	—	0.50	—	0.50
Lorig & Krause	1937	0.07	0.47	0.26	0.02	—	—	1.40
Cor-Ten	1937	0.10	0.10–0.30	0.10–0.20	0.50–1.00	—	0.50–1.50	0.30–0.50

[a]When specified.

decreased as strength and plate thickness increased and the transition temperature was raised.

The ensuing broad-scale research in the United States, Canada, Europe, and Japan on grain size effects, strengthening mechanisms, composition, and steel cleanliness resulted in the currently available tough and weldable HSLA steels. These steel products combine carefully controlled heating and rolling procedures with optimum use of alloying elements (Porter and Repas 1982).

Specifications

A242 Steel

In 1941, ASTM adopted for the first time a specification for weathering steel named "High-Strength Low-Alloy Structural Steel," designation A242. The material was at first limited to 5–50 mm in thickness and, in 1955, to 100 mm. In 1960, the atmospheric corrosion resistance was quantified as being "equal to or greater than" copper steel. In 1968, concurrently with the adoption of the new A588 specification, the atmospheric corrosion resistance was specified as being "at least" two times that of copper steel. That established A242 steel as having a higher corrosion resistance than A588 steel, which was specified to have corrosion resistance "approximately" two times that of copper steel.

A588 Steel

In 1968, ASTM adopted the "Specification for High-Strength Low-Alloy Steel with 345 MPa Minimum Yield Point to 100 mm Thick," designation A588. The steel was intended for use in welded bridges and buildings. Its atmospheric corrosion resistance was said to be approximately twice that of structural carbon steel with copper. It once was available in ten proprietary grades (A to K), each based on minor variations of the same basic chemical composition. By 1990, mergers and bankruptcies of steel producers have reduced the number of grades to the following four: A, Cor-Ten B supplied by US Steel; B, Mayari R-50 supplied by Bethlehem Steel; C, Stelcoloy 50 supplied by Stelco; and K, Dura Plate 50 supplied by Republic Steel. All four combine low nickel with medium silicon, chromium, and copper contents. According to the American Iron and Steel Institute, today eight companies in the United States and four in Canada produce shapes and plates made of weathering steel.

A709 Steel

In 1975, ASTM issued the "Standard Specification for Carbon and High-Strength Low-Alloy Structural Steel Shapes, Plates, and Bars and Quenched-and-Tempered Alloy Structural Steel Plates for Bridges," designation A709. In the 2000 edition, six grades are available in four strength levels: 250, 345, 485, and 690 MPa. Grades 345W, HPS 485W, and 690W are weathering steels.

Applications

The first major application of weathering steel, in 1933, was in the form of exterior painted coal hopper cars and barges. This met with instant success, and since then more than 1,000,000 such railway cars have been fabricated and placed in service worldwide. The next major application at that time resulted from the Great Depression. It was the standard design for an electrified railway or trolley car known as the President's Conference Car (PCC), a result of President Roosevelt's effort to improve the nation's transit system while providing an economic stimulus to the steel industry. Many such cars went into service from 1936 to 1939 with the remainder supplied from 1946 to 1949.

In 1948, an unpainted Cor-Ten steel angle member was installed in a galvanized carbon steel transmission tower in the Gary Steel Works of US Steel. Periodic inspections indicated that it was performing effectively. Encouraged by this example, a bare steel transmission tower line of Cor-Ten steel was built along Lake Michigan in the Gary Steel Works. General Electric in Massachusetts also erected an experimental tower. The first extended transmission line was erected by Vepco in Virginia followed by the erection of similar lines in Missouri and Georgia.

The John Deere Administration Center, located on the outskirts of Moline, Ill., includes a pedestrian bridge leading to a 2-story exhibition building, which is attached to the seven-story administration building. Additional buildings have since been added. This complex represents the first instance in which unpainted Cor-Ten steel was used for structural members and all exterior columns, posts, beams, girders, and sun-control devices.

The first application of weathering steel in a high-rise structure took place in the heart of a metropolitan business district, with the erection of the 32-story Chicago Civil Courts Building in 1964 and 1965. It was erected with uncoated steel curtain walls consisting of fascia, column covers, and window frames. Test panels exposed 800 m from the building had indicated that the environment was satisfactory and the terminal color would likely be reached in about three years. All exposed steel was fabricated from A242 (Cor-Ten A) steel. The Picasso sculpture on the plaza was fabricated from A588 Grade A (Cor-Ten B) steel.

The experience with transmission towers and buildings, coupled with the possibility of avoiding or minimizing costly maintenance painting of the structures during their service lives, led to the construction in 1964 and 1965 of the first four unpainted weathering steel bridges. Iowa erected a bridge on Iowa Rt. 28 over the Racoon River. The Michigan Highway Department built four structures at the crossing of the Eight-Mile road over US 10 in Detroit. Ohio replaced an old bridge with a new all-welded, three-span steel bridge across Brush Creek in Miami County, north of Dayton. New Jersey erected a four-span composite bridge as an adjunct to the New Jersey Turnpike near Morristown. Since then about 4,000 bridges on state highway systems have been built of bare, exposed weathering steel. This number does not include bridges owned by local government agencies.

Experience with bare, exposed weathering steel has not always been satisfactory. The Racoon and 8-Mile Road bridges corroded excessively and were remedially painted. The Brush Creek bridge was strengthened along the web-to-flange junction because corrosion had perforated the web. Crevice corrosion severely deformed bolted joints in transmission towers, also called rust pack-out. Corrugated siding and roofing sheets for buildings rusted through along the end laps and side laps. All of these problems can be traced to long periods of wetness, which prevent the formation of a dense, protective rust coat. They can be overcome by a better understanding of the material's limitation and greater attention to proper detailing.

Experimental Methods

Atmospheric Corrosion

Several excellent references describe the characteristics of atmospheric corrosion of steel (Kucera and Mattson 1987, Uhlig and

Revie 1985, and Schweitzer 1989). From an engineering viewpoint, the most succinct is that offered by Wranglen (1972, p. 62):

Atmospheric corrosion in general is the result of the conjoint action of two factors: oxygen and moisture (water in liquid form). If one of these factors is missing, corrosion does not occur. In dry air, as under the freezing point or at a relative humidity (RH) less than about 60%, steel does not rust. Corrosion is therefore negligible in polar regions and in hot deserts. Indoors in heated, i.e., dry localities, steel does not rust either. In unheated premises, however, the humidity may be so high that rusting can occur.

The atmospheric corrosion increases strongly if the air is polluted by smoke gases, particularly sulfur dioxide from fossil fuels, or aggressive salts, as in the vicinity of chimneys and marine environments. The atmospheric corrosion is therefore particularly strong in industrial and coastal areas. The corrosion is, furthermore, much higher if the metal surface is covered by solid particles, such as dust, dirt, and soot, because moisture and salts are then retained for longer time.

The degree to which steel will rust in the atmosphere is directly related to the so-called time of wetness, defined as the average number of hours per year during which the steel is wet. The longer a steel surface remains wet, the more it will corrode. The time of wetness varies with the climatic conditions at the site. It depends on the RH of the atmosphere, as well as the duration and frequency of fog, dew, rain, and snowfall.

Direct precipitation is not needed to wet a steel surface. Moisture can be deposited also by capillary action of the porous rust coating, and adsorption by corrosion products or salt deposits on the steel surface. Pores and cracks in the rust coating, crevices, and small pits all foster capillary action.

Such moist atmospheric corrosion, as opposed to wet atmospheric corrosion, takes place when a thin, invisible film of electrolyte forms on the surface at a relative humidity well below 100%. The critical RH, above which the atmospheric corrosion rate increases sharply, depends on the type of metal, surface contaminants, atmospheric pollutants, and the nature of the corrosion product.

Vernon (1931; 1933; 1935) first showed the effect of sulfur dioxide, in conjunction with relative humidity, on the corrosion of iron. He corroded carbon steel specimens in bell jar atmospheres with controlled humidity, SO_2 content, and temperature. In the absence of SO_2, even saturated humid air (100% RH) did not cause appreciable rusting. When the SO_2 content was raised to 0.01%, the critical RH dropped to 60%. At 0.01% SO_2 content and relative humidity of 70, 75, and 99%, iron corroded 7, 8, and 12 times more than in 100% humid air free of SO_2.

Tomashov (1966) determined the effect of steel surface condition on the critical RH. He found that the corrosion rate of iron, with a prior rust coating that formed in water or in 3% sodium chloride solution, increases sharply when the RH of the air rises above 65 and 55%, respectively. At these values, a continuous film of moisture forms by capillary condensation or by hydration of salts, corrosion products, and other films that might be present on the surface.

Environment Types

General Classification

Atmospheric corrosion tests are generally performed in three types of environments—rural, industrial, and marine. Industrial environments are generally more corrosive than rural environments, and marine environments are much more corrosive than rural and industrial environments. These broad classifications are not entirely accurate because many factors contribute to the corrosivity of a test site. For example, the corrosivity of a marine environment depends on the distance from the shore, prevailing wind direction, storm activity, and chloride concentration.

While exposure sites fall into one of the three broad categories, some distinctions are made. For example, the test site in Kure Beach, N.C., is called moderate marine when the specimens are exposed 250 m from the ocean and severe marine when exposed 25 m from the ocean. In other examples, South Bend, Pa., is classified as being semirural and Monroeville, Pa., as semiindustrial. While Potter County, Pa., is rural, the acidity of the rainfall makes it behave toward steel more like an industrial environment.

Rural Environments

The atmosphere of a rural environment is generally found at locations remote from urban centers and industrial plants. The corrosion rates of steel in these environments, which are usually free of aggressive agents, are generally lower than in the other types of environments. The corrosive elements that exist in rural environments include moisture, and relatively small amounts of sulfur oxides (SO_x) and carbon dioxide (CO_2). Ammonia (NH_3), resulting from the decomposition of farm fertilizers or animal excrement, may also be found. Since these elements are present in small quantities, rural environments are not generally aggressive toward steel.

Industrial Environments

Industrial environments contain pollutants produced by nearby industrial facilities, such as power plants or manufacturing plants. The principal corrosives found in these environments are sulfur oxides (SO_x) and nitrogen oxides (NO_x). Sulfur dioxide is emitted from the burning of coal and oil, or from smokestacks in refineries and chemical plants. When sulfur dioxide dissolves in moisture that is present on a steel surface, sulfuric acid forms and corrodes the steel. Levels of sulfur oxides in the air are highest during the heating season.

Government regulations on combustion emissions have reduced pollution significantly over the years. As an example of the steadily improving air quality, Gallagher (1976) reported that the first 6 months thickness losses of copper steel exposed at the Kearny, N.J., test site of US Steel decreased from 0.046 to 0.025 mm between 1934 and 1974.

Marine Environments

Marine environments are generally the most aggressive towards steel. Among the reasons for this are the salt spray produced by nearby bodies of water and the nightly temperature drops near the shore. Moisture retained by salt on steel surfaces, coupled with the presence of chloride ions, causes all structural steels to corrode severely. In the absence of direct sunlight, a salt-covered surface can stay moist most of the day. Other factors affecting corrosion in marine environments are wave action at the surf line, prevailing wind direction, shoreline topography, and relative humidity. Because these factors are related to bodies of water, corrosivity decreases with increased distance from the shore. Sea salt and deicing salt on highways are extremely corrosive to all structural steels.

Baker and Lee (1982) measured chloride by wet candle in both the 25 and 250 m Kure Beach sites, at a fixed height of 1.50 m above ground. The average annual levels from 1962 to 1979 remained relatively constant at about 100 mg/m^2/day at the 250 m site, but continuously declined from 650 to 250 mg/m^2/day at the 25 m site. Iron calibration specimens exposed vertically at the 25 m site for either 1 or 2 years showed a decline in corrosion rates consistent with the decline in chloride levels. Baker and Lee (1982, p. 262) attributed the decline "to the steadily increasing size of the sand dunes between the 25-m site and the ocean. This in turn would tend to shift the maximum chloride content in the atmosphere to a somewhat higher elevation and thereby reduce the corrosiveness of the 25-m site." Beginning in 1999, large two-story beach houses were being built on portions of the flat and wide-open 25 and 250 m sites, forever changing local wind patterns and, hence, the environmental characteristics.

Climatic and Pollution Data

The examples of industrial Kearny, N.J., and marine Kure Beach, N.C., illustrate how atmospheric test sites are changing with time. Therefore, steel specimens of identical steel composition exposed at the same site, but at different times, cannot be expected to corrode equally.

Climatic and pollution data should be collected during exposure tests of steels. These include air temperature, relative humidity, sulfur dioxide deposition, airborne chloride, time of wetness of steel surfaces, and surface temperature of steel specimens (Pourbaix and Miranda 1983). Such measurements are needed to evaluate the suitability of bare, exposed weathering steel at the site of a planned new structure based on its performance at atmospheric test sites. Many European studies have collected environmental data along with thickness loss data so that the results could be extrapolated to other sites and environments. Unfortunately, this has not been the rule in domestic studies, which have focused almost exclusively on product development, that is, the relative corrosion resistance of carbon, copper, and weathering steels.

Exposure Tests

Steel Types

Four types of steel were exposed in domestic atmospheric corrosion tests over the years: A242 weathering steel, A588 weathering steel, copper steel (0.20% Cu minimum), and carbon steel (0.02% Cu maximum). Compositions vary with grade and time. US Steel produces ASTM A242 and A588 grade A under the trade names Cor-Ten A and Cor-Ten B. Bethlehem Steel produces ASTM A242 and A588 grade B under the trade names Mayari R and Mayari R-50.

Ten of the 14 major atmospheric corrosion studies of weathering steel examined in this paper used copper steel as a reference. The copper content was 0.21 to 0.23% in six studies, 0.26% in three studies, and 0.32% in one study. As Fig. 1 shows, variations in copper content from 0.21 to 0.32% have only a small effect on thickness loss. Thus, differences in performance of copper steel can be used to determine the relative severity of exposure sites; this procedure is often called calibration of sites.

Specimens

Most specimens tested after the 1930s had standard dimensions of 152 mm length by 102 mm width. The *Standard Practice for Conducting Atmospheric Corrosion Tests on Metals, ASTM designation G50*, recommends a minimum thickness of 0.75 mm. One notable exception to this standard size is found in the study by Reed and Kendrick (1982), who exposed two types of specimens: one type consisted of a 305 by 305 mm plate containing two butt welds; the other type consisted of a 305 by 152 mm plate to which a 75-mm-long angle of 75 mm leg sizes was welded to each side of the plate.

Test Racks

Test specimens for determining the atmospheric corrosion resistance of carbon, copper, and weathering steels were mounted on racks constructed from a variety of materials, the only requirement being that the material does not alter the corrosion of the specimen in any way. Orientation and angle of exposure are the most important characteristics of a test rack. Both affect the corrosion rate significantly.

To obtain comparable results, the G50 practice recommends exposing specimens at an orientation and angle that yields the least amount of corrosion. This condition is met in the northern hemisphere when the sun is above the equator and the specimens face south at an angle equal to the latitude of the test site. Conversely, in the southern hemisphere, specimens should be exposed facing north at an angle equal to the latitude of the test site.

Most atmospheric corrosion tests in the United States were performed at an angle of 30° facing south. One consistent exception has been the 30° exposures facing the ocean in the severe marine site in Kure Beach. As an example, the following northern cities, beginning on the Pacific Coast of the Americas and moving eastward to Europe and Asia, are close to the 30° N latitude: Rosario, Baja California, Mexico; Houston, Tex; New Orleans, La.; Jacksonville, Fla.; Agadir, Morocco; Cairo, Egypt; Delhi, India; Chengdu, China; and Shanghai, China. Very few test data have been reported for sites south of the equator.

Although the 30° latitude passes through Houston, New Orleans, and Jacksonville in the deep south of the United States, most atmospheric exposure tests were performed at sites located farther north in the sates of Michigan, Ohio, Pennsylvania, Rhode Island, and North Carolina.

Exposures other than 30° facing south were chosen in some studies for various reasons. For example, at the severe marine site in Kure Beach, N.C., the specimens faced the ocean at 30° from the horizontal to maximize the exposure to thermal winds and salt spray from waves breaking on the beach.

In another study (Gallagher 1982b; Komp et al. 1993; Coburn et al. 1995), groups of four specimens were placed on the racks at 90° facing south, 90° facing north, 30° facing south, and 30° facing north. All specimens were painted on the back side.

Atmospheric exposure studies in Europe are usually done with specimens facing south at 45° from the horizontal, owing to the higher latitudes. The following northern cities, again beginning on the Pacific Coast of the Americas and moving eastward to Europe and Asia, are close to the 45° N latitude: Salem, Or.; Minneapolis, Minn.; Sherbrooke, Quebec; Halifax, Nova Scotia; Bordeaux, France; Turin, Italy; Belgrade, Yugoslavia; Sebastopol, Ukraine; Harbin, China; and Wakannai, Japan.

Exposure Times

Pairs of test specimens were removed from the racks at the predetermined intervals. The G50 practice recommends a removal schedule at 1, 2, 4, 8, and 16 years. Bethlehem Steel removed their specimens yearly in their early studies (Epstein and Horton 1949; Horton 1964, 1965) and adopted the G50 removal schedule in later studies. US Steel followed its own removal schedule of 1.5, 3.5, 7.5, and 15.5 years. Reed and Kendrick (1982) removed all specimens at one time, after 13 years of exposure.

Thickness Loss

The rust that had built up on the specimens was chemically removed, the mass loss was determined, and the specimens were discarded. According to the *Standard practice for preparing, cleaning and evaluating corrosion test specimens*, ASTM designation G1, the following equation yields the thickness:

$$C = K \frac{W}{AD} \qquad (1)$$

where C = average thickness loss per surface, mm; K = conversion constant; W = mass loss, grams; A = total exposed surface area, cm^2; and D = density, g/cm^3. When the units are used as specified above, the conversion constant is $K = 10$.

It is important to note that the thickness loss reported in corrosion studies is the average of the thickness losses of the exposed skyward and groundward surfaces of a specimen. Since the specimens are thin, their edges are often neglected in the calculation of exposed area. Larrabee (1953) reported that for specimens exposed at 30° from the horizontal, the skyward surface accounted for 38% of the total thickness loss, while the groundward surface accounted for 62%.

Corrosivity Bands

ISO Method

The International Standard *Corrosion of Metals and Alloys—Corrosivity of Atmospheres*, ISO designation 9224, specifies guiding values of corrosion rate for metals exposed to the atmosphere consisting of an average corrosion rate during the first 10 years of exposure, r_{av}, followed by a steady-state corrosion rate after 10 years of exposure, r_{lin}. Values are given for five corrosivity categories—C1, C2, C3, C4, and C5—the order being from the least to the most corrosive. Standard 9224 can be used to predict the service life of metals.

The category of most interest is C3, herein referred to as the medium corrosivity category because it defines the degree of thickness loss expected from well-performing weathering steel. Categories C1 and C2 can be characterized as being of very low and low corrosivity; categories C4 and C5 of high and very high corrosivity. The 9224 standard provides guiding values for the corrosion rate of carbon steel and weathering steel. Lacking guiding values for the corrosion rate of copper steel, the thickness loss data for copper steel are compared herein, by default, with the guiding values for weathering steel.

According to the 9224 standard specification, the upper and lower bounds on thickness loss for any corrosivity category are given by

$$C = t r_{av} \qquad (2)$$

for the first 10 years of exposure (dotted lines in Figs. 2 and 3) and

$$C = 10 r_{av} + (t - 10) r_{lin} \qquad (3)$$

for exposure beyond 10 years (solid lines in Figs. 2 and 3), where C = thickness loss, mm; t = time, years; r_{av} = average corrosion rate, mm/year (Table 3); and r_{lin} = steady-state corrosion rate, mm/year (Table 3).

Modified ISO Method

The 9224 standard acknowledges in a footnote of its Table 1, "the corrosion rate of carbon steel is not constant during the first 10

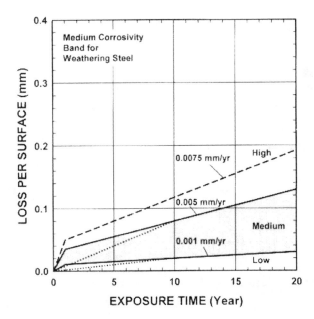

Fig. 2. Medium corrosivity band for weathering steel

years." Nor is that of weathering steel. To rectify this inaccuracy, the authors modified the bilinear thickness loss curves as follows. For the first year of exposure

$$C = 1 r_{av} \qquad (4)$$

and beyond the first year of exposure (solid lines in Figs. 2 and 3)

$$C = 1 r_{av} + (t - 1) r_{lin} \qquad (5)$$

Tables 3 and 4 list the corrosion rates and thickness losses at 1, 10, and 20 years of exposure, according to the 9224 standard and the herein modified ISO Standard 9224. The shaded space between the lower and upper bounds represents the medium corrosivity bands for weathering steel and carbon steel (Figs. 2 and 3).

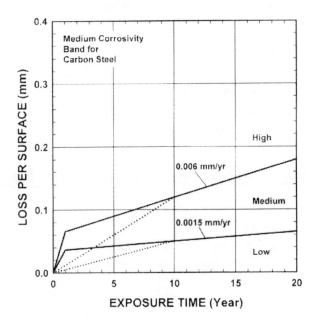

Fig. 3. Medium corrosivity band for carbon steel

Table 3. Corrosion Rates for Medium Corrosivity Category C3

Steel type	Bounds	CORROSION RATE (mm/year)			
		ISO Standard 9224		Modified ISO Standard 9224	
		Average 0 to 10 years r_{av}	Steady state >10 years r_{lin}	Modified 0 to 1 year	Steady state >1 year r_{lin}
Weathering	Upper	0.008	0.005	0.035	0.005
	Lower	0.002	0.001	0.011	0.001
Carbon	Upper	0.012	0.006	0.066	0.006
	Lower	0.005	0.0015	0.0365	0.0015

Three years before the 9224 standard was first issued, Albrecht et al. (1989) had reported that the thickness loss found for weathering steel in a variety of bridges in the contiguous United States had an upper bound of

$$C = 0.05 + (t-1)0.0075 \quad (6)$$

for exposure beyond the first year (dashed line in Fig. 2). They concluded that "weathering steel corroding at a rate higher than 0.0075 mm/year cannot be expected to develop a normal, protective rust coating. When such corrosion rates are anticipated, weathering steel should not be used in bare condition."

ISO's choice of an average 0.0075 mm/year corrosion rate for the high corrosivity category C4 in Standard 9224 is consistent with the upper bound rate for satisfactory performance proposed by Albrecht et al. (1989).

Thickness Loss Data

Introduction

The writers analyzed data from 14 domestic studies on atmospheric corrosion of structural steels that have been performed since the mid-1930s. In most studies, the specimens were made of A242, A588, copper, and carbon steels, had dimensions of 152 mm length and 52 mm width, and 0.25 to 5 mm thickness, and were exposed at 30° from the horizontal facing south for at least 8 years. Table 5 lists the following parameters in chronological order of the 14 studies: study number, start and length of exposure, exposure sites, exposure angle and orientation, steel types, and specimen dimensions.

The steel types include A242 (Cor-Ten A and Mayari R), A588 (Cor-Ten B and Mayari R-50), copper steel, and carbon steel.

Table 4. Thickness Loss for Medium Corrosivity Category C3

Steel type	Method	Bound	Thickness Loss (mm)		
			1 year	10 years	20 years
Weathering	ISO	Upper	0.008	0.08	0.13
	ISO	Lower	0.002	0.02	0.03
	ISO modified	Upper	0.035	0.08	0.13
	ISO modified	Lower	0.011	0.02	0.03
Carbon	ISO	Upper	0.012	0.12	0.18
	ISO	Lower	0.005	0.05	0.065
	ISO modified	Upper	0.066	0.12	0.18
	ISO modified	Lower	0.0365	0.05	0.065

Table 6 lists for each steel the content of the five elements known to contribute most to atmospheric corrosion resistance: P, S, Ni, Cr, and Cu. The last two columns of Table 6 give for each composition the predicted 15.5-year thickness loss that would have occurred if the steels had been exposed in industrial Kearny, N.J., from 1942 to 1958 (Larrabee and Coburn 1962).

The first value was calculated with Legault and Leckie's (1974) quadratic model, using 260 of the 270 compositions that Larrabee and Coburn (1962) had tested. This model became the basis for the corrosion index in the G101 guide. It is inaccurate because the quadratic model predicts an optimum level of copper at which thickness loss reaches a minimum within the compositional levels in the Larrabee and Coburn (1962) data set. McCuen and Albrecht (1994a) showed that the convex shape of the copper curves that Legault and Leckie (1974) reported in their Figs. 3 to 7 is an artifact of the quadratic model, not a representation of the data. The writers did not find any data in the literature showing an optimal level of copper at which thickness loss reaches a minimum. In all studies, weathering steels rust least at the highest contents of all five elements that contribute most to corrosion resistance—P, Si, Ni, Cr, and Cu.

The second value was calculated with McCuen and Albrecht's (1994a) exponential model using the 202 of the 270 compositions that had copper contents of 0.05% or greater. It describes the effect of the alloying elements more accurately. For example, the exponential model predicts a thickness loss of 0.255 mm for the 12 production heats of Cor-Ten steel at the top of Table 6 and a thickness loss of 0.033 mm for A242 Cor-Ten A steel in study No. 3 (Larrabee 1959). In contrast, the quadratic model predicts thickness gains of $(-)0.001$ and $(-)0.004$ mm.

The trade names of weathering steels are cited along with the corresponding ASTM specification numbers A242 and A588 for two reasons. First, the A242 steel specification does not identify the composition ranges for the various grades available from producers. Second, users are more familiar with the trade names than the letter grade designations in the A588 steel specification. In fact, both trade names have become synonymous with the term weathering steel.

The steel specimens were exposed in the rural, industrial, and marine environments described in the Appendix. Three sites are generally classified as being rural: Potter County, Saylorsburg, and South Bend—all three located in Pennsylvania.

Of the many industrial sites, the most commonly used are Kearny, N.J., and Bethlehem and Monroeville, Pa. Other industrial sites are Bayonne, N.J.; Columbus, Ohio; Pittsburgh, Pa.; Los Angeles and Sacramento, Calif.; and Detroit, Mich.

Kure Beach, located on the coast of North Carolina, is the major marine test site in the United States; Block Island, R.I., and Point Reyes, Calif., are rarely used today.

Differences between environments affect thickness loss much more than differences between grades of a steel type, be it A242 or A588 steel. Changes in the corrosivity of some environments have significantly affected thickness loss. Four examples are cited next. First, very tall exhaust stacks of fossil fuel power plants discharge sulphur dioxide that is carried by the west-to-east jet stream, giving rise to the acid rain phenomenon. This has increased the corrosivity of the rural Potter County, Pa., site, which is in the path of the jet stream. Second, "by contrast, the aggressive industrial site in Kearny, NJ, owing to the imposition of lower sulfur levels for fossil fuels, lost much of its corrosivity" (Komp et al. 1993). Third, steel plant closings in Bethlehem and

Table 5. Summary of Atmospheric Exposure Studies at Domestic Sites

Study/Curve Number	Exposure Time	Exposure Site	Exposure Angle and Orientation	Steel Type	Specimen Length, Width and Thickness (mm)
		(a) Epstein and Horton (1949), Horton (1964), Horton (1965), Madison (1966), Mayari (1967, 1983 or later)			
1	1934–1944	Pittsburgh, Pa. (I)	90° (vertical) tensile, specimen standing on edge	A242 Mayari R, test F Copper, carbon	254×13 to 19×0.8
	1938–1948	Bethlehem, Pa. (I)			
	10 years	Columbus, Ohio (I)			
		(b) Epstein and Horton (1949), Mayari (1967, 1983)			
2	1934–1951	Pittsburgh, Pa. (I)	90° (vertical) tensile, ,specimen standing on edge	A242 Mayari R, test H A242 Mayari R, Code LQ	254×13 to 19×0.8
	1938–1955				
	17 years				
		(c) Larrabee (1959), Larrabee and Coburn (1962), Cruz and Mullen (1962), U.S. Steel (1962, 1980 or later)			
3	1936–1959	South Bend, Pa (I) (semi-R)	30° facing south	A242 Cor-Ten A Copper, carbon	254×254×0.8, 1.6, 3.2, 4.8
	23 years				
	1936–1956	Kearny, N.J. (I)			
	20 years				
		(d) Copson (1960)			
4	1941–1959	Bayonne, N.J. (I)	30° facing south	A242, No. 14 Copper, No. 6, 41, 57 Carbon, No. 45	152×102×0.7 to 2.1
	18.1 years				
	1941–1956	Kure Beach, N.C. (moderate M)			
	15.5 years				
	1941–1958	Block Island, R.I. (M)	30° facing ocean		
	17.1 years				
		(e) Horton (1964)			
5	1956–1966	Bethlehem, Pa. (I)	30° facing south	A242 Mayari R Copper, carbon	152×102×0.25
	10.1 years	Kure Beach, N.C. (moderate M)			
		Kure Beach, N.C. (severe M)	30° facing ocean		
		(f) Schmitt and Mullen (1965), Gallagher (1976), Komp (1987)			
6	1957–1973	South Bend, Pa. (semi-R)	30° facing south	A242 Cor-Ten A A588 Cor-Ten B, E138 A588 Cor-Ten B, E158 Copper, carbon	152×102
	15.5 years	Monroeville, Pa. (semi-I)			
		Newark, N.J. (I)			
		Kure Beach, N.C. (moderate M)			
		Kure Beach, N.C. (severe M)	30° facing ocean		
		(g) Zoccola et al. (1970), Zoccola (1976)			
7	1966–1974	Detroit, Mich. (I)	Mean of 90°, 0° top painted, 0° bottom painted	A242 Mayari R	152×102
	8 years				

Table 5. (*Continued*)

Study/Curve Number	Exposure Time	Exposure Site	Exposure Angle and Orientation	Steel Type	Specimen Length, Width and Thickness (mm)
(h) Cosaboom et al. (1979), Townsend and Zoccola (1982), Mottola (1989)					
8	1968–1984 16 years	Newark, N.J. (I)	30° facing south	A242 Mayari R A588 Mayari R-50 Copper, carbon	152×102×2.5 to 5.0
(i) Townsend and Zoccola (1982), Shastry et al. (1986)					
9	1968–1984 16 years	Saylorsburg, Pa. (R) Bethlehem, Pa. (I) Kure Beach, N.C. (moderate M)	30° facing south	A242 Mayari R A242 Cor-Ten A A588 Mayari R-50 A588 Cor-Ten B Copper, carbon	152×102×2.5 to 5.0
(j) Reed and Kendrick (1982)					
10	1968–1981 13 years	Sacramento, Calif. (I) Los Angeles, Calif. (I) Point Reyes, Calif. (M)	30° facing S 30° facing E 30° facing W	A242 Cor-Ten A A242 Mayari R A588 Cor-Ten B Carbon	Butt-welded plate and plate with angle welded all around
(k) Gallagher (1982b), Komp et al. (1993), G101 Guide					
11	1972–1988 16 years	Potter County, Pa. (R) Kearny, N.J. (I) Kure Beach, N.C. (moderate M)	Mean of 30°S, 30°N, 90°S, 90°N; painted on back side	A242 Cor-Ten A A588 Cor-Ten B Copper, carbon	152×102×3.2 to 4.8
(l) Gallagher (1978, 1982a), Vrable (1985), Komp (personal communication, 1991)					
12	1976–1992 15.5 years	Potter County, Pa. (R) South Bend, Pa. (semi R) Monroeville, Pa. (Semi I) Kearny, N.J. (I) Kure Beach, N.C. (moderate M) Kure Beach, N.C. (severe M)	30° facing south	A242 Cor-Ten A A588 Cor-Ten B, pre-78 A588 Cor-Ten B post-78 Copper, carbon	152×102
(m) Tinklenberg (1986)					
13	1976–1993 16 6 years	Detroit, Mich. (I)	0° (horizontal)	A242 Mayari R A588 Mayari R-50 A588 Mayari R-50, low Ni Copper, carbon	152×102
(n) Townsend (2000)					
14	~ 1990–1998 8.05 years	Saylorsburg, Pa. (R) Bethlehem, Pa. (I) Kure Beach, N.C (moderate M)	30° facing south	A588 Mayari R-50 norm. A588 Mayari R-50 Q & T A588 Mayari R-50 lean A	152×102

Note: I, Industrial; M, Marine; R, Rural.

Table 6. Chemical Composition of Steels Exposed to Atmosphere at Domestic Sites

Study/Curve Number	Steel Type	Composition (%) P	Si	Ni	Cr	Cu	Predicted Thickness Loss per Surface[a] (mm) Legault and Leckie (1974)	McCuen and Albrecht (1994a)
	(a) Average composition of 12 production heats							
—	Cor-Ten	0.156	0.846	0.057	0.979	0.433	−0.001	0.255
	(b) Average composition of 36 production heats (Komp, personal communication, 1991; Townsend, personal communication, 1991)							
—	A242 Cor-Ten A	0.109	0.415	0.080	0.964	0.298	0.038	0.216
—	A242 Mayari R	0.076	0.292	0.292	0.549	0.274	0.066	0.188
—	A588 Cor-Ten B	0.017	0.433	0.210	0.552	0.329	0.083	0.171
—	A588 Mayari R-50	0.013	0.360	0.286	0.519	0.281	0.088	0.166
	(c) Epstein and Horton (1949), Horton (1964), Horton (1965), Madison (1966), Mayari (1967, 1983 or later)							
1	A242 Mayari R, test F	0.100	0.40	0.33	0.56	0.62	0.107	0.033
	Copper	0.046	0.03	—	—	0.22	0.128	0.162
	Carbon	0.047	0.03	—	—	0.02	0.219	0.226
	(d) Epstein and Horton (1949), Mayari (1967, 1983, or later)							
2	A242 Mayari R, Code LM	0.012	0.38	0.46	0.68	0.63	0.144	0.035
	A242 Mayari R, Code LN	0.037	0.36	0.42	0.63	0.59	0.118	0.035
	A242 Mayari R, Code LQ	0.094	0.40	0.40	0.62	0.58	0.093	0.033
	A242 Mayari R, test H	0.102	0.35	0.36	0.62	0.56	0.083	0.033
	(e) Larrabee (1959), Larrabee and Coburn (1962), Cruz and Mullen (1962), U.S. Steel (1962, 1980) (or later)							
3	A242 Cor-Ten A	0.15	0.89	0.04	1.00	0.40	−0.004	0.033
	Copper	0.024	0.025	0.02	0.02	0.32	0.117	0.146
	Carbon	0.019	0.043	0.02	0.02	0.05	0.211	0.235
	(f) Copson (1960)							
4	A242, No. 14	0.089	0.29	0.47	0.75	0.39	0.050	0.035
	Copper; No. 6, 41, 57	0.007	0.008	0.037	0.023	0.207	0.147	0.203
	Carbon, No. 45	0.006	0.003	0.05	—	0.02	0.234	0.247
	(g) Horton (1964), Mayari (1967, 1983)							
5	A242 Mayari R	0.093	0.29	0.72	0.55	0.27	0.049	0.037
	Copper	0.008	—	0.02	0.04	0.23	0.141	0.196
	Carbon	0.009	—	0.02	0.03	0.05	0.216	0.242
	(h) Schmitt and Mullen (1965), Gallagher (1976), Komp (1987)							
6	A242 Cor-Ten A	0.090	0.56	0.49	0.94	0.44	0.040	0.033
	A588 Cor-Ten B, E138	0.021	0.20	0.05	0.46	0.36	0.094	0.085
	A588 Cor-Ten B, E158	0.013	0.26	0.04	0.44	0.31	0.100	0.102
	Copper	0.013	0.029	0.04	0.05	0.23	0.136	0.185
	Carbon	0.017	0.028	0.03	0.03	0.03	0.223	0.240
	(i) Zoccola et al. (1970), Zoccola (1976)							
7	A242 Mayari R, test 1	0.077	0.24	0.75	0.56	0.24	0.057	0.041
	A242 Mayari R, test 2	0.080	0.36	0.72	0.64	0.34	0.047	0.035
	Carbon, test 2	0.011	0.05	—	—	0.015	0.238	0.246
	(j) Cosaboom et al. (1979), Townsend and Zoccola (1982), Mottola (1989)							
8	A242 Mayari R	0.110	0.29	0.66	0.52	0.27	0.047	0.037
	A588 Mayari R-50	0.008	0.22	0.27	0.64	0.21	0.106	0.104
	Copper	0.005	0.01	0.02	0.02	0.21	0.148	0.206
	Carbon	0.009	0.001	0.01	0.02	0.021	0.235	0.246
	(k) Townsend and Zoccola (1982), Shastry et al. (1986)							
9	A242 Mayari R	0.110	0.29	0.66	0.52	0.27	0.047	0.037
	A242 Cor-Ten A	0.090	0.44	0.36	0.67	0.24	0.056	0.041
	A588 Mayari R-50	0.008	0.22	0.27	0.64	0.21	0.106	0.104
	A242 Cor-Ten B	0.017	0.27	0.03	0.52	0.29	0.099	0.099
	Copper	0.005	0.01	0.02	0.02	0.21	0.148	0.206

Table 6. (*Continued*)

Study/Curve Number	Steel Type	Composition (%)					Predicted Thickness Loss per Surface[a] (mm)	
		P	Si	Ni	Cr	Cu	Legault and Leckie (1974)	McCuen and Albrecht (1994a)
	Carbon	0.009	0.001	0.01	0.02	0.021	0.235	0.246
(l) Reed and Kendrick (1982)								
10	A242 Cor-Ten A plate	0.130	0.57	0.48	1.00	0.36	0.016	0.033
	A242 Cor-Ten A angle	0.110	0.35	0.40	1.06	0.32	0.030	0.034
	A242 Mayari R plate	0.090	0.33	0.73	0.60	0.30	0.046	0.036
	A242 Mayari R angle	0.060	0.35	0.64	0.61	0.29	0.056	0.041
	A242 Cor-Ten B plate	0.028	0.23	0.028	0.60	0.32	0.089	0.082
	A242 Cor-Ten B angle	0.021	0.20	—	0.46	0.36	0.095	0.089
	Carbon plate	0.017	0.01	—	—	0.08	0.199	0.235
	Carbon angle	0.011	0.06	—	—	0.02	0.234	0.245
(m) Gallagher (1982b), Komp et al. (1993)								
11	A242 Cor-Ten A	0.092	0.42	0.31	0.82	0.30	0.044	0.038
	A242 Cor-Ten B, pre-78	0.006	0.25	0.015	0.56	0.33	0.099	0.098
	Copper	0.002	0.004	0.014	0.014	0.26	0.137	0.194
	Carbon	0.012	0.016	0.012	0.025	0.014	0.237	0.246
(n) Gallagher (1978, 1982a), Vrable (1985), Komp (personal communication, 1991)								
12	A242 Cor-Ten A	0.092	0.42	0.31	0.82	0.30	0.044	0.038
	A588 Cor-Ten B, pre-78	0.006	0.25	0.015	0.56	0.33	0.099	0.098
	A588 Cor-Ten B, post-78	0.008	0.43	0.21	0.62	0.31	0.084	0.070
	Copper	0.002	0.004	0.014	0.014	0.26	0.137	0.194
	Carbon	0.012	0.016	0.012	0.025	0.014	0.237	0.246
(o) Tinklenberg (1986)								
13	A242 Mayari R	0.110	0.29	0.66	0.52	0.27	0.047	0.037
	A588 Mayari R-50, No. 921	0.011	0.28	0.32	0.55	0.28	0.089	0.080
	A588 Mayari R-50, No. 914	0.013	0.24	0.14	0.50	0.29	0.098	0.097
	Copper	0.006	0.01	0.01	0.01	0.26	0.136	0.190
	Carbon	0.007	0.01	0.01	0.02	0.015	0.239	0.247
(p) Townsend (2000)								
14	A588 Mayari R-50	0.013	0.38	0.27	0.41	0.31	0.088	0.076
	A588 Mayari R-50, lean A	0.009	0.29	0.23	0.22	0.28	0.104	0.109

[a]Predicted thickness loss per surface of specimens exposed from 1942 to 1958 (15.5 years) in Kearny, N.J.

Pittsburgh, Pa., during the past 2 decades have greatly reduced the sulfur dioxide content in the atmosphere. Fourth, sand dune growth at the severe marine test site in Kure Beach, N.C., reduced the amount of airborne chloride, thus reducing the corrosivity of the site (Baker and Lee 1982).

Data from the 14 studies are presented hereafter in sets of thickness loss versus exposure time curves (to be shown in Figs. 4 to 19) and organized by type of environment and type of steel. Within an environment, data are then presented by type of steel in the order of decreasing atmospheric corrosion resistance: A242, A588, copper, and carbon steels. This performance ranking holds for all studies.

A242 and A588 steels are compared against the medium corrosivity band for weathering steel (Fig. 2); the carbon steel data are compared against the medium corrosivity band for carbon steel (Fig. 3). Since ISO standard 9224 does not have corrosivity categories for copper steel, the copper steel data are compared by default against the medium corrosivity band for weathering steel, knowing that copper steel corrodes more than weathering steel.

To avoid overcrowding the first segment in the plots of thickness loss versus exposure time, the lines from the origin to the first data point of all curves were left out.

Rural Environments

A242 Steel (Fig. 4)

All curves fall inside the medium corrosivity band for weathering steel, indicating that the rural environments of South Bend, Saylorsburg, and Potter County are not aggressive toward A242 steel. Curve 3 is lowest because, as Table 6 shows, this early Cor-Ten steel from the mid-1930s had levels of 0.15% P and 0.89% Si that are much higher than the values of 0.109% P and 0.415% Si found in 36 production heats of A242 Cor-Ten A reported by Komp (personal communication, 1991). The thickness losses for curve 11 were calculated as the mean of the four specimens facing 90°N, 90°S, 30°N, and 30°S.

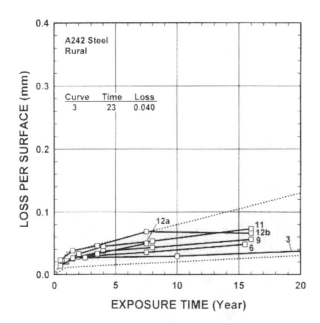

Fig. 4. A242 steel in rural environments

The corrosion rates of most curves (3, 6, 9, and 11) remained constant after 3.5 years of exposure. This observation is contrary to the G101 Guide, which recommends fitting thickness loss data with a power function that, by its nature, yields decreasing corrosion rates with exposure time and misrepresents the data (McCuen et al. 1992). Needed instead are composite functions that model the initial curved portion followed by a straight portion (McCuen et al. 1992).

In one case, curve 12b, the average thickness loss of the two specimens removed after 7.5 years was unexpectedly higher than the average of the two specimens removed after 16 years. Several other curves from study 12 exhibit unusually large kinks at 7.5

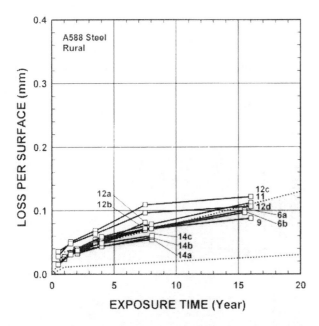

Fig. 5. A588 steel in rural environments

years of exposure (see for example Figs. 5 to 8, 10, 12, and 14). Because Vrable (1985) and Komp (personal communication, 1991) could not be reached, one is left to speculate that the problem arose when the rust was chemically removed from the specimens prior to calculating thickness loss from weight loss.

A588 Steel (Fig. 5)
Most curves for A588 steel exposed in rural environments lie on the upper half of the medium corrosivity band for weathering steel. Curves 11, 12c, and 12d exceed the upper bound. This is to be expected, given that A588 steel always has a leaner composition and, thus, rusts more than A242 steel (Fig. 4).

Because long-term atmospheric corrosion tests had not confirmed that Cor-Ten B steel of the pre-1978 composition had a corrosion rating of four relative to carbon steel (Albrecht and Naeemi 1984, Albrecht and Lee 1991), its composition was enriched by increasing the contents of silicon from 0.25 to 0.43% and nickel from 0.02 to 0.21% (Table 6). Accordingly, the post-1978 composition curves 12b and 12d lie lower than the pre-1978 composition curves 12a and 12c.

For the same reason, the former lean composition of Mayari R-50 was also enriched by increasing silicon from 0.29 to 0.38%, nickel from 0.23 to 0.27%, chromium from 0.22 to 0.31%, and copper from 0.28 to 0.31% (Table 6). Curves 14a and 14b in Fig. 5 represent the enriched composition, and curve 14c represents the former lean composition. The Mayari R-50 steels represented by curves 9 and 14 corroded least, having been exposed in the benign rural environment of Saylorsburg, Pa., located in the Pocono Mountains 50 km north of Bethlehem, Pa.

The acid rain effect in Potter County, Pa., and the proximity of semirural South Bend, Pa., to beehive coke ovens and burning coal-mine refuse several kilometers away (see Appendix) are the likely reason why the Cor-Ten B steel curves 6a, 6b, 12c, and 12d are relatively high. In comparison, curves 14a, 14b, and 14c are low because Saylorsburg (Pa.) is truly rural.

Copper Steel (Fig. 6)
As stated earlier, the copper steel data are compared with the medium corrosivity band for weathering steel because the 9224 standard does not have a band for copper steel. Copper steels rust more because they are not alloyed with P, Si, Ni, and Cr. Indeed, all curves lie on or above the weathering steel band. The copper content of the steels for the six curves listed in Table 6 varied from 0.21% (curve 9) to 0.32% Cu (curve 3). As Fig. 1 shows, such variations are not significant. Clearly, copper steel has insufficient corrosion resistance for use in bare, exposed steel structures with long service lives, such as bridges.

Carbon Steel (Fig. 7)
After 8 years of exposure, curve 11 was on the upper bound and curves 3, 6, 9, and 12b were well above the upper bound for medium corrosivity of carbon steel. The curve 12a test was discontinued at 7.5 years.

The carbon steels represented by curves 3 and 6 had copper contents of 0.05 and 0.03%. Both exceed the 0.02% Cu maximum for carbon steel, by definition. This explains the relatively low positions of these curves. The other curves represent steels that have equal to or less than 0.02% Cu. Considering that the rural environments in which the carbon steels were exposed are not severe, the ISO 9224 medium corrosivity band for carbon steel should be raised.

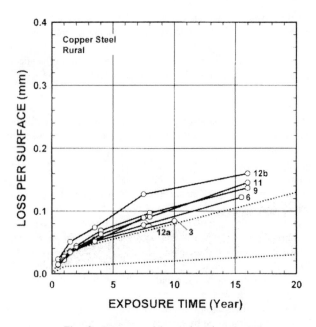

Fig. 6. Copper steel in rural environments

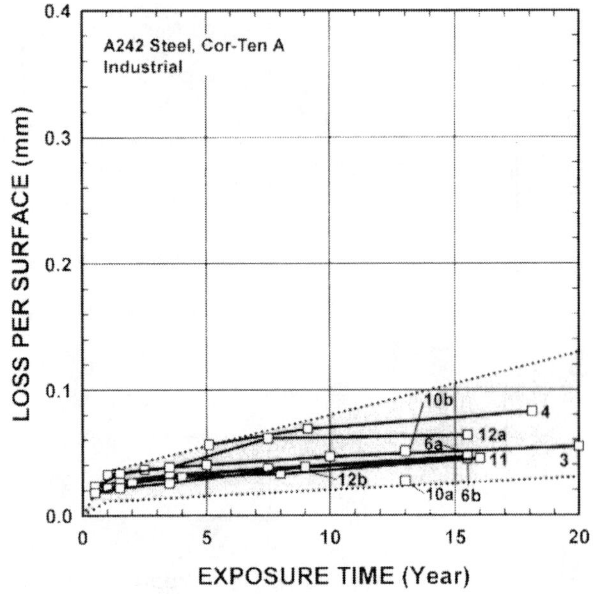

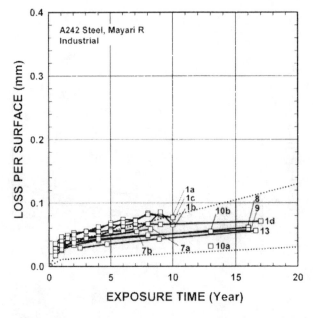

Fig. 8. A242 steel in industrial environments (Cor-Ten A)

Industrial Environments

The data for the four steels exposed to industrial environments are presented in four pairs of figures. The first figure in a pair contains data from studies on Cor-Ten steel, while the second figure contains data from studies on Mayari steel. Copson's data were included with the Cor-Ten steel data. The wealth of industrial exposure data made this separation necessary.

A242 Steel (Figs. 8 and 9)

All curves in Figs. 8 and 9 lie within the medium corrosivity band for weathering steel, a behavior similar to that of A242 steel in rural environments (Fig. 4). Curves 1a, 1b, and 1c in Fig. 9 os-

cillate because the specimens were removed yearly. Had they been removed at increasingly longer time intervals, as is now common, the oscillations would have vanished. Some curves had constant slopes after 4 years of exposure while others had decreasing slopes.

The single data points 10a and 10b are for butt-welded plates and plates with angles welded all around that were exposed for 13 years in Sacramento and Los Angeles, Calif. At both sites, the sun shines about 3,000 h per year, and precipitation is very low. The difference between the thickness losses can be ascribed to the proximity to the Pacific Ocean. The specimens rusted less in Sac-

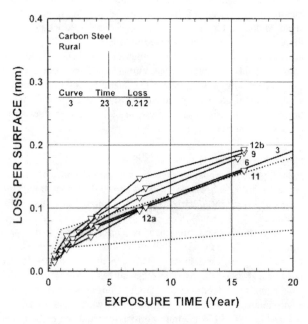

Fig. 7. Carbon steel in rural environments

Fig. 9. A242 steel in industrial environments (Mayari R)

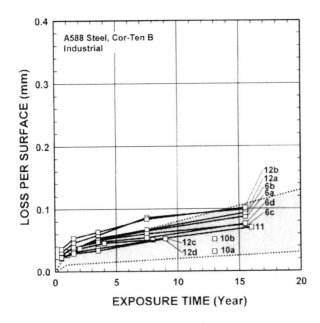

Fig. 10. A588 steel in industrial environments (Cor-Ten B)

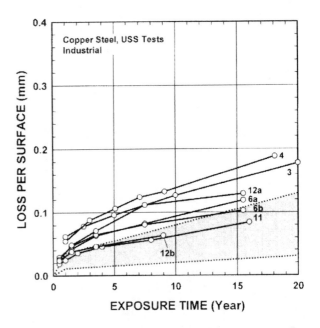

Fig. 12. Copper steel in industrial environments (USS tests)

ramento (point 10a) than in Los Angeles (point 10b), which are 160 and 27 km from the ocean, respectively.

A588 Steel (Figs. 10 and 11)
The curves for Cor-Ten B and Mayari R-50 steels exposed to industrial environments are remarkably similar. Except for curves 12a and 12b (Fig. 10) and curve 9 (Fig. 11), all lie in the upper half of the medium corrosivity band.

The Cor-Ten B steels exposed in Monroeville, Pa., curves 12a and 12b, rusted much more than the same steels, 12c and 12d, exposed in the milder industrial environment of Kearny, N.J.

As was the case for A242 steel, data points 10a and 10b are well below the other curves for A588 steel, owing to the dry and sunny climates of Sacramento and Los Angeles, Calif.

Two factors contribute to the differences in thickness losses of specimens made of Mayari R-50 steels that were exposed in Bethlehem, Pa. (Fig. 11). The steel for curves 14a and 14b had a richer composition, and the specimens were exposed during the 1990s when steel plants in Bethlehem were being closed, thus greatly reducing air pollution. In contrast, curve 9 represents specimens with a leaner composition that were exposed from 1968 to 1984, before Bethlehem Steel began to close plants.

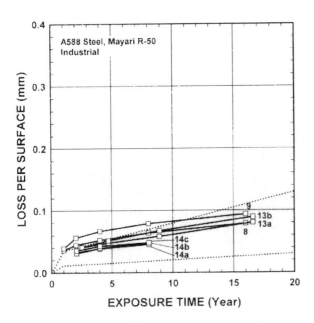

Fig. 11. A588 steel in industrial environments (Mayari R-50)

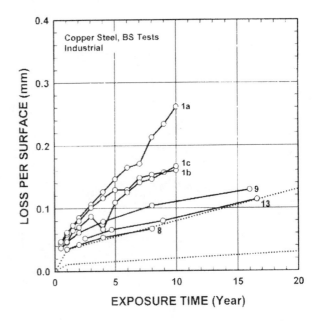

Fig. 13. Copper steel in industrial environments (BS tests)

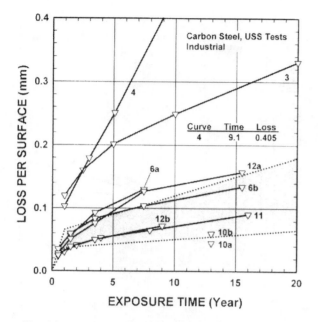

Fig. 14. Carbon steel in industrial environments (USS tests)

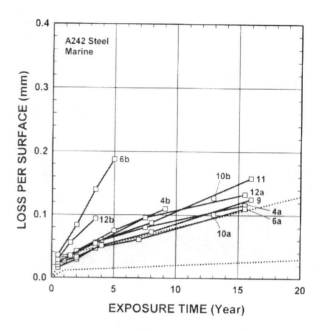

Fig. 16. A242 steel in marine environments

Several curves (6a, 6b, 6c, 6d, and 11) exhibit a nearly constant corrosion rate after 3.5 years.

Copper Steel (Figs. 12 and 13)

The older studies 1, 3, and 4 were conducted during the 1940s and 1950s (Table 5), well before pollution-control equipment was installed in steel plants. Accordingly, the copper steel curves 1, 3, and 4 lie highest in Figs. 12 and 13. From the studies performed since the early 1960s, only curves 11 and 12b fall inside the medium corrosivity band for weathering steel, both coming from steel exposed to the mild industrial environment of Kearny, N.J. All other curves are on or above the upper bound for weathering

steel. The highest curve 1a is for copper steel exposed in Pittsburgh, Pa., an environment that Horton (1965) had labeled severe industrial.

Carbon Steel (Figs. 14 and 15)

The thickness loss curves for carbon steel exposed to industrial environments are stratified much like those for copper steels. Curves 1, 3, and 4 lie highest; curves 11 and 12b are lowest. All other curves fall in between, either along the upper half or above the medium corrosivity band for carbon steel. Curves 3, 4, 6b, 8, 11, and 13 rise at constant corrosion rates.

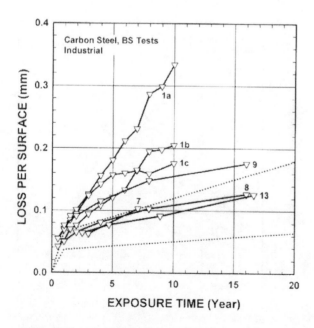

Fig. 15. Carbon steel in industrial environments (BS tests)

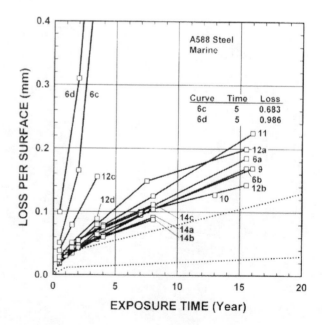

Fig. 17. A588 steel in marine environments

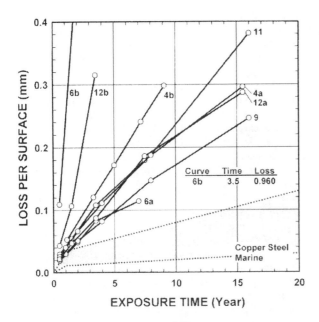

Fig. 18. Copper steel in marine environments

The data points 10a and 10b for carbon steels exposed in Sacramento and Los Angeles, Calif., are so low that they would even fall inside the weathering steel band. These two environments approach semidesert conditions.

Marine Environments

A242 Steel (Fig. 16)
The data for A242 steel exposed to moderate marine environments are represented by the following curves: 4b for Block Island, R.I.; 10a and 10b for Point Reyes, Calif.; and 4a, 6a, 9, 11,

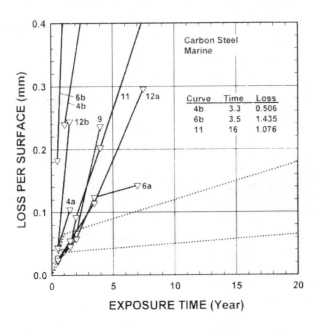

Fig. 19. Carbon steel in marine environments

and 12a for Kure Beach, N.C., 250 m from the shoreline. All curves lie on or above the upper bound of the medium corrosivity category for weathering steel.

Curves 6b and 12b represent specimens exposed to the severe marine environments in Kure Beach, N.C., 25 m from the shoreline. Most specimens rusted at a constant rate after 3.5 years of exposure. Clearly, boldly exposed and unpainted weathering structures should not be built in moderate marine environments and most certainly not in severe marine environments.

Bridges located in the snowbelt states are contaminated with deicing salts when runoff water from the bridge deck leaks through joints and cracks and runs down to the steel superstructure, or when trucks driving on a wet pavement under the bridge kick up a plume of moisture that settles on the overhead members. The resulting thickness losses can be more severe than those reported herein for A242 and A588 steels.

A588 Steel (Fig. 17)
The data for A588 steel exposed to moderate marine environments are represented by the following curves: 10 for Point Reyes, Calif., and 6a, 6b, 9, 10, 11, 12a, 12b, 14a, 14b, and 14c for moderate Kure Beach, N.C. Curves 6c, 6d, 12c, and 12d represent specimens exposed to severe Kure Beach, N.C. All curves lie much above the medium corrosivity band for weathering steel, illustrating the harshness of environments containing airborne salts.

As was observed for the A242 data, most curves for A588 steel rise at a constant corrosion rate after 3.5 years. A notable exception is curve 12a with its unusual kink at 7.5 years that can also be seen in several other curves from study 12.

Copper Steel (Fig. 18)
The curves that represent the copper steel data are as follows: 4b for Block Island, R.I.; 4a, 6a, 9, 11, and 12a for moderate Kure Beach; and 6b and 12b for severe Kure Beach, N.C. All curves lie far above the medium corrosivity band for weathering steels. Copper steel has little corrosion resistance in marine environments.

Carbon Steel (Fig. 19)
Last and least is carbon steel exposed to marine environments. The curve numbers are the same as those for the copper steel data shown in Fig. 18. The medium corrosivity band for carbon steel is not valid for marine environments. It is being used here only as a reference. For unexplained reasons, the corrosion rate of curve 6a declines dramatically after 4 years; this likely resulted from a data error.

Corrosion Allowance

Thickness loss due to corrosion increases the stress in structural members. In axially loaded members and in maximum moment regions of flexural members where the calculated stress approaches the allowable stress, a corrosion allowance should be added to the member thickness arrived by stress calculation. On the other hand, a certain decrease of member thickness may be allowed in low-stress members. When deciding how much corrosion allowance to add, it should be kept in mind that surfaces do not corrode uniformly as is implied by the average thickness loss data presented in Figs. 4 to 19. Albrecht and Shabshab (1994), for example, showed that the bottom flange of sheltered and deicing-salt contaminated W 14×30 beams exposed to the atmosphere corroded and pitted much more on the bottom surface than the top surface. Also, the lower 100 to 150 mm of the web corroded much more than the upper portion of the web. Some pits perforated the lower portion of the 7-mm-thick web in 6 years. The corrosion patterns cited above were also found in weathering steel bridges in Michigan (McCrum et al. 1985) and Ohio, and in a carbon steel bridge in Maryland (Albrecht and Xu 1988).

The addition of a corrosion allowance to member thickness is not intended to compensate for such corrosion losses as would occur in high corrosivity environments where weathering steel cannot develop a protective rust coating. It is meant for bare, exposed weathering steel structures that are expected to corrode at rates lower than 0.0075 mm/year (Fig. 2). Weathering steel corroding at a rate greater than 0.0075 mm/year should be painted.

Several ways of calculating the corrosion allowance have been proposed. Madison (1966) assumed an upper bound corrosion rate of 0.0013 mm/year based on thickness loss data reported by Horton, Larrabee, and Copson. He extrapolated the line to 100 years. "Since the data were obtained from small specimens rather than large structures, and for other reasons inherent in such tests," Madison assumed a safety factor of 3. The resulting corrosion allowance is then 0.0013 mm/year $\times 100$ years $\times 3 = 0.4$ mm per surface in semirural and industrial environments.

Albrecht et al. (1989, p. 26) recommended that the thickness of each component of a member should be increased, as follows, to allow for corrosion losses of 0.0075 mm/year over a 100-year service life:

- For weathering steel members less than 38 mm thick, the thickness should be increased by 0.8 mm per surface above the thickness arrived by stress calculations.
- For members 38 mm or greater in thickness, production tolerances will provide sufficient steel to compensate for these losses.

Other countries specify much greater increases in thickness than those recommended by Albrecht et al. (1989) to allow for corrosion during the service life of the structure. For example, the British Department of Transport calls for an increase of thickness of 1 mm of steel per surface in rural atmospheres and 2 mm per surface in industrial atmospheres.

The German guidelines call for the following increases in thickness per surface for a 60-year service life: 0.8 mm in rural, 1.2 mm in urban, and 1.5 mm in industrial and marine atmospheres.

The data presented herein provides the basis for developing more rational corrosion allowances that take into account the variability in composition and environment.

Summary

- The rural, industrial, and marine environments become increasingly severe in that order.
- The A242 (Cor-Ten A and Mayari R), A588 (Cor-Ten B and Mayari R-50), copper, and carbon steels become decreasingly corrosion resistant in that order.
- Thickness loss of a structural steel can vary significantly when it is exposed at different sites of the same environment type.
- Environment influences corrosion resistance more than does composition.
- Air quality standards mandated by the government have reduced the average sulfur dioxide content in the country by half in the last 20 years, resulting in smaller thickness losses for all steels exposed to rural and industrial environments.
- The thickness loss curves for A242 steel (Cor-Ten A and Mayari R) fall inside the medium corrosivity band for weathering

steel (ISO standard 9224) in rural and industrial environments. In marine environments, the curves lie on or above the upper bound of the band.
- One in five thickness loss curves for A588 steel (Cor-Ten B and Mayari R-50) exposed to rural and industrial environments lies above the medium corrosivity band for weathering steel; four in five lie on the upper half of the band. All curves representing A588 steel exposed to marine environments, without exception, are well above the band, suggesting that bare, exposed weathering steel should not be used in marine environments.
- All but three of 27 curves for copper steel in the rural, industrial, and marine environments fall above the medium corrosivity band for weathering steel. The 9224 standard does not provide a band for copper steel, and no direct comparison is possible in this case.
- Carbon steel should not be used in a bare condition in any environment, except perhaps in arid climates. Twenty-one of 30 curves lie above the medium corrosivity band for carbon steel, suggesting that ISO set the band too low. For example, even in rural environments two of the six carbon steels lie on the upper bound of the carbon band, while the other four curves lie above the band.
- Readers who are planning to build a bare, exposed steel structure can extrapolate to the end of the service life the thickness loss curve most applicable to the site. This corrosion allowance should then be added to member thicknesses arrived at by stress calculations.
- Thickness loss data obtained from simple test specimens that measure 150×100 mm—the size of an index card—must be applied carefully to complex structures such as a bridge, which has a variety of details with different exposure conditions.

It is appropriate to close the paper with a quotation from Larrabee (1953, p. 259), formerly from US Steel, to illustrate the difficulty of predicting the corrosivity of environments and the corrosion resistance of structural steels.

Data from tests in both industrial and marine atmospheres reveal that very great differences in corrosivity can exist at locations only a few miles apart, or—in some extreme cases—only a few hundred feet apart.

Zoccola (1976), formerly of Bethlehem Steel, provided corroborating evidence. He exposed specimens of Mayari R and carbon steels on the rooftop of a building and under the westbound service bridge of Eight-Mile Road where it crosses over Highway US 10. The bridge is located about 800 m from the building site. Fig. 20 shows the data for both sites. The Mayari R and carbon steel curves for the rooftop exposure compare well with most other data for Mayari R and carbon steels exposed in industrial environments (Figs. 9 and 15). However, the specimens exposed under the bridge rusted several times more than their counterparts at the nearby building (Fig. 20). In fact, they rusted more than Mayari R and carbon steel specimens exposed in marine environments (Figs. 16 and 19). In severe environments, where corrosion resistance is needed most, weathering steel has little advantage over carbon steel.

More information on the performance of structural steels in service can be found in Larrabee (1953), Albrecht and Naaemi (1984), and Albrecht et al. (1989).

Appendix. Atmospheric Test Sites

Specimens exposed on the rooftop of office buildings dried faster as a result of the specimens absorbing heat reflected by the roof

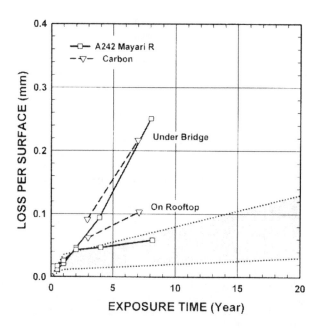

Fig. 20. Effect of deicing salt contamination on corrosion of weathering and carbon steels exposed in Detroit, Mich. (Zoccola 1976)

surface and interior heat emitted by the building. The average value of sulfur dioxide in the country has decreased by about one-half in the last 20 years (http://nadp.sws.uiuc.edu/nadpdata/; http://www.epa.gov/airtrends/data/aq.html). Specimens exposed 1.5 m above ground at the Kure Beach test sites corroded much less than specimens exposed at higher elevations at the same sites (Baker and Lee 1982).

Rural Sites
- Potter County, Pa. The exact location of the test site was not specified. Test racks were presumably at ground level. This rural test site is in the direct path of the jet stream carrying exhaust fumes emitted by very high stacks of fossil fuel power plants downwind, giving rise to the acid rain phenomenon (Komp et al. 1993).
- Saylorsburg, Pa. Bethlehem Steel's rural test site is located in the Pocono Mountains, 50 km north of Bethlehem (Townsend and Zoccola 1982). Test racks were presumably at ground level.
- South Bend, Pa. The exact location of the test site was not specified. Test racks were presumably at ground level. This semirural atmosphere is somewhat contaminated with sulfur fumes from beehive coke ovens and burning coal-mine refuse, both several miles away. However, with varying wind direction these emit sufficient pollution to cause about twice the corrosion that occurs at State College, Pa., which is designated as "rural" by ASTM (Larrabee 1959).

Industrial Sites
- Bayonne, N.J. The test site was located on the grounds of a laboratory (Copson 1960). The test racks were presumably at ground level.
- Bethlehem, Pa., first test site. The test site was located near blast furnaces and open-hearth furnaces at Bethlehem Steel. The test racks were on the rooftop of the 14-story main office building (Horton 1964). The tensile specimens were placed on

the racks with the lengthwise direction horizontal and the widthwise direction vertical (Epstein and Horton 1949).
- Bethlehem, Pa., second test site. The test site was located at the cinder dump of Bethlehem Steel, southeast of open-hearth furnaces and northeast of the coke plant (Horton 1964). The exact location of the test racks was not specified.
- Bethlehem, Pa., third test site. The test site was located at Bethlehem Steel's Homer Research Laboratory about 3 km from its integrated steel facility (Townsend and Zoccola 1982). The environment at the site is mild industrial. The exact location of the test racks was not specified.
- Columbus, Ohio. The test site was located at the Battelle Memorial Institute headquarters. The test racks were located on a rooftop (Horton 1964).
- Detroit, Mich. The specimens were exposed first on the rooftop of Northland Towers for 1.5 years; the remaining exposure took place at the National Guard Armory building (Zoccola 1976). The Armory is a large four-story building located about 800 m from the bridge exposure site. It is located on a major urban thoroughfare surrounded by a large shopping mall and mostly residential areas. There is some industry, particularly to the east, and the Armory has a large heating plant. These buildings are located near the crossing of Eight-Mile Road over highway US 10. Additional specimens were exposed under the westbound service bridge of Eight-Mile Road where it crosses over US 10.
- Kearny, N.J. The aggressive industrial site in Kearny, owing to the imposition of lower sulfur levels for fossil fuels, lost much of its corrosive nature and acted toward the test panels like a rural site (Komp et al. 1993). The exact location of the test racks was not specified, but presumably it was at ground level.
- Kearny/Newark/Kearny, N.J. Newark was US Steel's industrial test site from 1956 to 1970. The specimens were moved to the new Kearny test site in 1970. Copper steel specimens exposed for a period of six months at the original Kearny site (1932–1956), the Newark site (1956–1970) and the new Kearny site (1970–present) show a progressive decline in thickness loss per surface, from about 0.046 mm in 1934 to 0.023 mm in 1974. Clearly, the Kearny/Newark industrial site has become less corrosive to copper steel over the years (Gallagher 1976).
- Los Angeles, Calif. The test site was located on the grounds of the maintenance station of the California Department of Transportation close to the Santa Ana Freeway (Interstate 5) in the City of Commerce, an incorporated industrial area 27 km from the Pacific Ocean.
- Monroeville, Pa. US Steel's semi-industrial test site is located at its research center. The exact location of the test racks was not specified, but presumably it was at ground level.
- Newark, N.J. The test site was adjacent to heavily traveled expressways. The test racks were mounted on the rooftop of a two-story office building occupied by the New Jersey Department of Transportation (Cosaboom et al. 1979).
- Pittsburgh, Pa. The specimens were exposed in the Rankine area of Pittsburgh. The exact locations of the site and the test racks were not reported. The area was rated severe industrial (Horton 1965).
- Sacramento, Calif. The test site was located on Folsom Boulevard, between 59th and 65th Streets, on the southeastern outskirts of Sacramento. The test racks were mounted on the rooftop of the Caltrans Transportation Laboratory. The city has

no heavy industry and is located in the middle of a farming area about 160 km from the Pacific Ocean (Reed and Kendrick 1982).

Marine Sites

- Block Island, R.I. The test site was about 24 km from the mainland. The test racks were located on a bluff overlooking the open ocean (Copson 1960). The test racks were presumably at ground level.
- Kure Beach, N.C., 250 m from mean tide. This is ASTM's main marine test site on the Atlantic coast. The test site was owned previously by the International Nickel Company and the LaQue Center for Corrosion Technology. The test racks were at ground level. The atmosphere at 250 m from the surf is said to be moderate marine.
- Kure Beach, N.C., 25 m from mean tide. The atmosphere of the test site 25 m from the surf is said to be severe marine, although dune growth has reduced the severity of the atmosphere (Baker and Lee 1982).
- Point Reyes, Calif. ASTM's marine test site on the Point Reyes Peninsula is located 400 m from the Pacific Ocean and 56 km northwest of San Francisco. The test racks were presumably at ground level.

References

Albrecht, P., and Lee, H. Y. (1991). "Evaluation of rating numbers for atmospheric corrosion resistance of weathering steel." *J. Test. Eval.*, 19(6), 429–439.

Albrecht, P., and Naeemi, A. (1984). "Performance of weathering steel in bridges." *NCHRP Rep. 272*, Transportation Research Board, National Research Council, Washington, D.C.

Albrecht, P., and Shabshab, C. F. (1994). "Fatigue strength of weathered rolled beam made of A588 Steel." *J. Mater. Civ. Eng.*, 6(3), 407–428.

Albrecht, P., and Xu, G. P. (1988). "Fatigue strength of long-term weathered rolled beams." *Rep. No. FHWA/MD-89/11.* Dept. of Civil Engineering, Univ. of Maryland, College Park, Md.

Albrecht, P., Coburn, S. K., Wattar, F. M., Tinklenberg, G., and Gallagher, W. P. (1989). "Guidelines for the use of weathering steel in bridges." *NCHRP Rep. 314*, Transportation Research Board, National Research Council, Washington, D. C.

Baker, E. A., and Lee, T. S. (1982). "Calibration of atmospheric test sites." *Atmospheric Corrosion of Metals, STP 767*, ASTM West Conshohocken, Pa, 250–266.

Breuil. (1907). *J. Iron Steel Inst., London*, 2, 41 and 60.

Buck, D. M. (1913a). "Copper in steel—The influence on corrosion." *J. Ind. Eng. Chem.*, 5(6), 447–452.

Buck, D. M. (1913b). "Copper in steel—Its influence on corrosion." *Iron Age*, 91(16), 931–938.

Buck, D. M. (1915). "Recent progress in corrosion resistance." *Iron Age* 1231–1239.

Buck, D. M. (1919). "The influence of very low percentages of copper in retarding the corrosion of steel." *Proc. ASTM*, 224–235.

Buck, D. M. (1920). "A review of the development of copper steel." *Year Book of the American Iron and Steel Institute*, 373–379.

Buck, D. M., and Handy, J. O. (1916). "Research on the corrosion resistance of copper steel." *J. Ind. Eng. Chem.*, 3.

Burgess, C. F., and Aston, J. (1913). "Influence of various elements on the corrodibility of iron." *J. Ind. Eng. Chem.*

Coburn, S. K., Komp, M. E., and Lore, S. C. (1995). "Atmospheric corrosion rates of weathering steels at test sites in the eastern United States—Effect of environment and test panel." *Atmospheric Corrosion, STP 1239*, ASTM, West Conshohocken, 101–113.

Cone, E. F. (1934). "Low-alloy high tensile steels." *Steel*, 41–44.

Copson, H. R. (1960). "Long-time atmospheric corrosion tests on low-alloy steels." *ASTM* 60, 650–667.

Cosaboom, B., Mehalchick, G., and Zoccola, J. C. (1979). "Bridge construction with unpainted high-strength low-alloy steel: Eight-year progress report." New Jersey Department of Transportation, Division of Research and Development, Trenton, N.J.

Cruz, I. S., and Mullen, C. X. (1962). "Architectural application of bare USS Cor-Ten high-strength low-alloy steel." Applied Research Laboratory, United States Steel, Monroeville, Pa.

Daeves, K. (1926). "Atmospheric corrosion resistance of copper steel." Stahl und Eisen, 46, 1857–1863, (in German).

Epstein, S., and Horton, J. B. (1949). "Atmospheric corrosion tests of plain low carbon and low alloy steels." Research Department, Bethlehem Steel, Bethlehem, Pa.

Gallagher, W. P. (1976). "Long-term corrosion performance of USS Cor-Ten B steel in various atmospheres." *Technical Rep.*, United States Steel, Monroeville, Pa.

Gallagher, W. P. (1978). "One and one-half year atmospheric corrosion performance of modified Cor-Ten B steel." *Technical Rep.*, United States Steel, Monroeville, Pa.

Gallagher, W. P. (1982a). "Performance of USS Cor-Ten steels in chloride-contaminated atmospheres." *Technical Rep.*, United States Steel, Monroeville, Pa.

Gallagher, W. P. (1982b). "Eight-year results of the USS Cor-Ten licensees demonstration program." *Proc. 11th Cor-Ten Conf.*, U.S. Steel, Monroeville, Pa.

Horton, J. B. (1964). "The composition, structure and growth of atmospheric rust on various steels." PhD Dissertation, Lehigh University, Bethlehem, Pa.

Horton, J. B. (1965). "The rusting of low alloy steels in the atmosphere." Presented at the San Francisco Regional Technical Meeting, American Iron and Steel Institute.

Karsten, C. J. B. (1827). "*Handbook of iron-works practice.*" (in German).

Koch, G. H., Brongers, M. P. H., Thompson, N. G., Virmani, Y. P., and Payer, J. H. (2002). "Corrosion cost and preventive strategies in the United States," *Rep. No. FHWA-RD-01-156*, Office of Infrastructure Research and Development, Federal Highway Administration, McLean, Va.

Komp, M. E. (1987). "Atmospheric corrosion ratings of weathering steels—Calculation and significance." *Materials Performance*, National Association of Corrosion Engineers, Houston, Tex.

Komp, M. E., Coburn, S. K., and Lore, S. C. (1993). "Worldwide data on the atmospheric corrosion resistance of weathering steels." *Proc. 12th Int. Corrosion Congress*, Houston, Tex.

Kucera, V., and Mattson, E. (1987). "Atmospheric corrosion." Chapter 5, *Corrosion mechanisms*, F. Mansfeld, ed., Marcel Dekker, New York, 211–284.

Larrabee, C. P. (1953). "Corrosion resistance of high-strength low-alloy steels as influenced by composition and environment." *Corrosion*, 9(9), 259–271.

Larrabee, C. P. (1954). "Report of subcommittee XIV on inspection of black and galvanized sheets." *Proc. ASTM*, Vol. 54, West Conshohocken, Pa., 110–122.

Larrabee, C. P. (1959). "Twenty-year results of atmospheric corrosion tests on USS Cor-Ten steel, structural copper steel, and structural carbon steel." *Technical Rep.*, United States Steel, Monroeville, Pa.

Larrabee, C. P., and Coburn, S. K. (1962). "The atmospheric corrosion of steels as influenced by changes in chemical composition." *Proc. First Int. Congress on Metallic Corrosion*, Butterworths, London, 276–285.

Legault, R. A., and Leckie, H. P. (1974). "Effect of alloy composition on the atmospheric corrosion behavior of steels based on a statistical analysis of the Larrabee-Coburn data set." *Corrosion in natural environments, ASTM STP 558*, ASTM, West Conshohocken, Pa., 334–347.

Madison, R. B. (1966). "Unpainted steel for permanent structures." *Civ. Eng.*, 68–72.

"Mayari R Weathering Steel." (1967). *Booklet 2348*, Bethlehem Steel, Bethlehem, Pa.

"Mayari R Weathering Steel." (1983). *Booklet 3790*, Bethlehem Steel, Bethlehem, Pa.

McCrum, R. L., Arnold, C. J., and Dexter, R. P. (1985). "Current status report, effects of corrosion on unpainted weathering steel bridges." *Research Rep. No. R-1255*, Testing and Research Division, Michigan Dept. of Transportation, Lansing, Mich.

McCuen, R. H., Albrecht, P., and Cheng, J. G. (1992). "A new approach to power-model regression of corrosion penetration data." *Corrosion forms and control for infrastructure, ASTM STP 1137*, Victor Chaker, ed., ASTM, West Conshohocken, Pa., 46–76.

McCuen, R. H., and Albrecht, P. (1994a). "Predicting effect of chemical composition on atmospheric corrosion penetration of weathering steel." Dept. of Civil Engineering, Univ. of Maryland, College Park, Md.

McCuen, R. H., and Albrecht, P. (1994b). "Composite modeling of corrosion penetration data." *Application of accelerated corrosion tests to service life prediction of materials, STP 1194*, Gustavo Cragnolino, ed., ASTM, West Conshohocken, Pa., 65–102.

Mottola, V. E. (1989). "16-year corrosion penetration data for A242, A588 and carbon steels tested in Newark, New Jersey." New Jersey Dept. of Transportation, Trenton, N.J.

Porter, L. F., and Repas, P. E. (1982). "The evolution of HSLA steels." *J. Met.*, 14–21.

Pourbaix, A., and Miranda, L. (1983). "Weathering steels performance and the effect of copper." *Proc. Copper in Steel, Acta Technica Belgica, ATB Metallurgie* 23(4), 7.1–7.8.

Reed, F. O., and Kendrick, C. B. (1982). "Evaluation of weathering effects on structural steel." *Research Rep. No. FHWA/CA/TL-82/04*, Office of the Transportation Laboratory, California Dept. of Transportation, Sacramento, Calif.

Richardson, E. A., and Richardson, L. T. (1916). "Observations upon the atmospheric corrosion of commercial sheet iron." American Electro-Chemical Society.

Schmitt, R. J., and Mullen, C. X. (1965). "Corrosion performance of Mn-Cr-Cu-V type USS Cor-Ten high-strength low-alloy steel in various atmospheres." *Technical Rep.*, Applied Research Laboratory, United States Steel, Monroeville, Pa.

Schweitzer, P. A. (1989). "Atmospheric corrosion." *Corrosion and Corrosion Protection Handbook*, 2nd Ed., Marcel Dekker, New York.

Shastry, C. R., Friel, J. J., and Townsend, H. E. (1986). "Sixteen-year atmospheric corrosion performance of weathering steels in marine, rural and industrial environments." *Degradation of metals in the atmosphere, STP 965*, S. W. Dean and T. S. Lee, eds., ASTM, West Conshohocken, Pa., 5–15.

Stead and Wigham (1901). *J. Iron Steel Inst., London*, 2, 135.

Stengel, H. (1836). "On the effect of copper and sulfur on the quality of steel." *Arch. Bergbau und Hüttenwesen*, 9, 465–487, (in German).

Tinklenberg, G. L. (1986). "Evaluation of weathering steel in a Detroit freeway environment—Second eight-year study." *Research Rep. No. R-1277*, Materials and Technology Division, Michigan Dept. of State Highways and Transportation, Lansing, Mich.

Tomashov, N. D. (1966). "Atmospheric corrosion of metals." *Theory of corrosion and protection of metals*, Macmillan, New York.

Townsend, H. E. (2000). "Atmospheric corrosion performance of quenched-and-tempered, high-strength weathering steel." *Corrosion*, 56(9), 883–886.

Townsend, H. E., and Zoccola, J. C. (1982). "Eight-year atmospheric corrosion performance of weathering steel in industrial, rural, and marine environments." *Atmospheric corrosion of metals, STP 767*, S. W. Dean, Jr., and E. C. Rhea, eds., ASTM, West Conshohocken, Pa., 45–59.

Uhlig, H. H., and Revie, R. W. (1985). "Corrosion and corrosion control." 3rd Ed., Wiley, New York.

U.S. Steel. (1962). "Cor-Ten high-strength low-alloy steel for architectural applications." *ADUCO 02058*.

U.S. Steel. (1980). "Cor-Ten high-strength low-alloy steel." *ADUSS88-7888-02*.

Vernon, W. H. J. (1931). "A laboratory study of the atmospheric corrosion of metals, part 1." *Trans. Faraday Soc.*, 27, 255–277.

Vernon, W. H. J. (1933). "The role of the corrosion product in the atmospheric corrosion of iron." *Trans. Electrochem. Soc.*, 64, 31–33.

Vernon, W. H. J. (1935). "A laboratory study of the atmospheric corrosion of metals, part 2." *Trans. Faraday Soc.*, 31, 1668–1701.

Vrable, J. B. (1985). "Seven and one-half years atmospheric corrosion performance of improved USS Cor-Ten B steel." U.S. Steel, Monroeville, Pa.

Williams, F. H. (1900). "Influence of copper in retarding corrosion of soft steel and wrought iron," *Proc. Engineering Society of Western Pennsylvania*, 16, 231–233.

Wranglen, G. (1972). "An introduction to corrosion and protection of metals." Institut för Metallskydd, Fack 10041, Stockholm 26.

Zoccola, J. C. (1976). "Eight year corrosion test report—Eight Mile Road interchange." Bethlehem Steel, Bethlehem, Pa.

Zoccola, J. C., Permoda, A. J., Oehler, L. T., and Horton, J. B. (1970). "Performance of Mayari R weathering steel (ASTM A242) in bridges, Eight-Mile Road over John Lodge Expressway, Detroit, Michigan." Bethlehem Steel Corporation, Bethlehem, Pa.

American Society of
Civil Engineers **1852 – 2002**
150th Anniversary
Paper

Building a Better World

Mathematical Modeling of Watershed Hydrology

Vijay P. Singh, F.ASCE,[1] and David A. Woolhiser, M.ASCE[2]

Abstract: Mathematical modeling of watershed hydrology is employed to address a wide spectrum of environmental and water resources problems. A historical perspective of hydrologic modeling is provided, and new developments and challenges in watershed models are discussed. These include data acquisition by remote sensing and space technology, digital terrain and elevation models, chemical tracers, geographic information and data management systems, topographic representation, upscaling of hydrologic conservation equations, spatial variability of hydraulic roughness, infiltration and precipitation, spatial and temporal scaling, model calibration, and linking with water quality models. Model construction, calibration, and data processing have received a great deal of attention, while model validation, error propagation, and analyses of uncertainty, risk, and reliability have not been treated as thoroughly. Finally, some remarks are made regarding the future outlook for watershed hydrology modeling.

DOI: 10.1061/(ASCE)1084-0699(2002)7:4(270)

CE Database keywords: Hydrologic models; Watersheds; Geographic information systems; Remote sensing; Risk; Reliability.

Introduction

Hydrology was defined by Penman (1961) as the science that attempts to answer the question, "What happens to the rain"? This sounds like a simple enough question, but experience has shown that quantitative descriptions of the land phase of the hydrologic cycle may become very complicated and are subject to a great deal of uncertainty. The term "watershed hydrology" is defined as that branch of hydrology that deals with the integration of hydrologic processes at the watershed scale to determine the watershed response. The emphasis in this paper is on the models that accomplish this integration, not on the models of individual component processes.

A watershed may be as small as a flower bed or a parking lot or as large as hundreds of thousands of square kilometers as exemplified by the Mississippi River basin. Operative hydrologic processes and their spatial nonuniformity are defined by climate, topography, geology, soils, vegetation, and land use and are related to the basin size. The nonuniformity of hydrologic processes is also directly related to the watershed size.

Mathematical models of watershed hydrology are designed to answer Penman's question at a level of detail depending on the problem at hand and are employed in a wide spectrum of areas ranging from watershed management to engineering design (Singh 1995a). They are used in the planning, design, and opera-

tion of projects, to conserve water and soil resources and to protect their quality. At the field scale, models are used for varied purposes, such as planning and designing soil conservation practices, irrigation water management, wetland restoration, stream restoration, and water-table management. On a large scale, models are used for flood protection projects, rehabilitation of aging dams, floodplain management, water-quality evaluation, and water-supply forecasting.

Watershed models are fundamental to water resources assessment, development, and management. They are, for example, used to analyze the quantity and quality of streamflow, reservoir system operations, groundwater development and protection, surface water and groundwater conjunctive use management, water distribution systems, water use, and a range of water resources management activities (Wurbs 1998).

Watershed models are employed to understand dynamic interactions between climate and land-surface hydrology. For example, vegetation, snow cover, permafrost active layer, etc. are quite sensitive to the lower boundary of the atmospheric system. The water and heat transfer between the land surface and atmosphere significantly influences hydrologic characteristics and yield, and in turn, lower boundary conditions for climate modeling (Kavvas et al. 1998). An assessment of the impact of climate change on national water resources and agricultural productivity is made possible by the use of watershed models.

Water allocation requires an integration of watershed models with models of physical habitat, biological populations, and economic response. Estimating the value of instream water use allows recreational, ecological, and biological concerns to compete with traditional consumptive uses, i.e., agriculture, hydropower, municipality, and industry (Hickey and Diaz 1999). Watershed models are utilized to quantify the impacts of watershed management strategies, linking human activities within the watershed to water quantity and quality of the receiving stream or lake (Mankin et al. 1999; Rudra et al. 1999) for environmental and water resources protection.

[1]A. K. Barton Professor, Dept. Civil and Environmental Engineering, Louisiana State Univ., Baton Rouge, LA 70803-6405.

[2]Faculty Affiliate, Dept. Civil Engineering, Colorado State Univ., 1631 Barnwood Dr., Fort Collins, CO 80525.

Note. Discussion open until December 1, 2002. Separate discussions must be submitted for individual papers. To extend the closing date by one month, a written request must be filed with the ASCE Managing Editor. The manuscript for this paper was submitted for review and possible publication on January 5, 2002; approved on January 5, 2002. This paper is part of the *Journal of Hydrologic Engineering*, Vol. 7, No. 4, July 1, 2002. ©ASCE, ISSN 1084-0699/2002/4-270–292/$8.00+$.50 per page.

In summary, watershed models have become an essential tool for water resources planning, development, and management. In the years ahead, the models will become even more common and will play an increasing role in our day-to-day lives. The objective of this work is to provide a historical perspective of watershed modeling, provide a short synopsis of currently used models, reflect on new developments and challenges, and conclude with a personal view of what the future has in store for mathematical modeling of watershed hydrology.

Historical Perspective

Hydrological modeling has a long and colorful history. Its beginning can be traced to the development of civil engineering in the nineteenth century for the design of roads, canals, city sewers, drainage systems, dams, culverts, bridges, and water-supply systems. Until the middle of the 1960s, hydrologic modeling primarily involved the development of concepts, theories and models of individual components of the hydrologic cycle, such as overland flow, channel flow, infiltration, depression storage, evaporation, interception, subsurface flow, and base flow. The Hortonian mechanism, subsurface flow mechanism, and partial and source area contributions were recognized as contributors to runoff.

Development of Component Models

The origin of mathematical modeling dates back to the rational method developed by Mulvany (1850) and an "event" model by Imbeau (1892) for relating storm runoff peak to rainfall intensity. About four decades later, Sherman (1932) introduced the unit hydrograph concept for relating the direct runoff response to rainfall excess. About the same time, Horton (1933) developed a theory of infiltration to estimate rainfall excess and improve hydrograph separation techniques. Horton (1939) investigated overland flow and produced a semiempirical formula. Keulegan (1944) made a theoretical investigation of overland flow and suggested that simplifying the equations to what is now termed the kinematic wave form would be appropriate. Izzard (1944) followed with an experimental analysis. Horton (1945) developed a concept of erosional land-form development and streamflow generation dominated by overland flow. Presented in this pioneering work were a set of empirical laws, now known as Horton's laws, which constituted the foundation of quantitative geomorphology. In these contributions, evaporation and other abstractions were treated using coefficients or indices.

Concurrent with Horton's work, Lowdermilk (1934), Hursh (1936), and Hursh and Brater (1944) observed that subsurface water movement constituted one component of storm flow hydrographs in humid regions. Subsequently, Hoover and Hursh (1943) and Hursh (1944) reported significant storm-flow generation caused by a "dynamic form of subsurface flow." Roessel (1950) observed dynamic changes in streamside groundwater flow. Based on the works of Hewlett (1961a,b), Nielsen et al. (1959), Remson et al. (1960), among others, it is now accepted that downslope unsaturated flow can contribute to streamside saturated zones and thus generate streamflow. Through the years since the 1940s, this thinking culminated into what is now referred to as subsurface flow mechanism and has indeed expanded into a more integrated understanding of streamflow generation, of which Horton's theory is but a part.

One of the earliest attempts to develop a theory of infiltration was by Green and Ampt (1911) who, using simplified principles of physics, derived a formula that is still popular for computing the infiltration capacity rate. The empirical equations of Kostiakov (1932) and Horton (1933, 1935, 1939, 1940) are also used by some current watershed models. Early work describing evaporation from lakes was done by Richardson (1931) and Cummings (1935), while Thornthwaite (1948) and Penman (1948) made important contributions to models of evapotranspiration.

There were also attempts to quantify other abstractions, such as interception, depression storage, and detention storage. Horton (1919) derived a series of empirical formulas for estimating interception during a storm for various types of vegetal covers. The Soil Conservation Service (SCS) (1956), now called the Natural Resources Conservation Service of the U.S. Department of Agriculture, developed what is now referred to as the SCS-curve number method for computing the amount of storm runoff, taking abstractions into account. Although it was originally intended to model daily runoff as affected by land-use practices, it has been used to model infiltration as well as runoff hydrograph for continuous hydrologic simulation.

The underground phase of the hydrologic cycle was investigated by Fair and Hatch (1933), who derived a formula for computing the permeability of soil. Theis (1935) combined Darcy's law with the continuity equation to derive the relation between the lowering of the piezometric surface and the rate and duration of discharge of a well. This work laid the foundation of quantitative groundwater hydrology. Jacob (1943, 1944) correlated groundwater levels and precipitation on Long Island, N.Y. The study of groundwater and infiltration led to the development of techniques for separation of baseflow and interflow in a hydrograph (Barnes 1940).

Puls (1928), of the U.S. Army Corps of Engineers, Chattanooga District, developed a method for flow routing through reservoirs, assuming invariable storage-discharge relationships and neglecting the variable slope during flood propagation. This method, later modified by the U.S. Bureau of Reclamation (1949), is now referred to as the modified Puls method. Using the concept of wedge and prism storage, McCarthy and others developed the Muskingum method of flow routing in 1934–1935 (U.S Army Corps of Engineers 1936). This method is still used for flood routing in several watershed models.

After a lull of nearly a quarter century in the area of rainfall-runoff modeling, a flurry of modeling activity started around the middle of the 1950s. A major effort employed the theory of linear systems, which led to the theory of the instantaneous unit hydrograph by Nash (1957) and then the generalized unit hydrograph theory by Dooge (1959). Lighthill and Whitham (1955) developed kinematic wave theory for flow routing in long rivers. This theory is now accepted as a standard tool for modeling overland flow and a variety of other hydrologic processes.

Development of Watershed Models

The decade of the 1960s witnessed the digital revolution that made possible the integration of models of different components of the hydrologic cycle and simulation of virtually the entire watershed, as exemplified by the seminal contribution of the Stanford Watershed Model-SWM (now HSPF) by Crawford and Linsley (1966). This was probably the first attempt to model virtually the entire hydrologic cycle. Simultaneously, a number of somewhat less comprehensive models were developed. Examples of such models that became popular are the watershed models of Dawdy and O'Donnell (1965) and HEC-1 (Hydrologic Engineering Center 1968). Also, a number of semidistributed models capable of accounting for the spatial variability of hydrologic processes within the watershed were developed, as illustrated by tank models developed by Sugawara (1967) and Sugawara et al. (1974).

Indeed there has been a proliferation of watershed hydrology models since the development of SWM (or HSPF), with emphasis on physically based models. Examples of such watershed hydrology models are SWMM (Metcalf and Eddy et al. 1971), PRMS (Leavesley et al. 1983), NWS River Forecast System (Burnash et al. 1973), SSARR (Rockwood 1982), Systeme Hydrologique Europeen (SHE) (Abbott et al. 1986a,b), TOPMODEL (Beven and Kirkby 1979), IHDM (Morris 1980), and so on. All of these models have since been significantly improved. SWM, now called HSPF, is far more comprehensive than its original version. SHE has been extended to include sediment transport and is applicable at the scale of a river basin (Bathhurst et al. 1995). TOPMODEL has been extended to contain increased catchment information, more physically based processes, and improved parameter estimation.

The digital revolution also triggered two other revolutions, namely, numerical simulation and statistical simulation. The power of computers increased exponentially and, as a result, advances in watershed hydrology have occurred at an unprecedented pace during the past 35 years. During the decades of the 1970s and 1980s, a number of mathematical models were developed not only for simulation of watershed hydrology but also for their applications in other areas, such as environmental and ecosystems management. Development of new models or improvement of previously developed models continues today. Table 1 shows in chronological order a sample of popular hydrologic models from around the globe. These days virtually all federal agencies in the United States have their own models or some variants of models developed elsewhere.

In 1991, the Bureau of Reclamation (1991) prepared an inventory of 64 watershed hydrology models classified into four categories, and the inventory is currently being updated. Burton (1993) compiled the *Proceedings of the Federal Interagency Workshop on Hydrologic Modeling Demands for the 1990's*, which contains several important watershed hydrology models. Singh (1995b) edited a book that summarized 26 popular models from around the globe. The Subcommittee on Hydrology of the Interagency Advisory Committee on Water Data (USGS 1998) published the *Proceedings of the First Federal Interagency Hydrologic Modeling Conference*, which contains many popular watershed hydrology models developed by federal agencies in the United States. Wurbs (1998) listed a number of generalized water resources simulation models in seven categories and discussed their dissemination.

Currently used Watershed Models

There are several well known general watershed models in current use in the United States and elsewhere. These models vary significantly in the model construct of each individual component process, partly because these models serve somewhat different purposes. HEC-HMS is considered the standard model in the private sector in the United States for the design of drainage systems, quantifying the effect of land-use change on flooding, etc. The NWS model is the standard model for flood forecasting. HSPF and its extended water quality model are the standard models adopted by the Environmental Protection Agency (EPS). The MMS model of the USGS is the standard model for water resources planning and management works, especially those under

the purview of the U.S. Bureau of Reclamation. The UBC and WATFLOOD models are popular in Canada for hydrologic simulation. The RORB and WBN models are commonly employed for flood forecasting, drainage design, and evaluating the effect of land-use change in Australia. TOPMODEL and SHE are the standard models for hydrologic analysis in many European countries. The HBV model is the standard model for flow forecasting in Scandinavian countries. The ARNO, LCS, and TOPIKAPI models are popular in Italy. The Tank models are well accepted in Japan. The Xinanjiang model is a commonly used model in China.

Comparison of Watershed Models

The World Meteorological Organization (WMO) sponsored three studies on intercomparison of watershed hydrology models. The first study (World Meteorological Organization 1975) dealt with conceptual models used in hydrologic forecasting. The second study (WMO 1986) dealt with an intercomparison of models used for simulation of flow rates, including snowmelt. The third study (WMO 1992) dealt with models for forecasting streamflow in real time. Except for the WMO reports, no comprehensive effort has been made to compare most major watershed hydrology models. However, efforts have been made to compare models of some component processes. Also, developers of some models have compared their models with one or a few other models.

Review and Synthesis

During the period 1970–1995, several very instructive state-of-the-art papers dealing with watershed modeling appeared. It is beyond the scope of this paper to deal with a large sample of such papers, but it is interesting to compare modeling concepts and challenges expressed by Hornberger and Boyer (1995) with those considered to be important 22 years earlier (Clarke 1973; Woolhiser 1973). Clarke (1973) discussed important issues regarding model identification and diagnosis and parameter estimation and showed that interdependence between model parameters required extensive exploration of error objective function surfaces, particularly when the model is used to determine the likely effects of land-use change. Woolhiser (1973) pointed out the importance of estimates of initial conditions for nutrient transport models and also reasoned that model verification and estimation of model parameters needed more attention.

Several investigators reviewed hydrologic models developed up to the beginning of the 1980s and discussed model reliability and future directions (Linsley 1982; Dawdy 1982; James et al. 1982; James and Burges 1982a,b; Delleur 1982; Jackson 1982). Todini (1988a,b) reviewed the historical development of mathematical methods used in rainfall-runoff modeling and classified the models based on a priori knowledge and problem requirements. He foresaw the increasing role of distributed models, satellite, and radar technology in watershed hydrology and noted that techniques for model calibration and verification remained less than robust.

Goodrich and Woolhiser (1991) reviewed progress in catchment hydrology in the United States and emphasized that a detailed process-based understanding of hydrologic response over a range of catchment scales, 0.01–500 km^2, still eluded the hydrologic community. El-Kady (1989) reviewed numerous watershed models and concluded that the surface water-groundwater linkage needed improvement, while ensuring an integrated treatment of the complexity and scale of individual component processes.

Table 1. Sample of Popular Hydrologic Models

Model name/acronym	Author(s) (year)	Remarks
Stanford watershed Model (SWM)/Hydrologic Simulation Package-Fortran IV (HSPF)	Crawford and Linsley (1966), Bicknell et al. (1993)	Continuous, dynamic event or steady-state simulator of hydrologic and hydraulic and water quality processes
Catchment Model (CM)	Dawdy and O'Donnell (1965)	Lumped, event-based runoff model
Tennessee Valley Authority (TVA) Model	Tenn. Valley Authority (1972)	Lumped, event-based runoff model
U.S. Department of Agriculture Hydrograph Laboratory (USDAHL) Model	Holtan and Lopez (1971), Holtan et al. (1974)	Event-based, process-oriented, lumped hydrograph model
U.S. Geological Survey (USGS) Model	Dawdy et al. (1970, 1978)	Process-oriented, continuous/event-based runoff model
Utah State University (USU) Model	Andrews et al. (1978)	Process-oriented, event/continuous streamflow model
Purdue Model	Huggins and Monke (1970)	Process-oriented, physically based, event runoff model
Antecedent Precipitation Index (API) Model	Sittner et al. (1969)	Lumped, river flow forecast model
Hydrologic Engineering Center—Hydrologic Modeling System (HEC-HMS)	Feldman (1981), HEC (1981, 2000)	Physically-based, semidistributed, event-based, runoff model
Streamflow Synthesis and Reservoir regulation (SSARR) Model	Rockwood (1982), U.S. Army Corps of Engineers (1987), Speers (1995)	Lumped, continuous streamflow simulation model
National Weather service-River Forecast System (NWS-RFS)	Burnash et al. (1973a,b), Burnash (1975)	Lumped, continuous river forecast system
University of British Columbia (UBC) Model	Quick and Pipes (1977), Quick (1195)	Process-oriented, lumped parameter, continuous simulation model
Tank Model	Sugawara et al. (1974), Sugawara (1995)	Process-oriented, semidistributed or lumped continuous simulation model
Runoff Routing Model (RORB)	Laurenson (1964), Laurenson and Mein (1993, 1995)	Lumped, event-based runoff simulation model
Agricultural Runoff Model (ARM)	Donigian et al. (1977)	Process-oriented, lumped runoff simulation model
Storm Water Management Model (SWMM)	Metcalf and Eddy et al. (1971), Huber and Dickinson (1988), Huber (1995)	Process-oriented, semidistributed, continuous stormflow model
Xinanjiang Model	Zhao et al. (1980), Zhao and Liu (1195)	Process-oriented, lumped, continuous simulation model
Hydrological Simulation (HBV) Model	Bergstrom (1976, 1992, 1995)	Process-oriented, lumped, continuous streamflow simulation model
Great Lakes Environmental Research Laboratory (GLERL) Model	Croley (1982, 1983)	Physically based, semidistributed continuous simulation model
Pennsylvania State University—Urban Runoff Model (PSU-URM)	Aron and Lakatos (1980)	Lumped, event-based urban runoff model
Chemicals, Runoff, and Erosion from Agricultural Management Systems (CREAMS)	USDA (1980)	Process-oriented, lumped parameter, agricultural runoff and water quality model
Areal Non-point Source Watershed Environment Response Simulation (ANSWERS)	Beasley et al. (1977), Bouraoui et al. (2002)	Event-based or continuous, lumped parameter runoff and sediment yield simulation model
Erosion Productivity Impact Calculator (EPIC) Model	Williams et al. (1984), Williams (1995a,b)	Process-oriented, lumped-parameter, continuous water quantity and quality simulation model
Simulator for Water Resources in Rural Basins (SWRRB)	Williams et al. (1985), Williams (1995a,b)	Process-oriented, semidistributed, runoff and sediment yield simulation model
Simulation of Production and Utilization of Rangelands (SPUR)	Wight and Skiles (1987), Carlson and Thurow (1992), Carlson et al. (1995)	Physically based, lumped parameter ecosystem simulation model
National Hydrology Research Institute (NHRI) Model	Vandenberg (1989)	Physically based, lumped parameter, continuous hydrologic simulation model
Technical Report-20 (TR-20) Model	Soil Conservation Service (1965)	Lumped parameter, event based runoff simulation model

Table 1. *Continued*

Model name/acronym	Author(s) (year)	Remarks
Systeme Hydrologique Europeen/Systeme Hydrologique Europeen Sediment (SHE/SHESED)	Abbott et al. (1986a,b), Bathurst et al. (1995)	Physically based, distributed, continuous streamflow and sediment simulation
Institute of Hydrology Distributed Model (IHDM)	Beven et al. (1987), Calver and Wood (1995)	Physically based, distributed, continuous rainfall-runoff modeling system
Physically Based Runoff Production Model (TOPMODEL)	Beven and Kirkby (1976, 1979), Beven (1995)	Physically based, distributed, continuous hydrologic simulation model
Agricultural Non-Point Source Model (AGNPS)	Young et al. (1989, 1995)	Distributed parameter, event-based, water quantity and quality simulation model
Kinematic Runoff and Erosion Model (KINEROS)	Woolhiser et al. (1990), Smith et al. (1995)	Physically based, semidistributed, event-based, runoff and water quality simulation model
Groundwater Loading Effects of Agricultural Management Systems (GLEAMS)	Knisel et al. (1993), Knisel and Williams (1995)	Process-oriented, lumped parameter, event-based water quantity and quality simulation model
Generalized River Modeling Package—Systeme Hydroloque Europeen (MIKE-SHE)	Refsgaard and Storm (1195)	Physically based, distributed, continuous hydrologic and hydraulic simulation model
Simple Lumped Reservoir Parametric (SLURP) Model	Kite (1995)	Process-oriented, distributed, continuous simulation model
Snowmelt Runoff Model (SRM)	Rango (1995)	Lumped, continuous snowmelt-runoff simulation model
THALES	Grayson et al. (1195)	Process-oriented, distributed-parameter, terrain analysis-based, event-based runoff simulation model
Constrained Linear Simulation (CLS)	Natale and Todini (1976a,b, 1977)	Lumped parameter, event-based or continuous runoff simulation model
ARNO (Arno River) Model	Todini (1988a,b, 1996)	Semidistributed, continuous rainfall-runoff simulation model
Waterloo Flood System (WATFLOOD)	Kouwen et al. (1993), Kouwen (2000)	Process-oriented, semidistributed continuous flow simulation model
Topgraphic Kinematic Approximation and Integration (TOPIKAPI) Model	Todini (1995)	Distributed, physically based, continuous rainfall-runoff simulation model
Hydrological (CEQUEAU) Model	Morin et al. (1995, 1998)	Distributed, process-oriented, continuous runoff simulation model
Large Scale Catchment Model (LASCAM)	Sivapalan et al. (1996a,b,c)	Conceptual, semidistributed, large scale, continuous, runoff and water quality simulation model
Mathematical Model of Rainfall-Runoff Transformation System (WISTOO)	Ozga-Zielinska and Brzezinski (1994)	Process-oriented, semidistributed, event-based or continuous simulation model
Rainfall-Runoff (R-R) Model	Kokkonen et al. (1999)	Semidistributed, process-oriented, continuous streamflow simulation model
Soil-Vegetation-Atmosphere Transfer (SVAT) Model	Ma et al. (1999), Ma and Cheng (1998)	Macroscale, lumped parameter, streamflow simulation system
Hydrologic Model System (HMS)	Yu (1996), Yu and Schwartz (1998), Yu et al. (1999)	Physically based, distributed-parameter, continuous hydrologic simulation system
Hydrological Modeling System (ARC/EGMO)	Becker and Pfutzner (1987), Lahmer et al. (1999)	Process-oriented, distributed, continuous simulation system
Macroscale Hydrolgical Model-Land Surface Scheme (MODCOU-ISBA)	Ledoux et al. (1989), Noilhan and Mahfouf (1996)	Macroscale, physically based, distributed, continuous simulation model
Regional-Scale Hydroclimatic Model (RSHM)	Kavas et al. (1998)	Process-oriented, regional scale, continuous hydrologic simulation model
Global Hydrology Model (GHM)	Anderson and Kavvas (2002)	Process-oriented, semidistributed, large scale hydrologic simulation model
Distributed Hydrology Soil Vegetation Model (DHSVM)	Wigmosta et al. (1994)	Distributed, physically based, continuous hydrologic simulation model
Systeme Hydrologique Europeen Transport (SHETRAN)	Ewen et al. (2000)	Physically based, distributed, water quantity and quality simulation model
Cascade two dimensional Model (CASC2D)	Julien and Saghafian (1991), Ogden (1998)	Physically based, distributed, event-based runoff simulation model
Dynamic Watershed Simulation Model (DWSM)	Borah and Bera (2000), Borah et al. (1999)	Process-oriented, event-based, runoff and water quality simulation model
Surface Runoff, Infiltration, River Discharge and Groundwater Flow (SIRG)	Yoo (2002)	Physically based, lumped parameter, event-based streamflow simulation model

Table 1. *Continued*

Model name/acronym	Author(s) (year)	Remarks
Modular Kinematic Model for Runoff Simulation (Modular System)	Stephenson (1989) Stephenson and Randell (1999)	Physically based, lumped parameter, event-based runoff simulation model
Watershed Bounded Network Model (WBNM)	Boyd et al. (1979, 1996), Rigby et al. (1991)	Geomorphology-based, lumped parameter, event-based flood simulation model
Geomorphology-Based Hydrology Simulation Model (GBHM)	Yang et al. (1998)	Physically based, distributed, continuous hydrologic simulation model
Predicting Arable Resource Capture in Hostile Environments-The Harvesting of Incident Rainfall in Semi-arid Tropics (PARCHED-THIRST)	Young and Gowing (1996) Wyseure et al. (2002)	Process-oriented, lumped parameter, event-based agro-hydrologic model
Daily Conceptual Rainfall-Runoff Model (HYDROLOG)-Monash Model	Potter and McMahon (1976), Chiew and McMahon (1994)	Lumped, conceptual rainfall-runoff model
Simplified Hydrology Model (SIMHYD)	Chiew et al. (2002)	Conceptual, daily, lumped parameter rainfall-runoff model
Two Parameter Monthly Water Balance Model (TPMWBM)	Guo and Wang (1994)	Process-oriented, lumped parameter, monthly runoff simulation model
The Water and Snow Balance Modeling System (WASMOD)	Xu (1999)	Conceptual, lumped, continuous hydrologic model
Integrated Hydrometeorological Forecasting System (IHFS)	Georgakakos et al. (1999)	Process-oriented, distributed, rainfall and flow forecasting system
Stochastic Event Flood Model (SEFM)	Scaefer and Barker (1999)	Process-oriented, physically based event-based, flood simulation model
Distributed Hydrological Model (HYDROTEL)	Fortin et al. (2001a,b)	Physically based, distributed, continuous hydrologic simulation model
Agricultural Transport Model (ACTMO)	Frere et al. (1975)	Lumped, conceptual, event-based runoff and water quality simulation model
Soil Water Assessment Tool (SWAT)	Arnold et al. (1998)	Distributed, conceptual, continuous simulation model

While reviewing advances in watershed modeling, Hornberger and Boyer (1995) emphasized the need to deal with spatial variability and scaling and the need to explicitly consider linkages among hydrology, geochemistry, environmental biology, meteorology, and climatology. The most important recent advances in watershed modeling were noted to have been the employment of geographical information systems (GIS), remotely sensed data, and environmental tracers. The need for the acquisition of more data and more experimentation were emphasized for future progress of hydrology.

Advances in scientific understanding and subsequent engineering applications come about through new theoretical insights, unique observations, or by the development of new measurement or computational techniques. It appears that there have been few theoretical breakthroughs. For example, Freeze and Harlan (1969) laid out the blueprint for a three-dimensional watershed model, including precipitation, surface runoff, porous media flow, open channel flow, interaction of groundwater and channel flow, and transport of water to the atmosphere by evaporation and transpiration. The model could not be implemented at the time because of computational and data limitations. However, it is a conceptual forerunner of the watershed model SHE (Abbott et al. 1986a,b). The Stanford Watershed Model (Crawford and Linsley 1962, 1966) was considered to be the standard for applied models in 1973. Many current models have essentially the same fundamental structure. The modeling of water quality was just beginning, and models of dissolved oxygen in a reach of stream, transport of conservative and nonconservative pollutants, radioactive aerosols, and nutrients in streams and watersheds were under development.

Many of the advances after 1973 were due to improvements in computational facilities or new measurement techniques. Others were due to insights obtained by comparing model results with experimental data. For example, little thought was given to the problem of subgrid scale variability in 1973. It was assumed that it was only necessary for a distributed model to accommodate the variability of saturated hydraulic conductivity due to different soil types and vegetative cover as well as watershed topography and channel geometry. Small-scale spatial variability of saturated conductivity, i.e., within a computational element, was not considered important until analysis of data from rainfall simulator plots showed an increase in infiltration rate with rainfall intensity (e.g., Hawkins and Cundy 1987). Many of the challenges discussed by Hornberger and Boyer (1995) result from new technology; the use of digital elevation models (DEMs) and GIS raises the question of subgrid variability and the effect of pixel size on model calibration. One new concept that appeared is the use of topographic indices such as those used in TOPMODEL (Beven and Kirkby 1979; Binley et al. 1989a,b). Another new approach is the use of chemical tracers in conjunction with numerical models. Another new concept is one of upscaling of hydrologic conservation equations and subgrid spatial variability.

Classification of Watershed Hydrology Models

A watershed hydrology model is an assemblage of mathematical descriptions of components of the hydrologic cycle. The model structure and architecture are determined by the objective for which the model is built. For example, a hydrologic model for flood control is quite different from the one for hydropower generation or reservoir operation. Likewise, a model for water resources planning is significantly different from the one used for water resources design or ecological management. Singh (1995a) classified hydrologic models based on (1) process description; (2)

timescale; (3) space scale; (4) techniques of solution; (5) land use; and (6) model use. ASCE (1996) reviewed and categorized flood analysis models into (1) event-based precipitation-runoff models; (2) continuous precipitation-runoff models; (3) steady flow routing models; (4) unsteady-flow flood routing models; (5) reservoir regulation models; and (6) flood frequency analysis models.

Although the mathematical equations embedded in watershed models are continuous in time and often space, analytical solutions cannot be obtained except in very simple circumstances. Numerical methods (finite difference, finite element, boundary element, boundary fitted coordinate) must be used for practical cases. The most general formulation would involve partial differential equations in three space dimensions and time. If the spatial derivatives are ignored, the model is said to be "lumped"; otherwise, it is said to be "distributed," and the solution (output) is a function of space and time. Strictly speaking, if a model is truly distributed, all aspects of the model must be distributed including parameters, initial and boundary conditions, and sources and sinks. Practical limitations of data and discrete descriptions of watershed geometry and parameters to conform to the numerical solution grid or mesh do not permit a fully distributed characterization. Most watershed hydrology models are deterministic, but some consist of one or more stochastic components.

Several scientific disciplines have developed mathematical descriptions of components of the hydrological cycle, using basic physical principles in conjunction with experimental data. The physical fidelity of these models depends on the objective of the researcher and the tools available to solve the resulting equations. The watershed modeler has wide latitude in choosing the level of rigor or detail required of an individual component model, and the choices are affected by the objectives, watershed topography, geology, soils, land use, and the available information.

Although watershed models may be complicated with many parameters, frequently, the information that they are required to provide is very simple, as for example, the mean annual groundwater recharge rate over part of the basin or the 100-year flood. Statistical tools, including regression and correlation analysis, time-series analysis, stochastic processes, and probabilistic analysis are necessary to analyze the output to provide this type of information. Because of uncertainties in model structure, parameter values and precipitation, and other climatic inputs, uncertainty analysis and reliability analysis can be employed to examine their impact.

Wurbs (1998) highlighted the availability and role of generalized computer modeling packages and outlined the institutional setting within which the models are disseminated throughout the water community. Generalized water resources models were classified into (1) watershed models; (2) river hydraulics models; (3) river and reservoir water quality models; (4) reservoir/river system operation models; (5) groundwater models; (6) water distribution system hydraulic models; and (7) demand forecasting models.

Hydrologic Data Needs

Frequently, the type of a model to be built is dictated by the availability of data. In general, distributed models require more data than do lumped models. In most cases, needed data either do not exist or are not available in full. That is one reason why regionalization and synthetic techniques are useful. Even if the needed data are available, problems remain with regard to completeness, inaccuracy, and inhomogeneity of data. Then, of course, storage, handling, retrieval, analysis, and manipulation of data have to be dealt with. If the volume of data required is large, data processing and management can be quite a sophisticated undertaking.

The data needed for watershed hydrology modeling are hydrometeorologic, geomorphologic, agricultural, pedologic, geologic, and hydrologic. Hydrometeorologic data include rainfall, snowfall, temperature, radiation, humidity, vapor pressure, sunshine hours, wind velocity, and pan evaporation. Agricultural data include vegetative cover, land use, treatment, and fertilizer application. Pedologic data include soil type, texture and structure, soil condition, soil particle size diameter, porosity, moisture content and capillary pressure, steady-state infiltration, saturated hydraulic conductivity, and antecedent moisture content. Geologic data include data on stratigraphy, lithology, and structural controls. More specifically, data on the type, depth, and areal extent of aquifers are needed. Depending on the nature of aquifers, these data requirements vary. For confined aquifers, hydraulic conductivity, transmissivity, storativity, compressibility, and porosity are needed. For unconfined aquifers, data on specific yield, specific storage, hydraulic conductivity, porosity, water table, and recharge are needed. Each dataset is examined with respect to homogeneity, completeness, and accuracy. Geomorphologic data include topographic maps showing elevation contours, river networks, drainage areas, slopes and slope lengths, and watershed area. Hydrologic data include flow depth, streamflow discharge, base flow, interflow, stream-aquifer interaction, potential, water table, and drawdowns.

New Developments and Challenges in Watershed Models

Data Acquisition

Remote Sensing and Space Technology
New data collection techniques, especially remote sensing, satellites, and radar, received a great deal of attention in the 1980s and continue to do so. Major advances have been made in recent years in remote sensing and radar and satellite technology, which are going a long way in alleviating the scarcity of data that is one of the major difficulties in watershed hydrologic modeling. This technology provides synoptic data regarding spatial distribution of meteorological inputs; soil and land-use parameters; initial conditions; inventories of water bodies; such as dams, lakes, swamps, flooded areas, rivers, etc.; mapping of snow and ice conditions; water-quality parameters, etc. (Engman and Gurney 1991). Digital imagery provides mapping of spatially varying landscape attributes. Goodrich et al. (1994) employed remotely sensed soil wetness for modeling runoff in semiarid environments.

Radar is being employed for rainfall measurements. In contrast with point measurements provided by the usual rain-gauging techniques, the advantage of radar measurements is that they provide spatial mapping of rainfall, which is badly needed for distributed models. The Next Generation Weather Radar, Weather Surveillance Radars-88 Doppler, among others, are being employed to near real-time high-resolution precipitation volume and intensity over space and time. The Soil (now Natural Resources) Conservation Service collects real-time data on snowpacks from a network of about 500 snowpack telemetry sites located in remote mountainous areas of the western United States. These point measurements are augmented by satellite remote sensing to provide spatial and temporal distribution of snowpack properties. The National Operational Hydrologic Remote Sensing Center of the National Weather Service provides data on real-time snow-water equivalents for river basins in more than 25 states through its airborne gamma radiation measurements, and maps areal extent of snow cover for more than 4,000 river basins nationwide through satellite data from the Advanced Very High Resolution Radiometer and the Geostationary Operational Environmental Satellite.

The Landsat Thematic Mapper, *T* Multispectral Scanner, or Systeme Probatoire d' la Terre produce satellite imagery that, in conjunction with aerial photos and terrain data, has proved successful for providing data for mapping and classification of land use and vegetative land cover. Similary, the airborne light detection and ranging technology is being employed to provide accurate real-time flood inundation maps.

Nicks and Scheibe (1992) employed the Simulations for Water Resources in Rural Basins (SWRRB) model with NEXRAD radar information for rainfall data in modeling runoff from the Little Washita River watershed in southern Oklahoma. Duchon et al. (1992) employed these remotely sensed data in updating the SWRRB model parameters for analysis of the water budget of the Little Washita River watershed. Kite and Kouwen (1992) obtained improved estimates of hydrograph components from the Simple Lumped Reservoir Parametric (SLURP) model when they used Landsat-derived land-cover classes. Rango (1992) employed remotely sensed areal extent of snow-cover data in the SRM (Snowmelt Runoff Model) for 50 basins worldwide and discussed the potential use of this model to evaluate the effects of climate change scenarios.

With the vastly improved capability to observe hydrologic data, remote sensing and space technology are being increasingly coupled with watershed models for real-time flood forecasting, weather forecasting, forecasting of seasonal and/or short-term snowmelt runoff, evolution of watershed management strategies for conservation planning, development of reporting services for drought assessment/forecasting, mapping of groundwater potential to support the conjunctive use of surface water and groundwater, inventorying of coastal and marine processes, environmental impact assessment of large-scale water resource projects, flood-damage assessment, and the development of remote information matrix for irrigation development, to name but a few (Goodrich et al. 1991).

Digital Terrain and Elevation Models
Because physical characteristics of a watershed, such as soils, land use and topography, vary spatially, distributed watershed models may require huge volumes of data. The primary source of topographic information prior to the 1980s consisted of contour maps. Advances in digital mapping have provided essential tools to closely represent the 3D nature of natural landscapes. One such tool is the digital terrain (DTM) or DEM models. DEMs automatically extract topographic variables, such as basin geometry, stream networks, slope, aspect, flow direction, etc. from raster elevation data. Three schemes for structuring elevation data for DEMs are: triangulated irregular networks (TIN), grid networks, and vector or contour-based networks (Moore and Grayson 1991).

The most widely used data structures are grid networks. The ANSWERS (Beasley et al. 1980), AGNPS (Young et al. 1989), and SHE (Abbott et al. 1986a,b) are examples of hydrologic models that use a square grid or cell network as their basic structure. Although most efficient, Mark (1978) remarked that grid structures for spatially dividing watersheds are not appropriate for many hydrologic and geomorphologic applications. Moore and Grayson (1991) reported that computed flow paths took on zigzag shapes. Moore et al. (1988a,b) used contour-based DEMs for hydrologic and ecologic applications. O'Loughlin (1986) employed them to identify zones of saturation, and Moore and Burch (1986) used them to delineate zones of erosion and deposition. The grid and vector networks are useful for planning purposes. Silfer et al. (1987) used a kinematic wave model for computing overland flow, based on TIN-DEM representation. A similar concept was employed by Grayman et al. (1975), Vieux (1988), and Goodrich and Woolhiser (1991).

Hydrologic models with a spatial structure are being increasingly based on DEM or DTM (Moore et al. 1988a,b). Many of the existing models, such as SHE, TOPMODEL, etc., have been adapted to the new type of data. Integration of hydrologic models with remotely sensed, GIS, and DEM-based data has started to occur. Examples of newly developed or adapted models are those by Fortin et al. (2001a,b), Wigmosta et al. (1994), Julien et al. (1995), Desconnets et al. (1996), and Olivera and Maidment (1999), among others.

Chemical Tracers
Data on the chemical composition of water can be used for modeling the flow of water along different paths. These data help define surface, subsurface, and groundwater flows and thus help define hydrograph separation. Stable isotopes have been used for defining conceptual models of water flow (Stewart and McDonnell 1991). Radiogenic isotopes, both natural and anthropogenic, have been used as tracers (Rose 1992). Chloroflurocarbons have been employed to trace flow paths in groundwater systems (Dunkle et al. 1993). Chemical data can be used for model calibration, as was done by Robson et al. (1992) in the case of TOPMODEL. Adar and Neuman (1988) used environmental isotopes and hydrochemical data to estimate the spatial distribution of groundwater recharge. Tracers can provide a wealth of information on flow of water, its origin, source, and flow paths, etc.

Data Processing and Management: Geographic Information System and Database-Management Systems

For processing large quantities of data, GIS, database-management systems (DBMS), and graphic and visual design tools are some of the techniques available (Singh and Fiorentino 1996). Integration of these techniques with watershed hydrology models accomplishes a number of significant functions: designing, calibrating, modifying, evaluating, and comparing watershed hydrology models. For example, the use of GIS permits subdividing a watershed into hydrologically homogeneous subareas in both horizontal and vertical domains. Depending on the type of application requiring categorization of hydrologic properties, many combinations of spatial overlays can be performed. With the GIS technique, it is possible to delineate soil loss rates, identify potential areas of nonpoint source agricultural pollution, and map groundwater contamination susceptibility. GIS enhances the ability to incorporate spatial details beyond the existing capability of watershed hydrology models. With much better resolution of terrain-streams and drainage areas, the ability to delineate more appropriate grid layers for a finite-element or finite-difference watershed model is enhanced. The USGS Precipitation-Runoff Modeling System (PRMS) employs automated methods to derive required model parameters in which the hydrological response units (HRUs) are delineated using terrain analysis (Leavesley and Stannard 1990). Using a data-parameter interface, a GIS system computes the necessary model parameters within each HRU. Battaglin et al. (1993) found this interface concept to be useful in model parameterization and calibration on a series of basins. Vieux (1991) discussed several aspects of the use of GIS in watershed modeling.

Spatial Description of Topography

The various methods of simplifying watershed geometry can be divided into grid methods and conceptual methods (Singh 1996). Either method subdivides the watershed into subareas that are linked together by routing elements. Hromadka et al. (1988) attempted to quantify the effect of watershed subdivision on prediction accuracy of hydrologic models. When the watershed was divided into subareas, each subarea having identical parameters, the variance of peak flow estimates decreased significantly with increasing number of subareas. A grid method attempts to maintain model flow patterns similar to those in the prototype watershed response. This concept was introduced by Bernard (1937). Huggins and Monke (1968) used the same grid method to represent watershed geometry. Surkan (1969) developed a computer algorithm for numeric coding of natural geometry on a rectangular grid for hydrograph synthesis. These days, different types of grid structures, such as finite-element grid, rectangular grid, and boundary-fitted coordinate grid, etc. are used, depending on the numerical scheme of a model.

Conceptual methods represent watershed geometry using a network of elemental sections, including plane, triangular section, converging section, diverging section, and channel. Each element represents a particular portion of the watershed. These elements may be arranged to provide a detailed representation of the gross topographic features of a watershed, regardless of its geometric complexity. Many simplified geometric configurations that depend on the arrangement of these elements have been employed in hydrology. Examples of such configurations are V-shaped geometry, composite geometry, cascade of planes and channels, complex configurations of planes and channels, and so on. Harley et al. (1970) and Rovey et al. (1977) employed configurations of planes and channels. There have been many techniques for generating such configurations. Berod et al. (1995, 1999) employed a geomorphologically based method to define planes and channels. Boyd (1978) employed a watershed-bounded method to generate a network representation for his WBN model. This representation is commonly used these days. Lane and Woolhiser (1977) suggested a statistical procedure to select an appropriate geometric simplification of a watershed.

Scaling and Variability

Scale is normally defined as the sampling interval size at which hydrologic observations are made or as the grid size used for numerical computations. Thus, the size of a scale will correspond to the length in the spatial domain and to the duration in the time domain. Parameters and hydrologic processes controlling the watershed response operate at many different space and timescales. Using five field examples, Seyfried and Wilcox (1995) analyzed how the nature of spatial variability affects the hydrological response over a range of scales: (1) infiltration and surface runoff affected by shrub canopy; (2) groundwater recharge affected by soil depth; (3) groundwater recharge and streamflow affected by small-scale topography; (4) frozen soil runoff affected by elevation; and (5) snowfall distribution affected by large-scale topog-

raphy. Depending on the scale, the sources of variability can be stochastic or deterministic or both. It is not possible to describe watersheds in terms of a single deterministic length scale, independent of scale and watershed characteristics. For a consistent treatment of these hydrologic processes, observed and model scales should be commensurate. Morel-Seytoux (1988) reasoned that nature embodies both the elements of chance and the descriptive laws of physics. Therefore, excessive process description at one scale is lost through the processes of integration in time and space and through averaging. This justifies model simplification as long as the essential behavior is retained. He showed how simplifications can be made so that straightforward scaling integration is accomplished in a physically based stochastic framework.

Issues related to spatial and temporal scaling and variability started receiving much attention beginning in the 1980s. An assumption commonly employed in hydrologic modeling is one of homogeneity at the grid scale. Kavvas (1999) defined heterogeneity as the fluctuations in the values of hydrologic state variables, such as flow discharge, infiltration rate, and evapotranspiration rate, etc., in hydrologic parameters such as roughness, hydraulic conductivity, porosity, etc. and in boundary conditions and forcing functions such as rainfall, snowfall, wind, etc. Heterogeneity was further classified into stationary heterogeneity and nonstationary heterogeneity. If the mean, higher-order moments and probability density functions of the fluctuations in space/time remain constant with respect to all space/time origin locations, then the hydrologic process (or parameter) is stationary heterogeneous; otherwise, it is nonstationary heterogeneous. A stationary heterogeneous process (or parameter) is ergodic in the mean if the ensemble average of its fluctuations is equal to their spatial/time volumetric average or areal average. Kavvas (1999) showed that a hydrologic process (or parameter) that is nonstationary at one scale may become stationary at another scale. The fundamental reason for transformation from nonstationarity to stationarity with the increase in scale is the phenomenon of coarse-graining of hydrologic processes at increasing scales. The hydrologic equations, however, still remain parsimonious as the scales get larger.

Upscaling of Hydrologic Conservation Equations

The construction and complexity of a hydrologic model are greatly influenced by the domain in which it is built and the scale at which it is built. In the time domain, different scales are used, based on which models are classified as continuous, event-based, weekly, monthly, seasonal, or yearly models. Many hydrologic models employ equations based on conservation of mass, momentum, and energy. These equations are point-scale, and their averaging in space depends on the hydrologic process to be modeled. For example, for surface flow the St. Venant equations or their simplified forms are depth-averaged, but for subsurface flow, the governing equations are areally averaged. In either case, they require data at a scale much finer than is available. This means that the point-scale equations must be upscaled in order to conserve mass, momentum, and energy and to ensure compatibility between the scales of observed data and governing equations. Kavvas (1999) has shown that when a larger scale process is formed by averaging a small-scale process, the high frequency components of the smaller scale process are eliminated by averaging, and this leads to considerable simplification of the average hydrologic conservation equations. Indeed, there are evolving scales of heterogeneity with respect to space, and these scales influence the averaging of conservation equations as well as the removal of high-frequency components.

The upscaling of conservation equations plays an important role in dealing with subgrid variability and in parameter estimation. Because the parameters in the upscaled equations are also upscaled, the subgrid variability can be quantified by means of areal variance and covariance of the point-scale parameters. Chen et al. (1994a,b) treated spatially averaging of unsaturated flow equations under infiltration conditions over areally heterogeneous soils and presented a practical application. In these equations, areal median saturated hydraulic conductivity and areal variance of log-saturated hydraulic conductivity emerged as the main parameters when the saturated hydraulic conductivity was considered the main source of heterogeneity. Numerical experiments on unsaturated flow within a soil column with varying degrees of heterogeneity measured by the coefficient of the log-saturated hydraulic conductivity showed that the areally averaged Green-Ampt equation significantly outperformed the point-scale Richards equation incorporating areally-averaged, log-saturated hydraulic conductivity even in the 3D case.

For modeling overland flow over varying microtopographic surfaces, Tayfur and Kavvas (1994) showed that if these surfaces were replaced by smooth surfaces then the depth-averaged equations are indeed treated as large-scale averaged equations (Tayfur 1993). Such a treatment is mathematically not correct, and one should upscale the depth-averaged equations to conserve mass and momentum at a larger scale. Tayfur and Kavvas (1994) developed transectionally averaged flow equations for a hillslope transect. By assuming randomness in the flow variables due to randomness in parameters, Tayfur and Kavvas (1998) also obtained areally averaged flow equations for a hillslope surface. The resulting flow equations were only time-dependent and whose solution required a very simple numerical method. In the same vein, Horne and Kavvas (1997) averaged over the snowpack depth the energy and mass conservation equations that govern the snowmelt dynamics at a point location and obtained depth-averaged equations (DAE). By assuming the snowmelt process to be spatially ergodic, they then averaged the point-location DAE over the snowpack area. The areally averaged equations were obtained in terms of their corresponding ensemble averages.

The model parameters as normally determined these days are based on spatial variation of point-scale parameters obtained using GIS, and remote sensing, etc. In large-scale modeling of land-surface processes, the scales of upscaled hydrologic equations and upscaled parameters seem to be consistent with grid-area resolution. However, because of the subgrid scale variability within each grid area, there is a fundamental issue of the inconsistency of the point-scale parameter values with regard to the grid area they represent. Through regional scale land surface hydrologic modeling of California at 20 km grid resolution, Kavvas et al. (1998) have shown that this inconsistency can be removed by using the spatially averaged, upscaled conservation equations whose upscaled parameters are at the same scale as of the modeling grid areas.

Spatial Scaling

The spatial scale greatly influences the choice of a model. Hydrologic variables vary in space with respect to both direction and location. In case of terrestrial hydrology, one dimensional treatment is adequate in most cases. However, the variability is particularly high in the soil and aquifer environment in all three dimensions or at least in two dimensions. Thus, incompatibilities arise when the entire continuum is modeled and even more so when the model is coupled with a climatic model or an oceanic model, due to significantly different speeds of atmospheric processes, land-surface hydrologic processes, as well as oceanic processes caused by the significantly different time response characteristics of atmospheric, oceanic, and hydrologic processes (Kavvas et al. 1998).

Spatial heterogeneity in catchment response arises from three sources: variabilities, discontinuities, and processes. Spatial variabilities in climatic inputs such as rainfall and hydrometorological variables, in soil characteristics such as hydraulic conductivity and porosity, in topography, and land use, encompass a space-time continuum. The runoff from a watershed is governed by local combinations of these factors. Discontinuities encompass the boundaries separating soil types, geologic formations, or land covers. Physical properties control interception, surface retention, infiltration, overland flow, and evapotranspiration at different scales, and these processes control runoff. It has been observed empirically that the form of hydrologic response changes with the spatial scale of heterogeneities, usually considered to be simpler and more linear with increasing watershed size (Dooge 1981). This relation may be climate dependent because Goodrich et al. (1997) demonstrated that the response became more nonlinear in a semiarid watershed. When the spatial scale is extended from a point to larger areas, the runoff generation process becomes less sensitive to temporal variations of local precipitation or spatial variations of soil characteristics because of the averaging effect. However, the spatial extent is limited by differences in physical, vegetative, and topographic features. Sivapalan and Wood (1986) investigated the effect of spatial heterogeneity in soil and rainfall characteristics on the infiltration response of catchments. Eagleson and Qinliang (1987) found that both the first and second moments of peak streamflow decreased rapidly with increasing values of the catchment to storm scale ratio. Milly and Eagleson (1988) underscored the need to incorporate areal storm variability in large area hydrologic models. Osborn et al. (1993) found that runoff volumes calculated with input from a centrally located rain gauge on a 6.3 km^2 semiarid watershed was greater than runoff calculated using 10 recording rain gauges.

Investigating the impact of spatial rainfall and soil information on runoff prediction at the hillslope scale, Loague (1988) aggregated fine-scale realizations of rainfall fields and spatial hydraulic conductivity to coarser resolutions. He found that at the hillslope scale hydraulic conductivity was more critical than rainfall and that runoff peak, time to peak, and runoff volume required different information levels.

Physical Spatial Size

The minimum level of physical spatial scale to be used in watershed modeling, which would adequately represent the spatial heterogeneity of a watershed, has received considerable attention. Using the SHE model on the Wye watershed 10.55 km^2 in area, Bathurst (1986) suggested dividing the watershed into elements no larger than 1% of the total area to ensure that each element was more or less homogeneous. Introducing the concept of representative elementary area (REA), Wood et al. (1988) found that an REA of approximately 1 km^2 existed for hydrologic response of the Coweeta watershed and was more strongly influenced by basin topography than rainfall length scales.

Tao and Kouwen (1989) used 5×5 km and 10×10 km grid sizes on the 3,520 km^2 Grand River watershed containing four reservoirs in southwestern Ontario in Canada, and found that the two grid sizes had no significant effect on the model results. Pierson et al. (1994) employed a surface soil classification scheme to partition the spatial variability in hydrological and interrill erosion processes in a sagebrush plant community. Using a unit hydrograph model, Hromadka et al. (1988) found on 12 watersheds that the variance of model-simulated discharge decreased significantly with the level of discretization, but this decrease reflected a departure of the model results from the true watershed behavior. Using a length scale based on surface characteristics and excess rainfall duration, Julien and Moglen (1990) found that the influence of spatial variability of slope, roughness, width, and excess rainfall intensity on watershed runoff varied with the length scale.

Zhang and Montgomery (1994) examined the effect of digital elevation model (DEM) grid size on the portrayal of the land surface and hydrological simulations on two small watersheds in the western United States. They found that the DEM grid size significantly affected both the representation of the land surface and the results of hydrological simulation. A grid size smaller than the hillslope length was necessary to adequately simulate the processes controlled by land form. A 10 m grid size was proposed as a compromise between increasing spatial resolution and data handling requirements.

Using TOPMODEL on the 115.5 km^2 Sleepers River Research watershed in Vermont, Wolock (1995) found that a subwatershed should have an area of at least 5 km^2 before it is representative of larger watersheds along the same stream in terms of topographic characteristics and simulated flow paths. Wilgoose and Kuczera (1995) use subgrid approximations to provide an effective parameterization of the processes that occur on scales smaller than those that can be modeled. Using data from small plot experiments as well as large-scale watersheds, they found that infiltration parameters can be adequately calibrated from small-scale plots but not the kinematic parameters. Bruneau et al. (1995) analyzed the effect of space and time resolutions using TOPMODEL on the 12 km^2 Coetdan Experimental watershed in Britanny, France, with input derived from DEMs. An optimum region for modeling with a grid size of 50 m and a time step of about 1 h was found.

Vieux (1993) investigated the DEM aggregation and smoothing effects on surface runoff modeling and found that errors are propagated if the apparent slope is flattened or the flow path is shortened. Quinn et al. (1991) found that a grid-cell resolution larger than 5 m had a significant effect on soil moisture modeling. According to Tarboton et al. (1991), drainage network density and configurations are highly dependent on smoothing of elevations during the pit removal stage of network extraction. Low rainfall intensities produce proportionately larger errors than higher intensities for an extracted network.

Molnar and Julien (2000) evaluated the effects of square grid-cell size from 17 to 914 m on surface runoff modeling using a raster-based distributed CASC2D hydrologic model. For event-based simulation, their findings indicate that coarser grid-cell resolutions can be used for runoff simulations as long as parameters are appropriately calibrated, and the primary effect of increasing grid-cell size on simulation parameters is to require an increase in overland and channel roughness parameters. They found that they had to adjust overland and channel Manning's n values as grid size changed. Yao and Terakawa (1999) employed 1 km grids for a distributed model of the Fuji River basin (3,432 km^2) in Japan. Daily meteorological data were produced using GIS and step-wise regression. They found that it was possible to integrate daily and hourly scales to produce reasonable hydrologic response.

Winchell et al. (1998) investigated the effects of algorithm uncertainty and pixel aggregation on simulation of infiltration and saturation-excess runoff from a medium-sized (100 km^2) basin in northern Texas using radar-based rainfall estimates. Two types of uncertainty in precipitation estimates were considered: those aris-

ing from rainfall estimates and those due to spatial and temporal representation of the "true" rainfall field. The infiltration-excess runoff was more sensitive to both types of uncertainties than was the saturation-excess runoff. There was a significant reduction in infiltration-excess runoff volume when temporal and spatial resolution of the precipitation was reduced.

Mazion and Yen (1994) investigated the effect of computational spatial size on watershed runoff simulated by HEC-1, RORB, and a linear system. They found that the computational grid size had a significant effect on the model results if the physical scale was not finer, although the effects decreased with increasing rainfall duration. The effect of the computational size was about one order larger than the effect of the variability of surface conditions within the watershed, provided the overall watershed average runoff coefficient remained the same.

Recognizing the importance of spatial variability, the usual practice is to subdivide larger watersheds and then calibrate hydrologic models. However, a working concept of physical heterogeneity remains still elusive. For example, the methods of subdivision are governed more by data availability than by physical meaning. Song and James (1992) reviewed five scales used in hydrologic simulation: laboratory scale, hillslope scale, catchment scale, basin scale, and continental and global scale. They suggested a stochastic method in which a parametric-stochastic model can be formed from a parent parametric-deterministic model to find an optimal scale for its application.

The scaling issue assumes even a greater significance when developing regional or global hydrology models. There is a discrepancy in scale between regional climate models and hydrologic models. In fact there are incompatibilities among soil, surface water, and groundwater models attributed in part to oversimplifications of complex hydrologic processes in each of these models (Goodrich and Woolhiser 1991; Yu 1996). Thomas and Henderson-Sellers (1991) conclude that hydrologic and climatic models fail to represent day-to-day variability in streamflow and hypothesize that this variability could be accounted for by incorporating the spatial variability of different mechanisms of rainfall-runoff production (Wood et al. 1990).

Temporal Scaling

The timescale of model output (e.g., streamflow) greatly influences the type of the model or the details to be included in the model. For example, a monthly watershed hydrology model is quite different in its architecture and construct from, say, an hourly model. It remains an unresolved question as to the hydrologic laws operating at different timescales for different components of the hydrologic cycle. A solution to this question will greatly facilitate model construction and more clearly define data needs.

Many hydrologic simulation models employ more than one time interval in their computation. Diskin and Simon (1979) defined the time base as a combination of the interval used for input and internal computation and the time interval used for output and model calibration. They explored the relationship between the time bases of hydrologic models and their structure. Hughes (1993) suggested incorporation of variable time intervals in deterministic models. Woolhiser and Goodrich (1988) investigated the importance of time varying rainfall in a model of a small watershed and found that disaggregating total rainfall amounts into simple, constant, and triangular distributions caused significant distortion in the peak rate distributions for Hortonian runoff. Ormsbee (1989) found that uniform disaggregation grossly underestimated peak discharge frequencies from a continuous hydrologic model.

Spatial Variability of Hydraulic Roughness

Wu et al. (1982) examined the effects of spatial variability of roughness on runoff hydrographs from an experimental watershed facility and found that under certain conditions an equivalent uniform roughness could be used for a watershed with nonuniform roughness. Lehrsch et al. (1987, 1988) determined the spatial variation of eight physically significant roughness indices using a semivariogram analysis. Hairsine and Parlange (1986) demonstrated the formation of kinematic shocks on various surfaces with different degrees of roughness and analyzed the error incurred when a curved surface was represented by a kinematic cascade model. Vieux and Farajalla (1994) evaluated the error resulting from smoothing of the hydraulic roughness coefficients in modeling overland flow with a finite-element solution.

Spatial Variability of Infiltration

Spatial variability of infiltration has been amply documented (Sharma et al. 1980; Maller and Sharma 1984; Loague and Gander 1990; Sullivan et al. 1996; Turcke and Kueper 1996) and has been found to influence surface runoff characteristics, depending on rainfall and watershed characteristics. Milly and Eagleson (1988) showed that spatial variability in soil type and rainfall depth resulted in decreased cumulative infiltration and increased surface runoff. Smith and Hebbert (1979), Sivapalan and Wood (1986), and Woolhiser and Goodrich (1988), observed considerable differences in the infiltration rate when the average soil properties, as opposed to spatially varied properties, were used. Smith et al. (1990) incorporated small-scale spatial variability of soil saturated hydraulic conductivity into an infiltration model. This method has been enhanced by Smith and Goodrich (2000).

Using a 2D runoff model and a Monte Carlo methodology, Saghafian et al. (1995) examined the variability of Hortonian surface runoff discharge and volume produced by stationary storms on a watershed with spatially distributed soil saturated hydraulic conductivity. Greater peak flow was observed for spatially variable hydraulic conductivity than for uniform values. Woolhiser et al. (1996) showed that Hortonian runoff hydrographs were strongly affected by trends in hydraulic conductivity, especially for small runoff events. Using a 3D model of variably saturated flow on a hillslope, Binley et al. (1989a,b) found that the peak discharge and runoff volume generally increased with varying hydraulic conductivity, increasing with increasing variance and spatial dependence of the random saturated hydraulic conductivity field. For low permeability soils, they could not find an effective hydraulic conductivity parameter capable of reproducing surface and subsurface flow hydrographs.

Precipitation Variability

Singh (1997) reported on the effects of spatial and temporal variability in rainfall and watershed characteristics on the streamflow hydrograph. A short discussion of these effects is presented here.

Storm Movement
Yen and Chow (1968) and Marcus (1968) undertook laboratory studies to demonstrate the importance of rainstorm movement to the time distribution of surface runoff. Jensen (1984) determined the influence of storm movement and its direction on the shape,

peak, time to peak, and other characteristics of the runoff hydrograph. Maksimov (1964) showed that rainstorm movement altered peak discharge. Niemczynowicz (1984a,b) determined the influence of storm direction, intensity, velocity, and duration on the runoff hydrograph and peak discharge on a conceptual watershed and an actual watershed in the city of Lund in Sweden. Roberts and Klingman (1970) found that the direction of storm movement might augment or reduce flood peaks and modify the hydrograph recession. Surkan (1974) observed that peak flow rates and average flow rates were most sensitive to changes in the direction and speed of the rainstorms.

Sargent (1981, 1982) determined the effects of storm direction and speed on runoff peak, flood volume, and hydrograph shape. Stephenson (1984) simulated runoff hydrographs from a storm traveling down a watershed. Foroud et al. (1984) employed a 50-year hypothetical moving rainstorm to quantify the effect of its speed and direction on the runoff hydrograph. Ngirane-Katashaya and Wheater (1985) analyzed the effect of storm velocity on the runoff hydrograph. Ogden et al. (1995) investigated the influence of storm movement on runoff. Singh (1998) evaluated the effect of the direction of storm movement on planar flow and showed that the direction of storm movement exercised a significant influence on the peak flow, time to the peak flow, and the shape of the overland flow hydrograph.

Spatial Variability of Rainfall
The shape, timing, and peak flow of a stream-flow hydrograph are greatly influenced by spatial and temporal variability in rainfall. While examining the effects of spatially distributed rainfall for a conceptual watershed 100 km^2 in area, Watts and Calver (1991) found that an efficient resolution of rainfall data was around 2.5 km^2 along the storm path. Dawdy and Bergmann (1969) and Wilson et al. (1979) concluded that errors in rainfall volume and intensity over a watershed were likely to limit the accuracy of runoff simulation. Phanartzis (1972) stressed the importance of altitudinal pattern in runoff simulation on a watershed in the San Dimas Experimental Forest. Beven and Hornberger (1982) found that in a relatively homogeneous watershed the most important effect of rainfall variability was in the timing of the runoff hydrograph. The effect on peak flows was smaller but still significant, and the effect on storm volume was relatively minor.

Julien and Moglen (1990) found that for both correlated and uncorrelated spatial variability in rainfall excess the discharge hydrograph was quite sensitive to excess rainfall intensity, and the degree of sensitivity decreased with increasing rainfall duration. Ogden and Julien (1993) concurred with the findings of Julien and Moglen (1990). Stephenson (1984) noted that the time of concentration was nearly the same for uneven rainfall as for uniform distribution. Naden (1992) found that the effect of spatial variation in rainfall on the network channel response could be marked. Using a distributed model on a midsize catchment 150 km^2 in area, Michaud and Sorooshian (1994) found that errors in simulated peaks due to inadequate raingauge density (one gauge per 20 km^2) represented 58% of the observed peak flow. Rainfall sampling errors accounted for approximately half the difference between observed and simulated peaks. Faurés et al. (1995) found that spatial variability of rainfall can have significant effects on simulated Hortonian runoff, even at a very small scale.

Temporal Variability of Rainfall
In general, time-varying rainfall produces greater peak discharge than does constant rainfall. Southerland (1983) found that design

storms for flood estimation generally peaked in intensity in the first half of the storm. While evaluating the effect of maximum rainfall position, El-Jabi and Sarraf (1991) found that hydrograph timing was altered but not the hydrograph peak. Lambourne and Stephenson (1987) simulated runoff peaks and volumes for a series of synthetic 5-year storms having rectangular, triangular, and bimodal temporal distributions. The rectangular hyetograph underpredicted the peaks and volumes from an urbanized watershed. The triangular distribution overpredicted the peak discharge, and the bimodal distribution better predicted the runoff volumes and peaks than did the triangular distribution.

Ball (1994) employed 10 different rainfall excess patterns beginning with constant rainfall excess. The time of concentration for a watershed significantly changed with the pattern of rainfall excess. When compared with a constant rainfall excess pattern, hydrographs of design patterns of rainfall peaked early and were varied in shape. Stephenson (1984) noted that the peak runoff was approximately 10% greater for triangular distribution than for a uniform pattern of the same duration. Using weather radar for flood forecasting in the Sieve River basin in Italy, Pessoa et al. (1993) found no significant differences between hydrographs generated from 5, 15, and 30 min radar rainfall data. The hydrographs were generated from a distributed rainfall-runoff model (Cabral et al. 1990) that extracts topographic information from DEMs.

Model Calibration

Significant advances have been made in automated watershed model calibration during the past 2 decades, with focus on four main issues (Gupta et al. 1998): (1) development of specialized techniques for handling errors present in data; (2) search for a reliable parameter estimation algorithm; (3) determination of an appropriate quantity of and information-rich kind of data; and (4) efficient representation of the uncertainty of the calibrated model (structure and parameters) and translation of uncertainty into uncertainty in the model response. To account for data errors, maximum likelihood functions have been developed for measuring the closeness of the model and the data by Sorooshian and Dracup (1980), Sorooshian (1981), and Kuczera (1983a, b), among others.

Optimization methods have been developed for parameter estimation. A typical automatic parameter estimation methodology requires four elements: (1) objective function; (2) optimization algorithm; (3) termination criteria; and (4) calibration data. The choice of an objective function influences parameter estimates as well as the quality of model results. Rao and Han (1987) analyzed several objective functions in calibrating the urban watershed runoff model ILLUDAS and found the least-squares criterion to be the best. Servat and Dezetter (1991) employed five different objective functions for calibrating a rainfall-runoff model on a Sudanese savannah area in the Ivory Coast and found the Nash-Sutcliffe efficiency to be the best. Clarke (1973) noted that the assumptions underlying the use of a least-squares objective function for estimation of hydrologic model parameters were seldom valid and suggested basing the objective function on the stochastic properties of the errors in the model and the data. Investigating the effects of selecting different objective functions, Diskin and Simon (1977) proposed guidelines and made recommendations for selecting an objective function in model calibration.

Sorooshian and Gupta (1995) discussed several optimization methods, including direct search methods, gradient search methods, random search methods, multistart algorithms, and shuffled complex algorithms. The first two are local search methods and

the remaining are global search methods. Population-evolution-based search strategies have been popular (Brazil and Krajewski 1987; Brazil 1988; Wang 1991; Duan et al. 1992, 1993; Sorooshian et al. 1993). The shuffled complex evolution global optimization algorithm has, however, been found to be consistent, effective, and efficient in locating the globally optimum hydrologic model parameters (Duan et al. 1992, 1993; Sorooshian et al. 1993; Luce and Cundy 1994; Gan and Biftu 1996; Tanakamaru 1995; Tanakamaru and Burges 1997; Kuczera 1997).

Termination criteria are needed in an iterative search algorithm to determine when the slope of the function response surface is zero and the function value is minimum. Sorooshian and Gupta (1995) discussed several criteria, including the function convergence, parameter convergence, and maximum iterations and their limitations. In fact, none of these criteria are reliable in ascertaining the attainment of the global optimum, although parameter convergence was found to be most suitable for model calibration studies. The proper choice of calibration data may mitigate difficulties encountered in model calibration. Critical issues pertaining to calibration data are the amount of data necessary and sufficient for calibration and the quality of data resulting in the best parameter estimates. However, our understanding to address such issues is less than complete.

One of the main problems of optimization methods is the difficulty of finding a unique "best" parameter set. Another difficulty is the inadequacy of these methods for multi-input-ouput hydrologic models (Gupta and Sorooshian 1994a, b). To address these concerns, the generalized likelihood uncertainty estimation (Freer et al. 1996), Monte Carlo membership set procedure (van Straten and Keesman 1991), and the prediction uncertainty method (Klepper et al. 1991) have been proposed. These approaches are related to the generalized sensitivity analysis method developed by Spear and Hornberger (1980). These methods have weaknesses, however. Therefore, a more powerful calibration paradigm that considers the inherent multiobjective nature of the problem and recognizes the role of model error is needed. To that end, Gupta et al. (1998) proposed a new paradigm based on the multiobjective approach.

Artificial Neural Networks and Genetic Algorithms

Another fascinating area that has emerged in the 1990s is the application of artificial neural networks (ANNs) to hydrologic modeling. Because ANNs have the ability to recursively learn from data and can result in significant savings in time required for model development, they are particularly suited for modeling nonlinear systems where traditional parameter estimation techniques are not convenient. Preliminary concepts and hydrologic applications of ANNs have been detailed by ASCE (2000a, b). The book edited by Govindaraju and Rao (2000) contains a variety of applications of ANNs to hydrologic modeling. Lorrai and Sechi (1995) applied ANNs to evaluating rainfall-runoff models and river-flow forecasting. Hsu et al. (1995) employed ANNs to identify the model structure and concluded that ANNs provide a viable and effective alternative for input-output simulation and forecasting models that do not require modeling the internal structure of the watershed. Therefore, they are not a substitute for conceptual watershed modeling. Mason et al. (1996) suggested the use of radial basis functions for developing a neural network model of rainfall runoff, especially when a large database is involved. Minns and Hall (1996) used ANNs as rainfall-runoff models. Tokar and Markus (2000) compared ANNs with traditional models in predicting watershed runoff on three basins and found ANNs to yield higher accuracy. Gupta et al. (2000) proposed a multilayer feed-forward neural network for application to streamflow forecasting.

Wang (1991) developed a genetic algorithm for calibrating conceptual rainfall-runoff models. Savic et al. (1999) developed a genetic programming approach to structured system identification for rainfall-runoff modeling.

Global Hydrology Models

The decade of the 1990s started with an emphasis on regional and global hydrology that called for integration of hydrologic (terrestrial, pedologic, and lithologic), atmospheric, and hydrospheric models to evaluate the impact of climate change. The integration became possible because of the data being gathered by large-scale field experiments, such as STORM, GEWEX, HAPEX-MOBILHY, MAC-HYDRO, and so on. As a result, there exists a multitude of hydrologic models for application at the continental and global scale. The global hydrology model developed by Anderson and Kavvas (2002), the continental scale model, UMUS by Arnold et al. (1999), the regional-scale model developed by Yoshitani et al. (2002), and ISBA-MODCOU developed by Ledoux et al. (2002), among others are examples. One of the difficulties with such models is the lack of compatibility in scales at which data are available and the scales at which hydrologic, pedologic, atmospheric, and hydrospheric processes operate.

Model Error Analysis

Most models perform little to no error analysis. Thus, it is not clear what the model errors are and how different errors propagate through different model components and parameters. This is one of the major limitations of most current watershed hydrology models. Thus, from the standpoint of a user, it is not clear how reliable a particular model is. It is, therefore, no surprise that the user runs into difficulty when selecting a particular model.

Expert Systems

There was also some attention paid to the development of expert systems in hydrology. Gashing et al. (1981) probably were the first to develop a knowledge-based expert system for water resource problems. Underlying this system was SWM/HSPF. Simanovic (1990) described an expert system for selection of a suitable method for flow measurement in open channels. Although the area of artificial intelligence is very appealing, it somehow has not attracted much attention in the hydrologic community.

Linking of Water Quality

The decades of the 1980s and 1990s also witnessed the linking of hydrologic models with those of geochemistry, environmental biology, meteorology, and climatology. This linking became possible primarily for two reasons. First, there was increased understanding of spatial variability of hydrologic processes and the role of scaling. This was essential because different processes operate at different scales, and linking them to develop an integrated model is always challenging. Second, the digital revolution made possible the employment of GIS, remote sensing techniques, and database management systems. Currently, a number of watershed hydrology models have water-quality components built into their architecture, as seen in Table 1, as for example, HSPF, SHET-RAN, LASCAM, DVSM, DWSM, to name but a few.

Future Outlook

Mathematical models of watershed hydrology have now become accepted tools for water resources planning, development, design, operation, and management. It is anticipated that the future will witness even a greater and growing integration of these models with environmental and ecological management. With growing technologies triggered by the information revolution, remote sensing, satellite technology, geographic information systems, visual graphics, and data base management, the hydrologic models are getting increasingly more sophisticated and are being integrated with other process models.

The future of watershed hydrology models will be shaped by increasing societal demand for integrated environmental management; growing need for globalization by incorporation of biological, chemical, and physical aspects of the hydrological cycle; assessment of the impact of climate change; rapid advances in remote sensing and satellite technology, GIS, DBMS, and expert systems; enhanced role of models in planning and decision making; mounting pressure on transformation of models to user-friendly forms; and clearer statements of reliability and risk associated with model results.

The application of watershed hydrology models to environmental management will grow in the future. The models will be required to be practical tools—readily usable in planning and decision making. They will have to be interfaced with economic, social, political, administrative, and judicial models. Thus, watershed models will become a component in the larger management strategy. Furthermore, these models will become more global, not only in the sense of spatial scale but also in the sense of hydrologic details. Increasing fusion of biological and chemical courses in undergraduate curricula emphasizing hydrology is a healthy sign in that direction and will help achieve this goal.

Watershed hydrology models will have to embrace rapid advances occurring in remote sensing and satellite technology, geographical information systems, database management systems, error analysis, risk and reliability analysis, and expert systems. With the use of remote sensing, radar, and satellite technology, our ability to observe data over large and inaccessible areas and to map these areas spatially is vastly improved, making it possible to develop truly distributed models for both gauged and ungauged watersheds. Distributed models require large quantities of data that can be stored, retrieved, managed, and manipulated with the use of GIS and DBMS. This is possible because of literally unlimited computing capability available these days and will be even more so in the future. If watershed hydrology models are to become practical tools, then they have to be relatively easy to use, with a clear statement as to what they can and cannot do. They will need to assess the errors and determine how they propagate, define the reliability with which they accomplish their intended functions, and require the user to possess only a minimal amount of hydrologic training. Furthermore, the models will have to learn from the user as well as from empirical experience. Many of these functions can be performed by the use of expert systems in watershed hydrology modeling. Usually, the user is interested in what a model yields, its accuracy, and how easy it is to use, not the biology, chemistry, physics, and hydrology it is based on.

The models will have to be described in simple terms such that the interpretation of their results would not tax the ability of the user. They are designed to serve a practical end, and their constituency is one of users. After all, hydrologic models are to be used, not to be confined to academic shelves. Thus, model building will have to gravitate around the central theme of their eventual practical use in integrated environmental management. Although much progress has been made in mathematical modeling of watershed hydrology, there is still a long way to go before the models will be able to fully integrate rapidly evolving advances in information, computer, and space technology, and become "household" tools. Hydrologists are being challenged, but we have no doubt that they will meet the challenge.

Although much progress has been achieved in hydrology, there is a greater road ahead. A basic question is: What modeling technology is better? Because of the confusion, the technology developed decades ago is still in use in many parts of the world. This state of affairs is partly due to the lack of consensus as to the superiority of one type of technology to the other. Also, we have not been able to develop physically based models in a true sense and define their limitations. Thus, it is not always clear when and where to use which type of a model.

Conclusions

The following conclusions can be drawn from the foregoing discussion:

1. Many of the current watershed hydrology models are comprehensive, distributed, and physically based. They possess the capability to accurately simulate watershed hydrology and can be applied to address a wide range of environmental and water-resources problems.
2. The scope of mathematical models is growing, and the models are capable of simulating not only water quantity but also quality.
3. The technology of model calibration is much improved, although not all models have taken full advantage of it.
4. The models are becoming embedded in modeling systems whose mission is much larger, encompassing several disciplinary areas.
5. The technology of data collection, storage, retrieval, processing, and management has improved by leaps and bounds. In conjunction with literally limitless computing prowess, this technology has significantly contributed to the development of comprehensive distributed watershed models.

References

Abbott, M. B., Bathurst, J. C., Cunge, J. A., O'Connell, P. E., and Rasmussen, J. (1986a). "An introduction to the European Hydrologic System-Systeme Hydrologique Europeen, SHE, 1: History and philosophy of a physically-based, distributed modeling system." *J. Hydrol.,* 87, 45–59.

Abbott, M. B., Bathurst, J. C., Cunge, J. A., O'Connell, P. E., and Rasmussen, J. (1986b). "An introduction to the European Hydrologic System-Systeme Hydrologique Europeen, SHE, 2: Structure of a physically-based, distributed modeling system." *J. Hydrol.,* 87, 61–77.

Adar, E. M., and Neuman, S. P. (1988). "Estimates of spatial recharge distribution using environmental isotopes and hydrochemical data, II. Application to Aravaipa Valley in Southern Arizona, U.S.A." *J. Hydrol.,* 97, 279–302.

Anderson, M., and Kavvas, M. L. (2002). "Chapter 6: A global hydrology model." *Mathematical models of watershed hydrology,* V. P. Singh and D. K. Frevert, eds., Water Resources Publications, Littleton, Colo., in press.

Andrews, W. H., Riley, J. P., and Masteller, M. B. (1978). "Mathematical modeling of a sociological and hydrological system." *ISSR Research Monograph,* Utah Water Research Laboratory Utah State Univ., Logan, Utah.

Arnold, J. G., Srinivasan, R., Muttiah, R. S., and Williams, J. R. (1998). "Large area hydrologic modeling and assessment. Part I: Model development." *J. Am. Water Resour. Assoc.,* 34(1), 73–89.

Arnold, J. G., Srinivasan, R., Muttiah, R. S., and Allen, P. M. (1999). "Continental scale simulation of the hydrologic balance." *J. Am. Water Resour. Assoc.,* 35(5), 1037–1051.

Aron, G., and Lakatos, D. F. (1980). *Penn State urban runoff model: User's manual.* Institute for Research on Land and Water Resources Pennsylvania State Univ., University Park, Penn.

ASCE. (1996). *Handbook of hydrology, ASCE Manual and Rep. on Engineering Practice No. 28,* New York.

ASCE. (2000a). "Artificial neural networks in hydrology. 1: Preliminary concepts." *J. Hydrologic Eng.,* 5(2), 115–123.

ASCE. (2000b). "Artificial neural networks in hydrology. 2: Hydrology applications." *J. Hydrologic Eng.,* 5(2), 124–137.

Ball, J. B. (1994). "The influence of storm temporal pattern on catchment response." *J. Hydrol.,* 158, 285–303.

Barnes, B. S. (1940). "Discussion on analysis of runoff characteristics by O. H. Meyer." *Trans. Am. Soc. Civ. Eng.,* 105, 104–106.

Bathurst, J. C. (1986). "Sensitivity analysis of the Systeme Hydrologique European (SHE) for an upland catchment." *J. Hydrol.,* 87, 103–123.

Bathurst, J. C., Wicks, J. M., and O'Connell, P. E. (1995). "Chapter 16: The SHE/SHESED basin scale water flow and sediment transport modeling system." *Computer models of watershed hydrology,* V. P. Singh, ed., Water Resources Publications, Littleton, Colo. 563–594.

Battaglin, L. E., Parker, R. S., and Leavesley, G. H. (1993). "Applications of a GIS for modeling the sensitivity of water resources- to alterations in climate in the Gunnison River basin, Colorado." *Water Resour. Bull.,* 25(6), 1021–1028.

Beasley, D. B., Huggins, L. F., and Monke, E. J. (1980). "ANSWERS: a model for watershed planning." *Trans. ASAE,* 23, 938–944.

Beasley, D. B., Monke, E. J., and Huggins, L. F. (1977). "ANSWERS: A model for watershed planning." *Purdue Agricultural Experimental Station Paper No. 7038,* Purdue Univ., West Lafayette, Ind.

Becker, A., and Pfutzner, B. (1987). "EGMO-system approach and subroutines for river basin modeling." Acta Hydrophysica, 31.

Bergstrom, S. (1976). "Development and application of a conceptual runoff model for Scandinavian countries." *SMHI Rep. No. 7,* Norrkoping, Sweden.

Bergstrom, S. (1992). "The HBV model—its structure and applications." *SMHI Rep. RH, No. 4,* Norrkoping, Sweden.

Bergstrom, S. (1995). "Chapter 13: The HBV model." *Computer models of watershed hydrology,* V. P. Singh, ed., Water Resources Publications, Littleton, Colo.

Bernard, M. (1937). "Giving areal significance to hydrological research on small areas." *Headwaters control and use, Proc., Upstream Engineering Conf.*

Berod, D. D., Singh, V. P., Devrod, D., and Musy, A. (1995). "A geomorphologically nonlinear cascade (GNC) model for estimation of floods from small alpine watersheds." *J. Hydrol.,* 166, 147–170.

Berod, D. D., Singh, V. P., and Musy, A. (1999). "A geomorphologic kinematic-wave (GKW) model for estimation of floods from small alpine watersheds." *Hydrolog. Process.,* 13, 1391–1416.

Beven, K. J. (1995). "Chapter 18: TOPMODEL." *Computer models of watershed hydrology,* V. P. Singh, ed., Water Resources Publications, Littleton, Colo.

Beven, K. J., Calver, A., and Morris, E. (1987). "The Institute of Hydrology distributed model." *Institute of Hydrology Rep. No. 98,* Wallingford, U.K.

Beven, K. J., and Hornberger, G. M. (1982). "Assessing the effect of spatial pattern of precipitation in modeling streamflow hydrographs." *Water Resour. Bull.,* 18, 823–829.

Beven, K. J., and Kirkby, M. J. (1976). "Toward a simple physically-based variable contributing area of catchment hydrology." *Working Paper No. 154,* School of Geography, Univ. Leeds, U.K.

Beven, K. J., and Kirkby, M. J. (1979). "A physically-based variable contributing area model of basin hydrology." *Hydrol. Sci. Bull.*, 24(1), 43–69.

Bicknell, B. R., Imhoff, J. L., Kittle, J. L., Donigian, A. S., and Johanson, R. C. (1993). *Hydrologic simulation program—Fortran; User's manual for release 10*, U.S. EPA Environmental Research Laboratory, Athens, Ga.

Binley, A., Beven, K., and Elgy, J. (1989b). "A physically based model of heterogeneous hillslopes: 2. Effective hydraulic conductivities." *Water Resour. Res.*, 25, 1227–1233.

Binley, A., Elgy, J., and Beven, K. (1989a). "A physically based model of heterogeneous hillslopes: 1. Runoff production." *Water Resour. Res.*, 25, 1219–1226.

Borah, D. K., and Bera, M. (2000). "Hydrologic modeling of the Court Creek watershed." *Contract Rep. No. 2000-04*, Illinois State Water Survey, Champaign, Il.

Borah, D. K., Bera, M., Shaw, S., and Keefer, L. (1999). "Dynamic modeling and monitoring of water, sediment, nutrients and pesticides in agricultural watersheds during storm events." *Contract Rep. No. 655*, Illinois State Water Survey, Champaign, Ill.

Bouraoui, F., Braud, I., and Dillaha, T. A. (2002). "ANSWERS: A nonpoint source pollution model for water, sediment and nutrient losses." *Mathematical models of small watershed hydrology and applications*, V. P. Singh and D. K. Frevert, eds., Water Resources Publications, Littleton, Colo.

Boyd, M. J. (1978). "A storage-routing model relating drainage basin hydrology and geomorphology." *Water Resour. Res.*, 14(2), 921–928.

Boyd, M. J., Pilgrim, D. H., and Cordery, I. (1979). "A watershed bounded network model for flood estimation-computer programs and user guide." *Water Research Laboratory Rep. No. 154*, Univ. New South Wales, Sydney, Australia.

Boyd, M. J., Rigby, E. H., and vanDrie, R. (1996). "WBNM—a comprehensive flood model for natural and urban catchments." *Proc., 7th Int. Conf. on Urban Drainage*, Institution of Engineers, Sydney, Australia, 329–334.

Brazil, L. E. (1988). "Multilevel calibration strategy for complex hydrologic simulation models." PhD dissertation Colorado State Univ., Fort Collins, Colo.

Brazil, L. E., and Krajewski, W. F. (1987). "Optimization of complex hydrologic models using random search methods." *Proc., ASCE Conf. on Engineering Hydrology*, Williamsburg, New York, 726–731.

Bruneau, P., Gascuel-Odoux, C., Robin, P., Merot, P., and Beven, K. (1995). "Sensitivity to space and time resolution of a hydrological model using digital data." *Hydrolog. Process.*, 9, 69–81.

Bureau of Reclamation. (1991). *Inventory of Hydrologic Models*, U.S. Dept. of the Interior, Denver.

Burnash, R. J. C., Ferral, R. L., and McGuire, R. A. (1973a). "A generalized streamflow simulation system—conceptual modeling for digital computers." *Rep.*, U.S. Dept. of Commerce, National Weather Service, Silver Springs, Md., and State of California, Dept. of Water Resources, Sacramento, Calif.

Burnash, R. J. C., Ferral, R. L., and McGuire, R. A. (1973a). "A generalized streamflow simulation system—conceptual modeling for digital computers." *Rep.*, U.S. Dept. of Commerce, National Weather Service, Silver Springs, Md., and State of California, Dept. of Water Resources, Sacramento, Calif.

Burton, J. S., ed. (1993). "Proc., federal interagency workshop on hydrologic modeling demands for the 90's." *USGS Water Resources Investigations Rep. No. 93-4018*, Denver.

Burnash, R. J. C. (1975). "Chapter 10: The NWS river forecast system-catchment modeling." *Computer models of watershed hydrology*, V. P. Singh, ed., Water Resources Publications, Littleton, Colo.

Cabral, M. C., Bras, R. L., Tarboton, D., and Entekabhi, D. (1990). "A distributed physically based rainfall runoff model incorporating topography for real time flood forecasting." *R. M. Parsons Laboratory Rep. No. 332*, Massachusetts Institute of Technology, Cambridge, Mass.

Calver, A., and Wood, W. L. (1995). "Chapter 17: The Institute of Hydrology distributed model." *Computer models of watershed hydrology*, V. P. Singh, ed., Water Resources Publications, Littleton, Colo.

Carlson, D. H., and Thurow, T. L. (1992). *SPUR-91: Workbook and user guide*. Texas Agricultural Experimental Station MP-1743, College Station, Tex.

Carlson, D. H., Thurow, T. L., and Wight, J. R. (1995). "Chapter 27: SPUR-91: Simulation of production and utilization of rangelands." *Computer models of watershed hydrology*, V. P. Singh, ed., Water Resources Publications, Littleton, Colo.

Chen, Z. Q., Govindaraju, R. S., and Kavvas, M. L. (1994a). "Spatial averaging of unsaturated flow equations under infiltration conditions over areally heterogeneous fields: 1. Development of models." *Water Resour. Res.*, 30(2), 523–533.

Chen, Z. Q., Govindaraju, R. S., and Kavvas, M. L. (1994b). "Spatial averaging of unsaturated flow equations under infiltration conditions over areally heterogeneous fields: 2. Numerical simulations." *Water Resour. Res.*, 30(2), 535–548.

Chiew, F. H. S., and McMahon, T. A. (1994). "Application of the daily rainfall-runoff model MODHYDROLOG to twenty eight Australian catchments." *J. Hydrol.*, 153, 383–416.

Chiew, F. H. S., Peel, M. C., and Western, A. W. (2002). "Application and testing of the simple rainfall-runoff model SIMHYD." *Mathematical models of small watershed hydrology and applications*, V. P. Singh and D. K. Frevert, eds., Water Resources Publications, Littleton, Colo.

Clarke, R. T. (1973). "A review of some mathematical models used in hydrology, with observations on their calibration and use." *J. Hydrol.*, 19, 1–20.

Crawford, N. H., and Linsley, R. K. (1962). "The synthesis of continuous streamflow hydrographs on a digital computer." *Tech. Rep. No. 12*, Dept. of Civil Engineering, Stanford Univ., Palo Alto, Calif.

Crawford, N. H., and Linsley, R. K. (1966). "Digital simulation in hydrology: Stanford Watershed Model IV." *Tech. Rep. No. 39*, Stanford Univ., Palo Alto, Calif.

Croley, T. E. (1982). "Great Lake basins runoff modeling." *NOAA Tech. Memo No. EER GLERL-39*, National Technical Information Service, Springfield, Va.

Croley, T. E. (1983). "Lake Ontario basin runoff modeling," *NOAA Tech. Memo No. ERL GLERL-43*, Great Lakes Environmental Research Laboratory, Ann Arbor, Mich.

Cummings, N. W. (1935). "Evaporation from water surfaces: Status of present knowledge and need for further investigations." *Trans., Am. Geophys. Union*, 16(2), 507–510.

Dawdy, D. R. (1982). "A review of deterministic surface water routing in rainfall-runoff models." *Rainfall-runoff relationship*, V. P. Singh, ed., Water Resources Publications, Littleton, Colo., 23–36.

Dawdy, D. R., and Bergmann, J. M. (1969). "Effect of rainfall variability on streamflow simulation." *Water Resour. Res.*, 5, 958–966.

Dawdy, D. R., Litchy, R. W., and Bergmann, J. M. (1970). "Rainfall-runoff simulation model for estimation of flood peaks for small drainage basins." *USGS Open File Rep.*, Washington, D.C.

Dawdy, D. R., and O'Donnell, T. (1965). "Mathematical models of catchment behavior." *J. Hydraul. Div., Am. Soc. Civ. Eng.*, 91(HY4), 123–127.

Dawdy, D. R., Schaake, J. C., and Alley, W. M. (1978). "Users guide for distributed routing rainfall-runoff model." *USGS Water Resources Invest. Rep. No. 78-90*, Gulf Coast Hydroscience Center, NSTL, Miss.

Delleur, J. W. (1982). "Mathematical modeling in urban hydrology: Applied modeling." *Applied catchment hydrology*, V. P. Singh, ed., Water Resources Publications, Littleton, Colo., 399–420.

Desconnets, J. C., Vieux, B. E., Cappelaere, B., and Delclaux, F. (1996). "A GIS for hydrological modeling in the semi-arid, HAPEX-Sahel experiment area of Niger, Africa." *Trans. GIS*, 1(2), 82–94.

Diskin, M. H., and Simon, E. (1977). "A procedure for the selection of objective functions for hydrologic simulation models." *J. Hydrol.*, 34, 129–149.

Diskin, M. H., and Simon, E. (1979). "The relationship between the time bases of simulation models and their structure." *Water Resour. Bull.*, 15(6), 1716–1732.

Dooge, J. C. I. (1959). "A general theory of the unit hydrograph." *J. Geophys. Res.,* 64(2), 241–256.

Dooge, J. C. I. (1981). "Parameterization of hydrologic processes." *JSC Study Conf. on Land Surface Processes in Atmospheric General Circulation Models,* 243–284.

Donigian, A. S., Beyerlein, D. C., Davis, H. H., and Crawford, N. H. (1977). "Agricultural runoff management (ARM) model version II: Refinement and testing." *Rep. No. EPA-600/3-77-098,* U.S. EPA Environmental Research Laboratory, Athens, Ga.

Duan, Q., Gupta, V. K., and Sorooshian, S. (1992). "Effective and efficient global optimization for conceptual rainfall-runoff models." *Water Resour. Res.,* 28, 1014–1015.

Duan, Q., Gupta, V. K., and Sorooshian, S. (1993). "A shuffled complex evolution approach for effective and efficient global minimization." *J. Optim. Theory Appl.,* 76, 501–521.

Duchon, C. E., Salisbury, J. M., Williams, T. H. L., and Nicks, A. D. (1992). "An example of using Landsat and GOES data in a water budget model." *Water Resour. Res.,* 28(2), 527–538.

Dunkle, S. A., et al. (1993). "Chloroflurocarbons (CC13F and CC12F2) as dating tools and hydrologic tracers in shallow groundwater of the Delmarva Peninsula, Atlantic Coastal Plain, United States." *Water Resour. Res.,* 29, 3837–3860.

Eagleson, P. S., and Qinliang, W. (1987). "The role of uncertain catchment storm size in the moments of peak streamflow." *J. Hydrol.,* 96, 329–344.

El-Jabi, N., and Sarraf, S. (1991). "Effect of maximum rainfall position on rainfall-runoff relationship." *J. Hydraul. Eng.,* 117(5), 681–685.

El-Kady, A. I. (1989). "Watershed models and their applicability to conjunctive use management." *J. Am. Water Resour. Assoc.,* 25(1), 25–137.

Engman, E. T., and Gurney, R. J. (1991). *Remote sensing in hydrology,* Chapman and Hall, London.

Ewen, J., Parkin, G., and O'Connell, P. E. (2000). "SHETRAN: Distributed river basin flow and transport modeling system." *J. Hydrologic Eng.,* 5(3), 250–258.

Fair, G. M., and Hatch, L. P. (1933). "Fundamental factors governing the streamline flow of water through sand." *J. Am. Water Works Assoc.,* 25, 1551–1565.

Faurés, J. M., Goodrich, D. C., Woolhiser, D. A., and Sorooshian, S. (1995). "Impact of small scale-rainfall variability on runoff simulation." *J. Hydrol.,* 173, 309–326.

Feldman, A. D. (1981). "HEC models for water resources system simulation: Theory and experience." *Adv. Hydrosci.,* 12, 297–423.

Foroud, N., Broughton, R. S., and Austin, G. L. (1984). "The effect of a moving storm on direct runoff properties." *Water Resour. Bull.,* 20, 87–91.

Fortin, J. P., Turcotte, R., Massicotte, S., Moussa, R., Fitzback, J., and Villeneuve, J. P. (2001a). "A distributed watershed model compatible with remote sensing and GIS data. I: Description of model." *J. Hydrologic Eng.,* 6(2), 91–99.

Fortin, J. P., Turcotte, R., Massicotte, S., Moussa, R., Fitzback, J., and Villeneuve, J. P. (2001b). "A distributed watershed model compatible with remote sensing and GIS data II: Application to Chaudiere watershed." *J. Hydrologic Eng.,* 6(2), 100–108.

Freer, J., Beven, A. M., and Ambroise, B. (1996). "Bayesian estimation of uncertainty in runoff prediction and the value of data: An application of the GLUE approach." *Water Resour. Res.,* 32, 2161–2173.

Freeze, R. A., and Harlan, R. L. (1969). "Blueprint for a physically-based, digitally-simulated hydrologic response model." *J. Hydrol.,* 9, 237–258.

Frere, M. H., Onstad, C. A., and Holtan, H. N. (1975). "ACTMO, an agricultural chemical transport model." *Rep. No. ARS-H-3,* USDA, Washington, D.C.

Gan, T. Y., and Biftu, G. F. (1996). "Automatic calibration of conceptual rainfall-runoff models: Optimization algorithms, catchment conditions, and model structure." *Water Resour. Res.,* 32, 3513–3524.

Gashing, J., Reboh, R., and Rewiter, J. (1981). "Development of a knowledge-based expert system for water resource problems." *Final Rep. Stanford Research Institute,* Menlo Park, Calif.

Georgakakos, K. P., Sperfslage, J. A., Tsintikidis, D., Carpenter, T. M., Krajewski, W. F., and Kruger, A. (1999). "Design and tests of an integrated hydrometeorological forecast system for operational estimation and prediction of rainfall and streamflow in the mountainous Panama Canal watershed." *HRC Tech. Rep. No. 2,* Hydrologic Research Center, San Diego.

Goodrich, D. S., et al. (1994). "Runoff simulation sensitivity to remotely sensed initial soil water content." *Water Resour. Res.,* 30, 1393–1405.

Goodrich, D. C., Lane, L. J., Shillito, R. M., Miller, S. N., Syed, K. H., and Woolhiser, D. A. (1997). "Linearity of basin response as a function of scale in a semiarid basin." *Water Resour. Res.,* 33(12), 2951–2965.

Goodrich, D. C., and Woolhiser, D. A. (1991). "Catchment hydrology." *Rev. Geophys.,* 29, 202–209.

Goodrich, D. C., Woolhiser, D. A., and Keefer, T. O. (1991). "Kinematic routing using finite elements on a triangular irregular network (TIN)." *Water Resour. Res.,* 27(6), 995–1003.

Govindaraju, R. S., and Rao, A. R., ed. (2000). *Artificial Neural Networks in hydrology,* Kluwer Academic, Boston.

Grayman, W. M., Hadder, A. W., Gates, W. E., and Moles, R. M. (1975). "Land-based modeling system for water quality management studies." *J. Hydraul. Div., Am. Soc. Civ. Eng.,* 110(HY5), 567–580.

Grayson, R. B., Bloschl, G., and Moore, I. D. (1995). "Chapter 19: Distributed parameter hydrologic modeling using vector elevation data: THALES and TAPEC-C." *Computer models of watershed hydrology,* V. P. Singh, ed., Water Resources Publications, Littleton Colo.

Green, W. H., and Ampt, C. A. (1911). "Studies on soil physics: 1. Flow of water and air through soils." *J. Agric. Sci.,* 4, 1–24.

Guo, S., and Wang, G. (1994). "Water balance in the semi-arid regions." Yellow River, No. 12 (in Chinese).

Gupta, H. V., Hsu, K., and Sorooshian, S. (2000). "Chapter 1: Effective and efficient modeling for streamflow forecasting." *Artificial neural networks in hydrology,* R. S. Govindaraju and A. R. Rao, eds., Kluwer Academic, Dordrecht, The Netherlands.

Gupta, V. K., and Sorooshian, S. (1994a). "Calibration of hydrologic models: Past, present and future." *Trends in hydrology,* Council of Scientific Research Integration, ed., Research Trends, Kaithamukhu, India, 329–346.

Gupta, V. K., and Sorooshian, S. (1994b). "A new optimization strategy for global inverse solution of hydrologic models." *Numerical methods in water resources,* A. Peters et al., eds., Kluwer Academic, Boston.

Gupta, H. V., Sorooshian, S., and Yapo, P. O. (1998). "Toward improved calibration of hydrologic models: Multiple and noncommensurable measures of information." *Water Resour. Res.,* 34(4), 751–763.

Hairsine, S. Y., and Parlange, J. Y. (1986). "Kinematic shock waves on curved surfaces and application to the cascade approximation." *J. Hydrol.,* 87, 187–200.

Harley, B. M., Perkins, F. E., and Eagleson, P. S. (1970). "A modular distributed model of catchment dynamics." *R. M. Parsons Laboratory for Hydrodynamics and Water Resources Rep. No. 133,* Dept. of Civil Engineering, Massachesetts Institute of Technology, Cambridge, Mass.

Hawkins, R. H., and Cundy, T. W. (1987). "Steady-state analysis of infiltration and overland flow for spatially-varied hillslopes." *Water Resour. Bull.,* 32(2), 251–256.

Hewlett, J. D. (1961b). "Soil moisture as a source of base flow from steep mountain watersheds." *Southeast Forest Experimental Station Paper No. 132,* USDA Forest Service, Athens, Ga.

Hewlett, J. D. (1961a). "Some ideas about storm runoff and base flow." *Southeast Forest Experiment Station Annual Rep.,* USDA Forest Service, Athens, Ga., 62–66.

Hickey, J. T., and Diaz, G. E. (1999). "From flow to fish to dollars: An integrated approach to water allocation." *J. Am. Water Resour. Assoc.,* 35(5), 1053–1067.

Holtan, H. N., and Lopez, N. C. (1971). "USDAHL-70 model of watershed hydrology." *USDA-ARS Tech. Bull. No. 1435,* Agricultural Research Station, Beltsville, Md.

Holtan, H. N., Stilner, G. J., Henson, W. H., and Lopez, N. C. (1974). "USDAHL-74 model of watershed hydrology." *USDA-ARS Plant*

Physiology Research Rep. No. 4, Agricultural Research Station, Beltsville, Md.

Hornberger, G. M., and Boyer, E. W. (1995). "Recent advances in watershed modelling. U.S. national report to IUGG, 1991–1994." *Rev. Geophys.*, American Geophysical Union, Suppl., 949–957.

Hoover, M. D., and Hursh, C. R. (1943). "Influence of topography and soil-depth on runoff from forest land." *Trans., Am. Geophys. Union*, 24, 693–697.

Horne, F. E., and Kavvas, M. L. (1997). "Physics of the spatially averaged snowmelt process." *J. Hydrol.*, 191, 179–207.

Horton, R. E. (1919). "Rainfall interception." *Monthly Weather Rev.*, 147, 603–623.

Horton, R. E. (1933). "The role of infiltration in the hydrologic cycle." *Trans., Am. Geophys. Union*, 145, 446–460.

Horton, R. E. (1935). "Surface runoff phenomena, Part 1—Analysis of hydrograph." *Horton Hydrology Laboratory Publication No. 101*, Voorheesville, N.Y.

Horton, R. E. (1939). "Analysis of runoff plot experiments with varying infiltration capacities." *Trans., Am. Geophys. Union* 20(IV), 683–694.

Horton, R. E. (1940). "An approach toward a physical interpretation of infiltration capacity." *Soil Sci. Soc. Am. Proc.*, 5, 399–417.

Horton, R. E. (1945). "Erosional development of streams and their drainage basins: Hydrophysical approach to quantitative geomorphology." *Bull. Geol. Soc. Am.*, 56, 275–370.

Hromadka, T. V., McCuen, R. H., and Yen, C. C. (1988). "Effect of watershed subdivision on prediction accuracy of hydrologic models." *Hydrosoft*, 1, 19–28.

Hsu, K. L., Gupta, H. V., and Sorooshian, S. (1995). "Artificial neural network modeling of the rainfall-runoff process." *Water Resour. Res.*, 31(10), 2517–2531.

Huber, W. C. (1995). "Chapter 22: EPA storm water management model SWMM." *Computer models of watershed hydrology*, V. P. Singh, ed., Water Resources Publications, Littleton, Colo.

Huber, W. C., and Dickinson, R. E. (1988). "Storm water management model user's manual, version 4." *Rep. No. EPA/600/3-88/001a*, U.S. Environmental Protection Agency, Athens, Ga.

Huggins, L. F., and Monke, E J. (1968). "A mathematical model for simulating the hydrologic response of a watershed." *Water Resour. Res.*, 4(3), 529–539.

Huggins, L. F., and Monke, E. J. (1970). "Mathematical simulation of hydrologic events of ungaged watersheds." *Water Resources Research Center, Tech. Rep. No. 14*, Purdue Univ., West Lafayette, Ind.

Hughes, D. A. (1993). "Variable time intervals in deterministic hydrological models." *J. Hydrol.* 143, 217–232.

Hursh, C. R. (1936). "Storm water and absorption." *Trans., Am. Geophys. Union*, 17(II), 301–302.

Hursh, C. R. (1944). "Appendix B—Report of the subcommittee on subsurface flow." *Trans., Am. Geophys. Union*, 25, 743–746.

Hursh, C. R., and Brater, E. F. (1944). "Separating hydrographs into surface- and subsurface-flow." *Trans., Am. Geophys. Union*, 25, 863–867.

Hydrologic Engineering Center (HEC). (1968). "HEC-1 flood hydrograph package, user's manual." U.S. Army Corps of Engineers, Davis, Calif.

Hydrologic Engineering Center (HEC). (1981). "HEC-1 flood hydrograph package: Users manual." U.S. Army Corps of Engineers, Davis, Calif.

Hydrologic Engineering Center (HEC). (2000). "Hydrologic modeling system HEC-HMS user's manual, version 2." Engineering, U.S. Army Corps of Engineers, Davis, Calif.

Imbeau, M. E. (1892). "La Durance: Regime, crues et inundations." *Ann. Ponts Chaussees, Mem. Doc., Ser.*, 3(I), 5–18 (in French).

Izzard, C. F. (1944). "The surface profile of overland flow." *Trans., Am. Geophys. Union*, 25(VI), 959–968.

Jackson, T. J. (1982). "Chapter 12: Application and selection of hydrologic models." *Hydrologic modeling of small watersheds*, C. T. Haan, H. P. Johnson, and D. L. Brakensiek, eds., ASAE Monograph No. 5, American Society of Agricultural Engineers, St. Joseph, Mich.

Jacob, C. E. (1943). "Correlation of groundwater levels and precipitation on Long Island, New York: 1. Theory." *Trans., Am. Geophys. Union* 24, 564–573.

Jacob, C. E. (1944). "Correlation of groundwater levels and precipitation on Long Island, New York: 2. Correlation of data." *Trans., Am. Geophys. Union*, 24, 321–386.

James, L. D., Bowles, D. S., and Hawkins, R. H. (1982). "A taxonomy for evaluating surface water quantity model reliability." *Applied modeling in catchment hydrology*, V. P. Singh, ed., Water Resources Publications, Littleton, Colo., 189–228.

James, L. D., and Burges, S. J. (1982a.) "Precipitation-runoff modeling: Future directions." *Applied modeling in catchment hydrology*, V. P. Singh, ed., Water Resources Publications, Littleton, Colo., 291–312.

James, L. D., and Burges, S. J. (1982b). "Chapter 11: Selection, calibration and testing of hydrologic models." *Hydrologic modeling of small watersheds*, C. T. Haan, H. P. Johnson, and D. L. Brakensiek, eds., ASAE Monograph No. 5, American Society of Agricultural Engineers, St. Joseph, Mich.

Jensen, M. (1984). "Runoff pattern and peak flows from moving block rains based on a linear time-area curve." *Nord. Hydrol.*, 15, 155–168.

Julien, P. Y., and Moglen, G. E. (1990). "Similarity and length scale for spatially varied overland flow." *Water Resour. Res.*, 26(8), 1819–1832.

Julien, P. Y., and Saghafian, B. (1991). "CASC2D user's manual." *Dept. of Civil Engineering Rep.*, Colorado State Univ., Fort Collins, Colo.

Julien, P. Y., Saghafian, B., and Ogden, F. L. (1995). "Raster-based hydrologic modeling of spatially varied surface runoff." *Water Resour. Bull.*, 31(3), 523–536.

Kavvas, M. L. (1999). "On the coarse-graining of hydrologic processes with increasing scales." *J. Hydrol.* 27, 191–202.

Kavvas, M. L., et al. (1998). "A regional scale land surface parameterization based on areally-averaged hydrological conservation equation." *Hydrol. Sci. J.*, 43(4), 611–631.

Keulegan, G. H. (1944). "Spatially variable discharge over a sloping plane." *Trans., Am. Geophys. Union*, 25(VI), 959–965.

Kite, G. W., and Kouwen, N. (1992). "Watershed modeling using land classification." *Water Resour. Res.*, 28(12), 3193–3200.

Kite, G. W. (1995). "Chapter 15: The SLURP model." *Computer models of watershed hydrology*, V. P. Singh, ed., Water Resources Publications, Littleton, Colo.

Klepper, O., Scholten, H., and van de Kamer, J. P. G (1991). "Prediction uncertainty in an ecological model of the Oosterschelde estuary." *J. Forecast.*, 10, 191–209.

Knisel, W. G., Leonard, R. A., Davis, F. M., and Nicks, A. D. (1993). "GLEAMS version 2.10, Part III, user's manual." *Conservation Research Rep.*, USDA, Washington, D.C.

Knisel, W. G., and Williams, J. R. (1995). "Chapter 28: Hydrology components of CREAMS and GLEAMS models." *Computer models of watershed hydrology*, V. P. Singh, ed., Water Resources Publications, Littleton, Colo.

Kokkonen, T., Koivusalo, H., Karvonen, T., and Lepisto, A. (1999). "A semidistributed approach to rainfall-runoff modeling-aggregating responses from hydrologically similar areas." *MODSIM'99*, L. Oxley and F. Scrimgeour, eds., The Modelling and Simulation Society of Australia and New Zealand, Hamilton, New Zealand, 75–80.

Kostiakov, A. M. (1932). "On the dynamics of the coefficient of water percolation in soils and of the necessity of studying it from a dynamic point of view for purposes of amelioration." *Trans. 6th Communic., Int. Soil Science Society*, Part 1, 17–29. (Russia)

Kouwen, N. (2000). "WATFLOOD/SPL: Hydrological model and flood forecasting system." Dept. of Civil Engineering, Univ. of Waterloo, Waterloo, Ont.

Kouwen, N., Soulis, E. D., Pietroniro, A., Donald, J., and Harrington, R. A. (1993). "Grouped response units for distributed hydrologic modeling." *J. Water Resour. Plan. Manage.*, 119(3), 289–305.

Kuczera, G. (1983a). "Improved parameter inference in catchment models. 1: Evaluating parameter uncertainty." *Water Resour. Res.*, 19, 1151–1162.

Kuczera, G. (1983b). "Improved parameter inference in catchment models. 2: Combining different kinds of hydrologic data and testing their compatibility." *Water Resour. Res.,* 19, 1163–1172.

Kuczera, G. (1997). "Efficient subspace probabilistic parameter optimization for catchment models." *Water Resour. Res.,* 19, 1163–1172.

Lahmer, W., Becker, A., Muller-Wohlfelt, D.-I., and Pfutzner, B. (1999). "A GIS-based approach for regional hydrological modeling." *Regionalization in hydrology,* B. Diekkruger, M. J. Kirkby, and U. Schroder, eds., IAHS Publication No. 254, International Association of Hydrological Sciences, 33–43.

Lambourne, J. J., and Stephenson, D. (1987). "Model study of the effect of temporal storm distribution on peak discharges and volumes." *Hydrol. Sci. J.,* 32, 215–226.

Lane, L. J., and Woolhiser, D. A. (1977). "Simplifications of watershed geometry affecting simulation of surface runoff." *J. Hydrol.,* 35, 173–190.

Laurenson, E. M. (1964). "A catchment storage model for runoff routing." *J. Hydrol.,* 2, 141–163.

Laurenson, E. M., and Mein, R. G. (1993). "RORB version 4 runoff routing program: User's manual." Monash Univ., Dept. of Civil Engineering, Monash, Victoria, Australia

Laurenson, E. M., and Mein, R. G. (1995). "Chapter 5: RORB: Hydrograph synthesis by runoff routing." *Computer models of watershed hydrology,* V. P. Singh, ed., Water Resources Publications, Littleton, Colo.

Leavesley, G. H., Lichty, R. W., Troutman, B. M., and Saindon, L. G. (1983). "Precipitation-runoff modeling system user's manual." *USGS Water Resources Investigations Rep. No. 83-4238,* Denver.

Leavesley, G. H., and Stannard, L. G. (1990). "Application of remotely sensed data in a distributed parameter watershed model." *Proc., Workshop on Applications of Remote Sensing in Hydrology,* National Hydrologic Research Center, Environment Canada, Saskatoon, Sask.

Ledoux, E., Etchevers, P., Golaz, C., Habets, F., Noilhan, J., and Voirin, S. (2002). "Chapter 8: Regional simulation of the water budget and riverflows with ISBA-MODCOU coupled model: Application to the Adour and Rhone basins." *Mathematical models of large watershed hydrology,* V. P. Singh and D. K. Frevert, eds., Water Resources Publications, Littleton, Colo.

Ledoux, E., Girard, G., de Marsily, G., and Deschenes, J. (1989). "Spatially distributed modeling: Conceptual approach, coupling surface water and ground water." *Unsaturated flow hydrologic modeling—theory and practice,* H. J. Morel-Seytoux, ed., NATO ASI Series S 275, Kluwer Academic, Boston, 435–454.

Lehrsch, G. A., Whisler, F. D., and Romkens, M. J. M. (1987). "Soil surface as influenced by selected soil physical parameters." *Soil Till. Res.* 10, 197–212.

Lehrsch, G. A., Whisler, F. D., and Romkens, M. J. M. (1988). "Spatial variation of parameters describing soil surface roughness." *Soil Sci. Soc. Am. J.,* 52, 311–319.

Lighthill, M. J., and Whitham, G. B. (1955). "On kinematic waves: 1. Flood movement in long rivers." *Proc. R. Soc. London, Ser. A,* 229, 281–316.

Linsley, R. K. (1982). "Rainfall-runoff models-an overview." *Rainfall-Runoff relationship,* V. P. Singh, ed., Water Resources Publications, Littleton, Colo., 3–22.

Loague, K. M. (1988). "Impact of rainfall and soil hydraulic property information on runoff predictions at the hillslope scale." *Water Resour. Res.,* 24(9), 1501–1510.

Loage, K. M., and Gander, G. A. (1990). "R-5 revisited: 1. Spatial variability of infiltration on a small rangeland watershed." *Water Resour. Res.,* 26, 957–971.

Lorrai, M., and Sechi, G. M. (1995). "Neural nets for modeling rainfall-runoff transformations." *Water Resour. Manage.,* 9, 299–313.

Lowdermilk, W. C. (1934). "Forests and streamflow: A discussion of Hoyt-Trozell report." *J. Forestry,* 21, 296–307.

Luce, C. H., and Cundy, T. W. (1994). "Parameter identification for a runoff model for forest roads." *Water Resour. Res.,* 30, 1057–1069.

Ma, X., Fukushima, Y., Hashimoto, T., Hiyama, T., and Nakashima, T. (1999). "Application of a simple SVAT model in a mountain catchment model under temperate humid climate." *J. Jpn. Soc. Hydrol. Water Resour.,* 12, 285–294.

Ma, X., and Cheng, W. (1998). "A modeling of hydrological processes in a large low plain area including lakes and ponds." *J. Jpn. Soc. Hydrol. Water Resour.,* 9, 320–329.

Maksimov, V. A. (1964). "Computing runoff produced by a heavy rainstorm with a moving center." *Sov. Hydrol.,* 5, 510–513.

Maller, R. A., and Sharma, M. L. (1984). "Aspects of rainfall excess from spatial varying hydrologic parameters." *J. Hydrol.,* 67, 115–127.

Mankin, K. R., Koelliker, J. K., and Kalita, P. K. (1999). "Watershed and lake water quality assessment: An integrated modeling approach." *J. Am. Water Resour. Assoc.,* 35(5), 1069–1088.

Marcus, N. (1968). "A laboratory and analytical study of surface runoff under moving rainstorms." PhD dissertation, Univ. Illinois, Urbana, Ill.

Mark, D. M. (1978). "Concepts of 'data structure' for digital terrain models." *Proc., Symposium on DTMs,* American Society of Photogrammetry/American Congress of Surveying and Mapping, St. Louis.

Mason, J. C., Price, R. K., and Tem'me, A. (1996). "A neural network model of rainfall-runoff using radial basis functions." *J. Hydraul. Res.,* 34(4), 537–548.

Mazion, E., and Yen, B. C. (1994). "Computational discretization effect on rainfall-runoff simulation." *J. Water Resour. Plan. Manage.,* 120(5), 715–734.

Metcalf and Eddy, Inc., Univ. of Florida, and Water Resources Engineers, Inc. (1971). "Storm water management model, Vol. 1—Final report." *EPA Rep. No. 11024DOC07/71 (NITS PB-203289),* EPA, Washington, D.C.

Michaud, J. D., and Sorooshian, S. (1994). "Effect of rainfall sampling errors on simulation of desert flash floods." *Water Resour. Res.,* 30, 2765–2775.

Milly, P. C. D., and Eagleson, P. S. (1988). "Effect of storm scale on surface runoff volume." *Water Resour. Res.,* 24(4), 620–624.

Minns, A. W., and Hall, M. J. (1996). "Artificial neural networks as rainfall-runoff models." *Hydrol. Sci. J.,* 41(3), 399–417.

Molnar, D. K., and Julien, P. Y. (2000). "Grid-size effects on surface runoff modeling." *J. Hydrologic Eng.,* 5(1), 8–16.

Moore, I. D., and Burch, G. J. (1986). "Sediment transport capacity of sheet and rill flow: Application of unit stream power theory." *Water Resour. Res.,* 22, 1350–1360.

Moore, I. D., and Grayson, R. B. (1991). "Terrain-based catchment partitioning and runoff prediction using vector elevation data." *Water Resour. Res.,* 27(6), 1177–1191.

Moore, I. D., O'Loughlin, E. M., and Burch, G. J. (1988a). "A contour-based topographic model for hydrological and ecological applications." *Earth Surf. Processes Landforms,* 13, 305–320.

Moore, I. D., Pamuska, J. C., Grayson, R. B., and Srivastava, K. P. (1988b). "Application of digital topographic modeling in hydrology." *Proc., Int. Symposium on Modeling Agricultural, Forest, and Rangeland Hydrology, Publication No. 07-88,* American Society of Agricultural Engineers, St. Joseph, Mich., 447–461.

Morel-Seytoux, H. J. (1988). "Soil-aquifer-stream interactions—a reductionist attempt toward physical stochastic integration." *J. Hydrol.,* 102, 355–379.

Morin, G., Paquet, P., and Sochanski, W. (1995). "Le modele de simulation de quantite de qualite CEQUEAU, Manuel de references." *INRS Eau Rapport de Recherché No. 433,* Sainte-Foy, Que.

Morin, G., Sochanski, W., and Paquet, P. (1998). "Le modele de simulation de quantite CEQUEAU-ONU, Manuel de references." *Organisation des Nations-Unies et INRS Eau Rapport de Recherché No. 519,* Sainte-Foy, Que.

Morris, E. M. (1980). "Forecasting flood flows in grassy and forested basins using a deterministic distributed mathematical model." *IAHS Publication No. 129 (Hydrological Forecasting),* International Association of Hydrological Sciences, Wallingford, U.K., 247–255.

Mulvany, T. J. (1850). "On the use of self-registering rain and flood gauges." *Proc. Inst. Civ. Eng.,* 4(2), 1–8.

Naden, P. S. (1992). "Spatial variability in flood estimation for large catchments: The exploitation of channel network structure." *Hydrol. Sci. J.,* 37, 53–71.

Nash, J. E. (1957). "The form of the instantaneous unit hydrograph." *Hydrol. Sci. Bull.,* 3, 114–121.

Natale, L., and Todini, E. (1976a). "A stable estimator for large models 1: Theoretical development and Monte Carlo experiments." *Water Resour. Res.,* 12(4), 667–671.

Natale, L., and Todini, E. (1976b). "A stable estimator for large models 2: Real world hydrologic applications." *Water Resour. Res.,* 12(4), 672–675.

Natale, L., and Todini, E. (1977). "A constrained parameter estimation technique for linear models in hydrology." *Mathematical models of surface water hydrology,* T. A. Ciriani, U. Maione, and J. R. Wallis, eds., Wiley, London, 109–147.

Ngirane-Katashaya, G. G., and Wheater, H. S. (1985). "Hydrograph sensitivity to storm kinematics." *Water Resour. Res.,* 21, 337–345.

Nielsen, D. R., Kirkham, D., and van Wijk, W. K. (1959). "Measuring water stored temporarily above the field moisture capacity." *Soil Sci. Soc. Am. Proc.,* 23, 408–412.

Niemczynowicz, J. (1984a). "Investigation of the influence of rainfall movement on runoff hydrograph. Part I: Simulation of conceptual catchment." *Nord. Hydrol.,* 15, 57–70.

Niemczynowicz, J. (1984b). "Investigation of the influence of rainfall movement on runoff hydrograph. Part I: Simulation of real catchments in the city of Lund." *Nord. Hydrol.,* 15, 71–84.

Nicks, A. D., and Scheibe, F. R. (1992). "Using NEXRAD precipitation data as input to hydrologic models." *Proc., 28th AWRA Annual Conf. and Symposium on Managing Water Resources during Global Change,* AWRA Technical Publication Series No. TPS-92-4, American Water Resources Association Bethesda, Md., 75–83.

Noilhan, J., and Mahfouf, J. F. (1996). "The ISBA land surface parameterization scheme, global planet." *Climate Change,* 13, 145–159.

Ogden, F. L. (1998). CASC2D version 1.18 reference manual. *Dept. of Civil & Environmental Engineering Rep. U-37, CT1665-1679,* Univ. Connecticut, Storrs, Conn.

Ogden, F. L., and Julien, P. Y. (1993). "Runoff sensitivity to temporal and spatial rainfall variability at runoff plane and small basin scale." *Water Resour. Res.,* 29, 2589–2597.

Ogden, F. L., Richardson, J. R., and Julien, P. Y. (1995). "Similarity in catchment response: 2. Moving rainstorms." *Water Resour. Res.,* 31, 1543–1547.

Olivera, F., and Maidment, D. (1999). "Geographical information system (GIS)-based spatially distributed model for runoff routing." *Water Resour. Res.,* 35(4), 1155–1164.

O'Loughlin, E. N. (1986). "Prediction of surface saturation zones in natural catchments by topographic analysis." *Water Resour. Res.,* 22, 229–246.

Ormsbee, L. E. (1989). "Rainfall disaggregation model for continuous hydrologic modeling." *J. Hydraul. Eng.,* 115(4), 507–525.

Osborn, H. B., Goodrich, D. C., and Unkrich, C. L. (1993). "Characterization of thunderstorm rainfall for hydrologic modeling." *Proc., ASCE Special Conf. on Management of Irrigation and Drainage Systems,* ASCE, New York, 409–415.

Ozga-Zielinska, M., and Brzezinski, J. (1994). *Applied hydrology,* Wydawnictawa Naukowe, PWN, Warsaw, Poland (in Polish).

Penman, H. L. (1948). "Natural evaporation from open water, bare soil and grass." *Proc. R. Soc. London, Ser. A,* 193, 120–145.

Penman, H. L. (1961). "Weather, plant and soil factors in hydrology." *Weather,* 16, 207–219.

Pessoa, M., Bras, R. L., and Williams, E. R. (1993). "Use of weather radar for flood forecasting in the Sieve River basin: A sensitivity analysis." *J. Appl. Meteorol.,* 32, 462–475.

Phanartzis, C. A. (1972). "Spatial variation of precipitation in the San Dimas Experimental Forest and its effect on simulated streamflow." *Tech. Rep. No. 11,* Dept. Hydroly and Water Resources, Univ. Arizona, Tucson, Ariz.

Pierson, F. B., Blackburn, W. H., van Sactor, S. S., and Wood, J. C. (1994). "Partitioning small scale variability of runoff and erosion on sagebrush rangeland." *Water Resour. Bull.,* 30, 1081–1089.

Potter, J. W., and McMahon, T. A. (1976). "The Monash model: User manual for daily program HYDROLOG." *Resource Rep. 2/76,* Dept. of Civil Engineering, Monash Univ., Monash, Victoria, Australia.

Puls, L. G., (1928). "Flood regulation of the Tennessee River." *Proc., 70th Congress, 1st Session,* H. D. 185, Part 2, Appendix B.

Quick, M. C. (1995). "Chapter 8: The UBC watershed model." *Computer models of watershed hydrology,* V. P. Singh, ed., Water Resources Publications, Littleton, Colo.

Quick, M. C., and Pipes, A. (1977). "UBC watershed model." *Hydrol. Sci. Bull.,* XXI(1/3), 285–295.

Quinn, P., Beven, K., Chevalier, P., and Planchon, O. (1991). "The prediction of hillslope flow paths for distributed hydrological modelling using digital terrain models." *Hydrolog. Process.,* 5, 59–79.

Rango, A. (1992). "Worldwide testing of the snowmelt runoff model with applications for predicting the effects of climate change." *Nord. Hydrol.,* 23, 155–172.

Rango, A. (1995). "Chapter 14: The snowmelt runoff model (SRM)." *Computer models of watershed hydrology,* V. P. Singh, ed., Water Resources Publications, Littleton, Colo.

Rao, A. R., and Han, J. (1987). "Analysis of objective functions used in urban runoff models." *Adv. Water Resour.,* 10, 205–211.

Refsgaard, J. C., and Storm, B. (1995). "Chapter 23: MIKE SHE." *Computer models of watershed hydrology,* V. P. Singh, ed., Water Resources Publications, Littleton, Colo.

Remson, I., Randolf, J. R., and Barksdale, H. C. (1960). "The zone of aeration and ground water recharge in sandy sediments at Seabrook, New Jersey." *Soil Sci.,* 89, 145–156.

Richardson, B. (1931). "Evaporation as a function of insolation." *Trans. Am. Soc. Civ. Eng.,* 95, 996–1011.

Rigby, E. H., Boyd, M. J., and vanDrie, R. (1999). "Experiences in developing the hydrology model: WBNM2000." *Proc., 8th Int. Conf. on Urban Drainage,* Institution of Engineers, 3, 1374–1381.

Roberts, M. C., and Klingeman, P. C. (1970). "The influence of landform and precipitation parameters on flood hydrographs." *J. Hydrol.,* 11, 393–411.

Robson, A., Beven, K., and Neal, C. (1992). "Towards identifying sources of subsurface flow: A comparison of components identified by a physically based runoff model and those determined by chemical mixing techniques." *Hydrolog. Process.,* 6, 199–214.

Rockwood, D. M. (1982). "Theory and practice of the SSARR model as related to analyzing and forecasting the response of hydrologic systems." *Applied modeling in catchment hydrology,* V. P. Singh, ed., Water Resources Publications, Littleton, Colo., 87–106.

Roessel, B. W. P. (1950). "Hydrologic problems concerning the runoff in headwater regions." *Trans., Am. Geophys. Union,* 31, 431–442.

Rose, S. (1992). "Tritium in groundwater of the Georgia Piedmont: Implications for recharge and flow paths." *Hydrolog. Process.,* 6, 67–78.

Rovey, E. W., Woolhiser, D. A., and Smith, R. E. (1977). "A distributed kinematic model of upland watersheds." *Hydrology Paper No. 93,* Colorado State Univ., Fort Collins, Colo.

Rudra, R. P., Dickinson, W. T., Abedini, M. J., and Wall, G. J. (1999). "A multi-tier approach for agricultural watershed management." *J. Am. Water Resour. Assoc.,* 35(5), 1059–1070.

Saghafian, B., Julien, P. Y., and Zogden, F. L. (1995). "Similarity in catchment response:1. Similarity in rainstorms." *Water Resour. Res.,* 31, 1533–1541.

Sargent, D. M. (1981). "An investigation into the effect of storm movement on the design of urban drainage systems: Part I." *Public Health Eng.,* 9, 201–207.

Sargent, D. M. (1982). "An investigation into the effect of storm movement on the design of urban drainage systems: Part II: Probability analysis." *Public Health Eng.,* 10, 111–117.

Savic, D. A., Walters, G. A., and Davidson, J. W. (1999). "A genetic programming approach to rainfall-runoff modeling." *Water Resour. Manage.,* 13, 219–231.

Scaefer, M. G., and Barker, B. L. (1999). "Stochastic modeling of extreme floods for A. R. Bowman Dam." *MGS Engineering Consultants Rep.,* Olympia, Wash.

Servat, E., and Dezetter, A. (1991). "Selection of calibration objective functions in the context of rainfall-runoff modeling in a Sudanese savannah area." *Hydrol. Sci. J.,* 36(4/8), 307–330.

Seyfried, M. S., and Wilcox, B. P. (1995). "Scale and nature of spatial variability: Field examples having implications for hydrologic modeling." *Water Resour. Res.,* 31, 173–184.

Sharma, M. L., Gander, G. A., and Hunt, C. G. (1980). "Spatial variability of infiltration in a watershed." *J. Hydrol.,* 45, 101–122.

Sherman, L. K. (1932). "Stream flow from rainfall by the unit graph method." *Eng. News-Rec.,* 108, 501–505.

Silfer, A. T., Hassett, J. M., and Kinn, G. J. (1987). "Hydrologic runoff modeling of small watersheds: The Tinflow model." *Proc., National Symposium on Engineering Hydrology,* ASCE, New York, 545–550.

Simanovic, S. P. (1990). "An expert system for the selection of a suitable method for flow measurement in open channels." *J. Hydrol.,* 112, 237–256.

Singh, V. P. (1995a). "Chapter 1: Watershed modeling." *Computer models of watershed hydrology,* V. P. Singh, ed., Water Resources Publications, Littleton, Colo., 1–22.

Singh, V. P. ed. (1995b). *Computer models of watershed hydrology,* Water Resources Publications, Littleton, Colo.

Singh, V. P. (1996). *Kinematic wave modeling in water resources: Surface water hydrology,* Wiley, New York.

Singh, V. P. (1997). "Effect of spatial and temporal variability in rainfall and watershed characteristics on streamflow hydrograph." *Hydrolog. Process.,* 11, 1649–1669.

Singh, V. P. (1998). "Effect of the direction of storm movement on planar flow." *Hydrolog. Process.,* 12, 147–170.

Singh, V. P., and Fiorentino, M., eds. (1996). *Geographical information systems in hydrology,* Kluwer Academic, Boston.

Sittner, W. T., Scauss, C. E., and Munro, J. C. (1969). "Continuous hydrograph synthesis with an API-type hydrologic model." *Water Resour. Res.,* 5(5), 1007–1022.

Sivapalan, M., Ruprecht, J. K., and Viney, N. R. (1996a). "Water and salt balance modeling to predict the effects of land use changes in forested catchments: 1. Small catchment water balance model." *Hydrolog. Process.,* 10, 393–411.

Sivapalan, M., Viney, N. R., and Ruprecht, J. K. (1996b). "Water and salt balance modeling to predict the effects of land use changes in forested catchments: 2. Coupled model of water and salt balances." *Hydrolog. Process.,* 10, 413–428.

Sivapalan, M., Viney, N. R., and Jeevaraj, C. G. (1996c). "Water and salt balance modeling to predict the effects of land use changes in forested catchments: 3. The large catchment model." *Hydrolog. Process.,* 10, 429–4446.

Sivapalan, M., and Wood, E. F. (1986). "Chapter 5: Spatial heterogeneity and scale in the infiltration response of catchments." *Scale problems in hydrology,* V. K. Gupta, I. Rodriguez-Iturbe, and E. F. Wood, eds., Kluwer Academic, Dordrecht, The Netherlands, 81–106.

Smith, R. E., and Goodrich, D. C. (2000). "Model to simulate rainfall excess patterns on randomly heterogeneous areas." *J. Hydrologic Eng.,* 5(4), 365–362.

Smith, R. E., Goodrich, D. C., and Woolhiser, D. A. (1990). "Areal effective infiltration dynamics for runoff of small catchments." *Transactions of the 14th Int. Conf. Soil Sci.,* International Soil Science Society, I-22–I-27.

Smith, R. E., Goodrich, D. C., Woolhiser, D. A., and Unkrich, C. L. (1995). "Chapter 20: KINEROS—A kinematic runoff and erosion model." *Computer models of watershed hydrology,* V. P. Singh, ed., Water Resources Publications, Littleton, Colo.

Smith, R. E., and Hebbert, R. H. B. (1979). "A Monte Carlo analysis of the hydrologic effects of spatial variability of infiltration." *Water Resour. Res.,* 15, 419–429.

Soil Conservation Service (SCS). (1956). "Supplement A, Section 4, Chapter 10, Hydrology." *National engineering handbook,* USDA, Washington, D.C.

Soil Conservation Service (SCS). (1965). *Computer model for project formulation hydrology, Tech. Release No. 20,* USDA, Washington, D.C.

Song, Z., and James, L. D. (1992). "An objective test for hydrologic scale." *Water Resour. Bull.,* 28(5), 833–844.

Sorooshian, S. (1981). "Parameter estimation of rainfall-runoff models with heteroscedastic streamflow errors: The noninformative data case." *J. Hydrol.,* 52, 127–138.

Sorooshian, S., and Dracup, J. A. (1980). "Stochastic parameter estimation procedures for hydrologic rainfall-runoff models: Correlated and heteroscedastic error cases." *Water Resour. Res.,* 16, 430–442.

Sorooshian, S., Duan, Q., and Gupta, V. K. (1993). "Calibration of rainfall-runoff models: Application of global optimization to the Sacramento soil moisture accounting model." *Water Resour. Res.,* 29, 1185–1194.

Sorooshian, S., and Gupta V. K. (1995). "Chapter 2: Model calibration." *Computer models of watershed hydrology,* V. P. Singh, ed., Water Resources Publications, Littleton, Colo., 23–68.

Southerland, F. R. (1983). "An improved rainfall intensity distribution for hydrograph synthesis." *Water Systems Research Program Rep. No. 1/1983,* Univ. Witwatersrand, Johannesburg, South Africa.

Spear, R. C., and Hornberger, G. M. (1980). "Eutrophication in Peel Inlet II. Identification of critical uncertainties via generalized sensitivity analysis." *Cybernetics,* 14, 43–49.

Speers, D. D. (1995). "Chapter 11: SSARR model." *Computer models of watershed hydrology,* V. P. Singh, ed., Water Resources Publications, Littleton, Colo.

Stephenson, D. (1984). "Kinematic study of effects of storm dynamics on runoff hydrographs." *Water SA,* 10, 189–196.

Stephenson, D. (1989). "A modular model for simulating continuous or event runoff." *IAHS Publication No. 181,* International Association of Hydrological Sciences, Wallingford, U.K., 83–91.

Stephenson, D., and Randell, B. (1999). "Streamflow prediction model for the Caledon catchment." *ESKOM Rep. No. RES/RR/00171,* Cleveland.

Stewart, M. K., and MacDonnell, J. J. (1991). "Modeling baseflow soil water residence times from deuterium concentrations." *Water Resour. Res.,* 27, 2687–2689.

Sugawara, M. (1967). "The flood forecasting by a series storage type model." *Int. Symposium Floods and their Computation,* International Association of Hydrologic Sciences, 1–6.

Sugawara, M., et al. (1974). "Tank model and its application to Bird Creek, Wollombi Brook, Bikin River, Kitsu River, Sanga River and Nam Mune." Research Note, *National Research Center for Disaster Prevention, No. 11,* Kyoto, Japan, 1–64.

Sugawara, M. (1995). "Chapter 6: Tank model." *Computer models of watershed hydrology,* V. P. Singh, ed., Water Resources Publications, Littleton, Colo.

Sullivan, M., Warwick, J. J., and Tyler, S. W. (1996). "Quantifying and delineating spatial variations of surface infiltration in a small watershed." *J. Hydrol.,* 181, 149–168.

Surkan, A. J. (1969). "Synthetic hydrographs: Effects of channel geometry." *Water Resour. Res.,* 5(1), 112–128.

Surkan, A. J. (1974). "Simulation of storm velocity effects of flow from distributed channel network." *Water Resour. Res.,* 10, 1149–1160.

Tanakamaru, H. (1995). "Parameter estimation for the tank model using global optimization." *Trans. Jpn. Soc. Irrig., Drain. Reclam.,* 178, 103–112.

Tanakamaru, H., and Burges, S. J. (1996). "Application of global optimization to parameter estimation of the tank model." *Proc., Int. Conf. on Water Resources and Environmental Research: Towards the 21st Century,* Kyoto, Japan.

Tao, T., and Kouwen, N. (1989). "Spatial resolution in hydrologic modeling." *Proc., Int. Conf. on the Centennial of Manning's Formula and Kuichling's Rational Formula,* B. C. Yen, ed., 166–175.

Tarboton, D. G., Bras, R. L., and Rodriguez-Iturbe, I. (1991). "On extraction of channel networks from digital elevation data." *Hydrolog. Process.,* 5, 81–100.

Tayfur, G., and Kavvas, M. L. (1994). "Spatially averaged conservation equations for interacting rill-interrill area overland flows." *J. Hydraul. Eng.,* 120(12), 1426–1448.

Tayfur, G., and Kavvas, M. L. (1998). "Area averaged overland flow equations at hillslope scale." *Hydrol. Sci. J.,* 43(3), 361–378.

Tayfur, G., Kavvas, M. L., Govindaraju, R. S., and Storm, D. E. (1993). "Applicability of St. Venant equations for two dimensional overland flows over rough infiltrating surfaces." *J. Hydraul. Eng.,* 119(1), 51–63.

Tennessee Valley Authority (TVA). (1972). "A continuous daily-streamflow model: Upper Bear Creek." *Experimental Project Research Paper No. 8,* Knoxville, Tenn.

Theis, C. V. (1935). "The relation between the lowering of the piezometric surface and the rate and duration of discharge of a well using ground-water storage." *Trans., Am. Geophys. Union,* 16, 519–524.

Thomas, G., and Henderson-Sellers, A. (1991). "An evaluation of proposed representation of subgrid hydrologic processes in climate models." *J. Climate,* 4, 898–910.

Thornthwaite, C. W. (1948). "An approach toward a rational classification of climate." *Geogr. Rev.,* 38, 55–94.

Todini, E. (1988a). "Il modello afflussi deflussi del flume Arno. Relazione Generale dello studio per conto della Regione Toscana." *Tech. Rep.,* Univ. of Bologna, Bologna (in Italian).

Todini, E. (1988b). "Rainfall runoff modelling: Past, present and future." *J. Hydrol.,* 100, 341–352.

Todini, E. (1995). "New trends in modeling soil processes from hillslopes to GCM scales." *The role of water and hydrological cycle in global change,* H. R. Oliver and S. A. Oliver, eds., NATO Advanced Study Institute, Series 1: Global, Kluwer Academic, Dordrecht, The Netherlands.

Todini, E. (1996). "The ARNO rainfall-runoff model." *J. Hydrol.,* 175, 339–382.

Tokar, A. S., and Markus, M. (2000). "Precipitation-runoff modeling using artificial neural networks and conceptual models." *Water Resour. Res.,* 5(2), 156–161.

Turcke, M. A., and Kueper, B. H. (1996). "Geostatistical analysis of the border aquifer hydraulic conductivity field." *J. Hydrol.,* 178, 223–240.

U.S. Army Corps of Engineers. (1936). *Method of flow routing. Rep. on Survey for Flood Control, Connecticut River Valley,* Providence, R. I., Vol. 1, Section 1, Appendix.

U.S. Army Corps of Engineers. (1987). *SSARR user's manual,* North Pacific Division, Portland, Ore.

U.S. Bureau of Reclamation. (1949). "Chapter 6.10: Flood routing." Part. 6, *Water Studies, Part 6: Flood hydrology,* Washington, D.C., Vol. IV.

USDA. (1980). "CREAMS: A field scale model for chemicals, runoff and erosion from agricultural management systems." W. G. Knisel, ed., *Conservation Research Rep. No. 26,* Washington, D.C.

USGS. (1998). *Proc., 1st Federal Interagency Hydrologic Modeling Conf.,* Subcommittee on Hydrology, Interagency Advisory Committee on Water, Reston, Va.

Vandenberg, A. (1989). "A physical model of vertical integration, drain discharge and surface runoff from layered soils." *NHRI Paper No. 42, IWD Tech. Bull. No. 161,* National Hydrologic Research Institute, Saskatoon, Sask.

Van Straten, G., and Keesman, K. J. (1991). "Uncertain propagation and speculation in projective forecasts of environmental change: A lake eutrophication example." *J. Forecast.,* 10, 163–190.

Vieux, B. E. (1988). "Finite element analysis of hydrologic response areas using geographic information systems." PhD dissertation, Michigan State Univ., East Lansing, Mich.

Vieux, B. E. (1991). "Geographic information systems and non-point source water quality and quantity modeling." *Hydrolog. Process.,* 5, 101–113.

Vieux, B. E. (1993). "DEM aggregation and smoothing effects on surface runoff modeling." *J. Comput. Civ. Eng.,* 7(3), 310–338.

Vieux, B. E., and Farajalla, N. S. (1994). "Capturing the essential variability in distributed hydrological modeling: Hydraulic roughness." *Hydrolog. Process.,* 8, 221–236.

Wang, Q. J. (1991). "The genetic algorithm and its application to calibrating conceptual rainfall-runoff models." *Water Resour. Res.,* 27(9), 2467–2471.

Watts, L. G., and Calver, A. (1991). "Effects of spatially distributed rain-

fall on runoff for a conceptual catchment." *Nord. Hydrol.,* 22, 1–14.

Wight, J. R., and Skiles, J. W., eds. (1987). "SPUR-Simulation of production and utilization of rangelands: Documentation and user guide." *Rep. No. ARS-63,* USDA, Washington, D.C.

Wigmosta, M. S., Vail, L. W., and Lettenmaier, D. P. (1994). "A distributed hydrology-vegetation model for complex terrain." *Water Resour. Res.,* 30(6), 1665–1679.

Wilgoose, G., and Kuczera, G. (1995). "Estimation of subgrade scale kinematic wave parameters for hillslopes." *Hydrolog. Process.,* 9, 469–482.

Williams, J. R., Jones, C. A., and Dyke, P. T. (1984). "The EPIC model and its application." *Proc., ICRISAT-IBSNAT-SYSS Symposium on Minimum Data Sets for Agrotechnology Transfer,* 111–121.

Williams, J. R. (1995a). "Chapter 25: the EPIC model." *Computer models of watershed hydrology,* V. P. Singh, ed., Water Resources Publications, Littleton, Colo.

Williams, J. R. (1995b). "Chapter 24: SWRRB—A watershed scale model for soil and water resources management." *Computer models of watershed hydrology,* V. P. Singh, ed., Water Resources Publications, Littleton, Colo.

Williams, J. R., Nicks, A. D., and Arnold, J. G. (1985). "Simulator for water resources in rural basins." *J. Hydraul. Eng.,* 111(6), 970–986.

Wilson, C. B., Valdes, J. B., and Rodriguez-Iturbe, I. (1979). "On the influence of spatial distribution of rainfall on storm runoff." *Water Resour. Res.,* 15, 321–328.

Winchell, M., Gupta, H. V., and Sorooshian, S. (1998). "On the simulation of infiltration-and saturation-excess runoff using radar-based rainfall estimates: Effects of algorithm uncertainty and pixel aggregation." *Water Resour. Res.,* 34(10), 2655–2670.

Wolock, D. M. (1995). "Effects of subbasin size on topography characteristics and simulated flow paths in Sleepers River watershed, Vermont." *Water Resour. Res.,* 31, 1989–1997.

Wood, E. F., Sivapalan, M., and Beven, K. (1990). "Similarity and scale in catchment storm response." *Rev. Geophys.,* 28, 1–18.

Wood, E. F., Sivapalan, M., Beven, K. J., and Band, L. (1988). "Effects of spatial variability and scale with implications to hydrologic modeling." *J. Hydrol.,* 102, 29–47.

Woolhiser, D. A. (1973). "Hydrologic and watershed modeling—State of the art." *Trans. ASAE,* 16(3), 553–559.

Woolhiser, D. A., and Goodrich, D. C. (1988). "Effect of storm rainfall intensity patterns on surface runoff." *J. Hydrol.,* , 102, 29–47.

Woolhiser, D. A., Smith, R. E., and Giraldez, J. Y. (1996). "Effects of spatial variability of saturated hydraulic conductivity on Hortonian overland flow." *Water Resour. Res.,* 32, 671–678.

Woolhiser, D. A., Smith, R. E., and Goodrich, D. C. (1990). "KINEROS—A kinematic runoff and erosion model: Documentation and user manual." *Rep. No. ARS-77,* USDA, Washington, D.C.

World Meteorological Organization (WMO). (1975). "Intercomparison of conceptual models used in operational hydrological forecasting." *Operational Hydrology Paper No. 429,* Geneva.

World Meteorological Organization (WMO). (1986). "intercomparison of models of snowmelt runoff." *Operational Hydrology Paper No. 646,* Geneva.

World Meteorological Organization (WMO). (1992). "Simulated real-time intercomparison of hydrological models." *Operational Hydrology Paper No. 779,* Geneva.

Wu, Y., Woolhiser, D. A., and Yevjevich, V. (1982). "Effects of spatial variability of hydraulic roughness on runoff hydrographs." *Agric. Forest Meteorol.,* 59, 231–248.

Wurbs, R. A. (1998). "Dissemination of generalized water resources models in the United States." *Water Int.,* 23, 190–198.

Wyseure, G. C. L., Gowing, J. W., and Young, M. D. B. (2002). "Chapter 10: PARCHED-THIRST: An hydrological model for planning rainwater harvesting systems in semi-arid areas." *Mathematical models of small watershed hydrology and applications,* V. Singh and D. K. Frevert, eds., Water Resources Publications, Littleton, Colo.

Xu, C. Y. (1999). "Operational testing of a water balance model for predicting climate change impacts." *Agric. Forest Meteorol.,* 98–99(1–4), 295–304.

Yang, D., Herath, S., and Musiake, K. (1998). "Development of a geomorphology-based hydrological model for large catchments." *Ann. J. Hydraul. Eng.*, 42, 169–174.

Yao, H., and Terakawa, A. (1999). "Distributed hydrological model for Fuji River Basin." *J. Hydrologic Eng.*, 4(2), 106–116.

Yen, B. C., and Chow, V. T. (1968). "A study of surface runoff due to moving rainstorms." *Hydraulic Engineering Series No. 17*, Dept. Civil Engineering, Univ. Illinois, Urabana, Ill.

Yoo, D. H. (2002). "Numerical model of surface runoff, infiltration, river discharge and groundwater flow—SIRG." *Mathematical models of small watershed hydrology and applications*, V. P. Singh and D. K. Frevert, eds., Water Resources Publications, Littleton, Colo.

Yoshitani, J., Kavvas, M. L., and Chen, Z. Q. (2002). "Chapter 7: Regional-scale hydroclimate model." *Mathematical models of watershed hydrology*, V. P. Singh and D. K. Frevert, eds., Water Resources Publications, Littleton, Colo.

Young, M. D. B., and Gowing, J. W. (1996). "PARCHED-THIRST model user guide." *Rep.*, Univ. of Newcastle upon Tyne.

Young, R. A., Onstad, C. A., and Bosch, D. D. (1995). "Chapter 26: AGNPS: An agricultural nonpoint source model." *Computer models of watershed hydrology*, V. P. Singh, ed., Water Resources Publications, Littleton, Colo.

Young, R. A., Onstad, C. A., Bosch, D. D., and Anderson, W. P. (1989). "AGNPS: A nonpoint source pollution model for evaluating agricultural watershed." *J. Soil Water Conservat.*, 44, 168–173.

Yu, Z. (1996). "Development of a physically-based distributed-parameter watershed (basin-scale hydrologic model) and its application to Big Darby Creek watershed." PhD dissertation, Ohio State Univ., Columbus.

Yu, Z., et al. (1999). "Simulating the river-basin response to atmospheric forcing by linking a mesoscale meteorological model and a hydrologic model system." *J. Hydrol.*, 218, 72–91.

Yu, Z., and Schwartz, F. W. (1998). "Application of integrated basin-scale hydrologic model to simulate surface water and groundwater interactions in Big Darby Creek watershed, Ohio." *J. Am. Water Resour. Assoc.*, 34, 409–425.

Zhang, W., and Montgomery, D. R. (1994). "Digital elevation model grid size, landscape representation, and hydrologic simulations." *Water Resour. Res.*, 30, 1019–1028.

Zhao, R. J., and Liu, X. R. (1995). "Chapter 7: the Xinjiang model." *Computer models of watershed hydrology*, V. P. Singh, ed., Water Resources Publications, Littleton, Colo.

Zhao, R. J., Zhuang, Y.-L., Fang, L. R., Liu, X. R., and Zhang, Q. S. (1980). "The Xinanjiang model." *Proc., Oxford Symposium on Hydrological Forecasting, IAHS Publication No. 129*, International Association of Hydrological Sciences, Wallingford, U.K., 351–356.

American Society of Civil Engineers
1852–2002
Building a Better World

150th Anniversary Paper

Fiber-Reinforced Polymer Composites for Construction—State-of-the-Art Review

C. E. Bakis[1]; L. C. Bank, F.ASCE[2]; V. L. Brown, M.ASCE[3]; E. Cosenza[4]; J. F. Davalos, A.M.ASCE[5]; J. J. Lesko[6]; A. Machida[7]; S. H. Rizkalla, F.ASCE[8]; and T. C. Triantafillou, M.ASCE[9]

Abstract: A concise state-of-the-art survey of fiber-reinforced polymer (also known as fiber-reinforced plastic) composites for construction applications in civil engineering is presented. The paper is organized into separate sections on structural shapes, bridge decks, internal reinforcements, externally bonded reinforcements, and standards and codes. Each section includes a historical review, the current state of the art, and future challenges.

DOI: 10.1061/(ASCE)1090-0268(2002)6:2(73)

CE Database keywords: Composite materials; Fiber-reinforced materials; Polymers; Bridge decks; State-of-the-art reviews.

Introduction

In the last 200 years, rapid advances in construction materials technology have enabled civil engineers to achieve impressive gains in the safety, economy, and functionality of structures built to serve the common needs of society. Through such gains, the health and standard of living of individuals are improved. To

[1]Professor, Dept. of Engineering Science & Mechanics, 212 Earth-Engineering Sciences Building, Pennsylvania State Univ., University Park, PA 16802. E-mail: cbakis@psu.edu

[2]Professor, Dept. of Civil and Environmental Engineering, 2206 Engineering Hall, 1415 Engineering Dr., Univ. of Wisconsin—Madison, Madison, WI 53706. E-mail: bank@engr.wisc.edu

[3]Professor and Chair, Dept. of Civil Engineering, One University Pl., Widener Univ., Chester, PA 19013. E-mail: Vicki.L.Brown@widener.edu

[4]Professor, Dipartimento di Analisi e Progettazione Stutturale, Univ. di Napoli Federico II, Via Claudio 21, 80125 Naples, Italy. E-mail: cosenza@unina.it

[5]Professor, Dept. of Civil & Environmental Engineering, P.O. Box 6103, West Virginia Univ., Morgantown, WV 26505. E-mail: jdavalos@wvu.edu

[6]Associate Professor, Dept. of Engineering Science & Mechanics, Mail Code 0219, Virginia Tech, Blacksburg, VA 24061. E-mail: jlesko@vt.edu

[7]Professor, Dept. of Civil & Environmental Engineering, Saitama Univ., Urawa, Saitama 338, Japan. E-mail: machida@p.mtr.civil.saitama-u.ac.jp

[8]Distinguished Professor of Civil Engineering & Construction, 203 Constructed Facilities Laboratory, North Carolina State Univ., Raleigh, NC 27695. E-mail: sami_rizkalla@ncsu.edu

[9]Associate Professor, Dept. of Civil Engineering, Structures Division, Univ. of Patras, Patras 26500, Greece. E-mail: ttriant@upatras.gr

Note. Discussion open until October 1, 2002. Separate discussions must be submitted for individual papers. To extend the closing date by one month, a written request must be filed with the ASCE Managing Editor. The manuscript for this paper was submitted for review and possible publication on August 16, 2001; approved on November 15, 2001. This paper is part of the *Journal of Composites for Construction*, Vol. 6, No. 2, May 1, 2002. ©ASCE, ISSN 1090-0268/2002/2-73–87/$8.00+$.50 per page.

mark the occasion of the 150th anniversary of the American Society of Civil Engineers (ASCE), this paper reviews a class of structural materials that has been in use since the 1940s but only recently has won the attention of engineers involved in the construction of civil structures—fiber-reinforced polymer [or fiber-reinforced plastic (FRP)] composites.

The earliest FRP materials used glass fibers embedded in polymeric resins that were made available by the burgeoning petrochemical industry following World War II. The combination of high-strength, high-stiffness structural fibers with low-cost, lightweight, environmentally resistant polymers resulted in composite materials with mechanical properties and durability better than either of the constituents alone. Fiber materials with higher strength, higher stiffness, and lower density, such as boron, carbon, and aramid, were commercialized to meet the higher performance challenges of space exploration and air travel in the 1960s and 1970s. At first, composites made with these higher performing fibers were too expensive to make much impact beyond niche applications in the aerospace and defense industries. Work had already begun in the 1970s, however, to lower the cost of high-performance FRPs and promote substantial marketing opportunities in sporting goods. By the late 1980s and early 1990s, as the defense market waned, increased importance was placed by fiber and FRP manufacturers on cost reduction for the continued growth of the FRP industry. As the cost of FRP materials continues to decrease and the need for aggressive infrastructure renewal becomes increasingly evident in the developed world, pressure has mounted for the use of these new materials to meet higher public expectations in terms of infrastructure functionality. Aided by the growth in research and demonstration projects funded by industries and governments around the world during the late 1980s and throughout the 1990s, FRP materials are now finding wider acceptance in the characteristically conservative infrastructure construction industry. Hence, a brief review of the development, state of the art, and future of these promising construction materials is a timely and appropriate marker for the 150th anniversary of the ASCE.

369

For the purpose of writing this paper, a learned group of writers was recruited from the ranks of the editorial board of the ASCE *Journal of Composites for Construction*. The breadth and depth of topics covered in this work are necessarily constrained by space limitations. The breakdown of topics, which roughly follows the annual summary of topics in the journal, and the respective contributing writers are as follows:

- Editor—Professor Charles E. Bakis (The Pennsylvania State Univ.),
- Structural Shapes—Professors Lawrence C. Bank (University of Wisconsin—Madison) and Edoardo Cosenza (University of Naples, Italy),
- Highway Bridge Decks—Professors John J. Lesko (Virginia Polytechnic Institute and State University) and Julio F. Davalos (West Virginia University),
- Internal FRP Reinforcements—Professors Vicki L. Brown (Widener University) and Charles E. Bakis (The Pennsylvania State University),
- Externally Bonded Reinforcements—Professor Thanasis Triantafillou (University of Patras, Greece), and
- Standards and Codes—Professors Sami H. Rizkalla (North Carolina State University) and Atsuhiko Machida (Saitama University, Japan).

The style and content of each section vary according to the different historical developments and maturities of the respective subject matters. Pultruded shapes, for example, have been in wide use in a number of industries for approximately 30 years, whereas FRP decks have been under development for only about 10 years. Hence, the discussion of FRP decks is relatively biased toward methods of analysis (structural as well as economical) and design, whereas the discussion of pultrusions involves a more historical perspective. Similarly, design guides and codes for the use of internal and external FRP reinforcements for concrete structures are only recently being finalized. Therefore, a discussion of design approaches and remaining questions in these areas is of current interest.

Structural Shapes

Introduction

Constant cross-sectioned FRP composite structural shapes, also commonly referred to as structural profiles, produced for use in the construction industry for building and bridge superstructure applications are discussed in this section. Pultruded sections for highway bridge decks are described elsewhere. Pultrusion technology used for manufacturing FRP structural shapes is briefly reviewed. This is followed by a brief discussion of the development and evolution of structural shapes from nonstructural commodity applications to current structural applications. Significant building and bridge superstructure developments are described. Recent research on pultruded FRP shapes is described and comments on the future of pultruded structural shapes are provided.

Pultrusion Process

The manufacturing method of choice, for both product consistency and economy, for structural shapes is the pultrusion process. This continuous manufacturing process, which is highly automated, consists of "pulling" resin impregnated reinforcing fibers and fiber fabrics through a heated curing die at speeds of up to 3 m (120 in.)/min, depending on the size and complexity of the

profile shape. Both open-section and single or multicelled closed-section profiles can be produced. The common fiber reinforcement in pultruded shapes consists of fiber bundles (called rovings for glass fiber and tows for carbon fiber), continuous strand mat (also called continuous filament mat), and nonwoven surfacing veils. Typically, fiber volume fractions of 35–50% are utilized. In recent years, bidirectional and multidirectional woven, braided, and stitched fiber fabrics have been used to produce pultruded parts with enhanced mechanical properties. Filled thermosetting resins in the polyester and vinylester groups are generally used in pultrusion. Additionally, phenolic, epoxy, and thermoplastic resins have been developed for the pultrusion process. Meyer (1985) provides a thorough introduction to the pultrusion process, its early evolution, the key patents awarded, and the parameters controlling the design of a pultruded part. His book also provides an annotated bibliography of the key papers dealing with pultrusion technology from the late 1950s to the early 1980s.

Evolution of Fiber-Reinforced Polymer Structural Shapes

Large-sized pultruded FRP structural shapes [defined herein as having a cross-sectional envelope greater than 150×150 mm (6×6 in.)] for building and bridge superstructure construction applications were developed from earlier advances in pultrusion technology, which prior to the 1970s was primarily focused on developing small-sized commodity products for nonstructural building and electrical applications. A key application driver that led to significant developments and standardization of the technology was the FRP stepladder. The seminal paper by Werner (1979) describes the development of statistically reliable design property values for pultruded parts used in the pultruded ladder rail industry. B-basis allowable property values for design were developed for pultruded channels used as ladder rails following the procedures of the *U.S. military handbook 17* (*Composite* 2001). In the 1970s and early 1980s, advances in pultrusion technology led to the ability to produce larger pultruded parts capable of serving as structural members in load-bearing applications. Profile shapes were developed and first used in complete building structural systems to construct electromagnetic interference (EMI) test laboratory buildings that contained no metallic components (Smallowitz 1985). At the same time, a number of pultrusion companies in the United States began producing "standard" I-shaped beams for construction applications. In the pultrusion industry, then as now, the term standard does not imply anything more than the fact that the parts are produced on a regular basis by the company, are usually available off-the-shelf, have published dimensions, and meet minimum manufacturer-provided property values (Bank 1995). These standard profiles were, and still are, generally used for small structural units and parts of structural systems of nonprimary load carrying significance. Nonstandard shapes are called "custom" shapes. In the late 1980s and early 1990s, a customized building system of pultruded components for the construction of industrial cooling tower structures (Green et al. 1994) was developed. Cooling towers are currently constructed from either standard pultruded shapes (angles, tubes, and channel and I sections) produced by a variety of manufacturers (called "stick built") or customized components (called "modular") (i.e., www.strongwell.com or www.creativepultrusions.com). Pultruded cooling tower systems are designed according to existing building codes using property data supplied by the manufacturer and verified by the designer. In the absence of an American National Standards Institute approved design guide for pultruded structures, designers generally rely on engineering judgment, fundamental mechanics principles, experience, and manufacturer-produced "design guides" (Fiberline 1995; Creative Pultrusions 1999; Strongwell 1999). Since the early 1990s, the use of small pultruded FRP structural shapes [cross sections less than 100×100 mm (4×4 in.)] to build industrial platforms and walkways and to build relatively short single-span [9–18 m (30–60 ft)] pedestrian bridges has also increased significantly (Johansen et al. 1997).

Current Developments in Pultruded Structural Shapes

Since the 1990s, there has been a significant increase throughout the world in the use (albeit still on a demonstration project basis) of pultruded structural shapes in primary load-bearing systems for general construction. This excludes the cooling tower market, which is a well-established niche market for pultruded structural shapes. Significant bridge and building structures have been designed and constructed using pultruded profiles.

Major pedestrian bridges constructed of pultruded structural shapes include the 114-m (371-ft) long cable-stayed Aberfeldy Footbridge in Aberfeldy, Scotland, and the 40-m (130-ft) long cable-stayed Fiberline Bridge in Kolding, Denmark. The Aberfeldy Bridge, constructed in 1992, and designed by Maunsell Structural Plastics (Beckenham, U.K.), used a proprietary interlocking modular pultruded decking section produced by GEC Reinforced Plastics (now Fibreforce Composites, Runcorn, U.K.). The Kolding Bridge, constructed in 1997, and designed by the Danish firm RAMBØLL, uses standard pultruded shapes in both the cable tower and the decking system. For small highway bridge superstructure and building structural elements, Strongwell has recently developed a new standard pultruded FRP double web beam (DWB) in a 200×150 mm (8×6 in.) size. A design guide for the beam is available from Strongwell (1999). The beam has been used in a demonstration bridge (Hayes et al. 2000a). A 900×450 mm (36×18 in.) version of the beam that can be used as a highway bridge girder or a transfer girder in a building frame is also available. An eight-girder timber deck bridge [two lane, 11-m (38-ft) clear span] employing this 914-mm (36-in.) deep DWB was installed in Marion, Virginia, during the summer of 2001. Standard pultruded composites from Creative Pultrusions were recently used to design and construct a large [$7.4 \times 4.6 \times 1.5$ m ($24 \times 15 \times 5$ ft)] box-girder pultruded causeway structure (Bank et al. 2000).

Pultruded shapes, both standard and custom, have also been used in building and housing construction systems, and continue to be used in the EMI building market. A six-story 19.4-m (63-ft) high load-bearing stair-tower was recently constructed in Fort Story, Virginia, of 250×250 mm (10×10 in.) pultruded I-section shapes manufactured by Strongwell. In Avon, U.K., a two-story building was designed by Maunsell Structural Plastics using its proprietary pultruded interlocking panelized system (Raasch 1998). In Basel, Switzerland, a five-story 15 m (49 ft) high residential/office building was constructed with a primary load-bearing structural frame of FRP shapes for the 1999 Swissbau Fair. Named the "Eyecatcher Building," the building was designed using Fiberline shapes and the *Fiberline design manual* (Keller 1999). In Italy, the pultruder TopGlass S.p.A. produces a variety of structural shapes that have been used in small building constructions. In West Virginia, an experimental multicellular FRP building [$12.3 \times 6.5 \times 4.3$ m ($40 \times 21 \times 14$ ft)] was constructed for the West Virginia Department of Transportation using pultruded profiles manufactured by Creative Pultrusions. In non-

building structural product markets, custom structural shapes have been developed for latticed transmission towers, light poles, and highway luminaire supports and guardrails.

Recent Research

In the early 1990s, research on the use of pultruded structural shapes for civil engineering building and bridge applications increased significantly as the potential for the use of these products in nonproprietary load-bearing structural systems became evident. A search of the ASCE database (www.pubs.asce.org) reveals that archival journal papers published by the ASCE on the topic of FRP pultruded structures or materials date back only to 1990. Approximately 20 archival research papers on this topic were published since that time, and appear in the ASCE database. In 1997, the ASCE began publishing the *Journal of Composites for Construction*, which has provided an outlet for many of the research papers on the topic. In the United Kingdom, *The Structural Engineer*, published by the British Institution of Structural Engineers, has published a number of key research papers on pultruded structures for civil engineering.

Future for Pultruded Structural Shapes

The increased acceptance of pultruded structural shapes for mainstream building and bridge superstructure applications will depend on three key developments. The first is the development of an internationally accepted material specification for pultruded materials that will allow users to determine material properties of interest to designers in a rational and nonproprietary manner with well-known reliability. The second is the development of a design code for pultruded structures that is consensus based and incorporated into building and bridge codes such as the International Building Code and the American Association of State Highway and Transportation Officials (AASHTO) bridge code. The recent paper by Zureick and Steffen (2000) provides an example of what is needed to develop these two items. The third development required, as is to be expected, will be to reduce the cost of pultruded shapes, which are currently not competitive with shapes made from traditional materials for mainstream structural applications.

Highway Bridge Decks

Introduction

Within the field of highway structures, several new FRP structural systems have been proposed, designed, and experimentally implemented. These include bridge decks for rehabilitation and new construction, concrete filled FRP shells for drivable piles (Karbhari et al. 2000; Mirmiran et al. 2000), and wood FRP composite girders (Dagher et al. 1997). However, bridge decks have received the greatest amount of attention in the past few years, due to their inherent advantages in strength and stiffness per unit weight as compared to traditional steel reinforced concrete (RC) decks. Reducing the weight of replacement decks in rehabilitation projects presents the opportunity for rapid replacement and reduction in dead load, thus raising the live load rating of the structure. The New York State Department of Transportation and other organizations have successfully pursued this strategy in the rehabilitation of several short-span and through-truss bridges (Alampalli et al. 1999).

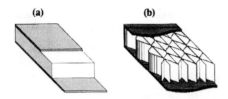

Fig. 1. Examples of two types of sandwich construction: (a) generic foam core; (b) proprietary (KSCI) corrugated core—taken from Davalos et al. (2001)

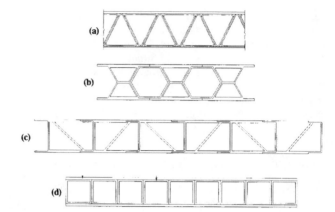

Fig. 2. FRP decks produced from adhesively bonded pultruded shapes: (a) EZSpan (Atlantic Research); (b) Superdeck (Creative Pultrusions); (c) DuraSpan (Martin Marietta Materials); (d) square tube and plate deck (Strongwell)

Thus, the focus of this discussion is on the efficacy of FRP bridge decks for rehabilitation and new construction. General performance and cost issues for the two major types of FRP decks currently in use—sandwich and adhesively bonded pultruded shapes—are outlined quantitatively.

Fiber-Reinforced Polymer Decks

A bridge deck in this discussion is defined as a structural element that transfers loads transversely to supports such as longitudinal running girders, cross beams, and/or stringers that bear on abutments. The connection of the deck to these underlying supports is typically made through the application of shear studs or a bolted connection in a simply supported condition that does not necessarily provide for composite action. Although decks have been designed and implemented bearing directly on abutments without underlying supports (e.g., the No-Name Creek Bridge in Russell, Kansas, and the Muddy Run Bridge on Delaware State Route 896), these types are not the focus of this discussion.

FRP decks commercially available at the present time can be classified according to two types of construction—sandwich and adhesively bonded pultruded shapes. In both cases, quality control of the product is enabled by standardized fabrication procedures within individual manufacturing facilities. The fabrication and performance of these types of decks are described next.

Sandwich Construction

Sandwich structures have been widely used for applications in the aerospace, marine, and automotive industries, where stiffness and strength requirements must be met with minimum weight, as explained in a number of textbooks on the subject (Vinson 1999). Sandwich construction implies the use of strong, stiff face sheets that carry flexural loads and a low-density, bonded core material that separates the face sheets and ensures composite action of the deck. Cellular materials are the most efficient core materials for weight-sensitive applications. Due to the ease with which face sheets and core materials can be changed in manufacturing, sandwich construction presents tremendous flexibility in designing for varied depths and deflection requirements. Face sheets of sandwich bridge decks are primarily composed of E-glass mats and/or rovings infused with a polyester or vinylester resin. Current core materials are rigid foams or thin-walled cellular FRP materials, such as those shown in Fig. 1.

Open or closed-mold composite fabrication processes whereby liquid resin is drawn through a dry fiber preform, typically with a vacuum, hold great promise for the economical manufacturing of bridge decks. In such processes, the core and face sheets can be impregnated with resin and cured at once. Changes in details related to materials, orientations, and thicknesses of the FRP face sheets or core can be determined analytically (Davalos et al. 2001) and are easily accommodated in many of these processes.

Individual sandwich deck panels are usually joined to each other by tongue-and-groove ends and are secured to the underlying structure using a clamp mechanism. Skew boundaries can be easily accommodated and edge panels can be delivered with preinstalled bridge railing systems. Special attention must be given to the design of connections due to the propensity of face sheets to delaminate from certain core materials in the presence of through-the-thickness tension. Connection points can be locally reinforced with thicker face sheets, a higher density core, and/or high-strength molded inserts. Thus far, the design process for sandwich decks is not in a code format. Rather, individual decks are designed on a job-by-job basis using finite-element techniques.

Adhesively Bonded Pultruded Shapes

Most currently available commercial decks are constructed using assemblies of adhesively bonded pultruded shapes. Such shapes can be economically produced in continuous lengths by numerous manufacturers using well-established processing methods (see the previous discussion of shapes in this paper). Secondary bonding operations are best done in the manufacturing plant for maximum quality control. Design flexibility in this type of deck is obtained by changing the constituents of the shapes (such as fibers and fiber orientations) and, to a lesser extent, by changing the cross section of the shapes. Due to the potentially high cost of pultrusion dies, however, variations in the cross section of shapes are feasible only if sufficiently high production warrants the tooling investment.

Closed-form, mechanics-based methods for designing section properties of a composite shape are detailed by Barbero et al. (1993). These mechanics concepts can be translated into approximate methods for estimating the equivalent orthotropic plate behavior of decks (Qiao et al. 2000). In this way, deck designs can be adjusted to derive structurally efficient and easy to manufacture pultruded sections. While systematic methods of optimizing pultruded shapes have been developed (Davalos et al. 1996), optimized deck designs are largely derived by trial and error (McGhee et al. 1991).

Several decks constructed with pultruded shapes are shown in Fig. 2. The pultruded shapes are typically aligned transverse to the traffic direction. Each deck design has advantages in terms of stiffness, strength, and field implementation. In laboratory testing,

Table 1. Summary of Deck Characteristics for Two Fabrication Methods

Deck system	Depth (mm)	kg/m² [a]	Dollars/m²	Deflection[b] (reported)	Deflection[c] (normalized)
(a) Sandwich Construction					
Hardcore composites	152–710	98–112	570–1,184	L/785[d]	L/1,120
KSCI	127–610	76[e]	700	L/1,300[f]	L/1,300
(b) Adhesively Bonded Pultrusions					
DuraSpan	194	90	700–807	L/450[g]	L/340
Superdeck	203	107	807	L/530[h]	L/530
EZSpan	229	98	861–1,076	L/950[i]	L/950
Strongwell	120–203	112	700[j]	L/605[k]	L/325

[a]Without wearing surface.

[b]Assumes plate action.

[c]Normalized to HS20+IM for a 2.4-m center-to-center span between supporting girders.

[d]HS20+IM loading of a 203-mm-deep section at a center-to-center span between girders of 2.7 m.

[e]For a 203-mm-deep deck targeted for RC bridge deck replacements.

[f]HS20+IM loading of a 203-mm-deep deck at a center-to-center span between girders of 2.4 m.

[g]HS20+IM loading of a 203-mm-deep deck at a center-to-center span between girders of 2.2 m.

[h]HS20+IM loading at a center-to-center span between girders of 2.4 m.

[i]HS20+IM loading at a center-to-center span between girders of 2.4 m.

[j]For a 171-mm-deep deck with a wearing surface under experimental fabrication processes.

[k]HS20+IM loading of a 171-mm-deep section at a center-to-center span between girders of 2 m.

the observed failures in such decks are generally by local punching shear and crushing or large-scale delamination of the shapes constituting the cross section. Local buckling, shearing, or delamination of internal stiffeners under concentrated wheel loads can also contribute to a loss in overall stiffness (GangaRao et al. 1999; Harik et al. 1999). Stitching and other forms of out-of-plane reinforcement are possible means of mitigating delamination.

Materials in FRP decks differ primarily in fiber architecture and resin type. Polyester resins are favored for their low cost, although vinylester resins are preferred in very moist environments (Pethrick et al. 2000). Woven and stitched fabrics are often employed (DuraSpan and Superdeck) for precise placement of multiaxial reinforcement for improved delamination resistance. EZSpan employs through-the-thickness braided preforms as the reinforcement for the triangular tubes.

The square plate tube and plate deck shown in Fig. 2 is a precommercial design composed of off-the-shelf shapes typical of many of the basic pultrusion companies (Strongwell 1999; Hayes et al. 2000b). These commonly stocked shapes are usually composed of continuous strand mat and roving, as it allows for efficient wet-out of the reinforcement and is a cost-effective means to develop a number of different shapes within a product line. Combining off-the-shelf plates and tubes of various sizes provides varied sections with performance suited to particular application requirements. A similar deck system was used by Bank et al. (2000) as the deck for a floating causeway system.

Comparison of Decks

A technical comparison of sandwich and pultruded decks is shown in Table 1. Although cost and stiffness are "system" dependent and a function of the application requirements, each manufacturer was asked *independently* to provide a representative value. Not surprisingly, there is greater flexibility with the sandwich-constructed decks to produce structures of varied depth and therefore stiffness. The mass per unit surface area is typically

near 98 kg/m² (20 lb/ft²), with the exception of the KSCI (Kansas Structural Composites, Inc.) corrugated core deck system, for which the reduced weight appears to suggest extra efficiency in the use of materials.

Cost Comparison

In terms of rough cost, $700/m² ($65/ft²) appears to be the lower bound for current FRP decks, which corresponds to about $7/kg ($3/lb) of material. This cost is greater than the roughly $322/m² ($30/ft²) typically quoted for the construction of a new bridge or a deck replacement with conventional materials (Lopez-Anido 2001). However, the higher costs of FRP decks can be absorbed in certain conditions, particularly when a complete reconstruction is necessary in the absence of a lightweight deck alternative. It remains to be determined if the higher initial cost of FRP decks can be justified based on other economic considerations.

Comparison of Deflections

Although limited field experience and concerns over costs have slowed the introduction of FRP decks into mainstream bridge applications, the specifications for deflections have presented the greatest number of questions in the design of these systems. As indicated by the normalized deflections in Table 1 for decks of about 203-mm depth, under HS2O+IM loading (AASHTO 1998), there is no current consensus on deck deflections by the manufacturers. The discrepancy can be attributed to the way in which FRP bridge decks are currently designed—i.e., on a case-by-case basis. However, the uncertainty in defining deflection limits is also present in existing design guidelines for conventional structures, as discussed next.

In section 2 of the AASHTO load and resistance factor design (LRFD) specifications (AASHTO 1998), deflection limits for steel, aluminum, and concrete constructions under live loads are provided as *optional*. The criteria are stated as optional because, at present, there are no definitive guidelines for the limits of tolerable static deflection or dynamic motion due to vehicular traffic (Taly 1998). Under general vehicular loads, the deflection of a structural element (i.e., a bridge deck) is limited to the span length divided by 800 (span/800). The deflection of an orthotropic (steel) plate deck is limited to the span length divided by 300 (span/300). For RC bridge decks, the approach for controlling deflections is indirect. In particular, a minimum depth [t_{min} (ft)] is stated in Eq. (1), where the span is in feet

$$t_{min} = \frac{1.2(\text{span} + 10)}{30} \tag{1}$$

The commentary to the AASHTO LRFD specification states that the purpose of the above criteria is twofold—(1) to prevent excessive deflections that may cause damage to the wearing surfaces applied to bridge decks; and (2) to provide for rider comfort. At present, no criteria are given for FRP composite constructions.

For the time being, FRP bridge deck designers quantify deck performance in terms of criteria developed for conventional materials. Further experience will determine if these criteria are appropriate and sufficient for design purposes or if other criteria unique to FRP decks will be needed.

Future Challenges

A relatively large number of FRP decks are already in service (GangaRao et al. 1999; Harik et al. 1999; Temeles et al. 2000) and several others are scheduled for installation in the near future (Davalos et al. 2001). Also, there exists growing interest in the future of these structures. However, several technical needs and questions must be addressed, as follows: (1) development of design standards and guidelines; (2) efficient design and characterization of panel-to-panel joints and attachment of decks to stringers; (3) fatigue behavior of panels and connections; (4) durability characteristics under combined mechanical and environmental loads; (5) failure mechanisms and ultimate strength, including local and global buckling modes; and (6) efficiency and durability of surface overlays (e.g., polymer concrete) and application of hot-mix asphalt in relation to glass-transition temperature of the polymer. In addition, the appropriate implementation of crash-tested guardrails remains an open question when considering the variability in the deck designs. Recently, progress has been made to address these concerns for specific systems (*Product* 2000).

The situation facing the FRP bridge deck industry is not dissimilar to that faced by previous industries—such as steel and concrete—upon the introduction of new materials to a well-entrenched marketplace. When iron was first introduced as a building material, it was fashioned into shapes that resembled timber. Perhaps the FRP decks of tomorrow will evolve to take fuller advantage of the material properties and manufacturing methods of FRP materials as experience and comfort with the material grow.

Internal Reinforcements

Introduction

Non-prestressed and prestressed internal FRP reinforcements for concrete have been under development since as early as the 1960s in the United States (Dolan 1999) and the 1970s in Europe (Taerwe and Matthys 1999) and Japan (Fukuyama 1999), although the overall level of research, demonstration, and commercialization has increased markedly since the 1980s. FRP reinforcements have been used primarily in concrete structures

requiring improved corrosion resistance or electromagnetic transparency. The scope of the remainder of this discussion on internal reinforcements includes products, structural behavior, applications, and future directions.

Products

The main differences between non-prestressed and prestressed FRP reinforcements are the level of stress and, correspondingly, the type of constituent FRP materials chosen for the application. Low-cost E-glass FRPs are generally chosen for non-prestressed applications, whereas high-strength carbon and aramid fiber FRPs are preferred for prestressed applications because of their capability of sustaining much higher stresses over the design life. Thermosetting resins such as vinylester and epoxy are the predominant polymers chosen for internal FRP reinforcements on account of their excellent environmental resistance, although affordable thermoplastic resins are recently gaining attention due to their potential for being heated and bent in the field.

Internal FRP reinforcements have been fabricated in a variety of one-dimensional and multidimensional shapes (Nanni 1993). To date, most commercially prefabricated multidimensional reinforcements are orthogonal, two-dimensional grids, although three-dimensional grids of various configurations have been proposed for certain precast structures. As with steel reinforcements, multidimensional FRP reinforcements can also be fabricated onsite by hand placement and tying of one-dimensional shapes.

One-dimensional FRP reinforcements are typically made by the pultrusion process or a close variant such as pull-forming. In such cases, the fibers are impregnated with resin, pulled through a forming die that compacts and hardens the material, and then coiled or cut to a prescribed length. To enhance the bond with concrete, surface deformations are applied to the bar before hardening by one or more of the following representative methods: (1) wrapping of one or more tows of fibers along the length of the bar; (2) molding of whisker reinforced ribs along the length of the bar; (3) wrapping of a textured release film along the length of the bar for the later removal and creation of a complementary impression in the bar; and (4) bonding of fine aggregate to the surface of the bar. In addition, bundles of tight-packed small-diameter FRP rods can be twisted to form a strand, as is commonly done with steel prestressing strands. Cosenza et al. (1997) review a number of one-dimensional FRP bar products. Although currently employed fibers in FRP reinforcements behave in a linear-elastic fashion to failure, developmental FRP reinforcements promise "pseudoductility," or graceful failure, by incorporating fibers of disparate ultimate strains or orientations in the reinforcement (Somboonsong et al. 1998). It is possible to incorporate an internal strain-sensing capability in pultruded FRP products as well (Benmokrane et al. 2000; Bakis et al. 2001).

Grid reinforcements have been made by winding resin-impregnated bundles of fibers into prescribed two- and three-dimensional shapes using a variety of manufacturing processes (Nanni 1993). The grids are often used as flat, two-dimensional flexural reinforcement in slabs or three-dimensional cages for combined shear and axial reinforcement in beams. Off-the-shelf pultruded shapes can also be mechanically joined with proprietary connection devices to create preformed, multidimensional grid reinforcements (Bank and Xi 1993). The joints of FRP grids dominate bond stiffness and strength, in effect providing a periodically bonded reinforcement system in cases where minimal bonding exists between the cross-over points (Matthys and Taerwe 2000).

Structural Behavior

Design Philosophy

Although analyses for flexural and shear capacities draw on many of the same assumptions used for steel reinforcement, significant differences between the material properties and mechanical behavior of FRP and steel necessitate a shift away from conventional concrete design philosophy. In particular, the linear-elastic stress-strain characteristic of most FRP composites (1–3% ultimate strain) implies that FRP-reinforced concrete design procedures must account for inherently less ductility than that exhibited by conventionally reinforced concrete.

Currently, FRP-reinforced concrete is designed using limit-states principles to ensure sufficient strength (typically based on some form of load and resistance factor design), to determine the governing failure mode, and to verify adequate bond strength. Serviceability limit states such as deflections and crack width, stress levels under fatigue or sustained loads, and relaxation losses (for prestressed concrete) are then checked. Although serviceability criteria are usually applied after strength design, relatively lower (especially for glass FRP and aramid FRP) elastic moduli mean that serviceability criteria will usually control the design.

Flexure

Flexural behavior is the best understood aspect of FRP-reinforced concrete, with basic principles applying regardless of member configuration, reinforcement geometry, or material type. Two possible flexural failure modes prevail. Sections with smaller amounts of reinforcement fail by FRP tensile rupture, while larger amounts of reinforcement result in failure by crushing of the compression-zone concrete prior to the attainment of ultimate tensile strain in the outermost layer of FRP reinforcement. The absence of plasticity in FRP materials implies that underreinforced flexural sections experience a sudden tensile rupture instead of a gradual yielding, as in the case with steel reinforcement. Thus, the concrete crushing failure mode of an overreinforced member is somewhat more desirable, due to enhanced energy absorption and greater deformability leading to a more gradual failure mode. Member recovery is essentially elastic with little or no energy dissipation resulting from large deformations.

Nominal flexural capacity is calculated from the constitutive behaviors of concrete and FRP reinforcement using strain compatibility and internal force equilibrium principles, assuming the tensile strength of the concrete is negligible, a perfect bond exists between the concrete and FRP, and strain is proportional to the distance from the neutral axis. The form of the analytical expression will depend upon the prevailing failure mode. The Whitney rectangular stress block is adequate for flexural capacity prediction when crushing of compression-zone concrete occurs in overreinforced sections, provided that strain compatibility is used to determine FRP tensile forces. If FRP tensile rupture controls failure, the Whitney stress block may not be applicable, unless compression-zone concrete is at near-ultimate conditions. An equivalent stress block that approximates the actual stress distribution in the concrete at FRP rupture can be used, or the moment capacity can be determined using an estimated tension force moment arm. Compression-zone FRP reinforcement is not considered effective for increasing moment capacity, enhancing ductility, or reducing long-term deflections.

For steel-reinforced concrete, ductility can be defined as the ratio of total deformation (curvature or deflection) at failure to deformation at yielding. Members with ductility ratios of four or more exhibit significant signs of distress prior to failure. As FRP reinforcement does not yield, alternate means of quantifying the warning signs of impending failure must be used. A variety of indices to measure pseudoductility have been proposed, including deformability indices, defined as the ratio of ultimate deflection to service-load deflection or ultimate curvature to service-load curvature (or curvature at a specified concrete strain level). Deformability ratios of about eight have been reported for overreinforced beams with glass FRP bars (GangaRao and Vijay 1997). The Canadian Highway Bridge Design Code uses an overall performance factor (J factor), computed as the ratio of the product of moment and curvature at ultimate to moment and curvature at a concrete strain of 0.001 (corresponding to the concrete proportional limit). The Canadian code specifies minimum acceptable values for this performance index of four for rectangular sections and six for T sections (Bakht et al. 2000). Another approach considers the magnitude of the net tensile strain in the outermost layer of FRP bars as the concrete compressive strain reaches the ultimate limit state. When the net tensile strain is 0.005 or greater, the section is "tension-controlled" and a lower resistance factor is required to compensate for the suddenness of FRP tensile rupture (ACI Committee 440 2001).

For overreinforced members, confinement of compression-zone concrete will increase the concrete ductility and thus the member deformability. Prestressed beams with unbonded carbon FRP tendons exhibit bilinear moment-curvature behavior with considerable rotation at failure. The large rotation capacity allows moments to redistribute and energy to be stored, but as energy is not dissipated, the members are not truly ductile. Typically, sections prestressed with bonded tendons do not achieve unbonded rotation capacities, although moment capacities may be greater. Possible alternatives are partial prestressing or partially bonded tendons (Lees and Burgoyne 1999). Anchors and connectors for FRP tendons differ from those used for steel tendons due to the anisotropic strength characteristics of FRP materials with highly oriented fibers. The most successful anchors for FRP tendons minimize localized transverse and shear-stress concentrations in the tendon by the use of gripping elements with tailored stiffness and geometry (Nanni et al. 1996).

Deflections and Cracking

Deflections and crack widths are typically larger in FRP-reinforced concrete beams and slabs (especially glass FRP) than in steel-reinforced concrete beams due to FRPs' lower elastic modulus. Limits on deflection or crack width frequently control designs and are usually satisfied by using overreinforced sections. Deflection prediction equations developed for steel-reinforced concrete typically underestimate immediate deflections, with disparities increasing as the load approaches ultimate. Such behavior correlates with observations that crack patterns at lower load levels are similar to those of steel-reinforced sections, but as loads increase beyond the service level, crack spacing decreases and crack width increases relative to steel reinforcement. Various modified expressions for the effective moment of inertia have been proposed for use with FRP (Masmoudi et al. 1998).

Creep and shrinkage behavior in FRP-reinforced members is similar to that in steel-reinforced members. American Concrete Institute (ACI) code equations for long-term deflection can be used for FRP reinforcement, with modifications to account for differences in concrete compressive stress and the particular elastic modulus and bond characteristics of the FRP reinforcement (ACI Committee 440 2001). Compression-zone FRP reinforcement does not reduce long-term deflections.

Shear

The concrete contribution to shear strength is reduced in beams with FRP longitudinal reinforcement because of smaller concrete compression zones, wider cracks, and smaller dowel forces. A reduction factor proportional to the modular ratio, E_{FRP}/E_{steel}, is typically applied to concrete shear contribution equations for conventional beams, although such an approach underestimates the shear strength in flexural members with larger amounts of FRP longitudinal reinforcement (Michaluk et al. 1998).

One-way deck slabs reinforced with glass FRP bars typically fail in diagonal tension shear, as opposed to flexure. Large deflections and crack widths provide an adequate warning of impending failure. The E_{FRP}/E_{steel} reduction factor underestimates the shear strength of deck test panels by a factor of three (Deitz et al. 1999). In two-way test slabs, punching shear failures have been observed. The shear capacity in such cases can be predicted using a nonlinear finite-element analysis.

In beams with FRP stirrups, shear failures occur either by FRP rupture at the bend points or by shear-compression failure in the shear span of the beam. Failure from stress concentrations at stirrup bends may limit the effective capacity to as little as 35% of the strength parallel to the fibers (Shehata et al. 2000). Multidirectional FRP grids can also be used as shear reinforcement (Bank and Ozel 1999; Razaqpur and Mostofinejad 1999).

Bond and Development of Reinforcement

Differences in FRP reinforcing products make bond characteristics quite variable. In some cases, the bond strength is comparable to or greater than that in steel reinforcement, while other products exhibit less bond strength (Cosenza et al. 1997). Bond strength is largely independent of concrete strength for smooth and sand-coated rods and twisted strands, provided there is adequate cover to prevent longitudinal cracking. Bars with molded deformations and helical wraps obtain bond from mechanical interlock, and thus have relatively good bond performance and more dependency on the concrete cover. Local bond-slip relations have been applied to glass FRP reinforcements of different diameters and embedment lengths (in pullout tests), and were found to match experimental results reasonably well while providing predictive capability for bars of arbitrary diameter and embedment length (Focacci et al. 2000). Development lengths for glass FRP reinforcement bars by pullout failure are generally in the range of 26–37 times the bar diameter (ACI Committee 440 2001).

In prestressed concrete, transfer lengths of FRP tendons have been noted to be less than approximately 50 times the tendon diameter. Development length in prestressed concrete beams depends on additional factors besides the diameter, such as the difference between the stresses in the tendon initially and at failure of the beam (Lu et al. 2000).

Applications

Bridges

During the period from 1980 through 1997, there were at least 32 documented bridge projects (20 with vehicular traffic) using concrete with FRP reinforcement (*A look* 1998). Of these, six bridges were constructed in Europe (primarily in Germany), seven in North America, and 19 in Japan. Prestressed applications predominated, including 11 bridges constructed with pretensioned FRP-reinforced concrete girders and 10 with posttensioned girders. Five bridges utilized FRP for prestressed slabs, and 11 had FRP rebar in the deck slab or beams. Several bridges utilized more than one type of FRP reinforcement. In general, carbon FRP was used for prestressed reinforcement, although there were also glass and aramid FRP applications. Non-prestressed reinforcement was typically glass FRP, although carbon FRP was also used.

Reinforced Concrete

FRP reinforcing bars have been used in magnetic resonance imaging facilities, an aircraft station compass calibration pad, tunnel boring operations, chemical plants, electrical substations, highway barriers, and a variety of seafront structures. Two-dimensional glass FRP grids have been used in RC decks and tunnel linings, while a demonstration project on FRP-reinforced underground chambers is currently under way in Canada. Glass FRP dowel bars appear feasible for load transfer across highway pavement joints, provided the dowel diameter increases and/or the dowel spacing decreases relative to designs with steel bars. Performance comparable to that of epoxy-coated steel dowels has been demonstrated (Davis and Porter 1999).

Prestressed Concrete

Bridges comprise the majority of prestressed FRP applications. Piles, piers, Maglev guideway beams, and airport pavements are other examples. Deck slab systems devoid of steel reinforcement and utilizing carbon FRP prestressed tendons in the transverse direction are under investigation. The system relies on compressive membrane action to increase punching shear failure loads and to limit deflections. Also under development are precast double-T panels reinforced with glass FRP bars and prestressed internally and externally with carbon FRP strands that require no shoring or forms for installation. The double-T panels are used in a multispan prestressed concrete bridge system utilizing internal bonded tendons and continuous externally draped tendons.

Future Directions

Applications of FRP reinforcement in major and innovative concrete structures have been facilitated by the development of "smart structure" sensor technology, thus alleviating the lack of performance data for FRP-reinforced structures. Canada's Intelligent Sensing for Innovative Structures (ISIS) project targets the development of smart sensor technology in combination with research on FRP infrastructure applications. As long-term data become available, designers will feel more comfortable specifying internal FRP reinforcements for concrete roadways and buildings. The brisk development of design guides and codes around the world will also speed the insertion of FRP reinforcements in construction practice.

Externally Bonded Reinforcements

Background

Due to the aging of infrastructure and the need for upgrading to meet more stringent design requirements, structural repair and strengthening have received considerable emphasis over the past two decades throughout the world. At the same time, seismic retrofit has become at least equally important, especially in earthquake-prone areas. State-of-the-art strengthening and retrofit techniques increasingly utilize externally bonded FRP composites, which offer unique properties in terms of strength, lightness, chemical resistance, and ease of application. Such techniques are most attractive for their fast execution and low labor costs.

Composites for structural strengthening are available today in the form of precured strips or uncured sheets. Precured shells meant to strengthen columns are also available, but are not treated further in this discussion. Precured strips are typically 0.5–1.5 mm (0.02–0.06 in.) thick and 50–200 mm (2–8 in.) wide, and made of unidirectional fibers (carbon, glass, aramid) in an epoxy matrix. Uncured sheets typically have a nominal thickness of less than 1 mm (0.04 in.), are made of unidirectional or bidirectional fibers (often called fabrics, in the latter case) that are either pre-impregnated or in situ impregnated with resin, and are highly conformable to the surface onto which they are bonded. Bonding is typically achieved with high-performance epoxy adhesives.

Historically, composites were first applied as flexural strengthening materials for RC bridges (Meier 1987; Rostasy 1987) and as confining reinforcement of RC columns (Fardis and Khalili 1981; Katsumata et al. 1987). Developments since the first research efforts in the mid-1980s have been tremendous. The range of applications has expanded to include masonry structures, timber, and even metals. The number of applications involving composites as strengthening/repair or retrofit materials worldwide has grown from just a few 10 years ago to several thousand today. Various types of structural elements have been strengthened, including beams, slabs, columns, shear walls, joints, chimneys, vaults, domes, and trusses.

Strengthening of Reinforced Concrete Structures with Fiber-Reinforced Polymer

General

Composites have gained widespread use as strengthening materials for RC structures in applications where conventional strengthening techniques may be problematic. For instance, one of the most popular techniques for upgrading RC elements has traditionally involved the use of externally epoxy-bonded steel plates. This technique is simple, cost-effective, and efficient, but it suffers from the following: deterioration of the bond at the steel-concrete interface caused by the corrosion of steel; difficulty in manipulating the heavy steel plates at the construction site; need for scaffolding; and limited delivery lengths of steel plates (in the case of flexural strengthening of long elements). As an alternative, the steel plates can be replaced by FRP strips or sheets. Another common strengthening technique involves the construction of RC, shotcrete, or steel jackets. Jacketing is quite effective as far as strength, stiffness, and ductility are concerned, but it increases the cross-sectional dimensions and dead loads of the structure, is labor intensive, obstructs occupancy, and provides RC elements with a potentially undesirable stiffness increase. Alternatively, FRP sheets may be wrapped around RC elements, resulting in considerable increases in strength and ductility without an excessive stiffness change. Furthermore, FRP wrapping may be tailored to meet specific structural requirements by adjusting the placement of fibers in various directions (ACI Committee 440 1996). An important point concerning the design of external FRP reinforcement is that, in order to maintain a sufficient safety factor in case of an accidental situation (e.g., FRP destruction due to fire), the degree of strengthening (ultimate capacity of the strengthened element divided by that of the unstrengthened element) should be limited, unless proper means of protecting the external reinforcement from loss are taken.

Flexural Strengthening

Flexural strengthening of RC elements using composites may be provided by epoxy bonding the materials to portions of the ele-

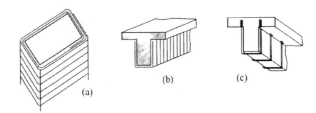

Fig. 3. Shear strengthening: (a) closed jacket applied to column; (b) open jacket applied to beam; (c) strips with end anchorage

ments in tension, with fibers parallel to the principal stress direction. Well-established strengthening procedures for RC structures may be followed, provided that special attention is paid to issues related to the linear-elastic nature of FRP materials and the bond between the concrete and FRP.

Central to the analysis and design of FRP-strengthened RC elements is the identification of all of the possible failure modes. These include the following modes: (1) steel yielding followed by FRP fracture; (2) steel yielding followed by concrete compressive crushing (while the FRP is intact); (3) concrete compressive crushing; (4) FRP peel-off at the termination or cutoff point, due to shear failure of the concrete; (5) FRP peel-off initiating far from the ends, due to inclined shear cracks in the concrete; (6) FRP peel-off at the termination point or at a flexural crack due to high tensile stresses in the adhesive; and (7) debonding at the FRP-concrete interface in areas of concrete surface unevenness or due to faulty bonding. Of the above, mode (2) is the most desirable. Modes (4) and (5) will be activated when the element's shear strength is approached; hence, they may be prevented by providing shear strengthening. Mode (6) can be suppressed by limiting the tensile strain in the FRP to a value of roughly 0.008. Finally, mode (7) may be avoided by proper quality control. Slip at the concrete-FRP interface may be ignored in design.

Shear Strengthening

Shear strengthening of RC elements may be provided by epoxy-bonding FRP materials with fibers as parallel as practically possible to the principal tensile stresses. Depending on accessibility, strengthening can be provided either by partial or by full wrapping of the elements, as illustrated in Fig. 3.

The effectiveness of the external FRP reinforcement and its contribution to the shear capacity of RC elements depend on the mode of failure, which may occur either by peeling-off through the concrete (near the concrete-FRP interface) or by FRP tensile fracture at a stress that may be lower than the FRP tensile strength (e.g., because of stress concentrations at rounded corners or at debonded areas). Whether peeling-off or fracture will occur first depends on the bond conditions, the available anchorage length and/or the type of attachment at the FRP termination point (full wrapping versus partial wrapping, with or without mechanical anchors), the axial rigidity of the FRP, and the strength of the concrete. In many cases, the actual mechanism is a combination of peeling-off at certain areas and fracture at others. In light of the above, the load carried by FRP at the ultimate limit state in shear of the RC element is extremely difficult to quantify based on rigorous analysis.

According to a simplified method of calculation of shear force in external FRP reinforcement, the FRP material is assumed to carry only normal stresses in the principal FRP material direction. It is also assumed that, in the ultimate limit state in shear, the FRP develops an *effective strain* (in the principal material direction),

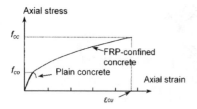

Fig. 4. Axial stress-strain response of FRP-confined concrete versus plain concrete

$\varepsilon_{f,e}$, which is generally less than the tensile failure strain. For RC elements with a rectangular cross section, the aforementioned effective strain decreases as the axial rigidity of the FRP (that is, the product of the FRP shear reinforcement ratio times its elastic modulus in the principal material direction) increases and as the concrete tensile strength decreases (Triantafillou and Antonopoulos 2000). Moreover, keeping in mind that large values of $\varepsilon_{f,e}$ correspond to considerable opening of diagonal cracks, to the extent that the contribution of concrete shear-resisting mechanisms is reduced by degraded aggregate interlock, $\varepsilon_{f,e}$ should be limited to a value on the order of 0.004–0.005 for the case of fibers perpendicular to the longitudinal axis of the RC element.

In the case of elements with a circular cross section (e.g., wrapped circular columns), experimental and analytical studies have demonstrated that FRP jackets with fibers in the circumferential direction significantly increase both the strength and the ductility in the presence of monotonic or cyclic lateral loads. The contribution of FRP to shear resistance in such cases may be estimated by taking $\varepsilon_{f,e}$ approximately equal to 0.004 (Priestley and Seible 1995).

Confinement

The enhancement of confinement in structurally deficient RC columns in seismically active regions of the world has proven to be one of the most significant early applications of FRP materials in infrastructure applications. Proper confinement increases the rotational capacity (and hence the ductility) in plastic hinge regions and prevents debonding of the internal reinforcement in lap splices. Confinement may be beneficial in nonseismic zones too, where, for instance, survivability of explosive attacks is required (Crawford et al. 1997) or the axial load capacity of a column must be increased due to higher vertical loads (e.g., increased traffic on a bridge). In any case, confinement may be provided by wrapping RC columns with FRP materials (prefabricated jackets or in situ cured sheets), in which the principal fiber direction is circumferential.

In *circular* columns, an FRP wrap effectively curtails the lateral expansion of concrete shortly after the unconfined strength is reached. It then reverses the direction of the volumetric response, and the concrete responds through large and stable volume contraction (this is not the case with steel confinement jackets, where yielding is associated with unstable volumetric expansion). As a result, the stress-strain response of FRP-confined concrete is characterized by a distinct bilinear response with a sharp softening at a stress level near the strength of unconfined concrete, f_{co} (Fig. 4). After this softening point, the tangent stiffness stabilizes at a nearly constant value. The ultimate state (f_{cc}, ε_{cu} in Fig. 4) is characterized by tensile failure of the wrap. At failure, the tensile stress in the FRP wrap is generally less than the uniaxial tensile strength of the FRP material due to triaxial stresses and variations in the quality of installation that could lead to unequal load shar-

ing among fibers, misalignment, and damaged fibers at sharp corners and local protrusions.

From the arguments discussed above, it is realized that reliable models for FRP-confined concrete must account for a number of parameters, including: (1) the circumferential stiffness of the FRP; (2) the continuous effect of the restraint provided by the FRP on the dilation tendency of the concrete; and (3) the composite action of the FRP-concrete column and the FRP-concrete interaction, based on micromechanics. As a simplified approach, one may assume a maximum compressive axial strain in the concrete at the ultimate limit state of approximately 0.01 and a fixed Poisson's ratio of approximately 0.5 to determine the confinement effect.

Confinement of RC columns is less effective if the cross section is rectangular. In this case, the confining stress is transmitted to the concrete at the four corners of the cross section and increases with the corner radius. The confinement model in this case must account for a reduced volume of fully confined concrete (Mirmiran et al. 1998).

Masonry

Recent years have seen proposals and practical applications that use composites as alternative strengthening materials for masonry structures, including those of considerable historical importance. The general approach is to epoxy-bond FRP strips to the surface of masonry in locations and directions dictated by the principal tensile stress field (Schwegler 1994).

In terms of design, masonry strengthened with FRP strips or sheets may be treated in the same manner as RC, following the procedures of modern design codes (Triantafillou 1998). The analysis of simple cases has led to the following conclusions: (1) When out-of-plane bending dominates (e.g., as in the case of upper levels of masonry buildings), horizontally applied FRP may offer a considerable strength increase; (2) in the (rather rare) case of in-plane bending, the amount and distribution of reinforcement are of high importance; high reinforcing ratios placed near the highly stressed zones give a significant strength increase; (3) the achievement of full in-plane flexural strength depends on the proper anchorage of the strips at their ends, in the sense that short anchorage lengths and/or the absence of clamping at the strips' curtailment positions may result in premature failures through peeling-off beneath the adhesive (as in the case of RC); and (4) the in-plane shear capacity of masonry walls strengthened with FRP may be quite high, too, especially in the case of low axial loads.

FRP composites can also be applied as confinement reinforcement to masonry using unbonded strips that are color-matched with the underlying masonry structure, and can be removed if necessary at a later time (Triantafillou and Fardis 1993). Recent applications include strengthening of vaults in old masonry buildings either from below, using transparent glass FRP fabrics, or from above, using epoxy-bonded FRP sheets in a gridlike pattern (Borri et al. 2000).

Timber and Metals

The high potential of FRP strips or sheets to increase the strength (flexural or shear), stiffness, and ductility of *timber* beams and/or columns has been demonstrated in various research studies and quite a few field applications (Plevris and Triantafillou 1992). Moreover, FRP wrapping has been used as an effective means of enhancing the durability of timber elements (Qiao et al. 1998).

Research and development related to FRP combined with *metals* used in construction (e.g., steel, cast iron, and wrought iron) have started relatively recently (Karbhari and Shulley 1995). High-stiffness sheets (such as carbon) may enhance the mechanical properties of metallic elements while offering certain other advantages, such as the low weight of bonded material, the easy applicability, and the ability to effectively cover areas with high bolt or rivet congestion.

Special Strengthening Techniques

A number of special techniques related to the application of composites as externally bonded reinforcement need to be mentioned, although space limitations prevent detailed descriptions.

- *Prestressed strips.* Prestressing of composite strips prior to the bonding procedure results in a more economical use of materials (Triantafillou et al. 1992), but requires special clamping devices.
- *Automated wrapping and curing.* Wrapping of columns (or other vertical elements, such as chimneys) with flexible FRP sheets is possible today by using automated machinery. The machinery can also apply heat and vacuum to assist curing.
- *Fusion-bonded pin-loaded straps.* A number of nonlaminated thermoplastic FRP layers that may move relative to each other when loaded are applied in a single, continuous, thin tape that is fusion-bonded (welded) to itself for anchorage (Winistoerfer and Mottram 1997).
- *Placement inside slits.* FRP strips or even rods may be bonded into slits, which are cut into the concrete or into masonry mortar joints (Blaschko and Zilch 1999; Tinazzi et al. 2000).
- *Prefabricated shapes.* Prefabricated angles or shells may be externally bonded to structures.
- *Mechanically fastened FRP strips.* Specially designed, pre-cured FRP strips can be rapidly attached to concrete beams for flexural strengthening using powder-actuated fasteners (Lamanna et al. 2001).

Concluding Remarks

The use of advanced composites as external reinforcement of concrete and other structures has progressed well in the past decade in selective applications where their cost disadvantage is outweighed by a number of benefits. There are clear indications that the FRP strengthening technique will increasingly continue to be the preferred choice for many repair and rehabilitation projects involving buildings, bridges, historic monuments, and other structures.

Codes and Standards

Introduction

Due to the importance of controlling risk in matters of public safety, standards and codes for FRP materials used in civil structures have been in development since the 1980s. FRP materials warrant separate treatment in standards and codes on account of their lower modulus and ductility in comparison with conventional materials such as metals. Without standards and codes, it is unlikely that FRP materials could make inroads beyond limited research and demonstration projects. Standardized test methods and material identification schemes minimize uncertainty in the performance and specification of FRP materials. Codes allow

structures containing FRP materials to be designed, built, and operated with safety and confidence. This section describes the standard and code development activities in Japan, Canada, the United States, and Europe. The main accomplishments of these activities, to date, pertain to the use of FRP materials for the reinforcement of new structures and for the repair and retrofit of existing structures.

Japan

Efforts to prescribe specifications for the design and construction of concrete structures with FRP reinforcements started in Japan in the 1980s. Examples of specifications for internal reinforcements completed by the middle of the 1990s are as follows:

1. Recommendation for Design and Construction of Concrete Structures Using Continuous Fiber Reinforcing Materials.
2. Guideline for Structural Design of FRP-Reinforced Concrete Buildings in Japan.
3. Design Methods for Prestressed FRP-Reinforced Concrete Building Structures.

Item 1, referred to here as the recommendation, was published by the Japan Society of Civil Engineers (JSCE) in 1997, and is intended for concrete structures other than buildings (Machida 1997b). The recommendation includes quality specifications and test methods for FRP materials, as well as recommendations for design and construction with FRP materials. The quality specifications for FRP reinforcements define the required characteristics and properties of the reinforcements, and serve to guide the development of new reinforcements for practical applications. Reinforcement characteristics addressed include fiber type and reinforcement configuration. Specified properties include the volume ratio of axial fibers, reinforcement cross-sectional area, guaranteed tensile strength, tensile modulus, elongation, creep rupture strength, relaxation rate, and durability, among others. Most of the specified properties are determined based on tests described in the recommendation. Further details are also given by Uomoto et al. (1997).

The design and construction recommendations in item 1 above are based on the JSCE Standard Specification for Design and Construction of Concrete Structures, which is for concrete structures in general (JSCE 1986a,b). The recommendations for construction in item 1 deal with issues such as FRP constituent materials, FRP storage and handling, assembly and placement of FRP reinforcements, precautions in concrete placing and tendon jacking, and quality control. Some details covered in the recommendation have also been presented elsewhere in the literature (Machida et al. 1995; Machida 1997a; Tsuji et al. 1997).

Items 2 and 3 listed above are intended for building structures. These specifications were developed in 1993 as the final output of the research and development project, "Effective Utilization of Advanced Composite on Construction," sponsored by the Ministry of Construction of the Japanese government. Item 2 adopts a limit state–based design method with specific provisions somewhat different from those of item 1. Details can be found in the English-language publications by Sonobe et al. (1995, 1997).

After the Hyogoken-Nanbu earthquake in 1995, the use of externally bonded carbon fiber sheets for seismic retrofitting of RC piers and columns greatly increased in Japan. Prior to this time, the use had been mainly for repair. Aramid fiber sheets for retrofit and repair were also developed at this time.

Design guidelines for the application of FRP sheets to highway bridge piers or railway viaduct columns are as follows:

1. Proposed Design and Construction Guidelines for Retrofitting of Reinforced Concrete Piers Using Carbon (Aramid) Fiber Sheets by the Japan Road Association and
2. Design and Construction Guidelines for Seismic Retrofitting of Railway Viaduct Columns Using Carbon (Aramid) Fiber Sheets by the Railway Technical Research Institute (1996a,b).

These guidelines include equations for evaluating the effects of FRP sheets on shear capacity and ductility. Similar guidelines have also been prescribed for building columns and center pillars of subway tunnels and bridge decks.

Standard test methods have been developed for FRP sheets by the Japan Concrete Institute (1998). The methods include a test for tensile properties of FRP sheets and a test for bond strength.

Canada

The use of FRP for civil engineering applications in Canada began in earnest in the late 1980s when the Canadian Society for Civil Engineers created a Technical Committee on the Use of Advanced Composite Materials in Bridges and Structures. Efforts of the committee were supported by the Canadian federal government, and led to the establishment of the Network on Advanced Composite Materials in Bridges and Structures in 1992. The network sponsored several missions in Japan, Europe, and the United States, and documented the findings in state-of-the-art reports in this field (Mufti et al. 1991a,b). In 1995, the Canadian federal government established the Network of Centers of Excellence on Intelligent Sensing for Innovative Structures. One area of focus of ISIS is the use of FRP materials for new structures and the rehabilitation of existing structures. ISIS published several design guidelines on externally bonded and internal FRP reinforcements, participated in several Code and Standards committees, and sponsored several national and international conferences.

In the year 2000, Canadian Highway Bridge Design Code section 16, "Fiber Reinforced Concrete," was completed (CSA 2000). The French translation is expected to be published in early 2001 (Bakht et al. 2000). The Canadian Standards Association also approved the code, "Design and Construction of Building Components with FRP in 2002" (CSA 2002).

United States

The United States has had a long and continuous interest in fiber-based reinforcement for concrete structures. Accelerated development and research activities on the use of these materials started in the 1980s through the initiatives and vision of the National Science Foundation and the Federal Highway Administration, who supported research at different universities and research institutions. In 1991, the ACI established Committee 440, "FRP Reinforcement." The committee published a state-of-the-art report on FRP reinforcement for concrete structures in 1996 (ACI Committee 440 1996). Committee 440 recently produced two documents approved by the Technical Activities Committee for publication in the year 2001. The documents are (1) "Guide for the design and construction of concrete reinforced with FRP bars" (ACI Committee 440 2001); and (2) "Guide for the design and construction of externally bonded FRP systems for strengthening concrete structures." The committee is currently working on the following documents: (1) "Stay-in place structural FRP forms;" (2) "Durability of FRP for concrete structures;" and (3) "Guide for the design and construction of concrete members prestressed with FRP," which are expected to be completed in 2002 and 2003.

Europe

Research on the use of FRP began in Europe in the 1960s. A Pan-European collaborative research program (EUROCRETE) was established in 1993 and ended in 1997. The program was aimed at developing FRP reinforcement for concrete, and included partners from the United Kingdom, Switzerland, France, Norway, and The Netherlands. The International Federation for Structural Concrete [Fédération Internationale du Béton (FIB) 2001] Task Group 9.3, "FRP Reinforcement for Concrete Structures," was convened in 1993 with an aim to establish design guidelines following the format of the Comité Euro-International du Béton–Fédération International de la Précontrainte Model Code and Eurocode 2 (http://allserv.rug.ac.be/~smatthys/fibTG9.3/). Task Group 9.3 is divided into subgroups on material testing and characterization, RC, prestressed concrete, externally bonded reinforcement, and marketing and applications. The task group consists of members representing most European universities, research institutes, and companies involved with FRP reinforcements for concrete. Membership includes representatives from Canada, Japan, and the United States. The task group has completed the development of an FIB bulletin on design guidelines for externally bonded FRP reinforcement for reinforced-concrete structures (FIB 2001). Supporting the work of Task Group 9.3 is a European Union Training and Mobility of Researchers (TMR) Network, "ConFibreCrete." More information about the TMR ConFibreCrete Network can be found at http://www.shef.ac.uk/~tmrnet.

In the United Kingdom, the Institution of Structural Engineers has published an interim guide on the design of RC structures with FRP reinforcement (Institution 1999). Prestressing and externally bonded reinforcements are not addressed in the guide. The guide is closely based on and refers to related British design codes (British 1985, 1990, 1997). The approaches adopted are similar to those under development in Japan, Canada, and the United States.

Future Work on Standards and Codes

From a technical standpoint, the need for specialized standards and codes for FRP materials arises from their substantially different mechanical and physical properties in comparison with conventional construction materials. As the preceding discussion points out, the development of standards and codes for the use of FRP reinforcement with concrete structures is ongoing and is expected to continue in the next several years. Much of this activity is motivated by immediate, obvious needs for improved, economical materials for the repair and retrofit of structures that are obsolete, degraded, or located in seismic zones. In other cases such as new construction, where the need for new materials is not always clear from a short-term economic standpoint, standards and codes will facilitate the use of FRP materials so that additional long-term experience can be accrued. This experience may eventually lead to the realization of promised life-cycle cost benefits of FRP materials by designers and owners of structures. Of the applications covered in this review paper, FRP shapes and bridge decks suffer from the least amount of development of standards and codes. Future research efforts on standards and codes should therefore be increasingly concentrated in these areas.

Conclusion

This paper attests to the many potential applications of FRP composite materials in construction, although the need for brevity prevents all topics from being fully addressed. It can be said that the amount of experience with various forms of FRP construction materials varies in accordance with the perceived near-term economic and safety benefits of the materials. In the case of externally bonded reinforcements, for example, the immediate cost and safety benefits are clear, and adoption of the material by industry is widespread. In other cases where FRP materials are considered to be primary load-bearing components of structures, field applications still maintain a research flavor while long-term experience with the material accumulates. A number of careful monitoring programs of structures with primary FRP reinforcement have been set up around the world and should provide this experience base in the coming years.

Standards and codes for FRP materials and their use in construction are either published or currently being written in Japan, Canada, the United States, and Europe. These official documents are typically similar in format to conventional standards and codes, which should ease their adoption by governing agencies and organizations. The most significant mechanical differences between FRP materials and conventional metallic materials are the higher strength, lower stiffness, and linear-elastic behavior to failure of the former. Other differences such as the thermal expansion coefficient, moisture absorption, and heat and fire resistance need to be considered as well.

The education and training of engineers, construction workers, inspectors, and owners of structures on the various relevant aspects of FRP technology and practice will be crucial in the successful application of FRP materials in construction. However, it should be emphasized that even with anticipated moderate decreases in the price of FRP materials, their use will be mainly restricted to those applications where their unique properties are crucially needed.

References

ACI Committee 440. (1996). "State-of-the-art report on fiber reinforced plastic (FRP) reinforcement for concrete structures." *ACI 440 R-96*, American Concrete Institute, Farmington Hills, Mich.

ACI Committee 440. (2001). "Guide for the design and construction of concrete reinforced with FRP bars." *ACI 440.1 R-01*, American Concrete Institute, Farmington Hills, Mich.

Alampalli, S., O'Connor, J., Yannotti, A. P., and Luu, K. T. (1999). "Fiber-reinforced plastics for bridge construction and rehabilitation in New York." *Materials and Construction: Exploring the connection, Proc., 5th Materials Engineering Congress*, L. C. Bank, ed., ASCE, Reston, Va., 344–350.

American Association of State Highway and Transportation Officials (AASHTO). (1998). *Load and resistance factor design (LRFD) bridge design specifications*, 2nd Ed., Washington, D.C.

Bakht, G. A., et al. (2000). "Canadian bridge design code provisions for fiber-reinforced structures." *J. Compos. Constr.*, 4(1), 3–15.

Bakis, C. E., Nanni, A., Terosky, J. A., and Koehler, S. W. (2001). "Self-monitoring, pseudo-ductile, hybrid FRP reinforcement rods for concrete applications." *Compos. Sci. Technol.*, 61(6), 815–823.

Bank, L. C. (1995). "How standards will develop markets in building construction." *Proc., 50th Annual SPI Con.*, Composites Institute, Society for the Plastics Industry, Cincinnati, 14.1–14.4.

Bank, L. C., et al. (2000). "Construction of a pultruded composite structure: Case study." *J. Compos. Constr.*, 4(3), 112–119.

Bank, L. C., and Ozel, M. (1999). "Shear failure of concrete beams reinforced with 3-D fiber reinforced plastic grids." *Proc., 4th Int.*

Symposium, Fiber Reinforced Polymer Reinforcement for Reinforced Concrete Structures, C. W. Dolan, S. H. Rizkalla, and A. Nanni, eds., *SP-188*, American Concrete Institute, Farmington Hills, Mich., 145–156.

Bank, L. C., and Xi, Z. (1993). "Pultruded FRP grating reinforced concrete slabs." *Fiber-Reinforced Plastic for Concrete Structures—Int. Symposium*, A. Nanni and C. W. Dolan, eds., *SP-138*, American Concrete Institute, Farmington Hills, Mich., 561–583.

Barbero, E. J., Lopez-Anido, R., and Davalos, J. F. (1993). "On the mechanics of thin-walled laminated composite beams." *J. Compos. Mater.*, 27(8), 806–829.

Benmokrane, B., et al. (2000). "Use of fibre reinforced polymer reinforcement integrated with fibre optic sensors for concrete bridge deck slab construction." *Can. J. Civ. Eng.*, 27(5), 928–940.

Blaschko, M., and Zilch, K. (1999). "Rehabilitation of concrete structures with CFRP strips glued into slits." *Proc., 12th Int. Conf. on Composite Materials*, T. Mussard, ed., International Committee on Composite Materials.

Borri, A., Avorio, A., and Bottardi, M. (2000). "Theoretical analysis and a case study of historical masonry vaults strengthened by using advanced FRP." *Proc., 3rd Int. Conf. on Advanced Composite Materials in Bridges and Structures*, J. L. Humar and A. G. Razaqpur, eds., *ACMBS-3*, Canadian Society for Civil Engineering, Montréal, Québec, 577–584.

British Standards Institute (BSI). (1985). "Structural use of concrete. Code of practice for special circumstances." *BS 8110-2*, London.

British Standards Institute (BSI). (1990). "Steel, concrete and composite bridges. Code of practice for design of concrete bridges." *BS 5400-4*, London.

British Standards Institute (BSI). (1997). "Structural use of concrete. Code of practice for design and construction." BS 8110-1, London.

Canadian Standards Association International (CSA). (2000). "Canadian Highway bridge design code." *CSA-S6-00*, Toronto.

Canadian Standards Association International (CSA). (2002). "Design and construction of building components with fibre reinforced polymers." *CSA-S8-06*, Toronto.

Composite materials handbook (MIL-HDBK-17). (2001). ⟨http://www.mil17.org⟩ (August 2001).

Cosenza, E., Manfredi, G., and Realfonzo, R. (1997). "Behavior and modeling of bond of FRP rebars to concrete." *J. Compos. Constr.*, 1(2), 40–51.

Crawford, J. E., Malvar, L. J., Wesevich, J. W., Valancius, J. R., and Aaron, D. (1997). "Retrofit of reinforced concrete structures to resist blast effects." *ACI Struct. J.*, 94(4), 371–377.

Creative Pultrusions. (1999). *Pultex design manual*, ⟨www.creativepultrusions.com⟩ (August 2001).

Dagher, H. J., Schmidt, A. L., Abdel-Magid, B., and Iyer, S. (1997). "FRP post-tensioning of laminated timber bridges." *Evolving Technologies for the Competitive Edge, Proc., 42nd Int. SAMPE Symposium*, Vol. 2, Society for the Advancement of Material and Process Engineering, Covina, Calif., 933–938.

Davalos, J. F., Qiao, P., and Barbero, E. J. (1996). "Multiobjective material architecture optimization of pultruded FRP I beams." *Compos. Struct.*, 35(3), 271–281.

Davalos, J. F., Qiao, P. Z., Xu, X. F., Robinson, J., and Barth, K. E. (2001). "Modeling and characterization of fiber-reinforced plastic honeycomb sandwich panels for highway bridge applications." *J. Compos. Constr.*, 52(3-4), 441–452.

Davis, D., and Porter, M. L. (1999). "Glass fiber reinforced polymer dowel bars for transverse pavement joints." *Proc., 4th Int. Symposium, Fiber Reinforced Polymer Reinforcement for Reinforced Concrete Structures*, C. W. Dolan, S. H. Rizkalla, and A. Nanni, eds., *SP-188*, American Concrete Institute, Farmington Hills, Mich., 297–304.

Deitz, D. H., Harik, I. E., and Gesund, H. (1999). "One-way slabs reinforced with glass fiber reinforced polymer reinforcing bars." *Proc., 4th Int. Symposium, Fiber Reinforced Polymer Reinforcement for Reinforced Concrete Structures*, C. W. Dolan, S. H. Rizkalla, and A.

Nanni, eds., *SP-188*, American Concrete Institute, Farmington Hills, Mich., 279–286.

Dolan, C. W. (1999). "FRP prestressing in the USA." *Concr. Int.,* 21(10), 21–24.

Fardis, M. N., and Khalili, H. (1981). "Concrete encased in fiberglass reinforced plastic." *ACI J.,* 78(6), 440–446.

Fédération Internationale du Béton (FIB). (2001). "Externally bonded FRP reinforcement for RC structures." *Bulletin 14,* Lausanne, Switzerland.

Fiberline. (1995). *Fiberline design manual,* ⟨www.fiberline.de⟩.

Focacci, F., Nanni, A., and Bakis, C. E. (2000). "Local bond-slip relationship for FRP reinforcement in concrete." *J. Compos. Constr.,* 4(1), 24–31.

Fukuyama, H. (1999). "FRP composites in Japan." *Concr. Int.,* 21(10), 29–32.

GangaRao, H. V. S., Thippeswamy, H. K., Shekar, V., and Craigo, C. (1999). "Development of glass fiber reinforced polymer composite bridge deck." *SAMPE J.,* 35(4), 12–24.

GangaRao, H. V. S., and Vijay, P. V. (1997). "Design of concrete members reinforced with GFRP bars." *Proc., 3rd Int. Symposium, Fiber Reinforced Polymer Reinforcement for Reinforced Concrete Structures,* Vol. 1, Japan Concrete Institute, Tokyo, 143–150.

Green, A., Bisarnsin, T., and Love, E. A. (1994). "Pultruded reinforced plastics for civil engineering structural applications." *J. Reinf. Plast. Compos.,* 13(10), 942–951.

Harik, I., et al. (1999). "Static testing on FRP bridge deck panels." *Proc., 44th Int. SAMPE Symposium and Exhibition,* Vol. 2, Society for the Advancement of Material and Process Engineering, Covina, Calif., 1643–1654.

Hayes, M. D., Lesko, J. J., Haramis, J., Cousins, T. E., Gomez, J., and Massarelli, P. (2000a). "Laboratory and field testing of composite bridge superstructure." *J. Compos. Constr.,* 4(3), 120–128.

Hayes, M. D., Ohanehi, D., Lesko, J. J., Cousins, T. E., and Witcher, D. (2000b). "Performance of tube and plate fiberglass composite bridge deck." *J. Compos. Constr.,* 4(2), 48–55.

Institution of Structural Engineers (ISE). (1999). "Interim guidance on the design of reinforced concrete structures using fiber composite reinforcement." *Reference No. 319,* London.

Japan Concrete Institute (JCI). (1998). "Technical report on continuous fiber reinforced concrete." *TC 952,* Committee on Continuous Fiber Reinforced Concrete, Tokyo.

Japan Society of Civil Engineers (JSCE). (1986a). "Standard specification for design and construction of concrete structures. Part 1 (design)." *SP1,* Tokyo.

Japan Society of Civil Engineers (JSCE). (1986b). "Standard specification for design and construction of concrete structures. Part 2 (construction)." *SP2,* Tokyo.

Johansen, G. E., Wilson, R. J., Roll, F., and Gaudini, G. (1997). "Design and construction of long span FRP pedestrian bridges." *Building to last, Proc., Structures Congress XV,* L. Kempner and C. B. Brown, eds., ASCE, New York, 46–50.

Karbhari, V. M., Seible, F., Burgueno, R., Davol, A., Wernli, M., and Zhao, L. (2000). "Structural characterization of fiber-reinforced composite short and medium-span bridge systems." *Appl. Compos. Mater.,* 7(2), 151–182.

Karbhari, V. M., and Shulley, S. B. (1995). "Use of composites for rehabilitation of steel structures—Determination of bond durability." *J. Mater. Civ. Eng.,* 7(4), 239–245.

Katsumata, H., Kobatake, Y., and Takeda, T. (1987). "A study on the strengthening with carbon fiber for earthquake-resistant capacity of existing reinforced concrete columns." *Proc., Workshop on Repair and Retrofit of Existing Structures,* U.S.-Japan Panel on Wind and Seismic Effects, U.S.-Japan Cooperative Program in Natural Resources, Tsukuba, Japan, 1816–1823.

Keller, T. (1999). "Towards structural forms for composite fibre materials." *Struct. Eng. Int. (IABSE, Zurich, Switzerland),* 9(4), 297–300.

Lamanna, A. J., Bank, L. C., and Scott, D. W. (2001). "Flexural strengthening of reinforced concrete beams using fasteners and fiber reinforced polymer strips." *ACI Struct. J.,* 98(3), 368–376.

Lees, J. M., and Burgoyne, C. J. (1999). "Design guidelines for concrete beams prestressed with partially-bonded fiber reinforced plastic tendons." *Proc., 4th Int. Symposium, Fiber Reinforced Polymer Reinforcement for Reinforced Concrete Structures,* C. W. Dolan, S. H. Rizkalla, and A. Nanni, eds., *SP-188,* American Concrete Institute, Farmington Hills, Mich., 807–816.

A look at the world's FRP composite bridges. (1998). Market Development Alliance, Society of Plastics Engineers Composites Institute, New York.

Lopez-Anido, R. (2001). "Life-cycle cost survey of concrete decks—A benchmark for FRP bridge deck replacement." (CD-Rom), *Proc., 80th Transportation Research Board Meeting.*

Lu, Z., Boothby, T. E., Bakis, C. E., and Nanni, A. (2000). "Transfer and development length of FRP prestressing tendons.," *Precast Concrete Institute J.,* 45(2), 84–95.

Machida, A. (1997a). "Issues in developing design code for concrete structures with FRP." *Proc., 3rd Int. Symposium on Non-Metallic (FRP) Reinforcement for Concrete Structures,* Vol. 1, Japan Concrete Institute, Tokyo, 129–140.

Machida, A., ed. (1997b). "Recommendation for design and construction of concrete structures using continuous fiber reinforcing materials." *Concrete Engineering Series 23,* Japan Society of Civil Engineers, Tokyo.

Machida, A., Kakuta, Y., Tsuji, Y., and Seki, H. (1995). "Initiatives in developing the specifications for design and construction of concrete structures using FRP in Japan." *Proc. 2nd Int. RILEM Symposium on Non-Metallic (FRP) Reinforcement for Concrete Structures,* L. Taerwe, ed., E & FN Spon, London, 627–635.

Masmoudi, R., Thériault, M., and Benmokrane, B. (1998). "Flexural behavior of concrete beams reinforced with deformed fiber reinforced plastic reinforcing rods." *ACI Struct. J.,* 95(6), 665–676.

Matthys, S., and Taerwe, L. R. (2000). "Concrete slabs reinforced with FRP grids. I: One-way bending." *J. Compos. Constr.,* 4(3), 145–153.

McGhee, K. K., Barton, F. W., and McKeel, W. T. (1991). "Optimum design of composite bridge deck panels." *Advanced Composite Materials in Civil Engineering Structures, Proc., Specialty Conf.,* S. I. Iyer and R. Sen, eds., ASCE, New York, 360–370.

Meier, U. (1987). "Bridge repair with high performance composite materials." *Mater. Tech. (Duebendorf, Switz.),* 4, 125–128 (in German).

Meyer, R. W. (1985). *Handbook of pultrusion technology,* Chapman & Hall, London.

Michaluk, C. R., Rizkalla, S., Tadros, G., and Benmokrane, B. (1998). "Flexural behavior of one-way concrete slabs reinforced by fiber reinforced plastic reinforcements." *ACI Struct. J.,* 95(3), 353–364.

Mirmiran, A., Naguib, W., and Shahawy, M. (2000). "Principles and analysis of concrete-filled composite tubes." *J. Adv. Mater.,* 32(4), 16–23.

Mirmiran, A., Shahawy, M., Samaan, M., El Echary, H., Mastrapa, J. C., and Pico, O. (1998). "Effect of column parameters on FRP-confined concrete." *J. Compos. Constr.,* 2(4), 175–185.

Mufti, A. A., Erki, M. A., and Jaeger, L. G., eds. (1991a). *Advanced composite materials in bridges and structures in Japan,* Canadian Society for Civil Engineering, Montréal, August.

Mufti, A. A., Erki, M. A., and Jaeger, L. G., eds. (1991b). *Advanced composite materials with application to bridges,* Canadian Society for Civil Engineering, Montréal, May.

Nanni, A., ed. (1993). *Fiber-reinforced-plastic for concrete structures: Properties and applications,* Elsevier Science, Amsterdam.

Nanni, A., Bakis, C. E., O'Neil, E. F., and Dixon, T. O. (1996). "Performance of FRP tendon-anchor systems for prestressed concrete structures." *PCI J.,* 41, 34–44.

Pethrick, R. A., Boinard, E., Dalzel-Job, J., and MacFarlane, C. J. (2000). "Influence of resin chemistry on water uptake and environmental aging in glass fibre reinforced composites—Polyester and vinyl ester laminates." *J. Mater. Sci.,* 35(8), 1931–1937.

Plevris, N., and Triantafillou, T. C. (1992). "FRP-reinforced wood as structural material." *J. Mater. Civ. Eng.,* 4(3), 300–317.

Priestley, M. J. N., and Seible, F. (1995). "Design of seismic retrofit measures for concrete and masonry structures." *Constr. Build. Mater.,* 9(6), 365–377.

Product selection guide: FRP composite products for bridge applications. (2000). 1st Ed., J. P. Busel and J. D. Lockwood, eds., Market Development Alliance, Harrison, N.Y.

Qiao, P., Davalos, J. F., and Brown, B. (2000). "A systematic approach for analysis and design of single-span FRP deck/stringer bridges." *Composites, Part B,* 31(6-7), 593–610.

Qiao, P., Davalos, J. F., and Zipfel, M. G. (1998). "Modeling and optimal design of composite-reinforced wood railroad crosstie." *Compos. Struct.,* 41(1), 87–96.

Raasch, J. E. (1998). "All-composite construction system provides flexible low-cost shelter." *Compos. Technol.,* 4(3), 56–58.

Railway Technical Research Institute (RTRI). (1996a). "Design and construction guidelines for seismic retrofitting of railway viaduct columns using aramid fiber sheets." Tokyo (in Japanese).

Railway Technical Research Institute (RTRI). (1996b). "Design and construction guidelines for seismic retrofitting of railway viaduct columns using carbon fiber sheets." Tokyo (in Japanese).

Razaqpur, A. G., and Mostofinejad, D. (1999). "Experimental study of shear behavior of continuous beams reinforced with carbon fiber reinforced polymer." *Proc., 4th Int. Symposium, Fiber Reinforced Polymer Reinforcement for Reinforced Concrete Structures,* C. W. Dolan, S. H. Rizkalla, and A. Nanni, eds., *SP-188,* American Concrete Institute, Farmington Hills, Mich. 169–178.

Rostasy, F. S. (1987). "Bonding of steel and GFRP plates in the area of coupling joints. Talbrucke Kattenbusch." *Research Report No. 3126/ 1429,* Federal Institute for Materials Testing, Braunschweig, Germany (in German).

Schwegler, G. (1994). "Masonry construction strengthened with fiber composites in seismically endangered zones." *Proc., 10th European Conf. on Earthquake Engineering,* Vol. 3, G. Duma, ed., Balkema, Rotterdam, The Netherlands, 2299–2304.

Shehata, E., Morphy, R., and Rizkalla, S. (2000). "Fiber reinforced polymer shear reinforcement for concrete members: Behavior and design guidelines." *Can. J. Civ. Eng.,* 27(5), 859–872.

Smallowitz, H. (1985). "Reshaping the future of plastic buildings." *Civ. Eng. (N.Y.),* 55(5), 38–41.

Somboonsong, W., Ko, F. K., and Harris, H. G. (1998). "Ductile hybrid fiber reinforced plastic reinforcing bar for concrete structures: Design methodology." *ACI Mater. J.,* 95(6), 655–666.

Sonobe, Y., et al. (1997). "Design guidelines of FRP reinforced concrete building structures." *J. Compos. Constr.,* 1(3), 90–115.

Sonobe, Y., Mochizuki, S., Matsuzaki, Y., Shimizu, A., Masuda, Y., and Fukuyama, H. (1995). "Guidelines for structural design of FRP reinforced concrete buildings in Japan." *Proc., 2nd Int. RILEM Symposium on Non-Metallic (FRP) Reinforcement for Concrete Structures,* L. Taerwe, ed., E&FN Spon, London, 636–645.

Strongwell. (1999). *EXTREN fiberglass structural shapes—Design manual,* Bristol, Va., ⟨www.strongwell.com⟩ (August).

Taerwe, L. R., and Matthys, S. (1999). "FRP for concrete construction." *Concrete Int.,* 21(10), 33–36.

Taly, N. (1998). *Design of modern highway bridges,* McGraw-Hill, New York.

Temeles, A. B., Cousins, T. E., and Lesko, J. J. (2000). "Composite plate and tube bridge deck design: Evaluation in the Troutville, Virginia weigh station test bed." *Proc., 3rd Int. Conf. on Advanced Composite Materials in Bridges and Structures, ACMBS-3,* J. L. Humar and A. G. Razaqpur, eds., Canadian Society for Civil Engineering, Montréal, 801–808.

Tinazzi, D., Arduini, M., Modena, C., and Nanni, A. (2000). "FRP-structural repointing of masonry assemblages." *Proc., 3rd Int. Conf. on Advanced Composite Materials in Bridges and Structures, ACMBS-3,* J. L. Humar and A. G. Razaqpur, eds., Canadian Society for Civil Engineering, Montréal, 585–592.

Triantafillou, T. C. (1998). "Strengthening of masonry structures using epoxy-bonded FRP laminates." *J. Compos. Constr.,* 2(2), 96–104.

Triantafillou, T. C., and Antonopoulos, C. P. (2000). "Design of concrete flexural members strengthened in shear with FRP." *J. Compos. Constr.,* 4(4), 198–205.

Triantafillou, T. C., Deskovic, N., and Deuring, M. (1992). "Strengthening of concrete structures with prestressed FRP sheets." *ACI Struct. J.,* 89(3), 235–244.

Triantafillou, T. C., and Fardis, M. N. (1993). "Advanced composites for strengthening historic structures." *Proc., Symposium on Structural Preservation of the Architectural Heritage,* International Association for Bridge and Structural Engineering, Zurich, Switzerland, 541–548.

Tsuji, Y., Umehara, H., Hattori, A., Uomoto, T., and Machida, A. (1997). "Japanese recommendation for design and construction of continuous fiber reinforced concrete in public works— Part 2: Construction." *Proc., 3rd Int. Symposium on Non-Metallic (FRP) Reinforcement for Concrete Structures,* Vol. 1, Japan Concrete Institute, Tokyo, 183–190.

Uomoto, T., Hattori, A., Maruyama, K., and Machida, A. (1997). "Notes on the proposed quality standards for continuous fiber reinforcing materials (*JSCE-E131-1995*)." *Proc., 3rd Int. Symposium on Non-Metallic (FRP) Reinforcement for Concrete Structures,* Vol. 1, Japan Concrete Institute, Tokyo, 167–173.

Vinson, J. R. (1999). *The behavior of sandwich structures of isotropic and composite materials,* Technomic, Lancaster, Pa.

Werner, R. I. (1979). "Properties of pultruded sections of interest to designers." *Proc., 34th Annual Technical Conf.,* Reinforced Plastics/Composites Institute, Society of the Plastics Industry, Inc., Washington, D.C., 1–7.

Winistoerfer, A., and Mottram, T. (1997). "The future of pin-loaded straps in civil engineering applications." *Recent Advances in Bridge Engineering, Proc., U.S.-Canada-Europe Workshop on Bridge Engineering,* U. Meier and R. Betti, eds., EMPA, Duebendorf, Switzerland, 115–120.

Zureick, A., and Steffen, R. (2000). "Behavior and design of concentrically loaded pultruded angle struts." *J. Struct. Eng.,* 126(3), 406–416.

American Society of Civil Engineers 1852 – 2002 *Building a Better World* *150th Anniversary Paper*

Tomorrow's Engineer Must Run Things, Not Just Make Them Run

Richard G. Weingardt, P.E., Hon.M.ASCE[1]

Abstract: Engineers make things run, but engineers rarely run things. Today's civil engineers, though highly respected for their technical expertise and problem-solving skills, hardly ever top the public's list of leaders. And they only rank in the second tier of professions when it comes to prestige in the general public's eye, mainly because few civil engineers get involved in highly visible public positions of power—and few people have a personal civil engineer they visit regularly as they do a doctor or dentist. In the distant past, civil engineers were involved as community leaders in developing history-altering projects and modernizing the growth of American cities and counties. Today, they typically are not. This paper discusses why this is and the strengths and weaknesses of today's civil engineers in public leadership—and emphasizes their reluctance to show up as leaders beyond their own industry. It highlights the inherent attributes members of the profession possess that recommend them for such leadership and discusses why civil engineers should be pacesetters in establishing public policy and direction. It addresses three monumental global developments currently impacting the industry and makes predictions about the future practice of civil engineering. Also covered is the need to better educate tomorrow's engineers, attract more bright youngsters into the profession, and keep them in it. Finally, it presents strategies outlining how, in addition to meaningful community involvement, civil engineering leaders can strengthen the stature of the profession through greater visibility and effective public relations.

CE Database subject headings: Leadership; Public policy; Engineers; Strategic planning; Education.

Introduction

The horrific terrorist acts of September 11, 2001, will be vividly etched in history forever. The terrible memories of that shocking day will be a constant reminder of the horrifying evil suicidal terrorists can do. The date will serve as the milestone when life as we Americans—and other civilized men and women throughout the world—once knew it changed for all time. It marks when the age of gullible innocence ended. Terrorism and homicidal fanaticism, however, did not end on that date but today continue—in some areas, virtually uncontrolled—on a daily basis around the globe.

The implications of terrorism for future engineering of the world's built environment have risen to the forefront as one of the three most momentous challenges civil engineers face. Wholehearted advancement into the first part of the new millennium will be drastically impaired until global terrorism is controlled—and the troubled Middle East is settled into a more peaceful existence.

Today's world situation is similar to that of the 1950s, with the Korean War and escalation of the Cold War between the United States and the U.S.S.R. These caused Kirby et al. (1956), the authors of *Engineering in History*, to comment, "Rapid developments in our ways of life are unsettling and confusing, but only primitive societies stand still and look backward to emulate the past. A civilization worthy of its name looks and moves forward. If knowledge of man and society can be increased and if ethical principles and human values can be bettered, there is no need to view the uncertain future with dark pessimism, although constant vigilance will be necessary if progress is to be achieved. It is

important for a nation of the world, as it is constituted today with its economic, political, and military rivalries, to know thoroughly its own *engineering* potential for security and welfare."

The events of September 11—and those that follow—have alerted Americans (and citizens worldwide) that ruthless fanatics and zealots will readily target innocent men, women, and children to advance their cause. This has underscored the need to design tomorrow's structures—and other infrastructure facilities—to repel terrorist actions and the like. Civil engineering of future projects and upgrading existing facilities will, increasingly, need to deal with expensive security to protect the safety of building occupants.

Sooner than later this nation's civil engineers need to step forward and become the stewards of America's infrastructure. Today's—and tomorrow's—civil engineers must do more than just make things run; they must get deeply involved in running things.

Crucial Worldwide Developments Underway

The challenges of the future concern much more than controlling terrorists and keeping warring peoples from destroying the world. As terrible as that probability is, terrorism is not the only concern. Two other global developments are impacting the world—and America's civil engineering industry. They are the world's rapidly expanding population and the fast-moving advances in computer usage and information technology (IT).

The effects of these avalanchelike developments—terrorism, demographics, and technology—will be far-reaching. And they drive home the point that this nation's best technologically trained minds and most skilled engineers must get involved with policy-

[1]CEO, Richard Weingardt Consultants, Inc., 9725 E. Hampden Ave., Ste. 200, Denver, CO 80231.

setting decisions to deal with the technical problems inherent to each of these developments.

Ever-increasing applications of technological advancements can either hinder or advance the quality of life in both industrial-power and developing countries. Individuals with in-depth knowledge of engineering nuances should be at the table helping make the rules for applying such advancements. Too often civil engineers have been used only as a resource to solve technical difficulties—and have not lent their leadership expertise in addressing broader problems. Few leaders in the industry believe this approach provides the best solution in the long term.

As reported by Bernstein and Lemer (1996), "The nation's built assets enable many of us to enjoy unprecedentedly high living standards, but we see all around us air pollution, traffic congestion, loss of open space, and other elements of environmental degradation. With the help of expanding transportation and water supply facilities, urban 'sprawl' has consumed land even faster than population has grown." They added, "Technological innovation in design and construction can help us achieve the goal of bringing sustainable development and improved living standards to all of the world's people. But this innovation will occur only if builders, bankers, owners and operators, users and neighbors, and all of the other stakeholders in our built environment agree that it should be." Civil engineers are part of the solution, not part of the problem, and they must be at the table when key decisions are being made.

Civil Engineers Underrepresented

The need to use the talents and leadership skills of civil engineers in resolving technological issues has never been more important than today. However, civil engineers continue to be woefully underrepresented on important policy-making bodies, such as decision-making regional boards and commissions, and *in* elected offices, from local levels to the halls of Congress. For instance, today there are only two professional engineers (PEs) in the U.S. Congress—neither is a civil engineer—and less than three dozen PEs among the 6,000-plus state legislators in the United States. It has been years since the position of U.S. secretary of transportation—a seat that cries out to be held by a leader with strong civil engineering credentials—has been filled by a professional engineer. Less than 40% of the 50 state secretaries of transportation are civil engineers. And none of America's 50 governors are civil engineers, nor are any of the mayors of major cities.

Civil engineering professionals in the recent past, by and large, have not been engrossed in consequential—and high-profile—public leadership opportunities. Notable exceptions are Andrew Card, chief of staff to President George W. Bush, and Lt. Governor Bill Ratliff of Texas. Both Card and Ratliff participate in making meaningful societal decisions—many with important engineering implications—that are altering history on a daily basis. Ratliff, a long-time Texas senator before his current appointment, has often stated, "It's amazing how much government needs the problem-solving skills of engineers. And not just on technical issues. Engineering skills and logic are needed for decisions on budgeting, investment, determining future goals—even on issues like education."

"The world is run by those who show up" (Weingardt 1997). Ratliff and Card—and other publicly active civil engineering leaders like them—validate this. As Ratliff noted, "Civil engineers can greatly motivate change. All we need to do is show up."

Engineering Legends

Because of their daring and stature in the community—and maybe because of their high level of education—civil engineers of the past were often the visionaries, instigators, and leaders of great projects. Many monumental edifices would have remained idle curiosities without their input and direction. For instance, the greatest building project of the nineteenth century—America's transcontinental railroad, the first of its kind in the world at the time—would never have been completed when it was without the vision, know-how, and lobbying skills of a young civil engineer, Theodore Judah (Weingardt 2002). The driving force behind the design and completion of cutting-edge American bridges of the 1800s—the Eads and the Brooklyn—were American engineers James Eads, John, Washington, and Emily Roebling.

Roswell Mason, the first president of the Western Society of Engineers (WSE) (1869–1870), served as mayor of Chicago, as did fellow civil engineer DeWitt Cregier, the sixth president of the WSE (1883–1885). Mason, who was also a major guiding force on the board of trustees of the University of Illinois for 10 years, "organized [near the close of his term as mayor] the relief and massive rebuilding of the city as a result of the great fire of October 9, 1871" (WSE 1970).

The Panama Canal—one of the world's greatest man-made wonders—was completed successfully under the leadership of civil engineers John Stevens and George Goethals, two giants in the profession, yet they are not household names in contemporary society. Why not? And who are today's counterparts to the likes of these legends of the past? Most people don't know!

More often than not, nonengineers are in charge of many of today's massive building ventures. The person heading up the international team of designers and constructors for the $18 billion Chunnel—the monumental tunnel project under the English Channel—was not a civil engineer but an architect, Jack Lemley. [Lemley has been decorated with honors from several engineering groups, ASCE and the American Council of Engineering Companies (ACEC) included. Apparently they believe he has the qualities and talents the civil and consulting engineering communities admire in someone responsible for great civil engineering feats.]

"The past is prologue to the future," said engineering historians Kirby et al. (1956). Will any of the civil engineering leaders of the future match the accomplishments made by the giants of the past? The answer lies in the hands of today's engineers.

Action Call to Civil Engineers

The public today perceives that engineers make things run but don't run things. Many in the industry may take exception to this, but perceptions are what they are—and often perception is more accurate than reality. Even though civil engineers are highly valuable in the advancement of any civilization, few in the media or the public arena recognize them. Even so, Kirby et al. (1956) reported, "An inescapable conclusion to be drawn from the story of engineering in history is that engineering has become an increasingly powerful factor in the development of civilization." Civil engineering, the historians stressed, "does not occur in a historical vacuum without reference to other human activities," and the fundamental changes stimulated by engineering developments "accelerate the rate of historical and social revolution."

Forecasts for tomorrow's world uncover many challenges—and opportunities—that professional engineers will encounter as America and the rest of the world quickly settle into the space-age twenty-first century. It's a century that began with America at peace and now finds the country at war in a far-reaching conflict against terrorists and their sponsors. To adequately deal with these factors, more and more civil engineering leaders will need to show up in leadership roles beyond the industry, at all levels of society.

If civil engineering leaders do reach the top of society's food chain of decision makers, the world—so dependent on civil engineering technology—will surely benefit. And civil engineers will increasingly be called upon not just to make things run, but to run things as well. The importance of engineers getting involved is underscored by Bernstein and Lemer (1996):

> The U.S. today possesses a physical infrastructure of extraordinary scale and scope. This civil infrastructure supports virtually all elements of our society, and the people and business that have produced it comprise a major segment of our economy. History indicates that the growth, flourishing and decline of any civilization are closely mirrored by the life cycle and performance of its civil infrastructure.

In the final analysis, how favorably the public perceives civil engineers—and their profession—will have a tremendous influence on how effectively engineers perform as leaders and even as technical experts. It will also influence whether civil engineers are thought of as professionals or technicians to be hired by low bid without regard to qualifications.

Public Perception Polls

Duke University professor Henry Petroski (2001a) stated:

> How any profession is perceived is very much under the control of its members and their collective surrogates, the professional societies. We should behave as we wish to be perceived. If we want to be shown the deference accorded doctors and lawyers, we should conduct ourselves accordingly. We should not want any stranger we might sit beside on an airplane or at a dinner party to express surprise that ladies and gentlemen can *also* be engineers.

Gallup Polls

In a 1945 Gallup survey, 13% of the adults polled in the United States said they would recommend engineering—either civil or electrical engineering—as a career choice for their children (and other bright youngsters); in contrast, 28% said they would recommend entering the medical profession (to be a doctor, nurse, dentist, or pharmacist); 9% said law (Gallup 1944–2001).

By 1950, in the rebuilding years after World War II, the percentage of Americans suggesting engineering (all disciplines) as a recommended profession was 16%. This figure jumped to an all time high of 20% in 1953, hovered in the mid-to-low teens until the late 1980s, then dropped like a rock. In 2001, less than 2% of Americans surveyed said they would advocate a career in engineering; 22% recommended the medical profession; 3% said law. A new category, computers, was added to the polls in the 1980s, and in 2001 18% of the respondents recommended "something in the computer field" as a preferred occupation.

In these same Gallup surveys, respondents rated the honesty and ethical standards of people engaged in engineering (all disciplines). In 2001, 60% rated engineers as "high to very high" in this category. Doctors were rated at 66% and police officers at 68%. From 1945 to the 1980s, when 12 to 20% of those surveyed were recommending engineering as a preferred occupation, only 48% (on average) ranked engineers' honesty and ethical standards as "high to very high."

Apparently today's public thinks highly of engineers and their standards but isn't as willing as in the distant past to recommend it as a profession for young people. Nor is the public clear about the different disciplines of engineering, nor do many know exactly what civil engineers actually do.

Harris Surveys

Harris surveys taken from 1977 to 2001 questioned the public about the prestige of different professions (Harris 1977–2001). In 2001, these surveys showed that 36% of American adults perceived engineering (all disciplines) as an occupation with "very great prestige." The Harris results show the public's favorable perception of engineers has ranged from a low of 30% (in 1982) to a high of 37% (in 1992) over the last two dozen years.

After an all-time high ranking in 1992, the public's favorable view of engineering dipped sharply to 32% by 1997, where it remained until 2000. Last year's 36% status—a jump of more than 10% from 2000 to 2001—still showed engineers ranking far behind doctors (a 61% rating in the most prestigious category, nearly double that of engineers), teachers (54%), and scientists (53%). Engineers also continued to rank below ministers (41%), military officers (40%), and police officers (37%). The 2001 Harris poll was taken before the horrific September 11, 2001, terrorist attacks. Most likely police officers and firefighters would have received an even higher ranking if the polling had occurred after 9/11.

In round numbers, the Harris polls indicate that only one out of three Americans currently think highly of engineers (all disciplines), while twice that many—two out of three—think highly of doctors. In addition to doctors, teachers, scientists, ministers, and military and police officers are perceived to be more prestigious than engineers.

ACEC Survey

In 2000, ACEC retained the public relations firm of Ogilvy Public Relations Worldwide to conduct studies on the public's perception of consulting engineers. Ogilvy reported back to ACEC that few people outside the engineering/construction industry knew who consulting engineers—consulting civil engineers included—were and what they did (Weingardt 2001).

Additionally, Ogilvy's ACEC study asserted that nonengineers such as media reporters, newscasters, and staffers for congressional leaders (at both federal and state levels) tend to group consulting engineers with consultants in general, and that the word "consulting" is a negative with government officials. One of the most significant findings of the ACEC-sponsored studies noted that large numbers of the public that consulting civil engineers want (and need) to reach do not know what engineers do.

Though the Gallup, Harris, and ACEC surveys—except for the 1945 Gallup survey—did not isolate civil engineering as such, they indicate the public's perception of engineering is not comprehensive nor as good as it could be. They show that much can be done to improve the profession's image and the public's understanding of the industry.

The managing editor of *New Civil Engineer*, Jackie Whitelaw (2000), suggested that a professional status for civil engineers is about "gaining and maintaining the trust and respect of people" who value the work civil engineers do. She also noted the profession is now experiencing "the dawning of the age of engineers"—and that the British public today appreciates that "civil engineers can also inspire, entertain and amaze." This "dawning" is something the mass media in the United States seem reluctant to acknowledge.

Ongoing American Public Relations Dilemmas

Public relations (PR) initiatives to enhance and increase the image, visibility, and relevance of civil engineers—and the profession of engineering—should stem from a basic desire to contribute to the betterment of society in a significant way and not merely to ballyhoo one's profession without regard to the impact one's work has on society. However, to be overly humble about one's contributions and talents—or stay isolated in ivory towers—serves no purpose.

Before any industrywide—or individual—PR effort can be truly effective, though, two deeply intertwined critical dilemmas plaguing the civil engineering profession must be resolved:
- The lack of respect for (and knowledge of) civil engineers' contributions to the built environment; and
- The public's (and the media's) failure to appreciate the significance of civil engineering accomplishments.

To resolve these two dilemmas, civil engineers should follow this advice: "Do good work and tell the world about it," not in an exaggerated or arrogant way, but truthfully and consistently—and not in violation of ASCE's code of ethics. Engineers need to take three steps to accomplish this:
- Stay technically competent;
- Be involved as leaders in society and contributors to the well-being of their communities; and
- Institute public relations programs that stress the relevance of civil engineering to everyday life and that highlight civil engineers as people (the person next door, not a faceless instigator of design or construction projects).

Civil engineers should understand that PR is about doing *all the things necessary* to let the people and/or groups they want to influence know who they are, what they do, and why they do it. Individuals and the civil engineering community should instigate PR programs that get them from behind closed doors to accomplish these objectives.

Future Education of Civil Engineers

Along with the need to solve the profession's PR dilemmas goes the requirement to improve the education of future civil engineers. ASCE's "Engineering the Future of Civil Engineering" by ASCE's Task Committee on the First Professional Degree outlined these concerns (ASCE 2001):
- The current four-year bachelor's degree is no longer adequate formal academic preparation for the practice of civil engineering at the professional level in the twenty-first century;
- Civil engineers are not being prepared to compete for leadership positions—their formal education is deficient in nontechnical knowledge and skills;
- Nonengineers are increasingly managing civil engineers, principally because they possess stronger leadership, communication, and business skills;
- Regardless of experience level, civil engineering salaries generally fall below those of other engineering professions; and
- By retaining the 200-year-old, 4-year basic education model, civil engineering has fallen behind accounting, architecture, dentistry, law, medicine, pharmacy, and veterinary medicine.

As stated by Norm Augustine, former head of aerospace giant Lockheed Martin, "One needs more training to give my neighbor's basset hound a vaccination than one needs to design a structure upon which the safety of thousands of people depends" (Weingardt 2000a).

In addition to calling the current American educational requirements for civil engineers lacking, the ASCE Task Committee compared American engineering education to others and reported, "The European educational system requires formal education beyond a baccalaureate degree as a condition for entering."

The Task Committee further reported that, "though maintaining the present civil engineering education model may meet the short-term needs of employers, its narrow focus does not serve the long-term interests of the public, employers and individual civil engineers." Stressing the implications, Walesh (2000) reported that in the future, principals of engineering firms and senior managers of government entities "will continue to complain about the inadequacies of entry-level and experienced civil engineers." The ASCE Task Committee summarized these inadequacies as:
- Poor communication skills;
- Inability to manage projects profitably;
- Lack of marketing interest and/or skill;
- Getting bogged down in technical matters;
- Failure to meet client expectations;
- Lack of visibility in the community;
- Inability to understand global context;
- Having little business sense.

The concern over the narrow focus of a civil engineer's education and training has been with the industry for more than 25 years. As expressed by past ASCE president Wallace Chadwick (1977), in *Celebrating ASCE's 125th Anniversary, Turning Points in U.S. Civil Engineering History*, "As engineering projects grow in size and complexity, men [and women] capable of expertly managing them are not being trained for such tasks. What we need is the man [or woman] who knows, not only engineering design and construction, but also client and public relations, economics, environmental considerations, finance and accounting, and particularly contracts."

Engineering as A Career

As crucial as having tomorrow's civil engineers properly educated is getting enough bright young people interested in engineering as a career in the first place. Many of today's industry leaders in both design and construction cite this as their main concern. To increase the credibility and maintain the lifeblood of the American civil engineering industry, substantial numbers of intelligent, U.S.-born youngsters need to be recruited into the field.

Attracting sharp young men and women today requires creating a much grander awareness of civil engineering than before—an awareness that vividly captures the imagination of those top students who believe they can become anything they want. Today's aspiring professionals want careers that have meaning and relate to the world around them—ones that will allow them to make a difference. In addition, they want role models—heroes and heroines—to look up to in the profession of their choice.

The civil engineering community needs to better celebrate its outstanding members—its stars—and highlight civil engineering accomplishments in ways that the average person as well as talented young people understand. The civil engineering community can benefit from studying successes in professions such as science, medicine, and architecture. As proposed by Weingardt (2000b) in a white paper, "Step Forward and Be Heard," for MIT's CEE New Millennium Colloquium, "Unlike architects who publish and widely distribute beautiful coffee-table books about architecture and star architects, civil engineers have hardly any such publications about their stars and notable projects. If our best and brightest young people want to find out about the heroes and heroines in the field, there are few places for them to look. We need more coffee-table books written *not* for engineers but *for* the non-engineer public, in a style they can relate to."

Additionally, individual civil engineers themselves must do everything possible—including being public figures—to assure would-be engineers that:

- There are heroes—and superstars—in the civil engineering industry;
- Civil engineering leaders are involved in shaping the nation's future; and
- Civil engineering is relevant to everyday events, the economy, and America's standard of living.

Demographics

Of the three major developments—world terrorism, runaway population growth, and increased reliance on computers and IT— demographics is having the most conspicuous impact on the practice of civil engineering in the United States and around the world.

The United States is in the midst of one of the most far-reaching changes in its history, comparable to the twentieth century's shift from agriculture to industry. Changing demographics are bringing about this momentous transformation, at a time when world commerce is becoming increasingly dependent on computer and information technology. The latter now allows ready access to different—and cheaper—labor markets around the globe and will continue to do so.

Demographically, members of the U.S. civil engineering profession and many of its customers have almost exclusively been white, middle-aged men over the years. But 20 to 30 years from now, both the design/construction industry's clients and its workforce will be very different.

Today's "minorities"—African-Americans, Asians, Hispanics, Native Americans, and so on—will make up the majority by 2060. "The U.S. engineering industry has not yet come to grips with this reality," according to Dorman (2000). "It [the civil engineering community] has not truly begun the internal transformation required to properly reflect this fact nor to adequately service the clients of the future. The need to do this, and to do it now, is urgent."

The predicted worldwide demographic changes, particularly in the first half of the twenty-first century, will dramatically alter the makeup of American engineering companies, governmental and educational entities, and their employment and customer base.

Population Growth

In 1991, the United Nations (UN) predicted the world's population would have grown 100% to 12 billion by the year 2100 (Snyder 2000). In 1999, the UN reduced this number by 40%, changing its projected growth from 100 to 60%. That means that, instead of the world's population doubling by the end of the twenty-first century, it would *only* grow by 60% to nearly 10 billion people (still a staggering number). This gradually slowing

population growth will start flattening out by the year 2060, and the population will stabilize at 10 billion around 2170.

International consulting futurist David Snyder (2000) believes the UN's reduced estimate of population growth rate was largely due to "the consequences of worldwide economic expansion and increasing prosperity in the majority of nations." The UN forecasts that more than 95% of the world's population growth will take place in developing countries. Europe, Russia, and Japan are expected to lose population. The United States is the only major industrial country in which large population increases are projected, mainly through immigration and higher-than-average birth rates among new immigrants.

Global Economic Growth

The current rapid growth of the global economy can largely be attributed to the efficiency of "frictionless transactions" made possible by IT and the Internet, and to new markets made available by free trade. Many futurists (Edwards and Snyder 1997) expect that "among the regional economic blocks around the world, NAFTA [North Atlantic Free Trade Agreement] will shoot past Europe to become the largest and most prosperous one by 2005—and will remain so throughout the 21st century."

One of the main reasons for North America's predicted economic dominance is tied to its population; North America's population will grow faster than that of all other continents except Africa during the next century. The North American population, including the United States, is expected to more than double by the twenty-second century.

"While policy makers in Europe and Japan are contemplating a century of labor shortfalls and increasing ratios of dependent-to-wage-earner, the U.S. will soon confront—as it has in the recent past—a temporary labor shortage that is the direct consequence of the low birth rate of the 1970s and early 1980s," reported Snyder (Weingardt 2001); "The combination of robust economic growth and a shrinking entry-level labor pool has indeed served to 'lift (almost) all the boats' in the U.S.; minority employment is at all-time highs and minority income at all-time highs, as is average income for women."

Older Workers

The average retirement age in the United States has been rising since the late 1980s, and the over-65 portion of the nation's workforce is expected to increase from 12 to 15% within the next 5 years. Snyder (Weingardt 2001) reported, "Human resources experts, however, point out that, while many seniors enjoy working and wish to remain productively employed, they often retire from their career employer because they are unhappy with their management or workplace environment, or they have encountered active age discrimination on the job."

School-Aged Children

The upward population growth in the United States is putting a major strain on our schools, which is one of the main reasons the 2001 ASCE infrastructure report card (ASCE 2002b) gave this nation's schools its lowest grade (a D−). The problem isn't just that school buildings and other facilities are inadequate and/or in poor repair; schools are clearly getting more and more overcrowded. The U.S. Department of Education reports that America's student enrollments are setting growth records and will continue to do so throughout the twenty-first century.

Upgrading and adding educational facilities—as well as building more elder-care facilities to serve the needs of increasing numbers of people living longer—will require extensive engineering and architectural services. How much of the required design and related work for modernizing America's educational system will be led by U.S.-based civil engineering companies or agencies will depend on how well civil engineers position themselves as pace setters and innovators in the near future.

Changes in Minority Status

The racial/ethnic/cultural enrichment of the United States over the past 20 years has set the stage for a transformation of American society in the century ahead. Snyder (2000) reported the following statistics: Hispanic-American immigration rates remain high, and their birthrates are 50% higher than those of all other ethnicities in the United States. Because of this, Hispanics will become the largest U.S. minority by 2010, rising to 17.6% of all Americans by 2025 and 34% by 2100. White non-Hispanics will constitute 40% of the U.S. population by 2100, while Asians will represent 13%, as will African-Americans.

By 2025–2030, over one-third of U.S. citizens will be of non-European descent, and by 2055 the percentage will be 50%. Because the United States will continue to become an increasingly diverse polyculture, all industries, trades, and professions will have to actively pursue cultural diversity in their recruitment, education, career development, and retention programs.

Changing Labor Practices in North America

Futurists predict that temporary but major labor shortages will occur over the first part of the twenty-first century (Edwards and Snyder 1997). To meet the staffing required by the future's infrastructure-driven, booming economy, employers of all types of human resources—from doctors, accountants, and engineers to construction workers, retail clerks, and general laborers—will find it necessary to employ older workers several years beyond typical retirement age.

Many employers are rapidly adopting "phased" retirement programs that allow senior personnel to retire gradually or get rehired after retirement. This situation, especially in the near future, could be magnified should the world find itself more deeply embroiled in an escalating war against terrorism and governments fostering the causes of terrorists.

Women in the Workplace

The number of women in the workplace will continue to increase and be a major factor in employment practices, especially in the United States, where three-quarters of working women now have school-aged children. In one-third of all two-income households, the woman is the principal wage earner in the family.

Currently both women and minorities are greatly underrepresented in the engineering professions. Because of the evolving demographics in this country, the makeup of civil engineering firms and groups will need to reflect the available workforce. Therefore the civil engineering industry must become aggressive in promoting women and minorities into its ranks.

With the increase in elderly and child-rearing-age women employees comes a stronger need for workplaces to be more family friendly than ever before. Supporters of family-friendly work environments not only suggest they promote greater productivity and make for better employer-employee relationships, but they will be extremely important in retaining workers in the future. Yet in spite of an increasing desire of workers for more family-friendly workplaces—and in spite of the projected tight labor market—most American employers have done little to accommodate their employees' growing sense of family obligation.

Impact of IT on U.S. Labor

Modern computer and communications technologies have allowed ready access to markets almost anywhere on the globe and spawned concern among U.S. engineers. They fear an overabundance of cheap labor—engineers and technicians working for considerably lower wages than Americans. At stake will be wholesale loss of American jobs to non-U.S. citizens.

In addition, the recent U.S. H1-b visa legislation increasing the numbers of "temporary" visas for foreigner workers has flooded the U.S. market in fields like software engineering. Because of H1-b, many electrical-computer-software-type engineers and technicians, in particular older workers, have lost their jobs, replaced by lower-salaried foreigners in the country on temporary visas. (Horror stories of the situation inundate the Internet on a daily basis.)

At the moment, U.S. civil engineering professionals remain aloof about H1-b implications. They also seem unconcerned about losing jobs to low-salaried, temporary non-U.S. citizens. (Some suggest it is not apathy but rather a feeling of powerlessness about the situation that is keeping civil engineers from taking action.) In any case, as reported by Weingardt (2000c) in "The Handwriting is on the Wall," what has happened to structural engineers in Germany is now transpiring in this country: much routine civil engineering and drafting work is being shipped out of the country, just like the German structural model. This trend—already a common practice with some of America's larger and more aggressive companies—will skyrocket in the coming years. Much of this country's less-challenging civil engineering work will be assumed by those beyond the boundaries of this country at substantially less cost.

Infrastructure

Infrastructure replacement and/or renewal may likely constitute the bulk of civil engineering work for American companies of tomorrow. The doubling of the U.S. population during the twenty-first century will require building an entire additional America within increasingly stringent environmental and land-use constraints. According to Bernstein and Lemer (1996), "The nation's buildings and physical infrastructure are valuable assets, estimated at some $20 trillion, and are a legacy left us by past generations." Doubling that number for a total asset of $40 trillion in national infrastructure facilities would be awesome. And accomplishing this, along with renewing the existing infrastructure, will pose a substantial civil engineering challenge, politically, financially, and technically.

Even today, many say the engineering and construction industry is at the heart of the U.S. economy. According to Bernstein and Lemer (1996), "The various enterprises involved in design, new construction, renovation and other construction-related activities, including equipment and materials manufacturing and supply, employ over 10 million people and account for roughly 13 percent of the nation's [current] economic activity, as measured by our gross domestic product (GDP). Taken as a whole, design and construction comprises the nation's largest manufacturing activity!"

ASCE Report Card for America's Infrastructure

The recent ASCE infrastructure report card (ASCE 2002b) gave the United States a dismal grade of D+ overall. This reflects a poor record and accentuates the need for comprehensive civil engineering solutions—a need that will escalate exponentially in America alone. (The 12 infrastructure categories reviewed in the report card were roads, bridges, transit, aviation, schools, drinking water, wastewater, dams, solid waste, hazardous waste, navigable waterways, and energy. These not only represent significant areas of work for civil engineers, but also areas of expertise for which they should be providing policy input and leadership.)

The nation's infrastructure systems have deteriorated from a grade of C in 1988, when the first national infrastructure report card was completed by a special commission appointed by the elder President Bush. That fact hardly endorses past public policy decisions as being optimum, but it does emphasize that more and more civil engineers must get involved in setting public priorities and direction.

International Markets

Not only will the American civil engineering industry feel the impact of this nation's future infrastructure needs, so will the international marketplace. As pointed out by Bernstein (2000):

> Forecasts of the future of the design and construction industry show a major shift in the future location of infrastructure projects over the next twenty years. In 1990, approximately two-thirds of infrastructure construction projects were located in industrialized countries; it is now estimated that by 2020, two-thirds of infrastructure will occur in developing countries.

If American civil engineers are to lead, according to Bernstein (2000), "We need to be positioned to understand the needs and requirements of working in developing countries, and better understand the policies on energy and sustainability in those countries, since they strongly influence construction-related issues."

U.S. Government: Friend or Foe?

Two recent propositions in California dealt with standing and proposed legislation and regulations restricting and/or curtailing the use of private-sector engineering and architectural services. Their wording emphasized that different goals often motivate those in the public versus the private sectors. The California case revealed a potentially ongoing conflict between public and private sectors and between union and nonunion engineering factions. In the mid-1990s, government ordinances similar to those recently voted on in California—in favor of the private sector, nonunion faction—were also proposed in other states, most notably in Massachusetts.

The concerns of private-sector engineering businesses about the unchecked power of certain government and special-interest groups were summarized in Bernstein (2000):

> Construction [and design] seems certain to continue its loss of control to government, environmental concerns, consumerism, the [impending] energy crisis, need for land-use policies, national growth policies and the whole matter of priorities. They will make construction [and design] ever more subservient to government—or in a democracy, to the will of the people.

More California-type legislation instigated by government union groups—as well as more government competition for private-sector work—will continue to surface. All-out competition, a prime force for change in both the civil engineering and construction industries, will not likely lessen, and private engineering businesses will have to continue dealing with competition from government and quasi-governmental bodies as a way of life. How successful future antiprivate business initiatives will be depends on how strongly America's private-sector civil engineering community positions itself, both from a leadership and a public image perspective. Consulting civil engineers have great potential to be perceived as part of the solution—not as part of the problem. The burden to make sure they are so viewed falls directly on the leaders in the engineering business community.

Government today has a direct impact on all publicly funded construction. In the future, it will have an even greater say in what and where private investors build. Private-sector civil engineers' responsibilities seem obvious. That is, the leaders in the fields of consulting engineering—design and construction—must participate in the formation of public policies and laws that regulate their activities. Likewise, preserving qualifications-based selection (QBS) procedures and legislation for selecting engineers and architects (E/As) for government projects—not for the benefit of E/As but because using QBS methods results in better undertakings at all levels—will require steadfast guidance from all leaders in the profession, from both the private and public sectors.

Unionization Efforts in the Private Sector

"An organizing drive by an operating engineers' local union aimed at engineering technicians is shooting off sparks in the Chicago area," began a May 2002 cover story in *Engineering News Record* (*ENR*) (Rubin et al. 2002). The incident is not isolated. Indications are that unions, much like they did in the 1960s, are targeting engineering firms—specifically those involved with construction—in efforts to expand union membership nationwide.

How far these efforts will go is currently not clear. That they will impact large numbers of consulting civil engineering firms, however, is not. "The owners of two major apartment projects have dropped an engineering firm whose workers rejected union membership ... and managers have reduced the role of an engineering firm whose workers have been targeted by Union Local 150," reported *ENR* (Rubin et al. 2002). The president of STS, one of the engineering firms caught in the middle of the fracas, claimed that his firm was caught unprepared. "We were novices in all this," he said. That will have to change.

Indications are that the current efforts by organized labor could reach epidemic portions in the coming months. Whether pro- or antiunion, private-sector engineering leaders will need to aggressively deal with the situation to ensure that the public's health, wealth, and safety are best served.

Sustainable Development

Given the current problems associated with a swelling global population—including urban sprawl, energy shortages, increasing clean water deficiencies, and air and surface transportation problems, as well as the crumbling infrastructure of many nations—sustainable development will be a major engineering issue. Sustainability will be a significant driver—if not explicitly, certainly in concept—for most engineering projects designed for the built environment. As summed up by Bernstein (2000), civil engineers will have to focus on balancing economic, environmental, and social benefits when designing their projects.

Henry Hatch, former head of the U.S. Army Corps of Engineers and a leading proponent of sustainable development practices globally, stated, "Though viewed by some in the early 1990s as a passing fad, the notion and goals of sustainability are clearly here to stay" (Weingardt 2001).

Hatch added:

Whatever you call it, pursuing our professions and businesses in ways that can be sustained without denying future generations their opportunities must become a bedrock principle. Today, nearly every U.S. and international professional and industry organization involved with the built environment has prominently included sustainability among its strategies, mission statements or ethics. From Presidential Executive Orders to a plethora of volunteer associations such as the rapidly growing U.S. Green Building Council, environmental, economic and social sustainability is gaining momentum as the driving set of principles for our industry in this new millennium (Weingardt 2001).

Opportunities and Imperatives

The United States model mirrors the civil engineering needs that all developed and developing countries have—and will continue to have—at an ever-increasing rate. In industrial nations, much of the infrastructure will include replacing and upgrading, while in developing nations more will be for new infrastructure. At the center of all this activity will be civil engineers serving society's need for expert, wise, cost-effective, and long-lasting engineered solutions.

How much will be provided by the private versus the public sector will depend on many things. One of the main factors is whether engineers in the private sector are perceived as being valuable problem solvers—and whether their work products are considered to be value-added services. As stated by Paul Zofnass (president of EFCG, Inc.), "Once you get away from being identified as a cost and you are considered a strategic value, that's something boards [and clients] are interested in—and is worth more" (Rubin and Powers 2001). "If we provide solutions, rather than the answers," added Bill Robertson (CEO of Roy Weston, Inc.), "we can redefine ourselves in our clients' eyes."

Snyder suggests clear "opportunities and imperatives for engineering firms" in the future (Weingardt 2001). The following six of these apply to civil engineers:

- Continued robust population growth in the United States throughout the twenty-first century will assure sustained long-term demands for civil engineering services and a sustained domestic supply of human resources to meet these demands;
- Continued robust economic growth in the United States will assure a sustained demand for sophisticated civil engineering services, plus the money to pay for them;
- America will become the world's first—and perhaps only true—polyculture, without a single dominant culture. Cultural diversity will become as American as Apple computers;
- The tight labor market will promote gender equity in the workplace, especially in fields involving professional and high-tech skills;
- The unbundling of vertically integrated U.S. industrial firms—through outsourcing administrative, off-line assembly, logistics, and in-house engineering services—will cut the proportion of lifetime career jobs in the American workplace in half over the next 10 to 20 years; and
- Rising birth rates among immigrants and older baby boomers are combining with the aging population to make boomers the

"sandwich generation," simultaneously responsible for both younger and older dependent family members. As a consequence, there is a growing movement in support of a family-friendly employment environment as a central issue in career planning and labor-management relations.

Predictions: Practice of Engineering

"The business of engineering and architecture is changing from a 'practice-centered business' to a 'business-centered practice'," stated FMI's Corey Hessen (2000) in "Looking at the Road Ahead for A/E Firms."

This concept and the predictions that follow paint a picture of what civil engineering might be like in the future. As with all predictions, they should be studied and updated continually.

Around the Clock Global Services

For several years, reports have been surfacing that the world is in the era of 24-h-around-the-world design services, 7 days a week. Engineering work done over an 8 h work period in the United States, for instance, is forwarded to Asia using today's IT tools. A project is worked on there for 8 h, then forwarded (using the latest and greatest IT tools) to Europe for an 8 h work session, then sent back to the United States. Many engineering leaders believe this process results in lower costs, more productivity, and shorter deadlines.

Emerging Technologies and Engineering Trends

The civil engineering industry will continue to expand its use of emerging advanced technologies and display the following characteristics:

- Creative uses of IT, involvement in 4D, paperless design, and so on, will increase;
- Engineering entities (firms and staffs) will be fully computer literate;
- Lean permanent core staffing with significant outsourcing will prevail; and
- Specialization in smaller firms, geographically and technically, will increase.

Virtually all drawing and engineering documents will be computer generated, and standard, uncomplicated engineering designs will be offered electronically on-line. A full array of engineering services, including funding from private sources, will be available on the Internet.

Engineering agencies and companies of all sizes and disciplines involved in engineering the built environment of the future will require a comprehensive understanding of human behavior, lifestyles, and social roles. To be successful, they will also have to exhibit a sound understanding of the economic potential and impact their work and projects have on communities. Firms will likewise need to employ innovative contracting and financial strategies for their projects and companies.

Prime design civil engineering firms, in particular, will have to show a greater knowledge about environmental consequences and materials use than in the past. Life-cycle costing, extensive virtual design procedures, affordable public housing knowledge, and so forth will require them to be flexible in their design practices—and in their skills leading and managing teams composed of members with diversified interests and objectives.

Some forecast small U.S. civil engineering design firms will continue to be healthy well into the next century. Many will become more specialized geographically and functionally and better able to provide superior services on request within tight time constraints. IT will allow such firms to become more productive and profitable at smaller sizes than today.

More small U.S. firms will work globally by merging with international partners on a project-by-project basis. "Mom and pop" operations will continue to spring up all over as a result of the IT evolution (or revolution) and will routinely contract specialized services, not only to large but also to small and medium-sized companies, stateside and overseas (Weingardt 2001).

A high percentage of U.S. engineering firms—large, medium, and small—will become multidisciplined, and the majority of them will increasingly participate in global work via the Internet. Because of IT, collaborations with overseas firms and universities will increase, and more geographically diverse teams will work on the same project. More partnering will occur among a wide array of firms, both foreign and domestic. Today's—and for sure, tomorrow's—global information systems will permit designs, drawings, and technology to be transferred across national boundaries instantaneously.

The Master Builder

Numerous industry leaders say civil engineering firms have the potential to become master builders. Many firms will lead in the return of the "master builder" concept: design and construction team leaders with a holistic view of projects who take on the ultimate responsibility for integrating all aspects of sustainability, resource productivity, and public and client service. They will also champion their projects through all barriers to their completion.

Plus, the focus of many engineering and construction activities will shift from just providing a product—with civil engineering regarded only as a commodity to be obtained by low price—into becoming full-service providers. "This may be the most important shift private-sector engineers can accomplish: from 'a commodity' to value," summarized Dorman (Weingardt 2001).

Civil engineering firms and their counterparts in construction will have to adopt the principle of "doing more with less" as a basis for effective business strategy. This will be done both for internal productivity and to deliver the "best bang for the buck" to the client. The aim? Providing the best overall project from a total resource productivity standpoint.

Computer-Driven Design Advancements

Design, material, and construction advances that will affect the practice of civil engineering include:

- Sophisticated sensors that detect early materials failures;
- Small implanted devices that direct materials to repair themselves;
- Earthquake detection devices that continually improve, thus providing better, more specific, and quicker warning systems than existing ones;
- Increased knowledge of advanced materials and their cost-effective use;
- Extensive and creative use of recycled materials and systems for construction projects;
- Advanced robots with intricate artificial brains capable of doing an increasing number of repetitive tasks;
- Widespread use of smoke-moisture-odor-light sensors that greatly improve public safety;
- Virtual design and 3D image-processing technology to optimize design of complex details and connections;

- Practical use of artificial intelligence to detect environmental changes on earth and in the atmosphere; and
- Advanced biotechnology leading to space-age-type innovations in environmental engineering.

Some forecast that "understanding the interaction of energy, information and infrastructure may bring about the biggest conceptual shift in urban infrastructure design in several hundred years" (Bernstein 2000).

The destruction of the World Trade Center towers by terrorists has brought forth predictions of all sorts about the future of super-tall high-rises. Historian and engineering professor Henry Petroski (2001b) reported:

Nontraditional structural material, such as ceramics, might someday provide the framework for new fire-resistant skyscrapers. And current research into nanotechnology—the manipulation of structures on the atomic scale—might in the distant future yield new materials suitable for building toward the sky. But ceramics are much more brittle than steel, more susceptible to snapping under impact, and nanotechnology is still in its infancy.

For now, Petroski suggested, "The era of the signature building may very well have ended on September 11, 2001, and America's skylines—as well as many others around the world—may remain for the next several decades as they are today."

America's Litigious Nature

One of the biggest detriments to the advancement of civil engineering innovations in the United States has been the threat of lawsuits if things don't go perfectly. Particularly troublesome has been that the civil engineering industry's standards of care are, in effect, established by trial lawyers rather than by the industry itself. Unless fair tort reform legislation is enacted in the United States, state-of-the-art advancements in American engineering will continue to be stymied.

As noted by Bernstein and Lemer (1996):

We have observed that many U.S. design and construction firms are finding it difficult and unprofitable to be as innovative as they might like. New technologies developed by U.S. industry and academic institutions are being commercialized overseas. Our global competitors are becoming more successful, not because they are necessarily more inventive, but because they operate in a setting more conducive to spreading innovation in the marketplace.

Competition from Nonengineering Companies

In recent years, some of the most aggressive recruiters of engineering graduates on college campuses have come from the big-five accounting firms, companies like Ernst and Young (E&Y) and Arthur Anderson (of the Enron Corporate Scandal shame). They, along with recruiters from the software industry, frequently offer signing bonuses and salaries much higher than those in the engineering industry. According to Rubin and Powers (2001), they are joining glitzy management consultants in making high-profile forays into the field of design and construction—and hiring the best talent around.

Mark Smith (partner in E&Y) said his firm's construction business has grown as owners downsized and focused more on core business (Rubin and Powers 2001). E&Y provides overall project management, including the selection of engineers and contractors; it currently oversees $1 billion a year in construction work. Said Smith, "Typically, we're involved in technology-related projects that have a higher degree of risk and more project controls."

Nonengineering project management consultants—and management consulting arms of big-five-type accounting firms—have come into demand because of government, business, and institutional entities' concerns about controlling construction costs, and the need to raise large sums of money quickly. For American A/E and construction firms to compete, they have to get more skilled—and diversified—at nonengineering activities such as financing, operations, locating funding sources, and so on.

Predictions: Projects and Customers

In the future, increasing numbers of projects will require a significant use of collaboration and partnering to be successful. Civil engineers—if they want to reach the top of the food chain of decision makers—will need broader knowledge about the planning, financing, and operating of facilities.

Lining up project funding may become a major part of the civil engineer's responsibility. Design/build/operate (DBO) as well as design/build/operate/transfer (DBOT) projects will increase.

Design/Build

Design/build (DB)—long a project-delivery mainstay in certain segments of the private sector—has made major inroads with federal and state governments in recent times. This trend will likely continue to escalate. "Declarations continue to be made that, by year 2005, approximately 50 percent of all new construction will be performed via design-build" (Hessen 2000).

Many civil engineering professionals have a serious concern about the wanton use of DB, not just because design firms often have to invest substantial monies upfront and take on the additional risk, but because of a tendency to ignore QBS selection procedures. Often engineering consultants for the DB team are selected by low bid, and engineers are rarely in the decision making or upper management layer of the team. On the other hand, when DB teams are properly structured with QBS-selected engineers in key leadership roles, design/build delivery procedures have worked well and projects were completed successfully.

Mass-Produced Products and Systems

The increased availability of mass-produced modular units/components for projects may eliminate the need for some design functions. Examples of this have been around for years: prefabricated precast concrete, steel, and timber structural systems. Their use will increase, as will the widespread use of mass-produced but efficient and inexpensive products such as on-site package water purification and treatment plants.

Their usage, say many experts, will create projects with hundreds of decentralized systems rather than designing and building just one central system. Such projects, to be successful, require high-quality customer service, market management, and project management skills. Environmental solutions will likely be built into the original design of a project rather than added on later in reducing the impact of construction on its surroundings.

Specific Project Types in the Future

Civil engineering professionals, while helping shape tomorrow's world, will be very much involved in the design and renewal of infrastructure projects (as outlined in the section on ASC's infrastructure report card). Constructing, refining, and upgrading the

nation's—and the world's—infrastructure systems will require massive engineering input and creative expertise.

Forks in the Road (Weingardt 1998) laid out additional project types on which civil engineers will have a major impact. They include the following:

- Public and private security systems and safeguards against terrorists and fanatics will be a major concern, and living with such will become a way of life for Americans;
- Recreational and health care needs for an aging population must be adequately thought out as well as society's ever-increasing demand for prisons, and so on. Finding lasting solutions for these must go beyond just building more new facilities;
- Exploring space and oceans will be ongoing and astonishing—and creative civil engineering solutions will be acutely needed for both;
- Mastering new technology and controlling the effects of faulty technology to prevent disasters such as Chernobyl, the Challenger, and Three Mile Island will be key challenges; and
- Environmental concerns, such as clean water and air, runaway waste issues, and global warming will require immense input from civil engineers. So will the increasing need for life-cycle engineering and costing, solutions for sustainable development, and saving the planet for future generations.

More important than knowing what types of projects the future holds, however, will be how the civil engineering industry positions itself: as experts who provide value-added services, as leaders rather than followers, and as professionals rather than technicians.

The public's perception of civil engineers will depend on three things: (1) the profession's ability to perform; (2) its willingness to be self-policing; and (3) its proactive PR efforts to convince others that civil engineers are critical to solving the problems in tomorrow's built environment.

Who Will Be the Customers?

As in the past, clients for private-sector civil engineering companies will come from a wide array of sources. For "interpro" firms—those who work mostly for other design firms, such as larger engineering firms, architects, A/Es, and so on—their core client base will change little, except they may become more global in their range of practice. Large prime designers will continue to serve public agencies and/or industrial and private-sector companies and businesses.

Plus, private-sector civil engineering consultants increasingly will work for nongovernmental organizations (NGOs), which fund their own projects and services. These NGOs include a wide range of groups, from local social service organizations to national or international environmental groups. Any financially stable group with ready funding and a passion for its cause will find resources for private-sector engineering services to build projects important to them.

In the coming decades, large companies in business, commerce, and industry will do more outsourcing. The reengineering of traditional company operations and staffing will continue, and many functions, such as in-house design and engineering, will be contracted to private sector and quasi-governmental consulting groups. Design/build delivery methods now in vogue will encourage one-stop-shopping type A/E service firms, those providing complete project services including planning, financing, design, construction, maintenance, and even operating expertise.

The increasing intrusion into the world of design and construction by accounting, financial, and management consulting firms is pushing traditional engineering firms to seek "high-end business in everything from program management to information technology consulting," according to Rubin and Powers (2001). "Engineering and construction firms are now positioning to deliver what owners want and need." This causes engineering and construction companies to actively work at influencing "decision-making at the earliest stages and highest levels of public and private-sector management."

Many large-firm engineering leaders, report Rubin and Powers (2001), basically say, "We can't just sit back and wait for the RFP to show up any more. In the old days, doing good business development and good engineering was enough. Today it is not. We must redefine ourselves in our clients' eyes, convincing them that we provide solutions rather than merely answers. Then we must go to them to do that."

Intercontinental Business Practices

IT, the Internet, and modern air travel have shrunk the world of business into a global marketplace. The meshing of foreign cultures—and intercontinental business practices, concepts, methods, activities, competition, and partnerships—will become a way of life for many more U.S. civil engineering firms than in the past.

Even locally focused firms who only practice within a small area will be exposed to international firms and/or projects. This will come either from foreign companies coming into their region or from ongoing clients doing ventures internationally (and using their U.S. engineering firms for the projects). In many cases, lasting partnerships/associations will develop between foreign and U.S. engineers after completing an international venture together.

Public Relations and Community Involvement

How U.S. civil engineers of tomorrow effectively address the trends that will greatly impact them will be influenced by how they are perceived by the public (which starts with how civil engineers perceive themselves). How successful such efforts will be essentially depends on whether civil engineers can indeed solve the two PR dilemmas—dealing with respect and appreciation (see the section on ongoing PR dilemmas)—that have plagued them for years. How well the U.S. engineering community can situate itself hinges, not just on civil engineers' technical talents, but on whether they are envisioned as societal leaders. Well-thought-out PR programs combined with the involvement of civil engineering leaders setting policy and direction in broader-based communities will benefit the profession enormously.

In addition to increasing visibility of civil engineers and creating a positive public perspective, persistent PR initiatives can be extremely valuable in the profession's quest to influence policy in both the private and public sectors. Neither PR by itself nor PR hype, however, substitute for sound reasoning and state-of-the-art engineering skills. As part of their overall PR plan, civil engineers must make sure their core message is being consistently communicated to key decision makers and community leaders. The message is that civil engineers can solve the problems of the built environment and must be included when key decisions are made. Said another way, civil engineers and engineering are not the problem; they are a vital part of the solution.

After the passage of California's Proposition 35 in 2000 (giving greater latitude to government contracting with private engineering firms), one of the activists behind its success, civil engineer John Baker, said, "Engineers are accustomed to merely being 'resources.' In the Prop 35 initiative, we initiated the movement and came across as decision makers and leaders. It's something we need to do more of to ensure that our businesses will survive successfully tomorrow" (Weingardt 2001).

Influencing Policy

The civil engineering profession's core message can be even more powerful if civil engineers partner with others as often as possible. Because the message concerning the public's perception of those responsible for designing, building, and maintaining the built environment is shared among dozens of industry and professional groups, it is imperative that civil engineering and construction groups closely collaborate with each other to be heard, understood, and believed.

Separate PR approaches in getting the message out are needed for federal, state, and local entities and for private-sector groups. For government groups, direct contacts between individual engineers and individual elected or appointed officials are extremely important, though lobbyists for engineering associations have an important role to play as well. But as stressed by civil engineer Brian Lewis, who served in both the House and Senate in the state of Washington, personal contacts are key. Said Lewis, "I listened to lobbyists, but those who really got my ear—and attention—were my constituents" (Weingardt 2001).

Influencing policy in the private-sector business community on issues dealing with procurement policies, design/build versus traditional design delivery, and so forth requires a totally different approach than does influencing public policy. In this area, personal contact can make an even more significant difference.

PR Is Not a Luxury

PR is not an isolated function—or merely a "feel-good" luxury item—within an organization or profession, but a vital part of the whole operation, especially in creating favorable perceptions. Perceptions significantly influence how the civil engineering profession comes across to its public and to fellow members: as technicians or professionals, as followers or leaders, and as technical resources or valuable decision makers who should be at the table when crucial decisions are made and public direction is set.

Community Involvement

Since the world is run by those who show up, civil engineers must show up as leaders on a regular basis—and become more involved and active in their communities—to have a consequential impact on the world around them. One effective way is to get appointed to—and serve on—local government boards and commissions, and so forth—and even run for elected public office.

Doing this will not only enhance the public's awareness and acceptance of civil engineers as community leaders, but will convince people that civil engineers care about more than engineering and construction projects—that they are serious about making their communities better places to work and live. To encourage those who might hesitate to get involved, Lewis suggested that "civil engineering societies [and associations] celebrate and publish stories of members who have been elected or appointed to decision-making bodies and explain the benefits accrued" (Weingardt 2001).

Conclusions

Looking into the future, certain trends become self-evident—such as the three current world developments discussed—while others remain elusive. For some issues, it comes down to reviewing the speculations of qualified people, making reasoned guesses, and then revising them as time dictates. Few people, including the most sophisticated futurists, predicted the fall of the Berlin Wall or the collapse of communism in the Soviet Union. So it is with detailed forecasts affecting the civil engineering industry.

That the future holds change is obvious. And clearly the rate of change, especially in the civil engineering arena, will keep accelerating. Engineers can anticipate and address change proactively, running the risk of being slightly wrong. Or they can ignore change until it is upon them. Then they may encounter the greater risk of being engulfed by events and conditions out of their control, forcing the profession to address them reactively. The choice belongs to engineers themselves.

As has been true throughout history, engineering and technology will lead most progress. Technological development will come fastest; change in political and social arenas will come more slowly. Paradoxically, it is in this area of social concerns about engineering and construction in the built environment that the industry's leaders can most increase their influence.

Civil Engineering of Tomorrow

As the world becomes more technologically dependent, the field of civil engineering will be at least as complex as it has ever been. As a result, greater numbers of tomorrow's civil engineering firms must become more diversified in the disciplines they practice and must add nonengineering expertise—finance, operations, political know-how, and so forth—to their arsenal of skills.

The growing advances in and use of IT will let more small U.S. firms compete on the international scene—and allow them to spring up anywhere and do well in niche markets by providing highly specialized services to others, often large engineering groups. Medium-sized engineering firms will either grow into bigger operations, merge with other like-sized firms, or be absorbed by megasized companies—or they will exist in strong niche markets and/or locales. Large-to-giant civil engineering firms will increasingly dominate regional and international markets, with numerous offices located everywhere, many staffed with partners indigenous to where the office is located. And they will become engrossed in 24-h-around-the-globe, design/build projects.

Influences of International Marketplace

Much U.S. project design work will be shipped to and done by inexpensive labor in foreign countries using the Internet and IT tools. North America will continue to be one of the most lucrative international marketplaces in the world, and foreign-based engineering and construction companies will make noticeable inroads into U.S. markets.

Even so, as shown in a December 2000 study by the University of Michigan, annual U.S. exports of services—including professional services—are expected to increase from $250 billion today to $650 billion in 2010. This is equal to today's annual exports of agricultural and industrial products (Weingardt 2001).

Government Bodies versus Private-Sector Groups

The future will see more efforts by certain government bodies, especially those with large unionized technical staffs, to do the engineering traditionally handled by private-sector consulting engineering companies. Recent California and Massachusetts legislative experiences will surface again in other forms and at other locations, as will efforts by unions to find work for unionized engineers and technicians. More than ever in history, U.S. private-sector civil engineers will need to become involved—actively and vocally—in the political process, ensuring their future existence and setting public policy and direction.

Much more could be accomplished if private-sector engineering firms would proactively strive to develop partnerships with government agencies rather than taking them on as adversaries after bad policy is enacted. Creating partnering agreements with bodies such as the U.S. Army Corps of Engineers and state departments of transportation, for instance, enhances the goals of both factions. For Americans to see public- and private-sector engineers doing battle with each other blackens the image and stature of the profession.

Attracting and Retaining Young Professionals

One of the major challenges facing the consulting civil engineering industry in the coming decades will continue to be attracting and retaining young professionals—young U.S. citizens—in the profession. With the pool of white males of European heritage rapidly declining as a percentage of the overall population, it is imperative that the profession hone its skills at bringing minorities and women into its ranks, immediately.

If adequate numbers of bright new professionals are not attracted into all the fields responsible for the built environment, managers of American-based civil engineering firms will have serious choices to make. They include:

- Ship civil engineering work overseas (something many large U.S. companies already do);
- Import technically skilled non-American workers into the United States (now under way, most visibly in the IT industry, that is, H1-b visa legislation); and
- Allow more foreign firms to enter the U.S. marketplace (get work and ship it "home").

Efforts to attract and retain new professionals will require enormous collaboration involving several professional societies (ASCE included), the National Academy of Sciences, the National Academy of Engineering, educators, and other members of the construction industry. Estimates indicate it will take a 5-to-10-year commitment to see results (Weingardt 2001).

U.S. society tends to educate more and more people—such as lawyers and stockbrokers—who are oriented toward dividing up the economic pie. At the same time, fewer and fewer are being educated to create and enlarge that pie. This needs to change, just as the education of tomorrow's civil engineers needs to change.

To keep up with the profession's technological and leadership needs, civil engineers of tomorrow will require a 5-year (minimum) degree plan to perform as professionals. Along with this, a requirement for life-long learning is essential for civil engineers to stay current with engineering advancements and to hone communication, leadership, and people skills. More education will broaden their outlook and allow civil engineers of the future to be leaders in society as well as in their industry.

QBS: Only the Beginning

In the past, much ado has been made about QBS as the favored A/E selection process for the federal and many local governments. In many cases, it has helped curtail low-balling fees and shoddy or inadequate design. Though a good concept, in reality, it often limits A/E profits by establishing fees based solely on hourly rates and overheads.

Civil engineering firms need to be selected on qualifications and rewarded (or paid) based on value added. In addition to the continued use of QBS, civil engineering fees and salaries need to be raised to levels more in line with the responsibilities engineers take on—and the value they add.

The Challenge

If engineering historians are right and the past is prologue to the future, American civil engineering legends in the 1800s—Judah, Eads, Roebling, Mason, Stevens, Goethals, and so forth—have set the bar high for today's engineers regarding leadership and community involvement. For them to be matched and surpassed, today's civil engineers need to have the will to do so *and* the dedication to make it happen.

ASCE's 2002 Vision

On the eve of its 150th anniversary celebration, ASCE presented its 2002 Vision Statement: "Engineers as global leaders building a better quality of life" (ASCE 2002a). This honorable and noble call to the society—and members of the profession—inspires them to set the pace in dealing with the challenges of the twenty-first century. The four key elements for accomplishing ASCE's vision are:

• Developing leadership;
• Advancing technology;
• Advocating lifelong learning; and
• Promoting the profession.

Individual civil engineers would be well advised to strive toward these goals themselves. Then they'll be able to leave a powerful legacy for those who come after them. The next group of engineers, in their turn, may then leave an even greater legacy for ongoing generations.

Technologically Complex World

As the world becomes more technologically complex, major public decisions cry out for insight from those with a thorough knowledge of engineering and technology. Because of this, the call for large numbers of savvy civil engineers to fill leadership roles—not just in the engineering industry but in the public arena—has never been greater. To be prepared to hold such crucial positions, tomorrow's civil engineers must both be well versed in emerging technologies and cutting-edge engineering developments and have political clout—to become Benjamin Franklin-type "citizens of the world." That means civil engineers will need to have their communication, leadership, and people talents honed to the highest levels possible.

If civil engineer leaders with high-level skills *don't* show up to provide the needed expertise concerning tomorrow's engineering nuances, who will? Who else is better suited to be the stewards of this nation's (and the world's) infrastructure systems—and to raise the public consciousness concerning the impact of civil engineering applications?

Global Demand for Engineers

Both developed and developing countries have always depended heavily on their national engineering base. Engineering will be an even stronger factor in the future. The strength of a nation's engineering talent determines its economic power and establishes its very standard of living. Engineers, say many think tank gurus, are the world's true wealth creators, those who help increase the size of the economic pie, not divide it up. They are the ones most often behind notable advances in progress.

In essence, the history of engineering has been the history of civilization. And tomorrow will be no different. The demand for civil engineers—and civil engineering leadership—will remain high well into the future. In response to these demands, civil engineers can come across as professionals or technicians—and be either leaders or followers, activists or laidback reactionaries. They can expand on the many opportunities being presented or stick their heads in the sand, waiting for others to tell them what to do. It is totally up to them.

The Fate of the Profession

If increasing numbers of engineers don't become proactive in helping make critical public judgment calls, it will be business as usual. And public direction—indeed, the very fate of the civil engineering profession—will continue to be in the hands of professional politicians and others with no engineering background.

In a perfect world, it would be wonderful if civil engineers could be community leaders and top engineers at the same time. This is not always possible. An engineer's years in college are so short, it is impossible to fulfill all the technical course work needed and still adequately study art, history, philosophy, and literature—the humanities—subjects that broaden a person's perspective. To accomplish those broadening objectives—and earn respect as a learned profession on a par with medicine, science, and architecture—civil engineers need to commit to lifelong learning, both at the university and afterward.

Finding Time for Community Service

Similarly, time available to working engineers for community involvement and public leadership is limited unless pursued after hours or unless employers allow time off. In the coming years, that will need to happen. Employers and upper management should be willing to provide matching paid time off for individuals willing to fill community leadership positions such as serving on public boards and commissions.

The Call to Show Up and Lead

The U.K. engineering press, it seems, would like to have Americans think that the world has entered the dawning of the age of the engineer. For that to happen, more top civil engineers—the cream of the crop—will need to become highly visible leaders outside of the profession. And they must unearth creative solutions to eliminate the profession's two PR dilemmas once and for all.

If civil engineers truly want to raise public awareness about the importance of their profession, interest the media, and influence public direction—plus attract and retain some of the brightest into their ranks—they will have to get actively involved as pacesetters beyond the field of engineering.

Tomorrow's civil engineers must *not just* show up, they must show up to lead. They *must* seize the moment to run things, not just make things run.

References

ASCE. (2001). *Engineering the future of civil engineering*, ASCE Task Committee on the First Professional Degree, Reston, Va.

ASCE. (2002a). *Building ASCE's future*, ASCE Strategic Plan Annual Element Fiscal Year 2002, Reston, Va.

ASCE. (2002b). "The 2001 report card for America's infrastructure." *ASCE* ⟨http://www.asce.org/reportcard/⟩

Bernstein, H. (2000). "The future of the design and construction industry." CERF ⟨http://www.cerf.org/PDFs/Abont/Future. PDF⟩.

Bernstein, H. M., and Lemer, A. C. (1996). *Solving the innovation puzzle: Challenges facing the U.S. design and construction industry*, ASCE, Reston, Va.

Chadwick, W. (1977). "How to overcome these threats to civil engineering." *Civil Engineering*, 47(10), 141.

Dorman, A. (2000). "Reaching out to Hispanics." White paper prepared for ACEC's 2000 Blue Ribbon Panel, American Council of Engineering Companies, Washington, D.C.

Edwards, G., and Snyder, D. (1997). "High tech and free trade in the 21st century: The outlook for international commerce at the transmillennium," *Foreign Service Journal*, May.

Gallup. (1944–2001). Gallup polls/surveys, Gallup Organization, Princeton, N.J.

Harris. (1977–2001). Harris Polls, Harris Interactive, New York.

Hessen, C. (2000). "Looking at the road ahead for A/E firms." *FMIdeas!*, 2(2).

Kirby, R., et al. (1956). *Engineering in history*, McGraw-Hill, New York.

Petroski, H. (2001a). "Expecting respect (refractions)." *Prism*, September.

Petroski, H. (2001b). "Onward but perhaps not upward (refractions)." *Prism*, November.

Rubin, D., et al. (2002). "Chicago operating engineers target testing technicians." *Eng. News Rec.*, May 20.

Rubin, D., and Powers, M. (2001). "Consultants vie for owners' hearts, minds and purses." *Eng. News Rec.*, February 12.

Snyder, D. (2000). "Demographic realities in the future of U.S. engineering companies." White paper prepared for ACEC's 2000 Blue Ribbon Panel, American Council of Engineering Companies, Washington, D.C.

Walesh, S. G. (2000). "Engineering a new education," *J. Manage. Eng.*, 16(2), 35–41.

Weingardt, R. (1997). "Leadership: The world is run by those that show up." *J. Manage. Eng.*, 13(4), 61–66.

Weingardt, R. (1998). *Forks in the road: Impacting the world around us*, Palamar Publishing, Denver.

Weingardt, R. (2000a). "Professional status: Debating the Five-Year Degree." *Structural Engineer*, 1(9).

Weingardt, R. (2000b). "Step forward and be heard." MIT's CEE New Millennium Colloquium, Cambridge, Mass.

Weingardt, R. (2000c). "The handwriting is on the wall." *Structural Engineer*, 1(6).

Weingardt, R. (2001). "Eye to the future." ACEC 2000 Blue Ribbon Panel Publication, American Council of Engineering Companies, Washington, D.C.

Weingardt, R. (2002). *Leadership Manage. Eng.*, 2(3), 53–55.

Western Society of Engineers (WSE). (1970). *The centennial of the engineer, the 100-year anniversary history of the Western Society of Engineers*, Chicago.

Whitelaw, J. (2000). "Is this the dawning of the age of engineers?" *New Civil Engineer*, September.

Subject Index

Page number refers to the first page of paper

Author Index

Page number refers to the first page of paper